Mycological Inventions for Sustainable Agriculture and Food Production

Arun Kumar Kashyap
Government E. Raghavendra Rao P.G. Science College, Bilaspur, India

Abhimanyu Jogawat
National Institute of Plant Genome Research, New Delhi, India

IGI Global
Scientific Publishing
Publishing Tomorrow's Research Today

Vice President of Editorial	Melissa Wagner
Managing Editor of Acquisitions	Mikaela Felty
Managing Editor of Book Development	Jocelynn Hessler
Production Manager	Mike Brehm
Cover Design	Phillip Shickler

Published in the United States of America by
IGI Global Scientific Publishing
701 East Chocolate Avenue
Hershey, PA, 17033, USA
Tel: 717-533-8845
Fax: 717-533-8661
Website: https://www.igi-global.com E-mail: cust@igi-global.com

Library of Congress Cataloging-in-Publication Data

Title: Mycological inventions for sustainable agriculture and food production / edited by Arun Kashyap, Abhimanyu Jogawat.
Description: Hershey, PA : IGI Global Scientific Publishing, [2025] | Includes bibliographical references and index. | Summary: "The objective of this book to collect latest advancement and application of fungi and their functional roles in improving agriculture and food systems. Further, this book will present vast picture of fungal community and their potential applications in agri-food systems which will further advance our knowledge in this field"-- Provided by publisher.
Identifiers: LCCN 2024055190 (print) | LCCN 2024055191 (ebook) | ISBN 9798369382011 hardcover | ISBN 9798369382028 paperback | ISBN 9798369382035 ebook
Subjects: LCSH: Fungi--Economic aspects | Fungi in agriculture | Fungi--Utilization | Mycology
Classification: LCC QK604.2.E27 M93 2025 (print) | LCC QK604.2.E27 (ebook) | DDC 579.5--dc23/eng/20250203
LC record available at https://lccn.loc.gov/2024055190
LC ebook record available at https://lccn.loc.gov/2024055191

British Cataloguing in Publication Data
A Cataloguing in Publication record for this book is available from the British Library.

Table of Contents

Preface xiv

Chapter 1
Role of Fungi in Sustainable Agriculture and Stress Mitigation 1
Lalit Kumar Sahu, Guru Ghasidas Vishwavidyalaya, India
Nikita Verma, Guru Ghasidas Vishwavidyalaya, India

Chapter 2
A Sustainable Approach to Unlocking Plant Potential With Fungal Biostimulants: Harnessing Fungal Powe to Sustainable Agriculture 25
Pooja Gangaram Malave, Dr. D.Y. Patil College of Arts, Commerce, and Science, India

Chapter 3
Fungal Biocontrol Agents for Sustainable Pest Management 45
Vikas Chandra, Guru Ghasidas Vishwavidyalaya, India
Abhishek Dadsena, Guru Ghasidas Vishwavidyalaya, India
Harsha Yadav, Guru Ghasidas Vishwavidyalaya, India
Gurucharan Sahu, Guru Ghasidas Vishwavidyalaya, India

Chapter 4
Arbuscular Mycorrhizal Fungal Diversity Under Abiotic Stress: How Does It Improve Plant Stress Tolerance and Nutrient Use Efficiency? 73
Soufiane Lahbouki, Chouaîb Doukkali University, Morocco
Naira Sbbar, Cadi Ayyad University, Marrakesh, Morocco
Salah Er-Raki, FST Marrakech, Morocco
Abdelilah Meddich, Cadi Ayyad University, Marrakesh, Morocco

Chapter 5
Fungal Biostimulants: Sustainable Tool to Boost Crop Yield Naturally 97
Upama Kalita, Sikkim University, India
Madhan Sanyasi, Sikkim University, India
Priyanka Kumari, Sikkim University, India
Anit Baidya, Sikkim University, India

Chapter 6
Enhancing Soil Fertility and Nutrient Cycling: Mycofertilizers 131
Khushboo Kumari, Mahatma Gandhi Central University, India
Khirabdhi Tanaya, Mahatma Gandhi Central University, India
Durgeshwer Singh, Mahatma Gandhi Central University, India

Chapter 7
Edible Mushrooms as Nutritious and Sustainable Superfoods 151
Dibyarupa Pal, JIS University, India
Sujan Samanta, JIS University, India

Chapter 8
Fungi as a Source of Single Cell Proteins: Current Progress and Future Prospects 163
Sharmistha Sharma, Assam University, India
Himanshu Kishore Prasad, Assam University, India
Anjali Roy, Assam University, India
Joydeb Dey, Assam University, India

Chapter 9
Mycoproteins: The Rise of Fungal-Based Meat Alternatives 191
Dipali Saxena, Shri Vaishnav Vidyapeeth Vishwavidyalaya, India
Neha Prasad, VLCC Ltd., India

Chapter 10
Arbuscular Mycorrhizal Fungi: The Metagenomics and the Circadian Rhythm Governed by the Environment 209
Fatima-Zahra Akensous, Faculty of Science Semlalia, Cadi Ayyad University, Morocco

Chapter 11
Fungi for Ameliorating Soil and Water Pollution: Mycoremediation 237
Khirabdhi Tanaya, Department of Botany, Mahatma Gandhi Central University, India
Durgeshwer Singh, Department of Botany, Mahatma Gandhi Central University, India

Chapter 12
Fungal Biodegradation: Breaking Down Agricultural Wastes 261
P. Selvakumar, Nehru Institute of Technology, India

Chapter 13
Mycoremediation: Leveraging Fungi for Ameliorating Soil and Water Pollution 283
Licon Kumar Acharya, National Research Centre for Integrated Pest Management, India
A. U. Akash, ICAR-Indian Agricultural Research Institute, India
Suryakant Manik, ICAR-Indian Institute of Agricultural Biotechnology, India
Amjada S. Khan, ICAR-Indian Agricultural Research Institute, India
Neha Singh, ICAR-Indian Agricultural Research Institute, India

Chapter 14
Antifungal Properties of Essential Oils From Rosmarinus Officinalis, Citrus Aurantium, and Origanum Majorana: In Vitro Evaluation for Biological Control Agents 309
Chaymae El-Hajjaji, Ibn Tofail University, Morocco
Soukaina Msairi, Mohammed V University, Morocco
Oumaima Ninich, Mohammed V University, Morocco
Yassine Mouniane, Ibn Tofail University, Morocco
Soumaya Choukri, Ibn Tofail University, Morocco
Rihab Hedhiri, Centre of Research and Water Technologies, Tunisia
Taoufiq Benali, Cadi Ayyad University, Morocco
Amir Souissi, Swift Current Research and Development Center, Canada
Amira Hachana, National Institute for Agricultural Research of Tunisia, Tunisia & Carthage University, Tunisia
Jamila Dahmani, Ibn Tofail University, Morocco

Chapter 15
Fungal Bioprospecting: Discovering Novel Bioactive Compounds 327
P. Selvakumar, Nehru Institute of Technology, India

Chapter 16
Edible Mushrooms: A Sustainable Superfood for the Future 349
Vishal Shridhar Moger, St. Aloysius University, India
Harshitha Gujaran, St. Aloysius University, India
Shreya Gowda, St. Aloysius University, India
Pallavi Baliga, St. Aloysius University, India

Chapter 17
Mycotoxins: Challenges and Strategies for Mitigation 377
Tarun Kumar Patel, Sant Guru Ghasidas Government P.G. College, India

Chapter 18
Hazards of Mycotoxin in Agriculture and Effective Mitigation Strategies 407
Suryakant Manik, ICAR-Indian Institute of Agricultural Biotechnology, India
S. Shreedevasena, Directorate of Medicinal and Aromatic Plant Research, India
Asharani Patel, ICAR-Indian Agricultural Research Institute, India
Sujit Kumar Bishi, ICAR-Indian Institute of Agricultural Biotechnology, India
Kuleshwar Prasad Sahu, College of Agriculture, Rani Lakshmi Bai Central Agricultural University, India
Licon Kumar Acharya, Division of Plant Pathology, ICAR-Indian Agriculture Research Institute, India, ICAR- National Research Centre for Integrated Pest Management, India

Compilation of References 439

About the Contributors 553

Index 559

Detailed Table of Contents

Preface..xiv

Chapter 1
Role of Fungi in Sustainable Agriculture and Stress Mitigation 1
Lalit Kumar Sahu, Guru Ghasidas Vishwavidyalaya, India
Nikita Verma, Guru Ghasidas Vishwavidyalaya, India

This chapter explores the critical role of fungi in sustainable agriculture and stress mitigation, emphasizing their contributions to enhancing soil health, improving crop resilience, and facilitating carbon sequestration. Fungi, particularly arbuscular mycorrhizal fungi (AMF) and various fungal biopesticides, offer innovative solutions for reducing reliance on synthetic fertilizers and pesticides. Recent advancements in fungal biofertilizers demonstrate their ability to improve nutrient uptake and disease resistance, while their potential for carbon storage aids in climate change mitigation. Despite the promising applications of fungi, challenges such as inconsistent efficacy, mass production limitations, and regulatory barriers remain. Addressing these challenges presents opportunities for further research, policy reform, and education to optimize the use of fungi in agricultural systems. By harnessing the diverse functions of fungi, we can create more resilient and sustainable agricultural practices.

Chapter 2
A Sustainable Approach to Unlocking Plant Potential With Fungal Biostimulants: Harnessing Fungal Powe to Sustainable Agriculture.. 25
Pooja Gangaram Malave, Dr. D.Y. Patil College of Arts, Commerce, and Science, India

The global population is projected to reach nearly 10 billion by 2050, necessitating a 60-70% increase in food production while minimizing environmental impacts (World Population Prospects 2019). This requires agriculture to produce more food with fewer resources and reduce reliance on unsustainable synthetic fertilizers and pesticides.Recent decades have shifted towards sustainable practices, particularly natural plant biostimulants that enhance growth, flowering, fruit set, and nutrient use efficiency (NUE). Fungal biostimulants effectively promote plant growth by forming symbiotic relationships with roots and improving nutrient uptake.This chapter explores the role of fungal biostimulants in agriculture, focusing on mycorrhizal, endophytic, and saprophytic fungi, with emphasis on arbuscular mycorrhizal biostimulants that enhance nutrient uptake and soil health.The chapter examines the sustainability and economic benefits of fungal biostimulants, addresses challenges related to consistency and regulation, and explores future research opportunities.

Chapter 3
Fungal Biocontrol Agents for Sustainable Pest Management 45
Vikas Chandra, Guru Ghasidas Vishwavidyalaya, India
Abhishek Dadsena, Guru Ghasidas Vishwavidyalaya, India
Harsha Yadav, Guru Ghasidas Vishwavidyalaya, India
Gurucharan Sahu, Guru Ghasidas Vishwavidyalaya, India

The rising demand for sustainable agriculture has increased interest in fungal biocontrol agents as eco-friendly alternatives to synthetic pesticides. This chapter examines fungi's role in pest management through mechanisms like mycoparasitism, resource competition, and insecticidal compound production. Key species, including Beauveria bassiana, Metarhizium anisopliae, and Trichoderma spp., effectively control insect pests, nematodes, and pathogens, reducing pest populations and enhancing crop yields. Benefits include minimal non-target effects, lower residues, improved soil health, and cost efficiency. Field studies demonstrate success in crops like cotton and vegetables. However, challenges such as field variability, limited shelf life, and regulatory constraints remain. Research on improved formulations and delivery methods is essential for wider adoption and long-term agricultural sustainability.

Chapter 4
Arbuscular Mycorrhizal Fungal Diversity Under Abiotic Stress: How Does It Improve Plant Stress Tolerance and Nutrient Use Efficiency? 73
Soufiane Lahbouki, Chouaîb Doukkali University, Morocco
Naira Sbbar, Cadi Ayyad University, Marrakesh, Morocco
Salah Er-Raki, FST Marrakech, Morocco
Abdelilah Meddich, Cadi Ayyad University, Marrakesh, Morocco

Arbuscular mycorrhizal fungi (AMF) play a key role in promoting plant growth and resilience to abiotic stresses like drought, salinity, and heat. By forming symbiotic relationships with plant roots, AMF significantly enhance the uptake of nutrients, especially phosphorus, often limited in stressed soils. This chapter explores how AMF help plants by improving water absorption, maintaining osmotic balance, and regulating ion homeostasis. It highlights the diversity of AMF species and their varying responses to stress, influencing plant support. The chapter also examines changes in AMF community structure under prolonged stress and their implications for plant productivity. Additionally, it delves into how AMF modify gene expression in host plants, impacting osmolyte production and stress-protective proteins like heat shock proteins. The strategic use of AMF offers sustainable agricultural solutions, reducing chemical inputs and enhancing crop resilience amid climate change.

Chapter 5
Fungal Biostimulants: Sustainable Tool to Boost Crop Yield Naturally 97
Upama Kalita, Sikkim University, India
Madhan Sanyasi, Sikkim University, India
Priyanka Kumari, Sikkim University, India
Anit Baidya, Sikkim University, India

From the beginning of humankind plants have been a great source of food. The increase in environmental pollution, elevated global temperature as well as widespread application of chemical fertilizers and insecticides have caused reduced soil fertility and decline in agricultural production. On the other hand, the need for food intensified with rise in the global population. The plant-microbe interface plays a vital role in moderating harsh environmental situation and improve crop strength. In the last few years plant-fungi interaction has become the most studied topic for production of ecofriendly substitutes to chemical products. Fungal biostimulants create a wide range of interactions with the host plant and encourage crop advancement, develop resilience, recover soil health and stimulate defense response. Overall, this chapter will provide extensive outlook on fungal biostimulants, role in plant stress tolerance, mechanism of action in plants and influence in organic agriculture. It also states their ecological role and challenges for advanced crop development in the near future.

Chapter 6
Enhancing Soil Fertility and Nutrient Cycling: Mycofertilizers.. 131
Khushboo Kumari, Mahatma Gandhi Central University, India
Khirabdhi Tanaya, Mahatma Gandhi Central University, India
Durgeshwer Singh, Mahatma Gandhi Central University, India

Soil fertility is essential for plant growth, but nutrient immobilization can lead to deficiencies. Although chemical fertilizers can address these issues, they pose environmental risks. Myco-fertilizers form symbiotic relationships with plants, extending the root network and enhancing nutrient absorption. They improve soil structure, nutrient cycling and enhance soil resilience against pathogens. Research indicates that myco-fertilizers can significantly reduce agriculture's environmental impact by limiting chemical runoff and soil degradation. This book chapter explores the use of myco-fertilizers in agriculture, focusing on their role in nutrient cycling. It also emphasizes how myco-fertilizers can reduce the need for chemical fertilizers and improve soil fertility. Incorporating myco-fertilizers into farming systems offers a path towards more eco-friendly agricultural practices, ensuring food security. The adoption of myco-fertilizers is a key element in sustainable agriculture, contributing to overall soil health and environmental sustainability.

Chapter 7
Edible Mushrooms as Nutritious and Sustainable Superfoods 151
Dibyarupa Pal, JIS University, India
Sujan Samanta, JIS University, India

The search for reforming diets has stemmed from the current invasion of Covid 19 pandemic and uprising of other viral diseases. People started relying more on plant protein source and the consumption of immunomodulatory bioactive compounds from herbal sources. The necessity for foods with several benefits, called as 'superfoods' has burgeoned to fulfil the desire for healthier lifestyle. Mushroom is one such cheap and versatile vegetable protein satisfying the role of superfood, has low-calorie, high-protein and loads of medicinal values to support diabetics and cardiovascular patients. It requires minimal investment and does not need arable land. This makes it an accessible business for unemployed youth, disabled people and women force providing a valuable income source. Biofortified mushroom products can lead to diversified income sources, benefitting the rural community. This book chapter discusses the miscellaneous applications of mushroom and its impending health-promoting and superfood qualities for its easy acceptance and popularization.

Chapter 8
Fungi as a Source of Single Cell Proteins: Current Progress and Future Prospects........................... 163
Sharmistha Sharma, Assam University, India
Himanshu Kishore Prasad, Assam University, India
Anjali Roy, Assam University, India
Joydeb Dey, Assam University, India

Fungi offer a promising solution for sustainable protein production and global food security as single-cell protein sources (SCPs). Their rapid growth, diverse substrate utilization, and high protein content make them attractive candidates. Key fungal species, including filamentous fungi and yeasts, are explored for their characteristics and applications. Various substrates, focusing on waste valorization and circular economy integration, are analyzed for fungal SCP production. Bioprocess engineering aspects, such as fermentation methods, bioreactor designs, and process optimization, are discussed. The nutritional value, safety, and diverse applications of fungal SCPs, ranging from food and feed to nutraceuticals, bioelectronics, and sustainable packaging materials, are examined. Challenges in scaling up fungal SCP production, including technical, economic, and consumer acceptance issues, are addressed, along with advances in genetic engineering, biorefinery integration, and novel extraction technologies. The economic and environmental impacts of fungal SCP production are evaluated.

Chapter 9
Mycoproteins: The Rise of Fungal-Based Meat Alternatives... 191
Dipali Saxena, Shri Vaishnav Vidyapeeth Vishwavidyalaya, India
Neha Prasad, VLCC Ltd., India

The rise of mycoproteins is contextualized within the broader movement towards sustainable food systems, driven by the growing global population, climate change, and the urgent need to reduce the environmental footprint of food production. This chapter delves into the historical development, production processes, nutritional benefits, and environmental impacts of mycoproteins. Mycoproteins offer several advantages, including lower greenhouse gas emissions, reduced land and water usage, and the ability to produce protein efficiently and at scale. The nutritional profile of mycoproteins is highlighted, emphasizing their high protein content, essential amino acids, and fiber, alongside low fat and cholesterol levels. The chapter also addresses consumer acceptance and market trends, noting the increasing popularity of mycoprotein-based products in mainstream diets. Mycoproteins serves as a viable and innovative solution to some of the most pressing challenges in sustainable food production, underscoring their potential to transform the future of food.

Chapter 10
Arbuscular Mycorrhizal Fungi: The Metagenomics and the Circadian Rhythm Governed by the Environment.. 209
Fatima-Zahra Akensous, Faculty of Science Semlalia, Cadi Ayyad University, Morocco

Arbuscular mycorrhizal fungi (AMFs) are behind the crucial symbiosis that occurs within over 80% of the root systems of vascular plants, reputed for their potentialities regarding the plant, soil, and conjecture with the existing microbiome. Researchers inclined toward studying AMFs in their occurrence in the environment are currently embracing investigation at the holobiont level within the sphere of the natural habitat. Hence, the present book division looks at how the metagenomics can provide important information in this respect. With a keen interest in studying microbial communities retrieved from their environment, relying on advanced sequencing techniques, the metagenomics can investigate the composition and diversity of AMFs from a close-up lens. In the fungal verse, the circadian rhythm, pioneeringly explored in the Sordariales (Neurospora), is slowly but steadily being investigated in the Glomerales (Rhizoglomus), providing genetic and molecular discernments worth spotlighting into the bargain.

Chapter 11
Fungi for Ameliorating Soil and Water Pollution: Mycoremediation 237
Khirabdhi Tanaya, Department of Botany, Mahatma Gandhi Central University, India
Durgeshwer Singh, Department of Botany, Mahatma Gandhi Central University, India

Pollution negatively impacts the environment due to human activities such as urbanization and industrialization. Pollutants that cause pollution can disrupt pH levels, nutrient balance, and microbial activity, resulting in lower crop yields and deteriorated soil and water quality. The removal of these pollutants is essential to ensure the sustainability of the environment. Many physicochemical treatment techniques are utilized. However, these methods often prove insufficient. Mycoremediation emerges as a promising alternative, harnessing fungi's natural abilities to degrade pollutants and sequester heavy metals. This process relies on fungal enzymatic activity and adsorption mechanisms, offering a more eco-friendly and cost-effective approach to environmental cleanup. This chapter delves into the various methods of mycoremediation, the specific enzymes responsible for breaking down pollutants, and the ways in which fungi accumulate heavy metals. The goal of this chapter is to develop more efficient, economically feasible, and eco-friendly pollution treatment strategies.

Chapter 12
Fungal Biodegradation: Breaking Down Agricultural Wastes 261
P. Selvakumar, Nehru Institute of Technology, India

Fungal biodegradation represents a promising approach in addressing environmental challenges posed by recalcitrant pollutants. With increasing global concern over pollution and its ramifications for ecosystems and human health, the search for sustainable solutions has intensified. Fungi, with their remarkable metabolic versatility and adaptive capabilities, offer a unique toolkit for bioremediation. This introductory exploration delves into the mechanisms underlying fungal biodegradation, highlighting its applications, benefits, and current research trends.Fungi have evolved sophisticated enzymatic systems capable of degrading a wide range of organic compounds, including complex pollutants such as pesticides, petroleum hydrocarbons, dyes, and pharmaceuticals. Unlike bacteria, fungi possess ligninolytic enzymes like lignin peroxidase, manganese peroxidase, and laccase, enabling them to break down recalcitrant substances found in industrial waste and contaminated environments.

Chapter 13
Mycoremediation: Leveraging Fungi for Ameliorating Soil and Water Pollution 283
Licon Kumar Acharya, National Research Centre for Integrated Pest Management, India
A. U. Akash, ICAR-Indian Agricultural Research Institute, India
Suryakant Manik, ICAR-Indian Institute of Agricultural Biotechnology, India
Amjada S. Khan, ICAR-Indian Agricultural Research Institute, India
Neha Singh, ICAR-Indian Agricultural Research Institute, India

The burgeoning population and increased demands for resources across the world have led to the industrial and agricultural revolution. As a result of anthropogenic activities, environmental pollutants are exacerbating day by day in nature leading to soil and water contamination. Mycoremediation, which uses fungi or their derivatives to ameliorate the pollutants has shown immense promise in the recent past. In addition, it is advantageous over other prevalent traditional methods due to its comparative cost-effectiveness, eco-friendliness, high efficiency, and its innate potential to degrade a wide range of contaminants across diverse environmental conditions. In this chapter, we explore the potential of fungi as effective agents for ameliorating soil and water pollution. We focus on the major classes of fungi that can be employed, the mechanisms involved, their applications, and real-world examples. Additionally, we address the major bottlenecks associated with mycoremediation and future directions to enhance this strategy.

Chapter 14
Antifungal Properties of Essential Oils From Rosmarinus Officinalis, Citrus Aurantium, and Origanum Majorana: In Vitro Evaluation for Biological Control Agents 309
Chaymae El-Hajjaji, Ibn Tofail University, Morocco
Soukaina Msairi, Mohammed V University, Morocco
Oumaima Ninich, Mohammed V University, Morocco
Yassine Mouniane, Ibn Tofail University, Morocco
Soumaya Choukri, Ibn Tofail University, Morocco
Rihab Hedhiri, Centre of Research and Water Technologies, Tunisia
Taoufiq Benali, Cadi Ayyad University, Morocco
Amir Souissi, Swift Current Research and Development Center, Canada
Amira Hachana, National Institute for Agricultural Research of Tunisia, Tunisia & Carthage University, Tunisia
Jamila Dahmani, Ibn Tofail University, Morocco

With the prohibition of many chemical pesticides due to their impact on the health of living resources, as well as on the quality of food, soil and water, it has become essential to look for alternatives of natural origin that promote sustainability. This study focuses on the antifungal activity of Rosmarinus officinalis, Origanum majorana and Citrus aurantium essential oils against Phytophthora palmivora and Alternaria alternata. The antifungal activity of six essential oil concentrations was tested in vitro by the direct contact method. After a seven-day follow-up, the results revealed a significant antifungal effect of Origanum majorana essential oil, with total inhibition of mycelial growth at a concentration of 1.5 µl/ml and 3.33 µl/ml on Phytophthora palmivora and Alternaria alternata respectively. While Rosmarinus officinalis and Citrus aurantium showed less antifungal activity, with MICs ranging from 6.66 µL/mL to 12.5 µL/mL. These promising results pave the way for in vivo efficacy tests to confirm its use in more sustainable agricultural practices.

Chapter 15
Fungal Bioprospecting: Discovering Novel Bioactive Compounds 327
P. Selvakumar, Nehru Institute of Technology, India

Introduction to fungal bioprospecting opens doors to exploring the vast diversity of fungi for discovering novel bioactive compounds, enzymes, and biotechnological applications. Fungi represent a diverse group of eukaryotic microorganisms found in various ecological niches, from terrestrial soils and plant surfaces to extreme environments like deep-sea vents and polar regions. Their metabolic diversity and evolutionary adaptations enable fungi to produce an array of secondary metabolites, enzymes, and biocatalysts with unique chemical structures and biological activities. The field of fungal bioprospecting focuses on systematically screening fungal species and isolates for their potential to produce valuable compounds and bioproducts beneficial to medicine, agriculture, biotechnology, and environmental management. Fungal enzymes, particularly lignocellulolytic enzymes like cellulases, hemicellulases, and ligninases, are essential in bioremediation, biofuel production, and biomass conversion industries.

Chapter 16
Edible Mushrooms: A Sustainable Superfood for the Future 349
Vishal Shridhar Moger, St. Aloysius University, India
Harshitha Gujaran, St. Aloysius University, India
Shreya Gowda, St. Aloysius University, India
Pallavi Baliga, St. Aloysius University, India

Edible mushrooms are termed as superfoods owing to their exceptional nutritional profile. This chapter highlights the dietary benefits of edible mushrooms derived from their high content of protein, fiber, complex carbohydrates, vitamins, minerals, and antioxidants. It details the health benefits of bioactive compounds in mushrooms, which include immune modulation, roles in chronic disease management, and reducing inflammation. In addition, this chapter explores various ways in which mushrooms might be exploited in food processing because of their culinary versatility and functional properties. Focus is laid on the environmentally sustainable aspects of mushroom farming, which requires little water and land resources and promotes waste recycling. This chapter discusses the challenges associated with mushroom cultivation and future research trends to improve cultivation and nutritional enhancement. Overall, this chapter provides a comprehensive overview of edible mushrooms, highlighting their nutritional and health benefits, ecofriendly nature, and future directions in this field.

Chapter 17
Mycotoxins: Challenges and Strategies for Mitigation 377
Tarun Kumar Patel, Sant Guru Ghasidas Government P.G. College, India

This chapter provides a comprehensive overview of mycotoxins, focusing on their occurrence, impact, detection, and mitigation strategies. It examines the major mycotoxins of concern, including aflatoxins, ochratoxins, fumonisins, trichothecenes, zearalenone, and patulin, highlighting their sources, health risks, and economic consequences. Key factors influencing mycotoxin contamination, such as environmental conditions and agricultural practices, are discussed. The chapter also explores various detection and quantification methods, including advanced technologies for accurate and efficient monitoring. A review of global regulatory frameworks, including those of the Codex Alimentarius Commission, the European Union, and the U.S. FDA, outlines current safety standards. Mitigation strategies, from pre-harvest practices to post-harvest and processing interventions, are detailed, alongside emerging technologies such as genetic engineering and biocontrol. The chapter concludes with future directions and policy recommendations to enhance global mycotoxin management efforts.

Chapter 18
Hazards of Mycotoxin in Agriculture and Effective Mitigation Strategies 407
Suryakant Manik, ICAR-Indian Institute of Agricultural Biotechnology, India
S. Shreedevasena, Directorate of Medicinal and Aromatic Plant Research, India
Asharani Patel, ICAR-Indian Agricultural Research Institute, India
Sujit Kumar Bishi, ICAR-Indian Institute of Agricultural Biotechnology, India
Kuleshwar Prasad Sahu, College of Agriculture, Rani Lakshmi Bai Central Agricultural University, India
Licon Kumar Acharya, Division of Plant Pathology, ICAR-Indian Agriculture Research Institute, India, ICAR- National Research Centre for Integrated Pest Management, India

Mycotoxins, toxic metabolites produced by fungi, represent a significant global health concern due to their prevalence and high toxicity. This chapter explores the major mycotoxins including aflatoxins, ochratoxin A, fumonisins, deoxynivalenol, T-2 toxins, zearalenone, etc. Both acute and chronic exposure to these contaminants can result in severe health outcomes demanding immediate and sustained attention. The importance of rapid and reliable detection of mycotoxin contamination in food and animal feed is emphasized as a cornerstone of health risk mitigation. The chapter highlights the role of advanced agricultural practices, encompassing both pre- and post-harvest stages, and the proper storage of agricultural commodities. Furthermore, it underscores the integration of routine monitoring with innovative technologies like biotechnological, biological and nanotechnological approaches as essential strategies to reduce mycotoxin risks. Together, these measures provide a comprehensive framework for addressing the complex challenges posed by mycotoxins to global health and food security.

Compilation of References 439

About the Contributors 553

Index 559

Preface

INTRODUCTION

Agriculture and food production are intricately connected to the vast and diverse world of microorganisms. While some microbes pose significant challenges to crop yields, others act as nature's allies, helping plants withstand environmental stressors, including both abiotic and biotic pressures. In recent years, the role of fungi in sustainable agriculture and food systems has garnered increasing attention. These mycological innovations offer promising solutions to enhance plant resilience, improve soil health, and optimize food production.

This edited volume, *Mycological Inventions for Sustainable Agriculture and Food Production*, aims to provide a comprehensive exploration of the latest advancements in fungal research and their applications in agriculture and food systems. The book delves Beneficial effect of fungi—while highlighting their potential to revolutionize modern farming practices and provide sustainable option for food. The book also discuss about how fungal interactions are helpful in solving problem like impact of climate change in agriculture to food production in sustainable way.

Our objective is to bring together cutting-edge research and emerging applications of fungi in agriculture, shedding light on their functional roles in improving crop yield, disease resistance, and sustainable food production. This book offers a holistic perspective on fungal communities and their vast potential, providing insights into novel biotechnological approaches and future directions in mycological interventions.

The intended audience for this book includes students, researchers, and scientists working in the fields of microbiology, agriculture, and food sciences. Additionally, industry professionals seeking to develop fungal-based solutions for sustainable agriculture and food security will find valuable insights within these pages. By bridging the gap between fundamental research and practical applications, this book serves as a crucial resource for harnessing the power of fungi in transforming agri-food systems.

We extend our gratitude to all contributors whose expertise and dedication have made this compilation possible. It is our hope that this book will serve as a valuable reference, inspiring further research and innovation in mycological applications for a more resilient and sustainable future.

ORGANIZATION OF THE BOOK

Chapter 1: Role of Fungi in Sustainable Agriculture and Stress Mitigation

This chapter delves into the indispensable role of fungi in fostering sustainable agriculture and mitigating environmental stressors. It highlights the contributions of fungal species, particularly arbuscular mycorrhizal fungi (AMF) and fungal biopesticides, in enhancing soil health, increasing crop resilience, and facilitating carbon sequestration. The discussion extends to recent advancements in fungal biofertilizers that improve nutrient uptake and disease resistance, presenting fungi as a viable alternative to synthetic fertilizers and pesticides. While the potential of fungi in agriculture is promising, challenges such as inconsistent efficacy, large-scale production limitations, and regulatory concerns persist. Addressing these issues requires continued research, policy adjustments, and widespread education to optimize the integration of fungi in sustainable farming practices.

Chapter 2: A Sustainable Approach to Unlocking Plant Potential with Fungal Biostimulants

With the global population projected to reach 10 billion by 2050, agricultural systems must adapt to meet food production demands while minimizing environmental impact. This chapter explores the transformative role of fungal biostimulants in sustainable agriculture, emphasizing their ability to enhance plant growth, nutrient uptake, and resilience. A particular focus is placed on mycorrhizal, endophytic, and saprophytic fungi, with special attention to arbuscular mycorrhizal biostimulants that improve soil health and nutrient efficiency. The chapter further examines the economic and sustainability benefits of fungal biostimulants while addressing the regulatory and consistency challenges that currently hinder widespread adoption. Future research directions are explored, offering insight into optimizing fungal biostimulant applications for enhanced agricultural productivity.

Chapter 3: Fungal Biostimulants: Sustainable Tools to Boost Crop Yield Naturally

As environmental pollution, rising temperatures, and excessive chemical inputs threaten soil fertility and crop production, alternative agricultural solutions have become a priority. This chapter underscores the critical role of fungal biostimulants in improving plant resilience, fostering soil health, and mitigating environmental stressors. The diverse interactions between fungi and plants offer eco-friendly substitutes to chemical fertilizers, enhancing crop growth and natural defense mechanisms. By providing an in-depth analysis of fungal biostimulants' mechanisms of action, their role in plant stress tolerance, and their contributions to organic farming, this chapter presents fungal biostimulants as a promising tool for sustainable agriculture. Challenges in large-scale implementation and future research prospects are also discussed.

Chapter 4: Arbuscular Mycorrhizal Fungal Diversity Under Abiotic Stress

Arbuscular mycorrhizal fungi (AMF) serve as key allies in supporting plant growth under abiotic stress conditions such as drought, salinity, and extreme temperatures. This chapter explores how AMF enhance plant stress tolerance by improving water absorption, regulating osmotic balance, and optimizing nutrient uptake, particularly phosphorus. It also examines the diversity of AMF species, their varying responses to environmental stressors, and their influence on plant productivity. Further discussion highlights the impact of AMF on gene expression in host plants, facilitating stress-protective responses. The strategic use of AMF presents a sustainable agricultural approach, reducing reliance on chemical inputs while bolstering crop resilience in the face of climate change.

Chapter 5: Fungal Biocontrol Agents for Sustainable Pest Management

With increasing concerns over the environmental and health impacts of synthetic pesticides, fungal biocontrol agents have emerged as a promising alternative in sustainable pest management. This chapter explores the mechanisms by which fungi, including *Beauveria bassiana, Metarhizium anisopliae*, and *Trichoderma* spp., control insect pests, nematodes, and plant pathogens. These fungi operate through mycoparasitism, resource competition, and insecticidal compound production, providing an eco-friendly solution to pest control. Field studies demonstrate their efficacy in various crops, yet challenges such as field variability, limited shelf life, and regulatory constraints hinder their widespread adoption. The chapter underscores the need for continued research to refine formulations and improve delivery methods to facilitate broader application in agriculture.

Chapter 6: Enhancing Soil Fertility and Nutrient Cycling with Mycofertilizers

Maintaining soil fertility is fundamental to agricultural productivity, yet nutrient immobilization and soil degradation remain persistent challenges. This chapter discusses the role of mycofertilizers in addressing these issues by forming symbiotic relationships with plants, enhancing root expansion, and improving nutrient absorption. Mycofertilizers contribute to soil structure improvement, nutrient cycling, and pathogen resilience, offering an environmentally friendly alternative to chemical fertilizers. The chapter emphasizes the benefits of incorporating mycofertilizers into farming systems, highlighting their potential to reduce agricultural pollution and ensure long-term soil sustainability.

Chapter 7: Edible Mushrooms as Nutritious and Sustainable Superfoods

The rising demand for nutrient-dense, sustainable food sources has positioned edible mushrooms as a valuable superfood. This chapter explores the nutritional and medicinal properties of mushrooms, emphasizing their low-calorie, high-protein content and health benefits for conditions such as diabetes and cardiovascular diseases. Additionally, the minimal resource requirements for mushroom cultivation make them an accessible economic opportunity for marginalized communities, including unemployed youth and women. The chapter highlights the potential of biofortified mushroom products in diversifying income sources and promoting food security, advocating for greater integration of mushrooms into global diets.

Chapter 8: Fungi as a Source of Single-Cell Proteins: Current Progress and Future Prospects

Fungi have emerged as a promising solution for sustainable protein production through single-cell protein (SCP) applications. This chapter explores key fungal species, such as filamentous fungi and yeasts, that exhibit rapid growth and efficient substrate utilization for protein production. The discussion extends to bioprocess engineering aspects, including fermentation methods, bioreactor designs, and optimization strategies. The potential of fungal SCPs in food, feed, nutraceuticals, and sustainable packaging is examined, alongside challenges in large-scale production, economic feasibility, and consumer acceptance. Advances in genetic engineering and biorefinery integration offer pathways for overcoming these challenges and expanding SCP applications.

Chapter 9: Mycoproteins: The Rise of Fungal-Based Meat Alternatives

As global food systems shift toward sustainability, mycoproteins have emerged as a viable alternative to conventional meat production. This chapter examines the development, production processes, and nutritional benefits of mycoproteins, emphasizing their low environmental footprint compared to traditional livestock farming. Mycoproteins offer high-quality protein with essential amino acids and fiber while minimizing land and water use. The chapter also explores market trends, consumer acceptance, and the increasing role of fungal-based meat substitutes in mainstream diets. By addressing pressing challenges in sustainable food production, mycoproteins present an innovative solution to meet future protein demands.

Chapter 10: Arbuscular Mycorrhizal Fungi (AMFs): Metagenomics and Environmental Circadian Rhythms

Arbuscular mycorrhizal fungi (AMFs) play a pivotal role in plant-soil-microbiome interactions, influencing plant growth and ecosystem stability. This chapter investigates the use of metagenomics to study AMF diversity, ecological functions, and interactions within plant root systems. Additionally, it explores how AMF activity is regulated by environmental circadian rhythms, affecting symbiotic efficiency and nutrient exchange. Understanding these dynamics provides valuable insights into optimizing AMF applications in agriculture, with implications for soil health, crop productivity, and ecological sustainability. The chapter underscores the importance of advancing research in AMF metagenomics to harness their full potential in sustainable agriculture.

Chapter 11: Fungi for Ameliorating Soil and Water Pollution: Mycoremediation

This chapter highlights the adverse environmental impacts of human activities such as urbanization and industrialization. Conventional physicochemical treatment methods often fall short in mitigating pollution effectively. This chapter introduces mycoremediation as an innovative and eco-friendly alternative, leveraging fungal enzymatic activity and adsorption mechanisms to degrade pollutants and sequester heavy metals. The discussion encompasses various mycoremediation techniques, the specific enzymes involved, and their role in developing efficient and sustainable pollution treatment strategies.

Chapter 12: Fungal Biodegradation: Breaking Down Agricultural Wastes

This chapter examines the critical role of fungi in addressing environmental challenges posed by persistent pollutants. Given their metabolic versatility, fungi possess enzymatic systems capable of decomposing complex organic compounds such as pesticides, petroleum hydrocarbons, dyes, and pharmaceuticals. This chapter delves into the mechanisms of fungal biodegradation, emphasizing their advantages over bacterial counterparts and showcasing their potential applications in bioremediation.

Chapter 13: Mycoremediation: Leveraging Fungi for Ameliorating Soil and Water Pollution

This chapter further expands on mycoremediation as a viable solution to environmental contamination resulting from industrial and agricultural activities. The chapter explores key fungal species, their mechanisms of action, and real-world applications of mycoremediation strategies. Additionally, it addresses major challenges in implementing this approach and suggests future directions for optimizing fungal remediation techniques.

Chapter 14: Antifungal Properties of Essential Oils from Rosmarinus Officinalis, Citrus Aurantium, and Origanum Majorana: In Vitro Evaluation for Biological Control Agents

This chapter focuses on the increasing need for sustainable alternatives to chemical pesticides. This study evaluates the antifungal efficacy of essential oils against plant pathogens, revealing promising results, particularly for *Origanum majorana.* The findings pave the way for further in vivo research to integrate essential oils into environmentally friendly agricultural practices.

Chapter 15: Fungal Bioprospecting: Discovering Novel Bioactive Compounds

This chapter explores the untapped potential of fungi in biotechnology. Fungi are a prolific source of bioactive compounds, enzymes, and bioproducts with applications in medicine, agriculture, and environmental management. The chapter highlights systematic screening approaches for identifying novel fungal metabolites, emphasizing their significance in industries such as biofuel production and bioremediation.

Chapter 16: Edible Mushrooms: A Sustainable Superfood for the Future

This chapter examines the nutritional and ecological benefits of edible mushrooms. With their rich composition of proteins, fiber, vitamins, and antioxidants, mushrooms offer numerous health advantages, including immune modulation and chronic disease prevention. The chapter also discusses sustainable mushroom cultivation methods that promote waste recycling and resource efficiency, positioning mushrooms as a key component of future food security strategies.

Chapter 17: Mycotoxins: Challenges and Strategies for Mitigation

This chapter provides an in-depth analysis of the threats posed by mycotoxins, toxic fungal metabolites that contaminate food and feed. The chapter covers the health risks, economic impact, and environmental factors contributing to mycotoxin proliferation. Detection methods and mitigation strategies, including regulatory frameworks and emerging technologies, are thoroughly examined to enhance global food safety efforts.

Chapter 18: Hazards of Mycotoxin in Agriculture and Effective Mitigation Strategies

The last chapter continues the discussion on mycotoxins, emphasizing their significant implications for human and animal health. The chapter explores detection techniques, storage solutions, and advanced mitigation strategies, including biotechnological and nanotechnological approaches. These insights contribute to a comprehensive understanding of mycotoxin management, highlighting the need for innovative interventions in global agriculture and food security.

Together, these chapters illustrate the diverse applications of fungi, from environmental remediation and sustainable agriculture to biotechnology and food safety. By harnessing the natural potential of fungi, researchers and practitioners can develop innovative solutions to some of the most pressing environmental and health challenges of our time.

CONCLUSION

The chapters in this volume highlight the transformative role of fungi in addressing critical environmental and agricultural challenges. From mycoremediation and fungal biodegradation to the discovery of bioactive compounds and the mitigation of mycotoxin hazards, this book underscores the vast potential of fungi as natural allies in sustainability. Through in-depth discussions on fungal-based biotechnologies, we have explored how these microorganisms contribute to soil and water remediation, waste decomposition, biological control, and food security.

As we continue to face mounting ecological and health concerns, the research and innovations presented in this book offer promising solutions that align with global sustainability goals. The interdisciplinary nature of fungal applications—spanning environmental science, agriculture, biotechnology, and food safety—demonstrates the need for continued collaboration and exploration in this field. We hope this volume serves as both a reference and an inspiration for researchers, practitioners, and policymakers committed to leveraging the potential of fungi for a more sustainable future.

The study of fungi remains an ever-evolving field with new discoveries continuously reshaping our understanding of their capabilities. Moving forward, the integration of cutting-edge technologies such as genomics, synthetic biology, and nanotechnology with fungal research will further enhance their applications. By fostering innovation and knowledge-sharing, we can harness the full power of fungi to create more resilient ecosystems and sustainable agricultural practices.

Chapter 1
Role of Fungi in Sustainable Agriculture and Stress Mitigation

Lalit Kumar Sahu
https://orcid.org/0009-0005-0632-0747
Guru Ghasidas Vishwavidyalaya, India

Nikita Verma
https://orcid.org/0009-0002-0558-8379
Guru Ghasidas Vishwavidyalaya, India

ABSTRACT

This chapter explores the critical role of fungi in sustainable agriculture and stress mitigation, emphasizing their contributions to enhancing soil health, improving crop resilience, and facilitating carbon sequestration. Fungi, particularly arbuscular mycorrhizal fungi (AMF) and various fungal biopesticides, offer innovative solutions for reducing reliance on synthetic fertilizers and pesticides. Recent advancements in fungal biofertilizers demonstrate their ability to improve nutrient uptake and disease resistance, while their potential for carbon storage aids in climate change mitigation. Despite the promising applications of fungi, challenges such as inconsistent efficacy, mass production limitations, and regulatory barriers remain. Addressing these challenges presents opportunities for further research, policy reform, and education to optimize the use of fungi in agricultural systems. By harnessing the diverse functions of fungi, we can create more resilient and sustainable agricultural practices.

1. INTRODUCTION

Developing sustainable agricultural practices is becoming more and more important as the world's population rises and climate change gets worse. Customary agricultural practices, which mainly depend on artificial fertilizers and pesticides, not only cause soil health to decline but also greatly increase the risk to biodiversity and ecosystem resilience. Overcoming these obstacles, fungi have become indispensable

DOI: 10.4018/979-8-3693-8201-1.ch001

to sustainable agriculture, providing creative and environmentally acceptable means of improving soil quality, encouraging plant development, and reducing stress.

The diverse role of fungi is explored in this chapter, with special attention paid to fungal biopesticides and biofertilizers, which have shown promise as efficient substitutes for synthetic inputs. In addition to enhancing nutrient availability and disease resistance, these fungi also increase resistance to environmental stressors like salinity, drought, and extreme temperatures. Fungi can contribute to more ecologically friendly farming methods by reducing dependency on chemical inputs through their capacity to establish symbiotic relationships with plants.

Recent years have seen an increase in research into the potential role that fungi may play in mitigating climate change. By enhancing soil structure and storing carbon in organic forms that stay in the soil for extended periods of time, fungi are essential to the sequestration of carbon. Fungi offer a viable way to mitigate climate change and improve agricultural sustainability by capturing carbon and lowering greenhouse gas emissions.

This chapter will focus on the latest developments in fungal biopesticides and biofertilizers, the increasing role that fungi play in sequestering carbon dioxide, and the ways in which creative uses of fungi have the potential to revolutionize contemporary agriculture. Fungal ecology research and genetic developments mean that fungi will play a significant role in the development of climate-resilient farming techniques in the future.

2. ROLE OF FUNGI IN SUSTAINABLE AGRICULTURE

Because they play a variety of roles in boosting plant growth, strengthening soil health, and increasing resistance to environmental stressors, fungi are essential to sustainable agriculture. Fungi provide environmentally friendly substitutes for synthetic pesticides and fertilizers, which can damage non-target organisms and degrade soil, thereby promoting long-term agricultural productivity. The most prominent fungi are saprophytic fungi, endophytic fungi, and arbuscular mycorrhizal fungi (AMF), each of which makes a distinct contribution to plant health and ecosystem balance(Fall et al., 2022; Frac et al., 2018).

2.1 Soil Health and Fertility

The contribution of fungi to soil health is one of their primary roles in agriculture. As decomposers, fungi break down organic matter and replenish the soil with nutrients like carbon, phosphorus, and nitrogen. By breaking down complex organic materials like cellulose and lignin, saprophytic fungi in particular contribute significantly to the cycling of nutrients and the development of soil organic matter. Consequently, these fungi improve the fertility and structure of the soil, creating an environment that sustains robust plant development. In addition to improving soil aggregate stability by breaking down dead plant matter, fungi also increase water retention capacity and decrease erosion, two important aspects of sustainable farming(Fall et al., 2022; Frac et al., 2018).

2.2 Plant Growth Promotion

Fungi promote plant growth through their symbiotic relationship with plant roots. Arbuscular mycorrhizal fungi (AMF) are particularly effective in improving nutrient availability, particularly phosphorus, which is often a limiting factor in soil. AMF's extensive hyphal networks extend far beyond the reach of plant roots, increasing surface area for nutrient and water uptake. This relationship not only increases plant growth, but also improves overall plant resilience, reducing the need for chemical fertilizers(Fall et al., 2022). Fungal symbiosis has been shown to increase plant biomass and yield, especially under stressful conditions such as drought and poor soil fertility(Baron & Rigobelo, 2022a).

2.3 Stress Tolerance and Crop Resilience

Fungi also contribute to the stress tolerance of plants and help them survive under abiotic stressors such as drought, salinity and extreme temperatures. AMF and endophytic fungi, for example, have been shown to improve a plant's ability to absorb water and nutrients from deeper soil layers, even under dry conditions. This increased access to resources supports plant health during times of water scarcity or high salinity. In addition, fungi produce bioactive compounds such as antioxidants and osmoprotectants that protect plants from oxidative stress, further improving plant resilience. As a result, plants inoculated with fungi often have higher survival rates and higher productivity in harsh environments(Fall et al., 2022; Frac et al., 2018).

2.4 Reduction of Chemical Inputs

By improving the efficiency of nutrient use and providing natural defenses against pests and diseases, fungi help reduce dependence on synthetic agrochemicals. Biofertilizers and biopesticides from fungi offer a sustainable alternative to chemical fertilizers and pesticides and reduce the ecological footprint of agriculture. For example, fungal biopesticides from species such as Beauveria bassiana and Trichoderma can effectively control pests and pathogens, reducing the need for harmful chemical inputs. This change not only benefits the environment by preventing chemical runoff and soil contamination but also improves the sustainability of agricultural practices(Frac et al., 2018; Yarzábal Rodríguez et al., 2024).

Fungi are crucial allies in sustainable agriculture, to sum up. They provide natural, environmentally friendly answers to the problems caused by conventional farming practices by enhancing soil fertility, encouraging plant growth, and boosting resistance to environmental stresses. Fungi will remain essential in the future for lowering chemical inputs and promoting a more resilient agricultural ecosystem.

3. FUNGI AND STRESS MITIGATION IN AGRICULTURE

Fungi are invaluable allies in enhancing crop resilience to both abiotic and biotic stresses, which are major limiting factors in agricultural productivity. Abiotic stresses such as drought, salinity, and heat, along with biotic stresses from pests and pathogens, can significantly reduce crop yields. Fungi, particularly arbuscular mycorrhizal fungi (AMF), saprophytic fungi, and endophytic fungi, have shown remarkable potential in mitigating these stresses and improving plant health(Begum et al., 2019; Morales-Vargas et al., 2024).

3.1 Abiotic Stress Tolerance

One of the most urgent problems in agriculture, particularly in arid and semi-arid areas, is abiotic stress, such as drought. Fungi, particularly endophytic and AMF fungi, improve water absorption and retention in plants, increasing their tolerance to drought. By extending the root system through their hyphal networks, AMF enable plants to reach water from previously unreachable deeper soil layers(Baron & Rigobelo, 2022b; Yarzábal Rodríguez et al., 2024). Furthermore, abscisic acid (ABA), a plant hormone that controls stomatal closure and lowers water loss during drought stress, can be modulated by AMF(Baron & Rigobelo, 2022b; Morales-Vargas et al., 2024).

Similarly, in saline environments, fungi help plants cope by reducing the uptake of toxic sodium ions while promoting the absorption of essential nutrients such as potassium and phosphorus. Endophytic fungi produce osmoprotectants like proline, which stabilize cellular structures and maintain cell turgor under saline conditions(Morales-Vargas et al., 2024; Yarzábal Rodríguez et al., 2024). This mechanism protects plant cells from salt-induced dehydration and oxidative stress, making fungi critical for enhancing crop resilience in saline soils.

Fungi also play a role in mitigating heat stress. Endophytic fungi can induce the production of heat shock proteins (HSPs) in plants, which protect cellular functions during high-temperature stress. These proteins act as molecular chaperones, preventing the denaturation of essential enzymes and helping plants maintain their metabolic processes during heat waves. Additionally, fungi enhance the production of antioxidants, which neutralize reactive oxygen species (ROS) generated under extreme temperatures, further safeguarding plant health(Begum et al., 2019; Yarzábal Rodríguez et al., 2024).

Another emerging area is the use of extremophilic fungi, which are capable of thriving in harsh environments. These fungi can help crops tolerate extreme conditions such as high salinity or prolonged drought. Extremophilic fungi produce unique metabolites that assist in maintaining cellular integrity under adverse conditions, allowing plants to survive in environments that would otherwise be unfavorable . This opens new possibilities for expanding agriculture into marginal lands where conventional crops struggle(Ripa et al., 2019; Yarzábal Rodríguez et al., 2024).

3.2 Biotic Stress Resistance

Fungi are also effective in controlling plant diseases and pests, offering a natural alternative to chemical pesticides. Fungal biopesticides such as *Beauveria bassiana*, *Metarhizium anisopliae*, and *Trichoderma* species provide an eco-friendly solution to managing biotic stress in agriculture. These fungi suppress pathogens through mechanisms like mycoparasitism, where one fungus parasitizes another, and the production of antifungal compounds that inhibit the growth of harmful pathogens(Baron & Rigobelo, 2022b).

Trichoderma species, for instance, have been extensively studied for their ability to control soil-borne pathogens like *Fusarium* and *Rhizoctonia*. These fungi colonize the root zone, outcompeting harmful pathogens for nutrients and space, while also inducing the plant's innate immune responses. This biological control method reduces the need for synthetic pesticides, thereby lowering the environmental impact of farming and promoting healthier ecosystems(Begum et al., 2019).

In addition to disease control, fungi can also enhance the natural defenses of plants against insect pests. Entomopathogenic fungi such as *Beauveria bassiana* and *Metarhizium anisopliae* infect insect pests, ultimately leading to their death. These fungi act by penetrating the insect's cuticle and proliferating within its body, causing the insect to die from fungal infection. This form of biocontrol is especially useful in organic farming systems, where chemical pesticides are limited. Moreover, it minimizes the ecological damage associated with conventional pesticides, such as the harm done to pollinators and beneficial insects(Baron & Rigobelo, 2022b; Begum et al., 2019; Verma et al., 2022).

3.3 Case Studies in Stress Mitigation

Several real-world examples underscore the effectiveness of fungi in mitigating agricultural stresses. Research has shown that crops inoculated with AMF demonstrate improved drought tolerance, with increased water use efficiency and nutrient uptake under water-limited conditions. For instance, maize and wheat inoculated with AMF have shown higher yields and better survival rates during periods of drought.

In saline soils, fungal inoculation has proven effective in managing salt stress in crops like rice and tomatoes. Studies have demonstrated that fungi not only improve nutrient uptake in saline conditions but also promote the accumulation of osmoprotectants, enhancing plant resilience to salt-induced stress. In rice, AMF have been shown to improve growth and yield, even under high salinity, by enhancing phosphorus availability and regulating ionic balance in plant tissues(Begum et al., 2019).

In viticulture, endophytic fungi have been used to mitigate heat stress in grapevines, improving their tolerance to rising temperatures caused by climate change. Fungi-induced heat shock proteins and antioxidants help grapevines maintain fruit quality and yields, even under extreme heat. This is particularly valuable in regions where climate change is leading to more frequent and intense heat waves, threatening the viability of crops like grapes(Baron & Rigobelo, 2022b; Gowtham et al., 2024).

3.4 Reduction of Chemical Pesticides

By promoting natural biocontrol, fungi significantly reduce the reliance on chemical pesticides, which have detrimental effects on the environment and human health. Fungal biopesticides, such as those derived from *Beauveria bassiana* and *Metarhizium anisopliae*, are particularly effective against insect pests. These entomopathogenic fungi infect and kill insect pests without harming beneficial organisms, making them an ideal solution for integrated pest management (IPM) strategies.

Furthermore, fungi-induced systemic resistance in plants offers long-term protection against a variety of pathogens. By boosting the plant's immune system, fungi help crops defend against future attacks, reducing the need for repeated pesticide applications. This leads to healthier crops, improved yields, and a reduced environmental impact.

Table 1. Role of fungi in stress mitigation

Stress Type	Fungal Mechanism	Example Fungi	References
Drought	Enhances water uptake through hyphal networks, regulates ABA levels	*Glomus* spp. (AMF), *Piriformospora indica*	(Ji et al., 2024; Raeisi Vanani et al., 2024; Ruiz-Lozano et al., 1995)
Salinity	Produces osmoprotectants (proline), regulates ion uptake	*Trichoderma harzianum*	(Wei et al., 2023)
Heat Stress	Induces heat shock proteins (HSPs)	*Trichoderma harzianum* (HSP70)	(Xiao et al., 2023)
Pathogen Resistance	Competes with pathogens, produces antimicrobial compounds	*Trichoderma* spp.	(Wei et al., 2023; Xiao et al., 2023)

The ability of fungi to mitigate both abiotic and biotic stresses provides a natural, sustainable solution for modern agriculture. Their use not only increases crop resilience but also reduces the environmental footprint of farming, offering a viable path toward more sustainable agricultural systems(Morales-Vargas et al., 2024; Yarzábal Rodríguez et al., 2024).

4. TYPES OF FUNGI IN AGRICULTURE

Fungi play diverse roles in agricultural ecosystems, and their applications in sustainable farming are vast. Different types of fungi interact with plants and soil in unique ways, offering various benefits such as nutrient acquisition, disease suppression, and enhanced plant resilience. In this section, we will explore the main categories of fungi used in agriculture: symbiotic fungi, saprophytic fungi, and endophytic fungi. Each of these groups plays a critical role in supporting sustainable farming practices.

4.1 Symbiotic Fungi

Symbiotic fungi, particularly arbuscular mycorrhizal fungi (AMF), are integral to plant health and soil fertility. AMF form mutualistic relationships with plant roots, extending their hyphal networks into the soil and significantly increasing the plant's ability to acquire essential nutrients like phosphorus, nitrogen, and other micronutrients. This symbiosis is crucial in nutrient-poor soils where plants would otherwise struggle to access these nutrients. The extended root network provided by AMF enhances not only nutrient uptake but also water absorption, which is especially beneficial in drought-prone regions.

There are two primary types of mycorrhizal fungi: arbuscular mycorrhizal fungi (AMF) and ectomycorrhizal fungi (EMF). AMF are the most widespread and are associated with a majority of agricultural crops, including cereals, legumes, and vegetables. They penetrate plant root cells to form structures called arbuscules, which facilitate the exchange of nutrients between the plant and the fungus. In return for providing essential nutrients like phosphorus, AMF receive carbohydrates from the plant, creating a mutually beneficial relationship (Kyslynska et al., 2023).

Ectomycorrhizal fungi (EMF), on the other hand, form external sheaths around plant roots and are typically associated with trees and shrubs. EMF play a vital role in forest ecosystems, where they assist in nutrient cycling and improve soil structure. Although EMF are less commonly used in conventional agriculture, their importance in agroforestry and tree-based systems highlights their potential for supporting sustainable land management.

AMF not only enhance nutrient acquisition but also contribute to improved soil health. Their hyphal networks improve soil structure by promoting soil aggregation, which increases water retention and reduces erosion. This makes AMF critical for maintaining soil fertility in the long term. Moreover, by reducing the need for chemical fertilizers, AMF help minimize environmental impacts, making them key players in sustainable agriculture.

4.2 Saprophytic Fungi

Saprophytic fungi are nature's decomposers, breaking down dead organic matter such as plant residues, animal remains, and organic waste. This decomposition process is vital for recycling nutrients within ecosystems, returning essential elements like nitrogen, phosphorus, and carbon to the soil. By doing so, saprophytic fungi support the nutrient cycles that sustain plant growth and maintain ecosystem balance.

One of the most notable genera of saprophytic fungi is *Trichoderma*, which not only excels at decomposing organic material but also promotes plant growth through its interactions with roots. *Trichoderma* is widely used in sustainable agriculture for its dual role as a decomposer and a growth promoter. It is known to improve seedling vigor, increase root biomass, and enhance nutrient uptake, making it an essential tool for organic farming systems(Philip et al., 2024).

Another important genus is *Aspergillus*, which contributes significantly to nutrient cycling, particularly by breaking down complex organic compounds like lignin and cellulose. This process improves soil fertility by converting organic matter into simpler forms that are more accessible to plants. By accelerating the decomposition process, saprophytic fungi play a crucial role in building and maintaining soil organic matter, which is vital for healthy, productive soils(Hu et al., 2021; Rodriguez & Redman, 1997).

Saprophytic fungi can also be utilized in composting, where they help break down agricultural waste into nutrient-rich compost. This practice reduces the need for synthetic fertilizers and supports a closed-loop agricultural system. By recycling nutrients through composting, farmers can reduce waste accumulation while maintaining soil fertility, contributing to more sustainable farming practices(Reverchon et al., 2010).

4.3 Endophytic Fungi

Endophytic fungi reside within plant tissues without causing harm to their host, forming mutualistic relationships that offer numerous benefits. These fungi enhance plant tolerance to a variety of biotic and abiotic stresses, including pests, diseases, drought, and salinity. Through the production of bioactive compounds such as phytohormones, antibiotics, and antioxidants, endophytic fungi bolster the plant's ability to resist pathogens and thrive under challenging environmental conditions.

One of the key mechanisms by which endophytic fungi support plant health is through improved nutrient acquisition. These fungi promote better root development and nutrient solubilization, which leads to healthier plants with increased productivity. In particular, endophytic fungi produce phytohormones like indole-3-acetic acid (IAA), a form of auxin that stimulates root growth, enabling plants to access more nutrients and water from the soil. This is especially crucial under stress conditions such as drought or nutrient deficiency(Sharma et al., 2023).

In addition to improving nutrient uptake, endophytic fungi help plants tolerate environmental stress through the production of protective compounds. For example, they produce antioxidants that mitigate the harmful effects of oxidative stress caused by drought, salinity, and temperature extremes. This enhanced stress tolerance enables plants to maintain growth and productivity under adverse conditions, making endophytic fungi an integral component of sustainable agriculture.

Endophytic fungi also contribute to plant defense by inducing systemic resistance, priming the plant to respond more effectively to future pathogen attacks or environmental stressors. This induced resistance allows plants to activate their defense mechanisms more rapidly and efficiently, providing protection against a wide range of threats. *Piriformospora indica* is one notable example of an endophytic fungus that has been shown to improve plant growth and stress tolerance across various crops, making it a valuable tool for promoting agricultural productivity(Varma et al., 2012).

4.4 Synergy Between Fungal Types

While each type of fungus has distinct roles, the combination of symbiotic, saprophytic, and endophytic fungi can lead to synergistic effects that further enhance plant growth and resilience. For instance, using AMF to improve nutrient uptake alongside saprophytic fungi for soil fertility management can create a robust, self-sustaining system. Endophytic fungi, with their stress-mitigating properties, complement this synergy by enhancing plant tolerance to environmental extremes(El-Metwally et al., 2023).

Table 2. Comparison of fungal types in sustainable agriculture

Type of Fungi	Function	Example Species	Key Benefits	References
Symbiotic Fungi	Form mutualistic associations with plant roots	*Arbuscular Mycorrhizal Fungi (AMF), Ectomycorrhizal Fungi*	Improve nutrient uptake, drought resistance, and soil structure	(Amaranthus, 1998; Begum et al., 2019)
Saprophytic Fungi	Decompose organic matter in soil	*Trichoderma spp., Aspergillus spp.*	Enhance soil fertility, increase organic matter content	(Nayak & Mishra, 2020; Yao et al., 2023a)
Endophytic Fungi	Live inside plant tissues without causing harm	*Piriformospora indica, Fusarium spp.*	Improve stress tolerance, enhance growth, and provide biocontrol	(Aleahmad & Ebrahimi*, 2023; Li et al., 2023)
Entomopathogenic Fungi	Act as natural biopesticides by infecting insect pests	*Beauveria bassiana, Metarhizium anisopliae*	Reduce pest populations naturally without harming beneficial organisms	(Bamisile et al., 2021; Bihal et al., 2023)

In future applications, fungal consortia—combinations of different fungal species—are being explored to provide more comprehensive benefits to crops. These consortia offer multiple advantages by addressing various aspects of plant health, from nutrient acquisition to pest resistance and stress tolerance. Such combinations represent an exciting frontier in sustainable agriculture, where the synergistic effects of multiple fungal types can be leveraged to optimize plant performance in diverse environments(Rodriguez & Redman, 1997).

5. RECENT ADVANCES IN FUNGAL BIOFERTILIZERS AND BIOPESTICIDES

Recent innovations in fungal biofertilizers and biopesticides have demonstrated their potential to revolutionize sustainable agriculture by improving soil health, boosting crop productivity, and reducing reliance on synthetic agrochemicals. Fungal biofertilizers enhance nutrient availability, while fungal biopesticides offer eco-friendly solutions for pest and pathogen control. As research advances, new formulations and technologies are making these fungal applications more efficient, scalable, and adaptable to diverse agricultural environments.

5.1 Fungal Biofertilizers

Fungal biofertilizers, especially those involving arbuscular mycorrhizal fungi (AMF), have seen significant advancements. These biofertilizers improve plant nutrient uptake by extending the root system through their extensive hyphal networks, which increases the absorption of phosphorus, nitrogen, and other essential nutrients. One of the most notable innovations in this area is the development of multifunctional biofertilizers that combine AMF with beneficial bacteria, creating synergistic effects that boost both nutrient acquisition and disease resistance.

Table 3. Benefits of fungal biofertilizers over chemical fertilizers

Parameter	Fungal Biofertilizers	Chemical Fertilizers	References
Nutrient Absorption	Improves efficiency through symbiotic nutrient exchange (N, P, K, Ca, Zn) via hyphal networks	Rapid availability but ~30-50% lost via leaching/volatilization	(Begum et al., 2019; Lehmann & Schroth, 2002; Y. Wang et al., 2023)
Soil Health	Enhances microbial diversity (+73-87 beneficial OTUs) and organic matter retention	Depletes organic matter (-40% over 10 yrs) and disrupts microbial communities	(Y. Wang et al., 2023)
Environmental Impact	Reduces GHG emissions (CO_2 fixation + 10-50x photosynthetic efficiency in microalgae)	Causes eutrophication (+300% N/P runoff) and soil acidification (pH ↓0.5-1.2)	(H. Wang et al., 2022)
Cost-Effectiveness	Long-term yield stability (+20-35%) with reduced input dependency	Requires 3-5x higher annual inputs to maintain yields	(Y. Wang et al., 2023)
Plant Growth	Stimulates root development (+200% hyphal density) and systemic stress resistance	Boosts short-term growth but increases heavy metal uptake (Zn/Cu +150%)	(Begum et al., 2019; Y. Wang et al., 2023)

AMF-based biofertilizers have been shown to enhance phosphorus availability, particularly in phosphorus-deficient soils, which is a critical factor for improving crop yields. New AMF formulations have been designed to improve their effectiveness in a variety of soil types, from nutrient-poor soils to those with high salinity. Moreover, these biofertilizers not only enhance plant productivity but also improve soil structure by increasing soil aggregation, which helps retain moisture and reduce erosion(Begum et al., 2019; Ebbisa, 2023).

Trichoderma species are another important group of fungi used in biofertilizers. These fungi are known for their ability to promote plant growth and suppress soil-borne pathogens. Commercial products like Promote® and RootShield® harness the beneficial effects of *Trichoderma* to improve root health, increase nutrient uptake, and protect plants from diseases (J. Kumar et al., 2021; Philip et al., 2024; Yao et al., 2023b). These biofertilizers are particularly effective in organic farming systems, where chemical inputs are minimized.

5.2 Fungal Biopesticides

Fungal biopesticides are gaining popularity as effective tools for managing agricultural pests and diseases while reducing the environmental and health risks associated with chemical pesticides. Biopesticides derived from fungi like *Beauveria bassiana*, *Metarhizium anisopliae*, and *Trichoderma* have shown remarkable efficacy in controlling a range of pests and pathogens.

One of the most notable fungal biopesticides is *Beauveria bassiana*, an entomopathogenic fungus that infects insect pests by penetrating their cuticle and proliferating inside the insect, eventually leading to its death. This fungus is effective against a wide range of insect pests, including aphids, whiteflies, and beetles. Similarly, *Metarhizium anisopliae* targets various insect pests, providing a natural and sustainable alternative to chemical insecticides. These fungi not only help control pest populations but also contribute to maintaining the ecological balance by minimizing harm to beneficial organisms, such as pollinators and predatory insects.

Fungal biopesticides like *Trichoderma* species are also highly effective against soil-borne pathogens. These fungi act through multiple mechanisms, including competition for space and nutrients, production of antifungal metabolites, and mycoparasitism, where the biocontrol fungus parasitizes the pathogenic fungus. *Trichoderma* species have been successfully used to control diseases caused by *Fusarium*, *Rhizoctonia*, and *Pythium*, which are major threats to crop production.

Recent advances in fungal biopesticides focus on improving their formulation and delivery methods. Nano-encapsulation techniques, for example, have been developed to protect fungal spores from environmental stresses such as UV radiation and desiccation, thus improving the shelf life and effectiveness of these products. This innovation enhances the viability of biopesticides during storage and application, making them more reliable for farmers in diverse climates(Tadesse Mawcha et al., 2024).

5.3 Emerging Technologies and Bioformulations

The future of fungal biofertilizers and biopesticides lies in the development of bioformulations that combine multiple microbial species, including fungi and bacteria, to provide a broader range of benefits to crops. These consortia leverage the strengths of each microorganism to enhance nutrient availability, improve disease resistance, and promote overall plant health. For example, combining AMF with nitrogen-fixing bacteria or phosphate-solubilizing bacteria can further optimize nutrient uptake and increase crop yields(Alam et al., 2021).

Additionally, gene editing technologies like CRISPR are opening new possibilities for enhancing the beneficial traits of fungi. By modifying specific genes, researchers are working to develop fungal strains that are more effective at solubilizing phosphorus, fixing nitrogen, or producing antifungal compounds. These advancements could lead to "designer fungi" that are tailored to meet the specific needs of different crops and environments(Nødvig et al., 2015; D. Wang et al., 2023).

The integration of fungal biofertilizers and biopesticides with precision agriculture technologies is another exciting development. Remote sensing tools, drones, and soil sensors can be used to monitor fungal activity and soil health in real-time, enabling farmers to apply fungal inoculants more precisely. This targeted approach improves the efficiency of fungal applications and ensures that crops receive the maximum benefits from these bioformulations.

5.4 Sustainability and Environmental Impact

One of the most significant advantages of fungal biofertilizers and biopesticides is their positive environmental impact. Unlike synthetic fertilizers and pesticides, which can lead to soil degradation, water pollution, and biodiversity loss, fungal-based products are biodegradable and do not accumulate harmful residues in the environment. By reducing the need for chemical inputs, fungal biofertilizers and biopesticides help preserve soil health, protect water resources, and promote biodiversity in agricultural ecosystems(A. Kumar et al., 2024).

Moreover, the use of fungi in agriculture supports sustainable farming practices by enhancing the resilience of crops to environmental stresses, reducing greenhouse gas emissions associated with chemical fertilizers, and improving the overall sustainability of farming systems. As global agriculture faces the dual challenges of feeding a growing population and mitigating climate change, fungi offer a natural and sustainable solution that aligns with the goals of environmental stewardship and food security(A. Kumar et al., 2024; Tadesse Mawcha et al., 2024).

6. INNOVATIVE APPLICATIONS OF FUNGI FOR CLIMATE CHANGE MITIGATION

Fungi are increasingly recognized for their crucial role in climate change mitigation through mechanisms like carbon sequestration, greenhouse gas reduction, and promoting plant resilience to environmental stresses. As the agricultural sector faces growing pressures from climate change, fungi offer innovative and sustainable solutions to both reduce the impact of farming on the environment and improve the resilience of crops to climate-induced stressors.

6.1 Fungi and Carbon Sequestration

One of the most significant contributions fungi make to climate change mitigation is their role in carbon sequestration. Fungi, particularly arbuscular mycorrhizal fungi (AMF), enhance soil carbon storage by promoting the formation of soil aggregates and increasing soil organic matter content. AMF facilitate the transfer of carbon from plants to soil, where it can be stored as stable organic matter for extended periods, reducing atmospheric carbon dioxide levels.

A key component of this process is the production of glomalin, a sticky glycoprotein secreted by AMF. Glomalin acts as a binding agent for soil particles, contributing to the formation of stable soil aggregates that protect organic carbon from microbial decomposition. Soils rich in glomalin are more effective at sequestering carbon, thus playing a vital role in mitigating climate change. Research indicates that up to 30% of the carbon stored in soils can be attributed to the presence of mycorrhizal fungi, underscoring their importance in maintaining soil carbon pools(Rillig et al., 2001).

In addition to AMF, saprophytic fungi also play a critical role in carbon cycling. These fungi break down dead plant material, converting it into simpler compounds that can be stored as organic matter in the soil. By decomposing organic material, saprophytic fungi help lock carbon into soil reservoirs, preventing it from being released back into the atmosphere as carbon dioxide. This not only contributes to carbon storage but also enhances soil fertility, creating a positive feedback loop that supports sustainable agriculture(Anthony et al., 2024; Rillig et al., 2001; Yang et al., 2022).

6.2 Reduction of Greenhouse Gas Emissions

Fungi can also help reduce greenhouse gas emissions from agricultural systems by improving nutrient use efficiency and reducing the need for synthetic fertilizers. Excessive use of nitrogen-based fertilizers is a major source of nitrous oxide (N_2O), a potent greenhouse gas. Fungal biofertilizers, particularly those that enhance nitrogen fixation and nutrient uptake, can reduce the need for these chemical inputs, thus lowering N_2O emissions.

By increasing the efficiency of nitrogen and phosphorus uptake, AMF and other fungal biofertilizers minimize the amount of fertilizer needed to sustain crop yields. This not only reduces the carbon footprint of fertilizer production but also prevents the runoff of excess nutrients into water bodies, which can lead to harmful algal blooms and the release of additional greenhouse gases. Moreover, fungal inoculants improve soil structure and promote water retention, reducing the need for irrigation and lowering the energy costs associated with pumping water for agricultural use.

Some fungi, such as extremophilic species, have demonstrated potential in mitigating methane (CH_4) emissions, especially in paddy fields where anaerobic conditions promote the production of methane. These fungi enhance oxygen availability in the soil, reducing the activity of methane-producing microbes and thereby lowering CH_4 emissions. This capacity of fungi to alter soil microbial communities and reduce methane emissions highlights their potential in addressing agricultural sources of greenhouse gases(Kalita et al., 2024).

6.3 Enhancing Crop Resilience to Climate-Induced Stresses

As climate change accelerates, agricultural systems are increasingly vulnerable to extreme weather events, such as droughts, floods, and heat waves. Fungi, particularly AMF and endophytic fungi, are emerging as key players in enhancing crop resilience to these climate-induced stresses. By improving water uptake, nutrient acquisition, and stress tolerance, fungi can help crops survive and thrive in increasingly unpredictable environments(Begum et al., 2019; Sharma et al., 2023; Verma et al., 2022).

In drought-prone regions, fungi enhance water retention and improve root architecture, allowing plants to access water from deeper soil layers. This makes crops more resilient to periods of water scarcity. Research has shown that crops inoculated with AMF demonstrate higher drought tolerance and yield compared to non-inoculated crops, even under severe water stress. This ability to increase water-use efficiency is particularly important as climate change exacerbates drought conditions in many agricultural regions.

Fungi also play a critical role in helping plants cope with salinity, which is becoming a more prevalent issue due to rising sea levels and soil degradation. AMF and endophytic fungi regulate the uptake of ions, such as sodium and potassium, helping plants maintain ionic balance in saline soils. This reduces the toxic effects of salt stress, allowing crops to continue growing in otherwise unfavorable environments. The use of fungal inoculants in saline soils has shown promise in improving the growth and yield of crops like rice, wheat, and tomatoes(Ripa et al., 2019).

In addition to drought and salinity, fungi enhance plant tolerance to heat stress, which is becoming more frequent with global temperature rises. Endophytic fungi help plants maintain cellular integrity under high temperatures by producing heat shock proteins and antioxidants. These compounds protect plant tissues from oxidative damage and prevent protein denaturation, allowing plants to survive and maintain productivity during heat waves. In regions where crops like grapes and wheat are vulnerable to heat-induced yield losses, fungal inoculants have been shown to mitigate the effects of extreme temperatures, ensuring food security in a changing climate.

6.4 Future Innovations: Fungi in Climate-Resilient Agriculture

Looking ahead, the integration of fungi into climate-resilient agricultural practices offers promising avenues for mitigating the effects of climate change while ensuring agricultural sustainability. One area of innovation is the development of bioformulations that combine fungi with other beneficial microorganisms, such as bacteria and archaea, to create synergistic effects that enhance crop resilience to multiple stressors. These formulations are designed to optimize nutrient uptake, improve soil structure, and boost plant immunity, providing a comprehensive solution to the challenges posed by climate change.

Advances in gene editing technologies, such as CRISPR, are also opening new possibilities for enhancing the capabilities of fungi. Researchers are exploring how to engineer fungi to improve their stress tolerance, nutrient acquisition, and carbon sequestration abilities. For instance, gene-edited fungi could be designed to produce higher levels of glomalin or more efficiently fix nitrogen, making them even more effective in mitigating the impacts of climate change(Nødvig et al., 2015; D. Wang et al., 2023; Yang et al., 2022).

The use of precision agriculture technologies to monitor fungal activity in real-time is another exciting development. Drones, soil sensors, and remote sensing tools can be used to assess soil health and fungal colonization, allowing farmers to apply fungal biofertilizers and biopesticides with greater precision(Guebsi et al., 2024; Sishodia et al., 2020). This targeted approach ensures that fungal inoculants are used where they are most needed, maximizing their effectiveness and minimizing waste.

By integrating fungi into sustainable farming practices, we can build more resilient agricultural systems that are better equipped to withstand the challenges of climate change. Fungi offer a natural and scalable solution to reducing greenhouse gas emissions, enhancing carbon sequestration, and improving crop resilience, making them indispensable tools in the global effort to mitigate climate change(Getahun et al., 2024; Traversari et al., 2021).

7. CHALLENGES AND OPPORTUNITIES IN FUNGAL APPLICATIONS

Despite the promising role of fungi in sustainable agriculture and climate change mitigation, several challenges hinder their widespread adoption. These challenges span from environmental variability and production issues to regulatory barriers and economic constraints. However, these hurdles also present opportunities for research advancements, technological innovations, and policy reforms that could unlock the full potential of fungal applications in agriculture.

7.1 Inconsistent Efficacy in Field Conditions

One of the primary challenges with fungal biofertilizers and biopesticides is their inconsistent efficacy under varying field conditions. The performance of fungal inoculants can vary significantly depending on factors such as soil type, climate, moisture levels, and temperature. For instance, while some AMF strains may thrive in specific soil types, they may struggle in others, leading to reduced effectiveness(Yadav &

Yadav, 2024). Additionally, the seasonal variations in temperature and humidity can affect the survival and activity of fungal spores in the soil, making their application unreliable in certain environments.

To address this challenge, researchers are working to develop more resilient fungal strains that can perform consistently across diverse environmental conditions. Strain improvement through selective breeding or genetic modification holds promise for creating fungi that can tolerate a broader range of soil types and climatic conditions. Field trials in different agroecological zones are also essential to identify the best-performing fungal species for specific regions, ensuring that farmers can use fungal products with confidence in their efficacy(Pavan et al., 2024; Tadesse Mawcha et al., 2025).

7.2 Mass Production and Formulation Challenges

Producing fungal biofertilizers and biopesticides at a commercial scale presents logistical and financial hurdles. The mass production of fungal spores requires specialized facilities and careful control of environmental conditions to maintain spore viability. Furthermore, during storage and transportation, fungal spores are vulnerable to desiccation, UV radiation, and temperature fluctuations, which can reduce their shelf life and effectiveness upon application(Fadiji et al., 2024; Tadesse Mawcha et al., 2025).

Recent advances in formulation technologies are addressing some of these challenges. Nano-encapsulation, for example, protects fungal spores from environmental stresses, improving their durability and extending shelf life. These encapsulation techniques involve coating the spores with protective materials that shield them from adverse conditions while ensuring their viability once applied to the field(Fadiji et al., 2024). Such innovations not only make fungal products more reliable but also reduce the need for frequent reapplications, lowering costs for farmers.

There is also ongoing research into liquid formulations that allow for more even distribution of fungal inoculants across the soil. These formulations offer greater flexibility in application methods, making it easier to integrate fungi into conventional and organic farming systems. Moreover, slow-release formulations that provide a steady supply of fungal inoculants over time are being developed to ensure long-lasting effects, further enhancing the practical utility of fungal biofertilizers and biopesticides(Fadiji et al., 2024).

7.3 Regulatory Barriers

The regulatory landscape for fungal biofertilizers and biopesticides can be complex and often acts as a bottleneck for their commercialization. In many regions, the approval process for microbial products is time-consuming and expensive, as it involves rigorous testing to ensure the safety and efficacy of the product. This can delay the introduction of new fungal-based products to the market and dissuade companies from investing in the development of fungal technologies(D. Kumar et al., 2023).

To overcome this barrier, there is a need for more streamlined regulatory processes that take into account the unique nature of biological products. Developing clear guidelines for the registration of fungal biofertilizers and biopesticides, along with support for fast-tracking approvals, could encourage more companies to invest in fungal innovations. Additionally, regulatory frameworks should be aligned with sustainability goals, recognizing the environmental benefits of fungi compared to synthetic chemicals.

International harmonization of regulations could also play a critical role in promoting the global adoption of fungal technologies. By creating standardized approval processes across countries, it would be easier to market fungal products internationally, reducing barriers to entry and fostering wider usage of these sustainable agricultural solutions(D. Kumar et al., 2023).

7.4 Economic Constraints for Smallholder Farmers

For smallholder farmers, the upfront costs associated with adopting fungal technologies can be a significant barrier. While fungal biofertilizers and biopesticides offer long-term benefits such as improved yields and reduced dependency on chemical inputs, the initial investment required to purchase these products may be prohibitive for farmers with limited financial resources. Additionally, smallholder farmers may lack access to the technical knowledge required to apply fungal inoculants effectively.

To address these economic constraints, there is an opportunity for governments and development organizations to provide financial incentives, such as subsidies or low-interest loans, to help smallholder farmers adopt fungal technologies. These incentives could be coupled with education and training programs that teach farmers how to integrate fungi into their farming practices, ensuring that they can maximize the benefits of fungal applications.

Furthermore, the development of cost-effective, locally produced fungal inoculants tailored to specific regional needs could reduce the price of these products, making them more accessible to smallholder farmers. Local production would also eliminate the high costs associated with importing microbial products, providing a sustainable supply of fungal biofertilizers and biopesticides to farmers in developing regions(Bisht et al., 2020; Yuan & Sun, 2024).

7.5 Knowledge Gaps and Research Needs

There are still significant gaps in our understanding of how fungi interact with plants and other soil microorganisms under different agricultural conditions. While research has established the beneficial effects of fungi in improving plant growth and stress tolerance, much remains unknown about the specific mechanisms behind these interactions. For instance, how do environmental stressors like drought or high salinity affect fungal colonization and efficacy? And what are the optimal conditions for fostering beneficial fungal communities in the soil?

Addressing these knowledge gaps requires interdisciplinary research that integrates plant biology, mycology, soil science, and agronomy. Collaborative efforts between academic institutions, agricultural extension services, and farmers can help generate the data needed to optimize fungal applications for various crops and environments(Kour et al., 2023a; Yarzábal & Chica, 2021). Such research will be crucial for developing tailored fungal solutions that are adaptable to local agricultural conditions.

In addition, more research is needed on the interactions between fungi and other microbial species. Understanding the synergies or antagonisms between fungi and bacteria, for example, could lead to the development of more effective microbial consortia for improving soil health and crop performance. Advances in metagenomics and microbiome studies will play a key role in uncovering these interactions, potentially leading to breakthroughs in fungal applications(Kour et al., 2023a).

7.6 Opportunities for Technological Innovations

While there are challenges, the opportunities for technological innovation in fungal applications are vast. As research into fungal ecology, genetics, and microbial interactions progresses, new techniques such as gene editing could be used to enhance the beneficial traits of fungi. For example, CRISPR-Cas9 technology could be employed to develop fungal strains with improved stress tolerance, nutrient uptake capabilities, or biocontrol properties(Kour et al., 2023a).

Precision agriculture tools, including drones, sensors, and satellite imagery, offer another exciting opportunity to optimize the use of fungi in agriculture. These technologies can provide real-time data on soil health, fungal activity, and plant stress, enabling farmers to apply fungal inoculants more efficiently. By ensuring that fungal products are applied only when and where they are needed, precision agriculture can help reduce input costs while maximizing the benefits of fungal biofertilizers and biopesticides(G-etahun et al., 2024; Guebsi et al., 2024; Traversari et al., 2021).

8. CONCLUSION: FUTURE DIRECTIONS FOR FUNGAL APPLICATIONS IN AGRICULTURE

Fungi are emerging as vital components of sustainable agriculture, offering a wide array of benefits ranging from improved soil health and nutrient cycling to enhanced plant resilience and biocontrol of pests. As global agriculture faces the dual challenges of climate change and the need to produce more food for a growing population, fungi provide natural, scalable, and eco-friendly solutions to improve crop productivity while reducing the environmental footprint of farming.

8.1 Integration with Advanced Agricultural Technologies

One of the most exciting future directions for fungal applications in agriculture is their integration with advanced technologies such as precision agriculture and gene editing. The use of **precision agriculture tools**—including drones, satellite imagery, and soil sensors—can help monitor fungal activity, soil conditions, and crop health in real time, enabling more targeted and efficient use of fungal biofertilizers and biopesticides(Vassilev & Mendes, 2024). This approach will help farmers optimize their input usage, reduce waste, and maximize the benefits of fungal inoculants.

Additionally, **gene editing technologies** like CRISPR-Cas9 present new possibilities for enhancing the performance of fungal species used in agriculture. By manipulating specific genes, researchers can develop fungal strains with improved traits such as enhanced stress tolerance, faster colonization of plant roots, or the production of higher levels of beneficial metabolites(Kour et al., 2023b). These "designer fungi" could be tailored to specific crops and environmental conditions, offering highly customized solutions for improving crop resilience and productivity.

Advances in **bioformulations** will also play a critical role in expanding the use of fungi in agriculture. Multifunctional bioformulations that combine fungi with other beneficial microbes—such as nitrogen-fixing bacteria or phosphate-solubilizing bacteria—will offer broader benefits to crops, improving both nutrient uptake and disease resistance(Vassilev & Mendes, 2024). These combinations of microbial species, often referred to as microbial consortia, are likely to be a cornerstone of future biofertilizer and biopesticide products.

8.2 Fungi and Climate Change Mitigation

As the impacts of climate change become more severe, fungi will play an increasingly important role in helping agriculture adapt to these challenges. Fungi, particularly arbuscular mycorrhizal fungi (AMF), can help sequester carbon in soils, contributing to global efforts to mitigate climate change. By enhancing soil organic matter and stabilizing carbon in long-lasting forms, fungi offer a natural way to capture and store carbon, thus helping to offset agricultural greenhouse gas emissions.

In addition to carbon sequestration, fungi can help reduce the need for chemical fertilizers, which are a significant source of nitrous oxide emissions—a potent greenhouse gas(Mwampashi et al., 2024b). By improving nutrient use efficiency, fungal biofertilizers reduce the amount of nitrogen and phosphorus required to sustain crop yields, minimizing both the environmental and economic costs associated with fertilizer use. Fungi can also contribute to reducing methane emissions in agricultural systems, particularly in paddy fields, by promoting microbial communities that suppress methane-producing organisms.

As climate change increases the frequency of extreme weather events, such as droughts and heat waves, fungi will also be critical in enhancing the resilience of crops to these stressors. Fungal inoculants, particularly endophytic fungi, have already demonstrated their ability to improve plant tolerance to abiotic stresses such as drought, salinity, and high temperatures. As research into extremophilic fungi—those that thrive in harsh environments—advances, new fungal species may be discovered that can further enhance crop resilience in areas affected by extreme climatic conditions.

8.3 Overcoming Barriers to Adoption

While the potential of fungi in agriculture is clear, significant barriers to their widespread adoption remain. Economic constraints, regulatory hurdles, and limited knowledge among farmers are some of the key challenges that need to be addressed. Moving forward, targeted efforts to educate farmers about the benefits of fungal biofertilizers and biopesticides will be essential. This can be achieved through agricultural extension services, demonstration projects, and training programs that show farmers how to integrate fungi into their farming practices effectively(Kour et al., 2023b).

Policymakers can also play a role by providing financial incentives, such as subsidies, to support the adoption of fungal technologies, particularly for smallholder farmers who may face financial barriers(Chaurasia & Bharati, 2021). Additionally, streamlining the regulatory approval process for fungal products will encourage more companies to invest in the development and commercialization of new fungal-based solutions. Aligning regulatory frameworks with sustainability goals—recognizing the environmental benefits of fungi—can further facilitate the integration of fungi into agricultural systems.

8.4 Research and Innovation

Continued research into fungal biology, ecology, and genetics will be critical for advancing the next generation of fungal biofertilizers and biopesticides. Exploring the diversity of fungal species in extreme environments, for instance, may lead to the discovery of new strains with unique abilities to enhance plant growth or withstand stress(Yarzábal Rodríguez et al., 2024). Research into microbial interactions—both between fungi and plants and between fungi and other microbes—will also provide valuable insights that can inform the development of more effective fungal inoculants(Vassilev & Mendes, 2024).

In addition to discovering new fungal species, biotechnological innovations, such as gene editing and synthetic biology, offer the potential to enhance the capabilities of existing fungal strains. These technologies could enable the creation of custom-engineered fungi that perform optimally under specific environmental conditions, offering highly targeted solutions for farmers(Kour et al., 2023b). For example, gene-edited fungi could be designed to produce higher levels of glomalin, a glycoprotein that contributes to soil carbon sequestration, or to fix nitrogen more efficiently in nutrient-poor soils.

Future innovations are also likely to focus on improving the formulation and delivery of fungal products. Slow-release formulations that provide a continuous supply of fungal inoculants over time, as well as liquid bioformulations that offer more uniform distribution in the soil, will enhance the practicality and effectiveness of fungal applications. These advancements will ensure that fungal products are more accessible, cost-effective, and adaptable to a wide range of agricultural systems.

8.5 A Path Toward Sustainable Agriculture

The future of agriculture will depend on the integration of sustainable practices that reduce environmental impacts while maintaining or even increasing food production. Fungi, with their ability to improve soil health, enhance plant resilience, and reduce reliance on chemical inputs, are set to play a central role in this transformation(Chaurasia & Bharati, 2021). As the global population continues to grow and the pressures of climate change intensify, the adoption of fungal technologies will be crucial for building agricultural systems that are resilient, productive, and environmentally sustainable.

By fostering partnerships between researchers, farmers, and policymakers, we can accelerate the adoption of fungal solutions and ensure that they are accessible to farmers worldwide(Chaurasia & Bharati, 2021). As we move toward a future where agriculture must balance productivity with sustainability, fungi offer a powerful tool for creating more resilient and climate-smart farming systems. The ongoing innovation in fungal research and the growing recognition of their environmental benefits suggest that fungi will be indispensable in the global effort to create a more sustainable food system.

REFERENCES

Alam, B., L , J., Gě, Q., Khan, M. A., Gōng, J., Mehmood, S., Yuán, Y., & G ng, W. (2021). Endophytic Fungi: From Symbiosis to Secondary Metabolite Communications or Vice Versa? *Frontiers in Plant Science*, *12*, 791033. Advance online publication. DOI: 10.3389/fpls.2021.791033 PMID: 34975976

Aleahmad, P., & Ebrahimi, L. (2023). The Possible Applications of Endophytic Fungi. *Research Journal of Pharmacognosy*, *10*(4), 81–94. DOI: 10.22127/rjp.2023.179187

Amaranthus, M. P. (1998). *The importance and conservation of ectomycorrizal fungal diversity in forest ecosystems: lessons from Europe and the Pacific Northwest.* DOI: 10.2737/PNW-GTR-431

Anthony, M. A., Tedersoo, L., De Vos, B., Croisé, L., Meesenburg, H., Wagner, M., Andreae, H., Jacob, F., Lech, P., Kowalska, A., Greve, M., Popova, G., Frey, B., Gessler, A., Schaub, M., Ferretti, M., Waldner, P., Calatayud, V., Canullo, R., & Averill, C. (2024). Fungal community composition predicts forest carbon storage at a continental scale. *Nature Communications*, *15*(1), 2385. DOI: 10.1038/s41467-024-46792-w PMID: 38493170

Bamisile, B. S., Akutse, K. S., Siddiqui, J. A., & Xu, Y. (2021). Model Application of Entomopathogenic Fungi as Alternatives to Chemical Pesticides: Prospects, Challenges, and Insights for Next-Generation Sustainable Agriculture. *Frontiers in Plant Science*, *12*, 741804. Advance online publication. DOI: 10.3389/fpls.2021.741804 PMID: 34659310

Baron, N. C., & Rigobelo, E. C. (2022a). Endophytic fungi: A tool for plant growth promotion and sustainable agriculture. *Mycology*, *13*(1), 39–55. DOI: 10.1080/21501203.2021.1945699 PMID: 35186412

Baron, N. C., & Rigobelo, E. C. (2022b). Endophytic fungi: A tool for plant growth promotion and sustainable agriculture. *Mycology*, *13*(1), 39–55. DOI: 10.1080/21501203.2021.1945699 PMID: 35186412

Begum, N., Qin, C., Ahanger, M. A., Raza, S., Khan, M. I., Ashraf, M., Ahmed, N., & Zhang, L. (2019). Role of Arbuscular Mycorrhizal Fungi in Plant Growth Regulation: Implications in Abiotic Stress Tolerance. *Frontiers in Plant Science*, *10*, 1068. Advance online publication. DOI: 10.3389/fpls.2019.01068 PMID: 31608075

Bihal, R., Al-Khayri, J. M., Banu, A. N., Kudesia, N., Ahmed, F. K., Sarkar, R., Arora, A., & Abd-Elsalam, K. A. (2023). Entomopathogenic Fungi: An Eco-Friendly Synthesis of Sustainable Nanoparticles and Their Nanopesticide Properties. *Microorganisms*, *11*(6), 1617. DOI: 10.3390/microorganisms11061617 PMID: 37375119

Bisht, I. S., Rana, J. C., & Pal Ahlawat, S. (2020). The Future of Smallholder Farming in India: Some Sustainability Considerations. *Sustainability (Basel)*, *12*(9), 3751. DOI: 10.3390/su12093751

Chaurasia, P. K., & Bharati, S. L. (2021). Applicability of fungi in agriculture and environmental sustainability. In *Microbes in Land Use Change Management* (pp. 155–172). Elsevier., DOI: 10.1016/B978-0-12-824448-7.00010-3

Ebbisa, A. (2023). Arbuscular Mycorrhizal Fungi (AMF) in Optimizing Nutrient Bioavailability and Reducing Agrochemicals for Maintaining Sustainable Agroecosystems. In *Arbuscular Mycorrhizal Fungi in Agriculture - New Insights*. IntechOpen. DOI: 10.5772/intechopen.106995

El-Metwally, M. M., Mekawey, A. A. I., El-Halmouch, Y., & Naga, N. G. (2023). Symbiotic Relationships with Fungi: From Mutualism to Parasitism. In *Plant Mycobiome* (pp. 375–413). Springer International Publishing. DOI: 10.1007/978-3-031-28307-9_15

Fadiji, A. E., Xiong, C., Egidi, E., & Singh, B. K. (2024). Formulation challenges associated with microbial biofertilizers in sustainable agriculture and paths forward. *Journal of Sustainable Agriculture and Environment, 3*(3), e70006. Advance online publication. DOI: 10.1002/sae2.70006

Fall, A. F., Nakabonge, G., Ssekandi, J., Founoune-Mboup, H., Apori, S. O., Ndiaye, A., Badji, A., & Ngom, K. (2022). Roles of Arbuscular Mycorrhizal Fungi on Soil Fertility: Contribution in the Improvement of Physical, Chemical, and Biological Properties of the Soil. *Frontiers in Fungal Biology, 3*, 723892. Advance online publication. DOI: 10.3389/ffunb.2022.723892 PMID: 37746193

Frac, M., Hannula, S., Bełka, M., & Jedryczka, M. (2018). Fungal Biodiversity and Their Role in Soil Health. *Frontiers in Microbiology, 9*, 707. Advance online publication. DOI: 10.3389/fmicb.2018.00707 PMID: 29755421

Getahun, S., Kefale, H., & Gelaye, Y. (2024). Application of Precision Agriculture Technologies for Sustainable Crop Production and Environmental Sustainability: A Systematic Review. *The Scientific World Journal, 2024*(1). DOI: 10.1155/2024/2126734

Gowtham, H. G., Hema, P., Murali, M., Shilpa, N., Nataraj, K., Basavaraj, G. L., Singh, S. B., Aiyaz, M., Udayashankar, A. C., & Amruthesh, K. N. (2024). Fungal Endophytes as Mitigators against Biotic and Abiotic Stresses in Crop Plants. *Journal of Fungi (Basel, Switzerland), 10*(2), 116. DOI: 10.3390/jof10020116 PMID: 38392787

Guebsi, R., Mami, S., & Chokmani, K. (2024). Drones in Precision Agriculture: A Comprehensive Review of Applications, Technologies, and Challenges. *Drones (Basel), 8*(11), 686. DOI: 10.3390/drones8110686

Hu, Y., Yesilonis, I., & Szlavecz, K. (2021). Microbial and environmental controls on wood decomposition in deciduous forests of different ages. *Applied Soil Ecology, 166*, 103986. DOI: 10.1016/j.apsoil.2021.103986

Ji, H., Zhang, M., Huang, C., Lin, W., Lu, Y., Wang, P., Dong, B., He, B., Wu, B., & Guo, L. (2024). The Impact of Piriformospora indica on plant heat and drought tolerance. *Frontiers in Plant Science, 15*, 1479561. Advance online publication. DOI: 10.3389/fpls.2024.1479561 PMID: 39711589

Kalita, B., Boro, P., Talukdar, S., & Borah, L. (2024). Greenhouse Gas Emission Reduction by Crop Management and Climate-Smart Agriculture. In *Agricultural Greenhouse Gas Emissions: Problems and Solutions* (pp. 1–31). Springer Nature Singapore. DOI: 10.1007/978-981-97-7554-5_1

Kour, D., Khan, S. S., Ramniwas, S., Kumar, S., Rai, A. K., Rustagi, S., Chaubey, K. K., Singh, S., Yadav, A. N., & Ahluwalia, A. S. (2023a). Beneficial fungal communities for sustainable development: Present scenario and future challenges. *Journal of Applied Biology and Biotechnology*. Advance online publication. DOI: 10.7324/JABB.2024.148256

Kour, D., Khan, S. S., Ramniwas, S., Kumar, S., Rai, A. K., Rustagi, S., Chaubey, K. K., Singh, S., Yadav, A. N., & Ahluwalia, A. S. (2023b). Beneficial fungal communities for sustainable development: Present scenario and future challenges. *Journal of Applied Biology and Biotechnology*. Advance online publication. DOI: 10.7324/JABB.2024.148256

Kumar, A., Saharan, B. S., Parshad, J., Gera, R., Choudhary, J., & Yadav, R. (2024). Revolutionizing Indian agriculture: The imperative of advanced biofertilizer technologies for sustainability. *Discover Agriculture*, *2*(1), 24. DOI: 10.1007/s44279-024-00037-y

Kumar, D., Singh, P. N., Willer, H., Sharma, S. K., Singh, U. B., Lagashetti, A. C., & Husain, R. (2023). Fungal Biofertilizers and Biopesticides and Their Roles in Sustainable Agriculture. In *Applied Mycology for Agriculture and Foods* (pp. 165–216). Apple Academic Press., DOI: 10.1201/9781003369868-9

Kumar, J., Ramlal, A., Mallick, D., & Mishra, V. (2021). An Overview of Some Biopesticides and Their Importance in Plant Protection for Commercial Acceptance. *Plants*, *10*(6), 1185. DOI: 10.3390/plants10061185 PMID: 34200860

Kyslynska, A. S., Nadkernychna, O. V., Kopylov, Y. P., & Tsekhmister, H. V. (2023). The relation between mutualistic mycorrhiza and endophytic plant-fungus associations and their effect on host plants. *Agricultural Science and Practice*, *10*(1), 54–75. DOI: 10.15407/agrisp10.01.054

Lehmann, J., & Schroth, G. (2002). Nutrient leaching. In *Trees, crops and soil fertility: concepts and research methods* (pp. 151–166). CABI Publishing., DOI: 10.1079/9780851995939.0151

Li, L., Feng, Y., Qi, F., & Hao, R. (2023). Research Progress of Piriformospora indica in Improving Plant Growth and Stress Resistance to Plant. *Journal of Fungi (Basel, Switzerland)*, *9*(10), 965. DOI: 10.3390/jof9100965 PMID: 37888222

Morales-Vargas, A. T., López-Ramírez, V., Álvarez-Mejía, C., & Vázquez-Martínez, J. (2024). Endophytic Fungi for Crops Adaptation to Abiotic Stresses. *Microorganisms*, *12*(7), 1357. DOI: 10.3390/microorganisms12071357 PMID: 39065124

Mwampashi, L. L., Magubika, A. J., Ringo, J. F., Theonest, D. J., Tryphone, G. M., Chilagane, L. A., & Nassary, E. K. (2024). Exploring agro-ecological significance, knowledge gaps, and research priorities in arbuscular mycorrhizal fungi. *Frontiers in Microbiology*, *15*, 1491861. Advance online publication. DOI: 10.3389/fmicb.2024.1491861 PMID: 39552643

Nayak, S. K., & Mishra, B. B. (2020). *Frontiers in Soil and Environmental Microbiology* (Nayak, S. K., & Mishra, B. B., Eds.). CRC Press., DOI: 10.1201/9780429485794

Nødvig, C. S., Nielsen, J. B., Kogle, M. E., & Mortensen, U. H. (2015). A CRISPR-Cas9 System for Genetic Engineering of Filamentous Fungi. *PLoS One*, *10*(7), e0133085. DOI: 10.1371/journal.pone.0133085 PMID: 26177455

Pavan, J., Vireesha, P., Chaitanya, M., Raja, G. P., & Priya, M. K. (2024). Biopesticides in Indian agriculture: Advances, challenges, and pathways to sustainable crop protection. *International Journal of Advanced Biochemistry Research*, *8*(8S), 936–943. DOI: 10.33545/26174693.2024.v8.i8Sm.1966

Philip, B., Behiry, S. I., Salem, M. Z. M., Amer, M. A., El-Samra, I. A., Abdelkhalek, A., & Heflish, A. (2024). Trichoderma afroharzianum TRI07 metabolites inhibit Alternaria alternata growth and induce tomato defense-related enzymes. *Scientific Reports*, *14*(1), 1874. DOI: 10.1038/s41598-024-52301-2 PMID: 38253713

Raeisi Vanani, A., Sheikhi Shahrivar, F., Nouri, A., & Sepehri, M. (2024). Exploring rice tolerance to salinity and drought stresses through Piriformospora indica inoculation: Understanding physiological and metabolic adaptations. *Frontiers in Plant Science*, *15*, 1428631. Advance online publication. DOI: 10.3389/fpls.2024.1428631 PMID: 39385986

Reverchon, F., María del Ortega-Larrocea, P., & Pérez-Moreno, J. (2010). Saprophytic fungal communities change in diversity and species composition across a volcanic soil chronosequence at Sierra del Chichinautzin, Mexico. *Annals of Microbiology*, *60*(2), 217–226. DOI: 10.1007/s13213-010-0030-7

Rillig, M. C., Wright, S. F., Nichols, K. A., Schmidt, W. F., & Torn, M. S. (2001). Large contribution of arbuscular mycorrhizal fungi to soil carbon pools in tropical forest soils. *Plant and Soil*, *233*(2), 167–177. DOI: 10.1023/A:1010364221169

Ripa, F. A., Cao, W., Tong, S., & Sun, J. (2019). Assessment of Plant Growth Promoting and Abiotic Stress Tolerance Properties of Wheat Endophytic Fungi. *BioMed Research International*, *2019*, 1–12. DOI: 10.1155/2019/6105865 PMID: 31032353

Rodriguez, R. J., & Redman, R. S. (1997). *Fungal Life-Styles and Ecosystem Dynamics: Biological Aspects of Plant Pathogens*. Plant Endophytes and Saprophytes., DOI: 10.1016/S0065-2296(08)60073-7

Ruiz-Lozano, J. M., Azcon, R., & Gomez, M. (1995). Effects of arbuscular-mycorrhizal glomus species on drought tolerance: Physiological and nutritional plant responses. *Applied and Environmental Microbiology*, *61*(2), 456–460. DOI: 10.1128/aem.61.2.456-460.1995 PMID: 16534929

Sharma, I., Raina, A., Choudhary, M., Apra, , Kaul, S., & Dhar, M. K. (2023). Fungal endophyte bioinoculants as a green alternative towards sustainable agriculture. *Heliyon*, *9*(9), e19487. DOI: 10.1016/j.heliyon.2023.e19487 PMID: 37662754

Sishodia, R. P., Ray, R. L., & Singh, S. K. (2020). Applications of Remote Sensing in Precision Agriculture: A Review. *Remote Sensing (Basel)*, *12*(19), 3136. DOI: 10.3390/rs12193136

Tadesse Mawcha, K., Malinga, L., Muir, D., Ge, J., & Ndolo, D. (2024). Recent Advances in Biopesticide Research and Development: A Focus on Microbial: A Review. *F1000 Research*, *13*, 1071. DOI: 10.12688/f1000research.154392.3 PMID: 39512238

Tadesse Mawcha, K., Malinga, L., Muir, D., Ge, J., & Ndolo, D. (2025). Recent Advances in Biopesticide Research and Development with a Focus on Microbials. *F1000 Research*, *13*, 1071. DOI: 10.12688/f1000research.154392.4 PMID: 39512238

Traversari, S., Cacini, S., Galieni, A., Nesi, B., Nicastro, N., & Pane, C. (2021). Precision Agriculture Digital Technologies for Sustainable Fungal Disease Management of Ornamental Plants. *Sustainability (Basel)*, *13*(7), 3707. DOI: 10.3390/su13073707

Varma, A., Sherameti, I., Tripathi, S., Prasad, R., Das, A., Sharma, M., Bakshi, M., Johnson, J. M., Bhardwaj, S., Arora, M., Rastogi, K., Agrawal, A., Kharkwal, A. C., Talukdar, S., Bagde, U. S., Bisaria, V. S., Upadhyaya, C. P., Won, P. S., Chen, Y., & Oelmüller, R. (2012). 13 The Symbiotic Fungus Piriformospora indica: Review. In *Fungal Associations* (pp. 231–254). Springer Berlin Heidelberg., DOI: 10.1007/978-3-642-30826-0_13

Vassilev, N., & Mendes, G. de O. (2024). Soil Fungi in Sustainable Agriculture. *Microorganisms*, *12*(1), 163. DOI: 10.3390/microorganisms12010163 PMID: 38257989

Verma, A., Shameem, N., Jatav, H. S., Sathyanarayana, E., Parray, J. A., Poczai, P., & Sayyed, R. Z. (2022). Fungal Endophytes to Combat Biotic and Abiotic Stresses for Climate-Smart and Sustainable Agriculture. *Frontiers in Plant Science*, *13*, 953836. Advance online publication. DOI: 10.3389/fpls.2022.953836 PMID: 35865289

Wang, D., Jin, S., Lu, Q., & Chen, Y. (2023). Advances and Challenges in CRISPR/Cas-Based Fungal Genome Engineering for Secondary Metabolite Production: A Review. *Journal of Fungi (Basel, Switzerland)*, *9*(3), 362. DOI: 10.3390/jof9030362 PMID: 36983530

Wang, H., Zhao, R., Zhao, D., Liu, S., Fu, J., Zhang, Y., Dai, N., Song, D., & Ding, H. (2022). Microbial-Mediated Emissions of Greenhouse Gas from Farmland Soils: A Review. *Processes (Basel, Switzerland)*, *10*(11), 2361. DOI: 10.3390/pr10112361

Wang, Y., Liu, Z., Hao, X., Wang, Z., Wang, Z., Liu, S., Tao, C., Wang, D., Wang, B., Shen, Z., Shen, Q., & Li, R. (2023). Biodiversity of the beneficial soil-borne fungi steered by Trichoderma-amended biofertilizers stimulates plant production. *NPJ Biofilms and Microbiomes*, *9*(1), 46. DOI: 10.1038/s41522-023-00416-1 PMID: 37407614

Wei, Y., Yang, H., Hu, J., Li, H., Zhao, Z., Wu, Y., Li, J., Zhou, Y., Yang, K., & Yang, H. (2023). Trichoderma harzianum inoculation promotes sweet sorghum growth in the saline soil by modulating rhizosphere available nutrients and bacterial community. *Frontiers in Plant Science*, *14*, 1258131. Advance online publication. DOI: 10.3389/fpls.2023.1258131 PMID: 37771481

Xiao, Z., Zhao, Q., Li, W., Gao, L., & Liu, G. (2023). Strain improvement of Trichoderma harzianum for enhanced biocontrol capacity: Strategies and prospects. *Frontiers in Microbiology*, *14*, 1146210. Advance online publication. DOI: 10.3389/fmicb.2023.1146210 PMID: 37125207

Yadav, A., & Yadav, K. (2024). Challenges and Opportunities in Biofertilizer Commercialization. *SVOA Microbiology*, *5*(1), 01–14. DOI: 10.58624/SVOAMB.2024.05.037

Yang, Y., Luo, W., Xu, J., Guan, P., Chang, L., Wu, X., & Wu, D. (2022). Fallow Land Enhances Carbon Sequestration in Glomalin and Soil Aggregates Through Regulating Diversity and Network Complexity of Arbuscular Mycorrhizal Fungi Under Climate Change in Relatively High-Latitude Regions. *Frontiers in Microbiology*, *13*, 930622. Advance online publication. DOI: 10.3389/fmicb.2022.930622 PMID: 35859742

Yao, X., Guo, H., Zhang, K., Zhao, M., Ruan, J., & Chen, J. (2023a). Trichoderma and its role in biological control of plant fungal and nematode disease. *Frontiers in Microbiology*, *14*, 1160551. Advance online publication. DOI: 10.3389/fmicb.2023.1160551 PMID: 37206337

Yao, X., Guo, H., Zhang, K., Zhao, M., Ruan, J., & Chen, J. (2023b). Trichoderma and its role in biological control of plant fungal and nematode disease. *Frontiers in Microbiology*, *14*, 1160551. Advance online publication. DOI: 10.3389/fmicb.2023.1160551 PMID: 37206337

Yarzábal, L. A., & Chica, E. J. (2021). Microbial-Based Technologies for Improving Smallholder Agriculture in the Ecuadorian Andes: Current Situation, Challenges, and Prospects. *Frontiers in Sustainable Food Systems*, *5*, 617444. Advance online publication. DOI: 10.3389/fsufs.2021.617444

Yarzábal Rodríguez, L. A., Álvarez Gutiérrez, P. E., Gunde-Cimerman, N., Ciancas Jiménez, J. C., Gutiérrez-Cepeda, A., Ocaña, A. M. F., & Batista-García, R. A. (2024). Exploring extremophilic fungi in soil mycobiome for sustainable agriculture amid global change. *Nature Communications*, *15*(1), 6951. DOI: 10.1038/s41467-024-51223-x PMID: 39138171

Yuan, Y., & Sun, Y. (2024). Practices, Challenges, and Future of Digital Transformation in Smallholder Agriculture: Insights from a Literature Review. *Agriculture*, *14*(12), 2193. DOI: 10.3390/agriculture14122193

Chapter 2
A Sustainable Approach to Unlocking Plant Potential With Fungal Biostimulants:
Harnessing Fungal Powe to Sustainable Agriculture

Pooja Gangaram Malave
Dr. D.Y. Patil College of Arts, Commerce, and Science, India

ABSTRACT

The global population is projected to reach nearly 10 billion by 2050, necessitating a 60-70% increase in food production while minimizing environmental impacts (World Population Prospects 2019). This requires agriculture to produce more food with fewer resources and reduce reliance on unsustainable synthetic fertilizers and pesticides.Recent decades have shifted towards sustainable practices, particularly natural plant biostimulants that enhance growth, flowering, fruit set, and nutrient use efficiency (NUE). Fungal biostimulants effectively promote plant growth by forming symbiotic relationships with roots and improving nutrient uptake.This chapter explores the role of fungal biostimulants in agriculture, focusing on mycorrhizal, endophytic, and saprophytic fungi, with emphasis on arbuscular mycorrhizal biostimulants that enhance nutrient uptake and soil health.The chapter examines the sustainability and economic benefits of fungal biostimulants, addresses challenges related to consistency and regulation, and explores future research opportunities.

1. HARNESSING FUNGAL POWER: AN INTRODUCTION TO SUSTAINABLE AGRICULTURE

The agricultural and horticultural sectors are facing a major challenge at this point; they need to meet the growing food requirements and at the same time minimize the ecological consequences (Duhamel & Vandenkoornhuyse, 2013). With a rise in the world population and a higher demand for food systems, the need to strike this balance has become a formidable challenge to sustainability(Owen et al., 2015). The increasing demand for sustainable farming is crucial due to rising environmental pollution and soil

DOI: 10.4018/979-8-3693-8201-1.ch002

degradation. Achieving this while maintaining crop yields and farmer income remains a major challenge in modern agriculture (Shahrajabian et al., 2021). In recent years, numerous technological advancements have been introduced to reduce dependence on synthetic materials while enhancing crop yields. Additionally, farmers face rising costs for pest management, which can represent over 40% of total production expenses, with pests and diseases contributing to significant yield losses(Fenu & Malloci, 2021).

Several technological advancements have been made throughout the previous decades to substantially reduce the number of synthetic agrochemicals like fertilizers and pesticides. An eco-friendly and viable novel alternative is to use natural **"Plant Biostimulants"** (PBs) (Rouphael et al., 2015). Biostimulants are defined as "Materials or substances, excluding nutrients and pesticides, that, when applied in minute quantities to plants, seeds, or growing substrates in specific formulations, can alter physiological processes to promote growth, development, and improve stress responses (Rouphael & Colla, 2020). As defined by the European Biostimulants Industry Council (EBIC) in 2012, plant biostimulants consist of substances and/or microorganisms that, when applied to plants or the rhizosphere in small amounts, stimulate natural processes to improve nutrient uptake, efficiency, tolerance to abiotic stress, and crop quality, without directly affecting pests (www.biostimulants.eu). Among these, fungal biostimulants form a distinct and highly effective category. These beneficial fungi enhance plant health by forming symbiotic relationships with plant roots, improving nutrient uptake, soil structure, and stress resistance. Fungal biostimulants offer a unique advantage over other plant biostimulants due to their ability to persist in the soil and establish long-term associations with host plants. Fungal biostimulants, such as arbuscular mycorrhizal fungi (*Rhizophagus irregularis*) and saprophytic fungi (*Trichoderma spp.),* have been shown to facilitate nutrient absorption, enhance plant resilience, and contribute to soil biodiversity. These fungi colonize plant roots, creating an extended hyphal network that increases water and nutrient uptake while improving soil aeration. By focusing on fungal biostimulants as a crucial subset of plant biostimulants, this chapter aims to explore their mechanisms, benefits, and potential applications in sustainable agriculture. The discussion will delve into the role of mycorrhizal and endophytic fungi in plant health, their contributions to soil ecosystems, and the challenges associated with their large-scale adoption. Understanding how these microorganisms interact with plants can provide valuable insights into developing innovative strategies for improving agricultural sustainability.

2. TYPES OF BIOSTIMULANTS

Biostimulants are heterogeneous materials and originate from biological and inorganic sources(Calvo et al., 2014). Plant biostimulants (PBs) can trigger a wide range of responses in plants, including changes in morpho-anatomical, biochemical, physiological, and molecular plant responses. These responses can enhance crop productivity, improve nitrogen use efficiency (NUE), and increase tolerance to abiotic stresses(Rouphael & Colla, 2020). *du Jardin (2015)*(du Jardin, 2015) categorized biostimulants into seven types (: (1) **humic and fulvic acids**, which are heterogeneous, highly acidic compounds formed through the decomposition of animal, microbial, and plant matter in the soil; (2) **protein hydrolysates** and other nitrogen (N)-containing substances, consisting of amino acids and peptides obtained via enzymatic or chemical hydrolysis of plant residues and animal by-products; (3) **seaweed extracts and botanicals**, derived from plants or seaweeds with biostimulant properties; (4) **chitosan and other biopolymers**, which are biodegradable and biocompatible polymers or oligomers produced from natural sources; (5) **inorganic compounds,** primarily beneficial elements like silicon, titanium, sodium, selenium, iodine, and aluminium ; (6) **beneficial fungi,** including plant-associated, endophytic, or symbiotic fungi with biostimulant effects; and (7) **beneficial bacteria**, which are plant-associated, endophytic, or symbiotic bacteria that promote growth.

Thus, biostimulants can generally be classified into two broad categories: non-microbial and microbial plant biostimulants(Bhupenchandra et al., 2022) Among these, fungal biostimulants stand out as a crucial category within microbial biostimulants. Unlike other biostimulants, fungi form complex associations with plant roots, enhancing water and nutrient uptake, improving soil structure, and increasing plant resilience against abiotic stress. Notably, arbuscular mycorrhizal fungi (*Rhizophagus irregularis*), endophytic fungi, and saprophytic fungi (*Trichoderma spp.*) have been extensively studied for their role in improving plant health and soil fertility. Given their profound impact on sustainable agriculture, this chapter will focus specifically on fungal biostimulants.

3. MECHANISMS AND PROPERTIES OF FUNGAL BIOSTIMULANTS

Fungal biostimulants promote plant growth and resilience by orchestrating a network of physiological, biochemical, and molecular interactions that enhance nutrient uptake, stress tolerance, and disease resistance. These fungi establish intricate symbiotic associations with plant roots, modulating rhizosphere dynamics to improve soil structure, microbial diversity, and nutrient bioavailability. Through mechanisms such as mycorrhizal symbiosis, secretion of bioactive compounds, and regulation of phytohormonal signaling, fungal biostimulants enable plants to withstand abiotic stressors like drought, salinity, and extreme temperatures. Additionally, they contribute to induced systemic resistance (ISR), suppressing pathogenic threats while fostering plant immunity. By optimizing root morphology and enhancing soil microbial activity, fungal biostimulants serve as a cornerstone of sustainable agriculture, reducing dependency on synthetic fertilizers and fostering long-term soil health. The detailed mechanisms underlying these benefits are discussed in the sections below.

3.1 Root System Expansion and Nutrient Uptake

Plants require a combination of essential nutrients, water, and favorable environmental conditions to achieve optimal growth and productivity. Nutrients such as nitrogen (N), phosphorus (P), potassium (K), and micronutrients like zinc (Zn) and iron (Fe) play vital roles in metabolic activities, enzyme function, and structural development. Water availability influences cell expansion, nutrient transport, and overall plant health, while environmental factors such as temperature, soil structure, and microbial interactions further determine growth efficiency(White et al., 2010).

One of the primary ways fungal biostimulants enhance plant growth is by modifying root structure and increasing nutrient availability. Arbuscular mycorrhizal fungi (AMF) form extensive hyphal networks that penetrate deep into the soil, increasing root surface area and enabling plants to access phosphorus (P), nitrogen (N), zinc (Zn), and copper (Cu) that would otherwise be unavailable (Smith, 2008; Wu et al., 2013). Similarly, endophytic fungi like *Piriformospora indica* enhance nutrient solubilization by breaking down minerals and facilitating their movement into plant cells. *Trichoderma spp.* further contribute by releasing organic acids and chelators, making iron (Fe) and zinc (Zn) more bioavailable (Sun & Shahrajabian, 2023).

These nutrient-enhancing mechanisms are crucial for plant growth, particularly in nutrient-deficient soils. Fungal endophytes significantly improve plant development in phosphorus-limited environments. For instance, inoculation with *Glomus mosseae* enhances phosphorus uptake and yield in grapevines. This occurs through a dual uptake system: (1) Direct root uptake via H+/Pi symporters from the PHT1/PT family, and (2) Mycorrhizal-assisted uptake, where fungal H+/Pi and Na+/Pi symporters scavenge phosphorus, store it as polyphosphates, and later release it through acid and alkaline phosphatases. Additionally, AMF-associated bacteria aid phosphate solubilization by secreting acid phosphatases, a process driven by carbon exchange between the plant, fungus, and microbes (Chiu & Paszkowski, 2019).

Beyond phosphorus, arbuscular fungal biostimulants also facilitate nitrogen assimilation, independently or through symbiotic nitrogen-fixing microbes. Studies show that *Glomus mosseae* inoculation enhances nitrogen fixation in chickpeas, improving plant growth in nitrogen-deficient soils. This is mediated by Nod factor signaling, promoting Rhizobium symbiosis and nitrogenase enzyme activity, leading to efficient nitrogen uptake and assimilation (Koide & Kabir, 2000). This intricate tripartite network of fungi, bacteria, and plant transporters optimizes nutrient availability, enhancing crop resilience in challenging soil conditions (Koide & Kabir, 2000).

Beyond improving nutrient availability, fungal biostimulants actively reshape root architecture, further amplifying nutrient uptake efficiency. These fungi induce lateral root formation and root hair development, optimizing root surface area for better absorption. Moreover, the formation of specialized fungal structures, such as arbuscules in fungal endophyte and hyphal networks in *Trichoderma spp.*, enhances root penetration and strengthens plant-fungal associations. To further facilitate nutrient uptake, these fungi secrete phosphatases, acid phosphatases, and organic acids, which break down insoluble phosphate and micronutrients, making them more accessible for plant absorption (Etesami et al., 2021; Sun & Shahrajabian, 2023).

Figure 1. Discusses the significance of the application of biostimulants on plant health

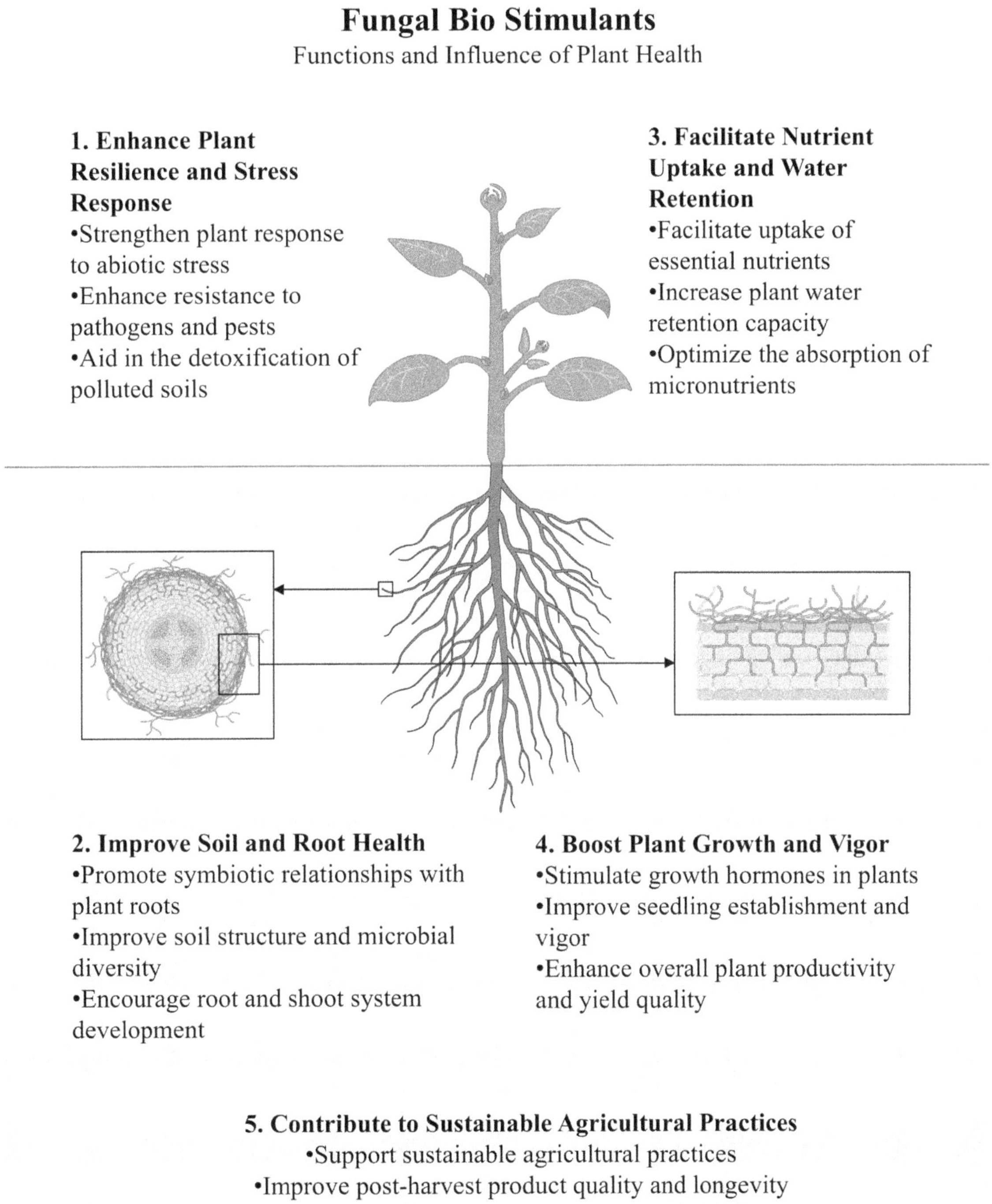

3.2 Enhanced Plant Metabolism and Crop Yield

Plant metabolism is a complex network of biochemical pathways that regulate growth, energy production, and stress adaptation. It encompasses primary metabolism, which includes photosynthesis, respiration, and nutrient assimilation, as well as secondary metabolism, which generates protective compounds such as phenolics and flavonoids. Efficient metabolic functioning is essential for plant development, particularly under stress conditions where energy allocation shifts toward defense and survival mechanisms (Jan et al., 2021). Fungal biostimulants have been shown to enhance plant metabolism, leading to higher yield potential and improved crop quality. By regulating hormonal balance and stimulating key metabolic pathways, fungi like arbuscular, *Trichoderma spp.,* and *Piriformospora indica* promote root and shoot development, photosynthesis, and nutrient assimilation (Yakhin et al., 2017). Fungal secondary metabolites such as auxins, gibberellins, and abscisic acid mimic plant hormones, helping modulate growth under biotic and abiotic stress conditions. These metabolites also activate **systemic acquired resistance** (SAR), equipping plants with stronger defence mechanisms against infections and environmental stressors (Pusztahelyi et al., 2015).

3.3 Drought Resistance and Induced Systemic Resistance

Drought stress is a major environmental challenge that significantly affects plant growth, development, and productivity. Limited water availability disrupts essential physiological processes, leading to reduced photosynthesis, impaired nutrient uptake, and oxidative stress. To cope with such adverse conditions, plants have evolved various adaptive mechanisms, including morphological, physiological, and biochemical responses (Zia et al., 2021). Fungal biostimulants contribute significantly to drought resistance by enhancing root hydraulic conductivity, optimizing water uptake, and improving osmotic adjustment. One of the critical mechanisms by which AMF aids in drought adaptation is through the regulation of stomatal function, which controls transpiration and minimizes excessive water loss. AMF colonization in *Cucumis sativus* has been shown to optimize stomatal regulation, reducing water loss under drought conditions (Huang & Lo, 2019). Similarly, *Rhizophagus irregularis* has been reported to enhance stomatal conductivity and increase shoot dry matter in *Digitaria eriantha* under water-deficit conditions, ensuring efficient water usage and maintaining physiological processes (Pedranzani et al., 2016).

Beyond stomatal regulation, some of the fungal endophytes improve nutrient uptake and osmotic adjustment, allowing plants to maintain cellular turgor and metabolic activity under drought stress. For instance, *Rhizophagus intraradices* have been shown to significantly enhance magnesium, zinc, copper, and fluorine accumulation in *Triticum durum*, improving overall plant nutrition during water stress (Goicoechea & Antolín, 2017). Additionally, this species plays a key role in increasing nitrogen, phosphorus, and potassium uptake in *Zea mays*, ensuring sustained growth even under moisture-limiting conditions (Zhao et al., 2015).

Several AMF species contribute to maintaining chlorophyll content, which is crucial for sustaining photosynthetic efficiency and energy production during drought stress. Under water-deficit conditions, chlorophyll degradation often leads to reduced carbon fixation, impairing plant growth and productivity. By preserving chlorophyll levels, AMF helps plants maintain metabolic activity, ensuring continued biomass accumulation and stress resilience. *Gigaspora decipiens* and *Glomus mosseae* have been found to enhance chlorophyll content in *Triticum aestivum*, thereby supporting photosynthetic performance even under moisture-limiting conditions (Goicoechea & Antolín, 2017). Similarly, *Glomus deserticola* aids in drought adaptation in *Antirrhinum majus* by not only preserving chlorophyll but also increasing proline accumulation, a critical osmoprotectant that stabilizes cellular structures under dehydration stress (A.-W. A. Asrar & Elhindi, 2011). In *Vigna subterranea*, *Gigaspora gregaria* has been shown to regulate mineral uptake while controlling excessive proline accumulation, preventing metabolic imbalances under drought stress (Esaïe, 2015). Meanwhile, *Glomus spp.* play a critical role in osmotic potential adjustment in *Ipomoea batatas*, while also boosting water uptake and antioxidant metabolite production, such as glutathione and phenolics, in *Saccharum arundinaceum* to counter oxidative stress caused by drought(Mirshad & Puthur, 2016).

In addition to their role in water conservation and nutrient assimilation, AMF and *Trichoderma spp.* activate plant defence mechanisms, priming systemic resistance against environmental stressors and pathogens. Studies on *Glomus intraradices* in tomato plants have demonstrated improved resistance against root-knot nematodes, leading to healthier root systems and increased crop productivity (Malviya et al., 2023). These fungi, alongside *Funneliformis mosseae,* also enhance plant growth parameters and mitigate the adverse effects of drought stress, reinforcing their importance in sustainable agriculture (Bozhinova, 2023).

Thus, we can say that fungal stimulants collectively contribute to stomatal regulation, nutrient uptake, osmotic balance, and induced systemic resistance highlight their essential role in plant stress adaptation, and can emerge as a powerful tool for promoting sustainable crop production under increasingly unpredictable climatic conditions.

3.4 Salinity Stress Adaptation and Ion Homeostasis

Salinity stress poses a significant challenge to plant growth by disrupting osmotic balance, reducing water uptake, and causing ion toxicity, ultimately leading to impaired metabolism and stunted development. High salt concentrations can interfere with nutrient absorption and induce oxidative stress, necessitating adaptive mechanisms to maintain cellular homeostasis (Balasubramaniam et al., 2023). In saline soils, fungi help plants regulate ion balance by reducing sodium (Na^+) and chloride (Cl^-) uptake while enhancing potassium (K^+) assimilation. Studies on *Rhizophagus irregularis* and maize have shown improved ion homeostasis, resulting in greater plant height and biomass under salt stress (Porras-Soriano et al., 2009). Furthermore, *Claroideoglomus etunicatum* has been reported to increase soluble sugar levels and improve dry mass accumulation in *Aleurites moluccanus*, thereby enhancing plant adaptability under high salinity conditions (Sun & Shahrajabian, 2023).

3.5 Heavy Metal Detoxification

Heavy metal contamination in soil poses a significant threat to plant growth by disrupting cellular metabolism, enzyme activity, and nutrient uptake. Toxic metals like cadmium (Cd), lead (Pb), and arsenic (As) induce oxidative stress, impair photosynthesis, and inhibit root development, ultimately reducing crop yield (Vasilachi et al., 2023). To mitigate these adverse effects, arbuscular mycorrhizal fungi (AMF) play a crucial role in metal sequestration, reducing bioavailability and enhancing plant tolerance through various mechanisms, including metal binding, compartmentalization, and improved antioxidant defence. Certain AMF, such as *Glomus versiforme* and *Glomus mosseae,* in celery experiments sequestered toxic heavy metals like cadmium (Cd), thereby reducing their bioavailability and minimizing toxicity effects on plants (Etesami & Glick, 2023) Additionally, *Glomus deserticola* has been shown to enhance proline accumulation and increase leaf count in *Antirrhinum majus*, further aiding in heavy metal stress tolerance by improving osmotic adjustment and overall plant health (Chun et al., 2018).

3.6 Increased Soil Fertility and Microbial Activities

Fungal biostimulants accelerate organic matter decomposition and nutrient cycling, enriching soil microbiota. By stimulating enzymatic activity, fungi enhance the breakdown of complex organic substrates, releasing essential nutrients into the soil. For example, AMF, particularly *Glomus species*, have been found to increase soil phosphatase activity, facilitating the release of phosphorus from organic matter (Huo et al., 2022). Additionally, *Trichoderma spp.* foster the proliferation of nitrogen-fixing bacteria, improving overall soil fertility and creating a more dynamic soil ecosystem (Calvo et al., 2014).

3.7 Biocontrol and Disease Suppression

Fungal biostimulants act as natural protectors, combating pathogens while boosting plant immunity. *Trichoderma spp.*, known for their **mycoparasitic** nature, actively hunt and invade harmful fungi by detecting, attaching, and coiling around their hyphae. These fungi also develop appressoria-like structures to penetrate the host, leading to its destruction(Tyśkiewicz et al., 2022).

Additionally, *Trichoderma spp.* release chitinases and glucanases, breaking down pathogen cell walls, while secreting antifungal compounds to further suppress infections (Druzhinina et al., 2011). Alongside this, AMF and Trichoderma enhance plant defense by activating systemic resistance, preparing plants to fend off future threats. This dual action of direct attack and immune priming makes them invaluable for sustainable disease management.

4. EXPLORING FUNGAL BIOSTIMULANTS: MYCORRHIZAL, ENDOPHYTIC, AND SAPROPHYTIC FUNGI

The rhizosphere, a region of soil that extends around the plant roots, is an active region that is teeming with microbial population and is rich in nutrient supra turnover in which the microorganisms and the plants work interactively to regulate the living and non-living factors of the environment (Thepbandit & Athinuwat, 2024). In this intricate region, microbial biostimulants play a crucial role in enhancing plant health and development. Microbial biostimulants can be broadly classified into two main categories. The first group comprises **beneficial fungi**, including species of endophytic mycorrhizal fungi such as arbuscular, exophytic fungi, and saprophytic fungi such as *Trichoderma,* yeast, etc (Ghorbanpour et al., 2018). The second group consists of **plant growth-promoting bacteria (PGPB),** including nitrogen-fixing bacteria from the genera *Rhizobium, Azotobacter*, and *Azospirillum* (Saeed et al., 2021).

4.1 Endophytic Fungi, the Arbuscular Mycorrhizal Fungi (AMF)

A wide variety of fungi known as mycorrhizal fungi are found populating the roots of nearly all land plants. These relationships are crucial for making some nutrients—phosphorus in particular—available to the plant in exchange for the carbon-based food produced during photosynthesis. Arbuscular mycorrhiza, ectomycorrhiza, ericoid mycorrhiza, and orchid mycorrhiza are among the multiple groups into which the various forms of mycorrhizae have historically been divided (Wei et al., 2022). These groupings have distinct functions in enhancing the health of plants and soil biomes.

With more than 90% of plant species, mycorrhizal fungi constitute a beneficial and diversified group of fungi. Mycorrhiza, are symbiotic associations of fungi and plant roots. **Arbuscular mycorrhizal fungi** (AMF) are a significant group of endomycorrhiza that form intracellular associations with plant roots. These fungi belong to the phylum *Glomeromycota* and are known for their widespread occurrence in various plant taxa, including crops and horticultural species (Bonfante & Genre, 2010). AMF forms symbiotic associations with wide range of essential horticultural crops, such as those from the Solanaceae family (e.g., tomato, eggplant, petunia), the Alliaceae family (e.g., onion, garlic, leek), various fruit trees (e.g., grapevine, citrus), as well as ornamental and herbal plants (e.g., basil, thyme, rosemary). With some exceptions, most arbuscular mycorrhizal fungi (AMF) can establish a mutualistic relationship with all mycorrhizal plants (Smith, 2008).

4.2 Life cycle and Arbuscular mycorrhizal fungi (AMF) symbiosis

Arbuscular mycorrhizal fungi (AMF) are characterized by an asymbiotic phase in their life cycle, during which asexual chlamydospores germinate in the soil. The germination is highly dependent on environmental conditions such as humidity and temperature. Since AMF are obligate biotrophs, they retract their cytoplasm and go dormant when there isn't a host plant present. The pre-symbiotic phase begins, with the formation of a germ tube, once the AMF approaches plant roots (Giovannetti, 1993). Root exudates have been shown to accelerate this process, as have some metabolites such as strigolactones (Akiyama et al., 2005; Tamasloukht et al., 2000).

The fungi create hyphopodia (appressoria) when they come into physical contact with the root surface. This causes the plant's underlying epidermal cells to react differently and create a pre-penetration apparatus, a temporary structure that is necessary for fungal entry (Genre et al., 2005). Fungal hyphae travel into or between plant cells, taking on linear or coiled shapes when they begin to colonize roots (Gianinazzi-Pearson & Gianinazzi, 1988). When the fungus reaches the inner cortex, it changes its approach to colonization and forms intricate, highly branching arbuscules inside plant cells called apoplasts. Furthermore, as storage organs, members of the *Glomineae* family create vesicles that are rich in lipids (Walker, 1995). At the same time, the fungus spreads its hyphae into the surrounding soil, where it colonizes the roots of neighboring plants of all sorts, taking nutrients and interacting with other microbes. As a result, a network of interconnected roots and hyphae is created, which helps plants and AM fungus communicate and exchange nutrients (Mikkelsen, 2008; Song et al., 2010).

4.3 Inoculum Production and Quality Considerations

Arbuscular mycorrhizal fungi (AMF) are the most widely used fungal biostimulants due to their ability to enhance plant growth, nutrient uptake, and stress tolerance. Since native arbuscular mycorrhizal fungi (AMF) in soils are becoming fewer in number, inoculating horticultural crops with AMF is becoming a normal procedure, especially in intensive cropping systems. High-quality inoculum is necessary for successful root colonization and should include the following: (1) a blend of different AMF species, as blends of two or more species perform better than single-species inocula; (2) a high density of infective AMF propagules; (3) the absence of pests and plant pathogens; (4) the addition of additives and beneficial bacteria that improve root mycorrhizal colonization and function; and (5) a dry solid form of inoculum with an extended shelf life (Rouphael et al., 2015).

As obligate biotrophs, AMF propagules can be produced on the roots of plants grown either in open fields ('on-farm inoculum')(Douds Jr. et al., 2012) or in controlled greenhouse conditions using containers (Feldmann & Schneider, 2008). Field propagation is the most cost-effective method, involving the cultivation of inoculated host plants in sandy soil, allowing the AMF to naturally develop and propagate. At the end of the growing cycle, mycorrhizal roots and the surrounding soil containing propagules are harvested, dried, and utilized as inoculum. Although this approach is straightforward, it presents challenges such as inconsistent production, difficulties in spore harvesting, and a heightened risk of contamination from pests, pathogens, and weeds. These issues may be mitigated through soilless AMF inoculum production in greenhouses using sterile materials like vermiculite(Feldmann & Schneider, 2008).

Currently, there is strong market interest in mycorrhizal inocula produced under sterile conditions, specifically those based on spores cultivated from plant roots in monoxenic systems (in vitro culture). Two primary sterile in vitro systems have been developed for monoxenic propagation of AMF: (1) propagation on genetically modified Ri T-DNA roots using *Agrobacterium rhizogenes*, which allows growth in root organ culture (ROC) within Petri dishes (Cranenbrouck et al., 2005)and (2) propagation on autotrophic plants where the shoot grows outside the Petri dish, either directly in the air or within a sterile tube connected vertically to the dish (Voets et al., 2005). Both systems are adapted for large-scale production of mycorrhizal inoculum for commercial use, utilizing small containers, airlift bioreactors, mist bioreactors container-based hydroponic cultures, or extended hydroponic systems (Akhtar & Abdullah, 2014). The advantages of in vitro culture include: (1) a pure, uncontaminated product due to sterile conditions; (2) easy traceability and monitoring; (3) straightforward concentration; and (4) the ability to produce mycorrhizal propagules throughout the year. However, in vitro propagation is currently limited to *Rhizophagus irregularis,* and the liquid form of the inoculum may result in a short shelf life, potentially hindering its commercial viability. Furthermore, there is still a scarcity of long-term studies directly comparing products from sterile versus unsterile production systems, although some negative impacts of sterile methods have been noted (Calvet et al., 2013). Ultimately, products that feature a high concentration of infective propagules and advanced inoculum forms (such as tablets or gels) could greatly facilitate their application in horticultural systems.

Table 1. Influence of Arbuscular mycorrhizal fungi (AMF) on the agronomic, physiological, and biochemical performance of horticultural crops

Fungal/Mycorrhizal species	Horticultural species	Effect on crop performance	Reference
R. intraradices	*Cucumis sativus*	AMF enhanced photosynthesis and optimized nutritional status, resulting in increased levels of P, K, Mg, Fe, Zn, and Mn, while reducing Na content.	(Rouphael et al., 2010)
Fusarium oxysporum Fusarium solani Fusarium sp.	*Senna Alata*	Improve seed germination rate	(Pradhan et al., 2023)
R. clarum, Claroideoglomus etunicatum, R. intraradices, G. etunicatum, F. mosseae, and mixture	Capsicum annuum	Plants that were inoculated showed a greater uptake of P and Zn compared to the control group.	(Kaya et al., 2009)

continued on following page

Table 1. Continued

Fungal/Mycorrhizal species	Horticultural species	Effect on crop performance	Reference
G. macrocarpum	*Apium graveolens*	AMF improved biomass production under Cd stress conditions. Inoculated plants also exhibited higher chlorophyll concentrations and increased photosynthate production	(Kapoor & Bhatnagar, 2007)
G. versiforme	*citrus sinensis and C. reticulate*	Inoculating two citrus cultivars with *G. versiforme* enhanced plant growth parameters, boosted photosynthesis, and increased magnesium concentration in plant tissues, particularly under low magnesium conditions.	(Xiao et al., 2014)
F. mosseae	*Capsicum annuum*	Under Cu-deficient conditions, inoculation enhanced plant growth, increased pigment biosynthesis, and improved the uptake of macronutrients such as phosphorus (P), potassium (K), calcium (Ca), and magnesium (Mg).	(Latef, 2011)
R. intraradices	*Dianthus caryophyllus*	Inoculation with AMF may have alleviated the negative effects of salinity on ornamental value, such as flower size and color, by increasing the levels of nitrogen (N), phosphorus (P), and calcium (Ca), while reducing the accumulation of toxic ions like sodium (Na) and chloride (Cl).	(Navarro et al., 2012)
R. clarum	*Capsicum annuum*	Inoculation improved key growth parameters of pepper under salt stress and reduced cell membrane leakage.	(Kaya et al., 2009)
G. deserticola	*Anthirhinum majus*	Inoculating plants resulted in higher flower yield, as well as increased shoot and root dry matter. The enhanced drought tolerance of mycorrhizal plants was attributed to improvements in water relations, chlorophyll content, and the levels of macronutrients	(A.-W. Asrar et al., 2012)
F. mosseae, F. geosporus and mixed inoculation	*Fragaria ananassa*	Inoculation with one or two fungal species increased strawberry growth sand yield	(Robinson Boyer et al., 2014)

5. ECTOMYCORRHIZA, ERICOID MYCORRHIZA, AND ORCHIDACEOUS MYCORRHIZA

In recent years, most studies have focused on the arbuscular mycorrhiza (AM) but, the other types of mycorrhizae – ectomycorhiza (EcM), ericoid mycorrhiza (ErM) and orchid mycorrhiza (OrM) – also help in widening the scope regarding the understanding of fungal biostimulants.

Ectomycorrhiza (EcM), on the other hand, is mainly found on many types of trees and shrubs plants including conifers and oak trees in which its symbiotic association aids in upeaking the nutrients due to scarcity in soils. This association is important in the forestry practice since it aids tree establishment and growth through nitrogen and phosphorus acquisition while enhancing the tolerance of the plants to environmental stresses.

Ericoid mycorrhiza (ErM) in contrast, is an expert in forming symbiotic partnerships with plants belonging to the Ericaceae family, which includes rhododendrons, blueberries, and cranberries. ErM fungi are adaptable in the breakdown of soil organic matter and nutrient cycling because they possess both saprotrophic and biotrophic traits. They are perfect for boosting the growth of commercially valuable ericaceous crops because they help plants acquire vital nutrients and increase their resistance to challenges including drought, heavy metal toxicity, and pathogen attacks. (Wei et al., 2022).

Orchidaceous mycorrhiza (OrM) symbiotically partners with orchids to facilitate the germination of their nearly nutrient-poor seeds. The fungus has a major impact on orchid propagation and conservation efforts because they provide the nutrients needed for seedling establishment.

6. EXOPHYTIC AND SAPROPHYTIC FUNGI

Ectophytic or exo-mycorrhizal fungi are commonly found on the surface of plant organs, such as roots, stems and leaves, but not on the plant cells themselves. These fungi, however, may not establish as deep associations as found with mycorrhizal and endophytic fungi, but can still sometimes benefit the plants that they inhabit. Here, ectophytic fungi appear more as a helping mechanism since they inhibit the activity of pathogenic organisms attacking the plants by actively colonizing the surface of the plant, and thereby starving out, other pathogenic organisms from that surface. Some of these ectophytic fungi such as penicillium, secrete bioactive substances which may serve the purpose of repellents to insects or act on the growth of other parasites. Even if they are not studied as much as the endophytes, they should not be underestimated since the ectophytes are biologically speaking a part of the plants defence system and can thus influence plant health (Gegenbauer et al., 2023).

Saprophytic fungi, in contrast, make use of dead organic matter and are capable of breaking down intricate materials such as cellulose lignin, and proteins. These types of fungi do not directly interact with healthy plants but help in the preservation of soils and their productivity. Because of their process of decomposition, saprophytic fungi are also considered as great recyclers of nutrients such as nitrogen, phosphorus, and potassium derived from organic wastes which are then made available for plant usage. This is a major aspect of enhancing soil productivity as well as the nutrient cycle in various ecosystems. In agriculture, saprophytic fungi can be strategically employed design to enhance soil structure and nutrient presence, thus their relevance in the context of sustainable farming. Furthermore, certain saprophytic fungi such as Trichoderma species have biocontrol activity as they inhibit soil-borne plant pathogens helping in crops protection naturally devoid of artificial insecticides (Abirami et al., 2022).

7. FUTURE PROSPECTS AND CHALLENGES IN FUNGAL BIOSTIMULANT ADOPTION: SCALABILITY, ENVIRONMENT, AND REGULATION

Fungal biostimulants have enormous potential to transform sustainable agriculture because of their substantial positive effects on plant resistance, nutrient uptake, and overall productivity. However, several issues must be resolved before their adoption is beneficial.

Scalability is one of the main areas of attention. Even though fungal biostimulants have been the subject of countless studies, the majority of these studies have been carried out in controlled environments like growth chambers or greenhouses. The intricate interconnections between biotic and abiotic elements that mushrooms encounter in their natural habitats make it difficult to translate these results into open-field applications. Further study is needed to understand how various fungal strains interact with various plant species and environmental factors to maximize their effects in the field.

Furthermore, it is imperative to improve the quality of fungal inoculants. For fungal biostimulants to be successfully and widely used, high-quality inocula with a high concentration of infectious propagules, a long shelf life, and easily manipulated formulations are necessary. This will guarantee that horticulturists and farmers can quickly implement these technologies and get the most out of them.

Combining fungal biostimulants with other advantageous microorganisms, including bacteria that stimulate plant development, is another exciting prospect. Increased productivity and plant health may result from these bacteria' synergistic interactions. Additional investigation into these microbial combinations has the potential to greatly improve the effectiveness of biostimulants.

Regulations must change as well to allow for the application of fungal biostimulants in contemporary farming practices. As of right now, the absence of uniform laws governing biostimulants may lead to misunderstandings about their application and marketing. To ensure the safe and efficient use of fungal biostimulants, governments, regulatory agencies, and industry partners must work together to establish clear rules and standards, including quality control procedures for both manufacturing and distribution. Thus, more research is needed to understand the molecular mechanisms behind the biostimulant effects of fungi on plants, particularly the augmentation of secondary metabolites and nutraceutical substances. This study may open up new avenues for enhancing the nutritional content and quality of agricultural products, hence benefiting human health.

One of the promising directions for advancing fungal biostimulants is the integration with omics technologies. Advances in metagenomics and transcriptomics can provide deeper insights into the microbial communities present in soil and plant environments. These approaches can help identify novel fungal strains with superior biostimulant properties, thereby improving strain selection for agricultural applications. The ability to analyze gene expression in fungi under different environmental conditions can also allow researchers to optimize their biostimulatory potential.

Precision agriculture applications offer another major opportunity for fungal biostimulants. By integrating sensor-based nutrient monitoring, farmers can determine the optimal application rates and timings for fungal inoculants, ensuring maximum efficiency. Smart agricultural technologies can be used to track fungal colonization and effectiveness in real-time, allowing adjustments based on environmental and crop-specific conditions.

Moreover, addressing environmental variability is crucial for the reliable use of fungal biostimulants in agriculture. Since fungi interact differently with plants depending on climate, soil composition, and moisture levels, extensive research is required to understand their performance under diverse agricultural settings. Studies that evaluate fungal biostimulants in a range of ecological conditions will help refine their application strategies, making them more adaptable to different geographic locations.

Metagenomics on the other hand serves as a powerful tool for monitoring and analyzing microbial associations in the rhizosphere. By providing comprehensive insights into microbial diversity and functional interactions, metagenomic approaches enable a deeper understanding of how fungal biostimulants influence plant health and soil ecosystems. Integrating these techniques with targeted inoculation of plant propagules and soils can enhance the effectiveness of fungal biostimulants, optimizing their application for improved crop resilience and sustainability.

Another key challenge lies in scaling fungal biostimulant production while ensuring storage viability. Many fungal inoculants suffer from limited shelf-life and reduced effectiveness over time. Advances in formulation technologies, such as microencapsulation and freeze-drying techniques, can enhance the longevity of fungal spores, improving their commercial viability. Research in this area will be crucial to making fungal biostimulants accessible and effective for a broader range of agricultural systems.

In conclusion, even though there are many benefits to using fungal biostimulants in agriculture, their effective integration into regular farming methods will depend on how well regulatory frameworks, microbial synergy, scalability, and inoculum quality are handled. A more resilient and sustainable agricultural future can be achieved by maximizing the potential of these biostimulants through increased cooperation between scientists, farmers, and policymakers.

REFERENCES

Abirami, S., Sree Gayathri, S., & Usha, C. (2022). Trichoderma as biostimulant - a plausible approach to alleviate abiotic stress for intensive production practices. In Singh, H. B., & Vaishnav, A. (Eds.), *New and Future Developments in Microbial Biotechnology and Bioengineering* (pp. 57–84). Elsevier., https://doi .org/https://doi.org/10.1016/B978-0-323-85577-8.00004-4 DOI: 10.1016/B978-0-323-85577-8.00004-4

Akhtar, M. S., & Abdullah, S. N. A. (2014). Mass production techniques of arbuscular mycorrhizal fungi: Major advantages and disadvantages: A review. In *Biosciences Biotechnology Research Asia* (Vol. 11, Issue 3, pp. 1199–1204). Oriental Scientific Publishing Company. DOI: 10.13005/bbra/1506

Akiyama, K., Matsuzaki, K., & Hayashi, H. (2005). Plant sesquiterpenes induce hyphal branching in arbuscular mycorrhizal fungi. *Nature*, *435*(7043), 824–827. DOI: 10.1038/nature03608 PMID: 15944706

Asrar, A.-W., Abdel-Fattah, G., & Elhindi, K. M. (2012). Improved growth, flower yield, and water relations of snapdragon (Antirhinum majus L.) plants grown under well-watered and water-stress conditions using arbuscular mycorrhizal fungi. *Photosynthetica*, *50*(2), 305–316. DOI: 10.1007/s11099-012-0024-8

Asrar, A.-W. A., & Elhindi, K. M. (2011). Alleviation of drought stress of marigold (Tagetes erecta) plants by using arbuscular mycorrhizal fungi. *Saudi Journal of Biological Sciences*, *18*(1), 93–98. https:// doi.org/https://doi.org/10.1016/j.sjbs.2010.06.007. DOI: 10.1016/j.sjbs.2010.06.007 PMID: 23961109

Balasubramaniam, T., Shen, G., Esmaeili, N., & Zhang, H. (2023). Plants' Response Mechanisms to Salinity Stress. *Plants*, *12*(12), 2253. Advance online publication. DOI: 10.3390/plants12122253 PMID: 37375879

Bhupenchandra, I., Chongtham, S. K., Devi, E. L., Choudhary, A. K., Salam, M. D., Sahoo, M. R., & Khaba, C. I. (2022). Role of biostimulants in mitigating the effects of climate change on crop performance. *Frontiers in Plant Science*, *13*, 967665.

Bonfante, P., & Genre, A. (2010). Mechanisms underlying beneficial plant–fungus interactions in mycorrhizal symbiosis. *Nature Communications*, *1*(1), 48. DOI: 10.1038/ncomms1046 PMID: 20975705

Bozhinova, R. (2023). Yield and chemical composition of oriental tobacco as affected by biostimulant application. [). https://cabidigitallibrary.org]. *Bulgarian Journal of Agricultural Science*, *29*(1), •••.

Calvet, C., Camprubí, A., Pérez-Hernández, A., & Lovato, P. (2013). Plant Growth Stimulation and Root Colonization Potential of In Vivo versus In Vitro Arbuscular Mycorrhizal Inocula. *HortScience: A Publication of the American Society for Horticultural Science, 48*, 897–901. DOI: 10.21273/HORTSCI.48.7.897

Calvo, P., Nelson, L., & Kloepper, J. W. (2014). Agricultural uses of plant biostimulants. *Plant and Soil*, *383*(1), 3–41. DOI: 10.1007/s11104-014-2131-8

Chiu, C. H., & Paszkowski, U. (2019). Mechanisms and impact of symbiotic phosphate acquisition. *Cold Spring Harbor Perspectives in Biology*, *11*(6), a034603. Advance online publication. DOI: 10.1101/cshperspect.a034603 PMID: 30910773

Chun, S. C., Paramasivan, M., & Chandrasekaran, M. (2018). Proline Accumulation Influenced by Osmotic Stress in Arbuscular Mycorrhizal Symbiotic Plants. *Frontiers in Microbiology*, *9*, 2525. Advance online publication. DOI: 10.3389/fmicb.2018.02525 PMID: 30459731

Cranenbrouck, S., Voets, L., Bivort, C., Renard, L., Strullu, D.-G., & Declerck, S. (2005). Methodologies for in Vitro Cultivation of Arbuscular Mycorrhizal Fungi with Root Organs. In Declerck, S., Fortin, J. A., & Strullu, D.-G. (Eds.), *In Vitro Culture of Mycorrhizas* (pp. 341–375). Springer Berlin Heidelberg., DOI: 10.1007/3-540-27331-X_18

Douds, D. D.Jr, Lee, J., Rogers, L., Lohman, M. E., Pinzon, N., & Ganser, S. (2012). Utilization of inoculum of AM fungi produced on-farm for the production of Capsicum annuum: A summary of seven years of field trials on a conventional vegetable farm. *Biological Agriculture and Horticulture*, *28*(2), 129–145. DOI: 10.1080/01448765.2012.693362

Druzhinina, I. S., Seidl-Seiboth, V., Herrera-Estrella, A., Horwitz, B. A., Kenerley, C. M., Monte, E., Mukherjee, P. K., Zeilinger, S., Grigoriev, I. V., & Kubicek, C. P. (2011). Trichoderma: The genomics of opportunistic success. In *Nature Reviews Microbiology* (Vol. 9, Issue 10, pp. 749–759). DOI: 10.1038/nrmicro2637

du Jardin, P. (2015). Plant biostimulants: Definition, concept, main categories and regulation. *Scientia Horticulturae*, *196*, 3–14. https://doi.org/https://doi.org/10.1016/j.scienta.2015.09.021. DOI: 10.1016/j.scienta.2015.09.021

Duhamel, M., & Vandenkoornhuyse, P. (2013). Sustainable agriculture: Possible trajectories from mutualistic symbiosis and plant neodomestication. *Trends in Plant Science*, *18*(11), 597–600. DOI: 10.1016/j.tplants.2013.08.010 PMID: 24055138

Esaïe, T. (2015). Early effects of water stress on some biochemical and mineral parameters of mycorrhizal Vigna subterranea (L.) Verdc. (Fabaceae) cultivated in Cameroon http://www.innspub.net Vol. 7, No. 2, p. 21-35, 2015. *IJAAR, 7*, 21–35.

Etesami, H., & Glick, B. R. (2023). Exploring the potential: Can mycorrhizal fungi and hyphosphere silicate-solubilizing bacteria synergistically alleviate cadmium stress in plants? *Current Research in Biotechnology*, *6*, 100158. https://doi.org/https://doi.org/10.1016/j.crbiot.2023.100158. DOI: 10.1016/j.crbiot.2023.100158

Etesami, H., Jeong, B. R., & Glick, B. R. (2021). Contribution of Arbuscular Mycorrhizal Fungi, Phosphate–Solubilizing Bacteria, and Silicon to P Uptake by Plant. *Frontiers in Plant Science*, *12*, 699618. Advance online publication. DOI: 10.3389/fpls.2021.699618 PMID: 34276750

European Biostimulants Industry Council (EBIC). (2012). *The European Biostimulants Industry Council*. EBIC.

Feldmann, F., & Schneider, C. (2008). How to produce arbuscular mycorrhizal inoculum with desired characteristics. *Mycorrhiza Work in Horticulture and Forestry*, 305–322.

Fenu, G., & Malloci, F. M. (2021). Forecasting Plant and Crop Disease: An Explorative Study on Current Algorithms. *Big Data and Cognitive Computing*, *5*(1), 2. Advance online publication. DOI: 10.3390/bdcc5010002

Gegenbauer, C., Bellaire, A., Schintlmeister, A., Schmid, M. C., Kubicek, M., Voglmayr, H., Zotz, G., Richter, A., & Mayer, V. E. (2023). Exo- and endophytic fungi enable rapid transfer of nutrients from ant waste to orchid tissue. *The New Phytologist*, *238*(5), 2210–2223. https://doi.org/https://doi.org/10.1111/nph.18761. DOI: 10.1111/nph.18761 PMID: 36683444

Genre, A., Chabaud, M., Timmers, T., Bonfante, P., & Barker, D. G. (2005). Arbuscular Mycorrhizal Fungi Elicit a Novel Intracellular Apparatus in Medicago truncatula Root Epidermal Cells before Infection [W]. *The Plant Cell*, *17*(12), 3489–3499. DOI: 10.1105/tpc.105.035410 PMID: 16284314

Ghorbanpour, M., Omidvari, M., Abbaszadeh-Dahaji, P., Omidvar, R., & Kariman, K. (2018). Mechanisms underlying the protective effects of beneficial fungi against plant diseases. *Biological Control*, *117*, 147–157. https://doi.org/https://doi.org/10.1016/j.biocontrol.2017.11.006. DOI: 10.1016/j.biocontrol.2017.11.006

Gianinazzi-Pearson, V., & Gianinazzi, S. (1988). Morphological Integration and Functional Compatibility Between Symbionts in Vesicular Arbuscular Endomycorrhizal Associations. In Scannerini, S., Smith, D., Bonfante-Fasolo, P., & Gianinazzi-Pearson, V. (Eds.), *Cell to Cell Signals in Plant, Animal and Microbial Symbiosis* (pp. 73–84). Springer Berlin Heidelberg. DOI: 10.1007/978-3-642-73154-9_5

Giovannetti, M., Sbrana, C., Avio, L., Citernesi, A. S., & Logi, C. (1993). Differential hyphal morphogenesis in arbuscular mycorrhizal fungi during pre-infection stages. *The New Phytologist*, *125*(3), 587–593. DOI: 10.1111/j.1469-8137.1993.tb03907.x PMID: 33874594

Goicoechea, N., & Antolín, M. C. (2017). Increased nutritional value in food crops. *Microbial Biotechnology*, *10*(5), 1004–1007. https://doi.org/https://doi.org/10.1111/1751-7915.12764. DOI: 10.1111/1751-7915.12764 PMID: 28696049

Huang, Y.-F., & Lo, S.-L. (2019). Predicting heating value of lignocellulosic biomass based on elemental analysis. *Energy*, *191*, 116501. DOI: 10.1016/j.energy.2019.116501

Huo, W. G., Chai, X. F., Wang, X. H., Batchelor, W. D., Kafle, A., & Gu, F. E. N. G. (2022). Indigenous arbuscular mycorrhizal fungi play a role in phosphorus depletion in organic manure amended high fertility soil. *Journal of Integrative Agriculture*, *21*(10), 3051–3066.

Jan, R., Asaf, S., Numan, M., Lubna, , & Kim, K.-M. (2021). Plant Secondary Metabolite Biosynthesis and Transcriptional Regulation in Response to Biotic and Abiotic Stress Conditions. *Agronomy (Basel)*, *11*(5), 968. Advance online publication. DOI: 10.3390/agronomy11050968

Kapoor, R., & Bhatnagar, A. (2007). Attenuation of cadmium toxicity in mycorrhizal celery (Apium graveolens L.). *World Journal of Microbiology & Biotechnology*, *23*(8), 1083–1089. DOI: 10.1007/s11274-006-9337-8

Kaya, C., Ashraf, M., Sonmez, O., Aydemir, S., Tuna, A., & Çullu, M. (2009). The influence of arbuscular mycorrhizal colonisation on key growth parameters and fruit yield of pepper plants grown at high salinity. *Scientia Horticulturae - SCI HORT-AMSTERDAM*, *121*, 1–6. DOI: 10.1016/j.scienta.2009.01.001

Koide, R. T., & Kabir, Z. (2000). Extraradical hyphae of the mycorrhizal fungus Glomus intraradices can hydrolyse organic phosphate. *The New Phytologist*, *148*(3), 511–517.

Latef, A. (2011). Influence of arbuscular mycorrhizal fungi and copper on growth, accumulation of osmolyte, mineral nutrition and antioxidant enzyme activity of pepper (Capsicum annuum L.). *Mycorrhiza*, *21*(6), 495–503. DOI: 10.1007/s00572-010-0360-0 PMID: 21221660

Malviya, D., Singh, P., Singh, U. B., Paul, S., Kumar Bisen, P., Rai, J. P., Verma, R. L., Fiyaz, R. A., Kumar, A., Kumari, P., Dei, S., Ahmed, M. R., Bagyaraj, D. J., & Singh, H. V. (2023). Arbuscular mycorrhizal fungi-mediated activation of plant defense responses in direct seeded rice (Oryza sativa L.) against root-knot nematode Meloidogyne graminicola. *Frontiers in Microbiology*, *14*, 1104490. Advance online publication. DOI: 10.3389/fmicb.2023.1104490 PMID: 37200920

Mikkelsen, B. L., Rosendahl, S., & Jakobsen, I. (2008). Underground resource allocation between individual networks of mycorrhizal fungi. *The New Phytologist*, *180*(4), 890–898. DOI: 10.1111/j.1469-8137.2008.02623.x PMID: 18801003

Mirshad, P. P., & Puthur, J. T. (2016). Arbuscular mycorrhizal association enhances drought tolerance potential of promising bioenergy grass (Saccharum arundinaceum retz.). *Environmental Monitoring and Assessment*, *188*(7), 425. DOI: 10.1007/s10661-016-5428-7 PMID: 27329476

Navarro, A., Elia, A., Conversa, G., Pasquale, C., & Mastrorilli, M. (2012). Potted mycorrhizal carnation plants and saline stress: Growth, quality and nutritional plant responses. *Scientia Horticulturae*, *140*, 131–139. DOI: 10.1016/j.scienta.2012.03.016

Owen, D., Williams, A. P., Griffith, G. W., & Withers, P. J. A. (2015). Use of commercial bio-inoculants to increase agricultural production through improved phosphrous acquisition. *Applied Soil Ecology*, *86*, 41–54. https://doi.org/https://doi.org/10.1016/j.apsoil.2014.09.012. DOI: 10.1016/j.apsoil.2014.09.012

Pedranzani, H., Rodríguez-Rivera, M., Gutiérrez, M., Porcel, R., Hause, B., & Ruiz-Lozano, J. M. (2016). Arbuscular mycorrhizal symbiosis regulates physiology and performance of Digitaria eriantha plants subjected to abiotic stresses by modulating antioxidant and jasmonate levels. *Mycorrhiza*, *26*(2), 141–152. DOI: 10.1007/s00572-015-0653-4 PMID: 26184604

Porras-Soriano, A., Soriano-Martín, M. L., Porras-Piedra, A., & Azcón, R. (2009). Arbuscular mycorrhizal fungi increased growth, nutrient uptake and tolerance to salinity in olive trees under nursery conditions. *Journal of Plant Physiology*, *166*(13), 1350–1359. https://doi.org/https://doi.org/10.1016/j.jplph.2009.02.010. DOI: 10.1016/j.jplph.2009.02.010 PMID: 19342122

Pradhan, J., Sarma, A., Kalita, S., Talukdar, R., & Tayung, K. (2023). Plant Growth-Promoting Effect of Seed-Borne Endophytic Fungi Isolated from Senna Alata (L.) Roxb. *Proceedings of the National Academy of Sciences. India. Section B, Biological Sciences*, *93*(4), 901–908. DOI: 10.1007/s40011-023-01483-2

Pusztahelyi, T., Holb, I. J., & Pócsi, I. (2015). Secondary metabolites in fungus-plant interactions. *Frontiers in Plant Science*, *6*. Advance online publication. DOI: 10.3389/fpls.2015.00573 PMID: 26300892

Robinson Boyer, L., Brain, P., Xu, X., & Jeffries, P. (2014). Inoculation of drought-stressed strawberry with a mixed inoculum of two arbuscular mycorrhizal fungi: Effects on population dynamics of fungal species in roots and consequential plant tolerance to water deficiency. *Mycorrhiza*, *25*(3), 215–227. Advance online publication. DOI: 10.1007/s00572-014-0603-6 PMID: 25186649

Rouphael, Y., & Colla, G. (2020). Editorial: Biostimulants in Agriculture. In *Frontiers in Plant Science* (Vol. 11). Frontiers Media S.A. DOI: 10.3389/fpls.2020.00040

Rouphael, Y., Franken, P., Schneider, C., Schwarz, D., Giovannetti, M., Agnolucci, M., De Pascale, S., Bonini, P., & Colla, G. (2015). Arbuscular mycorrhizal fungi act as biostimulants in horticultural crops. *Scientia Horticulturae*, *196*, 91–108. https://doi.org/https://doi.org/10.1016/j.scienta.2015.09.002. DOI: 10.1016/j.scienta.2015.09.002

Saeed, Q., Xiukang, W., Haider, F. U., Kučerik, J., Mumtaz, M. Z., Holatko, J., Naseem, M., Kintl, A., Ejaz, M., Naveed, M., Brtnicky, M., & Mustafa, A. (2021). Rhizosphere Bacteria in Plant Growth Promotion, Biocontrol, and Bioremediation of Contaminated Sites: A Comprehensive Review of Effects and Mechanisms. *International Journal of Molecular Sciences*, *22*(19), 10529. Advance online publication. DOI: 10.3390/ijms221910529 PMID: 34638870

Shahrajabian, M. H., Chaski, C., Polyzos, N., & Petropoulos, S. A. (2021). Biostimulants Application: A Low Input Cropping Management Tool for Sustainable Farming of Vegetables. *Biomolecules*, *11*(5), 698. Advance online publication. DOI: 10.3390/biom11050698 PMID: 34067181

Smith, S. E. R. D. J. (2008). *Mycorrhizal Symbiosis* (3rd ed.). Academic Press, London.

Song, Y. Y., Zeng, R. Sen, Xu, J. F., Li, J., Shen, X., & Yihdego, W. G. (2010). Interplant Communication of Tomato Plants through Underground Common Mycorrhizal Networks. *PLOS ONE, 5*(10), e13324-. DOI: 10.1371/journal.pone.0013324

Sun, W., & Shahrajabian, M. H. (2023). The Application of Arbuscular Mycorrhizal Fungi as Microbial Biostimulant, Sustainable Approaches in Modern Agriculture. *Plants*, *12*(17), 3101. Advance online publication. DOI: 10.3390/plants12173101 PMID: 37687348

Tamasloukht, B., Kuhn, B., Becard, G., & Franken, P. (2000). RNA accumulation patterns of arbuscular mycorrhizal fungi during presymbiotic stages. *Programs, Abstratcts and Papers of the Third International Congress on Symbiosis.*

Thepbandit, W., & Athinuwat, D. (2024). Rhizosphere Microorganisms Supply Availability of Soil Nutrients and Induce Plant Defense. *Microorganisms*, *12*(3), 558. Advance online publication. DOI: 10.3390/microorganisms12030558 PMID: 38543610

Tyśkiewicz, R., Nowak, A., Ozimek, E., & Jaroszuk-Ściseł, J. (2022). Trichoderma: The Current Status of Its Application in Agriculture for the Biocontrol of Fungal Phytopathogens and Stimulation of Plant Growth. *International Journal of Molecular Sciences*, *23*(4), 2329. Advance online publication. DOI: 10.3390/ijms23042329 PMID: 35216444

Vasilachi, I. C., Stoleru, V., & Gavrilescu, M. (2023). Analysis of Heavy Metal Impacts on Cereal Crop Growth and Development in Contaminated Soils. *Agriculture*, *13*(10), 1983. Advance online publication. DOI: 10.3390/agriculture13101983

Voets, L., Dupré de Boulois, H., Renard, L., Strullu, D.-G., & Declerck, S. (2005). Development of an autotrophic culture system for the in vitro mycorrhization of potato plantlets. *FEMS Microbiology Letters*, *248*(1), 111–118. DOI: 10.1016/j.femsle.2005.05.025 PMID: 15941625

Walker, C. (1995). AM or VAM: What's in a Word? In Varma, A., & Hock, B. (Eds.), *Mycorrhiza: Structure, Function, Molecular Biology and Biotechnology* (pp. 25–26). Springer Berlin Heidelberg., DOI: 10.1007/978-3-662-08897-5_2

Wei, X., Zhang, W., Zulfiqar, F., Zhang, C., & Chen, J. (2022). Ericoid mycorrhizal fungi as biostimulants for improving propagation and production of ericaceous plants. *Frontiers in Plant Science*, *13*, 1027390. Advance online publication. DOI: 10.3389/fpls.2022.1027390 PMID: 36466284

White, P. J., & Brown, P. H. (2010). Plant nutrition for sustainable development and global health. *Annals of Botany*, *105*(7), 1073–1080. DOI: 10.1093/aob/mcq085 PMID: 20430785

Wu, Q.-S., Srivastava, A. K., & Zou, Y.-N. (2013). AMF-induced tolerance to drought stress in citrus: A review. *Scientia Horticulturae*, *164*, 77–87. https://doi.org/https://doi.org/10.1016/j.scienta.2013.09.010. DOI: 10.1016/j.scienta.2013.09.010

Xiao, J., Hu, C., Chen, Y., Yang, B., & Hua, J. (2014). Effects of low magnesium and an arbuscular mycorrhizal fungus on the growth, magnesium distribution and photosynthesis of two citrus cultivars. *Scientia Horticulturae*, *177*, 14–20. DOI: 10.1016/j.scienta.2014.07.016

Yakhin, O. I., Lubyanov, A. A., Yakhin, I. A., & Brown, P. H. (2017). Biostimulants in Plant Science: A Global Perspective. *Frontiers in Plant Science*, *7*. Advance online publication. DOI: 10.3389/fpls.2016.02049 PMID: 28184225

Zhao, R., Guo, W., Bi, N., Guo, J., Wang, L., Zhao, J., & Zhang, J. (2015). Arbuscular mycorrhizal fungi affect the growth, nutrient uptake and water status of maize (Zea mays L.) grown in two types of coal mine spoils under drought stress. *Applied Soil Ecology*, *88*, 41–49. https://doi.org/https://doi.org/10.1016/j.apsoil.2014.11.016. DOI: 10.1016/j.apsoil.2014.11.016

Zia, R., Nawaz, M. S., Siddique, M. J., Hakim, S., & Imran, A. (2021). Plant survival under drought stress: Implications, adaptive responses, and integrated rhizosphere management strategy for stress mitigation. *Microbiological Research*, *242*, 126626. https://doi.org/https://doi.org/10.1016/j.micres.2020.126626. DOI: 10.1016/j.micres.2020.126626 PMID: 33189069

Chapter 3
Fungal Biocontrol Agents for Sustainable Pest Management

Vikas Chandra
https://orcid.org/0000-0003-2686-2373
Guru Ghasidas Vishwavidyalaya, India

Abhishek Dadsena
Guru Ghasidas Vishwavidyalaya, India

Harsha Yadav
Guru Ghasidas Vishwavidyalaya, India

Gurucharan Sahu
Guru Ghasidas Vishwavidyalaya, India

ABSTRACT

The rising demand for sustainable agriculture has increased interest in fungal biocontrol agents as eco-friendly alternatives to synthetic pesticides. This chapter examines fungi's role in pest management through mechanisms like mycoparasitism, resource competition, and insecticidal compound production. Key species, including Beauveria bassiana, Metarhizium anisopliae, and Trichoderma spp., effectively control insect pests, nematodes, and pathogens, reducing pest populations and enhancing crop yields. Benefits include minimal non-target effects, lower residues, improved soil health, and cost efficiency. Field studies demonstrate success in crops like cotton and vegetables. However, challenges such as field variability, limited shelf life, and regulatory constraints remain. Research on improved formulations and delivery methods is essential for wider adoption and long-term agricultural sustainability.

1. INTRODUCTION

Growing public concern over the misuse of agricultural chemicals and their environmental impact has driven research into safer, eco-friendly pest and disease control methods. Managing plant diseases is crucial for food, feed, and fiber production. While farmers traditionally use chemical fertilizers and pesticides, their environmental risks have led to stricter regulations and efforts to phase out harmful

DOI: 10.4018/979-8-3693-8201-1.ch003

chemicals. Consequently, researchers are developing alternative pest management strategies that reduce reliance on synthetic chemicals, promoting sustainable and environmentally friendly agricultural practices. Humans have utilized bacteria, algae, fungi, plants, animals, and their in vitro cultures to develop valuable applications and products (Chandra et al., 2022; Sheikh et al., 2022; Agrawal et al., 2023; Kosariya et al., 2023; Kumar et al., 2024). Biological pest and disease control is the most environmentally friendly strategy, relying on living organisms to manage insects, mites, weeds, and pathogens. Effective implementation requires understanding the complex interactions among plants, pests, microbes, and the environment. Biocontrol reduces disease-causing propagules naturally or by modifying the environment, host, or biocontrol agent. It involves parasites, predators, or microorganisms to lower pathogen populations. Broadly, it includes using natural compounds from various sources to control pests and diseases sustainably. Biocontrol depends on the goal, biological agent type, source, and intervention period. It involves using one or more organisms, often natural enemies, to control harmful organisms (Jyoti & Singh, 2016). Fungi, such as *Beauveria*, *Metarhizium*, *Cordyceps*, *Purpureocillium*, *Lecanicillium*, and *Trichoderma*, are effective biocontrol agents (Baron et al., 2019). Their use depends on virulence, environmental resistance, mass production feasibility, and host specificity (Pourseyed et al., 2010), making them valuable for pest and disease control. This chapter examines fungal biocontrol agents (FBAs), their mechanisms against insect pests and phytopathogens, and their role in sustainable agriculture. FBAs suppress plant diseases and pests through mycoparasitism, competition, antibiosis, entomopathogenesis, and systemic resistance induction. They aid in disease management, pest control, and weed suppression, reducing chemical pesticide use. Despite challenges like field variability, regulatory barriers, and environmental stress, FBAs enhance crop health, lower pest populations, and support agricultural sustainability. Figure 1 illustrates key aspects of FBAs.

Figure 1. Key aspects of Fungal Biocontrol Agents (FBAs), including their mechanisms of action, applications, advantages, challenges, and overall impact

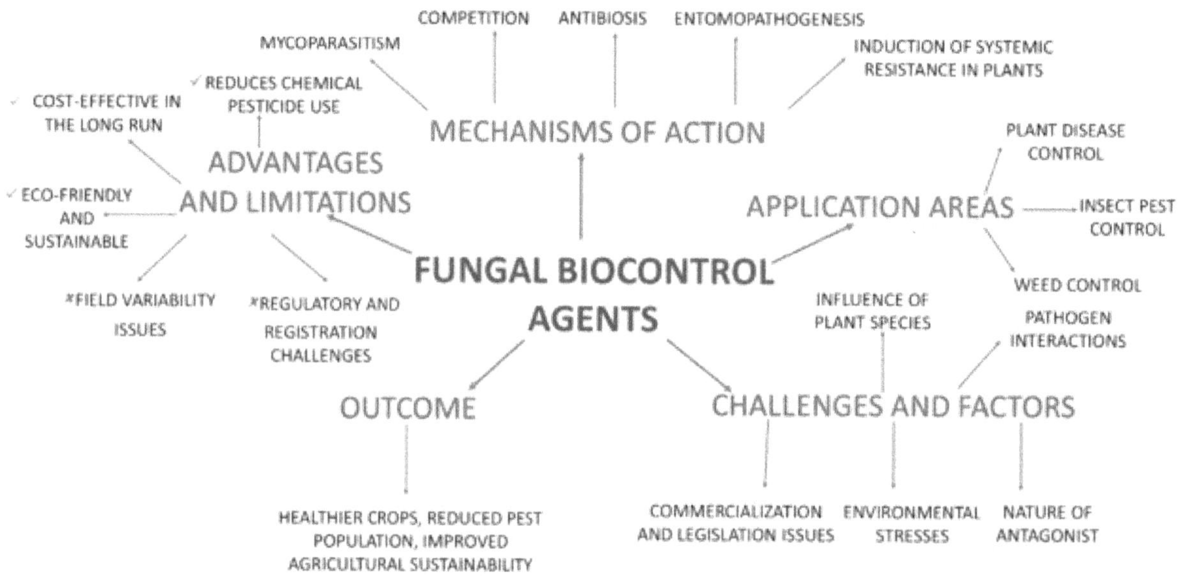

2. HISTORY OF BIOCONTROL AGENTS

The concept of harnessing living organisms to manage pest populations is not new. However, modern biological control techniques, which gained momentum in the late 19th century, have drastically evolved from their ancient origins, offering sustainable alternatives to synthetic chemicals. Some of the important landmarks in this area are highlighted in figure 2.

Figure 2. History of biocontrol agents

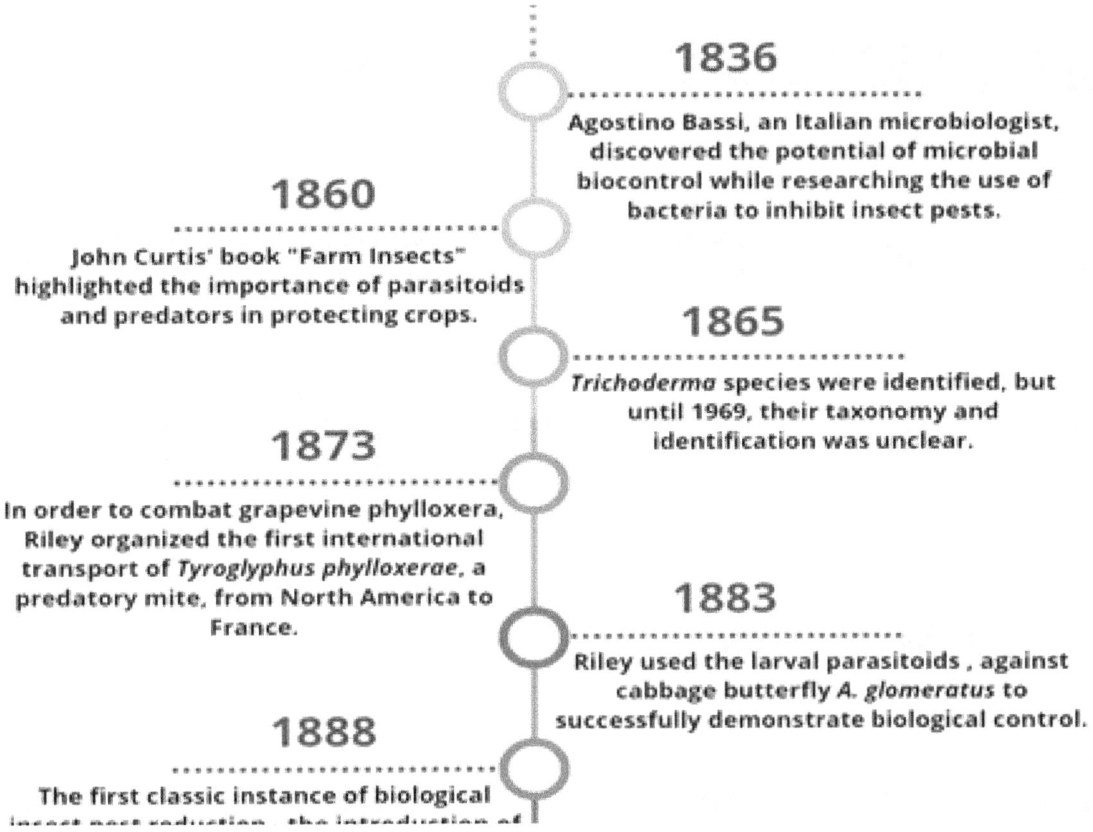

(Jyoti & Singh, 2016)

3. FUNGI AS BIOCONTROL AGENT

The investigation and application of fungal biocontrol agents (FBAs) against insect pests and plant diseases have gained significant attention due to fungi's unique biological attributes. These include high growth and replication rates, efficient spreading ability, short generation time, target specificity, and ease of laboratory cultivation with minimal maintenance. Moreover, in the absence of a suitable host, fungal biocontrol agents persist in the environment as saprophytes rather than acting as parasites. Since ancient times, humans have explored alternative agricultural strategies to enhance crop yield while minimizing pest and disease intensity (Singh & Chawla, 2012; Gupta & Sharma, 2014).

The potential of microorganisms in disease suppression led to the development of various biocontrol strategies using fungi. Hartley (1921) pioneered soil inoculation with antagonistic microbes, introducing fungal strains to reduce damping-off caused by *Pythium debaryanum* (Gupta & Sharma, 2014). In 1941, *Trichoderma* species were identified as antagonists against *Rhizoctonia solani* and *Sclerotium americana* (Weindling, 1941). *Trichoderma* produces gliotoxin, an antimycotic compound that suppresses plant pathogens, marking the first demonstration of fungal antimycotic antagonism in disease management (Howell, 2003).

Biotechnological advancements have expanded the potential of FBAs, leading to the validation of new fungal strains effective across various environmental conditions. FBAs serve as viable alternatives where chemical pesticides, such as organochlorines and methyl bromide, face restrictions or where pest resistance to conventional pesticides emerges. The use of *Puccinia chondrillina* to control rush skeleton weed (*Chondrilla juncea*) highlights fungi's effectiveness as biocontrol agents.

Entomopathogenic fungi (EPF) have attracted considerable interest due to their ability to infect insect pests at all life stages. The extent of infection depends on spore adhesion and penetration through the host exoskeleton. However, commercialization of fungal biocontrol agents faces challenges, including socioeconomic factors and biological complexities (Singh et al., 2017). Key EPF genera include *Beauveria*, *Metarhizium*, *Isaria*, *Lecanicillium*, and *Hirsutella*, which inhabit soil and interact with plant roots (Meyling & Eilenberg, 2007). Some EPF, such as *Beauveria bassiana* and *Metarhizium anisopliae*, function as endophytes, enhancing plant resistance by producing metabolites like phenolic acids, quinones, and chitinases (Akutse et al., 2013; Zaynab et al., 2018). Beyond agriculture, FBAs hold promise for controlling animal and human pests, including fire ants, tsetse flies, and mosquitoes (Sharma & Sharma, 2021).

4. MECHANISM OF ACTION

Biocontrol agents employ various mechanisms to suppress plant pathogens, insect pests, and weeds. Their efficacy depends on interspecies interactions, including mycoparasitism, competition, antibiosis, enzyme production, and induced resistance (activation of plant defense responses). Mycoparasitism, the most direct antagonism, involves attacking and degrading pathogens. Figure 3 illustrates these modes of action, highlighting their role in promoting plant health and disease suppression.

Figure 3. Modes of action of fungal biocontrol agents

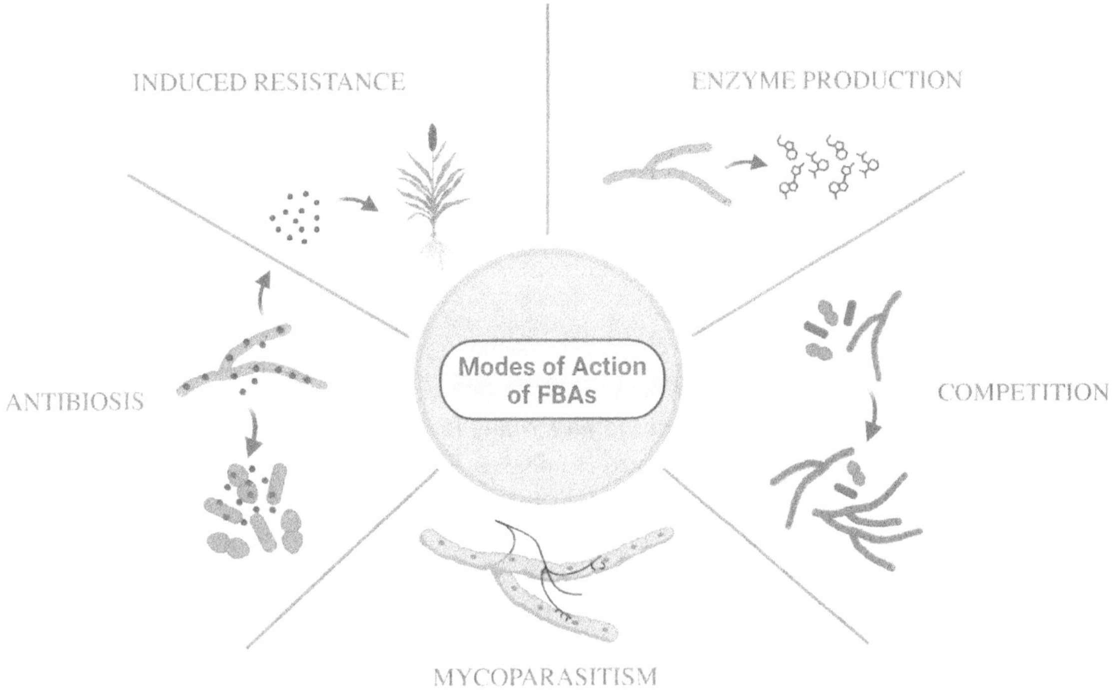

4.1. Mycoparasitism

Mycoparasitism, a form of hyperparasitism, occurs when one fungus parasitizes another, using its nutrients and body for growth. This ability makes fungi effective biocontrol agents, as they inhibit pathogenic fungi by feeding on them, ultimately causing their death (Atanasova, 2013). The process involves recognition, attachment, and penetration, often through specialized structures like appressoria. Once attached, the parasitic fungus secretes enzymes such as chitinases, cellulases, and glucanases, breaking down the host's cell wall (Rabea et al., 2003; Horbach et al., 2011), leading to tissue digestion and pathogen elimination. Ampelomyces quisqualis parasitizes powdery mildew (*Erysiphales*), invading its hyphae, conidiophores, and cleistothecia to form pycnidia within the host. This hyperparasitic interaction promotes *A. quisqualis* growth while suppressing powdery mildew, ultimately leading to its death (Jyoti & Singh, 2016). *Trichoderma lignorum* (now *Trichoderma viride*) was first reported in 1932 to parasitize *Rhizoctonia solani* hyphae (Weindling, 1941). In 1997, Lo suggested using *Trichoderma* spores to control citrus seedling damping-off. *Trichoderma* species effectively target diverse pathogens (Harman et al., 2004). Other fungal parasites like *Pythium oligandrum* and *Coniothyrium minitans* attack plant pathogens, while multiple hyperparasites, including *Acremonium alternatum*, *Acrodontium crateriforme*, and *Cladosporium oxysporum*, parasitize fungi responsible for powdery mildew (Kiss, 2003).

4.2. Competition

Competition for space and nutrients is essential in phytopathogen biocontrol. Starvation limits fungal pathogen survival, making resource competition a key suppression mechanism. Soil and plant-surface microbes compete when nutrients are scarce. *Trichoderma* spp. produces siderophores to extract iron, inhibiting pathogen growth (Eisendle et al., 2004). *Trichoderma harzianum* T35 suppresses *Fusarium oxysporum* through nutrient competition (Tjamos et al., 1992). Efficient iron uptake helps commensal microbes colonize roots, outcompeting pathogens like *Botrytis cinerea*, highlighting nutrient competition's role in biocontrol.

4.3. Antibiosis

Antibiosis is a biocontrol mechanism where antagonistic fungi secrete antimicrobial compounds to inhibit or eliminate pathogenic fungi in their vicinity. Many fungi produce secondary metabolites with antibiotic properties, often linked to specific growth phases. Some of these metabolites influence plant growth, while others target fungal processes such as hyphal elongation and sporulation (Keller et al., 2005). *Trichoderma* spp. are particularly effective in reducing pathogenic fungal sclerotia lifespan through antagonistic metabolites (Szekeres et al., 2005). These compounds disrupt lipid membranes, exhibit antimicrobial activity, and induce plant resistance. Certain *Trichoderma* species also produce enzymes that degrade fungal cell walls. Peptaibols and peptaibiotics, linear amphipathic polypeptides, function alongside cell-wall-degrading enzymes (CWDEs) to inhibit pathogens (Wiest et al., 2002; Szekeres et al., 2005). These non-ribosomal peptides form pores in lipid membranes, enhancing microbial inhibition. Exogenous peptaibol treatments in tobacco reduced susceptibility to the tobacco mosaic virus, triggering plant defense responses. *Trichoderma* species generate three primary secondary metabolites: (i) water-soluble compounds like koningic acid, (ii) volatile antibiotics like 6-pentyl-α-pyrone, and (iii) peptaibols containing amino-isobutyric acid (Mukherjee et al., 2012; Zeilinger et al., 2016). These metabolites play a crucial role in antagonism against *Rhizoctonia solani* and *Pythium ultimum* (Hanson & Howell, 2002).

4.4. Induction of Systemic Resistance

Studies have demonstrated that *Trichoderma* spp. produce diverse compounds within their interaction zone, inducing plant resistance mechanisms. These fungi secrete proteins like swollenin and xylanase, which exhibit enzymatic activity and trigger localized necrotic responses in plants (Martinez et al., 2001; Brotman et al., 2008). Additionally, *Trichoderma* endochitinase activates plant defense-related proteins, enhancing resistance (Harman & Shoresh, 2007). *T. virens* strains also release hydrophobin-like proteins that stimulate terpenoid phytoalexin synthesis (Hanson & Howell, 2004; Djonovic et al., 2007).

Many fungal biocontrol agents synthesize proteins encoded by avirulence-like (*Avr*) genes, which trigger plant defense responses (Woo et al., 2006). Peptabiols, a class of secondary metabolites, stimulate plant defenses and exhibit antibacterial properties (Szekeres et al., 2005). Certain *Trichoderma* metabolites enhance pathogenesis-related (PR) protein expression and reduce disease severity (Vinale et al., 2008a, b), modulating Ca^{2+} and H^{+} signaling (Vadassery et al., 2009). *Piriformospora indica* similarly induces systemic resistance (Stein et al., 2008). Plant-microbe interactions trigger systemic acquired resistance (SAR) and induced systemic resistance (ISR). SAR, activated by local infection, provides long-term defense and is regulated by salicylic acid (SA) signaling (Durrant & Dong, 2004).

4.5. Entomopathogenic Fungi (EPF)

Numerous insect species from the Lepidoptera, Coleoptera, Homoptera, Hymenoptera, and Diptera orders are susceptible to infection by EPF. The infection process begins when fungal conidia or spores adhere to the insect's cuticle. In Beauveria bassiana, dry spores interweave with the fascicles of hydrophobic rodlets in the outer layer. The conidial surface contains chemical components such as lectins and glycoproteins that facilitate adhesion by binding to carbohydrates. Once attached, conidia germinate and develop an infectious structure known as the appressorium. This structure, along with the penetration peg, helps breach the insect's exoskeleton. The penetration process is facilitated by extracellular enzymes, including proteases, chitinases, lipases, and esterases, which degrade the cuticle. Among these, endoprotease plays a crucial role. Biochemical and histochemical studies have identified these enzymes in *Metarhizium anisopliae* during infection of *Calliphora vomitoria* and *Manduca sexta*. Upon penetration, the fungus spreads into the hemocoel, forming blastospores that establish infection. EPF species employ different penetration strategies—*B. bassiana* and *M. anisopliae* develop specialized infection hyphae, while *Lecanicillium lecanii* penetrates using its germ tube. During mycosis, fungi release cytotoxins that disrupt cellular function, leading to behavioral symptoms such as sluggishness and paralysis. Neuromuscular toxins from *B. bassiana* and *M. anisopliae* cause these effects. Additionally, *M. anisopliae* produces destruxins (DTXs) and cytochalasins, while *B. bassiana* releases beauvericin, brassinolide, isarolide, and beauverolides, further weakening the host (Leger et al., 1987; Sandhu, 1995).

Figure 4 illustrates the infection process of a fungal pathogen, showing key structures involved. The infection process shown in the figure is relevant to EPF like *Beauveria bassiana* and *Metarhizium anisopliae*, which infect insect pests through similar steps (spore adhesion, germination, penetration, and colonization). It is also relevant to mycoparasitic fungi like *Trichoderma,* which invade other fungal pathogens. The role of entomopathogenic fungi as biocontrol agent is summaried in Table 1.

Figure 4. Infection process of a fungal pathogen on a suitable host

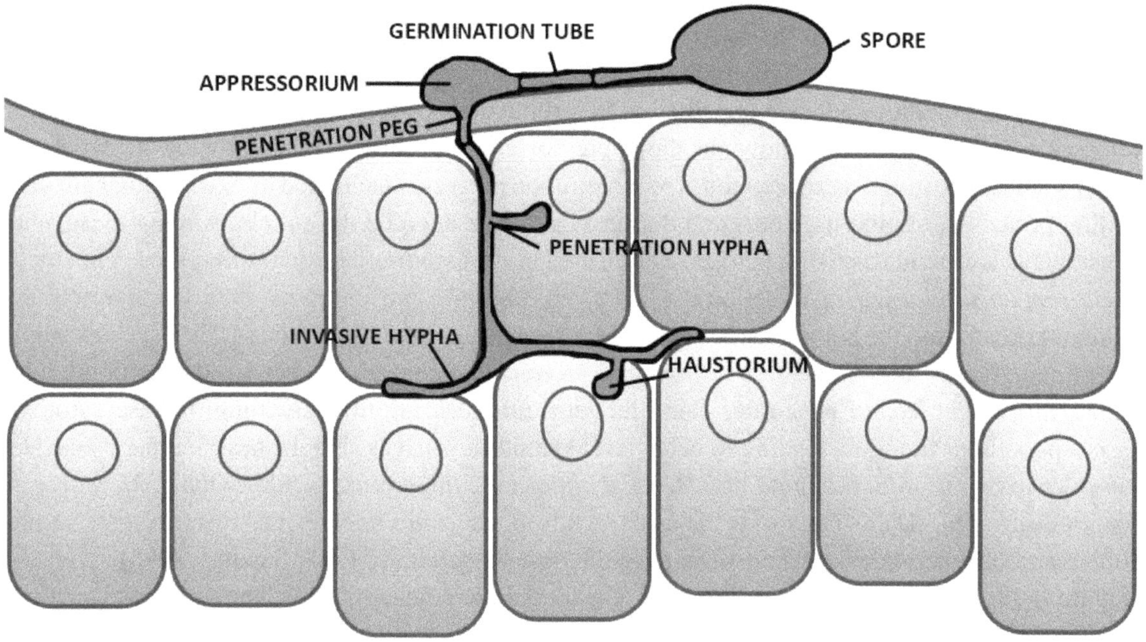

Table 1. Entomopathogenic fungi and their role in biocontrol

Fungal Species	Target Pests	Mode of Action	Application Methods	References
Beauveria bassiana	*Helicoverpa armigera*, Colorado potato beetle, codling moth, termites, whiteflies	Spore adhesion, cuticle penetration, colonization	Spore suspension sprays, emulsified formulations, bee-vectored application	Jain et al., 2008; Al Mazra'awi et al., 2006
Metarhizium anisopliae	Termites, locusts, grasshoppers, cockroaches, white grubs, leafhoppers	Cuticle penetration, tissue colonization	Oil formulations, direct soil application	Sandhu et al., 1993; Hasan et al., 2002
Lecanicillium lecanii	Aphids, whiteflies, *Trialeurodes vaporariorum*	Mycelial growth over host, epizootic spread	Greenhouse applications, foliar sprays	Kim et al., 2008; Hamlen, 1979
Nomuraea rileyi	Lepidopteran pests	Spore adhesion, host colonization	Spore suspension, direct foliar applications	Ignoffo, 1981; Mathew et al., 1998

continued on following page

Table 1. Continued

Fungal Species	Target Pests	Mode of Action	Application Methods	References
Paecilomyces lilacinus	Nematodes, root-knot nematodes, cyst nematodes	Egg parasitism, hyphal penetration	Soil treatment, direct application	Wraight et al., 2000; Faria & Wraight, 2001
Paecilomyces fumosoroseus	Whiteflies	Mycelial envelopment, spore germination	Greenhouse foliar sprays	Kim et al., 2002; Núñez del Prado et al., 2008

5. NEMATOPHAGOUS FUNGI (NF) AS BIOCONTROL AGENT

Nematophagous fungi have the ability to trap, kill, and metabolize nematodes using diverse mechanisms. They colonize both external and internal habitats of host organisms, utilizing these sites for growth. NF employs specialized structures, such as hyphal tips, conidia, and trapping devices, while also releasing toxins to infect nematodes. Predatory fungi primarily use hyphal structures to ensnare nematodes. Their trapping mechanisms include adhesive networks, constricting rings, adhesive branches, and adhesive knobs. These traps interact with the nematode cuticle, allowing fungal penetration and digestion. The process begins with trap formation, followed by nematode entrapment, penetration via a specialized structure called the penetration peg, and eventual digestion. Adhesive branches, or adhesive columns, consist of one to three interconnected cells that form anastomosing networks, efficiently capturing nematodes upon contact. This structure is predominant in *Dactylella cionopaga* and *Monacrosporium cionopagum* (Zhang et al., 2014). Adhesive networks, found in species like *Arthrobotrys oligospora*, arise from lateral hyphal branches and form three-dimensional traps, effectively capturing free-living nematodes. Adhesive knobs are specialized sticky-coated cells. When a nematode encounters a flattened sticky pad, it is firmly trapped, enabling fungal penetration and digestion. This enzymatic process involves the production of collagenase, which lyses the nematode cuticle. Species such as *Dactylellina arcuata* and *D. asthenopaga* utilize adhesive knobs. Constricting rings are intricate structures composed of three hyphal cells arranged in a circular formation. When a nematode enters, these cells rapidly expand, enclosing and trapping the prey, leading to fungal penetration and digestion (Liu et al., 2012). Non-constricting rings, another type of trap, consist of three-celled rings on a short stalk, used by species like *Dactylaria candida* and *D. lysipaga*. NF play a critical role in naturally controlling nematode populations through parasitism, trapping, and toxin production, making them valuable biocontrol agents in sustainable agriculture (Jiang et al., 2017). *Verticillium lecanii* significantly reduces cereal-cyst nematode populations, while *V. chlamydosporium* targets root-knot and cyst nematodes (Kerry et al., 1982). Additionally, *Paecilomyces* species infect and eliminate harmful nematodes. *Purpureocillium (Paecilomyces) lilacinum* is particularly effective against root-knot and cyst nematode eggs, making it a promising bionematicide for soil applications (Atkins et al., 2005).

6. FUNGAL MEDIATED BIOCONTROL OF PLANT DISEASES

The genus *Trichoderma* (Hypocreales) is a promising antagonist against various plant diseases, comprising numerous species commonly found in soil. Over 20 *Trichoderma* species serve as biocontrol agents against soil-borne and aerial phytopathogens. Key species include *T. harzianum, T. koningii, T. viride, T. atroviride, T. pseudokoningii, T. longibrachiatum, T. hamatum, T. polysporum,* and *T. reesei* (Monaco et al., 1991). *Trichoderma* grows rapidly in nature as a decomposer, primarily breaking down cellulosic materials (Jaklitsch, 2009; Kubicek et al., 2009). Due to their ability to produce antibiotics, industrially important enzymes, and act as biocontrol agents, *Trichoderma* species have gained commercial significance (Anees et al., 2010). These fungi suppress phytopathogenic fungi through multiple mechanisms, including direct antagonism, competition for nutrients and space, hyperparasitism (mycoparasitism), and induction of plant resistance (Howell, 2003). During mycoparasitism, *Trichoderma* forms appressoria or helical hyphae around the pathogen's hyphae. Some species also produce secondary metabolites and antimicrobial compounds that inhibit pathogens (Vinale et al., 2008a, b). *Trichoderma gamsii* effectively controls *Rhizoctonia solani*, with antibiosis being the primary mechanism (Anees et al., 2010).

Several studies have demonstrated the antagonistic potential of *Trichoderma* against pathogens such as *Sclerotium rolfsii, Fusarium ciceris, Macrophomina phaseolina, Rhizoctonia solani,* and plant-parasitic nematodes (Mukhopadhyay & Pan, 2012). *Trichoderma* and *Gliocladium virens* are crucial in controlling *S. rolfsii* infections in ginger rhizomes and stored vegetables (Mukherjee & Raghu, 1997). Additionally, *Trichoderma* produces volatile compounds that inhibit *Fusarium oxysporum* growth (Li et al., 2018). Beyond disease control, *Trichoderma* promotes plant growth, seed germination, and seedling emergence, enhancing overall plant health and yield (Joshi et al., 2010; Mukhopadhyay & Pan, 2012). Seed treatments with *Trichoderma* offer protection against *Pythium spp., R. solani, S. rolfsii, M. phaseolina,* and *Fusarium spp.* (Dubey et al., 2007; Nirmalkar et al., 2017). *Ampelomyces quisqualis* parasitizes powdery mildew hyphae, conidiophores, and cleistothecia, ultimately leading to mildew colony collapse (Németh et al., 2019). Over 64 species across genera like *Oidium, Oidiopsis, Brasilomyces, Erysiphe, Leveillula, Microsphaera, Phyllactinia, Podosphaera, Sphaerotheca,* and *Uncinula* have been identified as biocontrol agents (Falk et al., 1995a,b). While wettable sulfur and *A. quisqualis* are equally effective against powdery mildew, *A. quisqualis* is preferred due to its odorless, non-residual nature (Gopalakrishnan & Valluvaparidasan, 2009). Another beneficial biocontrol genus, *Purpureocillium*, enhances plant defense against biotic and abiotic stresses by releasing phytohormones like gibberellins and auxins (Khan et al., 2012).

Table 2 summarizes the fungal mediated biocontrol of plant diseases.

Table 2. Fungal mediated biocontrol of plant diseases

Fungal Species	Target Pathogens	Mode of Action	Key Benefits	References
Trichoderma spp. (T. harzianum, T. koningii, T. viride, T. atroviride, etc.)	*Sclerotium rolfsii, Fusarium spp., Macrophomina phaseolina, Rhizoctonia solani*	Antagonism, competition, hyperparasitism, resistance induction	Biocontrol, plant growth promotion, antibiotic production	Monaco et al. 1991; Anees et al. 2010, Howell 2003
Gliocladium virens	*Sclerotium rolfsii* (ginger rhizomes)	Antagonism, volatile compound production	Controls stored vegetable infections	Mukherjee & Raghu 1997
Ampelomyces quisqualis	Powdery mildew (*Erysiphe, Podosphaera, Sphaerotheca, Uncinula* spp.)	Hyperparasitism of mildew hyphae, conidiophores, cleistothecia	Non-residual, odorless, long survival on plant surfaces	Falk et al., 1995a,b; Németh et al., 2019
Purpureocillium lilacinum (Paecilomyces lilacinus)	Plant-parasitic nematodes	Parasitism, secondary metabolite production	Nematode control, plant growth stimulation	Atkins et al. 2005; Khan et al. 2012

7. FUNGAL MEDIATED BIOCONTROL OF INSECT PESTS

EPF are widely recognized for their ability to infect and cause disease in insect pests and other arthropods, making them crucial components of biological pest management (Vega et al., 2009; Pell et al., 2010). Among the well-characterized and extensively studied EPF, *Beauveria bassiana* and *Metarhizium anisopliae* stand out for their effectiveness in controlling insect pests. These fungi are facultative pathogens, meaning that in the absence of suitable hosts, they can survive as saprophytes in diverse habitats (Schrank & Vainstein, 2010; Ortiz-Urquiza et al., 2015). *Beauveria bassiana*, a filamentous soil fungus, causes white muscardine disease in arthropods. It has been isolated from numerous insect hosts of agricultural and medical importance (Jain et al., 2008). *Beauveria* species exhibit strong host specificity, enabling them to infect a wide range of insect pests, including *Helicoverpa armigera*, Colorado potato beetle, codling moth, termites, and whiteflies. The persistence of its spores in the environment can lead to epizootics, significantly reducing pest populations over time (Thakur & Sandhu, 2010). This fungus is widely used as a nonselective microbial pesticide, typically applied as an emulsified suspension or wettable powder on affected crops. To enhance its effectiveness, innovative spore dissemination strategies have been explored. Lin et al. (2017) suggested that invertebrates could act as EPF vectors, ensuring effective fungal spore delivery to targeted pests. Apivectory, involves using honeybees (*Apis mellifera*) to distribute *B. bassiana* spores efficiently. Al Mazra'awi et al. (2006) demonstrated this approach in *Brassica napus* (canola) fields to control the tarnished plant bug (*Lygus lineolaris*). Since honeybees are key pollinators, their movement ensures widespread fungal dispersion, leading to increased pest mortality. Parasitoids such as *Encarsia formosa* have been observed acting as passive EPF vectors during foraging, transferring spores from infected to uninfected hosts (Oreste et al., 2016). *Lecanicillium (Verticillium) lecanii*, thrives in warm, humid, tropical, and subtropical climates. It colonizes the underside of leaves using a filamentous mycelium and infects both nymphal and adult insect stages (Nunez et al., 2008). In South Korean greenhouses, *L. lecanii* has proven effective against *Trialeurodes vaporariorum* (Kim et al., 2002; 2008). Additionally, it has been used to control aphids and whiteflies in greenhouse chrysanthemums (Hamlen, 1979). *Metarhizium anisopliae* is another widely employed EPF, particularly for subterranean insect pests such as leafhoppers and white grubs (Sandhu et al., 1993). It is used to manage termites, locusts, grasshoppers, and cockroaches in countries including the U.S., Australia, and Africa. The fungus is highly effective when applied in oil formulations, ensuring spore contact with insect cuticle areas prone to high humidity. Depending on the spore concentration, *M. anisopliae* can kill up to 90% of locusts within 7–21 days post-application. Its effectiveness as a biocontrol agent has been validated against *Eutectona machaeralis* (teak skeletonizer) (Sandhu et al., 2000) and through solid-state fermentation for mass spore production (Hasan et al., 2002). *Nomuraea rileyi*, a dimorphic hyphomycete, induces epizootic outbreaks in several lepidopteran and coleopteran pests, including *Spodoptera litura*. Its host specificity and eco-friendly nature make it a valuable biocontrol agent (Ignoffo, 1981). Studies have demonstrated its efficacy against pests such as *Trichoplusia ni, Heliothis zea, Plathypena scabra, Bombyx mori,* and *Anticarsia gemmatalis* (Mathew et al., 1998). *Paecilomyces (Isaria) fumosoroseus* causes "Yellow Muscardine" disease in whiteflies (Nunez et al., 2008). It exhibits strong epizootic potential against *Bemisia* and *Trialeurodes* spp. in greenhouse and open-field conditions. It spreads rapidly in humid environments, drastically reducing *Bemisia tabaci* populations through widespread epizootics (Faria & Wraight, 2001). *Paecilomyces lilacinus* is an effective biocontrol agent in tropical and subtropical soils, particularly against whiteflies. Its ability to proliferate extensively on leaf surfaces under humid conditions facilitates its rapid spread in pest populations (Wraight et al., 2000). Beyond pest control, *Trichoderma* spp. have evolved into commercial biocontrol products used in both greenhouse and field agriculture (Harman et al., 2004; Harman 2006). Their ectomycorrhizal associations enable them to access nutrients from complex substrates like proteins, glucosamine, and tannins in soil. Additionally, *Trichoderma* plays a crucial role in bioremediation, improving soil mineral content and mitigating the overuse of herbicides.

8. FUNGAL MEDIATED BIOCONTROL OF WEEDS

Weeds are a major challenge to global agriculture, with herbicide resistance in crops like maize and wheat complicating management. Chemical herbicides are costly, environmentally harmful, or impractical in heavily infested areas. Fungal biological control agents (BCAs) offer a sustainable alternative by utilizing pathogenic fungi for weed control. Two main approaches exist: (i) classical biological control, which introduces coevolved pathogens from a weed's native habitat to manage invasive species, and (ii) inundative control, which mass-produces and applies native pathogens (myco-herbicides) to suppress endemic weeds. These fungal BCAs provide an eco-friendly, long-term solution for integrated weed management, reducing reliance on chemical herbicides and promoting sustainable agricultural practices. Table 3 lists some commercially available fungal bioherbicides along with their specifications.

Table 3. List of commercially available fungal bioherbicides and their specifications

Mycoherbicide	Fungal Species	Target Weed(s)	Mode of Action	References
Chontrol™	*Chondrostereum purpureum*	Hardwood and deciduous trees, shrubs	Induces wood rot, which leads to stump deterioration and inhibits resprouting.	Bailey et al., 2010; Charudattan, 2005
Collego™	*Colletotrichum gloeosporioides* f. sp. *aeschynomene*	*Aeschynomene virginica* (jointvetch), *Aeschynomene indica (Indian jointvetch), Sesbania exaltata (hemp sebania)*	Causes wilting and plant death by causing anthracnose sores.	Boyette et al., 2019; Cordeau et al., 2016
Di-Bak Parkinsonia™	*Lasiodiplodia pseudotheobromae, Macrophomina phaseolina, Neoscytalidium novaehollandiae*	*Parkinsonia aculeata* (parkinsonia)	Releases hydrolytic enzymes and toxins that break down plant tissues.	Galea, 2021
Phoma®	*Phoma macrostoma* Montagne 94–44B	*Taraxacum officinale* (dandelion), other broadleaf weeds	Interferes with root activity, blocking the uptake of nutrients and resulting in chlorosis.	Rai et al., 2021
Sarritor™	*Sclerotinia minor*	*Taraxacum officinale* (dandelion), other broadleaf weeds	Plant tissues are weakened by necrotrophic fungal infections, which lower their vigor.	Aneja et al., 2013
Stumpout™	*Cylindrobasidium laeve*	*Acacia mearnsii* (black wattle), *Acacia pycnantha* (golden wattle), *Poa annua* (winter grass)	Speeds up the breakdown of roots and stumps, preventing regrowth.	Green, 2003
Woad Warrior®	*Puccinia thalaspeos*	*Isatis tinctoria* (dyer's woad)	A systemic rust infection lowers seed viability by suppressing growth.	Bailey, 2014

9. COMMERCIAL ASPECT OF FBAS

EPF play a crucial role as rhizosphere colonizers, plant growth promoters, BCAs against pests, and antagonists of phytopathogens. EPF produce various bioactive metabolites, including antibiotics, volatile compounds (such as ammonia, hydrogen cyanide, alkyl pyrones, alcohols, esters, ketones, and lipids), and enzymes that help suppress plant pathogen-induced diseases (Ownley & Windham, 2007). Notably, some EPF, effective against insect pests, also exhibit antagonistic properties against plant diseases (Kim et al., 2008). Specific EPF species demonstrate efficacy in controlling major agricultural pests. For instance, *Beauveria bassiana* is reported to control the Colorado potato beetle (*Leptinotarsa decemlineata*) (Groden & Lockwood, 1991). Similarly, treating corn seeds with *Metarhizium anisopliae* conidia before sowing has been shown to increase plant stability and fresh weight while reducing wireworm damage (Kabaluk & Ericsson, 2007).

EPF not only protect against insect pests but also display an endophytic form of colonization in plants (Arnold & Lutzoni, 2007). Studies have identified endophytic colonization by EPF in various crops, including *Acremonium, Beauveria, Cladosporium, Clonostachys*, and *Isaria* (Vega et al., 2008). The growing interest in fungal pathogens as mycobiocontrol agents is due to their reliability, cost-effectiveness, and environmental safety (Wraight et al., 2001). *B. bassiana* effectively controls several hemipteran, coleopteran, and lepidopteran pests, and laboratory tests confirm its lethality against *Capnodis tenebrionis* larvae (El Khoury et al., 2020). Similarly, *M. anisopliae* demonstrates efficacy against termites, scarab beetle grubs, cutworms, and weevils, while *Lecanicillium lecanii* is used against Thysanoptera and Hemiptera insects.

Key EPF characteristics, such as enhanced virulence, specificity to target species, and resilience to environmental factors, are essential for successful field applications (Jackson et al., 2009). Common soil pest control methods include introducing barley kernels infected with EPF, which has effectively managed *Melolontha melolontha* populations (Vestergaard et al., 2002). Another effective method involves fungal-impregnated bands placed on tree branches or trunks, spreading spores to eliminate invasive pests. This approach was first applied to control the beetle *Monochamus alternatus*, the primary vector of pine wilt disease caused by *Bursaphelenchus xylophilus* (Shimazu, 2004). The fiber band technique has since been successfully employed against *Anoplophora glabripennis* and *Agrilus planipennis* (Augustyniuk-Kram & Kram, 2012).

The development of novel fungal biocontrol formulations through biotechnological advancements aims to improve efficiency, stability, and survivability (Heydari & Pessarakli, 2010). Various microbial-based solutions are globally marketed to combat insect pests and phytopathogens affecting economically significant crops. These products come in different formulations, including dust, granules, wettable powders, and aqueous or oil-based liquids (Ardakani et al., 2009; Nega, 2014). Many microbial-based insecticides and antagonists have been patented and evaluated for commercial application, providing effective crop protection (Schena et al., 2005; Nabi et al., 2017).

Table 4 lists some commercial fungal biocontrol agents. Table 5 list recent findings related to fungal biocontrol agents with their targeted pathogen and the the part of the plant utilized to isolate the antagonist. Formulations for mycoinsecticides rely on a limited number of fungal species, including *Paecilomyces fumosoroseus, B. bassiana, B. brongniartii, L. lecanii, L. longisporum*, and *M. anisopliae. B. bassiana* accounts for majority of mycopesticide formulations, with *M. anisopliae*, *Isaria fumosorosea*, and *B. brongniartii* following closely behind (Faria and Wraight, 2007). Stain improvement and genetic manipulations are essential to increase the survival and ecological fitness of these fungal biocontrol agents because their wild isolates lack adequate levels of virulence (Rangel et al., 2005).

Table 4. List of few commercialized fungal biocontrol agents

Biocontrol agents	Mode of Action	Target species	Phytopathology	Affected Crops
Cryptococcus laurentii	Increasing resistance enzymes, enhancing ROS tolerance, and secreting lytic enzymes.	*Penicillium expansum*	Gray mold	Strawberry, Apple, Orange, Tomato
Debaryomyces hansenii	Induces host resistance and compete for nutrients and space.	*Botrytis cynerea*	Rhizopus rot, Alternaria rot	Tomato
Trichosporon pullulans	Produces lytic enzymes	*Botrytis cinerea*	Alternaria rot, Gray rot	Cherry
Meyerozyma guilliermondii	Enhances host defense and regulate population density.	*Botrytis cinerea*	Penicillium rot	Orange, Tomato
Saccharomyces cerevisiae	Depletes iron.	*Penicillium expansum*	Penicillium rot	Grape
Trichoderma harzianum	Produces Antibiotics	*Collectotrichum gloeosporoides*	Anthracnose	Pear, Kiwi, Grape
Epicoccum nigram	Produces natural bioactive compounds	*Colletotrichum falcatum*	Late blight, Fusarium wilt	Tomato, Peas
Rodotorula glutinis	Competes for nutrients and space through site exclusion.	*Penicillium expansum*	Alternaria rot, Green mold	Banana, Pear

(Ramudingana et al., 2025)

Table 5. List of potential fungal biocontrol agents

Potential Antagonist	Target Pathogenic Fungus	Host Plant	Plant part for Isolation of Antagonist
Acrophialophora jodhpurensis	*Rhizoctonia solani*	Tomato (*Lycopersicon esculentum*)	Roots
Alternaria alternata	*Alternaria arborescens*	Japanese cornel dogwood (*Cornus officinalis*)	Biennial twigs
Alternaria destruens	*Fusarium graminearum*	Wheat (*Triticum aestivum*)	Stems or heads
Annulohypoxylon sp.	*Alternaria alternata*	Agarwood (*Aquilaria sinensis*)	Bark from branches and twigs
Botryosphaeria berengeriana	*Alternaria alternata*	Japanese cornel dogwood (*Cornus officinalis*)	Triennial twigs
Botryosphaeria dothidea	*Alternaria alternata*	Various plants	Various plant parts
Cladosporium sp.	*Fusarium oxysporum*	Tomato (*Lycopersicon esculentum*)	Stems
Diaporthe spp.	*Colletotrichum sp.*	*Avicennia nitida*	Branches
Eupenicillium javanicum	*Fusarium oxysporum*	Agarwood (*Aquilaria sinensis*)	Leaves
Guignardia mangiferae	*Fusarium oxysporum*	Mangrove (*Rhizophora stylosa*)	Stems
Trichoderma asperellum	*Corynespora cassiicola*	Lettuce (*Lactuca sativa*)	Leaves
Trichoderma longibrachiatum	*Rhizoctonia solani*	Soybean (*Glycine max*)	Roots
Xylaria feejeensis	*Alternaria solani*	Mangrove trees (*Ceriops decandra, Rhizophora apiculata*, etc.)	Leaves, petioles, and roots

(Priyashantha et al., 2023)

10. FACTORS AFFECTING BIOCONTROL OF PLANT DISEASES AND SELECTION OF POTENTIAL FBAS

BCA effectiveness depends on plant-pathogen interactions, environmental conditions, and native microbes. Fungal BCA, requires further research to optimize interactions and enhance agricultural efficacy (Thambugala et al., 2020; Howell, 2003).

Plant Species Influence the Biocontrol Activity: Pathogens interact uniquely with plants based on genetic and environmental factors. Their sensitivity to BCAs varies, and some develop resistance over time. Identifying vulnerable pathogen life cycle stages allows BCAs to exploit weaknesses, disrupt infection processes, and enhance disease suppression effectively.

Pathogen Influence the Biocontrol Activity: Pathogens interact with plants differently due to genetic and environmental adaptations (Thambugala et al., 2020). Their sensitivity to BCAs varies based on the mode of action. Some pathogens may develop resistance over time, reducing BCA effectiveness. Identifying vulnerable pathogen life cycle phase helps BCAs exploit weaknesses of pathogens and prevent disease spread.

Nature of the Antagonist: BCAs often fail in the field due to poor soil adaptation. Understanding their spread in the rhizosphere is crucial. Exploring native strains enhances biocontrol. Key factors include pathogen impact, survival in harsh environments, and optimal application timing. Traits like plant colonization, culture ease, metabolite production, and compatibility with other agents are also studied.

Impact of Environmental Stresses: Optimal conditions for BCAs vary, with soil pH, temperature, and moisture influencing their activity. Some thrive in acidic soils, while others prefer alkaline conditions. Ecological studies are essential to understand biotic and abiotic interactions affecting BCAs. Agricultural crop production is affected by a range of biotic and abiotic factors (Pant et al., 2025). Selecting stable agents resilient to environmental changes enhances plant disease management and crop resilience, especially amid climate change challenges.

11. CHALLENGES IN ESTABLISHING BIOCONTROL AGENTS

Researchers screen microbes to develop biological pesticides. Greenhouse studies assess BCA growth for field applications. Stability depends on formulation, storage, and transport. Optimizing production, extending shelf life, and refining storage conditions are crucial. Private sectors conduct safety evaluations, while governments oversee registration and risk assessment. Balancing efficacy, security, and commercialization remains a challenge.

Registering BCAs requires time, industry-academia collaboration, and navigating ethical and legal concerns. Some nations restrict new BCA species, limiting their application, though greenhouse use faces fewer risks. Regulatory challenges vary globally, hindering adoption. Farmers may see limited financial incentives compared to chemical pesticides. Awareness programs, workshops, and training can enhance understanding and encourage wider adoption.

BCAs undergo rigorous risk assessments before commercial approval. With growing interest in reducing synthetic pesticides and streamlining regulations for low-risk products, BCAs have the potential for global commercialization.

12. ADVANTAGES, DISADVANTAGES AND LIMITATIONS

Fungal BCAs reduce reliance on hazardous chemicals, offering a safer alternative for humans and the environment. They provide continuous protection as plants grow and are cost-effective, particularly hyphomycete fungi, which are easy to culture and produce in large quantities. Their spores, highly adaptable to modern agricultural applications like spray rigs, make them commercially viable. Additionally, their broad host range enables control of multiple pests with a single product. However, challenges exist. The presence of beneficial insects in crops raises concerns about non-target mortality, potentially affecting overall biological control success. Environmental factors also play a crucial role; high humidity is often required for effective control, and prolonged sunlight exposure can inactivate fungal spores, limiting persistence. Furthermore, the sporadic nature of fungal outbreaks can reduce their reliability in pest management. Bioinformatics approaches, such as molecular docking (Jaiswal et al., 2023), can help identify fungal molecules for targeted biocontrol.

13. CONCLUSION AND FUTURE DIRECTIONS

Fungal biocontrol agents are emerging as sustainable alternatives to synthetic pesticides, offering effective pest management solutions. Species such as *Beauveria*, *Metarhizium*, and *Trichoderma* control insect pests, nematodes, and plant pathogens while being biodegradable and compatible with Integrated Pest Management (IPM) and Integrated Disease Management (IDM). Their use reduces chemical residues, enhances soil health, and lowers production costs. A major advantage of these fungi is their host specificity, ensuring minimal impact on beneficial insects and non-target organisms. However, challenges such as inconsistent field performance, formulation stability, and regulatory barriers limit widespread adoption. Research should focus on improving fungal formulations through biotechnological and nanotechnology-based approaches. Novel delivery systems that enhance viability under diverse environmental conditions are essential for large-scale implementation. Extensive field trials and farmer education programs will further support their adoption. Advancements in genetic engineering have improved fungal biocontrol efficacy by enhancing traits such as pathogenicity, humidity tolerance, and resistance to abiotic stressors like UV radiation and temperature fluctuations. Genetic modifications targeting key genes strengthen fungal epizootic capabilities, while innovative inoculum delivery methods optimize field performance. Secondary metabolites produced by fungal antagonists play a crucial role in phytopathogen control. Enhancing beneficial traits through biotechnology can further optimize their agricultural effectiveness. Strain improvement and identification of entomopathogenic targets should be prioritized, alongside regulatory reforms to support commercialization while ensuring safety and efficacy. Understanding the metabolic and genetic potential of fungi is vital. Their association with plant roots strengthens crop resilience against biotic and abiotic stressors. Continued research, innovative delivery systems, and supportive policies will enable the seamless integration of fungal biocontrol agents into sustainable pest management strategies.

REFERENCES

Agrawal, P., Pant, K. K., Sonkar, M., & Chandra, V. (2023). Algal Biofuel. *Algae Refinery*, 187-208. Mehariya, S., & Verma, P. (eds.). (2023). Algae Refinery: Up- and Downstream Processes, CRC Press.

Akutse, K., Maniania, N., Fiaboe, K., Van den Berg, J., & Ekesi, S. (2013). Endophytic colonization of Vicia faba and Phaseolus vulgaris (Fabaceae) by fungal pathogens and their effects on the life-history parameters of Liriomyza huidobrensis (Diptera: Agromyzidae). *Fungal Ecology*, *6*(4), 293–301. DOI: 10.1016/j.funeco.2013.01.003

Al Mazra'awi, M. S., Shipp, J. L., Broadbent, A. B., & Kevan, P. G. (2006). Dissemination of Beauveria bassiana by honey bees (Hymenoptera: Apidae) for control of tarnished plant bug (Hemiptera: Miridae) on Canola. *Environmental Entomology*, *35*(6), 1569–1577. DOI: 10.1603/0046-225X(2006)35[1569:DOBB-BH]2.0.CO;2

Anees, M., Tronsmo, A., Edel-Hermann, V., Hjeljord, L. G., Héraud, C., & Steinberg, C. (2010). Characterization of field isolates of Trichoderma antagonistic against Rhizoctonia solani. *Fungal Biology*, *114*(9), 691–701. DOI: 10.1016/j.funbio.2010.05.007 PMID: 20943179

Aneja, K.R., Kumar, V., Jiloha, P., Kaur, M., Sharma, C., Surain, P., Dhiman, R., & Aneja, A. (2013). Potential Bioherbicides: Indian perspectives. *Biotechnology: Prospects and Applications*, 197-215.

Ardakani, S., Heydari, A., Khorasani, N., Arjmandi, R., & Ehteshami, M. (2009). Preparation of new Biofungicides using antagonistic bacteria and mineral compounds for controlling cotton seedling damping-off disease. *Journal of Plant Protection Research*, *49*(1). Advance online publication. DOI: 10.2478/v10045-009-0007-3

Arnold, A. E., & Lutzoni, F. (2007). Diversity and host range of foliar fungal endophytes: Are tropical leaves biodiversity hotspots? *Ecology*, *88*(3), 541–549. DOI: 10.1890/05-1459 PMID: 17503580

Atanasova, L., Crom, S. L., Gruber, S., Coulpier, F., Seidl-Seiboth, V., Kubicek, C. P., & Druzhinina, I. S. (2013). Comparative transcriptomics reveals different strategies of Trichodermamycoparasitism. *BMC Genomics*, *14*(1), 121. DOI: 10.1186/1471-2164-14-121 PMID: 23432824

Atkins, S. D., Clark, I. M., Pande, S., Hirsch, P. R., & Kerry, B. R. (2005). The use of real-time PCR and species-specific primers for the identification and monitoring of Paecilomyces lilacinus. *FEMS Microbiology Ecology*, *51*(2), 257–264. DOI: 10.1016/j.femsec.2004.09.002 PMID: 16329874

Augustyniuk-Kram, A., & Kram K.J. (2012). Entomopathogenic fungi as an important natural regulator of insect outbreaks in forests (Review). *Forest Ecosystems - More than Just Trees.*

Bailey, K., Boyetchko, S., & Längle, T. (2010). Social and economic drivers shaping the future of biological control: A Canadian perspective on the factors affecting the development and use of microbial biopesticides. *Biological Control*, *52*(3), 221–229. DOI: 10.1016/j.biocontrol.2009.05.003

Bailey, K.L. (2014). The Bioherbicide approach to weed control using plant pathogens. *Integrated Pest Management*, 245-266.

Baron, N. C., Rigobelo, E. C., & Zied, D. C. (2019). Filamentous fungi in biological control: Current status and future perspectives. *Chilean Journal of Agricultural Research*, *79*(2), 307–315. DOI: 10.4067/S0718-58392019000200307

Boyette, C. D., Hoagland, R. E., & Stetina, K. C. (2019). Extending the host range of the bioherbicidal fungus *Colletotrichum gloeosporioides* f. sp. *aeschynomene.Biocontrol Science and Technology*, *29*(7), 720–726. DOI: 10.1080/09583157.2019.1581130

Brotman, Y., Briff, E., Viterbo, A., & Chet, I. (2008). Role of Swollenin, an expansin-like protein from-*Trichoderma*, in plant root colonization. *Plant Physiology*, *147*(2), 779–789. DOI: 10.1104/pp.108.116293 PMID: 18400936

Chandra, V., Tiwari, A., Pant, K. K., & Bhatt, R. (2022). Animal Cell Culture: Basics and Applications. In Verma, P. (Ed.), *Industrial Microbiology and Biotechnology*. Springer., DOI: 10.1007/978-981-16-5214-1_24

Charudattan, R. (2005). Ecological, practical, and political inputs into selection of weed targets: What makes a good biological control target? *Biological Control*, *35*(3), 183–196. DOI: 10.1016/j.biocontrol.2005.07.009

Cordeau, S., Triolet, M., Wayman, S., Steinberg, C., & Guillemin, J. (2016). Bioherbicides: Dead in the water? A review of the existing products for integrated weed management. *Crop Protection (Guildford, Surrey)*, *87*, 44–49. DOI: 10.1016/j.cropro.2016.04.016

Djonovic, S., Vargas, W. A., Kolomiets, M. V., Horndeski, M., Wiest, A., & Kenerley, C. M. (2007). A proteinaceous elicitor Sm1 from the beneficial fungus Trichoderma virens is required for induced systemic resistance in maize. *Plant Physiology*, *145*(3), 875–889. DOI: 10.1104/pp.107.103689 PMID: 17885089

Dubey, S. C., Suresh, M., & Singh, B. (2007). Evaluation of Trichoderma species against Fusarium oxysporum F. Sp. ciceris for integrated management of chickpea wilt. *Biological Control*, *40*(1), 118–127. DOI: 10.1016/j.biocontrol.2006.06.006

Durrant, W., & Dong, X. (2004). Systemic acquired resistance. *Annual Review of Phytopathology*, *42*(1), 185–209. DOI: 10.1146/annurev.phyto.42.040803.140421 PMID: 15283665

Eisendle, M., Oberegger, H., Buttinger, R., Illmer, P., & Haas, H. (2004). Biosynthesis and uptake of Siderophores is controlled by the pacc-mediated Ambient-pH regulatory system in *Aspergillus nidulans. Eukaryotic Cell*, *3*(2), 561–563. DOI: 10.1128/EC.3.2.561-563.2004 PMID: 15075286

El Khoury, Y., Noujeim, E., Ravlić, J., Oreste, M., Addante, R., Nemer, N., & Tarasco, E. (2020). The effect of entomopathogenic nematodes and fungi against four xylophagous pests. *Biocontrol Science and Technology*, *30*(9), 983–995. DOI: 10.1080/09583157.2020.1781059

Falk, S. P., Gadoury, D. M., Cortesi, P., Pearson, R. C., & Seem, R. C. (1995a). Parasitism of Uncinula necator cleistothecia by the mycoparasite Ampelomyces quisqualis. *Phytopathology*, *85*(7), 794–800. DOI: 10.1094/Phyto-85-794

Falk, S. P., Gadoury, D. M., Pearson, R. C., & Seem, R. C. (1995b). Partial control of grape powdery mildew by the mycoparasite *Ampelomyces quisqualis.Plant Disease*, *79*(5), 483–490. DOI: 10.1094/PD-79-0483

Faria, M., & Wraight, S. P. (2001). Biological control of Bemisia tabaci with fungi. *Crop Protection (Guildford, Surrey)*, *20*(9), 767–778. DOI: 10.1016/S0261-2194(01)00110-7

Faria, M. R., & Wraight, S. P. (2007). Mycoinsecticides and Mycoacaricides: A comprehensive list with worldwide coverage and international classification of formulation types. *Biological Control*, *43*(3), 237–256. DOI: 10.1016/j.biocontrol.2007.08.001

Galea, V. J. (2021). Use of Stem Implanted Bioherbicide Capsules to Manage an Infestation of *Parkinsonia aculeata* in Northern Australia. *Plants*, *10*(9), 1909. DOI: 10.3390/plants10091909 PMID: 34579440

Gohel, N. M., Raghunandan, B. L., Patel, N. B., Parmar, H. V., & Raval, D. B. (2022). Role of fungal biocontrol agents for sustainable agriculture. *Fungal Biology*, •••, 577–606. DOI: 10.1007/978-981-16-8877-5_28

Gopalakrishnan, C., & Valluvaparidasan, V. (2009). Management of okra powdery mildew using Ampelomyces quisqualis. *Journal of Biological Control*, *23*(3), 325–327.

Green, S. (2003). A review of the potential for the use of bioherbicides to control forest weeds in the UK. *Forestry*, *76*(3), 285–298. DOI: 10.1093/forestry/76.3.285

Groden, E., & Lockwood, J. (1991). Effects of soil fungistasis on Beauveria bassiana and its relationship to disease incidence in the Colorado potato beetle, Leptinotarsa decemlineata, in Michigan and Rhode Island soils. *Journal of Invertebrate Pathology*, *57*(1), 7–16. DOI: 10.1016/0022-2011(91)90035-O

Gupta, S. K., & Sharma, M. (2014). *Approaches and trends in plant disease management*. Scientific Publishers.

Hajji, L., Hlaoua, W., Regaieg, H., & Horrigue-Raouani, N. (2016). Biocontrol potential of *Verticillium leptobactrum* and *Purpureocillium lilacinum* against Meloidogyne javanica and Globodera pallida on potato (Solanum tuberosum). *American Journal of Potato Research*, *94*(2), 178–183. DOI: 10.1007/s12230-016-9554-0

Hamlen, R. A. (1979). Biological control of insects and mites on European greenhouse crops: Research and commercial implementation. *Proceedings of the Annual Meeting of the Florida State Horticultural Society*, *92*, 367–368.

Hanson, L. E., & Howell, C. R. (2002). Biocontrol efficacy and other characteristics of protoplast fusants between Trichoderma koningii and T. virens. *Mycological Research*, *106*(3), 321–328. DOI: 10.1017/S0953756202005592

Hanson, L. E., & Howell, C. R. (2004). Elicitors of Plant Defense Responses from Biocontrol Strains of Trichoderma viren. *Phytopathology*, *94*(2), 171–176. DOI: 10.1094/PHYTO.2004.94.2.171 PMID: 18943540

Harman, G. E. (2006). Overview of mechanisms and uses of *Trichoderma* spp. *Phytopathology*, *96*(2), 190–194. DOI: 10.1094/PHYTO-96-0190 PMID: 18943924

Harman, G. E., Howell, C. R., Viterbo, A., Chet, I., & Lorito, M. (2004). Trichoderma species—Opportunistic, avirulent plant symbionts. *Nature Reviews. Microbiology*, *2*(1), 43–56. DOI: 10.1038/nrmicro797 PMID: 15035008

Harman, G. E., & Shoresh, M. (2007). The mechanisms and applications of symbiotic opportunistic plant symbionts. In Vurro, M., & Gressel, J. (Eds.), *Novel Biotechnologies for Biocontrol Agent Enhancement and Management. NATO Security through Science Series* (pp. 131–155). Springer., DOI: 10.1007/978-1-4020-5799-1_7

Hartley, C. (1921). Damping-off in forest nurseries /. https://doi.org/DOI: 10.5962/bhl.title.108260

Hasan, S., Bhamra, A. K., Sil, K., Rajak, R. C., & Sandhu, S. S. (2002). Spore production of Metarhizium anisopliaem (ENT-12) by Solid State fermentation. *Journal of the Indian Botanical Society*, *8*, 85–88.

Heydari, A., & Pessarakli, M. (2010). A review on biological control of fungal plant pathogens using microbial antagonists. *Journal of Biological Sciences (Faisalabad, Pakistan)*, *10*(4), 273–290. DOI: 10.3923/jbs.2010.273.290

Horbach, R., Navarro-Quesada, A. R., Knogge, W., & Deising, H. B. (2011). When and how to kill a plant cell: Infection strategies of plant pathogenic fungi. *Journal of Plant Physiology*, *168*(1), 51–62. DOI: 10.1016/j.jplph.2010.06.014 PMID: 20674079

Howell, C. R. (2003). Mechanisms Employed by Trichoderma Species in the Biological Control of Plant Diseases: The History and Evolution of Current Concepts. *Plant Disease*, *87*(1), 4–10. DOI: 10.1094/PDIS.2003.87.1.4 PMID: 30812698

Ignoffo, C. M. (1981). The fungus Nomuraea rileyi as a microbial insecticide. In Burges, H. D. (Ed.), *Microbial Control of Pests and Plant Diseases* (pp. 513–538). Academic Press.

Jackson, M. A., Dunlap, C. A., & Jaronski, S. T. (2009). Ecological considerations in producing and formulating fungal entomopathogens for use in insect biocontrol. *BioControl*, *55*(1), 129–145. DOI: 10.1007/s10526-009-9240-y

Jain, N., Rana, I. S., Kanojiya, A., & Sandhu, S. S. (2008). Characterization of Beauveria bassiana strains based on protease and lipase activity and their role in pathogenicity. *Journal of Basic and Applied Mycology*, *I-II*, 18–22.

Jaiswal, C., Pant, K. K., Behera, R. K. S., Bhatt, R., & Chandra, V. (2023). Development of New Molecules Through Molecular Docking. In Verma, P. (Ed.), *Industrial Microbiology and Biotechnology*. Springer., DOI: 10.1007/978-981-99-2816-3_22

Jaklitsch, W. M. (2009). European species of Hypocrea Part I. The green-spored species. *Studies in Mycology*, *63*, 1–91. DOI: 10.3114/sim.2009.63.01 PMID: 19826500

Jiang, X., Xiang, M., & Liu, X. (2017). Nematode-Trapping Fungi. *Microbiology spectrum*, *5*(1), . https://doi.org/DOI: 10.1128/microbiolspec.FUNK-0022-2016

Joshi, B. B., Bhatt, R. P., & Bahukhandi, D. (2010). Antagonistic and plant growth activity of Trichoderma isolates of Western Himalayas. *Journal of Environmental Biology*, *31*(6), 921–928. PMID: 21506476

Jyoti and Singh. D. P. (2016). Fungi as biocontrol agents in sustainable agriculture. In J. S. Singh & D. P. Singh (Eds.), Microbes and Environmental Management (pp. 172–194). Studium Press, New Delhi.

Kabaluk, J. T., & Ericsson, J. D. (2007). Metarhizium anisopliae seed treatment increases yield of Field corn when applied for Wireworm control. *Agronomy Journal*, *99*(5), 1377–1381. DOI: 10.2134/agronj2007.0017N

Keller, N. P., Turner, G., & Bennett, J. W. (2005). Fungal secondary metabolism - from biochemistry to genomics. *Nature Reviews. Microbiology*, *3*(12), 937–947. DOI: 10.1038/nrmicro1286 PMID: 16322742

Kenis, M., Hurley, B. P., Hajek, A. E., & Cock, M. J. (2017). Classical biological control of insect pests of trees: Facts and figures. *Biological Invasions*, *19*(11), 3401–3417. DOI: 10.1007/s10530-017-1414-4

Kerry, B. R., Crump, D. H., & Mullen, L. A. (1982). Studies of the cereal cyst-nematode, *Heterodera avenae* under continuous cereals, 1975–1978. II. Fungal parasitism of nematode females and eggs. *Annals of Applied Biology*, *100*(3), 489–499. DOI: 10.1111/j.1744-7348.1982.tb01415.x

Khan, A. L., Hamayun, M., Kang, S. M., Kim, Y. H., Jung, H. Y., Lee, J. H., & Lee, I. J. (2012). Endophytic fungal association via gibberellins and indole acetic acid can improve plant growth under abiotic stress: An example of Paecilomyces formosus LHL10. *BMC Microbiology*, *12*(1), 3. DOI: 10.1186/1471-2180-12-3 PMID: 22235902

Kim, J. J., Goettel, M. S., & Gillespie, D. R. (2008). Evaluation of Lecanicillium longisporum, Vertalec® for simultaneous suppression of cotton aphid, Aphis gossypii, and cucumber powdery mildew, Sphaerotheca fuliginea, on potted cucumbers. *Biological Control*, *45*(3), 404–409. DOI: 10.1016/j.biocontrol.2008.02.003

Kim, J. J., Lee, M. H., Yoon, C. S., Kim, H. S., Yoo, J. K., & Kim, K. C. (2002). Control of cotton aphid and greenhouse whitefly with a fungal pathogen. Journal of Natural Institute of Agricultural Science and Technology, 7–14.

Kiss, L. (2003). A review of fungal antagonists of powdery mildews and their potential as biocontrol agents. *Pest Management Science*, *59*(4), 475–483. DOI: 10.1002/ps.689 PMID: 12701710

Kosariya, A., Kumar, D., Pant, K. K., Bhatt, R., & Chandra, V. (2023). Strategies for Strain Improvement of Economically Important Microorganisms. In Verma, P. (Ed.), *Industrial Microbiology and Biotechnology*. Springer., DOI: 10.1007/978-981-99-2816-3_24

Kubicek, C. P., Mikus, M., Schuster, A., Schmoll, M., & Seiboth, B. (2009). Metabolic engineering strategies for the improvement of cellulase production by Hypocrea jecorina. *Biotechnology for Biofuels*, *2*(1), 19. DOI: 10.1186/1754-6834-2-19 PMID: 19723296

Kumar, A. V., B, P., Chandra, V., & Singam, P. (2024). In silico analysis and characterization of the F5H gene family in sorghum bicolor and its role in lignin production. *Plant Cell Biotechnology and Molecular Biology*, *25*(11-12), 84–106. DOI: 10.56557/pcbmb/2024/v25i11-128885

Leger, R. J., Charnley, A. K., & Cooper, R. M. (1987). Characterization of cuticle-degrading proteases produced by the entomopathogen Metarhizium anisopliae. *Archives of Biochemistry and Biophysics*, *253*(1), 221–232. DOI: 10.1016/0003-9861(87)90655-2 PMID: 3545084

Li, N., Alfiky, A., Wang, W., Islam, M., Nourollahi, K., Liu, X., & Kang, S. (2018). Volatile Compound-Mediated Recognition and Inhibition between *Trichoderma* Biocontrol Agents and *Fusarium oxysporum*. *Frontiers in Microbiology*, *9*, 2614. DOI: 10.3389/fmicb.2018.02614 PMID: 30455673

Lin, G., Tanguay, A., Guertin, C., Todorova, S., & Brodeur, J. (2017). A new method for loading predatory mites with entomopathogenic fungi for biological control of their prey. *Biological Control*, *115*, 105–111. DOI: 10.1016/j.biocontrol.2017.09.012

Liu, K., Tian, J., Xiang, M., & Liu, X. (2012). How carnivorous fungi use three-celled constricting rings to trap nematodes. *Protein & Cell*, *3*(5), 325–328. DOI: 10.1007/s13238-012-2031-8 PMID: 22528749

Lo, C. T. (1997). Biological control of turf diseases using Trichoderma harzianum. Plant Protection. *Bulletin.*, *39*, 207–225.

Martinez, C., Blanc, F., Le Claire, E., Besnard, O., Nicole, M., & Baccou, J. C. (2001). Salicylic acid and ethylene pathways are differentially activated in melon cotyledons by active or heat-denatured cellulase from Trichoderma longibrachiatum. *Plant Physiology*, *127*(1), 334–344. DOI: 10.1104/pp.127.1.334 PMID: 11553761

Mathew, S. O., Sandhu, S. S., & Rajak, R. C. (1998). Bioactivity of Nomuraea rileyi against Spilosoma obliqua: Effect of dosage, temperature, and relative humidity. *Journal of the Indian Botanical Society*, *77*, 23–25.

Meyling, N. V., & Eilenberg, J. (2007). Ecology of the entomopathogenic fungi Beauveria bassiana and Metarhizium anisopliae in temperate agroecosystems: Potential for conservation biological control. *Biological Control*, *43*(2), 145–155. DOI: 10.1016/j.biocontrol.2007.07.007

Mohamed, H., & Saad, A. (2009). The biocontrol of postharvest disease (Botryodiplodia theobromae) of guava (Psidium guajava L.) by the application of yeast strains. *Postharvest Biology and Technology*, *53*(3), 123–130. DOI: 10.1016/j.postharvbio.2009.04.001

Monaco, C., Perello, A., Alippi, H. E., & Pasquare, A. O. (1991). Trichoderma spp.: A biocontrol agent of Fusarium spp. and Sclerotium rolfsii by seed treatment. *Advances in Horticultural Science*, *5*(3), 92–95.

Mukherjee, P. K., Horwitz, B. A., & Kenerley, C. M. (2012). Secondary metabolism in Trichoderma—A genomic perspective. *Microbiology (Reading, England)*, *158*(Pt 1), 35–45. DOI: 10.1099/mic.0.053629-0 PMID: 21998165

Mukherjee, P. K., & Raghu, K. (1997). Effect of temperature on antagonistic and biocontrol potential of Trichoderma sp. on Sclerotium rolfsii. *Mycopathologia*, *139*(3), 151–155. DOI: 10.1023/A:1006868009184 PMID: 16283454

Mukhopadhyay, R., & Kumar, D. (2020). Trichoderma: A beneficial antifungal agent and insights into its mechanism of biocontrol potential. *Egyptian Journal of Biological Pest Control*, *30*(1), 133. DOI: 10.1186/s41938-020-00333-x

Mukhopadhyay, R., & Pan, S. K. (2012). Isolation and selection of some antagonistic Trichoderma species from different new alluvial zones of Nadia district, West Bengal. *Journal of the Botanical Society of Bengal*, *66*(2), 149–152.

Nabi, S. U., Raja, W. H., Kumawat, K. L., Mir, J. I., Sharma, O. C., Singh, D. B., & Sheikh, M. A. (2017). Post-harvest diseases of temperate fruits and their management strategies—A review. *International Journal of Pure and Applied Biosciences*, *5*(3), 885–898. DOI: 10.18782/2320-7051.2981

Nega, A. (2014). Review on concepts in biological control of plant pathogens. *Journal of Biology, Agriculture and Healthcare*, *4*(27), 33–54.

Németh, M. Z., Pintye, A., Horváth, Á. N., Vági, P., Kovács, G. M., Gorfer, M., & Kiss, L. (2019). Green Fluorescent Protein Transformation Sheds More Light on a Widespread Mycoparasitic Interaction. *Phytopathology*, *109*(8), 1404–1416. DOI: 10.1094/PHYTO-01-19-0013-R PMID: 30900938

Nirmalkar, V. K., Singh, S., Tiwari, R. S., Said, P. P., & Kaushik, D. K. (2017). Field efficacy of Trichoderma harzianum and rhizobium against wilt complex of chickpea. *International Journal of Current Microbiology and Applied Sciences*, *6*(7), 1421–1429. DOI: 10.20546/ijcmas.2017.607.170

Noma, T., & Strickler, K. (2000). Effects of Beauveria bassiana on Lygus hesperus (Hemiptera: Miridae) feeding and oviposition. *Environmental Entomology*, *29*(2), 394–402. DOI: 10.1093/ee/29.2.394

Núñez del Prado, E., Iannacone, J., & Gómez, H. (2008). Effect of two Entomopathogenic fungi in controlling Aleurodicus cocois (Curtis, 1846) (Hemiptera: Aleyrodidae). *Chilean Journal of Agricultural Research*, *68*(1), 21–30. DOI: 10.4067/S0718-58392008000100003

Oreste, M., Bubici, G., Poliseno, M., & Tarasco, E. (2016). Effect of *Beauveria bassiana* and *Metarhizium anisopliae* on the *Trialeurodes vaporariorum-Encarsia Formosa* system. *Journal of Pest Science*, *89*(1), 153–160. DOI: 10.1007/s10340-015-0660-4

Ortiz-Urquiza, A., Luo, Z., & Keyhani, N. O. (2015). Improving mycoinsecticides for insect biological control. *Applied Microbiology and Biotechnology*, *99*(3), 1057–1068. DOI: 10.1007/s00253-014-6270-x PMID: 25503318

Ownley, B. H., & Windham, M. (2007). Biological control of plant pathogens. In Trigiano, R. N., Windham, M. T., & Windham, A. S. (Eds.), *Plant pathology: Concepts and laboratory exercises* (2nd ed., pp. 423–436). CRC Press.

Pant, K. K., Naik, J., Barthakur, S., & Chandra, V. (2025). High-temperature stress in wheat (*Triticum aestivum L.*): Unfolding the impacts, tolerance and methods to mitigate the detrimental effects. *Cereal Research Communications*. Advance online publication. DOI: 10.1007/s42976-025-00634-7

Pell, J. K., Hannam, J. J., & Steinkraus, D. C. (2010). Conservation biological control using fungal entomopathogens. *BioControl*, *55*(1), 187–198. DOI: 10.1007/s10526-009-9245-6

Pourseyed, S. H., Tavassoli, M., Bernousi, I., & Mardani, K. (2010). Metarhizium anisopliae (Ascomycota: Hypocreales): an effective alternative to chemical acaricides against different developmental stages of fowl tick Argas persicus (Acari: Argasidae). *Veterinary Parasitology*, *172*(3-4), 305–310. DOI: 10.1016/j.vetpar.2010.05.014 PMID: 20541868

Priyashantha, A. K. H., Karunarathna, S. C., Lu, L., & Tibpromma, S. (2023). Fungal Endophytes: An Alternative Biocontrol Agent against Phytopathogenic Fungi. *Encyclopedia*, *3*(2), 759–780. DOI: 10.3390/encyclopedia3020055

Rabea, E. I., Badawy, M. E., Stevens, C. V., Smagghe, G., & Steurbaut, W. (2003). Chitosan as antimicrobial agent: Applications and mode of action. *Biomacromolecules*, *4*(6), 1457–1465. DOI: 10.1021/bm034130m PMID: 14606868

Rai, M., Zimowska, B., Shinde, S., & Tres, M. V. (2021). Bioherbicidal potential of different species of Phoma: Opportunities and challenges. *Applied Microbiology and Biotechnology*, *105*(8), 3009–3018. DOI: 10.1007/s00253-021-11234-w PMID: 33770245

Ramudingana, P., Makhado, N., Kamutando, C. N., Thantsha, M. S., & Mamphogoro, T. P. (2025). Fungal Biocontrol Agents in the Management of Postharvest Losses of Fresh Produce-A Comprehensive Review. *Journal of Fungi (Basel, Switzerland)*, *11*(1), 82. DOI: 10.3390/jof11010082 PMID: 39852501

Rangel, D. E., Braga, G. U., Anderson, A. J., & Roberts, D. W. (2005). Variability in conidial thermotolerance of Metarhizium anisopliae isolates from different geographic origins. *Journal of Invertebrate Pathology*, *88*(2), 116–125. DOI: 10.1016/j.jip.2004.11.007 PMID: 15766928

Roy, H. E., & Cottrell, T. E. (2008). Forgotten natural enemies: Interactions between coccinellids and insect-parasitic fungi. *European Journal of Entomology*, *105*(3), 391–398. DOI: 10.14411/eje.2008.049

Sandhu, S. S. (1995). Effect of physical factors on germination of entomopathogenic fungus Beauveria bassiana conidia. *National Academy Science Letters*, *18*(1–2), 1–5.

Sandhu, S. S., Rajak, R. C., & Agarwal, G. P. (1993). Studies on prolonged storage of Beauveria bassiana conidia: Effects of temperature and relative humidity on conidial viability and virulence against chickpea borer Helicoverpa armigera. *Biocontrol Science and Technology*, *3*(1), 47–53. DOI: 10.1080/09583159309355258

Sandhu, S. S., Rajak, R. C., & Hasija, S. K. (2000). Potential of entomopathogens for the biological management of medically important pests: Progress and prospect. Glimpses in Plant Sciences, 110–117.

Sandhu, S. S., Sharma, A. K., Beniwal, V., Goel, G., Batra, P., Kumar, A., Jaglan, S., Sharma, A. K., & Malhotra, S. (2012). Myco-biocontrol of insect pests: Factors involved, mechanism, and regulation. *Journal of Pathogens*, *126819*, 1–10. Advance online publication. DOI: 10.1155/2012/126819 PMID: 22567344

Schena, L., Nigro, F., Soleti Ligorio, V., Yaseen, T., Ippolito, A., & El Ghaouth, A. (2005). Biocontrol activity of Bio-Coat and Biocure against postharvest rots of table grapes and sweet cherries. *Acta Horticulturae*, (682), 2115–2120. DOI: 10.17660/ActaHortic.2005.682.288

Schrank, A., & Vainstein, M. H. (2010). Metarhizium anisopliae enzymes and toxins. *Toxicon: official journal of the International Society on Toxinology, 56*(7), 1267–1274. https://doi.org/DOI: 10.1016/j.toxicon.2010.03.008

Sharma, R., & Sharma, P. (2021). Fungal entomopathogens: A systematic review. *Egyptian Journal of Biological Pest Control*, *31*(1), 57. DOI: 10.1186/s41938-021-00404-7

Sheikh, A. A., Loveleen, L., Pant, K. K., & Chandra, V. (2022). Cherries: A Treasure Of Bioactive Compounds With Promising Health Promoting Functional Potential. *IJFANS International Journal of Food and Nutritional Sciences*, *11*(8), 794–821.

Shimazu, M. (2004). A novel technique to inoculate conidia of entomopathogenic fungi and its application for investigation of susceptibility of the Japanese pine sawyer, Monochamus alternatus, to Beauveria bassiana. *Applied Entomology and Zoology*, *39*(3), 485–490. DOI: 10.1303/aez.2004.485

Singh, D., Raina, T. K., & Singh, J. (2017). Entomopathogenic fungi: An effective biocontrol agent for management of insect populations naturally. *Journal of Pharmaceutical Sciences and Research.*, *9*(6), 830–839.

Singh, H. B. (2014). Management of plant pathogens with microorganisms. *Proceedings of the Indian National Science Academy. Part A, Physical Sciences*, *80*(2), 443–454. DOI: 10.16943/ptinsa/2014/v80i2/55120

Singh, V., & Chawla, S. (2012). Cultural Practices: An Eco-friendly Innovative Approach in Plant Disease Management. DOI: 10.13140/RG.2.1.4216.2327

Stein, E., Molitor, A., Kogel, K. H., & Waller, F. (2008). Systemic resistance in Arabidopsis conferred by the mycorrhizal fungus Piriformospora indica requires jasmonic acid signaling and the cytoplasmic function of NPR1. *Plant & Cell Physiology*, *49*(11), 1747–1751. DOI: 10.1093/pcp/pcn147 PMID: 18842596

Szekeres, A., Leitgeb, B., Kredics, L., Antal, Z., Hatvani, L., Manczinger, L., & Vágvölgyi, C. (2005). Peptaibols and related peptaibiotics of Trichoderma. A review. *Acta Microbiologica et Immunologica Hungarica*, *52*(2), 137–168. DOI: 10.1556/AMicr.52.2005.2.2 PMID: 16003936

Thakur, R., Rajak, R., & Sandhu, S. S. (2005). Biochemical and molecular characteristics of indigenous strains of the entomopathogenic fungus Beauveria bassiana of Central India. *Biocontrol Science and Technology*, *15*(7), 733–744. DOI: 10.1080/09583150500136584

Thakur, R., & Sandhu, S. S. (2010). Distribution, occurrence and natural invertebrate hosts of indigenous entomopathogenic fungi of Central India. *Indian Journal of Microbiology*, *50*(1), 89–96. DOI: 10.1007/s12088-010-0007-z PMID: 23100813

Thambugala, K. M., Daranagama, D. A., Phillips, A. J. L., Kannangara, S. D., & Promputtha, I. (2020). Fungi vs. Fungi in Biocontrol: An Overview of Fungal Antagonists Applied Against Fungal Plant Pathogens. *Frontiers in Cellular and Infection Microbiology*, *10*, 604923. DOI: 10.3389/fcimb.2020.604923 PMID: 33330142

Tjamos, E. C., Papavizas, G. C., & Cook, R. J. (1992). *Biological control of plant diseases: Progress and challenges for the future*. Plenum Press. DOI: 10.1007/978-1-4757-9468-7

Vadassery, J., Tripathi, S., Prasad, R., Varma, A., & Oelmüller, R. (2009). Monodehydroascorbate reductase 2 and dehydroascorbate reductase 5 are crucial for a mutualistic interaction between Piriformospora indica and Arabidopsis. *Journal of Plant Physiology*, *166*(12), 1263–1274. DOI: 10.1016/j.jplph.2008.12.016 PMID: 19386380

Vega, F. E., Goettel, M. S., Blackwell, M., Chandler, D., Jackson, M. A., Keller, S., Koike, M., Maniania, N. K., Monzón, A., Ownley, B. H., Pell, J. K., Rangel, D. E., & Roy, H. E. (2009). Fungal entomopathogens: New insights on their ecology. *Fungal Ecology*, *2*(4), 149–159. DOI: 10.1016/j.funeco.2009.05.001

Vega, F. E., Posada, F., Aime, M. C., Pava-Ripoll, M., Infante, F., & Rehner, S. A. (2008). Entomopathogenic fungal endophytes. *Biological Control*, *46*(1), 72–82. DOI: 10.1016/j.biocontrol.2008.01.008

Vestergaard, S., Nielsen, C., Harding, S., & Eilenberg, J. (2002). First field trials to control Melolontha melolontha with Beauveria brongniartii in christmas trees in Denmark. *IOBC/WPRS Bulletin*, *25*(7), 51–58.

Vinale, F., Sivasithamparam, K., Ghisalberti, E., Marra, R., Barbetti, M., Li, H., Woo, S. L., & Lorito, M. (2008a). A novel role for Trichoderma secondary metabolites in the interactions with plants. *Physiological and Molecular Plant Pathology*, *72*(1-3), 80–86. DOI: 10.1016/j.pmpp.2008.05.005

Vinale, F., Sivasithamparam, K., Ghisalberti, E. L., Marra, R., Woo, S. L., & Lorito, M. (2008b). Trichoderma-plant-pathogen interactions. *Soil Biology & Biochemistry*, *40*(1), 1–10. DOI: 10.1016/j.soilbio.2007.07.002

Wang, C., & St Leger, R. J. (2005). Developmental and transcriptional responses to host and nonhost cuticles by the specific locust pathogen Metarhizium anisopliae var. acridum. *Eukaryotic Cell*, *4*(5), 937–947. DOI: 10.1128/EC.4.5.937-947.2005 PMID: 15879528

Weindling, R. (1941). Experimental consideration of the mold-toxins of Gliocladium and Trichoderma. *Phytopathology*, *31*(11), 991–1003.

Wiest, A., Grzegorski, D., Xu, B. W., Goulard, C., Rebuffat, S., Ebbole, D. J., Bodo, B., & Kenerley, C. (2002). Identification of peptaibols from Trichoderma virens and cloning of a peptaibol synthetase. *The Journal of Biological Chemistry*, *277*(23), 20862–20868. DOI: 10.1074/jbc.M201654200 PMID: 11909873

Wisniewski, M., Droby, S., Norelli, J., Liu, J., & Schena, L. (2016). Alternative management technologies for postharvest disease control: The journey from simplicity to complexity. *Postharvest Biology and Technology*, *122*, 3–10. DOI: 10.1016/j.postharvbio.2016.05.012

Woo, S. L., Scala, F., Ruocco, M., & Lorito, M. (2006). The molecular biology of the interactions between Trichoderma spp., phytopathogenic fungi, and plants. *Phytopathology*, *96*(2), 181–185. DOI: 10.1094/PHYTO-96-0181 PMID: 18943922

Wraight, S., Carruthers, R., Jaronski, S., Bradley, C., Garza, C., & Galaini-Wraight, S. (2000). Evaluation of the Entomopathogenic fungi Beauveria bassiana and Paecilomyces fumosoroseus for microbial control of the Silverleaf Whitefly, Bemisia argentifolii. *Biological Control*, *17*(3), 203–217. DOI: 10.1006/bcon.1999.0799

Wraight, S. P., Jackson, M. A., & Kokc, S. L. (2001). Production, stabilization and formulation of fungal biocontrol agents. Fungi as biocontrol agents: progress, problems and potential, 253-287. DOI: 10.1079/9780851993560.0253

Zaynab, M., Fatima, M., Abbas, S., Sharif, Y., Umair, M., Zafar, M. H., & Bahadar, K. (2018). Role of secondary metabolites in plant defense against pathogens. *Microbial Pathogenesis*, *124*, 198–202. DOI: 10.1016/j.micpath.2018.08.034 PMID: 30145251

Zeilinger, S., Gruber, S., Bansal, R., & Mukherjee, P. K. (2016). Secondary metabolism in Trichoderma – Chemistry meets genomics. *Fungal Biology Reviews*, *30*(2), 74–90. DOI: 10.1016/j.fbr.2016.05.001

Zhang, Q., Zhang, J., Yang, L., Zhang, L., Jiang, D., Chen, W., & Li, G. (2014). Diversity and biocontrol potential of endophytic fungi in Brassica napus. *Biological Control*, *72*, 98–108. DOI: 10.1016/j.biocontrol.2014.02.018

Zhang, Y., Zhang, K., & Hyde, K. (2014). The ecology of Nematophagous fungi in natural environments. Fungal Diversity Research Series, 211-229. DOI: 10.1007/978-94-017-8730-7_4

Chapter 4
Arbuscular Mycorrhizal Fungal Diversity Under Abiotic Stress:
How Does It Improve Plant Stress Tolerance and Nutrient Use Efficiency?

Soufiane Lahbouki
Chouaîb Doukkali University, Morocco

Naira Sbbar
Cadi Ayyad University, Marrakesh, Morocco

Salah Er-Raki
FST Marrakech, Morocco

Abdelilah Meddich
https://orcid.org/0000-0001-9590-4405
Cadi Ayyad University, Marrakesh, Morocco

ABSTRACT

Arbuscular mycorrhizal fungi (AMF) play a key role in promoting plant growth and resilience to abiotic stresses like drought, salinity, and heat. By forming symbiotic relationships with plant roots, AMF significantly enhance the uptake of nutrients, especially phosphorus, often limited in stressed soils. This chapter explores how AMF help plants by improving water absorption, maintaining osmotic balance, and regulating ion homeostasis. It highlights the diversity of AMF species and their varying responses to stress, influencing plant support. The chapter also examines changes in AMF community structure under prolonged stress and their implications for plant productivity. Additionally, it delves into how AMF modify gene expression in host plants, impacting osmolyte production and stress-protective proteins like heat shock proteins. The strategic use of AMF offers sustainable agricultural solutions, reducing chemical inputs and enhancing crop resilience amid climate change.

DOI: 10.4018/979-8-3693-8201-1.ch004

INTRODUCTION

Arbuscular mycorrhizal fungi (AMF) play a pivotal role in supporting plants and helping them adapt to their surrounding environmental conditions, whether in natural or agricultural ecosystems (Madouh et al., 2023; Xing et 2019). It is estimated that AMF which form symbiotic partnerships with about 80% of vascular terrestrial plants, contribute to enhancing plant growth and development (Raja et al., 2022). These improving their ability to absorb essential water and nutrients, particularly phosphorus (P), which is a key nutrient for plant growth but is often insufficiently available in the soil (Smith et al., 2010). This efficient absorption occurs thanks to the specialized fungal structures, such as arbuscules, which act as bridges for nutrient transfer between the soil and the plant.

Moreover, AMF enhance the uptake of essential nutrients beyond P, including nitrogen (N), calcium (Ca), copper (Cu), magnesium (Mg), manganese (Mn), sulfate (SO_4^{2-}), and zinc (Zn) (Stover et al., 2012). They improve nutrient availability by modifying soil microbial composition and influencing denitrifying and nitrifying bacteria, thereby enhancing nitrogen use efficiency (NUE) (Tang et al., 2023). Additionally, AMF regulate potassium (K) and P use efficiency, especially under salinity and drought stress (Mulyadi et al., 2023). They also facilitate the direct uptake of ammonium, nitrate, and amino acids, boosting plant biomass and productivity (Stover et al., 2012). These stresses, exacerbated by climate change, reduce soil water availability and increase toxic salt accumulation, hindering plant growth (Raja et al., 2022; Auge, 2001). AMF mitigate these effects by enhancing physiological and biochemical adaptations, such as osmotic regulation and reducing toxic ion accumulation (Das and Roychoudhury, 2014).

.Moreover, AMF play a critical role in boosting plant resilience to environmental changes, whether by improving root structure or influencing the production of protective molecules such as osmolytes. These fungi help plants maintain water and nutrient balance even in the harshest conditions (Ruiz-Lozano, 2003). Studying the effects of AMF under extreme environmental conditions is essential for understanding how to improve plant productivity and ensure the sustainability of agricultural systems.

AMF diversity under the influence of these stresses represents a crucial research area, as these fungi interact differently depending on species, family, and the surrounding environment.

Understanding AMF ecological diversity requires examining community shifts under stress gradients, considering species-specific adaptations. Some taxa thrive in extreme conditions, while others decline, yet rare species enhance ecosystem resilience through nutrient cycling and plant support. AMF populations respond dynamically to drought, salinity, and soil degradation, favoring species with adaptive traits like drought tolerance or efficient phosphorus acquisition (Krishnamoorthy et al., 2015). Exploring AMF diversity under abiotic stress is crucial for sustainable agriculture and ecological restoration.

Some AMF species, like Glomus and Rhizophagus, withstand harsh conditions, while others decline, altering fungal diversity and ecosystem function (Meena et al., 2022. Their role extends beyond nutrient uptake, as they influence plant community stability through host specificity, affecting ecosystem resilience. Understanding AMF diversity in an ecological context, rather than solely through functional traits, is crucial for assessing their contributions to plant adaptation and sustainable agriculture (Lin, 2015).

In this context, AMF are one of the promising natural tools that can be exploited to enhance crop productivity in harsh environments. In addition to their traditional role in improving nutrient absorption, they also provide natural solutions to increase plant resistance to abiotic stresses, making them a key partner in sustainable agriculture (Kumar et al., 2010). A deeper understanding of the effects and diversity of these AMF under these stresses will contribute to developing sustainable strategies that ensure the preservation of natural resources and increase crop productivity amid accelerating climate changes.

Therefore, this chapter explores AMF diversity and its response to abiotic stresses, focusing on drought, salinity, and high temperatures. It examines how enhancing AMF diversity improves plant adaptation and tolerance, contributing to innovative agricultural strategies for sustainable productivity and ecological balance.

INFLUENCE OF ABIOTIC STRESS ON AMF DIVERSITY

Impact of Drought on AMF Communities

Drought stress significantly reduces soil moisture, impacting agricultural and natural ecosystems. As drought progresses, shifts in AMF species richness and phylogenetic diversity are observed. Some AMF species, such as Glomus sp. and *Rhizophagus intraradices*, exhibit resilience under water scarcity due to their ability to enhance osmotic adjustments and water uptake efficiency (Meena et al., 2022). In contrast, species that lack such adaptations decline under prolonged drought conditions (Krishnamoorthy et al., 2015).

Diversity indices such as the Shannon-Weaver and Simpson indices have been widely used to quantify shifts in AMF communities under abiotic stress conditions (Davison et al., 2020). Studies indicate that under increasing drought stress, AMF diversity tends to decline, with only highly stress-tolerant species persisting (Van der Heijden et al., 2015). However, rare AMF taxa, which typically exist in low abundance, may play a critical role in stabilizing ecosystem functions by contributing to nutrient cycling and stress resilience (Powell et al., 2009). These rare species can act as reservoirs of functional traits that facilitate plant survival under fluctuating environmental conditions.

Over the past decades, pioneering research has significantly shaped our understanding of AMF and their role in plant nutrient acquisition and stress tolerance. Early study by Lahbouki et al. (2023) and Smith et al. (2010) established the fundamental mechanisms by which AMF enhance phosphorus uptake and facilitate nutrient exchange between soil and plants. Subsequent research extended these findings to nitrogen uptake, with Zeng et al. (2021) demonstrating that AMF influence nitrogen cycling in the rhizosphere, affecting both nitrate and ammonium availability. Recent advances, such as those by Yang et al. (2022), have highlighted the temporal complementarity between plant roots and AMF in optimizing NUE in wheat systems. These foundational studies provide historical context to the role of AMF in plant nutrition and set the stage for current investigations into their broader ecological functions.

These changes cause significant modifications in the physical and chemical properties of the soil, directly and indirectly influencing AMF communities and their interactions with plants (Y. Liu et al., 2024; Preece et al., 2019). When soil moisture decreases due to drought, the soil structure hardens, increasing its density and reducing the pore spaces essential for gas exchange, water movement, and nutrient flow (Y. Liu et al., 2024; Kang et al., 2022). This reduction limits the spread and colonization of fungal hyphae on roots, ultimately hindering the biological activity of AMF (Preece et al., 2019).

In addition, drought causes significant chemical alterations in the soil. A key impact is the reduction of soil moisture, which decreases the solubility of essential nutrients such as P, K, and Mg, making them less available to plants and AMF (Fischer et al., 2019; Ortas et al., 2021). Drought can also alter the soil P cycle by decreasing the solubility of calcium phosphate and increasing the binding of phosphorus with secondary minerals like Fe/Al oxides, further reducing its bioavailability (Fischer et al., 2019). This reduction affects the symbiotic relationship between plants and AMF (Ortas et al., 2021).

Moreover, drought influences soil pH, which further affects the availability of P, a critical nutrient for both fungal and plant growth (Fan et al., 2021). As drought progresses, salt concentrations increase due to evaporation, altering Ca and Mg levels in the soil (Qadeer et al., 2022). Despite these challenges, certain AMF genrea such as Glomus, Funneliformis, and Rhizophagus are capable of dominating under these conditions because they can access P even in its less soluble forms (Sery et al., 2018). These genrea adapt to the chemical changes in the soil, supporting plant resilience in drought-affected ecosystems.

Drought conditions significantly impact photosynthesis in plants (Lahbouki et al., 2022). In response to water scarcity, plants close their stomata to conserve moisture, leading to reduced carbon dioxide uptake and decreased carbohydrate production (Qiao et al., 2024; Lahbouki et al., 2022). This is an important energy source for AMF (Lahbouki et al., 2023; Bogati & Walczak, 2022). Under these stressed conditions, some AMF species heavily dependent on high carbohydrate levels, such as *Funneliformis mosseae*, *Rhizophagus intraradices* and *Paraglomus occultum* enhance carbohydrate metabolism to support their survival and sustain symbiotic relationships with their host plants (Zheng et al., 2024; Liu et al., 2022; Zhang et al., 2018).

As drought progresses, the most water-deficient-adapted genera, such as *Glomus sp.*, *R. intraradices*, *Claroideoglomus etunicatum*, tend to dominate due to their physiological and morphological flexibility (Madouh et al., 2023; Meena et al., 2022; Lahbouki et al., 2021;). These species form highly resilient spores that can withstand prolonged drought, remaining dormant until favorable conditions return (Madouh et al., 2023; Meena et al. 2022). In addition, these species are also efficient in forming extensive hyphal networks that reach deeper soil layers to access scarce water and nutrients, including phosphorus, helping to sustain plant growth even in dry environments (Meena et al., 2022). These adaptations make them more suited to arid conditions compared to less drought-tolerant AMF species (Meena et al., 2022).

Among other critical adaptations, drought-resistant species such as *R.intraradices*, *C.etunicatum* accumulate osmolytes like trehalose and mannitol (Pooja et al., 2024; Ocón et al., 2007; . These compounds help stabilize proteins and cell membranes, enabling the fungi to withstand water scarcity. Studies have shown that this species can increase its trehalose concentration by up to 60% under drought conditions, enhancing its survival and continued support of host plants (Pooja et al., 2024). In addition, in drought conditions, these AMF species adapt to the chemical changes in the soil, such as increased salt concentration and altered nutrient availability (Pooja et al., 2024). These adaptations allow AMF species to survive in rapidly changing environments (Meena et al., 2022).

However, As drought intensifies, less-resistant AMF species decline, while adaptable species like *F. mosseae*, *Glomus sp.*, and *R.irregularis* dominate (Sery et al., 2018; Ocón et al., 2007). Although these species maintain nutrient uptake, including P, the levels are lower due to the reduced availability of water and nutrients (Pooja et al., 2024). Despite these limitations, they help plants cope with drought by enhancing nutrient absorption and promoting root hair formation, which helps maintain plant growth rates during water stress (Lahbouki et al. 2024; Zheng et al., 2024).

Effects of saline soils on AMF communities

High salt concentrations in soil increase osmotic pressure, directly impacting the ability of plants and fungi to absorb water and essential nutrients (Raja et al., 2022). High salinity reduces AMF diversity, favoring salt-tolerant species like *Funneliformis mosseae* while limiting sensitive ones like *Scutellospora calospora* (Dorostkar et al., 2014). Rare AMF species support ecosystem stability under stress, with dominant taxa aiding nutrient uptake and low-abundance species contributing to niche differentiation and stress resilience through unique metabolic adaptations.

Actually, salinty conditions alter the soil's chemical and physical properties, such as pH and ion concentration, which affects AMF diversity and changes their community structure and function (Qadeer et al., 2022; Dorostkar et al., 2014). Severl studies show that salt-tolerant AMF species, such as *C. etunicatum* and *F. mosseae*, maintain root colonization and functionality in soils with salinity levels up to 10 dS/m, which is considered moderate to high salinity (Zhou et al., 2022). However, at salinity levels above 15 dS/m, even these species struggle with effective symbiosis. As salinity increases, overall AMF diversity declines, and species like *C. etunicatum*, *F. mosseae*, and *Rhizophagus irregularis* dominate due to their adaptive mechanisms (Bogati & Walczak, 2022; Dorostkar et al., 2014). In fact, these species depend on high carbohydrate availability to sustain symbiosis, especially under stress, with elevated sugar concentrations aiding in salt tolerance and biomass production (Dorostkar et al., 2014; Bogati & Walczak, 2022).

They produce osmolytes like glycine betaine, proline, and trehalose to maintain osmotic balance and prevent dehydration (Zhang et al., 2018; Ocón et al., 2017). Under saline conditions, they increase osmolyte concentrations, adjust membrane composition to limit ion influx, and express transporter proteins to expel or sequester sodium ions, reducing cytotoxicity (Raja et al., 2022; Pooja et al., 2024). These adaptations enable them to sustain symbiotic functions even under high salinity.

Less salt-tolerant species like *S. calospora* and *Gigaspora margarita* struggle in saline environments due to inefficient sodium transporters and weak osmoprotective mechanisms, leading to ion toxicity and metabolic disruption (Juniper & Abbott, 1993; Raja et al., 2022). Their compromised hyphal structures reduce root colonization and symbiotic efficiency under high salinity (Dorostkar et al., 2014). As salinity rises, AMF communities shift towards more resilient species, altering functional dynamics and reducing diversity (Dorostkar et al., 2014). Salt-tolerant AMF mitigate salinity effects by secreting glomalin, enhancing soil structure and stability (Pooja et al., 2024; Benages et al., 2022).

Impact of high temperatures on AMF diversity

High temperatures are a significant abiotic stress affecting soil ecosystems, particularly AMF communities (Xu et al., 2020; Oliveira et al., 2013). Heat stress induces notable shifts in AMF community composition, favoring species with physiological adaptations such as increased heat shock protein expression and osmolyte accumulation (Tian et al., 2023). While species like Rhizophagus irregularis exhibit moderate tolerance to high temperatures, prolonged heat exposure reduces their colonization efficiency and symbiotic functionality (Cornell et al., 2022).

Community-level shifts in AMF diversity under heat stress indicate a decline in overall richness, with thermotolerant taxa such as *Funneliformis* and *Rhizophagus* becoming dominant (Nouh et al., 2021). However, these changes may lead to functional trade-offs, where reduced species diversity limits the resilience of ecosystems to additional stressors like nutrient depletion. Rare AMF species, often overlooked in biodiversity assessments, may provide essential stress-adaptive functions, such as secondary metabolite production and enhanced soil aggregation, which buffer plants against extreme temperature fluctuations.

High temperatures conditions lead to changes in soil moisture, nutrient availability, and microbial activity, which directly influence the diversity and functionality of AMF (Xu et al., 2020). They can also cause soil desiccation, reducing moisture levels and increasing soil temperature beyond the optimal range for many AMF species (Radujković et al., 2018). This results in shifts in community composition and a decline in species diversity.

Additionally, high temperatures impact photosynthesis in plants, causing them to close their stomata to minimize water loss, which reduces carbon dioxide uptake and carbohydrate production (Lahbouki et al.; 2022; Rodrigues et al., 2018). Since AMF heavily rely on carbohydrates from their host plants, a decrease in carbohydrate availability affects the growth and survival of AMF, especially those species that require significant energy inputs (Rodrigues et al., 2018). This decline in photosynthetic activity limits AMF's ability to develop extensive hyphal networks, and species that cannot efficiently utilize the reduced carbohydrate supply may diminish (Gavito et al., 2019).

Furthermore, high temperatures influence soil pH and nutrient dynamics, further affecting AMF diversity (Wei et al., 2022). Elevated temperatures can accelerate mineralization processes, such as N and P cycling, temporarily increasing nutrient availability (Wei et al., 2022). However, rapid decomposition and loss of organic matter in the soil may eventually reduce nutrient availability, making it difficult for AMF to access the resources needed for growth and symbiotic functions (Toleikiene et al., 2020; Qin et al., 2015). As nutrient availability changes, AMF community composition may shift, favoring species that can quickly adapt to these new conditions (Leal et al., 2013).

Research indicates that high temperatures, especially those exceeding 35°C, can negatively affect AMF survival and root colonization capacity (Ndeko et al., 2022). However, many AMF species are sensitive to extreme heat, and sustained high temperatures can inhibit spore germination, hyphal growth, and colonization ability (Ndeko et al., 2022; A. Liu et al., 2004). This reduction in AMF colonization directly impacts their ability to establish and maintain symbiotic relationships with plants, affecting nutrient exchange and plant growth (Ndeko et al., 2022; A. Liu et al., 2004).

Species like *R. intraradices* and *Glomus mosseae* tolerate temperatures up to 35°C by regulating osmolyte-related genes, such as proline and trehalose, which stabilize proteins and membranes under thermal stress (Cornell et al., 2022; Xing et al., 2019; Ma et al., 2015). However, prolonged exposure above 40°C reduces even heat-tolerant AMF efficiency. *Diversispora versiformis* experiences decreased root colonization and mycelium length but activates heat-responsive genes, including Hsp70 and PIPs, aiding thermal adaptation (Zhao et al., 2024; Tian et al., 2023; Cornell et al., 2022). *D. versiformis* upregulates 11 of 12 *Hsp70* genes under heat stress, demonstrating direct gene regulation in response to high temperatures (Tian et al., 2023). Thus, AMF employ osmolyte production and heat shock protein activation to withstand heat stress, though prolonged exposure over 40°C impairs colonization and symbiotic function.

In such environments, species like *C. etunicatum* tend to dominate, as they possess physiological mechanisms that enable them to withstand high temperatures (Xu et al., 2024). These mechanisms include modifications in membrane lipid composition to maintain fluidity and stability under heat stress, as well as increased production of osmolytes like proline and mannitol, which help maintain osmotic balance and protect cellular components (Zhang et al., 2023; Zhou et al., 2022).

Heat stress also interacts with other abiotic stressors, such as drought and salinity, compounding the challenges faced by AMF (Sousa et al., 2024; Suzuki et al., 2014). When high temperatures coincide with drought conditions, soil moisture levels decrease drastically, exacerbating the negative effects on AMF colonization and diversity (Albracht et al., 2023). The combined impact can further reduce AMF species richness, as only the most resilient species capable of withstanding both heat and low moisture levels persist (Diagne et al., 2020). For example, species like *R. irregularis* and *F. mosseae* demonstrate some ability to endure the combined stresses of drought and high temperatures, but their efficiency is significantly reduced compared to when each stressor occurs independently (Duc et al., 2023; Musyoka et al., 2020).

SYMBIOTIC MECHANISMS AND CHEMICAL SIGNALING

The symbiosis between AMF and plants relies on precise biochemical interactions and chemical signals that regulate communication between roots and AMF (Wani et al., 2020; Lanfranco et al., 2018;). This symbiotic relationship typically begins when the fungi detect suitable moisture levels in the soil, which is crucial for spore germination (Lanfranco et al., 2018). The germination of spores heavily depends on soil moisture levels, providing the ideal environment for fungal growth (Khan et al., 2024). Once the spores germinate and the AMF begin to grow, they become ready to receive chemical signals from the plant (Boyno & Demir, 2022).

Before colonization, AMF recognize potential hosts, while non-host plants do not stimulate hyphal growth and may even inhibit it (Requena et al., 2007). The colonization process is tightly regulated by both partners, beginning with presymbiotic communication. Host plants secrete flavonoids into the soil, attracting AMF and triggering cellular reprogramming for infection. This promotes hyphal elongation and sustains fungal growth even before root contact (Choi et al., 2018; Nadal et al., 2013).

Host root exudates activate AM fungal metabolism and, probably as a consequence, distinguish hyphal branching patterns that increase the chances of encountering the host by secreting 2-hydroxy fatty acids released by the plant and this triggers a low degree of hyphal branching from the primary hypha. Typically, application of the branching factor to a hypha results in gene activation in *Gigaspora rosea, as exemple,* after 0.5–1 h, physiological alterations after 1.5–3 h, and the onset of hyphal branching could be observed after 5 h (Tamasloukht et al., 2003). More than 10 years ago, strigolactones (SLs) were identified as key components. The high degree of hyphal branching mediated by these SLs that only occurs when the AMF is in close proximity to the root. Secretion of SLs into the rhizosphere involves the ABC transporter Pleiotropic drug resistance1 (PDR1), which is located in the non-suberized hypodermal passage cell and produces a concentration gradient that provides a positional signal to guide the beneficial fungus toward the host root. Perception of the cutin monomers 1,16-hexadecanediol and 16-hydroxyhexadecanoic acid triggers the development of hyphopodium on the root surface.

Chemical communication in AMF frequently involves oligosaccharide signals. Germinated spores of AM fungi secrete lipochitooligosaccharides, which are two types, lipochitooligosaccharides and short-chain chitooligosaccharides (COs), namely chitotetraose and chitopentaose that trigger lateral root formation, which is partially mediated by the nod factor receptor NFP. These root hairs, located in the exodermis layer, interact with the surrounding soil environment, where strigolactones—typically released when the plant experiences phosphorus deficiency—stimulate hyphal branching in the fungi, promoting root colonization and initiating the symbiotic relationship.

The AMF respond to these signals by increasing their activity near the root, producing more branched hyphae, which facilitates attachment to the roots and initiates the symbiotic interaction (Pons et al., 2020). This initial phase is crucial for the successful establishment of the symbiotic relationship between the AMF and the plant. Subsequently, only specific cells promote the formation of an appressorium and the prepenetration apparatus (Genre et al., 2005), enabling the fungus to penetrate the root through the epidermis. In the root, AMF hyphae form foot-like attachment structures, called hyphopodia, on the surface of the root epidermis. At no point does the AMF penetrate the host plasmalemma. In addition to arbuscules, some AMF can form vesicles, thought to store lipids. To complete its life cycle, the AMF emerges from the root and, after extensive growth of the outer hyphae, sporulates in the soil. Throughout the life cycle, from successful colonization of the root cortex to the subsequent formation of new fungal spores, signals must be exchanged between symbiotic partners (Genre et al. 2008; Parniske 2008), inducing stage-specific gene expression patterns in host roots and AMF.

At this stage, Ca signaling plays a key role in regulating the plant's response to the symbiotic interaction (Luginbuehl & Oldroyd, 2016). This pathway triggers Ca pulses within the root cells, which serve as vital signals for activating genes associated with symbiosis and the formation of arbuscules, the branched fungal structures formed inside plant root cells (Gutjahr et al., 2008). In addition, Ca pulses decode the signals from the fungi and activate genes that regulate arbuscule formation, enhancing the effectiveness of the symbiotic interaction and nutrient exchange between fungi and plants.

The Vapyrin protein facilitates periarbuscular membrane (PAM) formation by interacting with *VAMP721 R-SNARE* proteins, essential for vesicle fusion and symbiotic stability (Zhang et al., 2015). This process is regulated by transcription factors WRI5a and MYB1, ensuring efficient nutrient exchange (Floss et al., 2017; Jiang et al., 2018; Bapaume et al., 2019). Within the PAM, phosphate transporters like MtPT4 in Medicago truncatula enhance P uptake at arbuscular branches, while carbon transfer sustains fungal interaction (Pumplin et al., 2012; Wang et al., 2014; Hammer et al., 2011). Cytoskeletal reorganization, involving microtubules and actin, supports fungal penetration and arbuscule formation, coordinated with calcium signaling (Genre et al., 2005; Gutjahr et al., 2008). DMI3 interprets Ca signals, activating genes crucial for arbuscule development and nutrient exchange (Gutjahr et al., 2008).

Sugars play a pivotal role in supporting the symbiotic relationship. Sucrose, produced by plants through photosynthesis, is transported to the fungi to support their growth and development (Courty et al., 2016; Doidy et al., 2012). Sugars are broken down into smaller molecules that fungi can use as an energy source (Courty et al., 2016; Doidy et al., 2012). Then, in plants, specific sugar transporters, such as the *SWEET* family, are involved in transporting sugars to AMF (An et al., 2019; Lahmidi et al., 2016). These transporters are up-regulated during fungal colonization, supporting symbiosis by sustaining metabolism and facilitating heme-dependent root penetration and dendritic root formation (Roth & Paszkowski, 2017).

In addition, lipids play a key role in this process. AMF depend on lipids produced by the plants and transferred to the fungi to provide the necessary energy for growth (Keymer et al., 2017). The production and transport of lipids in plants are regulated by *RAM2*, a gene involved in lipid biosynthesis, which is crucial for maintaining symbiosis with AMF. *RAM1* is necessary for the activation of the lipid biosynthetic pathway, including the induction of *RAM2* expression, which supports symbiotic interactions (Dai et al., 2022; Xue et al., 2018; Luginbuehl et al., 2017). These lipids are vital for the fungi to sustain their activity and form arbuscules, structures that facilitate nutrient exchange between the plant and the AMF (Feng et al., 2020; Luginbuehl et al., 2017).

Once the fungi fully colonize the roots, arbuscules are formed within the root cells. These structures serve as hubs for nutrient exchange between AMF and plants, with the AMF supplying phosphorus and other minerals absorbed from the soil to the plant (Bisht et al., 2024; Wang et al., 2014; Gutjahr et al., 2008). In return, plants provide the AMF with the sugars and lipids needed to support fungal growth (An et al., 2019; Keymer et al., 2017).

In the later stages of arbuscule formation, cytoskeletal proteins such as actin play a crucial role in stabilizing the arbuscules and enhancing the transport process between AMF and plants (Bisht et al., 2024;). Actin forms a dynamic network that helps regulate and direct the flow of nutrients within both plant and fungal cells. This supporting structure ensures the continuity of arbuscules within plant root cells, facilitating complete nutrient exchange between fungi and plants and strengthening the sustainability of the symbiotic relationship (Bisht et al., 2024).

To ensure the continued success of the symbiotic interaction between fungi and plants, transcription factors such as *RAM1* play a critical role in regulating this process (Dai et al., 2022). *RAM1* regulates the expression of *RAM2*, a glycerol-3-phosphate acyltransferase (Gobbato et al., 2012, 2013). These factors activate the genes responsible for arbuscule formation. and ensure balanced nutrient exchange between fungi and plants (Feng et al., 2020; Luginbuehl et al., 2017). This genetic regulation supports the plants' uptake of essential nutrients and enhances their ability to grow in nutrient-poor environments, ensuring the continuity of the symbiotic relationship between the plant and the fungi (Salwan et al., 2023; Fischer et al., 2019).

ROLE OF AMF IN ENHANCING PLANT STRESS TOLERANCE

AMF-Induced Changes in Plant Physiology and Soil Aggregation

Abiotic stresses, such as drought, salinity and extreme temperature regimes, are interconnected, as they all reduce plant cells' access to water and nutrients. AMF community composition influences plant stress tolerance, as certain AMF species are better suited to improve plant physiological responses under specific stress conditions. For example, drought-resistant species such as *R. intraradices* accumulate osmolytes like trehalose and mannitol, enabling the fungi to withstand water scarcity while supporting their host plants (Pooja et al., 2024).

These stresses impact crop productivity and sustainability, exacerbated by nutrient deficiencies, organic contaminants, and heavy metals. Mycorrhizal inoculation enhances plant growth under these conditions. Lahbouki et al. (2023) demonstrated that cactus regulates metabolism and osmotic balance to combat water loss, with decreased stomatal conductance in response to drought. However, AMF application increased this parameter. Similarly, Ouhaddou et al. (2023) found that AMF improved chlorophyll fluorescence and stomatal conductance under salt stress by up-regulating chloroplastic genes (RppsbA, RppsbD), enhancing PSII efficiency and photosynthesis. Ikan et al. (2023) reported that AMF mitigated salinity effects in wheat by reducing Na and Cl uptake while improving nutrient absorption. Regarding heavy metal stress, AMF strengthen plant defense mechanisms, promoting growth in contaminated soils. Zhu et al. (2022) showed that Glomus versiforme facilitated upland rice growth in Cd-contaminated soil, increasing P uptake and photosynthesis through enhanced phosphate transport protein expression and small molecule organic secretion.

AMF enhance plant stress tolerance and contribute to carbon sequestration by producing glomalin, a resistant glycoprotein that stabilizes soil aggregates (Wang et al., 2023). Released into the soil, glomalin improves porosity, water retention, and soil fertility, promoting sustainable carbon storage (Wang et al., 2023). AMF hyphae protect organic matter from decomposition, further stabilizing soil structure (Solaiman et al., 2024).

AMF also produces glomalin-related proteins, a major component of soil organic carbon. These proteins, rich in carbon (20–59%) and N (3-5%), can account for up to 40% of total organic carbon (Wang et al., 2023). Due to their resistant chemical structure, they remain present in the soil for several decades, ensuring prolonged carbon storage.

AMF-induced Hormonal balance

Arbuscular mycorrhizae play a key role in regulating the hormonal balance of plants, promoting growth and resistance to abiotic stresses. The symbiotic association between AMF and plant roots leads to changes in the levels of several plant hormones, which are essential for growth, development, and response to stress factors. Under water stress, Asensio et al. 2012, working on *Solanum lycopersicum* L. found that the application of AMF altered the isoprenoid content in the plant's leaves, increasing the level of essential isoprenoids instead of non-essential isoprenoids. This is explained by the increased need for and biosynthesis of carotenoid-derived compounds such as ABA and SLs in mycorrhizal plants. This is in line with the study by Ludwig-Müller, 2010, who observed a change in certain plant hormones such as ABA, Sls, and jasmonic acid in mycorrhizal plants under water-deficit conditions (Goyal et al., 2020). As under salt and metal stress, AMF symbiosis could enhance the plant's potential to withstand these abiotic stresses by producing plant growth hormones.

AMF-induced biochemical changes

The symbiotic association between plants and AMF induces biochemical changes that enhance tolerance to abiotic stress by regulating redox balance and sugar and protein metabolism. These adaptations, driven by enzymatic activity and osmoprotectant accumulation, strengthen plant resistance. Soluble sugars, polyols, and proline play a key role in stress mitigation, with proline scavenging reactive oxygen species (ROS), reducing oxidative damage, and maintaining cell turgor. Synthesized via the glutamate pathway, proline accumulation signals enhanced osmoprotection and stress adaptation. Anli et al. (2020) found that biostimulant application in *Phoenix dactylifera* increased osmolytes, particularly soluble sugars and proline, while reducing superoxyde dismutase (SOD), peroxidase, and catalase activity, highlighting AMF's role in stress resilience.

In parallel, stress indicators such as malondialdehyde and hydrogen peroxide also decreased, signaling an improved plant response to stress conditions. Under metal stress, AMF inoculation also led to an increase in nitrate reductase activity and protein content, with similar results reported in *Triticum aestivum*, *M. truncatula*, and *Vigna unguiculata* (Shaher et al., 2024) (**Figure 1**).

Figure 1. Mycorrhizal functions to regulate various physiological, hormonal, biochemical, and molecular processes of plants under abiotic stress conditions.

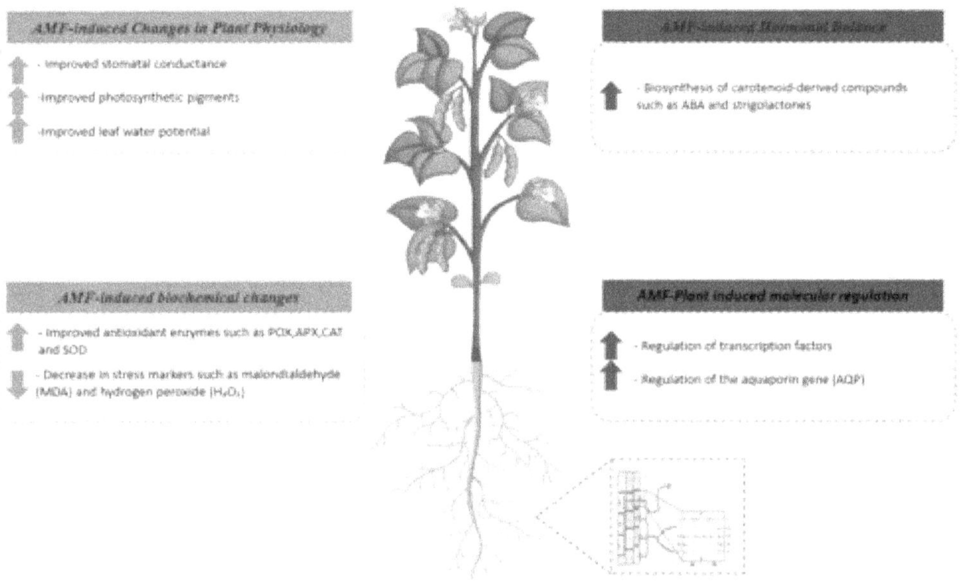

Molecular regulation and gene expression induced by AMF-plant symbiosis

Plants activate avoidance and tolerance mechanisms to cope with abiotic stress, regulated by transcription factors and signaling pathways. In M. truncatula, AMF colonization and arbuscular development depend on MtbHLHm1;1 and ammonium facilitator MtAMF1;3 (Ovchinnikova et al., 2023).

AMF also modulates root development genes, including CEP2 downregulation, promoting lateral root formation via auxin pathways (Hsieh et al., 2022; Mendoza-Soto et al., 2022). AMF enhances stress tolerance by influencing molecular responses, particularly aquaporins regulating water transport (Sharma et al., 2021). Under drought, AMF boosts SOD genes for antioxidant defense and MAPK genes for stress response, as seen in Bombax ceiba, where AMF increased BcCSD and BcMSD expression to mitigate oxidative stress (Luo et al., 2024).

R. irregularis' RiMsn2 regulates osmotic balance and trehalose, enhancing drought tolerance (Fan et al., 2023). Against salt stress, AMF in maize upregulated ZmAKT2 and ZmSKOR to maintain a high $K+/Na+$ ratio. In rice, AMF regulated ion homeostasis genes (OsSOS1, OsNHX3, OsHKT1;5, OsHKT2;1), reducing $Na+$ translocation and improving salt tolerance (Porcel et al., 2016). AMF also mitigates heavy metal stress; C. etunicatum and R. intraradices reduced molybdenum transport, enhancing biomass and nutrient uptake (Zhang et al., 2023). Additionally, genetic modifications improve metal tolerance by altering transport and detoxification genes, including metal tolerance proteins (Grabsztunowicz et al., 2024).

Table 1 illustrates the effect of AMF on the physiology, biochemistry, hormonal, and molecular balance of plants subjected to various abiotic stresses.

Table 1. The effect of AMF on the physiology, biochemistry, hormonal, and molecular balance of plants subjected to various abiotic stresses

Responses observed on	Abiotic conditions	Plant species	Consequences on Plants	References
Morphological and physiological traits	Drought	*Opuntia ficus-indica*	-Increased stomatal conductance -Increased stomata density -Increased chlorophyll fluorescence	Lahbouki et al., 2023
	Salinity	*Lactuca sativa*	-Increased chlorophyll fluorescence -Increased stomatal conductance	Ouhaddou et al., 2023
	Heavy metal	*Oryza sativa*	-Improved rice growth -Improved phosphorus uptake -Improved photosynthesis.	Zhu et al., 2022
Hormonal regulation	Drought	*Solanum lycopersicum*	-Changes in isoprenoid content -Increased levels of essential isoprenoids	Asensio et al., 2012
	Salinity	*Medicago truncatula*	-Changes in certain plant hormones such as ABA, strigolactones and jasmonic acid	Ludwig-Müller et al., 2019
	Heavy metal	*Cucumis sativus*	-Increases levels of stress-related hormones such as jasmonic acid and salicylic acid	Hashem et al., 2018
Biochemical properties	Drought	*Opuntia ficus-indica*	-Increased osmolyte content, in particular soluble sugars and protein content. -Reduced stress indicators such as malondialdehyde (MDA) and hydrogen peroxide (H_2O_2)	Lahbouki et al., 2023
	Heavy metal	*Robinia pseudoacacia*	-Increased activity of antioxidant enzymes, notably superoxide dismutase, ascorbate peroxidase and glutathione peroxidase -Reduced oxidative damage	Yang et al., 2015
Molecular regulation	Drought	*Zea mays*	-Overaccumulation of TSS and expression of aquaporin genes (ZmPIP1;1, ZmPIP1;3, ZmPIP1;4, ZmPIP1;6, ZmPIP2;2, ZmPIP2;4, ZmTIP1;1, and ZmTIP2;3 down-regulated and ZmTIP4;1 up-regulated).	Quiroga et al., 2017
	Salinity	*Oryza sativa*	-Up-regulated OsNHX3, OsSOS1, OsHKT2;1, and OsHKT1;5	Porcel et al., 2016
	Heavy metal	*Oryza sativa*	- Nramp5 and HMA3 expression is induced in the roots.	Chen et al., 2019

INTERACTIONS BETWEEN AMF AND OTHER SOIL MICROBES

AMF interact with a wide range of soil microorganisms, including bacteria and other fungal species, which collectively influence their functionality and impact on plant health. The presence of nitrogen-fixing bacteria, such as Rhizobium and Azospirillum, enhances nutrient availability by increasing nitrogen content in the rhizosphere, thereby complementing the phosphorus acquisition facilitated by AMF

(Masrahi et al., 2032). Similarly, phosphate-solubilizing bacteria improve phosphorus bioavailability, further strengthening the mutualistic benefits of AMF-plant symbioses (Sngwan et al., 2022).

Beyond mutualistic interactions, AMF also engage in competitive and antagonistic relationships with other soil microbes, including pathogenic fungi and bacteria. Certain soil-borne pathogens, such as *Fusarium* spp. and *Pythium* spp., negatively affect AMF colonization by disrupting root structures and altering root exudates essential for fungal establishment (Wahab et al., 2020).

APPLICATION OF AMF IN SUSTAINABLE AGRICULTURE

Microbial inoculants consist of free-living bacteria, fungi, and AMF sourced from soil, plants, plant waste, water, and compost (Gu et al., 2014). AMF exhibit high affinity for diverse plant families (Wang & Qiu, 2006) and form mutualistic relationships that enhance K, N, and micronutrient uptake, such as Zn, Cu, and Fe (Diagne et al., 2020; Varghese et al., 2023). They improve chlorophyll pigments, hormone production, secondary metabolites, and plant resistance to root pathogens and abiotic stresses, including heat, salinity, drought, and heavy metals (Varghese et al., 2023). AMF mitigate stress due to their ubiquity, genetic diversity, and adaptability to harsh environments (Sharma et al., 2021). Acting as extended plant roots, they enhance redox homeostasis, osmoregulation, water absorption, nutrient uptake, photosynthetic efficiency, and antioxidant metabolism (Meddich, 2023). Accumulation of osmoprotectants like proline, sugars, and polyols plays a key role in stress alleviation, with proline aiding in ROS scavenging (Meddich, 2023). AMF improve drought resistance by facilitating water and nutrient uptake, regulating transcriptional and aquaporin activity, optimizing photosynthesis, adjusting osmotic pressure, modulating phytohormones, and strengthening oxidative stress defense (Diagne et al., 2020). Additionally, they produce Glomalin, a stable glycoprotein that enhances soil aggregation and water retention (Wang et al., 2023). Their symbiotic role in sustainable agriculture reduces agrochemical dependency, particularly chemical fertilizers (Varghese et al., 2023) (**Figure 2**).

The mycorrhizal product market is growing but faces regulatory, quality, and adoption challenges. Europe lacks unified regulations, and awareness among farmers remains low. Efforts focus on field trials, while the EU works on market harmonization. Key organizations and research aid adoption (Chen et al., 2018).

Figure 2. European additive manufacturing market growth: Analysis of the number of companies selling amf inoculants. based on the year of creation, the number of companies was assessed over three distinct periods: 2000, 2010, and 2017

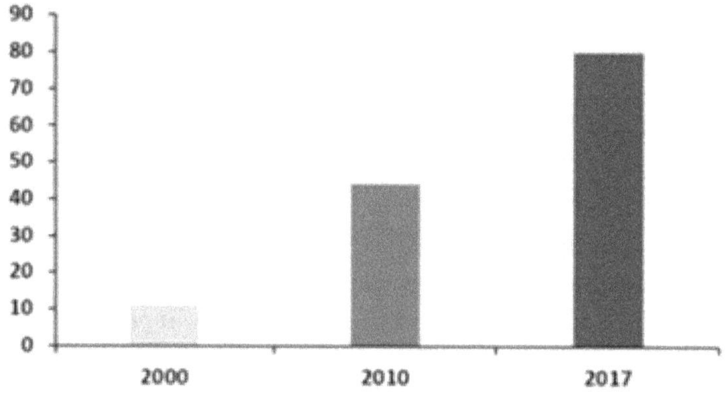

(Chen et al., 2018).

CONCLUSION AND FUTURE PROSPECTS

Arbuscular mycorrhizal fungi represent a promising tool for improving plant resilience to abiotic stresses, offering a natural solution to enhance water and nutrient absorption. Their ability to thrive in extreme environmental conditions while supporting plant growth makes them invaluable in sustainable agriculture. Looking forward, further research is needed to explore the molecular and genetic mechanisms behind AMF-mediated stress tolerance, including their role in regulating gene expression and nutrient transport. Advancing commercial applications of AMF inoculants can provide farmers with eco-friendly alternatives to chemical fertilizers, helping to reduce environmental impacts. Continued exploration of AMF biodiversity will be key to enhancing plant productivity in an era of increasing environmental challenges. Sustainable strategies that integrate AMF into modern agriculture will be vital in ensuring food security and the long-term viability of agricultural systems.

REFERENCES

Albracht, C., Eisenhauer, N., Vogel, A., Wagg, C., Buscot, F., & Heintz-Buschart, A. (2023). Effects of recurrent summer droughts on arbuscular mycorrhizal and total fungal communities in experimental grasslands differing in plant diversity and community composition. *Frontiers in Soil Science*, *3*, 1129845. DOI: 10.3389/fsoil.2023.1129845

An, J., Zeng, T., Ji, C., de Graaf, S., Zheng, Z., Xiao, T. T., Deng, X., Xiao, S., Bisseling, T., Limpens, E., & Pan, Z. (2019). A *Medicago truncatula* SWEET transporter implicated in arbuscule maintenance during arbuscular mycorrhizal symbiosis. *The New Phytologist*, *224*(1), 396–408. DOI: 10.1111/nph.15975 PMID: 31148173

Anli, M., Baslam, M., Tahiri, A., Raklami, A., Symanczik, S., Boutasknit, A., Ait-El-Mokhtar, M., Ben-Laouane, R., Toubali, S., Ait Rahou, Y., Ait Chitt, M., Oufdou, K., Mitsui, T., Hafidi, M., & Meddich, A. (2020). Biofertilizers as Strategies to Improve Photosynthetic Apparatus, Growth, and Drought Stress Tolerance in the Date Palm. *Frontiers in Plant Science*, *11*, 516818. Advance online publication. DOI: 10.3389/fpls.2020.516818 PMID: 33193464

Asensio D, Rapparini F, Peñuelas J (2012) AM fungi root colonization increases the production of essential isoprenoids vs. nonessential isoprenoids especially under drought stress conditions or after jasmonic acid application. Phytochemistry 77:149 161.DOI: 10.1016/j.phytochem.2011.12.012

Bapaume, L., Laukamm, S., Darbon, G., Monney, C., Meyenhofer, F., Feddermann, N., Chen, M., & Reinhardt, D. (2019). VAPYRIN marks an endosomal trafficking compartment involved in arbuscular mycorrhizal symbiosis. *Frontiers in Plant Science*, *10*, 666. DOI: 10.3389/fpls.2019.00666 PMID: 31231402

Bisht, A., Lather, S., Goyal, P., & Garg, N. (2024). Molecular Determinants and Regulatory Mechanisms of Nutrient Exchange Between Plant and AMF. In *Arbuscular Mycorrhizal Fungi in Sustainable Agriculture: Nutrient and Crop Management* (pp. 77–97). Springer Nature Singapore. DOI: 10.1007/978-981-97-0300-5_3

Bogati, K., & Walczak, M. (2022). The impact of drought stress on soil microbial community, enzyme activities and plants. *Agronomy (Basel)*, *12*(1), 189. DOI: 10.3390/agronomy12010189

Boyno, G., & Demir, S. (2022). Plant-mycorrhiza communication and mycorrhizae in inter-plant communication. *Symbiosis*, *86*(2), 155–168. DOI: 10.1007/s13199-022-00837-0

Chen, M., Arato, M., Borghi, L., Nouri, E., & Reinhardt, D. (2018). Beneficial services of arbuscular mycorrhizal fungi–from ecology to application. *Frontiers in Plant Science*, *9*, 1270. DOI: 10.3389/fpls.2018.01270 PMID: 30233616

Chen, X. W., Wu, L., Luo, N., Mo, C. H., Wong, M. H., & Li, H. (2019). Arbuscular mycorrhizal fungi and the associated bacterial community influence the uptake of cadmium in rice. *Geoderma*, *337*, 749–757. DOI: 10.1016/j.geoderma.2018.10.029

Cornell, C., Kokkoris, V., Turcu, B., Dettman, J., Stefani, F., & Corradi, N. (2022). The arbuscular mycorrhizal fungus *Rhizophagus irregularis* harmonizes nuclear dynamics in the presence of distinct abiotic factors. *Fungal Genetics and Biology*, *158*, 103639. DOI: 10.1016/j.fgb.2021.103639 PMID: 34800644

Courty, P., Doidy, J., Garcia, K., Wipf, D., & Zimmermann, S. D. (2016). The transportome of mycorrhizal systems. *Molecular Mycorrhizal Symbiosis*, 239–256.

Dai, H., Zhang, X., Zhao, B., Shi, J., Zhang, C., Wang, G., Yu, N., & Wang, E. (2022). Colonization of mutualistic mycorrhizal and parasitic blast fungi requires *OsRAM2*-regulated fatty acid biosynthesis in rice. *Molecular Plant-Microbe Interactions*, *35*(3), 178–186. DOI: 10.1094/MPMI-11-21-0270-R PMID: 34941378

Diagne, N., Ngom, M., Djighaly, P. I., Fall, D., Hocher, V., & Svistoonoff, S. (2020). Roles of arbuscular mycorrhizal fungi on plant growth and performance: Importance in biotic and abiotic stressed regulation. *Diversity*, *12*(10), 370. DOI: 10.3390/d12100370

Doidy, J., Grace, E., Kühn, C., Simon-Plas, F., Casieri, L., & Wipf, D. (2012). Sugar transporters in plants and in their interactions with fungi. *Trends in Plant Science*, *17*(7), 413–422. DOI: 10.1016/j.tplants.2012.03.009 PMID: 22513109

Dorostkar, V., Afyuni, M., Khoshgoftarmanesh, A. H., & Rejali, F. (2014). Effects of salinity on exudates of Glomus species in wheat rhizosphere. *Advances in Environmental Biology*, •••, 219–226.

Duc, N. H., Szentpéteri, V., Mayer, Z., & Posta, K. (2023). Distinct impact of arbuscular mycorrhizal isolates on tomato plant tolerance to drought combined with chronic and acute heat stress. *Plant Physiology and Biochemistry*, *201*, 107892. DOI: 10.1016/j.plaphy.2023.107892 PMID: 37490823

Fan, D., Ji, M., Wu, J., Chen, H., Jia, H., Zhang, X., Zhuang, X., & Kong, W. (2023). Grazing does not influence soil arbuscular mycorrhizal fungal diversity, but increases their interaction complexity with plants in dry grasslands on the Tibetan Plateau. *Ecological Indicators*, *148*, 110065. DOI: 10.1016/j.ecolind.2023.110065

Fan, Y., Lu, S., He, M., Yang, L., Hu, W., Yang, Z., Liu, X., Hui, D., Guo, J., & Yang, Y. (2021). Long-term throughfall exclusion decreases soil organic phosphorus associated with reduced plant roots and soil microbial biomass in a subtropical forest. *Geoderma*, *404*, 115309. DOI: 10.1016/j.geoderma.2021.115309

Feng, Z., Liu, X., Feng, G., Zhu, H., & Yao, Q. (2020). Linking lipid transfer with reduced arbuscule formation in tomato roots colonized by arbuscular mycorrhizal fungus under low pH stress. *Environmental Microbiology*, *22*(3), 1036–1051. DOI: 10.1111/1462-2920.14810 PMID: 31608569

Fischer, S., Hilger, T., Piepho, H.-P., Jordan, I., & Cadisch, G. (2019). Do we need more drought for better nutrition? The effect of precipitation on nutrient concentration in East African food crops. *The Science of the Total Environment*, *658*, 405–415. DOI: 10.1016/j.scitotenv.2018.12.181 PMID: 30579198

Floss, D. S., Gomez, S. K., Park, H.-J., MacLean, A. M., Müller, L. M., Bhattarai, K. K., Lévesque-Tremblay, V., Maldonado-Mendoza, I. E., & Harrison, M. J. (2017). A transcriptional program for arbuscule degeneration during AM symbiosis is regulated by MYB1. *Current Biology*, *27*(8), 1206–1212. DOI: 10.1016/j.cub.2017.03.003 PMID: 28392110

Gavito, M. E., Jakobsen, I., Mikkelsen, T. N., & Mora, F. (2019). Direct evidence for modulation of photosynthesis by an arbuscular mycorrhiza-induced carbon sink strength. *The New Phytologist*, *223*(2), 896–907. DOI: 10.1111/nph.15806 PMID: 30891762

Gobbato, E., Marsh, J. F., Vernié, T., Wang, E., Maillet, F., Kim, J., Miller, J. B., Sun, J., Bano, S. A., Ratet, P., Mysore, K. S., Dénarié, J., Schultze, M., & Oldroyd, G. E. D. (2012). A GRAS-type transcription factor with a specific function in mycorrhizal signaling. *Current Biology*, *22*(23), 2236–2241. DOI: 10.1016/j.cub.2012.09.044 PMID: 23122845

Gobbato, E., Wang, E., Higgins, G., Bano, S. A., Henry, C., Schultze, M., & Oldroyd, G. E. D. (2013). RAM1 and RAM2 function and expression during arbuscular mycorrhizal symbiosis and Aphanomyces euteiches colonization. *Plant Signaling & Behavior*, *8*(10), e26049. DOI: 10.4161/psb.26049 PMID: 24270627

Goyal, D., Misera, S., & Dantu, P. K. (2020). *Arbuscular mycorrhizae associations and role in mitigation of drought stress in plants. Agriculturally Important Fungi for Sustainable Agriculture* (Vol. 2). Functional Annotation for Crop Protection.

Grabsztunowicz, M., Stolarska, E., Tanwar, U. K., Arasimowicz-Jelonek, M., & Sobieszczuk-Nowicka, E. (2024). Heavy metal tolerance in crop plants: An in silico comprehensive analysis of Metal-tolerance gene family in barley. *Postepy Biochemii*, *70*(3), 358–370. DOI: 10.18388/pb.2021_559 PMID: 39365567

Gutjahr, C., Banba, M., Croset, V., An, K., Miyao, A., & Paszkowski, U. (2008). Arbuscular mycorrhiza–specific signaling in rice transcends the common symbiosis signaling pathway. *The Plant Cell*, *20*(11), 2989–3005. DOI: 10.1105/tpc.108.062414 PMID: 19033527

Hammer, E. C., Pallon, J., Wallander, H., & Olsson, P. A. (2011). Tit for tat? A mycorrhizal fungus accumulates phosphorus under low plant carbon availability. *FEMS Microbiology Ecology*, *76*(2), 236–244. DOI: 10.1111/j.1574-6941.2011.01043.x PMID: 21223336

Hashem, A., Alqarawi, A. A., Radhakrishnan, R., Al-Arjani, A. B. F., Aldehaish, H. A., Egamberdieva, D., & Abd_Allah, , E. F. (2018). Arbuscular mycorrhizal fungi regulate the oxidative system, hormones and ionic equilibrium to trigger salt stress tolerance in Cucumis sativus L. *Saudi Journal of Biological Sciences*, *25*(6), 1102–1114. DOI: 10.1016/j.sjbs.2018.03.009 PMID: 30174509

Hsieh, Y. H., Wei, Y. H., Lo, J. C., Pan, H. Y., & Yang, S. Y. (2022). Arbuscular mycorrhizal symbiosis enhances tomato lateral root formation by modulating CEP2 peptide expression. *The New Phytologist*, *235*(1), 292–305. DOI: 10.1111/nph.18128 PMID: 35358343

Ikan, C., Ben-Laouane, R., Ouhaddou, R., Anli, M., Boutasknit, A., Lahbouki, S., & Meddich, A. (2023). Interactions between arbuscular mycorrhizal fungus and indigenous compost improve salt stress tolerance in wheat (*Triticum durum*). *South African Journal of Botany*, *158*, 417–428. DOI: 10.1016/j.sajb.2023.05.038

Jiang, Y., Xie, Q., Wang, W., Yang, J., Zhang, X., Yu, N., Zhou, Y., & Wang, E. (2018). Medicago AP2-domain transcription factor WRI5a is a master regulator of lipid biosynthesis and transfer during mycorrhizal symbiosis. *Molecular Plant*, *11*(11), 1344–1359. DOI: 10.1016/j.molp.2018.09.006 PMID: 30292683

Juniper, S., & Abbott, L. (1993). Vesicular-arbuscular mycorrhizas and soil salinity. *Mycorrhiza*, *4*(2), 45–57. DOI: 10.1007/BF00204058

Kang, E., Li, Y., Zhang, X., Yan, Z., Zhang, W., Zhang, K., Yan, L., Wu, H., Li, M., Niu, Y., Yang, A., Wang, J., & Kang, X. (2022). Extreme drought decreases soil heterotrophic respiration but not methane flux by modifying the abundance of soil microbial functional groups in alpine peatland. *Catena*, *212*, 106043. DOI: 10.1016/j.catena.2022.106043

Keymer, A., Pimprikar, P., Wewer, V., Huber, C., Brands, M., Bucerius, S. L., Delaux, P.-M., Klingl, V., von Röpenack-Lahaye, E., & Wang, T. L. (2017). Lipid transfer from plants to arbuscular mycorrhiza fungi. *eLife*, *6*, e29107. DOI: 10.7554/eLife.29107 PMID: 28726631

Khan, W., Zhu, Y., Khan, A., Zhao, L., Yang, Y.-M., Wang, N., Hao, M., Ma, Y., Nepal, J., Ullah, F., Rehman, M. M. U., Abrar, M., & Xiong, Y.-C. (2024). Above-and below-ground feedback loop of maize is jointly enhanced by plant growth-promoting rhizobacteria and arbuscular mycorrhizal fungi in drier soil. *The Science of the Total Environment*, *917*, 170417. DOI: 10.1016/j.scitotenv.2024.170417 PMID: 38280611

Krishnamoorthy, R., Kim, C. G., Subramanian, P., Kim, K. Y., Selvakumar, G., & Sa, T. M. (2015). Arbuscular mycorrhizal fungi community structure, abundance and species richness changes in soil by different levels of heavy metal and metalloid concentration. *PLoS One*, *10*(6), e0128784. DOI: 10.1371/journal.pone.0128784 PMID: 26035444

Lahbouki, S., Anli, M., El Gabardi, S., Ait-El-Mokhtar, M., Ben-Laouane, R., Boutasknit, A., Ait-Rahou, Y., Outzourhit, A., Wahbi, S., Douira, A., & Meddich, A. (2022). Evaluation of arbuscular mycorrhizal fungi and vermicompost supplementation on growth, phenolic content and antioxidant activity of prickly pear cactus (*Opuntia ficus-indica*). *Plant Biosystems*, *156*(4), 882–892. DOI: 10.1080/11263504.2021.1947408

Lahbouki, S., Ech-chatir, L., Er-Raki, S., Outzourhit, A., & Meddich, A. (2022). Improving drought tolerance of Opuntia ficus-indica under field using Subsurface Water Retention Technology: Changes in physiological and biochemical parameters. *Canadian Journal of Soil Science, ja.*

Lahbouki, S., Fernando, A. L., Rodrigues, C., Ben-Laouane, R., Ait-El-Mokhtar, M., Outzourhit, A., & Meddich, A. (2023). Effects of Humic Substances and Mycorrhizal Fungi on Drought-Stressed Cactus: Focus on Growth, Physiology, and Biochemistry. *Plants*, *12*(24), 4156. DOI: 10.3390/plants12244156 PMID: 38140483

Lahmidi, N. A., Courty, P.-E., Brulé, D., Chatagnier, O., Arnould, C., Doidy, J., Berta, G., Lingua, G., Wipf, D., & Bonneau, L. (2016). Sugar exchanges in arbuscular mycorrhiza: *RiMST5* and *RiMST6*, two novel Rhizophagus irregularis monosaccharide transporters, are involved in both sugar uptake from the soil and from the plant partner. *Plant Physiology and Biochemistry*, *107*, 354–363. DOI: 10.1016/j.plaphy.2016.06.023 PMID: 27362299

Lanfranco, L., Fiorilli, V., Venice, F., & Bonfante, P. (2018). Strigolactones cross the kingdoms: Plants, fungi, and bacteria in the arbuscular mycorrhizal symbiosis. *Journal of Experimental Botany*, *69*(9), 2175–2188. DOI: 10.1093/jxb/erx432 PMID: 29309622

Leal, P. L., Siqueira, J. O., & Stuermer, S. L. (2013). Switch of tropical Amazon forest to pasture affects taxonomic composition but not species abundance and diversity of arbuscular mycorrhizal fungal community. *Applied Soil Ecology*, *71*, 72–80. DOI: 10.1016/j.apsoil.2013.05.010

Liu, A., Wang, B., & Hamel, C. (2004). Arbuscular mycorrhiza colonization and development at suboptimal root zone temperature. *Mycorrhiza*, *14*(2), 93–101. DOI: 10.1007/s00572-003-0242-9 PMID: 12748840

Liu, C. Y., Guo, X. N., Wu, X. L., Dai, F. J., & Wu, Q. S. (2022). The comprehensive effects of Rhizophagus intraradices and P on root system architecture and P transportation in Citrus limon L. *Agriculture*, *12*(3), 317. DOI: 10.3390/agriculture12030317

Liu, Y., Feng, S., & Liu, H. (2024). Measurements of Drying and Wetting Gas Diffusion Coefficients and Gas Permeability of Unsaturated Soils Using a New Flexible-Wall Device. *Journal of Geotechnical and Geoenvironmental Engineering*, *150*(11), 6024006. DOI: 10.1061/JGGEFK.GTENG-12475

Ludwig-Müller, J. (2010). Hormonal responses in host plants triggered by arbuscular mycorrhizal fungi. Arbuscular mycorrhizas: physiology and function, 169-190

Luginbuehl, L., & Oldroyd, G. E. D. (2016). Calcium signaling and transcriptional regulation in arbuscular mycorrhizal symbiosis. *Molecular Mycorrhizal Symbiosis*, 125–140.

Luginbuehl, L. H., Menard, G. N., Kurup, S., Van Erp, H., Radhakrishnan, G. V., Breakspear, A., Oldroyd, G. E. D., & Eastmond, P. J. (2017). Fatty acids in arbuscular mycorrhizal fungi are synthesized by the host plant. *Science*, *356*(6343), 1175–1178. DOI: 10.1126/science.aan0081 PMID: 28596311

Ma, J., Janoušková, M., Li, Y., Yu, X., Yan, Y., Zou, Z., & He, C. (2015). Impact of arbuscular mycorrhizal fungi (AMF) on cucumber growth and phosphorus uptake under cold stress. *Functional Plant Biology*, *42*(12), 1158–1167. DOI: 10.1071/FP15106 PMID: 32480753

Madouh, T. A., & Quoreshi, A. M. (2023). The function of arbuscular mycorrhizal fungi associated with drought stress resistance in native plants of arid desert ecosystems: A review. *Diversity*, *15*(3), 391. DOI: 10.3390/d15030391

Masrahi, A. S., Alasmari, A., Shahin, M. G., Qumsani, A. T., Oraby, H. F., & Awad-Allah, M. M. (2023). Role of arbuscular mycorrhizal fungi and phosphate solubilizing bacteria in improving yield, yield components, and nutrients uptake of barley under salinity soil. *Agriculture*, *13*(3), 537. DOI: 10.3390/agriculture13030537

Meddich, A. (2023). Biostimulants for resilient agriculture—Improving plant tolerance to abiotic stress: A concise review. *Gesunde Pflanzen*, *75*(4), 709–727. DOI: 10.1007/s10343-022-00784-2

Meena, K. K., Kumar, P., Sorty, A. M., Bitla, U., & Pathak, H. (2022). Ecology of Arbuscular Mycorrhizae and Influence on Drought Tolerance in Crop Plants. In *Microbial BioTechnology for Sustainable Agriculture* (Vol. 1, pp. 261–285). Springer Nature Singapore. DOI: 10.1007/978-981-16-4843-4_8

Mendoza-Soto, A. B., Rodríguez-Corral, A. Z., Bojórquez-López, A., Cervantes-Rojo, M., Castro-Martínez, C., & Lopez-Meyer, M. (2022). Arbuscular mycorrhizal symbiosis leads to differential regulation of genes and mirnas associated with the cell wall in tomato leaves. *Biology (Basel)*, *11*(6), 854. DOI: 10.3390/biology11060854 PMID: 35741375

Menezes-Oliveira, V. B., Scott-Fordsmand, J. J., Soares, A. M. V. M., & Amorim, M. J. B. (2013). Effects of temperature and copper pollution on soil community—Extreme temperature events can lead to community extinction. *Environmental Toxicology and Chemistry, 32*(12), 2678–2685. DOI: 10.1002/etc.2345 PMID: 23939831

Mulyadi, & Jiang, L. (2023). Combined application of arbuscular mycorrhizal fungi (AMF) and nitrogen fertilizer alters the physicochemical soil properties, nitrogen uptake, and rice yield in a polybag experiment. *Agriculture, 13*(7), 1364.

Musyoka, D. M., Njeru, E. M., Nyamwange, M. M., & Maingi, J. M. (2020). Arbuscular mycorrhizal fungi and Bradyrhizobium co-inoculation enhances nitrogen fixation and growth of green grams (*Vigna radiata* L.) under water stress. *Journal of Plant Nutrition, 43*(7), 1036–1047. DOI: 10.1080/01904167.2020.1711940

Ndeko, A. B., Founoune-Mboup, H., Kane, A., & Cournac, L. (2022). Arbuscular mycorrhizal fungi alleviate the negative effect of temperature stress in millet lines with contrasting soil aggregation potential. *Gesunde Pflanzen, 74*(1), 53–67. DOI: 10.1007/s10343-021-00588-w

Nouh, F. A. A., & Abdel-Azeem, A. M. (2025). Climatic changes and drought against forest fungi. In *Forest Fungi* (pp. 47–62). Academic Press. DOI: 10.1016/B978-0-443-18870-1.00008-1

Ocón, A., Hampp, R., & Requena, N. (2007). Trehalose turnover during abiotic stress in arbuscular mycorrhizal fungi. *The New Phytologist, 174*(4), 879–891. DOI: 10.1111/j.1469-8137.2007.02048.x PMID: 17504469

Ortas, I., Rafique, M., & Çekiç, F. Ö. (2021). Do Mycorrhizal Fungi enable plants to cope with abiotic stresses by overcoming the detrimental effects of salinity and improving drought tolerance? *Symbiotic Soil Microorganisms: Biology and Applications*, 391–428.

Ouhaddou, R., Meddich, A., Ikan, C., Lahlali, R., Ait Barka, E., Hajirezaei, M. R., Duponnois, R., & Baslam, M. (2023). Enhancing Maize Productivity and Soil Health under Salt Stress through Physiological Adaptation and Metabolic Regulation Using Indigenous Biostimulants. *Plants, 12*(21), 3703. DOI: 10.3390/plants12213703 PMID: 37960059

Ovchinnikova, E., Chiasson, D., Wen, Z., Wu, Y., Tahaei, H., Smith, P. M., & Kaiser, B. N. (2023). Arbuscular-Mycorrhizal Symbiosis in Medicago Regulated by the Transcription Factor *MtbHLHm1*; 1 and the Ammonium Facilitator Protein *MtAMF1*; 3. *International Journal of Molecular Sciences, 24*(18), 14263. DOI: 10.3390/ijms241814263 PMID: 37762569

Pons, S., Fournier, S., Chervin, C., Bécard, G., Rochange, S., Frei Dit Frey, N., & Puech Pagès, V. (2020). Phytohormone production by the arbuscular mycorrhizal fungus Rhizophagus irregularis. *PLoS One, 15*(10), e0240886. DOI: 10.1371/journal.pone.0240886 PMID: 33064769

Pooja, P., Tallapragada, S., Lamba, A., & Punia, S. (2024). Role played by arbuscular mycorrhizal fungi in amelioration of salinity stress: A review. *Plant and Soil*, •••, 1–16.

Porcel, R., Aroca, R., Azcon, R., & Ruiz-Lozano, J. M. (2016). Regulation of cation transporter genes by the arbuscular mycorrhizal symbiosis in rice plants subjected to salinity suggests improved salt tolerance due to reduced Na+ root-to-shoot distribution. *Mycorrhiza, 26*(7), 673–684. DOI: 10.1007/s00572-016-0704-5 PMID: 27113587

Preece, C., Verbruggen, E., Liu, L., Weedon, J. T., & Peñuelas, J. (2019). Effects of past and current drought on the composition and diversity of soil microbial communities. *Soil Biology & Biochemistry, 131*, 28–39. DOI: 10.1016/j.soilbio.2018.12.022

Pumplin, N., Zhang, X., Noar, R. D., & Harrison, M. J. (2012). Polar localization of a symbiosis-specific phosphate transporter is mediated by a transient reorientation of secretion. *Proceedings of the National Academy of Sciences of the United States of America, 109*(11), E665–E672. DOI: 10.1073/pnas.1110215109 PMID: 22355114

Qadeer, A., Wakeel, A., Cheema, S. A., & Sanaullah, M. (2022). Interactive effects of salinity and drought stresses on soil respiration and microbial activities. *Pakistan Journal of Agricultural Sciences, 59*(1), 35–42. DOI: 10.21162/PAKJAS/22.1263

Qiao, M., Hong, C., Jiao, Y., Hou, S., & Gao, H. (2024). Impacts of drought on photosynthesis in major food crops and the related mechanisms of plant responses to drought. *Plants, 13*(13), 1808. DOI: 10.3390/plants13131808 PMID: 38999648

Qin, H., Lu, K., Strong, P. J., Xu, Q., Wu, Q., Xu, Z., Xu, J., & Wang, H. (2015). Long-term fertilizer application effects on the soil, root arbuscular mycorrhizal fungi and community composition in rotation agriculture. *Applied Soil Ecology, 89*, 35–43. DOI: 10.1016/j.apsoil.2015.01.008

Quiroga, G., Erice, G., Aroca, R., Chaumont, F., & Ruiz-Lozano, J. M. (2017). Enhanced drought stress tolerance by the arbuscular mycorrhizal symbiosis in a drought-sensitive maize cultivar is related to a broader and differential regulation of host plant aquaporins than in a drought-tolerant cultivar. *Frontiers in Plant Science, 8*, 1056. DOI: 10.3389/fpls.2017.01056 PMID: 28674550

Radujković, D., Verbruggen, E., Sigurdsson, B. D., Leblans, N. I. W., Janssens, I. A., Vicca, S., & Weedon, J. T. (2018). Prolonged exposure does not increase soil microbial community compositional response to warming along geothermal gradients. *FEMS Microbiology Ecology, 94*(2), fix174. DOI: 10.1093/femsec/fix174 PMID: 29228354

Raja, B.-L., Soufian, L., Salma, T., Wissal, B., Ouissame, R., Said, W., Cherkaoui, E. M., Marouane, B., & Abdelilah, M. (2022). Use of biostimulants to improve salinity tolerance in cereals. In *Sustainable Remedies for Abiotic Stress in Cereals* (pp. 471–517). Springer. DOI: 10.1007/978-981-19-5121-3_19

Rodrigues, W. P., Silva, J. R., Ferreira, L. S., Machado Filho, J. A., Figueiredo, F. A., Ferraz, T. M., Bernado, W. P., Bezerra, L. B. S., de Abreu, D. P., & Cespom, L. (2018). Stomatal and photochemical limitations of photosynthesis in coffee (*Coffea* spp.) plants subjected to elevated temperatures. *Crop & Pasture Science, 69*(3), 317–325. DOI: 10.1071/CP17044

Roth, R., & Paszkowski, U. (2017). Plant carbon nourishment of arbuscular mycorrhizal fungi. *Current Opinion in Plant Biology, 39*, 50–56. DOI: 10.1016/j.pbi.2017.05.008 PMID: 28601651

Salwan, R., Sharma, A., Kaur, R., Sharma, R., & Sharma, V. (2023). Signaling in arbuscular mycorrhizal association. In *The Chemical Dialogue Between Plants and Beneficial Microorganisms* (pp. 127–135). Elsevier. DOI: 10.1016/B978-0-323-91734-6.00005-3

Sangwan, S., & Prasanna, R. (2022). Mycorrhizae helper bacteria: Unlocking their potential as bioenhancers of plant–arbuscular mycorrhizal fungal associations. *Microbial Ecology*, *84*(1), 1–10. DOI: 10.1007/s00248-021-01831-7 PMID: 34417849

Sery, D. J.-M., Van Tuinen, D., Drain, A., Mounier, A., & Zézé, A. (2018). The genus Rhizophagus dominates arbuscular mycorrhizal fungi communities in contrasted cassava field soils in Côte d'Ivoire. *Rhizosphere*, *7*, 8–17. DOI: 10.1016/j.rhisph.2018.06.007

Shaher, H., Naushin, F., Hasan, M., & Bagyaraj, D. J. (2024). Synergistic impact of Rhizobium and arbuscular mycorrhizal fungi on lentil plant tolerance to heavy metal-rich fly ash amended soil. *Discover Plants*, *1*(1), 9. DOI: 10.1007/s44372-024-00010-5

Sharma, K., Gupta, S., Thokchom, S. D., Jangir, P., & Kapoor, R. (2021). Arbuscular mycorrhiza-mediated regulation of polyamines and aquaporins during abiotic stress: Deep insights on the recondite players. *Frontiers in Plant Science*, *12*, 642101. DOI: 10.3389/fpls.2021.642101 PMID: 34220878

Solaiman, Z. (2024). Role of AMF in Organic Matter Decomposition, Carbon Sequestration and Climate Change Mitigation. In *Arbuscular mycorrhizal fungi in sustainable agriculture: Nutrient and crop management* (pp. 131–141). Springer Nature Singapore. DOI: 10.1007/978-981-97-0300-5_6

Son, Y., Martínez, C. E., & Kao-Kniffin, J. (2024). Three important roles and chemical properties of glomalin-related soil protein. *Frontiers in Soil Science*, *4*, 1418072. DOI: 10.3389/fsoil.2024.1418072

Stover, H. J., Thorn, R. G., Bowles, J. M., Bernards, M. A., & Jacobs, C. R. (2012). Arbuscular mycorrhizal fungi and vascular plant species abundance and community structure in tallgrass prairies with varying agricultural disturbance histories. *Applied Soil Ecology*, *60*, 61–70. DOI: 10.1016/j.apsoil.2012.02.016

Tang, H., Hassan, M. U., Feng, L., Nawaz, M., Shah, A. N., Qari, S. H., & Miao, J. (2022). The critical role of arbuscular mycorrhizal fungi to improve drought tolerance and nitrogen use efficiency in crops. *Frontiers in Plant Science*, *13*, 919166. DOI: 10.3389/fpls.2022.919166 PMID: 35873982

Tian, X., Liu, X. Q., Liu, X. R., & Li, Q. S., Abd_Allah, E. F., & Wu, Q. S. (. (2023). Mycorrhizal cucumber with *Diversispora versiformis* has active heat stress tolerance by up-regulating expression of both *CsHsp70s* and *CsPIPs* genes. *Scientia Horticulturae*, *319*, 112194. DOI: 10.1016/j.scienta.2023.112194

Toleikiene, M., Arlauskiene, A., Fliessbach, A., Iqbal, S., Sarunaite, L., & Kadziuliene, Z. (2020). The decomposition of standardised organic materials in loam and clay loam arable soils during a non-vegetation period. *Soil and Water Research*, *15*(3), 181–190. DOI: 10.17221/31/2019-SWR

Varghese, R., & Ray, J. G. (2023). Sustainable agriculture of tropical spices: Arbuscular mycorrhizal fungi as an ecotechnological tool-a critical review. *Journal of Applied Research on Medicinal and Aromatic Plants*, *36*, 100507. DOI: 10.1016/j.jarmap.2023.100507

Wahab, A., Muhammad, M., Munir, A., Abdi, G., Zaman, W., Ayaz, A., & Reddy, S. P. P. (2023). Role of arbuscular mycorrhizal fungi in regulating growth, enhancing productivity, and potentially influencing ecosystems under abiotic and biotic stresses. *Plants*, *12*(17), 3102. DOI: 10.3390/plants12173102 PMID: 37687353

Wang, B., & Qiu, Y. L. (2006). Phylogenetic distribution and evolution of mycorrhizas in land plants. *Mycorrhiza*, *16*(5), 299–363. DOI: 10.1007/s00572-005-0033-6 PMID: 16845554

Wang, E., Yu, N., Bano, S. A., Liu, C., Miller, A. J., Cousins, D., Zhang, X., Ratet, P., Tadege, M., Mysore, K. S., Downie, J. A., Murray, J. D., Oldroyd, G. E. D., & Schultze, M. (2014). A H+-ATPase that energizes nutrient uptake during mycorrhizal symbioses in rice and *Medicago truncatula.The Plant Cell*, *26*(4), 1818–1830. DOI: 10.1105/tpc.113.120527 PMID: 24781115

Wang, F., Sun, Y., & Shi, Z. (2019). Arbuscular mycorrhiza enhances biomass production and salt tolerance of sweet sorghum. *Microorganisms*, *7*(9), 289. DOI: 10.3390/microorganisms7090289 PMID: 31450847

Wang, Y. J., He, X. H., Meng, L. L., Zou, Y. N., & Wu, Q. S. (2023). Extraradical mycorrhizal hyphae promote soil carbon sequestration through difficultly extractable glomalin-related soil protein in response to soil water stress. *Microbial Ecology*, *86*(2), 1023–1034. DOI: 10.1007/s00248-022-02153-y PMID: 36471016

Wani, K. I., Zehra, A., Choudhary, S., Naeem, M., Khan, M. M. A., Castroverde, C. D. M., & Aftab, T. (2020). Mechanistic insights into strigolactone biosynthesis, signaling, and regulation during plant growth and development. *Journal of Plant Growth Regulation*, •••, 1–17.

Wei, D., Wei, S., Peng, A., Yang, C., & Chen, C. (2022). Responses of bacterial communities in soils under winter wheat to nightly warming and nitrogen addition. *Agronomy (Basel)*, *12*(7), 1616. DOI: 10.3390/agronomy12071616

Xing, H.-S., & Sun, P.-F., LI, F., LI, W., & Guo, S.-X. (2019). Effects of arbuscular mycorrhizal fungi on heat tolerance of *Lavandula angustifolia.Junwu Xuebao*, *38*(5), 698–706.

Xu, G., Pan, J., Rehman, M., Li, X., Cao, S., Wang, C., Wang, X., Chen, C., Nie, J., Wang, M., Li, R., Li, R., & Chen, P. (2024). Arbuscular mycorrhizal fungi-mediated drought stress tolerance in kenaf (*Hibiscus cannabinus* L.): A mechanistic approach. *Plant Growth Regulation*, *103*(3), 1–22. DOI: 10.1007/s10725-024-01150-0

Xu, X., Wang, X., Cleary, M., Wang, P., Lu, N., Sun, Y., & Rönnberg, J. (2020). Slope position rather than thinning intensity affects arbuscular mycorrhizal fungi (AMF) community in Chinese fir plantations. *Forests*, *11*(3), 273. DOI: 10.3390/f11030273

Xue, L., Klinnawee, L., Zhou, Y., Saridis, G., Vijayakumar, V., Brands, M., Dörmann, P., Gigolashvili, T., Turck, F., & Bucher, M. (2018). *AP2* transcription factor *CBX1* with a specific function in symbiotic exchange of nutrients in mycorrhizal Lotus japonicus. *Proceedings of the National Academy of Sciences of the United States of America*, *115*(39), E9239–E9246. DOI: 10.1073/pnas.1812275115 PMID: 30209216

Yang, H., Fang, C., Li, Y., Wu, Y., Fransson, P., Rillig, M. C., & Weih, M. (2022). Temporal complementarity between roots and mycorrhizal fungi drives wheat nitrogen use efficiency. *The New Phytologist*, *236*(3), 1168–1181. DOI: 10.1111/nph.18419 PMID: 35927946

Yang, Y., Han, X., Liang, Y., Ghosh, A., Chen, J., & Tang, M. (2015). The combined effects of arbuscular mycorrhizal fungi (AMF) and lead (Pb) stress on Pb accumulation, plant growth parameters, photosynthesis, and antioxidant enzymes in Robinia pseudoacacia L. *PLoS One*, *10*(12), e0145726. DOI: 10.1371/journal.pone.0145726 PMID: 26698576

Zeng, H., Yu, L., Liu, P., Wang, Z., Chen, Y., & Wang, J. (2021). Nitrogen fertilization has a stronger influence than cropping pattern on AMF community in maize/soybean strip intercropping systems. *Applied Soil Ecology*, *167*, 104034. DOI: 10.1016/j.apsoil.2021.104034

Zhang, F., Jia-Dong, H. E., Qiu-Dan, N. I., Qiang-Sheng, W. U., & Ying-Ning, Z. O. U. (2018). Enhancement of drought tolerance in trifoliate orange by mycorrhiza: Changes in root sucrose and proline metabolisms. *Notulae Botanicae Horti Agrobotanici Cluj-Napoca*, *46*(1), 270–276. DOI: 10.15835/nbha46110983

Zhang, J., Diao, F., Hao, B., Xu, L., Jia, B., Hou, Y., Ding, S., & Guo, W. (2023). Multiomics reveals Claroideoglomus etunicatum regulates plant hormone signal transduction, photosynthesis and La compartmentalization in maize to promote growth under La stress. *Ecotoxicology and Environmental Safety*, *262*, 115128. DOI: 10.1016/j.ecoenv.2023.115128 PMID: 37315361

Zhang, X., Pumplin, N., Ivanov, S., & Harrison, M. J. (2015). EXO70I is required for development of a sub-domain of the periarbuscular membrane during arbuscular mycorrhizal symbiosis. *Current Biology*, *25*(16), 2189–2195. DOI: 10.1016/j.cub.2015.06.075 PMID: 26234213

Zhao, J., Xie, X., Jiang, Y., Li, J., Fu, Q., Qiu, Y., Fu, X., Yao, Z., Dai, Z., Qiu, Y., & Chen, H. (2024). Effects of simulated warming on soil microbial community diversity and composition across diverse ecosystems. *The Science of the Total Environment*, *911*, 168793. DOI: 10.1016/j.scitotenv.2023.168793 PMID: 37996030

Zheng, F. L., Wang, Y. J., & Hashem, A., Abd_Allah, E. F., & Wu, Q. S. (. (2024). Mycorrhizae with *Funneliformis mosseae* regulate the trehalose synthesis and sucrose cleavage for enhancing drought tolerance in trifoliate orange. *Scientia Horticulturae*, *337*, 113486. DOI: 10.1016/j.scienta.2024.113486

Zhou, Y. H., Liu, Y. J., Wang, Y. D., & Yang, C. X. (2022). *Claroideoglomus etunicatum* improved the growth and saline–alkaline tolerance of *Potentilla anserina* by altering physiological and biochemical properties. *Biocell*, *46*(8), 1967–1978. DOI: 10.32604/biocell.2022.019304

Zhu, B., Gao, T., Zhang, D., Ding, K., Li, C., & Ma, F. (2022). Functions of arbuscular mycorrhizal fungi in horticultural crops. *Scientia Horticulturae*, *303*, 111219. DOI: 10.1016/j.scienta.2022.111219

Zhu, C., Tian, G., Luo, G., Kong, Y., Guo, J., Wang, M., & Shen, Q. (2018). N-fertilizer-driven association between the arbuscular mycorrhizal fungal community and diazotrophic community impacts wheat yield. *Agriculture, Ecosystems & Environment*, *254*, 191–201. DOI: 10.1016/j.agee.2017.11.029

Chapter 5
Fungal Biostimulants:
Sustainable Tool to Boost Crop Yield Naturally

Upama Kalita
Sikkim University, India

Madhan Sanyasi
https://orcid.org/0009-0001-4786-3300
Sikkim University, India

Priyanka Kumari
https://orcid.org/0009-0006-4528-2912
Sikkim University, India

Anit Baidya
Sikkim University, India

ABSTRACT

From the beginning of humankind plants have been a great source of food. The increase in environmental pollution, elevated global temperature as well as widespread application of chemical fertilizers and insecticides have caused reduced soil fertility and decline in agricultural production. On the other hand, the need for food intensified with rise in the global population. The plant-microbe interface plays a vital role in moderating harsh environmental situation and improve crop strength. In the last few years plant-fungi interaction has become the most studied topic for production of ecofriendly substitutes to chemical products. Fungal biostimulants create a wide range of interactions with the host plant and encourage crop advancement, develop resilience, recover soil health and stimulate defense response. Overall, this chapter will provide extensive outlook on fungal biostimulants, role in plant stress tolerance, mechanism of action in plants and influence in organic agriculture. It also states their ecological role and challenges for advanced crop development in the near future.

DOI: 10.4018/979-8-3693-8201-1.ch005

INTRODUCTION

Plant biostimulants are the 'materials that in minute quantities promote plant growth' (Du Jardin, 2015) or according to European Biostimulants Industry Council (EBIC), "Plant biostimulants include compound(s) and/or microorganisms whose role is to enhance natural processes like nutrient uptake, nutrient efficacy, resistance to abiotic stress and improve crop quality when applied to plants or the rhizosphere. Biostimulants are the key element in sustainable agriculture because of their ability to improve soil health. These are derived from natural sources hence eco-friendly. However, the concept of biostimulants is not new to agriculture systems. Early farmers from various civilizations had been using biostimulants like seaweed extract, plant residues as mulch and manures to enhance crop quality and soil fertility. In modern agriculture biostimulants are getting a lot of attention due to its multiple benefits from a single source. Biostimulants are not categorized under plant protectants as they do not provide nutrients directly to plants, but increase their nutrient use efficacy, hence decreasing the reliability on synthetic fertilizers (Baltazar et al., 2021). They are mainly divided into two categories – microbial and non-microbial. Non microbial biostimulants include humic acid, fulvic acid, plant and seaweed extract, vermicompost, chitosan and other biopolymers, protein hydrolases and amino acids etc. Microbial biostimulants mainly include beneficial bacteria such as phosphorus solubilizing bacteria, ACC deaminase bacteria, non-pathogenic fungi, plant growth promoting rhizobacteria (PGPR) and microalgae (Zulfiqar et al., 2024). These biostimulants have been tested and implemented and have led to better nutritional composition, improved water-use capability, enhanced resistance and recovery from biotic and abiotic stress of the subject plant (Albrecht, n.d.). The effectiveness of biostimulants vary depending on their formulation, climate conditions and crop types. They play a vital role in all aspects of plant development from seed germination to fruit production. Biostimulants go by numerous other terms such as bioregulator, plant conditioners, soil improvers, plant strengtheners, metabolic enhancers etc. Some commercially available biostimulants till date are Activewave®, Wuxai-Ascofol®, Maxicrop® and GoëmarBM86® which are sourced from *Ascophyllum nodum*. Others include SM3(Sea Magic 3), originating from *Laminariaceae* and *Fucaceae* species, Bio-Algen® from *phaeophyceae,* Comcat® from *lychnis Viscaria* and Aminoplant (Siapton®) which is derived from animal origin (Yakhin et al., 2017).

Humic and fulvic substances are derived from decomposed organic matters, peatland soils and organic lignite and they are known to enhance physical as well as chemical characteristics of soil, nutrient uptake by the plant roots and the development of lateral roots. Seaweed extracts contain specific polysaccharides, plant hormones and betaines which are known to have antioxidant properties, hence helping in crops development and defense against several environmental stresses (Canellas et al., 2015). Chitosans are naturally occurring biopolymers found in nematodes and crustaceans which is also known to trigger defensive response & stress tolerance in plants (Abdelaziz et al., 2022).

Recent advancements in biochemical and molecular field have enabled the commercialization of microorganisms as biostimulants. The interaction between microbial biostimulant and the host plant can be mutualistic symbiotic (beneficial relation for both the partner, living in direct contact with the host) to parasitism. Beneficial bacteria, particularly plant growth promoting bacteria (PGPBs) help resist pathogens found in rhizosphere by producing siderophores and also form root-nodule which are helpful in nitrogen fixation (Albrecht, n.d.; Ruzzi & Aroca, 2015). Beneficial fungi include mycorrhizal and non-mycorrhizal fungi, mycorrhizal fungi help in improving root area, stress tolerance & water and nutrient use efficiency of plants (López-Bucio et al., 2015). Symbiotic fungi like arbuscular mycorrhizal fungi (AMF) are known to enhance nutrient and water uptake as well as defense against various stresses (Jogawat et al., 2016). In this chapter we will discuss the benefits and limitations of using fungal biostimulants in modern day farming as well as their application for enhancing crop productivity and pathogen control.

What are Fungal Biostimulants?

Fungi are eukaryotic organisms that act as a great source of food, enzymes, organic acids and several secondary metabolites. Beneficial fungi include yeasts, rusts, smuts, mildews, molds, mushrooms etc. and they play a crucial role in decomposition and nutrient cycling. They are ubiquitous in nature and present in soil, water, air, dead and decaying matters, plants and animals. Fungi can tolerate a diverse range of pH from acidic to extreme alkaline, hence they are able to withstand extreme environmental stresses. The diversity of fungal species differs depending on soil type, microbial interaction and climate. Fungal biostimulants enhance agricultural productivity, nutrition, yield and stress tolerance either by forming a symbiotic or non-symbiotic relation with the host plant (Du Jardin, 2015). Several fungal species act against pests and abiotic stress conditions, most commonly used are *Rhizophagus irregularis*, *Trichoderma harzianum*, *T. asperellum*, *Glomus mosseae*, *Phanerochaete chrysosporium*, *Saccharomyces cerevisiae*, *Aspergillus niger*, *A. oryzae*, *Trichoderma viride*, and *Beauveria bassiana* (Tarigan et al., 2022). They offer synergetic relation with most of the crop plants such as rice, wheat, maize, barley, potato, corn, tomato acting as biofertilizer, maintaining phytohormone balance and promoting ecosystem stability.

Mycorrhiza is a symbiotic association between fungi and green plants that occurs at the rhizosphere level, where the fungus permanently colonizes the plant roots. Being a diverse group, mycorrhizal fungi establish mutual relationship with about 90% of plant species (Calvo et al., 2014). Mycorrhizal fungi include ectomycorrhiza and endomycorrhiza;

- Ectomycorrhiza – These fungi associate with plant roots externally and can thrive without the host but in order to complete their life cycle they from symbiotic association. They mainly colonize woody plant species forming a sheath around the root where nutrient is exchanged through a structure called 'hartig net.' Example includes *Boletus* sp., *Amanita* sp. etc. (Nasslahsen et al., 2022).
- Endomycorrhiza – These fungi form internal association with host cells, especially herbaceous plants, penetrates into the root cortex and forms a nutrient exchange structure called 'arbuscules or vesicles.' Fungi belonging to *Glomeromycota* phylum is known to form this symbiotic association. Examples include *Rhizophagus* sp. and *Funneliformis* sp. (Tedersoo et al., 2018; Nasslahsen et al., 2022).

AMF (Arbuscular mycorrhizal fungi) is one of the most common plant-fungal associations, it colonizes the plant root cortex and provide better nutrient acquisition from soil through their hyphal extension and in return plants provide them with carbohydrates. Recent research suggests that AMF also have synergistic relationship with soil bacteria which can boost soil health and crop production in near future (Sible et al., 2021).

Endophytic fungi are systemic in nature, hence live within the healthy plant tissue in above ground parts without causing any negative impacts (Stępniewska & Kuźniar, 2013). They provide positive effects to their host while gaining beneficial outcome for themselves. They play a vital role by providing resistance against plant disease, secondary metabolite synthesis, plant growth promotion (Hardoim et al., 2015). Most fungal endophytes reported till now belong to *Ascomycota* and *Basidiomycota* and they include genus *Fusarium, Curvularia, Cladosporium, Phomopsis, Chaetomium, Trichoderma, Alternaria,* and *Penicillium etc. (Sharma et al., 2018).* The bioactive compounds produced by these endophytes relieve stress by boosting photosynthesis, respiration and stomatal conductance. A pseudomycetous fungi *Serendipita indica* is used as a model organism which can colonize wide range of plant species including monocot and dicots and their association can enhance growth as well as resistance to biotic & abiotic stress (Varma et al., 1999; Jogawat et al., 2016). *Pichia guilliermondii* is another important endophyte that provide resistance to several crops such as tomato, citrus and chili paper against biotic stress. It produces an antibiotic, helvolic acid, which successfully restricts *Magnaporthe oryzae*, a prominent pathogen from rice field (Zhao et al., n.d.). *Acromonium* spp. such as *Acremonium zeae*, *Acremonium ochraceum* & *Acremonium strictum* are known to reduce fusarium wilt in tomato, reduce pest attack in date palm and increase resilience in corn and gingseng by enhancing jasmonic acid production (Han et al., 2020) .

Fungal endophytes can be classified into clavicipitaceous and non-clavicipitaceous. The former usually associate with grasses whereas the latter colonize vascular and non-vascular plants. Rodriguez et al. (2009), classified clavicipitaceous and non-clavicipitaceous fungi into 4 classes:

I. Class I: These are clavicipitaceous fungi that mainly colonize grasses and can be found in aerial parts and roots. They have narrow host range but occupy its host extensively. They help plants tolerate stress, provide resistance against nematode and produce anti-fungal compound. (Verma et al., 2022; Aamir, et al., 2020; Baron et al., 2018).
II. Class II: non-clavicipitaceous - they can colonize shoots, roots and rhizomes. They are less diversified and have a broader host range. These endophytes increase root and shoot biomass of its host and confer resistance to diseases, drought, heat and salinity. (Verma et al., 2022; Aamir et al., 2020; Baron et al., 2018).
III. Class III: non-clavicipitaceous - these are divergent groups of fungus that inhabit aerial parts (leaves) and are non-habitat specific. They are helpful in interaction with other microorganisms. (Verma et al., 2022; Aamir et al., 2020; Baron et al., 2018).
IV. Class IV: non-clavicipitaceous – these are called dark septate fungi and have melanin in their septate. They are found mostly in the roots of their host and in high stress environment. (Verma et al., 2022; Aamir et al., 2020; Baron et al., 2018).

DSE (dark septate endophytic) fungi play an important role as bioremediator, biocontrol agent, biofertilizer and bioregulator contributing to plant health and ecosystem stability. They mainly belong from *Ascomycota* family and have dark septate hyphae which colonize plant without causing any negative symptoms. Though less studied than mycorrhizal fungi, they have an impactful role on sustainable agriculture due to their widespread host range. They can colonize both mycorrhizal and nonmycorrhizal plants including members of Brassicaceae, Caryophyllaceae, and Chenopodiaceae (Usuki & Narisawa, 2007). They can be cultivated independently, also they can establish early colonization in seedlings, which make them very useful for field applications.

The prime role of filamentous fungi in the soil are to degrade organic materials and help in soil structure formation. In addition, bound species of *Alternaria, Aspergillus, Cladosporium, Dematium, Gliocladium, Humicola* and *Metarhizium* produce some compounds in the soil which are essential to maintain soil organic content (Yuvaraj & Ramasamy, 2020). *Trichoderma*, a genus of cosmopolitan filamentous endophytic fungus, has gained immense popularity in the past years. They form a symbiotic relationship with plants, release phytohormones like auxins, volatiles & active metabolites, small peptides that promotes root branching, plants increase nutrient uptake, therefore boosting plant growth

& yield, along with that it eliminates the competing pathogenic fungus by means of parasitism, competition & antagonism, helps to control root diseases. (López-Bucio et al., 2015). Other species like *T. virens* GV41 improves nitrogen efficiency and antioxidant activity in lettuce and *T. saturnisporum* promotes seed germination, yield, and fruit quality in melon (Baltazar et al., 2021). *T. asperellum* targets *Helopeltis theivora*, a tea pest, on the other hand, *T. virens* affects *Gryllotalpa,* a major pest from cowpea and soybean plants. T*richoderma harzianum* is one of the most potent biocontrol agents that suppresses plant pathogens like *Pythium aphanidermatum*, *Rhizoctonia solani*, and *Fusarium oxysporum and* also provide defense against various leaf, root, and fruit diseases (Tarigan et al., 2022).

In summary, fungal species have the potential to enhance global food security and promote eco-friendly farming. However, further research is needed to develop effective inoculation methods, ensure their viability under field conditions, and sustain microbial diversity.

FUNCTIONAL MECHANISM AND INTERACTIONS

The utilization of biostimulants presents a viable and sustainable approach to managing crop nutrition, perhaps mitigating the unfavorable consequences of excessive chemical fertilizer application. The most common ways that endophytic associations help plants are by increasing their ability to absorb nutrients, reducing stress, secreting phytohormones and other bioactive metabolites, and strengthening their ability to fight against pests and illness (Parthasarathi et al.,2012). Enhanced water uptake, osmolyte accumulation and mineral nutrition absorption under stressful conditions are a few of the processes that drive plant growth. Together, these processes enhance the host ability to metabolize nutrients, which promotes growth and resilience to stress. Alkaloids, aromatic ketones, phenol carboxylic acids, acetogenin compounds, acyclic compounds, corticosteroids, isoprenoids, and glycosides or the products generated from them, are the most often created or discovered metabolites (Aamir et al., 2020). It has been shown that these products possess antibacterial, anticancer, anti-insect, and/or antipathogenic qualities. Here, we have covered a few particular interactions and mechanisms pertaining to the generation of phytohormones, siderophores, nitrogen fixation, biofertilizers and volatile chemicals that promote plant growth and development.

Nitrogen Fixation

Symbiotic nitrogen fixation ecologists are very interested in symbiotic plant-fungal interactions as they can impact important ecological functioning such as crop yield, species diversity and plant-pathogen relationship (B. Yang et al., 2015). According to B. Yang et al. (2014) and H. Yang et al. (2014), rice crop infected with *P. liquidambari* have markedly changed N metabolism. In the roots or stems of the rice seedlings, *P. liquidambari* infection increased the expression of particular members of the OsAMT and OsNRT gene families. This implies that rice's intake of N may be improved by the fungal endophyte. By raising the number of soluble sugars, total free amino acids and plant organic acids in rice plant root transudes, *P. liquidambari* promotes plant growth (B. Yang et al., 2015). In a variety of settings, endophytic soil fungi *Metarhizium* and *Beauveria*, which are widely distributed, help plants absorb nitrogen from insects. These fungi infect both plants and insects breaking down nitrogen from the former and releasing it in a form that may be absorbed by the latter. This mechanism, called endophytic, insect pathogenic fungi (EIPF)-mediated nitrogen transfer, can give plants a substantial supply of nitrogen, particularly in environments with low nitrogen availability. This is the new process through which the EIPF *Metarhizium* and *Beauveria* create a symbiotic connection with all four plant species (maize, rice, soybean, and tomato) and reacquire this lost nitrogen (Behie & Bidochka, 2014). Arbuscular mycorrhizal fungi (AMF) can also have an effect on this vital component, which is required for crop production, plant development, and growth. AMF affects the health of ecosystems, the nitrogen cycle, and the growth of plants. A study by McFarland et al. (n.d.) found that over half of the nitrogen required by plants comes from mycorrhizal connections. Unlike non-arbuscular mycorrhizal plants, arbuscular mycorrhizal plants have simple access to all types of nitrogen. AMF exterior hyphae are capable of absorbing inorganic nitrogen such as ammonium ($NH4^{+)}$ and $NO3^{-}$. Fungal hyphae have a lower capacity to absorb $NH4^{+}$ than do roots (Lone et al., 2017).

Biofertilizers

The processes of atmospheric nitrogen fixation, nutrient mobilization, absorption, and solubilization are significantly aided by biofertilizers, which comprise bacteria, algae, and fungi (Mishra et al., 2015; Nath et al., 2015). Aamir et al. (2020) claim that symbiotic fungi can act as biofertilizers by growing within the underground tissues of many plant species without harming them and boosting the tolerance of the plants to biological and physical stresses. Chadha et al. (2015) studied 30 inhabiting endophytes separated from tomato roots in the central Himalayan region of India. They found that all of them boosted plant development, with *Trichoderma pseudokoningii* and *Fusarium fusarioides* having the most noteworthy effects. It is possible to use these fungi as biofertilizers to support sustainable agriculture. Under varied salt (NaCl) stress conditions, Hamayun et al. (2017) explored the growth enhancing characteristics of the endophytic fungus *P. spadiceum AGH786* on seedlings of *Glycine max* L. Merril. According to their research, *P. spadiceum AGH786* effectively reduces salt stress, which makes it a viable option for use in saline soil farming. Similarly, *Brassica campestris* spp. *chinensis* L. was studied by Khalid et al. (2017) for the effects of *Piriformospora indica* in combination with bio-fertilizer. They discovered that the introduction of endophytic organisms into plants resulted in improved antioxidant pathways, higher concentrations of carotenoids, phenolic acids and flavonoids, as well as improved plant development. Furthermore, it was shown by Doni et al. (2018) that treated plants exhibited considerably higher levels of tiller and particle counts, photosynthetic activity, chlorophyll a and b concentration, stomatal permeability and plant stature when treated with biofertilizers based on *Trichoderma* compared to untreated plants. *Trichoderma harzianum* (T22) was applied as an organic fertilizer to maize, and this improved a number of growth metrics, such as production, total protein, chlorophyll amount, complex glucose and phytohormone levels (Aamir et al., 2020).

Siderophore Production

Hydroxamate siderophores, which range in concentration from 2 to 12 nM in soil solutions, are detectable and suggest microbial activity in iron uptake. In situations where iron availability is a constraint, siderophore production can be used to measure the proliferation of microorganisms in the soil. Because both pathogenic and mycorrhizal fungi excrete siderophores, they are significant in plant-microbe interactions (Kajula et al., 2010). According to numerous analyses, siderophores are necessary for various fungal species pathogenicity, tolerance to oxidative damage, non-sexual and sexual development, iron accumulation, and defense against iron-induced toxicity (Johnson, 2008). Furthermore, siderophores are advantageous to plants in two important ways. Firstly, they have the ability to solubilize iron that would otherwise be unavailable, providing vital nutrients to a variety of plants, such as sorghum, sunflower, cucumber, and maize. Second, the generation of siderophores by nonpathogenic bacteria which can inhibit the development of harmful microbes caused by iron deficiency. This phenomenon demonstrates siderophores dual function of preventing infections and fostering plant development. By restricting the amount of iron that pathogens can access, this iron competition mechanism offers an environmentally benign method of biocontrol (Renshaw et al., 2002). And few are discussed below in details.

Iron Acquisition

Numerous hydroxamate siderophores, mainly used as iron transport chemicals, are produced by fungi. Internal iron concentrations control siderophore production and uptake system expression. These are transported in a stereoselective and energy-dependent manner that is dependent on the identification of the metal ions spatial arrangements. Apart from their role in iron transport, siderophores also play crucial roles in pathogenicity enhancement, accumulation of iron inside cells and prevention of the growth of other microbes. Aside from iron, they can also complex other metals, especially the actinides. Most of fungal siderophores are hydroxamates, with the exception of zygomycetes' carboxylate-type siderophore rhizoferrin. Aerobic microorganisms, such as ericoid mycorrhizal fungi, produce hydroxamate siderophores when iron levels are low (Haselwandter et al., 1992). There is also evidence that ectomycorrhizal and arbuscular mycorrhizal fungi may release these siderophores (Bartholdy et al., n.d.). Similar to mycorrhizal fungi, dark-septate endophytes also penetrate plant roots and may contribute to iron feeding (Kajula et al., 2010). Numerous mycorrhizal fungal strains generate siderophores, and the capacity of disease-causing fungi to produce siderophores is often linked to their virulence. In one study, the synthesis of extracellular siderophores and antioxidant and antibacterial compounds was examined in foliar endophytic fungi of Labrador tea (*Rhododendron tomentosum Harmaja*) and Scots pine (*Pinus sylvestris* L.). Ferricrocin was found in Labrador tea leaves, which indicates that it might be involved in the endophyte-plant host interaction in vivo (Aamir et al., 2020). Plants benefit from fungal siderophores because they facilitate iron absorption and mobilization. Plants find it difficult to absorb enough iron in settings low in iron because Fe (III) is mostly insoluble. Siderophores are chemicals that chelate iron and bind to Fe (III) to make it soluble which are produced by fungi. The iron-siderophore complexes are subsequently conveyed to the roots of plants, where the iron is absorbed directly via the siderophore complex or following the enzymatic reduction of Fe(III) to Fe(II). This procedure is essential for plant growth in soils lacking in nutrients because it increases the availability of iron, which boosts the production of chlorophyll and overall plant vigor. Furthermore, by sequestering iron and depriving pathogens of this vital nutrient, certain fungal siderophores aid in protecting plants against harmful organisms (Renshaw et al., 2002).

Pathogen Suppression

In addition to being essential for iron uptake, fungus siderophores are essential for ecological interactions, particularly in the suppression of pathogenicity. Because pathogens require iron for many essential functions, their development and pathogenicity can be severely restricted by their capacity to sequester iron. In the competition for this limited resource, fungi that produce siderophores can efficiently suppress the growth of harmful microbes (Weinberg, 1999). According to Lemanceau et al. (1986), non-pathogenic *Fusarium* species, for instance, have been demonstrated to produce more siderophores than their pathogenic counterparts. This superior iron acquisition leads to the pathogen's suppression. This behavior is also observed in other systems, where the creation of siderophores by helpful fungi increases plant development by increasing the availability of iron while also inhibiting the spread of diseases. For example, numerous studies have shown that competition for iron suppresses pathogenic fungi in the rhizosphere, demonstrating the dual function of siderophores in fostering plant development and thwarting infections (Renshaw et al., 2002). By restricting the amount of iron available to pathogens and hence decreasing their pathogenicity, this iron competition mechanism offers a natural and environmentally beneficial type of biocontrol.

Volatile Compounds

VOCs are low polarity, having a high room pressure vapor pressure and low molecular weight (less than 300 Da) chemicals. Particularly attractive for use in agriculture are volatile substances such as aromatic compounds, heterocyclic compounds, aliphatic aldehydes, ketones, aliphatic alcohols, aromatic alcohols, thiols, thiol esters, and their derivatives, as well as small molecules with a high diffusible ability in the atmosphere. (Morath et al., 2012; Oro et al., 2018). Microorganism growth phase, environmental conditions (including pH and temperature), and microbial species all affect the synthesis and chemical makeup of the volatiles (Korpi et al., 2009). According to Aamir et al. (2020), numerous studies have been reported on the synthesis of fuel-grade hydrocarbons and their derivatives, known as mycodiesel from some fungal endophytes, including *Gliocladium* sp., *Ascocoryne sarcodies*, *Phoma*, and *Hypoxylon* sp. Thines et al. (2004) claim that inhabiting fungi are a rich producer of specialized metabolites and a variety of chemicals with antibiotic actions. In a different investigation, *Lentinula* mycelia treated to volatile organic compounds (VOCs) from *M. pulcherrima* showed increased lipid and protein content suggesting an improvement in certain metabolic characteristics. The principal organic volatile molecules (VOCs) generated by the *M. pulcherrima* Mc9 strain were esters. A vast pool of volatile compounds, known as volatilomes, can be produced by *Metshnikowia pulcherrima*, *Aureobasidium pullulans*, *Wickerhamomyces anomalus*, and *Saccharomyces cerevisiae*. These substances has the capacity to impede the development of several fungal infections, such as *Botrytis cinerea, Penicillium* spp., *Monilinia* spp., and *Trichoderma* spp. (Di Francesco et al., 2024).

Phytohormones Production

The ability of endophytic bacteria and fungi to produce physiologically active metabolites and substances which not only help the growth of their host plants but also sustain their own metabolism has long been recognized as one of their beneficial impacts (A. L. Khan et al., 2016). The mutualistic association between non-pathogenic fungi and host crops has garnered significant interest, especially due to its function in augmenting plant growth and assisting plants in withstanding environmental stressors (A. L. Khan et al., 2011). Plant growth-promoting fungi (PGPF) are widely known for helping plants grow and develop by helping them deal with a variety of biological and environmental challenges. A frequent auxin produced from the metabolism of L-tryptophan is indole-3-acetic acid (IAA), which is produced by a variety of fungi associated with plant roots and rhizosphere soil. As per Khan et al. (2016), IAA is produced by endophytic fungi which stimulates numerous activities including cell expansion, root development, seed sprouting, seedling development and light-directed growth. IAA generated by the ectomycorrhizal fungus *Paxillus involutus* and *Pisolithus tinctorius* boosted root development and growth in Scots pine, as reported by Niemi et al. (2002). A study evaluating the growth-promoting and stress-mitigating effects of *Penicillium* species RDA01, NICS01 and DFC01 on sesame (*Sesamum indicum* L.) revealed that NICS01 and DFC01 significantly increased shoot and root length and seedling biomass, which was attributed to the secretion of various amino acids (Radhakrishnan et al., 2014). Additionally, thirteen endophytic fungi that were separated from the roots of three orchid species- *Paphiopedelum bellatulum*, *Phaius tankervilleae*, and *Spathoglottis affinis*, produced substantial amounts of IAA in culture. These fungi improved the development of the kidney beans (*Phaseolus vulgaris*) aided in the germination of seeds and the elongation of roots in *Zea mays* and initiates the growth of *Oryza sativa* (Chutima & Lumyong, 2012). Gibberellins (GAs), which are known to promote growth, are produced by the endophytic strain GMH-1a, which has been found to enhance several aspects of soybean plants under salt stress, including shoot length, biomass, foliage area, chlorophyll amount, and photosynthetic efficiency. Three fungal endophytes were identified by Fouda et al. (2015) from the medicinal plant *Asclepias sinaica: Penicillium chrysogenum* Pc_25, *Alternaria alternata* Aa_27, and *sterile hyphae* Sh_26. Further corroborating their function in augmenting plant growth parameters, these isolates were shown to promote the growth of plants by producing ammonia and IAA (Aamir et al., 2020).

FUNGAL BIOSTIMULANTS IN PLANT STRESS TOLERANCE

Conditions such as raised salinity, drought, high temperature, pests & diseases are liable for about 50 to 70 percentage of productivity loss by affecting crop performance (Rai et al., 2021). In coming future a more sustainable and environment friendly approach is required to enhance food production and combat harsh environmental conditions (López-Bucio et al., 2015). Fungal biostimulants provide resistance and sustainability to crops against several biotic as well as abiotic stresses. These stresses instigate numerous physical, biochemical and molecular changes in plants such as ROS generation, lipid peroxidation, reduced photosynthesis and hormonal imbalance. Biostimulants help plants to adapt by inducing their defense responses and enhancing osmoprotection, membrane stability, metal chelation and ROS scavenging activity (Rai et al., 2021).

Biotic Stress Tolerance

Some non-pathogenic fungi are known to save crops from various insect herbivore and parasitic fungi that modify the host plant to obtain nutrition. Damage causing pathogens of plants are usually inhibited or controlled by the upregulation or production of compounds having antifungal, antibacterial, nematocidal, and insecticidal properties (Sharma et al., 2018). Tactics such as competition, antibiosis, mycoparasitism, induced resistance, antibiotic production and enzyme production are used by non-pathogenic fungi in order to protect host plants. They colonize inside host plant and competes for nutrition and space, induce resistance through physical and chemical barriers and stimulates synthesis of signaling molecule. In order to avoid illness, induced resistance is one of the vital tactics used by endophytes. Species of fungi like *Aspergillus* sp., *Colletotrichum* sp., *Gliocladium* sp., *Fusarium* sp. etc., are some of the endophytes that have the ability to promote tolerance against biotic stress as well as helping plants in primary growth and yield (Verma et al., 2022). *Piriformospora indica,* a symbiotic-fungi colonizes the plant roots and protect them from necrotrophic pathogens. Enhanced antioxidative enzymic activities lead to increase crop yields which is the result of defense response in the host through activation the glutathione-ascorbate cycle. In mycoparasitism, an endophyte parasitizes on pathogenic fungi through direct attachment, puncture their hyphae and breakdown the cell wall. Endophytic fungi do not cause harm to the related host plants and produce compounds such as flavonoids, alkaloids, phenols, terpenoids, VOCs and upregulate defense related genes which results in decreased penetration and capability of the pathogen (Aamir et al., 2020). *Trichoderma*, a ubiquitous endophyte can strongly colonize any part of its host plant providing resistance against wide range of pests and disease-causing organisms. Several species of *Trichoderma* including *T. botryosum*, *T. caeruloviride*, *T. lentissimum*, *T. harzianum* and *T. koningii* were proven to be efficient when utilized as biofertilizer and pathogen control agents in agricultural field (Verma et al., 2022). *Aureobasidium pullalans*, an endophyte was effective against pathogens such as *Penicillium expansum* and *B. cinerea*, which suppresses apples growth and development. It also controls the scab disease of wheat and protects kernel contaminated with pathogens like *Fusarium culmorum*, *Acremonium*, *Cladosporium*, and *Penicillium* (Aamir et al., 2020).

Entomopathogenic fungi (EPF) carries out a significant role in pest management through a sustainable manner. Various endophytes including *Purpureocillium lilacinum*, *Isaria fumosorosea* and *Metarhizium robertsii* act as biocontrol agents in several crop species for instance cotton, maize, tomato, barley and sorghum (Lugtenberg et al., 2016). They reduce pest population and crop damage by controlling reproduction rate, food consumption and development rate of the pests. *Beauveria bassiana*, the most studied EPF also known for reducing the larval population of *Helicoverpa armigera*, *Liriomyza huidobrensis* and *Acyrthosiphon pisum* in faba bean. *Metarhizium anisopliae*, a universal endophyte reduces feeding and reproduction rate of many pests such as *Acyrthosiphon pisum*, *Myzus persicae*, *Aphis fabae* and *Plutella xylostella* in faba bean, pepper, rapeseed and strawberry (Mantzoukas & Eliopoulos, 2020).

Abiotic Stress Tolerance

Salt Stress

In regions where the soil is adversely affected by salinity, fungal biostimulants have emerged to be a potent and sustainable mechanism for boosting crop tolerance. Salinity stress generates excess amount of ROS in plant cells which results in high osmotic pressure and ionic disturbance that causes eventual apoptosis (Rai et al., 2021). To overcome salt toxicity endophytes stimulates antioxidant enzymes such as ascorbate peroxidase (APX), superoxide dismutase (SOD) and catalase (CAT), increase osmo-protectants like as L-proline and glycine-betaine, while regulating the level of phytohormones (López-Bucio et al., 2015). With the conditions of high salinity, the levels of stress hormones like jasmonic acid (JA), abscisic acid (ABA) increases resulting in stunted plant growth. Presence of fungal biostimulants enhance the levels of gibberellic acid (GA) in salt stressed plants which enrich the isoflavones content, thereby inhibiting the biosynthesis of JA and ABA (Aamir et al., 2020). Crops having symbiotic relationship with endophytes shows salinity tolerance with higher chlorophyll content, increased rate of photosynthesis, better yield, increased stomatal conductance, low oxidative damage and enhanced nutrient uptake than non-symbiotic crop. For example, when rice plants are associated with a root endophyte *Piriformospora indica*, it significantly increase their yield, protect PSII and reduce salt induced toxicity (Verma et al., 2022). Beneficial fungi like *Trichoderma* spp. and *S. vermifera* are known to enhance physiological parameters of crops, upregulate antioxidant biosynthesis genes, regenerate ascorbate-glutathione in the cells, maintain water potential and the Na^+ and K^+ efflux in crop under salinity stress. *R.intraradices*, a potential candidate for biostimulant enhanced nutrient accumulation, reduced transport of toxic ion (Na^+ and Cl^-) to shoot and increased sink strength in grapevine and some leafy vegetables such as lettuce and zucchini squash (Rouphael et al., 2015).

Drought Stress

Drought being a sensitive environmental stress, cause an annual reduction of 30-70% in crop yield and productivity worldwide. When water is unavailable in the soil surface, plant cells lose its turgidity resulting in withering and parching of the plants (Rai et al., 2021). To prevent water loss under these situations endophytic fungi cause some alteration in the plant mechanism such as reduction in leaf conductance, delaying stomatal opening, slowing down the transpiration stream and thick cuticle formation (Verma et al., 2022). Endophytes develop drought resilience in crops by accumulating amino acids, inducing antioxidant enzymes, enhancing water efficiency of plants, decreasing ROS production and increasing growth hormones such as IAA and GA (Aamir et al., 2020). Crops in association with symbiotic fungi such as *Fusarium* sp., *Trichoderma* sp., *Penicillium* sp., *Aspergillus* sp. and *Alternaria* sp. use significantly less water, gather more compatible solutes, increase photosynthetic rate and produce more biomass than non-symbiotic crops (Verma et al., 2022). Symbiotic fungi such as *Penicillium minioluteum* and *P. indica* get firmly established in the roots of crop plants and change root structure by altering their length, diameter and density for improved drought endurance. These alteration helps the fungal hyphae to penetrate deeper into root rhizosphere and cause better water and mineral uptake by the host plant. *P. indica* is a broadly functional mycorrhizal fungi that reallocate resources in the crop plant, maintain the aquaporin water channels and alter the plant proteomic profile to overcome drought stress (Toppo et al., 2023). Under water deficient environment AMF like *Glomus* spp., *F. mosseae* and *R. intraradices* showed improved leaf water content, enriched nutritional status, enhanced productivity and increased root water capacity in several crops like tomato, watermelon, strawberry and chili pepper (Rouphael et al., 2015).

Temperature Stress

Heat stress is a serious risk to crop plants that can hamper any developmental stages from seedling to maturity. It is considered as an escalation in temperature ranging from 10 ^{0}C -15 ^{0}C above the ambient level. (Shaffique et al., 2022). Thermal stress causes several physiological and cellular changes in plants such as disruption of thylakoid ultrastructure, leaf pigment reduction, membrane disruption, protein denaturation, ROS generation and chromatin modification (Rai et al., 2021). Application of endophytic fungi on heat stressed crops enhance osmolyte production in the plant, accumulate several metabolites such as phenol, sugars and flavonoids, induces heat shock proteins (HSPs) and increase crop biomass and yield. HSPs play a major role in heat tolerance as it maintains protein homeostasis by binding to denatured and misfolded proteins (Morales-Vargas et al., 2024). *Rhizophagus irregularis*, a most widely propagated AMF is known to promote GA levels, photochemical efficiency, relative water content and melatonin levels and decrease ABA levels, electrolyte leakage and senescence-associated genes (SAGs) in heat stressed plants (Sun & Shahrajabian, 2023). *Paraphaeosphaeri aquadriseptata* and *Stenotrophomonas maltophilia* are potential candidates for heat tolerant endophytes that enhances growth of plant during stressed conditions (Verma et al., 2022). In wheat and tomato plants, endophytic fungus such as *Curvularia protuberate, Stenotrophomonas maltophilia* and *Paraphaeosphaeri* sp. are known to provide heat tolerance by regulating biological, biochemical and metabolic activities of their host. (Lata et al., 2018).

Like heat stress, chilling stress also causes numerous functional and chemical modifications in plants like decreased seed germination, altered membrane fluidity, change in intracellular calcium and poor yield. At low temperature the plasma membrane becomes very rigid and it modifies the function of various channels, receptors, and kinases which causes a rapid increase in ROS within the cells (Godara & Ramakrishna, 2023). In blueberry, endophytes like *Penicillium rubens* and *P. bialowienzense* induce cold tolerance by enhancing photochemical efficiency, upregulating cold regulated genes (COR) and lowering oxidative stress (Morales-Vargas et al., 2024).

Heavy Metal Stress

This is a serious problem in many parts of the world which not only impacts crops but also causes severe health hazard to humans as well as animals upon consumption. Because of natural and human activities these toxic metalloids leach out to soil water system and get accumulated in edible parts of the plants. Heavy metals cause inhibition of seed germination, reduce plant growth, interferes with photosynthesis, hinder nutrient uptake and induce oxidative damage ultimately leading to the plants death (Nagajyoti et al., 2010). Endophytic fungi act as a solution in mitigating heavy metal toxicity by enhancing crop tolerance and performance under stress conditions. They limit heavy metal uptake and translocation, relocate HM to non-edible parts and promote crop growth (Sun & Shahrajabian, 2023). Endophytes like *Penicillium* sp., *Mucor* sp., *Alternaria* sp. And *Microsphaeropsis* sp are known to involve in biosorption and detoxification of lethal metals, enhance antioxidant enzyme, reduce ROS production and improve crop yield (Verma et al., 2022). Fungal endophytes *Fusarium tricinctum* and *Alternaria alternata* reduced Cadmium uptake by roots and shoots, increased content of chlorophyll, leaf area and dry biomass in *Solanum nigrum* (Khan et al., 2017). *Exophiala pisciphila*, a drak septate endophyte (DSE) changes the expression of 24 Glutathione S-transferases (GSTs) genes, upregulate tryptophan decarboxylase and serotonin N-acetyltransferase genes involved in metal detoxification. Endophytes such as *Phialocephala fortinii*, *Rhizodermea veluwensis*, and *Rhizoscyphus* sp. are involved in phytoremediation which shows growth enhancement and promote K uptake in plants growing in heavy metal contaminated areas. When inoculated with endophytic fungi, *Lactuca sativa* showed better seedling growth in Cr contamined soil. *P. indica* belonging from Sebacinaceae family was found to be effective in enhancing seedling growth and improviing Cr tolerance in sunflower (Toppo et al., 2023). AMF like *R. intraradices, G. constrictum*, and *R. mosseae* are known to enhance antioxidant enzyme activity and reduce root shoot movement of Cd in marigold and Pea under Cd stress conditions. In *Limonium sinuatum* (sea lavender), mixture of two endophytes *G. mosseae* and *G. intraradices* have been identified to subdue the harmful effects of Cd and Pb (Rouphael et al., 2015).

RECENT ADVANCEMENTS IN THE FIELD OF FUNGAL BIOSTIMULANTS

Fungal biostimulants are a fast-developing field of sustainable agriculture characterized by their capacity to enhance plant vigor, maximize nutrient uptake, and strengthen plants against environmental stress. Recent advances in fungal biostimulants involve a wide range of innovations ranging from the exploitation of extremophilic fungi and the understanding of synergistic microbial interactions to incorporating nanotechnology which have broadened our knowledge about the diverse functions of fungi to enhance plant well-being. Overall, these innovations contribute to the development of sustainable and resilient agriculture systems. The complexity of biostimulant research is further added to by the different sources of such compounds, such as bacteria, fungi, algae, higher plants, animals, and humate-containing raw materials, and also the variety of industrial processes that are employed during their preparation (Yakhin et al., 2017). Endophytic fungi, living symbiotically inside plant tissue, have also shown vast promise in improving the tolerance of the plant to several abiotic stresses. One prominent study emphasized the contribution of *Aspergillus flavus* towards the endurance of wheat plants under lead-stress. The fungus and exopolysaccharides were proved to promote growth factors of plants and stimulate antioxidant defense in order to counteract the deleterious influences of heavy metal toxicity (El-Khawaga et al.,2024). The search for more durable farming techniques has prompted studies into extremophilic fungi—organisms capable of growing in extreme environmental settings. These fungi have specialized adaptations that can be utilized to produce biofertilizers and biocontrol agents that can function under stressful conditions. Their potential to enhance soil fertility while helping to eliminate chemical use makes them more useful in times of worldwide climate change (Rodríguez et al., 2024). The antagonistic activity of specific fungi has been used to fight plant diseases. Studies aimed at Fusarium wilt suppression in tomatoes have isolated endophytic fungi with the ability to suppress the growth of *Fusarium oxysporum* f. sp. *lycopersici*. Not only did these fungi hinder pathogen growth but also enhanced overall plant health, highlighting their dual functionality as biostimulants and biocontrol agents (Muhorakeye et al.,2024). Progress has also been achieved in elucidating the synergistic interactions among various fungal species. Research has indicated that the integration of *Trichoderma guizhouense* with indigenous soil fungi, e.g., Humicola spp., can be used to increase disease suppression in plants such as bananas (Tao et al., 2023). Not only does this synergistic interaction decrease pathogen burden but also triggers plant immune systems, providing an integrated approach to disease control. The marriage of nanotechnology and fungal biostimulant is a revolutionary development. For example, the use of biogenic α-Fe_2O_3 nanoparticles has been found to enhance the antagonistic activity of Trichoderma species against diseases such as *Sclerotinia sclerotiorum*. The method increases the efficiency of biocontrol agents, opening the way for more powerful plant protection mechanisms (Bilesky-Jose et al.,2021).

ECOLOGICAL ROLE AND SUSTAINABILITY

The main objective of using biostimulants is sustainable agriculture with minimum or no harm to the environment, since the use of chemical fertilizers and pesticides can interfere with animals, plants as well as human wellbeing. These toxic chemicals can get incorporated into the food chain leading to health disorders, also hampering the aquatic and terrestrial habitat (Van Lexmond et al., 2015; Baron et al., 2018). Fungi plays an important role in the ecosystem by maintaining soil structure, nurturing valuable plant microbe relation, improving nutrient movement and dissolution which not only profits crops but also larger ecological communities. It can withstand a wide range of soil environment ranging from acidic to alkaline making them a promising candidate for adverse environmental conditions (Gianinazzi et al., 2010). Mycoremediation is one of the key functions carried out by fungi which results in enriched soil fertility and plant diversity. Some fungi play a role of pathogen control against harmful bacteria, fungi, for example, *Trichoderma* sp. can control soil borne diseases as well as panicle disease in different plants (Yao et al., 2023). Arbuscular mycorrhiza (AM) symbiosis can help ecosystem in different ways like improvement in root morphology of host plant by ramifying fungal mycelial network and by secreting 'glomalin, (sticky proteinaceous substance) into the soil that promotes soil stability. They are an excellent choice for reforestation as it is cost effective, increase survival rate of plants and help in preserving the natural system. Fungal endophytes are a vital element in the ecosystem, breaking down organic waste materials and continuing the geochemical cycle (Gianinazzi et al., 2010).

CHALLENGES AND LIMITATIONS

Over the past few years endophytic fungi have gained immense popularity in the field of agricultural sustainability. In many European countries fungal bio stimulants have seen successful adoption in the global market, however legal recognition of these bio stimulants is still limited on a broader level. Although derived from natural sources, they have to undergo toxicological and environmental assessment by regulatory bodies to avoid long term risk concerns. Due to complex nature of biostimulants it is extremely challenging to understand their mode of action, while external factor like environmental conditions, crop type, method of inoculation further complicates their standardization (Yakhin et al., 2017) Their compatibility with edaphic factors, climatic conditions, crop variety, shelf life and farmers lack of awareness and trust make commercialization relatively tough (du Jardin, 2015; Calvo et al., 2014; Yakhin et al., 2017)

Difficulties in practical implementation:

i. *Dependency on external factors*:

Microbes require optimal conditions to reach their maximum potential. They are affected by various environmental factors (temperature, moisture) and edaphic factors (soil pH, Soil type, soil structure, salinity) which make their survival and functionality under unfavorable situations very difficult (Ansari et al., 2023). Additionally, their effectiveness also depends on type of host crops, and some of them may fail to establish proper symbiotic relation resulting in little to no benefit.

ii. Competition

Soil is known to harbor a wide range and variety of microorganisms; this may lead to competition between the native communities and the introduced strain. Native microbe may outcompete the introduced strain leading to inconsistencies in their performance. If non-native the introduced microbe can become invasive threatening the local microbial diversity and disturbing the ecosystem (Ansari et al., 2023). Similarly, AMF has found to enhance nutrient availability in many crops with adequate fertile condition, however these beneficial fungi also require adequate macronutrient for their survival. When the fertility level is low, especially in low N condition, they were found to compete with the host crops. (Sible et al., 2021).

iii. Storage Risks

There are significant challenges in maintaining microbial biostimulant viability during preservation, transport and storage. In order to tackle this hurdle protective carriers, encapsulation methods should be used to prevent unsuitable circumstances like desiccation and temperature fluctuation (Fadiji et al., 2024).

iv. Slow Response and establishment

In order to observe the effects of microbial biostimulant on crop plant, it requires time and patience. Because prior to that there has to be a successful interaction which may manifest gradually as there is a process of colonization followed by a slower plant response. Being an obligate biotroph AMF require living host for propagation making large scale production and availability challenging. Use of highly effective strain and the optimization of application can be the solution to overcome this problem (Castiglione et al., 2021).

v. Risk of contamination

Although AMF show beneficial effects but some strains may become pathogenic or parasitic reducing plant growth. They the ability to absorb heavy metal which can lead to accumulation of toxic elements in plants and potential food contamination (Fadiji et al., 2024; Kumari et al., 2022).

FUTURE PROSPECTS

Several researches are going on to understand the host-endophyte relation, joint action of endophytes-fertilisers on crops, effect of adverse environmental condition and more field studies is necessary in this regard. Incorporation of fungal bio stimulants as biofertilizer and pest management remedy in agricultural field can promote ecofriendly food production. Future possibilities include discovery of suitable endophytic strains from local regions and detailed analysis on molecular mechanism of fungal bio stimulants to produce genetically modified endophytes. However, due to the complicated nature of interaction, it is hard to pinpoint the exact mechanism and pathways responsible for positive effects biostimulants. Recent advancements in the field of molecular biology can help overcome this issue. Transcriptome sequencing of the microbial populations can help understand the complex interaction of introduced strain with other microbes assisting in identification of key genes and signalling pathways. This can help standardize the inoculation method and increase their shelf life in different environmental conditions. Cutting edge technologies such as genomics, proteomics, metabolomics, RT-PCR and cytogenetic methods can help identify genetic makeup, differential gene expression, and protein profiles. Advanced tools like GC-MS/MS, HPLC-MS/MS, UPLC-MS/MS and NMR can detect secondary metabolites, phytohormones and enzymes responsible for beneficial traits. Metagenomics and NGS technology can help identify host crop specific fungal strain to increase long term crop productivity. Numerous research has proved the efficiency of endophytic fungi in controlling pathogens and pests invitro but their role in the host plant is yet to be explored. Addressing these gaps will enhance organic food production, combat escalating food demands and help these endophytes execute complex biological role.

CONCLUSION

In conclusion this chapter highlights that crop plants host diverse communities of symbiotic fungal endophytes which is beneficial in several ways. They provide resistance against biotic and abiotic stress factors thereby improving the crop production and yield stability in a manner which is environmentally sustainable. These fungal endophytes boost plant resistance by mitigating the adverse effects of salinity, drought and heavy metal toxicity and also combating the pathogens which are becoming pesticide-resistant with time ensuring a huge boost in growth for different agricultural crops. They are known to enhance crop yield, nutrient use efficiency, soil fertility and maintain soil microbial diversity. The use of these symbiotic associations as means of biostimulation has become popular in last two decades but significant challenges still remain. One such concern is regarding the nature of scientific studies performed in understanding the role of biostimulants. As these studies that report positive effects of fungal biostimulants have been operated either in laboratory, growth chamber or greenhouse provisions, the final effect of them might be different in real field situation as presence of different stressors will affect the fungi too altering their performance. Different crops react differently in presence of various biostimulants, alongside different product formulations, agricultural practices and environmental conditions may add another layer of complication. Also, commercialization of these fungal strains remains a prominent challenge due to lack of in-depth study on their longterm effect on human health and environment. Widespread use of these fungal biostimulants without proper research can lead to pathogen resistance, disruption of local microbial population and ecosystem imbalance. Therefore, locally and holistically adapted solutions are recommended for sustainable agricultural and horticultural use of fungal biostimulants. Overall, these biostimulants hold huge potential in crop protection and dealing with climatic change in a sustainable way.

ACKNOWLEDGEMENT

The authors are indebted to Department of Botany, Sikkim University, Gangtok, India for continuous support and encouragement.

REFERENCES

Aamir, M., Rai, K. K., Zehra, A., Kumar, S., Yadav, M., Shukla, V., & Upadhyay, R. S. (2020). Fungal endophytes: Classification, diversity, ecological role, and their relevance in sustainable agriculture. In Microbial Endophytes (pp. 291–323). Elsevier., 0-12-818734-0.00012-7.DOI: 10.1016/B978-0-12-818734-0.00012-7

Abdelaziz, A. M., Kalaba, M. H., Hashem, A. H., Sharaf, M. H., & Attia, M. S. (2022). Biostimulation of tomato growth and biocontrol of Fusarium wilt disease using certain endophytic fungi. *Botanical Studies*, *63*(1), 34. DOI: 10.1186/s40529-022-00364-7 PMID: 36484866

Ahmad, I., Jiménez-Gasco, M. D. M., Luthe, D. S., Shakeel, S. N., & Barbercheck, M. E. (2020). Endophytic Metarhizium robertsii promotes maize growth, suppresses insect growth, and alters plant defense gene expression. *Biological Control*, *144*, 104167. DOI: 10.1016/j.biocontrol.2019.104167

Albrecht, U. (n.d.). *Plant Biostimulants: Definiition and Overview of Categories and Effects.*

Ansari, M., Devi, B. M., Sarkar, A., Chattopadhyay, A., Satnami, L., Balu, P., Choudhary, M., Shahid, M. A., & Jailani, A. A. K. (2023). Microbial Exudates as Biostimulants: Role in Plant Growth Promotion and Stress Mitigation. *Journal of Xenobiotics*, *13*(4), 572–603. DOI: 10.3390/jox13040037 PMID: 37873814

Baltazar, M., Correia, S., Guinan, K. J., Sujeeth, N., Bragança, R., & Gonçalves, B. (2021). Recent Advances in the Molecular Effects of Biostimulants in Plants: An Overview. *Biomolecules*, *11*(8), 1096. DOI: 10.3390/biom11081096 PMID: 34439763

Baron, N. C., Costa, N. T. A., Mochi, D. A., & Rigobelo, E. C. (2018). First report of Aspergillus sydowii and Aspergillus brasiliensis as phosphorus solubilizers in maize. *Annals of Microbiology*, *68*(12), 863–870. DOI: 10.1007/s13213-018-1392-5

Bartholdy, B. A., Berreck, M., & Haselwandter, K. (n.d.). *Hydroxamate siderophore synthesis by Phialocephala fortinii, a typical dark septate fungal root endophyte.*

Behie, S. W., & Bidochka, M. J. (2014). Ubiquity of Insect-Derived Nitrogen Transfer to Plants by Endophytic Insect-Pathogenic Fungi: An Additional Branch of the Soil Nitrogen Cycle. *Applied and Environmental Microbiology*, *80*(5), 1553–1560. DOI: 10.1128/AEM.03338-13 PMID: 24334669

Bilesky-Jose, N., Maruyama, C., Germano-Costa, T., Campos, E., Carvalho, L., Grillo, R., ... & De Lima, R. (2021). Biogenic α-Fe_2O_3 nanoparticles enhance the biological activity of Trichoderma against the plant pathogen Sclerotinia sclerotiorum.

Canellas, L. P., Olivares, F. L., Aguiar, N. O., Jones, D. L., Nebbioso, A., Mazzei, P., & Piccolo, A. (2015). Humic and fulvic acids as biostimulants in horticulture. *Scientia Horticulturae*, *196*, 15–27. DOI: 10.1016/j.scienta.2015.09.013

Castiglione, A. M., Mannino, G., Contartese, V., Bertea, C. M., & Ertani, A. (2021). Microbial Biostimulants as Response to Modern Agriculture Needs: Composition, Role and Application of These Innovative Products. *Plants*, *10*(8), 1533. DOI: 10.3390/plants10081533 PMID: 34451578

Chutima, R., & Lumyong, S. (2012). Production of indole-3-acetic acid by Thai native orchid-associated fungi. *Symbiosis*, *56*(1), 35–44. DOI: 10.1007/s13199-012-0158-2

Di Francesco, A., Moret, E., Cignola, R., Garagozzo, L., Torelli, E., & Di Foggia, M. (2024). Yeasts volatile organic compounds (VOCs) as potential growth enhancers and molds biocontrol agents of mushrooms mycelia. *Fungal Biology*, *128*(4), 1859–1867. DOI: 10.1016/j.funbio.2024.05.007 PMID: 38876538

Doni, F., Zain, C. R. C. M., Isahak, A., Fathurrahman, F., Anhar, A., Mohamad, W. N. W., Yusoff, W. M. W., & Uphoff, N. (2018). A simple, efficient, and farmer-friendly Trichoderma-based biofertilizer evaluated with the SRI Rice Management System. *Organic Agriculture*, *8*(3), 207–223. DOI: 10.1007/s13165-017-0185-7

Du Jardin, P. (2015). Plant biostimulants: Definition, concept, main categories and regulation. *Scientia Horticulturae*, *196*, 3–14. DOI: 10.1016/j.scienta.2015.09.021

El-Khawaga, H. A., Mustafa, A. E., El Khawaga, M. A., Mahfouz, A. Y., & Daigham, G. E. (2024). Biostimulating effect of endophytic Aspergillus flavus AUMC 16068 and its respective ex-polysaccharides in lead stress tolerance of Triticum aestivum plant. *Scientific Reports*, *14*(1), 11952. PMID: 38796501

Fadiji, A. E., Xiong, C., Egidi, E., & Singh, B. K. (2024). Formulation challenges associated with microbial biofertilizers in sustainable agriculture and paths forward. *Journal of Sustainable Agriculture and Environment*.

Fouda, A. H., Hassan, S. E.-D., Eid, A. M., & Ewais, E. E.-D. (2015). Biotechnological applications of fungal endophytes associated with medicinal plant Asclepias sinaica (Bioss.). *Annals of Agricultural Science*, *60*(1), 95–104. DOI: 10.1016/j.aoas.2015.04.001

Gajc-Wolska, J., Kowalczyk, K., Nowecka, M., Mazur, K., & Metera, A. (n.d.). *EFFECT OF ORGANIC-MINERAL FERTILIZERS ON THE YIELD AND QUALITY OF ENDIVE*.

Gianinazzi, S., Gollotte, A., Binet, M.-N., Van Tuinen, D., Redecker, D., & Wipf, D. (2010). Agroecology: The key role of arbuscular mycorrhizas in ecosystem services. *Mycorrhiza*, *20*(8), 519–530. DOI: 10.1007/s00572-010-0333-3 PMID: 20697748

Godara, H., & Ramakrishna, W. (2023). Endophytes as nature's gift to plants to combat abiotic stresses. *Letters in Applied Microbiology*, *76*(2), ovac067. Advance online publication. DOI: 10.1093/lambio/ovac067 PMID: 36763787

Hamayun, M., Hussain, A., Khan, S. A., Kim, H.-Y., Khan, A. L., Waqas, M., Irshad, M., Iqbal, A., Rehman, G., Jan, S., & Lee, I.-J. (2017). Gibberellins Producing Endophytic Fungus Porostereum spadiceum AGH786 Rescues Growth of Salt Affected Soybean. *Frontiers in Microbiology*, *8*, 686. DOI: 10.3389/fmicb.2017.00686 PMID: 28473818

Hamayun, M., Khan, S. A., Khan, A. L., Rehman, G., Kim, Y.-H., Iqbal, I., Hussain, J., Sohn, E.-Y., & Lee, I.-J. (2010). Gibberellin production and plant growth promotion from pure cultures of *Cladosporium* sp. MH-6 isolated from cucumber (*Cucumis sativus* L.). *Mycologia*, *102*(5), 989–995. DOI: 10.3852/09-261 PMID: 20943499

Han, Y., Feng, J., Han, M., & Zhu, B. (2020). Responses of arbuscular mycorrhizal fungi to nitrogen addition: A meta-analysis. *Global Change Biology*, *26*(12), 7229–7241. DOI: 10.1111/gcb.15369 PMID: 32981218

Hardoim, P. R., Van Overbeek, L. S., Berg, G., Pirttilä, A. M., Compant, S., Campisano, A., Döring, M., & Sessitsch, A. (2015). The Hidden World within Plants: Ecological and Evolutionary Considerations for Defining Functioning of Microbial Endophytes. *Microbiology and Molecular Biology Reviews : MMBR*, *79*(3), 293–320. DOI: 10.1128/MMBR.00050-14 PMID: 26136581

Harrach, B. D., Baltruschat, H., Barna, B., Fodor, J., & Kogel, K.-H. (2013). The Mutualistic Fungus *Piriformospora indica* Protects Barley Roots from a Loss of Antioxidant Capacity Caused by the Necrotrophic Pathogen *Fusarium culmorum.Molecular Plant-Microbe Interactions*, *26*(5), 599–605. DOI: 10.1094/MPMI-09-12-0216-R PMID: 23405867

Harwoko, H., Hartmann, R., Daletos, G., Ancheeva, E., Frank, M., Liu, Z., & Proksch, P. (2019). Biotransformation of Host Plant Flavonoids by the Fungal Endophyte *Epicoccum nigrum.ChemistrySelect*, *4*(45), 13054–13057. DOI: 10.1002/slct.201903168

Haselwandter, K., Dobernigg, B., Beck, W., Jung, G., Cansier, A., & Winkelmann, G. (1992). Isolation and identification of hydroxamate siderophores of ericoid mycorrhizal fungi. *Biometals*, *5*(1), 51–56. DOI: 10.1007/BF01079698

Imran, A., Saadalla, M. J. A., Khan, S.-U., Mirza, M. S., Malik, K. A., & Hafeez, F. Y. (2014). Ochrobactrum sp. Pv2Z2 exhibits multiple traits of plant growth promotion, biodegradation and N-acyl-homoserine-lactone quorum sensing. *Annals of Microbiology*, *64*(4), 1797–1806. DOI: 10.1007/s13213-014-0824-0

Jaber, L. R., & Araj, S.-E. (2018). Interactions among endophytic fungal entomopathogens (Ascomycota: Hypocreales), the green peach aphid Myzus persicae Sulzer (Homoptera: Aphididae), and the aphid endoparasitoid Aphidius colemani Viereck (Hymenoptera: Braconidae). *Biological Control*, *116*, 53–61. DOI: 10.1016/j.biocontrol.2017.04.005

Jogawat, A., Vadassery, J., Verma, N., Oelmüller, R., Dua, M., Nevo, E., & Johri, A. K. (2016). PiHOG1, a stress regulator MAP kinase from the root endophyte fungus Piriformospora indica, confers salinity stress tolerance in rice plants. *Scientific Reports*, *6*(1), 36765. DOI: 10.1038/srep36765 PMID: 27849025

Johnson, L. (2008). Iron and siderophores in fungal–host interactions. *Mycological Research*, *112*(2), 170–183. DOI: 10.1016/j.mycres.2007.11.012 PMID: 18280720

Kajula, M., Tejesvi, M. V., Kolehmainen, S., Mäkinen, A., Hokkanen, J., Mattila, S., & Pirttilä, A. M. (2010). The siderophore ferricrocin produced by specific foliar endophytic fungi in vitro. *Fungal Biology*, *114*(2–3), 248–254. DOI: 10.1016/j.funbio.2010.01.004 PMID: 20943135

Khalid, M., Hassani, D., Bilal, M., Liao, J., & Huang, D. (2017). Elevation of secondary metabolites synthesis in Brassica campestris ssp. Chinensis L. via exogenous inoculation of Piriformospora indica with appropriate fertilizer. *PLoS One*, *12*(5), e0177185. DOI: 10.1371/journal.pone.0177185 PMID: 28493970

Khan, A. L., Halo, B. A., Elyassi, A., Ali, S., Al-Hosni, K., Hussain, J., Al-Harrasi, A., & Lee, I.-J. (2016). Indole acetic acid and ACC deaminase from endophytic bacteria improves the growth of Solanum lycopersicum. *Electronic Journal of Biotechnology*, *21*, 58–64. DOI: 10.1016/j.ejbt.2016.02.001

Khan, A. L., Hamayun, M., Kim, Y.-H., Kang, S.-M., Lee, J.-H., & Lee, I.-J. (2011). Gibberellins producing endophytic Aspergillus fumigatus sp. LH02 influenced endogenous phytohormonal levels, isoflavonoids production and plant growth in salinity stress. *Process Biochemistry (Barking, London, England)*, *46*(2), 440–447. DOI: 10.1016/j.procbio.2010.09.013

Khan, A. R., Ullah, I., Waqas, M., Park, G.-S., Khan, A. L., Hong, S.-J., Ullah, R., Jung, B. K., Park, C. E., Ur-Rehman, S., Lee, I.-J., & Shin, J.-H. (2017). Host plant growth promotion and cadmium detoxification in Solanum nigrum, mediated by endophytic fungi. *Ecotoxicology and Environmental Safety*, *136*, 180–188. DOI: 10.1016/j.ecoenv.2016.03.014 PMID: 27931714

Khan, S. A., Hamayun, M., Yoon, H., Kim, H.-Y., Suh, S.-J., Hwang, S.-K., Kim, J.-M., Lee, I.-J., Choo, Y.-S., Yoon, U.-H., Kong, W.-S., Lee, B.-M., & Kim, J.-G. (2008). Plant growth promotion and Penicillium citrinum. *BMC Microbiology*, *8*(1), 231. DOI: 10.1186/1471-2180-8-231 PMID: 19099608

Korpi, A., Järnberg, J., & Pasanen, A.-L. (2009). Microbial Volatile Organic Compounds. *Critical Reviews in Toxicology*, *39*(2), 139–193. DOI: 10.1080/10408440802291497 PMID: 19204852

Kumari, M., Swarupa, P., Kesari, K. K., & Kumar, A. (2022). Microbial Inoculants as Plant Biostimulants: A Review on Risk Status. *Life (Chicago, Ill.)*, *13*(1), 12. DOI: 10.3390/life13010012 PMID: 36675961

Lahlali, R., & Hijri, M. (2010). Screening, identification and evaluation of potential biocontrol fungal endophytes against Rhizoctonia solani AG3 on potato plants: Biocontrol of Rhizoctonia solani using fungal endophytes. *FEMS Microbiology Letters*, *311*(2), 152–159. DOI: 10.1111/j.1574-6968.2010.02084.x PMID: 20738401

Lahlali, R., McGregor, L., Song, T., Gossen, B. D., Narisawa, K., & Peng, G. (2014). Heteroconium chaetospira Induces Resistance to Clubroot via Upregulation of Host Genes Involved in Jasmonic Acid, Ethylene, and Auxin Biosynthesis. *PLoS One*, *9*(4), e94144. DOI: 10.1371/journal.pone.0094144 PMID: 24714177

Lemanceau, P., Alabouvette, C., & Meyer, J. M. (1986). Production of Fusarinine and Iron Assimilation by Pathogenic and Non-Pathogenic Fusarium. In Swinburne, T. R. (Ed.), *Iron, Siderophores, and Plant Diseases* (pp. 251–259). Springer US., DOI: 10.1007/978-1-4615-9480-2_27

Lone, R., Shuab, R., Khan, S., Ahmad, J., & Koul, K. K. (2017). Arbuscular Mycorrhizal Fungi for Sustainable Agriculture. In Kumar, V., Kumar, M., Sharma, S., & Prasad, R. (Eds.), *Probiotics and Plant Health* (pp. 553–577). Springer Singapore., DOI: 10.1007/978-981-10-3473-2_25

López-Bucio, J., Pelagio-Flores, R., & Herrera-Estrella, A. (2015). Trichoderma as biostimulant: Exploiting the multilevel properties of a plant beneficial fungus. *Scientia Horticulturae, 196*, 109–123. DOI: 10.1016/j.scienta.2015.08.043

Lugtenberg, B. J. J., Caradus, J. R., & Johnson, L. J. (2016). Fungal endophytes for sustainable crop production. *FEMS Microbiology Ecology, 92*(12), fiw194. Advance online publication. DOI: 10.1093/femsec/fiw194 PMID: 27624083

Manoussopoulos, Y., Mantzoukas, S., Lagogiannis, I., Goudoudaki, S., & Kambouris, M. (2019). Effects of Three Strawberry Entomopathogenic Fungi on the Prefeeding Behavior of the Aphid Myzus persicae. *Journal of Insect Behavior, 32*(2), 99–108. DOI: 10.1007/s10905-019-09709-w

Mantzoukas, S., & Eliopoulos, P. A. (2020). Endophytic Entomopathogenic Fungi: A Valuable Biological Control Tool against Plant Pests. *Applied Sciences (Basel, Switzerland), 10*(1), 360. DOI: 10.3390/app10010360

McFarland, J. W., Ruess, R. W., Kielland, K., Pregitzer, K., Hendrick, R., & Allen, M. (n.d.). *Cross-Ecosystem Comparisons of In Situ Plant Uptake of Amino Acid-N and NH4+.*

Mishra, S., Singh, A., Keswani, C., Saxena, A., Sarma, B. K., & Singh, H. B. (2015). Harnessing Plant-Microbe Interactions for Enhanced Protection Against Phytopathogens. In Arora, N. K. (Ed.), *Plant Microbes Symbiosis: Applied Facets* (pp. 111–125). Springer India., DOI: 10.1007/978-81-322-2068-8_5

Morales-Payan, J. P., & Stall, W. M. (n.d.). *PAPAYA (CARICA PAPAYA) RESPONSE TO FOLIAR TREATMENTS WITH ORGANIC COMPLEXES OF PEPTIDES AND AMINO ACIDS. 14.*

Morales-Vargas, A. T., López-Ramírez, V., Álvarez-Mejía, C., & Vázquez-Martínez, J. (2024). Endophytic Fungi for Crops Adaptation to Abiotic Stresses. *Microorganisms, 12*(7), 1357. DOI: 10.3390/microorganisms12071357 PMID: 39065124

Morath, S. U., Hung, R., & Bennett, J. W. (2012). Fungal volatile organic compounds: A review with emphasis on their biotechnological potential. *Fungal Biology Reviews, 26*(2–3), 73–83. DOI: 10.1016/j.fbr.2012.07.001

Muhorakeye, M. C., Namikoye, E. S., Khamis, F. M., Wanjohi, W., & Akutse, K. S. (2024). Biostimulant and antagonistic potential of endophytic fungi against fusarium wilt pathogen of tomato Fusarium oxysporum f. sp. lycopersici. *Scientific Reports, 14*(1), 15365. DOI: 10.1038/s41598-024-66101-1 PMID: 38965302

Nagajyoti, P. C., Lee, K. D., & Sreekanth, T. V. M. (2010). Heavy metals, occurrence and toxicity for plants: A review. *Environmental Chemistry Letters, 8*(3), 199–216. DOI: 10.1007/s10311-010-0297-8

Nassimi, Z., & Taheri, P. (2017). Endophytic fungus *Piriformospora indica* induced systemic resistance against rice sheath blight via affecting hydrogen peroxide and antioxidants. *Biocontrol Science and Technology, 27*(2), 252–267. DOI: 10.1080/09583157.2016.1277690

Nasslahsen, B., Prin, Y., Ferhout, H., Smouni, A., & Duponnois, R. (2022). Management of Plant Beneficial Fungal Endophytes to Improve the Performance of Agroecological Practices. *Journal of Fungi (Basel, Switzerland)*, *8*(10), 1087. DOI: 10.3390/jof8101087 PMID: 36294652

Nath, A., Chattopadhyay, A., & Joshi, S. R. (2015). Biological Activity of Endophytic Fungi of Rauwolfia serpentina Benth: An Ethnomedicinal Plant Used in Folk Medicines in Northeast India. *Proceedings of the National Academy of Sciences. India. Section B, Biological Sciences*, *85*(1), 233–240. DOI: 10.1007/s40011-013-0184-8

Niemi, K., Vuorinen, T., & Ernstsen, A. (2002). Ectomycorrhizal fungi and exogenous auxins influence root and mycorrhiza formation of Scots pine hypocotyl cuttings in vitro. *Tree Physiology*, *22*(17), 1231–1239. DOI: 10.1093/treephys/22.17.1231 PMID: 12464576

Oro, L., Feliziani, E., Ciani, M., Romanazzi, G., & Comitini, F. (2018). Volatile organic compounds from Wickerhamomyces anomalus, Metschnikowia pulcherrima and Saccharomyces cerevisiae inhibit growth of decay causing fungi and control postharvest diseases of strawberries. *International Journal of Food Microbiology*, *265*, 18–22. DOI: 10.1016/j.ijfoodmicro.2017.10.027 PMID: 29107842

Parrado, J., Bautista, J., Romero, E. J., García-Martínez, A. M., Friaza, V., & Tejada, M. (2008). Production of a carob enzymatic extract: Potential use as a biofertilizer. *Bioresource Technology*, *99*(7), 2312–2318. DOI: 10.1016/j.biortech.2007.05.029 PMID: 17601731

Pinos, N. Q., Berbara, R. L. L., Tavares, O. C. H., & García, A. C. (2019). Different Structures in Humic Substances Lead to Impaired Germination but Increased Protection against Saline Stress in Corn. *Communications in Soil Science and Plant Analysis*, *50*(17), 2209–2225. DOI: 10.1080/00103624.2019.1659294

Prado, M. R. V., Weber, O. L. D. S., Moraes, M. F. D., Santos, C. L. R. D., Santos Tunes, M., & Ramos, F. T. (2016). Humic Substances on Soybeans Grown Under Water Stress. *Communications in Soil Science and Plant Analysis*, *47*(21), 2405–2413. DOI: 10.1080/00103624.2016.1243715

Radhakrishnan, R., Kang, S.-M., Baek, I.-Y., & Lee, I.-J. (2014). Characterization of plant growth-promoting traits of *Penicillium* species against the effects of high soil salinity and root disease. *Journal of Plant Interactions*, *9*(1), 754–762. DOI: 10.1080/17429145.2014.930524

Rai, N., Rai, S. P., & Sarma, B. K. (2021). Prospects for Abiotic Stress Tolerance in Crops Utilizing Phyto- and Bio-Stimulants. *Frontiers in Sustainable Food Systems*, *5*, 754853. DOI: 10.3389/fsufs.2021.754853

Renshaw, J. C., Robson, G. D., Trinci, A. P. J., Wiebe, M. G., Livens, F. R., Collison, D., & Taylor, R. J. (2002). Fungal siderophores: Structures, functions and applications. *Mycological Research*, *106*(10), 1123–1142. DOI: 10.1017/S0953756202006548

Rétif, F., Kunz, C., Calabro, K., Duval, C., Prado, S., Bailly, C., & Baudouin, E. (2023). Seed fungal endophytes as biostimulants and biocontrol agents to improve seed performance. *Frontiers in Plant Science*, *14*, 1260292. DOI: 10.3389/fpls.2023.1260292 PMID: 37941673

Rose, M. T., Patti, A. F., Little, K. R., Brown, A. L., Jackson, W. R., & Cavagnaro, T. R. (2014). A Meta-Analysis and Review of Plant-Growth Response to Humic Substances. In *Advances in Agronomy* (Vol. 124, pp. 37–89). Elsevier., DOI: 10.1016/B978-0-12-800138-7.00002-4

Rouphael, Y., Franken, P., Schneider, C., Schwarz, D., Giovannetti, M., Agnolucci, M., Pascale, S. D., Bonini, P., & Colla, G. (2015). Arbuscular mycorrhizal fungi act as biostimulants in horticultural crops. *Scientia Horticulturae*, *196*, 91–108. DOI: 10.1016/j.scienta.2015.09.002

Ruzzi, M., & Aroca, R. (2015). Plant growth-promoting rhizobacteria act as biostimulants in horticulture. *Scientia Horticulturae*, *196*, 124–134. DOI: 10.1016/j.scienta.2015.08.042

Shaffique, S., Khan, M. A., Wani, S. H., Pande, A., Imran, M., Kang, S.-M., Rahim, W., Khan, S. A., Bhatta, D., Kwon, E.-H., & Lee, I.-J. (2022). A Review on the Role of Endophytes and Plant Growth Promoting Rhizobacteria in Mitigating Heat Stress in Plants. *Microorganisms*, *10*(7), 1286. DOI: 10.3390/microorganisms10071286 PMID: 35889005

Sharma, M., Kansal, R., & Singh, D. (2018). Endophytic Microorganisms: Their Role in Plant Growth and Crop Improvement. In *Crop Improvement Through Microbial Biotechnology* (pp. 391–413). Elsevier. DOI: 10.1016/B978-0-444-63987-5.00020-7

Sible, C. N., Seebauer, J. R., & Below, F. E. (2021). Plant Biostimulants: A Categorical Review, Their Implications for Row Crop Production, and Relation to Soil Health Indicators. *Agronomy (Basel)*, *11*(7), 1297. DOI: 10.3390/agronomy11071297

Stępniewska, Z., & Kuźniar, A. (2013). Endophytic microorganisms—Promising applications in bioremediation of greenhouse gases. *Applied Microbiology and Biotechnology*, *97*(22), 9589–9596. DOI: 10.1007/s00253-013-5235-9 PMID: 24048641

Sun, W., & Shahrajabian, M. H. (2023). The Application of Arbuscular Mycorrhizal Fungi as Microbial Biostimulant, Sustainable Approaches in Modern Agriculture. *Plants*, *12*(17), 3101. DOI: 10.3390/plants12173101 PMID: 37687348

Tao, C., Wang, Z., Liu, S., Lv, N., Deng, X., Xiong, W., Shen, Z., Zhang, N., Geisen, S., Li, R., Shen, Q., & Kowalchuk, G. A. (2023). Additive fungal interactions drive biocontrol of Fusarium wilt disease. *The New Phytologist*, *238*(3), 1198–1214. DOI: 10.1111/nph.18793 PMID: 36740577

Tarigan, S. I., Toth, S., Szalai, M., Kiss, J., Turoczi, G., & Toepfer, S. (2022). Biological control properties of microbial plant biostimulants. A review. *Biocontrol Science and Technology*, *32*(12), 1351–1371. DOI: 10.1080/09583157.2022.2129589

Tedersoo, L., Sánchez-Ramírez, S., Kõljalg, U., Bahram, M., Döring, M., Schigel, D., May, T., Ryberg, M., & Abarenkov, K. (2018). High-level classification of the Fungi and a tool for evolutionary ecological analyses. *Fungal Diversity*, *90*(1), 135–159. DOI: 10.1007/s13225-018-0401-0

Thines, E., Anke, H., & Weber, R. W. S. (2004). Fungal secondary metabolites as inhibitors of infection-related morphogenesis in phytopathogenic fungi. *Mycological Research*, *108*(1), 14–25. DOI: 10.1017/S0953756203008943 PMID: 15035501

Ting, A. S. Y., Mah, S. W., & Tee, C. S. (2012). Evaluating the feasibility of induced host resistance by endophytic isolate Penicillium citrinum BTF08 as a control mechanism for Fusarium wilt in banana plantlets. *Biological Control*, *61*(2), 155–159. DOI: 10.1016/j.biocontrol.2012.01.010

Toppo, P., Subba, R., Roy, K., Mukherjee, S., & Mathur, P. (2023). Elucidating the Strategies for Isolation of Endophytic Fungi and Their Functional Attributes for the Regulation of Plant Growth and Resilience to Stress. *Journal of Plant Growth Regulation*, *42*(3), 1342–1363. DOI: 10.1007/s00344-022-10638-w

Usuki, F., & Narisawa, K. (2007). A mutualistic symbiosis between a dark septate endophytic fungus, *Heteroconium chaetospira,* and a nonmycorrhizal plant, Chinese cabbage. *Mycologia*, *99*(2), 175–184. DOI: 10.1080/15572536.2007.11832577 PMID: 17682770

Vadassery, J., Ritter, C., Venus, Y., Camehl, I., Varma, A., Shahollari, B., Novák, O., Strnad, M., Ludwig-Müller, J., & Oelmüller, R. (2008). The Role of Auxins and Cytokinins in the Mutualistic Interaction Between *Arabidopsis* and *Piriformospora indica.Molecular Plant-Microbe Interactions*, *21*(10), 1371–1383. DOI: 10.1094/MPMI-21-10-1371 PMID: 18785832

Van Lexmond, M. B., Bonmatin, J.-M., Goulson, D., & Noome, D. A. (2015). Worldwide integrated assessment on systemic pesticides: Global collapse of the entomofauna: exploring the role of systemic insecticides. *Environmental Science and Pollution Research International*, *22*(1), 1–4. DOI: 10.1007/s11356-014-3220-1 PMID: 25146115

Van Oosten, M. J., Pepe, O., De Pascale, S., Silletti, S., & Maggio, A. (2017). The role of biostimulants and bioeffectors as alleviators of abiotic stress in crop plants. *Chemical and Biological Technologies in Agriculture*, *4*(1), 5. DOI: 10.1186/s40538-017-0089-5

Varma, A., Savita Verma, , Sudha, , Sahay, N., Bütehorn, B., & Franken, P. (1999). *Piriformospora indica*, a Cultivable Plant-Growth-Promoting Root Endophyte. *Applied and Environmental Microbiology*, *65*(6), 2741–2744. DOI: 10.1128/AEM.65.6.2741-2744.1999 PMID: 10347070

Verma, A., Shameem, N., Jatav, H. S., Sathyanarayana, E., Parray, J. A., Poczai, P., & Sayyed, R. Z. (2022). Fungal Endophytes to Combat Biotic and Abiotic Stresses for Climate-Smart and Sustainable Agriculture. *Frontiers in Plant Science*, *13*, 953836. DOI: 10.3389/fpls.2022.953836 PMID: 35865289

Waller, F., Achatz, B., Baltruschat, H., Fodor, J., Becker, K., Fischer, M., Heier, T., Hückelhoven, R., Neumann, C., Von Wettstein, D., Franken, P., & Kogel, K.-H. (2005). The endophytic fungus Piriformospora indica reprograms barley to salt-stress tolerance, disease resistance, and higher yield. *Proceedings of the National Academy of Sciences of the United States of America*, *102*(38), 13386–13391. DOI: 10.1073/pnas.0504423102 PMID: 16174735

Weinberg, E. D. (1999). The Role of Iron In Protozoan and Fungal Infectious Diseases. *The Journal of Eukaryotic Microbiology*, *46*(3), 231–238. DOI: 10.1111/j.1550-7408.1999.tb05119.x PMID: 10377984

Yakhin, O. I., Lubyanov, A. A., Yakhin, I. A., & Brown, P. H. (2017). Biostimulants in Plant Science: A Global Perspective. *Frontiers in Plant Science*, *7*. Advance online publication. DOI: 10.3389/fpls.2016.02049 PMID: 28184225

Yang, B., Ma, H.-Y., Wang, X.-M., Jia, Y., Hu, J., Li, X., & Dai, C.-C. (2014). Improvement of nitrogen accumulation and metabolism in rice (Oryza sativa L.) by the endophyte Phomopsis liquidambari. *Plant Physiology and Biochemistry*, *82*, 172–182. DOI: 10.1016/j.plaphy.2014.06.002 PMID: 24972305

Yang, B., Wang, X.-M., Ma, H.-Y., Yang, T., Jia, Y., Zhou, J., & Dai, C.-C. (2015). Fungal endophyte Phomopsis liquidambari affects nitrogen transformation processes and related microorganisms in the rice rhizosphere. *Frontiers in Microbiology*, *6*. Advance online publication. DOI: 10.3389/fmicb.2015.00982 PMID: 26441912

Yang, H., Wang, Y., Zhang, Z., Yan, R., & Zhu, D. (2014). Whole-Genome Shotgun Assembly and Analysis of the Genome of *Shiraia* sp. Strain Slf14, a Novel Endophytic Fungus Producing Huperzine A and Hypocrellin A. *Genome Announcements*, *2*(1), e00011–e00014. DOI: 10.1128/genomeA.00011-14 PMID: 24503982

Yao, X., Guo, H., Zhang, K., Zhao, M., Ruan, J., & Chen, J. (2023). Trichoderma and its role in biological control of plant fungal and nematode disease. *Frontiers in Microbiology*, *14*, 1160551. DOI: 10.3389/fmicb.2023.1160551 PMID: 37206337

Yarzábal Rodríguez, L. A., Álvarez Gutiérrez, P. E., Gunde-Cimerman, N., Ciancas Jiménez, J. C., Gutiérrez-Cepeda, A., Ocaña, A. M. F., & Batista-García, R. A. (2024). Exploring extremophilic fungi in soil mycobiome for sustainable agriculture amid global change. *Nature Communications*, *15*(1), 6951. DOI: 10.1038/s41467-024-51223-x PMID: 39138171

Yuan, Y., Feng, H., Wang, L., Li, Z., Shi, Y., Zhao, L., Feng, Z., & Zhu, H. (2017). Potential of Endophytic Fungi Isolated from Cotton Roots for Biological Control against Verticillium Wilt Disease. *PLoS One*, *12*(1), e0170557. DOI: 10.1371/journal.pone.0170557 PMID: 28107448

Yuvaraj, M., & Ramasamy, M. (2020). Role of Fungi in Agriculture. In Mahyar Mirmajlessi, S., & Radhakrishnan, R. (Eds.), *Biostimulants in Plant Science*. IntechOpen., DOI: 10.5772/intechopen.89718

Zhao, J., Zhou, L., Wang, J., Shan, T., Zhong, L., Liu, X., & Gao, X. (n.d.). *Endophytic fungi for producing bioactive compounds originally from their host plants.*

Zulfiqar, F., Moosa, A., Ali, H. M., Bermejo, N. F., & Munné-Bosch, S. (2024). Biostimulants: A sufficiently effective tool for sustainable agriculture in the era of climate change? *Plant Physiology and Biochemistry*, *211*, 108699. DOI: 10.1016/j.plaphy.2024.108699 PMID: 38749375

KEY TERMS AND DEFINITIONS

Symbiotic Relationship: A close, long-term, mutually beneficial biological relationship between two different species.

Endophyte: A group of microorganisms, generally fungi or bacteria, that colonize intra or intracellular spaces inside plants. They complete all or part of their life cycle within the host plants and do not harm them by causing any disease.

Sustainability: Environment friendly practice to meet the requirement of the society without jeopardizing the environmental health by the best use of the non-renewable resources and increasing nature's capacity.

Stress Tolerance: Ability of plants to tolerate and withstand adverse environmental conditions (biotic and abiotic) through several tolerance and resistance mechanisms.

Nutrient Use Efficiency: Nutrient use efficiency is defined as amount of plant growth per unit of nutrient absorbed by the plants.

Crop Improvement: Improvement in yield, disease resistance, performance against abiotic stress factors as well as improving nutritional value per unit yield.

Defense Response: The physiological adaptations of an organism to withstand harmful elements such as pests, structural damage or disease.

Soil Health: The ability and the continued capacity of soil to function as a living ecosystem to support all forms of life within the environmental and land-use boundaries to sustain normal functioning of organisms supporting their productivity.

APPENDIX I

Table 1. List of some biostimulants, host and their activity on crop plants

Class	*Name/source*	*Crop*	*Activity*	*Reference*
Beneficial fungi	*Piriformospora indica* *Fusarium verticillioides*	Maize (*Zea mays*), Barley (*Hordeum vulgare*), Rice (*Oryza sativa*), Tomato (*Solanum lycopersicum* L.), etc. Soybean (*Glycine max*)	Increase nutrient uptake, imparts biotic and abiotic stress tolerance, plant growth promotion, increase fruit biomass. Improve seed germination under salt stress with low ABA content and high protein content.	Jogawat et al., 2016; Vadassery et al., 2008; Waller et al., 2005; Varma et al., 1999 Retif et al., 2023
Beneficial Bacteria	*Ochrobacterium* sp. Pv2Z2 *Pseudomonas syringae* pv. *Phaseolicola*	Common bean (*Phaseolus vulgaris* L.) Barley (*Hordeum vulgare*)	Improves nitrogen uptake, plants height, dry and fresh weight. It can greatly increase the fresh weight and dry weight of roots and shoots as well as the shoot length.	Imran et al., 2014
Humic & Fulvic substances	Degradation of dead biota in soil	Cucumber (*Cucumis sativa*), Rice (*Oryza sativa*), Soybean (*Glycine max*), Maize (*Zea mays*)	Positive effects on the growth under stress conditions.	*Demir1999*, n.d.; Pinos et al., 2019; Prado et al., 2016; Rose et al., 2014; Van Oosten et al., 2017
Protein hydrolysates	Animal or plant origin	Spinach (*Spinacia oleracea*), Papaya (*Carica papaya*), Tomato (*Solanum lycopersicum*)	Tolerance to biotic & abiotic stress, can increase soil fertility and soil microbial activities, thereby indirectly affecting plant growth and productivity, increase in number of flowers and yield	Gajc-Wolska et al., n.d.; Morales-Payan & Stall, n.d.; Parrado et al., 2008
Seaweed extracts	Mostly brown seaweeds *Macrocystis pyrifera, Fucus* spp., *Ascophyllum nodusum*	Lettuce, Grape, Tomato	Improves fruit production, plant development, yield, seed germination and resistance to abiotic and biotic stresses.	Calvo et al., 2014 & Sible et al., 2021

Table 2. Some fungal biostimulants and their activity on crop plants

Name	**Host**	**Activity**	**Reference**
Aspergillus sydowii	maize (*Zea mays*)	Phosphorous accumulation	Baron et al., 2018
Penicillium citrinum IR-3-3	Rice (*Oryza sativa*), sandy plant (*Atriplex gmelinii*)	Promote growth, improves length of seedlings.	S. A. Khan et al., 2008
Cladosporium sphaerospermum	Cucumber (*Cucumis sativus*), Soyabean (*Glycine max*)	Improvement in plant growth	Hamayun et al., 2010

continued on following page

Table 2. Continued

Name	Host	Activity	Reference
Aspergillus niger	Sunflower (*Helianthus annuus*), soybean (*Glycine max*)	Tolerance to high temperature, increase plant height, biomass and chlorophyll content	Imran et al., 2014
Metarhizium aestivum	Broad bean (*Vicia faba*)	promote plant growth (shoot height, root length, and fresh root and shoot weights)	Lr J. et al., 2018
Metarhizium robertsii	Maize (*Zea mays*)	Promotes plant growth and biomass, modulate the expression of defense genes	Ahmad et al., 2020
Chaetomium globosum	Chili pepper (*Capsicum annuum*)	Plant growth promotion and increased photosynthetic response under water stress condition	Rai et al., 2022
R. irregularis	Black gram (*Vigna mungo*)	Induce phenols, lignin, antioxidant, reduce feeding activity of *Spodoptera*	Tarigan et al., 2022
T. harzianum	Tomato (*Solanum lycopersicum*), Common bean (*Phaseolus vulgaris*)	Reduce Nezera viridula, increase jasmonic acid production, impacts *Acanthoscelides obtectus*	Tarigan et al., 2022
Penicillium simplicissimum & *Leptosphaeria* sp.	Cotton (*Gossypium* spp.)	Suppress verticillium wilt in, reduce disease occurance, and increase yield	Yuan et al., 2017
Epicoccum nigrum	Potato (*Solanum tubersoum*), sugarcane (*Saccharum officinarum*), ginger (*Zingiber officinale*)	Enhance tuber production, resist blackleg disease, suppress pathogen, produce antimicrobial compound	Harwoko et al., 2019

Table 3. List of fungal biostimulants and their activity in disease/pest prone crop plants

Crop Plant	Endophytic Fungi	Pests/ Causal agent	Activity	References
Rice	*P. indica*	*R.solani*	increase plant biomass and antioxidant enzyme activity, decrease sheath blight of rice	Nassimi & Taheri, 2017
Barley	*P. indica*	*Fusarium culmorum*	prevent root rot, decrease lipid peroxidation, increase antioxidant enzymes, regulate ascorbate-glutathione cycle	Harrach et al., 2013
Canola	*Heteroconium chaetospira*	*Plasmodiophora brassicae*	upregulate defense related gene like JA, ethylene, auxin, synthesize secondary metabolites such as flavones, phenols, antioxidants, prevent clubroot disease	Lahlali et al., 2014
Potato	*Alternaria longipes, Epicoccum nigrum,Phomopsis* sp., and *Trichoderma atroviride*	*R.solani*	improve yield, decrease stem canker and tuber blemish, produce secondary metabolites	Lahlali & Hijri, 2010
Banana	*Penicillium citrinum*	*Fusarium oxysporum*	elevate peroxidase, PPO, PAL, produce antimicrobial compound, prevent wilt	Ting et al., 2012
Cherry tomato	*Nodulisporium sp*	*Fusarium oxysporum*	increase VOC in host plant, promote plant growth	Toppo et al., 2023

continued on following page

Table 3. Continued

Crop Plant	Endophytic Fungi	Pests/ Causal agent	Activity	References
Lemon mint	*Muscodor yucatanensis,Fusarium chlamydosporum, Fusarium oxysporum, and Fusarium redolens*	*Sclerotinia* sp., *Colletotrichum capsici, Aspergillus flavus, Aspergillus fumigatus, and Fusarium solani*	inhibit pathogen growth, enhance plant defense	Aamir et al., 2020
Sugarcane, cotton, apple, peach	*Epicoccum nigrum*	*Sclerotinia sclerotiorum, Pythium, phytoplasma, Monilinia* spp.	increase plant biomass, control pathogen attack	Lugtenberg et al., 2016
Barley	*Penicillium brevicompactum*	*Cochiolobus, Fusarium, Pyrenophora Rhynchosporium*	suppress seed borne infection	Lugtenberg et al., 2016
Coffee	*Beauveria bassiana* and *Clonostachys rosea*	*Hypothenemus hampei*	regulate nitrogen cycling in plants, control pest population	Lugtenberg et al., 2016
Common bean	*Beauveria, Lecanicillium, Isaria*	*Tetranychus urticae,* spider mite	increased plant biomass, reduction in larval development	Mantzoukas & Eliopoulos, 2020
Wild barley	*Neotyphodium*	*Mayetiola destructor, Metopolophium dirhodum*	hamper development and survival of insects	Sharma et al., 2018
Strawberry	*Beauveria bassiana, Isaria fumosorosea, Metarhizium anisopliae* var. *robertsii*	*M. persicae*	effect feeding behavior, induce secondary metabolite production	Manoussopoulos et al., 2019
Rapeseed	*Metarhizium anisopliae*	*Plutella xylostella*	increase larval mortality	Manoussopoulos et al., 2019
Bell pepper	*Beauveria bassiana, Metarhizium brunneum*	*Myzus persicae, Aphidius colemani*	increased plant growth parameters, reduces reproduction and controls birth rate of aphids	Jaber & Araj, 2018

Table 4. A list of fungal biostimulants and their role in abiotic stress tolerance in crop plants

Crop Plant	Fungal Species	Environmental Stresses	Activity	References
Cucumber	*Exophiala* sp., *Paecilomyces formosus*	Salinity and drought	Promotes plant growth	Aamir et al., 2020
Barley, rice	*Piriformospora indica*	Salinity	Enhance antioxidant activity and biomass, decrease oxidative stress	Verma et al, 2022
Rice	*Fusarium culmorum*	Salinity and drought	Increase plant height, root length, biomass and osmolyte content	Verma et al, 2022

continued on following page

Table 4. Continued

Crop Plant	Fungal Species	Environmental Stresses	Activity	References
Cucumber, tomato	*Trichoderma harzianum*	Salinity and Cold	Increase content of metabolites like soluble sugar, protein, enhance nutrient uptake, root growth and proline accumulation, reduce electrolyte leakage	Toppo et al.,2023
Cowpea	*Sordariomycetes* sp., *Melanconiella elegans*	Salinity	Enhance growth, photosynthetic rate and nutrition status	Toppo et al.,2023
Wild barley	*Epichloe bromicola*	Salinity	Improve seed germination, enhance polyamines like putrescine, spermidine and spermine	Retif et al., 2023
Tomato	*Alternaria* spp., and *Trichoderma harzianum*	Salinity and drought	increase photosynthesis, plant biomass, water-use efficacy and lower ROS content	Toppo et al.,2023
Orange	*F.mosseae, Paraglomus occultum, Glomus versiforme*	Salinity and drought	increased shoot length, stem size, shoot and root biomass, low Na and Cl, high K-Na ratio, increased leaf mineral content	Rouphael et al., 2015
Olive, pistachio	*F. mosseae, R. intraradices, Claroideoglomus claroideum*	Salinity and drought	increased shoot and root total weight, K uptake, increase in leaf water content	Rouphael et al., 2015
Lettuce, tomato	*R. irregularis, G.iranicum, Tenuihypharum* sp., *R. intraradices*	Salinity	Control phytohormone production, photosynthesis and stomatal conductance	Rouphael et al., 2015
Tomato	*Pyrrhopappus carolinianus, Acer negundo, N. haematococca*	Salinity and drought	enhance yield, fruit production, leaf area and number of leaflet per leaf	Toppo et al.,2023
Maize	*P.indica, Trichoderma atroviride*	Drought	enhance proline and drought related genes, prevent membrane damage, decrease MDA content	Toppo et al.,2023
Finger millet	*Rhizophagus intraradices, P. indica*	Drought	improves antioxidant system, chlorophyll content, and osmoregulation	Toppo et al.,2023
Bell pepper, melon	*G. Diaphanum, G. claroides, G. albidium, R. intraradices*	Drought	Maintain water potential and antioxidant activity, Increase net rate of photosynthesis and efficiency in use of nutrient	Rai et al., 2021
Cucumber	*Thermomyces* sp	Heat	Improvement in efficiency of PS II (photosystem II), Increase rate of photosynthesis, and efficiency of water use.	Morales-Vargas et al., 2024
Soybean, sunflower, rice	*Aspergillus niger, A.japonica, Rhizopus oryzae*	Heat	Increase height of the plant, total weight, chlorophyll content, lipid peroxidation and ROS content reduction	Morales-Vargas et al., 2024
Soybean	*Paecilomyces formosus, Penicillium funiculosum*	Heat and drought	Improvement in height of plant, biomass, chlorophyll content; decrease electrolyte leakage	Morales-Vargas et al., 2024
Barley, banana, grapevine	*P. indica*	cold	reduce MDA and H202 content, increase antioxidant enzyme, proline and soluble sugar	Morales-Vargas et al., 2024

continued on following page

Table 4. Continued

Crop Plant	Fungal Species	Environmental Stresses	Activity	References
Lettuce, chard, and spinach	*Penicillium fuscoglaucum, P. glabrum*	cold	increase nutritional status and antioxidant enzyme	Morales-Vargas et al., 2024
Wheat	*penicillium* spp	Heat, drought and heavy metal	improve seed germination and development, increase IAA content	Retif et al., 2023
Mexican tea	*Epicoccum nigrum*	heavy metal	increase seed production, IAA, GA and JA level	Rétif et al., 2023
Corn	*Paraphaeosphaeria* sp	heavy metal	promote HM detoxification and plant growth	Toppo et al.,2023
Soyabean	*P. funiculosum*	heavy metal	phytohormone regulation, downregulate stress related genes	Toppo et al.,2023
Tomato, Rice	*Aspergillus flavus, Aspergillus niger*	heavy metal	induce metal chelation, stress responsive gene and secondary metabolite production	Toppo et al.,2023
Celery	*G. macrocarpum*	heavy metal	improved nutrient uptake and chlorophyll content, enhanced photosynthetic rate and biomass	Rouphael et al., 2015
Cucumber	*F. mosseae*	heavy metal	translocation of Cadmium from root to shoot is inhibited	Rouphael et al., 2015

Figure 1. Mechanism of plant growth promotion by fungal endophytes. This figure illustrates how fungal biostimulants are involved in both direct and indirect growth of plants through multiple vital mechanisms such as phytohormone production, phosphorus solubilization, siderophore and secondary metabolite production etc.

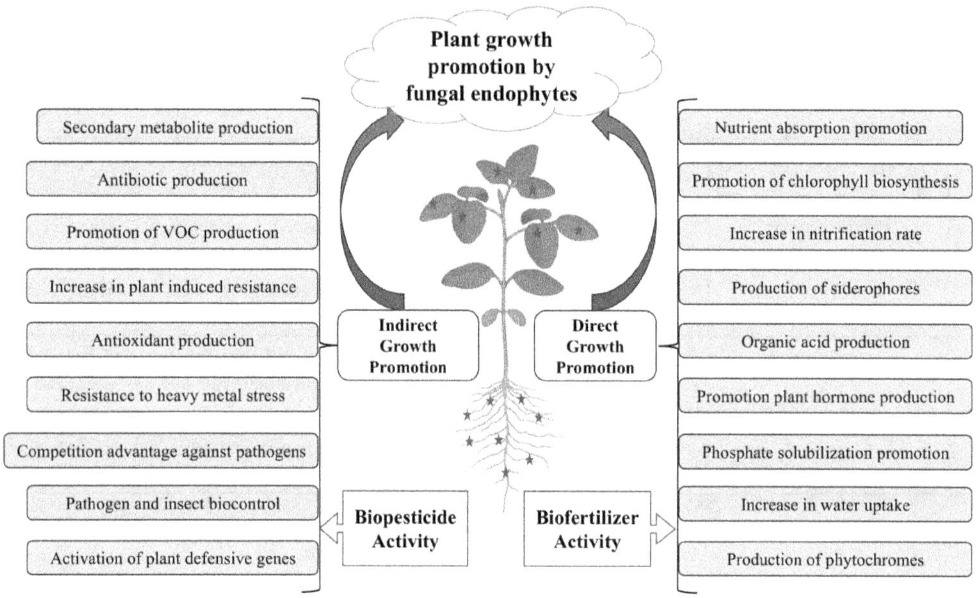

Figure 2. Role of fungal biostimulants under different environmental stress conditions. This figure demonstrates how fungal biostimulants overcome both biotic stress (bacteria, virus, fungi, pathogen, insects) and abiotic stress (heat, cold, drought, heavy metal) by inducing defense response genes, scavenging ROS and maintaining nutrient balance in the host plants.

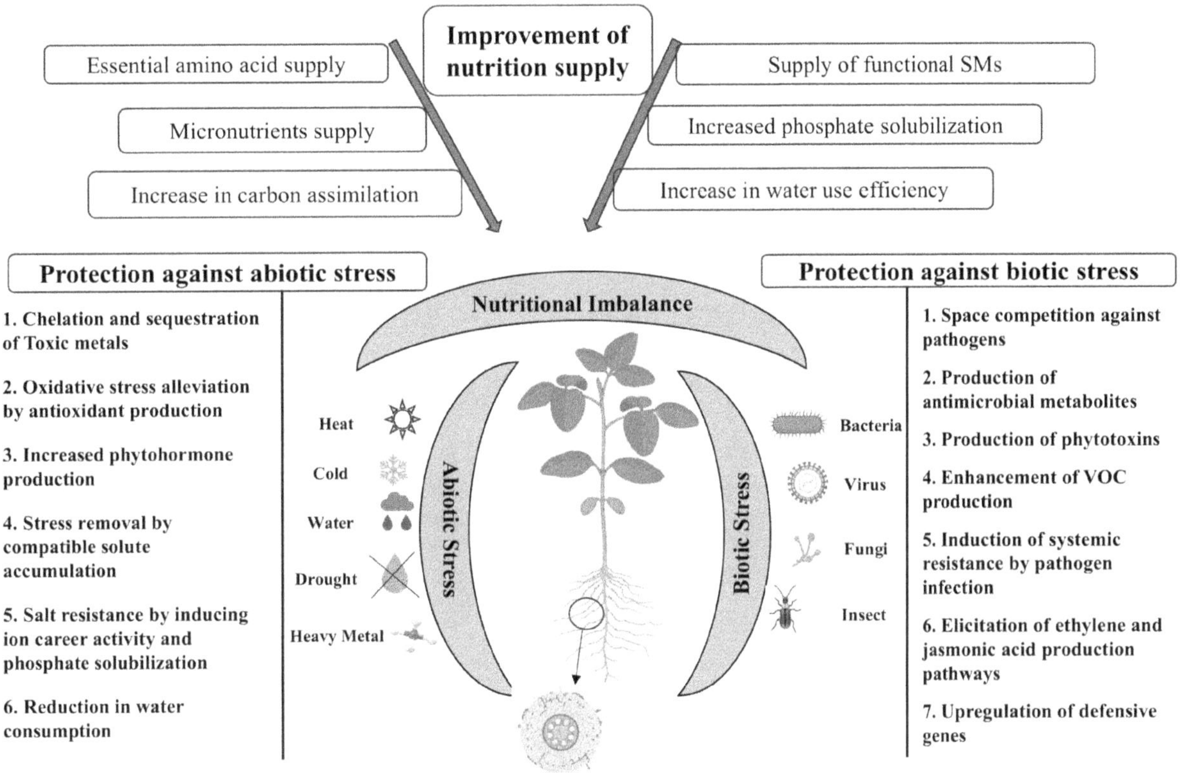

Chapter 6
Enhancing Soil Fertility and Nutrient Cycling:
Mycofertilizers

Khushboo Kumari
https://orcid.org/0009-0002-1494-0303
Mahatma Gandhi Central University, India

Khirabdhi Tanaya
https://orcid.org/0009-0005-6875-2035
Mahatma Gandhi Central University, India

Durgeshwer Singh
Mahatma Gandhi Central University, India

ABSTRACT

Soil fertility is essential for plant growth, but nutrient immobilization can lead to deficiencies. Although chemical fertilizers can address these issues, they pose environmental risks. Myco-fertilizers form symbiotic relationships with plants, extending the root network and enhancing nutrient absorption. They improve soil structure, nutrient cycling and enhance soil resilience against pathogens. Research indicates that myco-fertilizers can significantly reduce agriculture's environmental impact by limiting chemical runoff and soil degradation. This book chapter explores the use of myco-fertilizers in agriculture, focusing on their role in nutrient cycling. It also emphasizes how myco-fertilizers can reduce the need for chemical fertilizers and improve soil fertility. Incorporating myco-fertilizers into farming systems offers a path towards more eco-friendly agricultural practices, ensuring food security. The adoption of myco-fertilizers is a key element in sustainable agriculture, contributing to overall soil health and environmental sustainability.

INTRODUCTION

Soil functions as an essential repository for water and vital nutrients, playing a key role in the growth and development of plants. While soil inherently contains macronutrients Potassium (K), Nitrogen (N), Phosphorus (P), and micronutrients that are required for the growth of plant and soil biota, these elements

DOI: 10.4018/979-8-3693-8201-1.ch006

can become immobilized over time, leading to deficiencies that hinder plant absorption. These inaccessible nutrient complexes can be rendered to make them bioavailable through various processes, including weathering (influenced by biological, climatic, geological, and topographical factors), organic matter decomposition, and microbial activities. Globally, chemical fertilizers have been employed to address nutrient deficiencies. However, their use is associated with numerous drawbacks, including soil fertility degradation, high economic costs, eutrophication of water bodies, and environmental risks (Bhardwaj et al., 2014). In response to these challenges, focus has been increased on cultivating nutrient-dense, high-quality crops using sustainable practices that prioritize environmental safety.

The use of biofertilizers is a noteworthy alternative to harmful synthetic fertilizers. They are cost-effective, improve soil health, are environment-friendly, non-hazardous, and have role in crop yield enhancement (Iqbal et al., 2023). The extensive utilization of cyanobacteria, mycorrhizal fungi, root-associated beneficial bacteria, and other favorable microflora helps with enhanced plant growth by absorption of sufficient amount of water and nutrients by interacting with plant roots. The myco-fertilizers contain beneficial fungi that convert insoluble cation complexes of soil into plant-accessible form. These fertilizers are formulated to integrate mycorrhizal fungi into the soil, establishing a mutually beneficial relationship with the plant roots. This association with fungi increases surface area by extending the roots and fungal hyphae, which enhances nutrient uptake, allowing plants to uptake essential nutrients and water more efficiently. The utilization of myco-fertilizers in agricultural practices was largely overshadowed by chemical fertilizers because of their immediate and observable impact on crop productivity (Smith and Read, 2010). In recent years, the recognition of the hazardous effects of chemical fertilizers on the environment has sparked a renewed interest in myco-fertilizers, which are of natural origin, cost-effective, non-hazardous, and eco-friendly. Myco-fertilizers have provided significant contributions to sustainable practices by limiting the use of chemical inputs and improving nutrient uptake and soil structure (Jansa and Bukovská, 2024). Ongoing research continues to reveal the benefits of mycorrhizal fungi used as a myco-fertilizer, suggesting their utilization offers a viable alternate method to chemical fertilizers.

The present book chapter imparts a review on integrating myco-fertilizers in agriculture, emphasizing their critical roles in the improvement of soil nutrient availability, plant growth promotion, soil sustainability and the environmental protection. It also evaluates their interaction with other biofertilizers and examines how various agro-practices affect their functioning. Myco-fertilizers in farming practices are both economically and ecologically friendly. As a result, myco-fertilizers can be utilized to convert insoluble cation complexes into a soluble state, which is necessary for proper growth and metabolism of plant as well as microbes, making it an optimal choice for organic farming in sustainable agriculture. Incorporating myco-fertilizer not only increases crop yield but also lessens dependence on synthetic fertilizer, promoting diversity and fertility of soil, offering it as a viable strategy for environmentally friendly farming methods.

INFLUENCE OF CHEMICAL FERTILIZERS ON ENVIRONMENT

Chemical fertilizers have been commonly employed since the 20th century and offer quick access to nutrients for growth of the plant; they have been associated with environmental drawbacks, including degradation of soil, adverse impacts on biodiversity, and pollution of water bodies (Tilman et al., 2002). Intensive reliance on chemical fertilizers can disturb the nutrient dynamics of agricultural land and can be harmful to soil structure. Applying enormous amounts of nitrogen leads to imbalances in soil pH, reduces water retention capability, limit plant growth, and result in deficiencies of other essential macronutrients (P and K) (Baweja et al.,2020; Pahalvi et al., 2021). Chemical fertilizers remain in the soil and environment for long durations, impacting various physical and biological factors. It adversely affects soil quality, soil microbial populations, other organisms, nutrient runoff (describes eutrophication of water bodies, which limits oxygen supply resulting in harm to aquatic biota), environment, and human health (Schindler et al., 2016). As a result, these concerns have driven alternative interests, like sustainable agricultural practices such as utilization of biofertilizer (myco-fertilizers and other organic fertilizers), which are becoming more attractive to farmers and receiving increasing focus in the present scenario. Myco-fertilizers, which involve the utilization of mycorrhizal fungi, provide a sustainable farming option by improving plant absorption of nutrients and soil health.

Biofertilizer

As the global population has rapidly increased, the demand for agricultural crop production has also risen, leading to substantial utilization of synthetic fertilizers. The use of synthetic fertilizers is expensive and has decreased beneficial soil microbes and insects, lost soil nutrient content, caused excessive nutrient buildup in water bodies, harmed the environment, and increased the susceptibility of crops to disease. To overcome this alarming situation, biofertilizers has emerged as an alternate method to chemical fertilizers. Biofertilizers contain live microbes like algae (cyanobacteria), fungi (myco-fertilizers), or bacteria (Plant-growth promoting bacteria (PGPB) and Plant Growth Promoting Rhizobacteria (PGPR)). It enhances soil fertility by increasing their number, accelerating their performances, and aiding in processes like Nitrogen (N) fixation, Phosphorus (P) solubilization and mineralization, and Potassium (K) mobilization and dissolution of organic matter. Hence, it improves nutrient availability for uptake by plants (Khan et al., 2023). Biofertilizers offer a viable and eco-friendly alternative approach by improving soil health and minimizing reliance on chemical fertilizers (Bhattacharyya and Jha, 2012). The microorganisms involved as biofertilizers have produced various potent extracellular enzymes that effectively break down different organic materials. Some of the commonly reported myco-fertilizers utilized under greenhouse and field conditions are *Trichoderma* spp., *Chaetomium* spp., *Gliocladium* spp., *Penicillium* spp., *Aspergillus* spp. (Elkhateeb et al., 2023).

Mycorrhiza-based fungi, used as myco-fertilizer and combined with other pragmatic microorganisms, are gaining popularity in agricultural and life science research. The bacteria and fungi interact and play a synergistic role in ecosystems, as it is important for the biochemical cycle and can influence the health of the plants, including disease dynamics (Deveau et al., 2018). Fungi and bacteria are integral to processes of fermentation in food production contributing to preservation and flavors. Moreover, their application in biotechnology, antibiotics or other metabolites synthesis and food production accentuates their importance in various fields. They possess the capability to synthesize metabolites vital for biotechnological applications, making them invaluable to ecological balance. Previous studies have shown a positive impact of mycorrhizal fungi on crop yields, soil fertility, and resistance to pathogens. This symbiotic relationship is especially valuable for improving tolerance to plant abiotic stresses. The consortia of beneficial rhizo-bacteria and myco-fertilizer can serve as effective alternatives to chemical fertilizers in sustainable organic farming.

Myco-fertilizers

Fungi are a broad category of organisms that includes yeasts, mushrooms, and molds. Ascomycota (also known as Sac fungi), Basidiomycota (Club fungi) and Deuteromycota (Imperfect fungi) are amongst the major groups of fungi. They are ubiquitous in nature, having advantageous or detrimental consequences on the growth of plants. They rely on absorption for obtaining nutrients rather than synthesizing through photosynthesis due to lack of chlorophyll. Fungi act as decomposers (saprotrophic), forming symbiotic relationships with plants (mycorrhizal), living within host tissues (endophytic), and acting as pathogens (parasitic). The study of fungi has revealed their important contribution to nutrient cycling, as well as their role in decomposing organic substances. The use of fungi on agricultural land is now being acknowledged as a way for the improvement of soil quality and enhancement of plant growth. The majority of fungi, such as *Aspergillus*, *Penicillium*, etc., have been observed as significant organisms for solubilizing insoluble phosphorus, particularly in agricultural scenarios (Bagyaraj and Revanna, 2017). Various fungi are reported for their impact on plant growth by performing diverse mechanisms (Table 1). These fungi synthesize organic acids, which lowers their surrounding pH or makes soil acidic and causes enhancement in the availability of phosphorus by breaking down mineral complexes into soluble forms. Inoculating seedlings or the plant surface with fungi, or using it directly in the soil promotes maturation and proliferation of plants (Kumar et al., 2018). Moreover, they function as useful biocontrol agents in controlling plant diseases as part of their physiological activities. For instance, *Trichoderma harzianum*, recognized as a fungal antagonist showed effectiveness against *Macrophomina phaseolina*, the causal organism of charcoal rot, in various plant species (Powlson et al., 2014) .

Table 1. Fungal contributions to plant growth enhancement through diverse mechanisms

Strain Name	Mechanisms involved	Role in plant growth	References
Alternaria alternate	Induction of defense-related genes	Induces systemic resistance	Prins et al., (2000)
Aspergillus niger	Secretes organic acids	Increases phosphate solubilization	Rudresh et al., (2005)
Beauveria bassiana	Enhances plant defense against pests	Endophytic colonization	Ownley et al., (2008)
Chaetomium globosum	Antagonistic action against pathogens	Enhances seedling vigor	Sarrocco and Vannacci, (2018)
Cladosporium cladosporioides	Secretes elicitors	Induces plant defense responses	Matsumoto et al., (2015)
Colletotrichum tofieldiae	Endophytic colonization	Promotes nutrient assimilation	Hiruma et al., (2016)
Curvularia lunata	Symbiotic relationship	Increases drought tolerance	Redman et al., (2001)
Fusarium oxysporum	Symbiotic relationship and production of osmo-protectants	Enhances drought tolerance	Al-Karaki et al., (2004)
Funneliformis mosseae, and *Rhizophagus intraradices*	Symbiotic association with host plant	Promotes plant growth, yield and P uptake efficiency	Adeyemi et al., (2021)
Glomus intraradices	Mycorrhizal association	Improves phosphorus uptake	Smith and Read, (2010)

continued on following page

Table 1. Continued

Strain Name	Mechanisms involved	Role in plant growth	References
Laccaria bicolor	Release of polygalacturonase enzyme, ectomycorrhizal association	Improves nutrient assimilation and promotes plant growth	Zhang et al., (2022)
Mortierella hyaline	Symbiosis with roots	Increases nitrogen uptake	Zhang et al., (2018)
Myrothecium verrucaria	Production of growth-promoting enzymes	Promotes root elongation	Khan et al., (2014)
Penicillium bilaii	Phosphatase enzyme production	Improves phosphorus availability	Whitelaw, (2000)
Phoma spp.	Endophytic interaction	Improves root colonization	Khan et al., (2012)
Piriformospora indica	Maintenance and modulation of defense enzyme activity, hormone metabolism pathways	Enhances plant growth under stress conditions	Li et al., (2023)
Pleurotus ostreatus	Lignin peroxidase secretion	Enhances decomposition of lignin	Das et al., (2020)
Rhizopus stolonifera	Decomposition of organic matter	Increases soil health	Zhao et al., (2015)
Rhizoctonia solani	Pathogen *antagonism*	Improves nutrient absorption	Gonzalez et al., (2021)
Serendipita indica	Enhances root surface area	Improves salinity tolerance	Franken, (2012)
Talaromyces flavus	Production of enzymes	Pathogen antagonism	Duo-Chuan et al., (2005)
Trichoderma harzianum	Induced systemic resistance, nutrient solubilization	Enhances root growth and plant resistance to phytopathogens	Harman et al., (2004)
Verticillium lecanii	Endophytic colonization and synthesis of degrading enzymes	Promotes plant defense against pathogens	Hasan et al., (2013)

Our dependence on agriculture is profound, as it provides the essential basic needs of food, clothing, and shelter. In other words, agriculture is considered the practice of plant cultivation and management of natural resources to synthesize food, fiber, and other essential products. Agriculture and myco-fertilizers are closely connected through the role that fungi play in soil health and plant growth promotion. In 1885, The term mycorrhiza was introduced by a forest pathologist, Albert Bernhard Frank. Mycorrhizal fungi, diverse groups of biological organisms found in soil interact with a wide range of plants, including those in grasslands, croplands, tropical forests, and alpine. Mycorrhizal fungi have been applied as biofertilizers since the late 1950s (Domínguez-Núñez et al., 2019). They are among the most prevalent living organisms in soil, significantly contributing to the proliferation of plants, enhancing plant stress tolerance, sequestering carbon from plants, cleaning up soil contaminants, and supporting food security and agricultural sustainability. The extended root systems (hyphae), of mycorrhizal fungi play a critical role in soil health by forming soil particles aggregates. This binding effect improves aggregation of soil, resulting in more stable and organized soil framework. As a result, soil structure is improved, which facilitates better aeration and enhanced water absorption and retention capacity. Improved soil structure lowers erosion risk by stabilization of soil, creating a healthy environment for root proliferation (Powlson et al., 2011) Mycorrhizae secretes organic acids, glomalin-related soil proteins (GRSP), and polyamines that can serve as adhesion agents, promoting the cohesion of soil particles. This process improves aeration, and aggregation of soil, promotes water stability, and improves soil pH. Furthermore, these interactions can enhance the soil redox potential, contributing to optimal growth of plants and strengthening resistance to diseases (Weng et al., 2022). These fungi are linked to a wide range of agricultural species, with the exception of those from the families *Juncaceae*, *Cyperaceae*, *Amaranthaceae*, *Commelinaceae*,

Brassicaceae, *Polygonaceae Caryophyllaceae*, and *Chenopodiaceae* (Mishra et al., 2013). Mycorrhizal interaction with plants improves the absorption of nutrients including N, P, Iron (Fe), Copper (Cu), Sulfur (S) Zinc (Zn) and Boron (B). Xerophytes, trees, shrubs, hydrophytes, herbs, and epiphytes are various plant types that form symbiotic relationships with mycorrhizal species (El-Ghwas et al., 2023). *Serendipita indica,* an endophytic fungus serves beneficial roles in modulating immune responses, providing protection against diverse kinds of biotic and abiotic challenges imposed on the host plants, reducing stress levels, enhancing growth in plants, and hence, acting as a biofertilizer. A commercial formulation designed for agricultural use, called "Rootonic," has been developed to make it easier for farmers to handle and store. This formulation incorporates the mycelium of *S. indica* with magnesium sulfate as a carrier. In addition, a new tool called Nanoembedded, *S. indica* has been developed by incorporating carbon nanotubes, zinc oxide, and titanium dioxide (Jangir et al., 2021).

Inoculation of mycorrhiza changes the roots' framework, and research indicates that the capability of inoculated roots for nutrient absorption is significantly higher than that of non-inoculated ones. In return for nutrients, mycorrhizal fungi receive approximately 20% of photosynthesized carbon from the plant. These photosynthesized carbon compounds are essential for mycorrhizal fungi's growth and metabolism, as the life cycle of majority of these fungi cannot complete without them. It is more complex than a simple nutrient trade, involving biochemical signals and biochemical pathways that regulate the mutualistic interaction (Bücking et al., 2012). Synthesis and exudation of root secretions and plant growth regulators from mycorrhiza also impact the proper growth of plants. Based on their mycorrhizal interactions with roots of the plant and ecological roles various types of mycorrhizal fungi are found namely, ericoid mycorrhiza fungi (ERM), ectomycorrhizal (ECM) fungi, arbuscular mycorrhizal fungi (AMF), orchid mycorrhizal (OM) fungi, etc. (Pandey et al., 2019).

Ectomycorrhizal Fungi (ECM)

ECM, a higher class of fungi, also known as macro-mycetes forms a symbiotic relationship with higher plant roots. ECM is predominantly associated with woody plants and gymnosperms (Owiny and Dusengemungu, 2024). It is beneficial for soil aggregation-stabilization and forest ecosystem restoration. They are considered inexpensive, eco-friendly biofertilizers which are utilized as biotechnological tools in agricultural practices. It is beneficial in bioremediation, manages forest ecosystems, and acts as an inhibitor for pathogenic soil microbes. ECM are capable of converting insoluble compounds into simpler forms or plants' available forms. ECM establishes a mutual symbiotic association with host roots, resulting in structural and biochemical changes in both plants and fungi. They surround the root tips with a defensive sheath by penetrating the roots intercellularly into the root cortex via secreting enzymes which could degrade cell wall (Zhang et al., 2022).

In the process of root colonization, ECM hyphae firstly helps create a pseudoparenchymatous mantle that covers root apices and later, the hyphae form an intraradical, apoplastic hyphal network named Hartig net intercellularly in plants roots and extend into the soil. The Hartig net maturation starts from the mantle sheath's innermost layer, with finger like hyphae expanding in a lobed hyphal front known as labyrinthine hyphal structure (essential for functional ectomycorrhiza). In the Hartig net, at symbiotic interface the metabolites are bidirectionally translocated between the symbionts, facilitated by the fusion of fungal and plant cell wall polysaccharide and proteins (Balestrini and Kottke, 2016). LbGh28A- symbiotic induced pectinase helps in formation of the Hartig net and is a key factor in achieving successful colonization (Zhang et al., 2022).

ECM fungi form mutually beneficial partnerships with plants by enhancing nutrient uptake, especially phosphorus and nitrogen, improving water absorption and drought resistance, providing protection against soil pathogens, and supporting seedling establishment and survival. Improvement in soil structure, plant nutrient uptake, enhancement of edible fruit biomass, and stress tolerance are the reported significant characteristics of ECM fungi. These fungi are associated with various trees and shrubs in forest ecosystems, which is vital for the growth and health of forest plants. ECM can significantly contribute to phytoremediation, aiding in the restoration of contaminated soil and improving the establishment of plants in revegetation initiatives (Bücking et al., 2012). By fostering a healthier ecosystem and supporting plant resilience, these fungi are crucial for sustaining the balance and integrity of forest habitats. According to Smith and Read (2010), in soils with limited nutrient content, ECM maximizes the surface area for absorption, crucial for phosphorus uptake by extending the root system through hyphal networks. ECM fungi, such as *Suillus luteus* and *Hebeloma crustuliniforme* play an essential role in promoting the growth of commercially valuable conifers like *Picea spp.* (spruce) and *Abies spp.* (fir) (Smith and Read 2008). Menkis et al. (2007) demonstrated that ECM fungi, specifically *Cenococcum geophilum, Piceirhiza bicolorata*, and *Hebeloma crustuliniforme*, were used to inoculate seedlings of *Pinus sylvestris* L. and *Picea abies* (L.) planted on abandoned agricultural land. Over a three-year period, the inoculated seedlings exhibited higher survival rates and better growth. This demonstrates the practical benefits of using mycorrhizal fungi in reforestation projects. A field experiment conducted in copper mine tailings investigated the effects of ectomycorrhizal fungi (*Pisolithus sp., Cenococcum geophilum, Laccaria laccata*) on seedlings of *Pinus densiflora* and *Quercus variabilis*. Inoculation with ectomycorrhizal fungi significantly improved the survival, growth, and nutrient uptake of pine seedlings, while also reducing the translocation of heavy metals to the shoots. In contrast, oak seedlings exhibited only an increase in biomass. The study demonstrates that the inoculation of ectomycorrhizal fungi can greatly enhance seedling performance in mine waste areas, highlighting its potential for reforesting degraded lands (Zong et al., 2015).

Arbuscular mycorrhizal fungi (AMF)

Arbuscular mycorrhizal fungi, formerly known as vesicular arbuscular mycorrhizal fungi (VAM fungi), are incredibly prevalent, occurring in about 80-90% of plant species globally. They thrive in various terrestrial ecosystems, including associations with metal-polluted soil and aquatic plants. This extensive prevalence highlights their critical role in agricultural practices and ecological systems (Owiny and Dusengemungu, 2024). AMF improves water absorption, boosts nutrient utilization, enhances growth of plants and increases resilience to environmental stresses. The host plant in return provides fixed carbon source to the fungus. AMF produces two specialized structures that are finely branched hyphal tip arbuscules, which contributes in nutrients exchange, and the structure that resides within the root cortical cells of the host, forming balloon-like vesicles for the storage of nutrients (Pepe et al., 2016). AMF associations help obtain sufficient phosphorus (P) from the soil, particularly in P-deficient conditions. Indeed, this symbiotic associations significantly increase plant maturation and P content. AMF produces acids, which makes their surroundings acidic and enhances phosphorus availability for uptake by plants along with root surface area increment. Plants inoculated with combination of PGPR and AMF can exhibit a synergistic effect, to augment growth indices of plant while maintaining natural resources, including P stocks. Besides this, alkaline phosphatase (ALP) has been identified as one of the primary enzymes involved in this process, with its activity being analyzed through ultra-cytochemistry and gel electrophoresis. The localization of ALP in the well-developed arbuscules of AMF within the vacuole points to its role in catalyzing P uptake in plants. In a pot experiment, inoculation of *Glomus mosseae*, *G. deserticola* and *Gigaspora gergaria* (AMF) on wheat plant (*Triticum aestivium* L. cv. Henta), were evaluated. Inoculation of AMF highly enhanced the growth of plants, activities of phosphatases, nutrient uptake, total soluble protein, and proline content as compared to control (Abdel-Fattah and Asrar, 2012). Arbuscular mycorrhizal fungi have gained recognition as an effective biological control method for a number of fungi that prove to be phytopathogenic. Schonbeck and Dehne (1977) showed that cotton plants that had been mycorrhized showed more resistance to infection of the phytopathogen *Thielaviopsis basicola*. This highlights the potential of AMF as a complement to defensive mechanisms in the plant in relation to overall resistance to certain pathogens. According to Lin et al. (2021), AMF inoculation enhances the growth of *Musa acuminate* Cavendish cv. Brail (seedlings of Brazilian banana) and activate defense-related genes that may help to inhibit wilt disease. A study evaluated three strains of AMF *Funneliformis mosseae*, *Rhizophagus irregularis*, and *Claroideoglomus etunicatum* on cotton under saline-alkali stress. The inoculation of AMF improved chlorophyll levels, photosynthesis, plant growth, and the salt tolerance of the cotton plants (Peng et al., 2024). A study conducted by Navarro and Morte (2024) evaluated the effectiveness of two AMF *Rhizophagus irregularis* and *Funneliformis mosseae* as biofertilizers for *Citrus aurantium* seedlings in south-eastern Spain. Both AMF species significantly enhanced the growth, phosphorus nutrition, and chlorophyll levels of the seedlings. These findings emphasize the potential of AMF inoculation to reduce the reliance on chemical fertilizers, thereby promoting sustainable agriculture in regions vulnerable to climate change.

ROLES OF MYCO-FERTILIZERS IN NUTRIENT AND WATER ABSORPTION

The maturation of plants depends on nutrients and water availability in the surrounding soil. Critical elements such as nitrogen, phosphorus, and potassium are required for growth of the plant and can be influenced by factors such as soil pH, moisture content, soil texture, presence of soil microbes, organic matter content, and soil fertility. The nutrient and water absorption mechanism performed by mycorrhizal fungi involves mainly two methods that facilitates nutrient and water transport from soil to plant. The plant root releases root exudates, which is beneficial for hyphal morphogenesis of germinated arbuscular mycorrhiza in asymbiotic phase (Coelho et al., 2019). The hyphae grow as germling hyphae into long and branched structures, and upon contact with the host roots, they penetrate the intercellular spaces with appressoria, which are characteristic penetration structures that develop arbuscules in the root cortex (Giovannini et al., 2020). The developing extracellular hyphal network enhances the rate of absorption by increasing the surface to volume ratio, which highly improves the uptake and diffusion of water and nutrients such as phosphorus and nitrogen (Baum et al., 2015). It is facilitated by the activity of nutrient transporter genes within the hyphae. Furthermore, the hyphae that emanate from the roots may be able to uptake nutrients and water from the finer crevices in the soil that the roots cannot reach (Püschel et al., 2020). The high order of mycelial network ensures absorption at a much higher surface area than the root system and also facilitates retention of moisture in the soil (Augé et al., 2015). Absorption plays a very significant role in this mutualistic relationship as the mycelium further exhausts nutrients in the vicinity. This can be achieved through the absorption network set by the mycelia in other areas spread across the depleted zone (Johri et al., 2015; Kobae, 2019). In addition, AMF involved in phosphate solubilization. They releases low-molecular weight organic acids, inorganic acids, alkaline phosphatases, and organic acids to mineralize and release nutrients through ligand exchange mechanisms (Sato et al., 2015; Andrino et al., 2021). The physiological interaction between the plant and the fungus determines the efficiency of phosphorus uptake. Loading and unloading of nutrient transport by the extracellular hyphae and cortical arbuscules determine translocation. The amount of nutrients needed by the plant determines the overall uptake process, affected by the dynamic of particular species of AMF and environmental conditions (Cao et al., 2020). AMF plays a very significant role in phosphate assimilation from poor soils where it mobilizes phosphorus stored in root phosphate (Etesami, 2020). There is evidence that the nitrogen from organic matter is transferred by AMF hyphae into the plant tissues which enhances the biomass of the plants (Thirkell et al., 2017). Apart from that, the AM hyphal network can favor the uptake of potassium and other micronutrients such as magnesium, zinc, copper, calcium, sulfur, sodium, manganese, boron, molybdenum, and iron, among others, which are essential for the growth of a plant (Hajiboland et al., 2015). The function of AMF in nutrient cycling is of prime importance to ensure a continued supply of nutrients in nutrient-scarce or low-fertility soils (Johnson et al., 2015). Myc factors (similar to Nod factors, which are involved in root nodulation) from arbuscular mycorrhizal (AM) fungi, mediate the expression of genes related to plant symbiosis and make ready the root for arbuscular mycorrhizal symbiosis. Myc factors, particularly lipo-chito-oligosaccharides, induce root responses to prepare for AM association and have a related structure to Nod factors, suggesting a shared evolutionary origin. AM and rhizobial symbioses also share the same signal transduction pathway, the common symbiosis pathway, which involves seven key SYM genes, which are essential for both types of root symbiosis (Bücking et al., 2012).

Role in Enhancement of Soil Structure and Organic Matter

Soil structure is the configuration of soil particles along with the gaps in between, which play a key role in the health and functioning of soil ecosystem. It significantly influences various essential soil properties, including water retention, aeration and availability of nutrients. In order to support agricultural productivity, well-structured soil encourages growth of roots, microbes functioning enhancement and proper circulation of air and water. Organic matter is a crucial nutrient source that is essential for shaping the structure of soil. Soil structure is influenced by the decomposition of organic matter. Presence of organic matter in soil promotes plant growth and the functioning of the ecosystem. The organic matter that is comprised of decayed microbes, plant and animal parts, provides soil additional nutrients enriches its structure, and improves water retention. Myco-fertilizers, which are enriched with beneficial mycorrhizal fungi improves soil health by enhancing structure of soil and organic matter content. Symbiotic relationships are established in mycorrhizal fungi with roots of higher plants, improving water and nutrient uptake. Mycorrhiza colonizes around 70-80% of plants root across various species (Owiny and Dusengemungu, 2024). For improvement of soil aeration and retention of water ability, myco-fertilizers help in soil aggregation and facilitate better growth of roots. Additionally, myco-fertilizers aid in the formation of organic matter by decomposing residues of plants and promoting buildup of humus. This process improves soil quality and enhances fertility of soil.

Role in Disease Resistance

Myco-fertilizers that harbor beneficial mycorrhizal fungi enhances resistance against diseases among plants. The mycorrhizal fungi live symbiotically with the plant roots, that leads to a variety of positive impacts such as boost in the Plant's immunity and health. The mechanism of myco-fertilizers, involves disease resistance, promotes root health, improvement in nutrient uptake, induction of systemic resistance, competition with surrounding pathogenic microflora, making plant environmental stress tolerant and improve soil health (Brar et al., 2024). A robust system of roots which helps in nutrient uptake also enables enhancement of plant resistance against diseases (Ghorbanpour et al., 2018). In general, nutrient-balanced and healthy plants are considered more resilient to stress and pathogen. Mycorrhizal fungi have the potency to trigger plant systemic resistance. When these fungi colonize plant roots, they can stimulate the production of defensive compounds namely, pathogenesis related proteins, phytoalexins, etc., which enhance resistance against pathogens. They can also outcompete advantageously with harmful pathogens for space and resources in the soil. They create a barrier that limits pathogens from successful invasion and formation of specialized structures that may prevent maturation and multiplication of disease-causing organisms, by occupying the root zone (Larbi-Koranteng et al., 2022). Mycorrhizal fungi provide plant, skill of enhanced tolerant from environmental stresses. This tolerance can indirectly enhance the capability of plants to resist diseases because plants that are under stress are more susceptible to infections (El-Saadony et al., 2022). These fungi also contribute to enhance structure of soil and diversity of microflora and hence, supports a good ecological balance that can suppress the growth of harmful pathogens.

Role in environmental stresses

In general, plants with AMF tend to have an improved environmental stresses tolerance such as soil stress from water and salt levels in the medium and are generally healthier and harder (Wahab et al., 2023). For instance, Chandrasekaran, (2022), revealed that under varying degree of drought stress, inoculating plants with AMF significantly impacts on root morphology, enhanced plant biomass and nutrient absorption. Inoculation of myco-fertilizers would stimulate nutrient uptake, resulting in reduced chemical fertilizer use, which in turn leads to cost reductions for those fertilizers while at the same time causing a reduction in the harms to the environment. It would stimulate nutrient uptake, resulting in reduced chemical fertilizer use, which in turn leads to cost reductions for those fertilizers while at the same time causing a reduction in the harm to the environment. Myco-fertilizers enhance nutrient assimilation; they improve the health of the soils, and they are suited to sustainable agricultural practice because of harnessing the natural benefits associated with AMF.

MYCO-FERTILIZERS IN SUSTAINABLE AGRICULTURE

Myco-fertilizers has been popular due to its harmonious relationship between agricultural methods and ecological benefits offered by mycorrhizal fungi for the integration of mycorrhizae into sustainable and organic farming. Myco-fertilizers, containing beneficial mycorrhizal fungi, are crucial for sustainable farming. Methods such as crop rotation, intercropping, and minimum tillage promote the growth of mycorrhiza. Crop rotation is for the prevention of pest and disease buildup, nutrient demand diversity, and improvement of the mycorrhizal networks in the soil. In intercropping, several species of plant are farmed together, which helps facilitate mycorrhizal relationships due to the availability of different exudates from roots. Another good practice that will ensure healthy populations of mycorrhizal includes cover cropping. Seeding of crops that are facilitative for mycorrhizal fungi will ensure that these beneficial fungi can grow in the soil with health, hence improving crop success. Apart from supporting fungal health, utilization of cover cropping promotes other forms of improvement in structure and fertility of soils. Organic additions, such as composts and green manures, also represent a food source rich in nutrients for mycorrhizal fungi as well as other microorganisms of similar nature in the soil. Increased organic inputs improve soil health and fertility, therefore establishing a healthy ecosystem that fosters plant growth. Organic matter improves the texture of the soil and increases its moisture retention capacity; hence, it indirectly strengthens mycorrhizal activity. Mycorrhizae improves soil health and aids in the plant growth through various essential functions. Such as enhanced nutrient and water uptake, improvement in soil structure, enhancement of soil organic matter content, resistance from disease and pest and drought tolerance. Myco-fertilizers enhance nutrient and water absorption, especially P, by extending root systems with fungal hyphae that access nutrients inaccessible to roots of plants alone. This can limit the dependency on synthetic fertilizers, promoting more environmental-friendly method of nutrient management. *Aspergillus niger* and *Penicillium chrysogenum*, were reported as best phosphate solubilizer myco-fertilizer for growth of plant (David et al., 2023). Some studies reported that inoculated plants with mycorrhizal fungi acquired higher phosphorus uptake than the non-inoculated plants. Smith et al. (2020) report that maize plants treated with myco-fertilizers contain up to 50% more phosphorus uptake as compared to others who did not show much development and yield. Myco-fertilizers contribute positively towards soil health by improving the soil structure and promoting biodiversity. Hyphae of mycorrhizal fungi join soil particles into clusters, which create porosity in soil, thus enhancing water retention, better aeration and reduced soil erosion, which are essential for eco-friendly land utilization (Khanday et al., 2016). These are preferred conditions for growth by roots and microbes. Moreover, through their interaction with many other microorganisms of the soil ecosystem, the network formed by mycorrhizal fungi allows nutrient cycling. According to Wei et al., (2024), myco-fertilizers improved soil organic matter content and enhanced microbial diversity, thus enhancing soil fertility.

Myco-fertilizers also act as an enhancer that helps tolerate the environmental stress in plants. It has been found that mycorrhizal fungi improve drought and salinity tolerance. For example, Zhao et al., (2021), found that myco-fertilized plants presented higher drought resistance and enhancement in water use efficiency compared to the non-treated ones. Resilience goes a long way in this context of climate change, where unpredictable patterns of weather threaten agricultural productivity hugely. There is sustainability in the use of myco-fertilizers since they assist in reducing the application rate of chemical fertilizers and prevent negative environmental impacts caused by chemical fertilizers. There is a possibility of easily overcoming the various disadvantages posed by chemical fertilizers, which include high quantities of ammonia that cause degradation in soils, polluted water, and loss of biodiversity. Various studies being carried out, revealed that the utilization of myco-fertilizers gave the opportunity to minimize the use of synthetic fertilizers, thereby saving farmers usage costs and improving soil health conditions. Generally, the reason why mycorrhizae are included in sustainable and organic agriculture systems is because this is definitely the epitome of the holistic nature of agriculture. Through these fungi, ecological benefits are being drawn upon by farmers to conduct healthier soil farming, enhance crop resilience, and diminish reliance on chemical inputs, all the while maintaining a sustainable agriculture system.

CONCLUSIONS AND FUTURE PROSPECTS

In conclusion, myco-fertilizers have emerged as a significant innovation in sustainable agriculture, offering valuable advantages for improving nutrient cycling and soil fertility. This chapter has explored the different mechanisms in which mycorrhizal fungi contribute to agricultural yield by enhancement of nutrient uptake, inducing systemic acquired resistance activity, promoting soil health and ultimately led to protection of environmental health. These capability of mycorrhizal fungi highlights their role as a valuable tool in contemporary farming methods. Mycorrhizal fungi facilitate nutrients uptake, particularly phosphorus. These fungi enable access to essential nutrients from inaccessible compounds more effectively by solubilization and mineralization processes, resulting in proper growth of plants and higher crop yields. Furthermore, improves soil structure and soil organic matter content, which in turn promotes better water retention and aeration. The environmental benefits of myco-fertilizers are significant, making them attractive alternate approach to traditional chemical fertilizer. Unlike chemical fertilizers, which is detrimental to soil and environment, myco-fertilizers promote a more balanced and resilient soil-environment system. They limit the need of chemical inputs by adding myco-fertilizers to soil by improving the nutrient availability naturally, thus mitigating the harmful impacts of excessive utilization of chemical fertilizers. Furthermore, these fertilizers can enhance plant resistance to various pathogens and stress conditions. Recent research has advanced our understanding of the mechanisms through which myco-fertilizers influence fertility of soil and plant health. Studies have also revealed their production and secretion of bioactive compounds needed for growth of plant and resistance against soil-borne pathogens. As research continues, the utilization pf myco-fertilizers is expected to evolve and become more prevalent, integrating seamlessly into various agricultural practices to optimize their benefits.

In the light of the above mention insights, myco-fertilizers have the ability to hold the future for sustainable agriculture. Ongoing research and innovation in this field and the complete molecular mechanism deciphering involved in the solubilization and mineralization processes of complexes are expected to further advancements, addressing the challenges to meet the ever-growing population demands while preserving environmental health.

CONFLICTS OF INTEREST

The authors declare that they have no financial or personal relationships that could be perceived as influencing the content of this article.

ACKNOWLEDGEMENT

Authors are thankful to honorable vice-chancellor, Head, Department of Botany, Mahatma Gandhi Central University, Bihar for their support and encouragement.

REFERENCES

Abdel-Fattah, G. M., & Asrar, A.-W. A. (2012). Arbuscular mycorrhizal fungal application to improve growth and tolerance of wheat (Triticum aestivum L.) plants grown in saline soil. *Acta Physiologiae Plantarum*, *34*(1), 267–277. DOI: 10.1007/s11738-011-0825-6

Adeyemi, N. O., Atayese, M. O., Sakariyawo, O. S., Azeez, J. O., Olubode, A. A., Ridwan, M., Adebiyi, A., Oni, O., & Ibrahim, I. (2021). Influence of different arbuscular mycorrhizal fungi isolates in enhancing growth, phosphorus uptake and grain yield of soybean in a phosphorus deficient soil under field conditions. *Communications in Soil Science and Plant Analysis*, *52*(10), 1171–1183. DOI: 10.1080/00103624.2021.1879117

Al-Karaki, G. N., & Al-Raddad, A. (2004). Effects of arbuscular mycorrhizal fungi and drought stress on growth and nutrient uptake of two wheat genotypes differing in drought resistance. *Mycorrhiza*, *14*(4), 263–269. DOI: 10.1007/s00572-003-0265-2 PMID: 12942358

Andrino, A., Guggenberger, G., Kernchen, S., Mikutta, R., Sauheitl, L., & Boy, J. (2021). Production of organic acids by arbuscular mycorrhizal fungi and their contribution in the mobilization of phosphorus bound to iron oxides. *Frontiers in Plant Science*, *12*, 661842. DOI: 10.3389/fpls.2021.661842 PMID: 34335645

Augé, R. M., Toler, H. D., & Saxton, A. M. (2015). Arbuscular mycorrhizal symbiosis alters stomatal conductance of host plants more under drought than under amply watered conditions: A meta-analysis. *Mycorrhiza*, *25*(1), 13–24. DOI: 10.1007/s00572-014-0585-4 PMID: 24831020

Bagyaraj, D., & Revanna, A. (2017). *Soil Biodiversity: Role in Sustainable Horticulture* (pp. 1 18).

Balestrini, R., and Kottke, I. (2016). Structure and development of ectomycorrhizal roots. *Molecular Mycorrhizal Symbiosis*, 47–61. https://doi.org/DOI: 10.1002/9781118951446.ch4

Baum, C., El-Tohamy, W., and Gruda, N. (2015). Increasing the productivity and product quality of vegetable crops using arbuscular mycorrhizal fungi: a review. *Scientia Horticulturae, 187*, 131 141. https://doi.org/DOI: 10.1016/j.scienta.2015.03.002

Bhardwaj, D., Ansari, M. W., Sahoo, R. K., & Tuteja, N. (2014). Biofertilizers function as key player in sustainable agriculture by improving soil fertility, plant tolerance and crop productivity. *Microbial Cell Factories*, *13*(1), 66. DOI: 10.1186/1475-2859-13-66 PMID: 24885352

Bhattacharyya, P. N., & Jha, D. K. (2012). Plant growth-promoting rhizobacteria (PGPR): Emergence in agriculture. *World Journal of Microbiology & Biotechnology*, *28*(4), 1327–1350. DOI: 10.1007/s11274-011-0979-9 PMID: 22805914

Brar, B., Bala, K., Saharan, B. S., Sadh, P. K., & Duhan, J. S. (2024). Bio-boosting agriculture: Harnessing the potential of fungi-bacteria-plant synergies for crop improvement. *Discover Plants*, *1*(1), 21. DOI: 10.1007/s44372-024-00023-0

Bücking, H., Liepold, E., & Ambilwade, P. (2012). *The Role of the Mycorrhizal Symbiosis in Nutrient Uptake of Plants and the Regulatory Mechanisms Underlying These Transport Processes* (N. K. Dhal and S. C. Sahu (eds.); p. Ch. 4). IntechOpen. DOI: 10.5772/52570

Cao, Y., Wu, X., Zhukova, A., Tang, Z., Weng, Y., Li, Z., & Yang, Y. (2020). Arbuscular mycorrhizal fungi (AMF) species and abundance exhibit different effects on saline-alkaline tolerance in Leymus chinensis. *Journal of Plant Interactions*, *15*(1), 266–279. DOI: 10.1080/17429145.2020.1802524

Chandrasekaran, M. (2022). Arbuscular Mycorrhizal Fungi Mediated Enhanced Biomass, Root Morphological Traits and Nutrient Uptake under Drought Stress: A Meta-Analysis. In *Journal of Fungi* (Vol. 8, Issue 7). DOI: 10.3390/jof8070660

Coelho, L. C. S., Mignoni, D. S. B., Silva, F. S. B., & Braga, M. R. (2019). Seed exudates of Sesbania virgata (Cav.) Pers. stimulate the asymbiotic phase of the arbuscular mycorrhizal fungus Gigaspora albida Becker & Hall. *Hoehnea*, *46*(1), e272018. DOI: 10.1590/2236-8906-27/2018

Das, A., & Sakthivel, K. M. (2020). Anthraquinone from Edible Fungi Pleurotus ostreatus Protects Human SH-SY5Y Neuroblastoma Cells Against 6-Hydroxydopamine-Induced Cell Death—Preclinical Validation of Gene Knockout Possibilities of PARK7, PINK1, and SNCA1 Using CRISPR SpCas9. *Applied Biochemistry and Biotechnology*, *191*(2), 555–566. DOI: 10.1007/s12010-019-03188-7 PMID: 31820379

David, O. M., Olawusi, A. C., Oluwole, O. A., Adeola, P. O., & Odeyemi, A. T. (2023). Isolation, Molecular Characterization and Application of Aspergillus niger and Penicillium chrysogenum with Biofertilizer Potentials to Enhance Rice Growth. *Tropical Journal of Natural Product Research*, *7*(4), 2790–2795. DOI: 10.26538/tjnpr/v7i4.20

Deveau, A., Bonito, G., Uehling, J., Paoletti, M., Becker, M., Bindschedler, S., Hacquard, S., Herve, V., Labbé, J., Lastovetsky, O. A., Mieszkin, S., Millet, L. J., Vajna, B., Junier, P., Bonfante, P., Krom, B. P., Olsson, S., van Elsas, J. D., & Wick, L. Y. (2018). Bacterial–fungal interactions: Ecology, mechanisms and challenges. *FEMS Microbiology Reviews*, *42*(3), 335–352. DOI: 10.1093/femsre/fuy008 PMID: 29471481

Domínguez-Núñez, J. A., Berrocal-Lobo, M., & Albanesi, A. S. (2019). *Ectomycorrhizal Fungi: Role as Biofertilizers in Forestry BT - Biofertilizers for Sustainable Agriculture and Environment* (B. Giri, R. Prasad, Q.-S. Wu, & A. Varma (eds.); pp. 67–82). Springer International Publishing. DOI: 10.1007/978-3-030-18933-4_4

Duo-Chuan, L. I., Chen, S., & Jing, L. U. (2005). Purification and partial characterization of two chitinases from the mycoparasitic fungus Talaromyces flavus. *Mycopathologia*, *159*(2), 223–229. DOI: 10.1007/s11046-004-9096-8 PMID: 15770448

El-Ghwas, D. E., Elkhateeb, W. A., & Daba, G. M. (2023). Fungi: The next generation of biofertilizers. *Environ Sci Arch*, *2*(1), 34–41. DOI: 10.5281/zenodo.7500741

El-Saadony, M. T., Saad, A. M., Soliman, S. M., Salem, H. M., Ahmed, A. I., Mahmood, M., El-Tahan, A. M., Ebrahim, A. A. M., Abd El-Mageed, T. A., Negm, S. H., Selim, S., & Babalghith, A.

Elkhateeb, W., Elghwas, D., & Daba, G. (2023). Fungi. *The Next Generation of Biofertilizers.*, *2*, 34–41. DOI: 10.5281/zenodo.7500741

Etesami, H. (2020). In Meena, R. S. (Ed.), *Enhanced Phosphorus Fertilizer Use Efficiency with Microorganisms BT -Nutrient Dynamics for Sustainable Crop Production* (pp. 215–245). Springer Singapore., DOI: 10.1007/978-981-13-8660-2_8

Franken, P. (2012). The *Piriformospora indica* story: From laboratory to field. *Mycorrhiza*, *22*(8), 631–639. DOI: 10.1007/s00572-012-0459-x PMID: 22476582

Ghorbanpour, M., Omidvari, M., Abbaszadeh-Dahaji, P., Omidvar, R., & Kariman, K. (2018). Mechanisms underlying the protective effects of beneficial fungi against plant diseases. *Biological Control*, *117*, 147–157. https://doi.org/https://doi.org/10.1016/j.biocontrol.2017.11.006. DOI: 10.1016/j.biocontrol.2017.11.006

Gill, S. S., Gill, R., Trivedi, D. K., Anjum, N. A., Sharma, K. K., Ansari, M. W., Ansari, A. A., Johri, A. K., Prasad, R., Pereira, E., Varma, A., & Tuteja, N. (2016). Piriformospora indica: Potential and significance in plant stress tolerance. *Frontiers in Microbiology*, *7*, 332. DOI: 10.3389/fmicb.2016.00332 PMID: 27047458

Giovannini, L., Sbrana, C., Avio, L., & Turrini, A. (2020). Diversity of a phosphate transporter gene among species and isolates of arbuscular mycorrhizal fungi. *FEMS Microbiology Letters*, *367*(2), fnaa024. Advance online publication. DOI: 10.1093/femsle/fnaa024 PMID: 32043113

González-Hernández, A. I., Suárez-Fernández, M. B., Pérez-Sánchez, R., Gómez-Sánchez, M. Á., & Morales-Corts, M. R. (2021). Compost Tea Induces Growth and Resistance against *Rhizoctonia solani* and *Phytophthora capsici* in Pepper. *Agronomy (Basel)*, *11*(4), 781. DOI: 10.3390/agronomy11040781

Hajiboland, R., Dashtebani, F., & Aliasgharzad, N. (2015). Physiological responses of halophytic C 4 grass Aeluropus littoralis to salinity and arbuscular mycorrhizal fungi colonization. *Photosynthetica*, *53*(4), 572–584. DOI: 10.1007/s11099-015-0131-4

Harman, G. E., Howell, C. R., Viterbo, A., Chet, I., & Lorito, M. (2004). Trichoderma species opportunistic, avirulent plant symbionts. *Nature Reviews. Microbiology*, *2*(1), 43–56. DOI: 10.1038/nrmicro797 PMID: 15035008

Hasan, S., Ahmad, A., Purwar, A., Khan, N., Kundan, R., & Gupta, G. (2013). Production of extracellular enzymes in the entomopathogenic fungus *Verticillium lecanii.Bioinformation*, *9*(5), 238–242. DOI: 10.6026/97320630009238 PMID: 23515428

Hiruma, K., Gerlach, N., Sacristán, S., Nakano, R. T., Hacquard, S., Kracher, B., & Schulze-Lefert, P. (2016). Root endophyte Colletotrichum tofieldiae confers plant fitness benefits that are phosphate status dependent. *Cell*, *165*(3), 464–474. DOI: 10.1016/j.cell.2016.02.028 PMID: 26997485

Iqbal, A., Shafi, M. I., Rafique, M., Jabeen, A., Asif, S., Zaman, M., & Jiang, L. (2023). Biofertilizers to improve soil health and crop yields. In *Sustainable Agriculture Reviews 61: Biochar to Improve Crop Production and Decrease Plant Stress under a Changing Climate* (pp. 247–272). Springer International Publishing.

Jangir, P., Shekhawat, P. K., Bishnoi, A., Ram, H., & Soni, P. (2021). Role of Serendipita indica in enhancing drought tolerance in crops. *Physiological and Molecular Plant Pathology*, *116*, 101691. https://doi.org/https://doi.org/10.1016/j.pmpp.2021.101691. DOI: 10.1016/j.pmpp.2021.101691

Jansa, J., & Bukovská, P. (2024). In Parihar, M., Rakshit, A., Adholeya, A., & Chen, Y. (Eds.), *Role of Arbuscular Mycorrhizal Fungi in Nitrogen and Phosphorus Cycling Within Terrestrial Ecosystems BT - Arbuscular Mycorrhizal Fungi in Sustainable Agriculture: Nutrient and Crop Management* (pp. 143–166). Springer Nature Singapore., DOI: 10.1007/978-981-97-0300-5_7

Johnson, N. C., Wilson, G. W. T., Wilson, J. A., Miller, R. M., & Bowker, M. A. (2015). Mycorrhizal phenotypes and the l aw of the m inimum. *The New Phytologist*, *205*(4), 1473–1484. DOI: 10.1111/nph.13172 PMID: 25417818

Johri, A. K., Oelmüller, R., Dua, M., Yadav, V., Kumar, M., Tuteja, N., Varma, A., Bonfante, P., Persson, B. L., & Stroud, R. M. (2015). Fungal association and utilization of phosphate by plants: Success, limitations, and future prospects. *Frontiers in Microbiology*, *6*, 984. DOI: 10.3389/fmicb.2015.00984 PMID: 26528243

Khan, A., Panthari, D., Sharma, R. S., Punetha, A., Singh, A. V., & Upadhayay, V. K. (2023). *Biofertilizers: a microbial-assisted strategy to improve plant growth and soil health.*

Khan, A. L., Hamayun, M., Kang, S. M., Kim, Y. H., Joo, G. J., & Lee, I. J. (2014). Myrothecium verrucaria-mediated growth promotion of soybean plants and its potential to regulate plant endogenous hormones. *Biological Control*, *68*(1), 14–21. DOI: 10.1016/j.biocontrol.2013.10.004

Khan, S. A., Hamayun, M., Khan, A. L., Lee, I. J., Shinwari, Z. K., Kamran, M., & Rehman, G. (2012). Isolation of plant growth promoting endophytic fungi from Phoma sp., native to saline soil. *African Journal of Biotechnology*, *11*(21), 4833–4842.

Khanday, M., Bhat, R. A., Haq, S., Dervash, M. A., Bhatti, A. A., Nissa, M., & Mir, M. R. (2016). In Hakeem, K. R., Akhtar, J., & Sabir, M. (Eds.), *Arbuscular Mycorrhizal Fungi Boon for Plant Nutrition and Soil Health BT - Soil Science: Agricultural and Environmental Prospectives* (pp. 317–332). Springer International Publishing., DOI: 10.1007/978-3-319-34451-5_14

Kobae, Y. (2019). Dynamic phosphate uptake in arbuscular mycorrhizal roots under field conditions. *Frontiers in Environmental Science*, *6*, 159. DOI: 10.3389/fenvs.2018.00159

Kumar, M. S., Reddy, G. C., Phogat, M., & Korav, S. (2018). Role of bio-fertilizers towards sustainable agricultural development: A review. *Journal of Pharmacognosy and Phytochemistry*, *7*(6), 1915–1921.

Larbi-Koranteng, S., Kankam, F., Adomako, J., & Abdulai, M. (2022). Role of Mycorrhizae in Crop Protection. In *Arbuscular Mycorrhizal Fungi in Agriculture-New Insights*. IntechOpen., DOI: 10.5772/intechopen.109020

Li, L., Feng, Y., Qi, F., & Hao, R. (2023). Research Progress of Piriformospora indica in Improving Plant Growth and Stress Resistance to Plant. In *Journal of Fungi* (Vol. 9, Issue 10). DOI: 10.3390/jof9100965

Lin, P., Zhang, M., Wang, M., Li, Y., Liu, J., & Chen, Y. (2021). Inoculation with arbuscular mycorrhizal fungus modulates defense-related genes expression in banana seedlings susceptible to wilt disease. *Plant Signaling & Behavior, 16*(5), 1884782. DOI: 10.1080/15592324.2021.1884782 PMID: 33793381

Matsumoto, M., Takahashi, Y., Fujimoto, H., Shindo, K., Okada, K., & Ikeda, H. (2015). *Cladosporium* cladosporioides promotes plant defense responses via secreted elicitors. *Bioscience, Biotechnology, and Biochemistry, 79*(1), 149–157. DOI: 10.1080/09168451.2014.949272 PMID: 25175163

Menkis, A., Vasiliauskas, R., Taylor, A. F. S., Stenlid, J., & Finlay, R. (2007). Afforestation of abandoned farmland with conifer seedlings inoculated with three ectomycorrhizal fungi—Impact on plant performance and ectomycorrhizal community. *Mycorrhiza, 17*(4), 337–348. DOI: 10.1007/s00572-007-0110-0 PMID: 17277941

Mishra, D., Rajvir, S., Mishra, U., & Kumar, S. S. (2013). Role of bio-fertilizer in organic agriculture: A review. *Research Journal of Recent Sciences ISSN, 2277*, 2502.

Navarro, J. M., & Morte, A. (2024). Arbuscular Mycorrhizal Fungi as Biofertilizers to Increase the Plant Quality of Sour-Orange Seedlings. *Agronomy (Basel), 14*(1), 230. DOI: 10.3390/agronomy14010230

O., Elrys, A. S., El-Tarabily, K. A., and Abu Qamar, S. F. (2022). Plant growth-promoting microorganisms as biocontrol agents of plant diseases: Mechanisms, challenges and future perspectives. *Frontiers in Plant Science, 13.* https://www.frontiersin.org/journals/plant-science/articles/10.3389/fpls.2022.923880

Owiny, A. A., & Dusengemungu, L. (2024). Mycorrhizae in mine wasteland reclamation. *Heliyon, 10*(13), e33141. https://doi.org/https://doi.org/10.1016/j.heliyon.2024.e33141. DOI: 10.1016/j.heliyon.2024.e33141 PMID: 39035525

Ownley, B. H., Griffin, M. R., Klingeman, W. E., Gwinn, K. D., Moulton, J. K., & Pereira, R. M. (2008). Beauveria bassiana: Endophytic colonization and plant disease control. *Journal of Invertebrate Pathology, 98*(3), 267–270. DOI: 10.1016/j.jip.2008.01.010 PMID: 18442830

Pahalvi, H. N., Rafiya, L., Rashid, S., Nisar, B., & Kamili, A. N. (2021). *Chemical Fertilizers and Their Impact on Soil Health BT - Microbiota and Biofertilizers, Vol 2: Ecofriendly Tools for Reclamation of Degraded Soil Environs* (G. H. Dar, R. A. Bhat, M. A. Mehmood, and K. R. Hakeem (eds.); pp. 1–20). Springer International Publishing. DOI: 10.1007/978-3-030-61010-4_1

Pandey, D., Kehri, H. K., Zoomi, I., Akhtar, O., & Singh, A. K. (2019). *Mycorrhizal Fungi: Biodiversity, Ecological Significance, and Industrial Applications BT - Recent Advancement in White Biotechnology Through Fungi: Volume 1: Diversity and Enzymes Perspectives* (A. N. Yadav, S. Mishra, S. Singh, and A. Gupta (eds.); pp. 181–199). Springer International Publishing. DOI: 10.1007/978-3-030-10480-1_5

Peng, Z., Zulfiqar, T., Yang, H., Wang, M., & Zhang, F. (2024). Effect of Arbuscular Mycorrhizal Fungi (AMF) on photosynthetic characteristics of cotton seedlings under saline-alkali stress. *Scientific Reports, 14*(1), 8633. DOI: 10.1038/s41598-024-58979-8 PMID: 38622240

Pepe, A., Giovannetti, M., & Sbrana, C. (2016). Different levels of hyphal self-incompatibility modulate interconnectedness of mycorrhizal networks in three arbuscular mycorrhizal fungi within the Glomeraceae. *Mycorrhiza*, *26*(4), 325–332. DOI: 10.1007/s00572-015-0671-2 PMID: 26630971

Powlson, D. S., Gregory, P. J., Whalley, W. R., & Quinton, J. N. (2014). Soil management in relation to sustainable agriculture and ecosystem services. *Food Policy*, *36*, 572–587.

Powlson, D. S., Gregory, P. J., Whalley, W. R., Quinton, J. N., Hopkins, D. W., Whitmore, A. P., Hirsch, P. R., & Goulding, K. W. T. (2011). Soil management in relation to sustainable agriculture and ecosystem services. *Food Policy*, *36*, S72–S87. https://doi.org/https://doi.org/10.1016/j.foodpol.2010.11.025. DOI: 10.1016/j.foodpol.2010.11.025

Prins, T. W., Tudzynski, P., von Tiedemann, A., Tudzynski, B., ten Have, A., Hansen, M. E., & van Kan, J. A. L. (2000). Infection strategies of Botrytis cinerea and related necrotrophic pathogens. *Fungal Genetics and Biology*, *30*(2), 1–23. DOI: 10.1006/fgbi.2000.1224

Püschel, D., Bitterlich, M., Rydlová, J., & Jansa, J. (2020). Facilitation of plant water uptake by an arbuscular mycorrhizal fungus: A Gordian knot of roots and hyphae. *Mycorrhiza*, *30*(2), 299–313. DOI: 10.1007/s00572-020-00949-9 PMID: 32253570

Redman, R. S., Sheehan, K. B., Stout, R. G., Rodriguez, R. J., & Henson, J. M. (2001). Thermotolerance generated by plant/fungal symbiosis. *Science*, *298*(5598), 1581–1583. DOI: 10.1126/science.1078055 PMID: 12446900

Rudresh, D. L., Shivaprakash, M. K., & Prasad, R. D. (2005). Tricalcium phosphate solubilizing abilities of Aspergillus niger and Penicillium sp. in maize rhizosphere. *Brazilian Journal of Microbiology : Publication of the Brazilian Society for Microbiology*, *36*(1), 9–12. DOI: 10.1590/S1517-83822005000100002

Sarrocco, S., & Vannacci, G. (2018). Preharvest application of beneficial fungi as a strategy to prevent mycotoxin contamination. *World Mycotoxin Journal*, *11*(1), 227–238. DOI: 10.3920/WMJ2017.2276

Sato, T., Ezawa, T., Cheng, W., & Tawaraya, K. (2015). Release of acid phosphatase from extraradical hyphae of arbuscular mycorrhizal fungus Rhizophagus clarus. *Soil Science and Plant Nutrition*, *61*(2), 269–274. DOI: 10.1080/00380768.2014.993298

Schindler, D. W., Carpenter, S. R., Chapra, S. C., Hecky, R. E., & Orihel, D. M. (2016). Reducing phosphorus to curb lake eutrophication is a success. *Environmental Science & Technology*, *50*(17), 8923–8929. DOI: 10.1021/acs.est.6b02204 PMID: 27494041

Schonbeck, F., & Dehne, H. W. (1977). Damage to mycorrhizal and nonmycorrhizal cotton seedlings by Thielaviopsis basicola. *The Plant Disease Reporter*, *61*(4), 266–267.

Smith, J. M., Whiteside, M. D., & Jones, M. D. (2020). Rapid nitrogen loss from ectomycorrhizal pine germinants signaled by their fungal symbiont. *Mycorrhiza*, *30*(4), 407–417. DOI: 10.1007/s00572-020-00959-7 PMID: 32363468

Smith, S. E., & Read, D. J. (2010). *Mycorrhizal symbiosis*. Academic press.

Smith, S. E., & Smith, F. A. (2011). Roles of arbuscular mycorrhizas in plant nutrition and growth: New paradigms from cellular to ecosystem scales. *Annual Review of Plant Biology*, *62*(1), 227–250. DOI: 10.1146/annurev-arplant-042110-103846 PMID: 21391813

Thirkell, T. J., Charters, M. D., Elliott, A. J., Sait, S. M., & Field, K. J. (2017). Are mycorrhizal fungi our sustainable saviours? Considerations for achieving food security. *Journal of Ecology*, *105*(4), 921–929. DOI: 10.1111/1365-2745.12788

Tilman, D., Cassman, K. G., Matson, P. A., Naylor, R., & Polasky, S. (2002). Agricultural sustainability and intensive production practices. *Nature*, *418*(6898), 671–677. DOI: 10.1038/nature01014 PMID: 12167873

Wahab, A., Muhammad, M., Munir, A., Abdi, G., Zaman, W., Ayaz, A., Khizar, C., & Reddy, S. P. P. (2023). Role of Arbuscular Mycorrhizal Fungi in Regulating Growth, Enhancing Productivity, and Potentially Influencing Ecosystems under Abiotic and Biotic Stresses. *Plants*, *12*(17), 3102. Advance online publication. DOI: 10.3390/plants12173102 PMID: 37687353

Wei, X., Xie, B., Wan, C., Song, R., Zhong, W., Xin, S., & Song, K. (2024). Enhancing Soil Health and Plant Growth through Microbial Fertilizers: Mechanisms, Benefits, and Sustainable Agricultural Practices. In *Agronomy* (Vol. 14, Issue 3). DOI: 10.3390/agronomy14030609

Weng, W., Yan, J., Zhou, M., Yao, X., Gao, A., Ma, C., Cheng, J., & Ruan, J. (2022). Roles of Arbuscular mycorrhizal Fungi as a Biocontrol Agent in the Control of Plant Diseases. *Microorganisms*, *10*(7), 1266. Advance online publication. DOI: 10.3390/microorganisms10071266 PMID: 35888985

Whitelaw, M. A. (2000). Growth promotion of plants inoculated with phosphate-solubilizing fungi. *Advances in Agronomy*, *69*, 99–151. DOI: 10.1016/S0065-2113(08)60948-7

Zhang, F., Labourel, A., Haon, M., Kemppainen, M., Da Silva Machado, E., Brouilly, N., Veneault-Fourrey, C., Kohler, A., Rosso, M., Pardo, A., Henrissat, B., Berrin, J.-G., & Martin, F. (2022). The ectomycorrhizal basidiomycete *Laccaria bicolo*r releases a GH28 polygalacturonase that plays a key role in symbiosis establishment. *The New Phytologist*, *233*(6), 2534–2547. DOI: 10.1111/nph.17940 PMID: 34942023

Zhang, L., Xu, X., Li, Y., Zhang, F., & Shen, Q. (2018). *Mortierella hyalina* increases the effectiveness of nitrogen fertilization by improving the soil nitrogen status. *Applied Microbiology and Biotechnology*, *102*(3), 1459–1469. DOI: 10.1007/s00253-017-8647-9 PMID: 29274059

Zhao, S., Zhang, Q., Xu, Z., and Zhang, Y. (2015). Effects of *Rhizopus stolonifer* on soil microbial community and organic matter decomposition. Journal of Environmental Sciences, 37, 92 99. https://doi.org/DOI: 10.1016/j.jes.2015.04.002

Zhao, Z., Chen, L., & Xiao, Y. (2021). The combined use of arbuscular mycorrhizal fungi, biochar and nitrogen fertilizer is most beneficial to cultivate *Cichorium intybus* L. in Cd-contaminated soil. *Ecotoxicology and Environmental Safety*, *217*, 112154. DOI: 10.1016/j.ecoenv.2021.112154 PMID: 33901784

Zong, K., Huang, J., Nara, K., Chen, Y., Shen, Z., & Lian, C. (2015). Inoculation of ectomycorrhizal fungi contributes to the survival of tree seedlings in a copper mine tailing. *Journal of Forest Research*, *20*(6), 493–500. DOI: 10.1007/s10310-015-0506-1

Chapter 7
Edible Mushrooms as Nutritious and Sustainable Superfoods

Dibyarupa Pal
JIS University, India

Sujan Samanta
JIS University, India

ABSTRACT

The search for reforming diets has stemmed from the current invasion of Covid 19 pandemic and uprising of other viral diseases. People started relying more on plant protein source and the consumption of immunomodulatory bioactive compounds from herbal sources. The necessity for foods with several benefits, called as 'superfoods' has burgeoned to fulfil the desire for healthier lifestyle. Mushroom is one such cheap and versatile vegetable protein satisfying the role of superfood, has low-calorie, high-protein and loads of medicinal values to support diabetics and cardiovascular patients. It requires minimal investment and does not need arable land. This makes it an accessible business for unemployed youth, disabled people and women force providing a valuable income source. Biofortified mushroom products can lead to diversified income sources, benefitting the rural community. This book chapter discusses the miscellaneous applications of mushroom and its impending health-promoting and superfood qualities for its easy acceptance and popularization.

INTRODUCTION

Mushroom is a type of fungi of Basidiomycetes group, saprophytic in nature, cultivable easily with zero requirement of arable land with emphasis on sustainable production. Among the 12,000 species of mushrooms discovered, only 2000 species are edible and can be considered in the category of superfood (Cerletti *et al.*, 2021; Das *et al.*, 2021). Some of the chief cultivated mushrooms in demand are *Agaricus bisporus* (button mushroom), *Lentinula edodes* (shiitake), *Pleurotus ostreatus* (oyster mushroom), *Flammulina velutipes* (enoki or winter mushroom), *Pleurotus eryngii* (king trumpet mushroom), *Volvariella volvacea* (paddy straw mushroom), *Calocybe indica* (milky mushroom). Not only as a culinary delicacy mushrooms are preferred by the population worldwide, but nutritionally also have many positive attributes like high in protein and dietary fibre, with packed source of vitamins, minerals and less fat

DOI: 10.4018/979-8-3693-8201-1.ch007

(Das *et al.*, 2021). Moreover mushroom cultivation can be attained on the agricultural waste which gets piled up in the field after each harvest season is over. Approximately 63.6% of paddy and wheat refuse accumulated can be utilized efficiently as substrate for mushroom cultivation instead of burning it openly creating extensive air pollution (IARI, 2012). Thus growing mushroom on these substrates can act in a sustainable way of agricultural waste clean-up and lessen the problem of crop refuse management (Dubey *et al.*, 2019; Paul *et al.*, 2024). Agricultural wastes like wheat/paddy straw, sugarcane bagasse, saw dust, cotton seed meal, soyabean meal and others can act as substrates for mushroom (Besufekad, 2020; Ray *et al.*, 2023).

The supply demand graph of mushroom in India shows that there is growing trend for demand of mushroom to Canada, US, Israel, Mexico both in fresh, processed or preserved form (Raman *et al.*, 2020; Ghosh *et al.*, 2024). The rising demand is not only due to its dietary value but also medicinal value like its therapeutics in Diabetes treatment and cure for high blood pressure, antitumor, cancer, cardiovascular disease, depression and modulators of immune system, (Chong *et al.*, 2019). The paradigm shift in the culinary habits globally towards vegetarianism is noticeable, where mushroom protein can be good alternative for meat replacement products (ICAR, Solan; Leroy *et al.*, 2022). Edible mushrooms consist of high-quality proteins, widely available, cheap to cultivate and meet ecological standards, making them suitable candidates for alternative proteins (Ayimbila et al., 2023). Availability of mushroom-based meat analogues can lead to improvement in nutritional status of people with vegetarian diets. Indian diet is mainly cereal based and lacks adequate protein like Lysine and Methionine which can be supplemented through diet. Mushrooms uphold 20-30% protein (dry weight) along with essential amino acids necessary for human consumption. The current trend shows more towards 'plant based' protein rich food, and in this regard consuming mushroom vegetable has the potential to bridge the protein gap in meals of rural people (Ayimbila et al., 2023). Mushroom is considered as 'cholesterol-free food' and the presence of antioxidants and the absence of starch makes it an ideal food for people suffering from cardiac problems. Even the high potassium: sodium ratio in mushroom makes it suitable diet for lactating and pregnant ladies. It is rich source of valuable enzymes with therapeutic properties. Component analysis of Total fatty acids content of mushroom exhibits 72-85% polyunsaturated fatty acids (PUFA) which significantly has positive health value impact on humans. Among the fatty acids found in mushrooms; linoleic acid has the highest percentage (63-74%), and also it is the only essential amino acid needed in human diet. Mushroom is an identical nutraceutical with immense health benefits and immunomodulatory properties, increased body resistance and reduce proneness to disease.

Global nutrition targets for 2025 encompasses the overall improvement in health and nutrition of the world's inhabitants (WHO, 2019). The burden of infectious disease with emerging and re-emerging infections have persevered throughout the twenty-first century and is also on the rise. Additionally, with fluctuating weather conditions, situations get worsened and affects people with compromised immune systems and those suffering from disease. Thus diets formulated with superfoods can only be the prime choice in such conditions. Mushroom cultivation is a sustainable process as it requires the least diversity of substrates to grow ranging from agricultural waste to compost, including the spent mushroom substrate. They are super efficient in utilizing lignocellulosic waste as their food with minimum requirement of water, fertilizer and other nutritional resources to grow. Mushrooms can be cultivated in a 'vertical' way or under controlled indoor conditions depending on the variety grown. Thus requirement of large arable land is not required like traditional crops and the agricultural waste are utilized which would otherwise 'go to landfills' or burnt openly causing massive air pollution. Thus cultivating mushrooms not only ameliorate income of different strata of people but improves waste reduction in a sustainable way minimizing the environmental effect associated with food production. Due to the ease of operation, unskilled labourers and women folks can carry mushroom cultivation with bare minimum investment (Das et al., 2020; Dey *et al.*, 2020). Mushroom also has a short life cycle of few months and can be grown year-round, providing a consistent source of fresh, 'protein pumped' local diet even in areas with extreme climates.

SUPERFOOD COMPONENTS OF MUSHROOM

The definition of superfood comprise foods which are nutrient packed with one or more distinctive properties and which enhance vitality of the human body by reinforcing the immune system (Franco Lucas *et al.*, 2022; Fernández-Ríos *et al.*, 2022). Presently, mushrooms are in demand not for their palatability but has expanded into wider areas like nutraceuticals and pharmaceuticals, used for the benefit of mankind. Existing investigations have authenticated and documented much of the traditional knowledge on mushrooms as functional foods and revealed their nutritional and medicinal attributes. For example different components like polyunsaturated fatty acids (ω-3, ω-6), glucans, phenolic compounds, organic acids, present in mushrooms assists human health. The tethering superfood components of different commonly consumed mushrooms are reported below.

Carbohydrate, amino acid, PUFA, vitamin and minerals

The proximate composition of *Pleurotus ostreatus* revealed 43.42% carbohydrate, 17.06% crude protein, 23.63% crude fiber, 1.21% lipid and a high moisture content (Effiong *et al.*, 2024). The saccharide profile indicates the presence of varied sugar groups like glucose, xylose, fructose, galactose, trehalose, maltose, sucrose and others. Abundant mineral like potassium, iron and magnesium are present along with high concentration of vitamin B (99.96 mg/kg). The vitamin profile revealed presence of other vitamin also like vitamin A, C and E where the highest concentration of 92.97 mg/kg vitamin B2 (riboflavin) was present. Vitamins are indispensable constituents of outmost necessary to our bodies to function effectively. They are essential for growth, development, and maintaining healthy body systems. On the other hand, amino acids are the building blocks of proteins, essential for body growth, tissue repairing and other function. They are categorized into essential and non-essential amino acids based on their nutritional needs and sources. Essential amino acids are needed but bodies cannot produce them and obtained these through diet. In *Pleurotus ostreatus,* non-essential amino acids (89.27%) are more than essential amino acid (10.73%). Some specific amino acid belonging to different categories are aspartic acid (only acidic amino acid), Lysine (only basic amino acid), 11 neutral amino acids, Phenylalanine (only aromatic amino acid) which are prominent in this mushroom. These amino acids are essential for various bodily functions, like metabolism, immune health, structural protein synthesis, detoxification, neurotransmission, energy production, blood sugar control, wound curing, hormone production, enzyme synthesis and haemoglobin production. Mature fruiting bodies are *C. indica* has high protein content (19.72%) of which glycine is predominant and moderate quantity of essential amino acids like tryptophan, methionine and Lysine are present (Chelladurai *et al.*, 2021). Analysis of Fatty acid profiles, shows that polyunsaturated fatty acid (PUFA) like linoleic (18:2n-6) existed in *P. ostreatus* mushroom. PUFAs are a type of unsaturated fat that our bodies cannot produce on their own, so must be obtain from diet. Mushrooms are a nutritional powerhouse, with nearly 75% of their fatty acid consist of health-promoting PUFAs. Consuming mushrooms (70 grams fresh or 5.6 grams dehydrated) can prevent cardiovascular diseases and other prolonged illnesses associated with oxidative stress (Nyong and Njenga, 2017; Chelladurai *et al.*, 2021).

Proximate composition analysis of other edible variety common mushroom like *Agaricus bisporus* (button mushroom), *Volvariella volvacea* (paddy straw mushroom), *Lentinula edodes* (shiitake), *Pleurotus eous* (pink oyster mushroom) and *Calocybe indica* (milky mushroom) revealed the following nutritional composition. *V. volvacea* had the highest protein content 38.10%), followed by *Agaricus bisporus, Pleurotus eous* and *Lentinula edodes*. *Agaricus bisporus* contained higher levels of fat, vitamin D, Sodium, potassium, and Selenium. *Pleurotus eous* was particularly rich in iron and zinc, with levels nearly double those found in the other mushrooms. *Lentinula edodes* had the highest potassium-to-sodium ratio and higher levels of manganese.

Vitamin D is available rarely through vegetarian sources and is synthesized in human skin when exposed to sunlight (Cashman, 2022). Vitamin D is essential for several physiological bodily function as it stimulates absorption of dietary calcium through synthesis of calcium transport proteins in the small intestine. This function helps prevent bone-related disorders, such as osteomalacia in adults and rickets in children, caused due to insufficient mineralization of calcium in the bones. Sufficient concentrations of 25-hydroxyvitamin D is required to uphold healthy body obtained through dietary sources. Wild mushrooms has been reported to have 3–30 µg D2/100 g fresh weight compared to market bought mushrooms, but sun-dried and UV radiation-exposed mushrooms enhance the bioavailability of vitamin D2. UV radiation triggers a photochemical reaction where the fungal sterol ergosterol is converted into vitamin D through a chain of reactions. Exposing the fruiting bodies of *Agaricus bisporus* and *Volvariella volvacea* to UV-C light (254 nm) for various durations resulted in a 2-4 fold increase in vitamin D content (Ahlawat *et al.*, 2016; Cardwell *et al.*, 2018; Starck *et al.*, 2024). Deficiency of vitamin D is prevalent in all age group as it play pivotal role in calcium homeostasis, bone strength and immunomodulation. Vitamin D receptor are widespread in internal organs and immune cells with it impact on cardiovascular, neurodegenerative and inflammatory diseases among others (Koduah *et al.*, 2017). Inadequate vitamin D levels influence formation of multiple sclerosis, Parkinson's disease and Alzheimer's disease.

MEDICINAL COMPONENT OF MUSHROOM

Mushroom is rich in digestible proteins and many essential amino acids, carbohydrates, fibre, vitamins, minerals like selenium (Raman *et al.*, 2020). The 'magical food' also harbours high dietary fibre which improves the gut microbiota of the intestine and improves digestive power and beneficial to people suffering from cardiovascular disease, diabetes and cancer (Kumar *et al.*, 2020; Li *et al.*, 2021). Rich in phenolics, proteoglycans, prebiotic sources and immunomodulatory substances, the edible mushrooms prevents tumor growth, reduces inflammation and improves heart conditions (Fernández-Ríos et al., 2022; Ayimbila *et al.*, 2023). *Pleurotus* sp. has several natural endowments of bioactive compounds out of which pleurans significantly decreases blood cholesterol level and can act as superfood for Diabetics people. *Pleurotus sajor caju* powder contains protein (22.41%), fat (2.30%) and carbohydrate (60.47%), with calorific value of 451.60 cal/g (Han *et al.*, 2016). Mushroom composition shows negligible amount of sucrose and high amount of Total dietary fibre (mainly β-glucan), making it an excellent choice in diet formulation for diabetic patient. Mushroom fortified diets like incorporation of Maize flour increases the protein content of *Agaricus bisporus* and *Pleurotus ostreatus* significantly (Ishara *et al.*, 2018). Mushroom polysaccharides are used in various health related products and it also acts as prebiotic which stimulates immune system and colon function (Sawangwan *et al.*, 2018). Immunomodulators like lectins, terpenoids, proteins, and polysaccharides are naturally found in mushrooms. A brief account of this is given below:

Phenolic compounds

These are important secondary metabolites of mushrooms, which include phenolic acids, exhibiting antioxidative and anti-inflammatory properties. The phenolic acids act against cell damage by neutralizing the harmful reactive oxygen species (ROS) or by chelation of those components which are responsible for generate ROS like Fe and copper ions. Antioxidants of muhroom neutralize the ROS to nonreactive oxygen species, interrupt the propagation of free radicals and hinder their formation. They break the radical chain reactions and helps to prevent oxidative stress-related damage and protect cells from damaging action of free radicals. Even events like interaction with oxidative enzymes like lipoxygenase, cyclooxygenase, and NADH oxygenase can prevent ROS generation. Several investigations were carried on the white oyster mushroom (*Pleurotus ostreatus*) evaluated its antioxidant properties, revealing a total phenol content ranging from 5.45 to 8.03 mg GAE (gallic acid equivalent)/g at different concentrations. The study also measured the mushroom's radical scavenging abilities, with IC_{50} values for ABTS cation radical reduction and hydrogen peroxide scavenging assays indicating remarkable antioxidant potential. Thus mushroom consumption supports the body' defence system by maintaining a sufficient level of antioxidants. Similarly, the medicinal mushroom *Ganoderma lucidum*, commonly known as the Reishi mushroom, is well-known for antitumor, anti-inflammatory, antidiabetic, antioxidant and antimicrobial activities due to its inherent polysaccharides and triterpenoids components. These are complex β-1,3-glucans and lucidenic acids which exert an immune stimulatory functions (Cör *et al.*, 2018; Bristy *et al.*, 2022). These findings emphasize the significant antioxidant capacity of edible mushroom, attributed to its high levels of phenols and flavonoids, making it a promising natural source of antioxidants for promoting health and preventing oxidative stress-related diseases.

Terpenes and terpenoids

Terpenes are naturally occurring compounds in plants, serving as major constituents of essential oils and possess significant medicinal properties like the Sterols. Several researcher monitored the presence of sterol in various species of mushroom in varied proportion, for instance ergosta-7-enol, ergosta-5,7-dienol and ergosta-7,22-dienol applying techniques of gas chromatography-mass spectrometry or thin-layer chromatography (Tagkouli, 2020; Devi *et al.*, 2024). These agents have been reported copiously in *P. ostreatus* with significant potential as an antioxidant, antitumor, antiinflammatory, and anti-microbial agent, specifically against lymphoma cell line and methicillin-resistant *Staphylococcus aureus* (MRSA) strains. These properties make *P. ostreatus* a valuable candidate for medicinal applications, particularly in cancer and infectious disease treatments. Similarly, lanostane triterpenes, were isolated from the fruiting bodies of *Ganoderma lucidum* responsible for anti-inflammatory activity. Triterpenes are the subtype of terpenes and contribute to the bioactive ability of *G. lucidum* for several biological properties, including anti-inflammatory, anti-tumourigenic and hypocholesterolemic, hepatoprotective and anti-histaminic activity. For instance the ganodermic acid lead to decrease in the proliferation of some cancer cells demonstrated both in vitro and in vivo studies. (Cör *et al.*, 2018; Bristy *et al.*, 2022). Thus mushroom consumption contribute to Triterpenoids which have tremendous biological activities.

Polysaccharides

Polysaccharides such as α-glucans, β-glucans, heteroglycans and proteoglycan are present in mushrooms and has significance health benefit like ROS scavenger, anti-tumour, anti-inflammatory effects. Some specific examples of polysaccharides from mushrooms like *Coriolus versicolor*, (commonly tunzhi in China) with potential health benefits include: Polysaccharide peptide (PSP) and polysaccharide-K (PSK) with potent immunomodulatory properties. Here, the glycoprotein PSK (krestin), are mainly composed of β-glucans, derived from a particular strain CM101, widely popular and most studied mushroom in Japan. PSK is a 100 kDa proteoglycan with carbohydrates composition of sugars like mannose, xylose, galactose, arabinose and rhamnose. Krestin displayed both direct and indirect cytotoxic effects on cancer cells in vitro through stimulation of several key immune cells such as cytotoxic T cells, B lymphocytes, dendritic cells, macrophages, and natural killer (NK) cells (Saleh *et al.*, 2017; Venturella *et al.*, 2022). Immunomodulatory action also involves a critical pathway like the Toll-like receptor 4 (TLR-4) which plays an essential role in recognizing pathogens and initiating immune responses. Furthermore, these immune cells release a range of signalling molecules, like tumor necrosis factor-alpha (TNF-α) and interleukins (IL-1, IL-2, IL-3 and IL-6) to regulate the immune response. This complex immunomodulatory process helps to strengthen the body's defence mechanisms along with overall health being. The medicinal mushroom, *Ganoderma lucidum*, as demonstrated by the works of Xie *et al.*, 2016, displayed cardio protective activity. The mushroom has been extensively used in Asian provinces to treat not only cancer but diabetes as well. Polysaccharide from the fruiting bodies of grey oyster mushroom, (1→3)-β-d-glucan, has exhibited number of medicinal effects like hypolipidemic effects in animal models (Jayasuria *et al.*, 2015). Similarly, immune modulation and antibacterial activities were demonstrated in humans by pleuran isolated from various other mushroom species like *P. sajor-caju, P. eryngii, P. ostreatus, P. pulmonarius* and *P. florida* (Baeva *et al.*, 2019; Patninan *et al.*, 2023). Glucan isolated from gray oyster mushroom showed antidiabetic expression by demonstrating the active pathway of glucose uptake in the myotube. The experiment carried by Patninan *et al.*, 2023, proved the isolated polysaccharide as an antidiabetic agent. This polysaccharide has also shown to progress neutrophil-killing activity against *Candida albicans* and to increase glucose absorption activity in L6 myotubes. Polysaccharide from *P. ostreatus* demonstrated inhibition of tumor growth factor by elevating the levels of tumor necrosis factor α (TNF-α) and nitric oxide (NO). Whereas sulfated polysaccharides from the same mushroom exhibited improved plasma clot activity, anti-tumor properties by inhibiting cancer cell growth, promoting apoptosis (programmed cell death) and enhancing the body's immune response to target cancerous cells. Antioxidative, neuroprotective and immunostimulating properties of quite a few polysaccharide isolated from several mushrooms like *Cordiceps militaris, Pleurotus nebrodensis, Hericium erinaceus, Agaricus brasiliensis* and *Ganoderma* are highlighted in several work of various authors. The working mechanism, visible 'in vivo' in model organisms seem to suppress growth of melanoma cell, inhibit superoxide and hydroxyl radicals with reduction in oedema by 56% in anti-inflammatory assay (Friedman, 2016; Cör *et al.*, 2018).

Diabetes is a metabolic disorder characterized by hyperglycemia (high blood sugar) due to defects in insulin metabolism. Insulin, a hormone produced and secreted by β-cells in the pancreatic islets, primarily functions to stimulate cells to absorb glucose from the bloodstream. Insulin also plays a crucial role in the metabolism of carbohydrates, fats, and proteins. Type 2 diabetes is a condition, caused by acquired resistance to insulin-stimulated glucose uptake hera Type 1 diabetes is caused by insufficient insulin production by the pancreas β-cells. Hypoglycaemic effects have been detected in several work on *Agaricus bisporus, Cordyceps sinensis, Ganoderma lucidum, Inonotus obliqus, Phellinus linteus* and *Pleurotus* spp., among other. Mushrooms are rich in alpha-amylase inhibitors, which are enzymes involved in the breakdown of carbohydrates. Mushrooms produce a wide variety of protein inhibitors that target alpha-amylase. Certain polysaccharides have been shown to regulate blood glucose levels, making them useful in preventing and managing type 2 diabetes by improving insulin sensitivity and reducing blood sugar spikes (Shamim *et al.*, 2023)

HEALTH BENEFITS THROUGH BIOFORTIFICATION OF MUSHROOMS

Biofortification helps in increasing the nutraceutical properties of the mushroom (Fernández-Ríos *et al.*, 2022; Madaan *et al.*, 2024). Budzyńska *et al.*, 2022 biofortified three mushroom species of mushroom (*Pleurotus eryngii, P. ostreatus,* andr *Pholiota nameko*) with iron salts of different composition. Maize flour fortified with powder of *Agaricus bisporus* and *Pleurotus ostreatus* increased protein content by 15.87% and 19.32% significantly. Selenium and Zinc content are biofortified from several *Pleurotus* spp. fruiting bodies (Zieba *et al.*, 2020; Maddan *et al.*, 2024). Since there is negligible amount of sucrose and high amount of total dietary fibre (56.99%), with 3.32% of β-glucan, it is an excellent choice in diet formulation of diabetic patients. Post-harvest, the spent mushroom substrate can act as fortified animal feed as they are good source of protein, minerals, microbes and extracellular enzymes.

PROSPECTS OF MUSHROOM PRODUCTION

Effective training programmes and post harvesting technologies can sustain these untapped potential. The challenges and probable solutions for popularizing the superfood can be done by haring the success stories of others. Dissemination of knowledge on potentiality and profitability of mushroom through training programmes holds huge significance. Not only success stories inspire others but also the neglected section of the society, the women folk can earn consistently sitting in their homes through this business. Notably, in Birbhum district of West Bengal, after training, number of mushroom growers increased (88%) and total mushroom productivity (89%) also increased (Fernández-Ríos *et al.*, 2022). The women folk of Manipur after receiving training from ICAR, gathered knowledge on mushroom cultivation method, its nutritional value and profitability. Profit margin and involvement in mushroom production escalated after such workshops. Current reports from north east India shows rise of the 'Mushroom-preneurs' as it is considered 'Roadmap for Economic Development- Post COVID-19 Pandemic' to rebuild the current economy.

Even the substrate left after mushroom harvesting are rich in lignocellulosic enzymes, with numerous secondary uses. It can act a 'fresh substrate' supplemented with wheat, rice bran and nitrogen-rich manure for growth of second cycle of mushroom. The spent substrate from mushroom cultivation, can be converted into compost, biofertilizer, biopesticide, cattle feed, fish feed and used in various industrial applications. The spent straw can be used for biogas and biofuel production and the remaining slurry can be used as manure encouraging a circular economy prospect. Excess mushrooms can be transformed into value-added products through post-harvest technologies. Mushroom cultivation provides opportunity to ameliorate rural income and combat poverty and simultaneously bridging the nutrition gap with protein-rich food. This process contributes to a 'zero pollution' model, integrating effortlessly into a circular economy (Raut *et al.*, 2021).

To encourage the sustainable superfood, new robust strains of different mushroom variety which can acclimatize through temperature differences are introduced consistently by Directorate of Mushroom Research (ICAR-DMR). Despite all efforts, India is still not graded among the leading mushroom producer among other countries. Change in perception is also needed which can be only given through proper training programmes. Creating awareness of the potentiality of this 'white vegetable' and closing the information gap can change the perception of people in their purchasing behaviour.

CHALLENGES

Despite all the credibility, mushroom is yet to find a permanent place in commoners table. Education, occupation and income level are the major criteria associated with mushroom consumption. Whole hearted acceptance of the 'toadstool' is still yet to be achieved. To overcome, technical and financial assistance from government is highly needed. Conduction of awareness programs on its versatile potentiality, workshops in villages, success stories through media can only bridge the gap of underutilization. Animal meat is also a major competitor of Mushroom. But the conversion from animal to plant meat can only be achieved if the versatile functions of mushroom are popularized. Short shelf life of mushroom makes it easily perishable; browning in colour loses the palatability of mushroom and leads to microbial spoilage. Hence post-harvest storage conditions and processing are the primary requisites.

Thus summarizing the major points which can improve the prospect of income generation, food security are the following components.

- Development of proper technologies for mushroom preservation is needed because due to its high moisture content it becomes easily perishable.
- Proper market chain from farmers household to market to the table of consumers.
- More workshops and training programmes with field level demonstration to be conducted to increase awareness, its nutritional value, profit percent etc.

SUMMARY AND CONCLUSION

Mushroom cultivation has bright prospect in Indian economy for uplifting the standards of people in terms of economy, livelihood and nutrition. The increasing trend of mushroom consumption in India, could be sustained through proper demand supply, storage and processing. Value added products derived from mushroom can be utilized in soups, pickles, noodles, pappads, ketchup, cookies which can elicit the nutritional status of the commonly consumed food especially by children. The ease of mushroom cultivation can motivate, encourage and channelize the young entrepreneur and enterprising farmers to grow the superfood. Home growers can be encouraged to initiate mushroom production for nutritional benefits and ameliorating their incomes.

REFERENCES

Ahlawat, O. P., Manikandan, K., & Singh, M. (2016). Proximate composition of different mushroom varieties and effect of UV light exposure on vitamin D content in Agaricus bisporus and Volvariella volvacea. *Mushroom Research*, *25*(1).

Ayimbila, F., & Keawsompong, S. (2023). Nutritional Quality and Biological Application of Mushroom Protein as a Novel Protein Alternative. *Current Nutrition Reports*, *12*(2), 290–307. DOI: 10.1007/s13668-023-00468-x PMID: 37032416

Baeva, E., Bleha, R., Lavrova, E., Sushytskyi, L., Čopíková, J., Jablonsky, I., Klouček, P., & Synytsya, A. (2019). Polysaccharides from Basidiocarps of Cultivating Mushroom *Pleurotus ostreatus*: Isolation and Structural Characterization. *Molecules (Basel, Switzerland)*, *24*(15), 2740. DOI: 10.3390/molecules24152740 PMID: 31357717

Baker, R. E., Mahmud, A. S., Miller, I. F., Rajeev, M., Rasambainarivo, F., Rice, B. L., Takahashi, S., Tatem, A. J., Wagner, C. E., Wang, L.-F., Wesolowski, A., & Metcalf, C. J. E. (2022). Infectious disease in an era of global change. *Nature Reviews. Microbiology*, *20*(4), 193–205. DOI: 10.1038/s41579-021-00639-z PMID: 34646006

Besufekad, Y., Mekonnen, A., Girma, B., Daniel, R., Tassema, G., Melkamu, J., Asefa, M., Fikiru, T., & Denboba, L. (2020). Selection of appropriate substrate for production of oyster mushroom (Pleurotus ostreatus). *Journal of Yeast and Fungal Research*, *11*(1), 15–25. DOI: 10.5897/JYFR2019.0187

Bristy, A. T., Islam, T., Ahmed, R., Hossain, J., Reza, H. M., & Jain, P. (2022). Evaluation of Total Phenolic Content, HPLC Analysis, and Antioxidant Potential of Three Local Varieties of Mushroom: A Comparative Study. *International Journal of Food Sciences*, *2022*, 3834936. DOI: 10.1155/2022/3834936 PMID: 36310853

Budzyńska, S., Siwulski, M., Gąsecka, M., Magdziak, Z., Kalač, P., Niedzielski, P., & Mleczek, M. (2022). Biofortification of Three Cultivated Mushroom Species with Three Iron Salts—Potential for a New Iron-Rich Superfood. *Molecules (Basel, Switzerland)*, *27*(7), 2328. DOI: 10.3390/molecules27072328 PMID: 35408727

Cardwell, G., Bornman, J. F., James, A. P., & Black, L. J. (2018). A Review of Mushrooms as a Potential Source of Dietary Vitamin D. *Nutrients*, *10*(10), 1498. DOI: 10.3390/nu10101498 PMID: 30322118

Cashman, K. D. (2022). Global differences in vitamin D status and dietary intake: A review of the data. *Endocrine Connections*, *11*(1), 11. DOI: 10.1530/EC-21-0282 PMID: 34860171

Cerletti, C., Esposito, S., & Iacoviello, L. (2021). Edible mushrooms and beta-glucans: Impact on human health. *Nutrients*, *13*(7), 2195. DOI: 10.3390/nu13072195 PMID: 34202377

Chelladurai, G., Yadav, T. K., & Pathak, R. K. (2021). Chemical composition and nutritional value of paddy straw milky mushroom (Calocybe indica). *Nature Environment and Pollution Technology*, *20*(3), 1157–1164. DOI: 10.46488/NEPT.2021.v20i03.023

Chong, P. S., Fung, M. L., Wong, K. H., & Lim, L. W. (2019). Therapeutic potential of Hericiumerinaceus for depressive disorder. *International Journal of Molecular Sciences*, *21*(1), 163. DOI: 10.3390/ijms21010163 PMID: 31881712

Cör, D., Knez, Ž., & Hrnčič, M. K. (2018). Antitumour, antimicrobial, antioxidant and antiacetylcholinesterase effect of Ganoderma lucidum terpenoids and polysaccharides: A review. *Molecules (Basel, Switzerland)*, *23*(3), 649. DOI: 10.3390/molecules23030649 PMID: 29534044

Das, A. K., Nanda, P. K., Dandapat, P., Bandyopadhyay, S., Gullón, P., Sivaraman, G. K., McClements, D. J., Gullón, B., & Lorenzo, J. M. (2021). Edible mushrooms as functional ingredients for development of healthier and more sustainable Muscle Foods: A Flexitarian approach. *Molecules (Basel, Switzerland)*, *26*(9), 2463. DOI: 10.3390/molecules26092463 PMID: 33922630

Devi, P. V., Islam, J., Narzary, P., Sharma, D., & Sultana, F. (2024). Bioactive compounds, nutraceutical values and its application in food product development of oyster mushroom. *Journal of Future Foods*, *4*(4), 335–342. DOI: 10.1016/j.jfutfo.2023.11.005

Dey, A., Patel, S., & Rashid, M. M. (2020). Small-scale mushroom production unit for the upliftment of rural economy and women empowerment in India: A Review. *Current Journal of Applied Science and Technology*, *39*, 38–46. DOI: 10.9734/cjast/2020/v39i230495

Dubey, D., Dhakal, B., Dhami, K., Sapkota, P., Rana, M., Poudel, N. S., & Aryal, L. (2019). Comparative study on effect of different substrates on yield performance of oyster mushroom. *Global Journal of Biology, Agriculture and Health Sciences*, •••, 14.

Effiong, M. E., Umeokwochi, C. P., Afolabi, I. S., & Chinedu, S. N. (2024). Assessing the nutritional quality of Pleurotus ostreatus (oyster mushroom). *Frontiers in Nutrition*, *10*, 1279208. DOI: 10.3389/fnut.2023.1279208 PMID: 38292699

Fernández-Ríos, A., Laso, J., Hoehn, D., Amo-Setién, F. J., Abajas-Bustillo, R., Ortego, C., Fullana-i-Palmer, P., Bala, A., Batlle-Bayer, L., Balcells, M., Puig, R., Aldaco, R., & Margallo, M. (2022). A critical review of superfoods from a holistic nutritional and environmental approach. *Journal of Cleaner Production*, *379*, 134491. DOI: 10.1016/j.jclepro.2022.134491

Franco Lucas, B., Götze, F., Vieira Costa, J. A., & Brunner, T. A. (2022). Consumer Perception Toward "Superfoods": A Segmentation Study. *Journal of International Food & Agribusiness Marketing*, *35*(5), 603–621. DOI: 10.1080/08974438.2022.2044955

Friedman, M. (2016). Mushroom Polysaccharides: Chemistry and Antiobesity, Antidiabetes, Anticancer, and Antibiotic Properties in Cells, Rodents, and Humans. *Foods*, *5*(4), 80. DOI: 10.3390/foods5040080 PMID: 28231175

Ghosh, D., Siddique Md, A., & Pal, D. (2024). *AI driven approach for Precision Agriculture in the Book 'AI in Agriculture for Sustainable and Economic Management.* Taylor and Francis.

Han, N. S., Ahmad, W. A. N. W., & Ishak, W. W. (2016). Quality characteristics of Pleurotus sajor-caju powder: Study on nutritional compositions, functional properties and storage stability. *Sains Malaysiana*, *45*, 1617–1623.

ICAR-Directorate of Mushroom Research. Solan https://dmrsolan.icar.gov.in/Binding-Agent-mushroom-meat-Dr.Anuradha-1.pdf. Accessed 30.07.2024.

Ishara, J. R., Sila, D. N., Kenji, G. M., & Buzera, A. K. (2018). Nutritional and functional properties of mushroom (Agaricus bisporus & Pleurotus ostreatus) and their blends with maize flour. *American Journal of Food Science and Technology*, *6*, 33–41. DOI: 10.12691/ajfst-6-1-6

Jayasuriya, W. J. A. B. N., Wanigatunge, C. A., Fernando, G. H., Abeytunga, D. T. U., & Suresh, T. S. (2015). Hypoglycaemic activity of culinary Pleurotus ostreatus and P. cystidiosus mushrooms in healthy volunteers and type 2 diabetic patients on diet control and the possible mechanisms of action. *Phytotherapy Research*, *29*(2), 303–309. DOI: 10.1002/ptr.5255 PMID: 25382404

Koduah, P., Paul, F., & Dörr, J. M. (2017). Vitamin D in the prevention, prediction and treatment of neurodegenerative and neuroinflammatory diseases. *The EPMA Journal*, *8*(4), 313–325. DOI: 10.1007/s13167-017-0120-8 PMID: 29209434

Kumari, K. (2020). Mushrooms as source of dietary fiber and its medicinal value: A review article. *Journal of Pharmacognosy and Phytochemistry*, *9*, 2075–2078.

Leroy, F., Beal, T., Gregorini, P., McAuliffe, G. A., & van Vliet, S. (2022). Nutritionism in a food policy context: The case of 'animal protein'. *Animal Production Science*, *62*(8), 712–720. DOI: 10.1071/AN21237

Li, M., Yu, L., Zhao, J., Zhang, H., Chen, W., Zhai, Q., & Tian, F. (2021). Role of dietary edible mushrooms in the modulation of gut microbiota. *Journal of Functional Foods*, *83*, 104538. DOI: 10.1016/j.jff.2021.104538

Madaan, K., Sharma, S., & Kalia, A. (2024). Effect of selenium and zinc biofortification on the biochemical parameters of Pleurotus spp. under submerged and solid-state fermentation. *Journal of Trace Elements in Medicine and Biology*, *82*, 127365. DOI: 10.1016/j.jtemb.2023.127365 PMID: 38171269

Nyong, T. M., & Njenga, P. K. (2017). Comparative Analysis of Agaricus bisporus Under Different Substrates And Agaricus bitorquis initiation, Incubation, And Casing Period. *IOSR Journal of Pharmacy and Biological Sciences*, *12*, 82–88.

Patninan, K., Sermwittayawong, D., Nuinamwong, S., Khimmaktong, W., Noipha, K., & Nongporn, H.-T. (2023). Purified Polysaccharides Extracted from Gray Oyster Mushroom (Pleurotus sajor-caju (Fr.) Sing.) Stimulate Glucose Uptake in C2C12 Myotubes through the activation of AMP-Activated Protein Kinase (AMPK) and Glucose Transporter 1 (GLUT1) Proteins. *Sains Malaysiana*, *52*(10), 2815–2827. DOI: 10.17576/jsm-2023-5210-07

Paul, B., Das, D., Aich, T., & Pal, D. (2024). Plant Based Biocoagulants from Cucurbita pepo and Cicer arietinum for Improving Water Quality. *International Journal of Agriculture Environment and Biotechnology*, *17*(1), 29–36. DOI: 10.30954/0974-1712.01.2024.5

Raman, J., Jang, K. Y., Oh, Y. L., Oh, M., Im, J.-H., Lakshmanan, H., & Sabaratnam, V. (2020). Cultivation and nutritional value of prominent Pleurotusspp: An overview. *Mycobiology*, *49*(1), 1–14. DOI: 10.1080/12298093.2020.1835142 PMID: 33536808

Raut J.K., Adhikari M.K. (2021) Mushroom: a true super food. Comprehensive insights in vegetables of Nepal. Nepal academy of science and technology (NAST), 3323: 201-234.

Ray, P., Pal, D., & Pandey, S. (2023). Influence of Sustainable Biocoagulants Trigonella foenumgraecum and Moringa oleifera for Improving Water Potability. [IJTSRD]. *International Journal of Trend in Scientific Research and Development*, *7*, 1239–1246.

Saleh, M. H., Rashedi, I., & Keating, A. (2017). Immunonodulatory properties of Coriuolus versicolor: The role of Polysaccharopeptide. *Frontiers in Immunology*, *8*, 1087. DOI: 10.3389/fimmu.2017.01087 PMID: 28932226

Sawangwan, T., Wansanit, W., Pattani, L., & Noysang, C. (2018). Study of prebiotic properties from edible mushroom extraction. *Agriculture and Natural Resources (Bangkok)*, *52*(6), 519–524. DOI: 10.1016/j.anres.2018.11.020

Shamim, M. Z., Mishra, A. K., Kausar, T., Mahanta, S., Sarma, B., Kumar, V., Mishra, P. K., Panda, J., Baek, K. H., & Mohanta, Y. K. (2023). Exploring Edible Mushrooms for Diabetes: Unveiling Their Role in Prevention and Treatment. *Molecules (Basel, Switzerland)*, *28*(6), 2837. DOI: 10.3390/molecules28062837 PMID: 36985818

Siegrist, M., Michel, F., & Hartmann, C. (2024). The shift from meat to plant-based proteins: Consumers and public policy. *Current Opinion in Food Science*, *58*, 101182. DOI: 10.1016/j.cofs.2024.101182

Starck, C., Cassettari, T., Wright, J., Petocz, P., Beckett, E., & Fayet-Moore, F. (2024). Mushrooms: A food-based solution to vitamin D deficiency to include in dietary guidelines. *Frontiers in Nutrition*, *11*, 1384273. DOI: 10.3389/fnut.2024.1384273 PMID: 38660061

Tagkouli, D., Kaliora, A., Bekiaris, G., Koutrotsios, G., Christea, M., Zervakis, G. I., & Kalogeropoulos, N. (2020). Free amino acids in three Pleurotus species cultivated on agricultural and agro-industrial by-products. *Molecules (Basel, Switzerland)*, *25*(17), 4015. DOI: 10.3390/molecules25174015 PMID: 32887476

Venturella, G., Ferraro, V., Cirlincione, F., & Gargano, M. L. (2021). Medicinal Mushrooms: Bioactive Compounds, Use, and Clinical Trials. *International Journal of Molecular Sciences*, *2*(2), 634. DOI: 10.3390/ijms22020634 PMID: 33435246

Xie, Y., Yang, F., Tan, W., Li, X., Jiao, C., Huang, R., & Yang, B. B. (2016). The anti-cancer components of Ganoderma lucidum possesses cardiovascular protective effect by regulating circular RNA expression. *Oncoscience*, *3*(7–8), 203–207. DOI: 10.18632/oncoscience.316 PMID: 27713910

Yu, Q., Guo, M., Zhang, B., Wu, H., Zhang, Y., & Zhang, L. (2020). Analysis of nutritional composition in 23 kinds of edible fungi. *Journal of Food Quality*, *2020*, 8821315. DOI: 10.1155/2020/8821315

Zięba, P., Kała, K., Włodarczyk, A., Szewczyk, A., Kunicki, E., Sękara, A., & Muszyńska, B. (2020). Selenium and Zinc Biofortification of Pleurotus eryngii Mycelium and Fruiting Bodies as a Tool for Controlling Their Biological Activity. *Molecules (Basel, Switzerland)*, *25*(4), 889. DOI: 10.3390/molecules25040889 PMID: 32079328

Chapter 8
Fungi as a Source of Single Cell Proteins:
Current Progress and Future Prospects

Sharmistha Sharma
https://orcid.org/0009-0006-1101-504X
Assam University, India

Himanshu Kishore Prasad
https://orcid.org/0000-0003-0215-1323
Assam University, India

Anjali Roy
https://orcid.org/0009-0001-2759-7201
Assam University, India

Joydeb Dey
https://orcid.org/0009-0000-8968-475X
Assam University, India

ABSTRACT

Fungi offer a promising solution for sustainable protein production and global food security as single-cell protein sources (SCPs). Their rapid growth, diverse substrate utilization, and high protein content make them attractive candidates. Key fungal species, including filamentous fungi and yeasts, are explored for their characteristics and applications. Various substrates, focusing on waste valorization and circular economy integration, are analyzed for fungal SCP production. Bioprocess engineering aspects, such as fermentation methods, bioreactor designs, and process optimization, are discussed. The nutritional value, safety, and diverse applications of fungal SCPs, ranging from food and feed to nutraceuticals, bioelectronics, and sustainable packaging materials, are examined. Challenges in scaling up fungal SCP production, including technical, economic, and consumer acceptance issues, are addressed, along with advances in genetic engineering, biorefinery integration, and novel extraction technologies. The economic and environmental impacts of fungal SCP production are evaluated.

DOI: 10.4018/979-8-3693-8201-1.ch008

1. INTRODUCTION

Single cell protein (SCP) encompasses protein-rich biomass derived from microbial sources, specifically cultured algae, bacteria, fungi, and yeast (Ritala et al., 2017). These microorganisms represent a significant advancement in addressing contemporary food security challenges while promoting sustainable protein production methodologies. The nomenclature "single cell" reflects the unicellular morphology of these organisms or the distinct cellular architecture of filamentous fungi, characteristics that facilitate their rapid proliferation and efficient biomass accumulation (Linder, 2024; Onyeaka et al., 2022). The biochemical composition of SCPs demonstrates remarkable complexity, including essential nutritional constituents such as lipids, carbohydrates, vitamins, nucleic acids, and minerals (Onyeaka et al., 2022).

The imperative for alternative protein sources arises from demographic projections from the United Nations, indicating a global population trajectory that will reach 9.7 billion by 2050 (Cleland, 2013; United Nations, Department of Economic and Social Affairs, Population Division, 2017). This demographic transformation requires increased production of protein, particularly in developing regions. Contemporary animal-based protein production systems exhibit numerous sustainability constraints, with livestock operations contributing approximately 14.5% of global anthropogenic emissions. The expansion of agricultural areas to produce animal feed has resulted in extensive deforestation and diminished biodiversity. Although aquaculture presents an alternative methodology, it faces significant challenges, including the depletion of wild fish populations for feed production.

Within this context, fungal SCPs (FSCPs) have emerged as a significant focus of research due to their distinctive attributes and sustainable production capabilities. These organisms exhibit remarkable metabolic plasticity, enabling the utilization of diverse substrates, including agricultural by-products and industrial waste materials (Figure 1). This metabolic versatility facilitates waste valorization while adhering to the principles of circular economy. The historical development of fungal SCPs, originating in the 1960s and 1970s, initially focused on nutritional efficacy and safety parameters (Ugalde & Castrillo, 2002). A significant advancement in this field was marked by the commercialization of Quorn™, a mycoprotein derived from *Fusarium venenatum* (Wiebe, 2002). Contemporary biotechnological innovations have facilitated the development of fungal strains that exhibit a higher protein content, optimized amino acid profiles, and improved growth kinetics.

The applications of FSCPs demonstrate remarkable versatility, covering human nutrition, animal feed formulations, aquaculture systems, and the synthesis of nutraceuticals and functional food components. Through fungal fermentation processes, these systems yield various metabolites, including citric acid, vitamin B12, riboflavin, and gluconic acid (Sharma et al., 2020). The nutritional integrity of SCP requires the presence of essential amino acids, specifically lysine, phenylalanine, leucine, methionine, isoleucine, valine, and threonine (Pobiega et al., 2024). Production protocols require a rigorous evaluation of the potential toxigenic and carcinogenic compounds generated during bioprocessing. The production methodology incorporates waste substrate utilization, producing products characterized by elevated protein content and reduced lipid concentrations (Raziq et al., 2020). Through genetic engineering, SCPs can be modulated to enhance amino acid profiles and other nutritional parameters.

This chapter presents a comprehensive examination of the selection of fungal species, substrate optimization, bioprocessing methodologies, nutritional assessment protocols, safety parameters, scale-up challenges, and economic feasibility within global protein production frameworks.

Figure 1. An overview of the single cell protein (SCP) producers and substrates used. SCP-producing microbes broadly include algae, bacteria, fungi, and yeast. Fungal SCP producers thrive on substrates such as (A) Carbon sources (Methanol, ethanol, carbon dioxide) (B) Nonrenewable sources (Wood, coal, industrial exhaust), (C) Agro by-products (Cellulose, hemicellulose and lignin), and (D) Industrial waste products (Bagasse, molasses, whey).

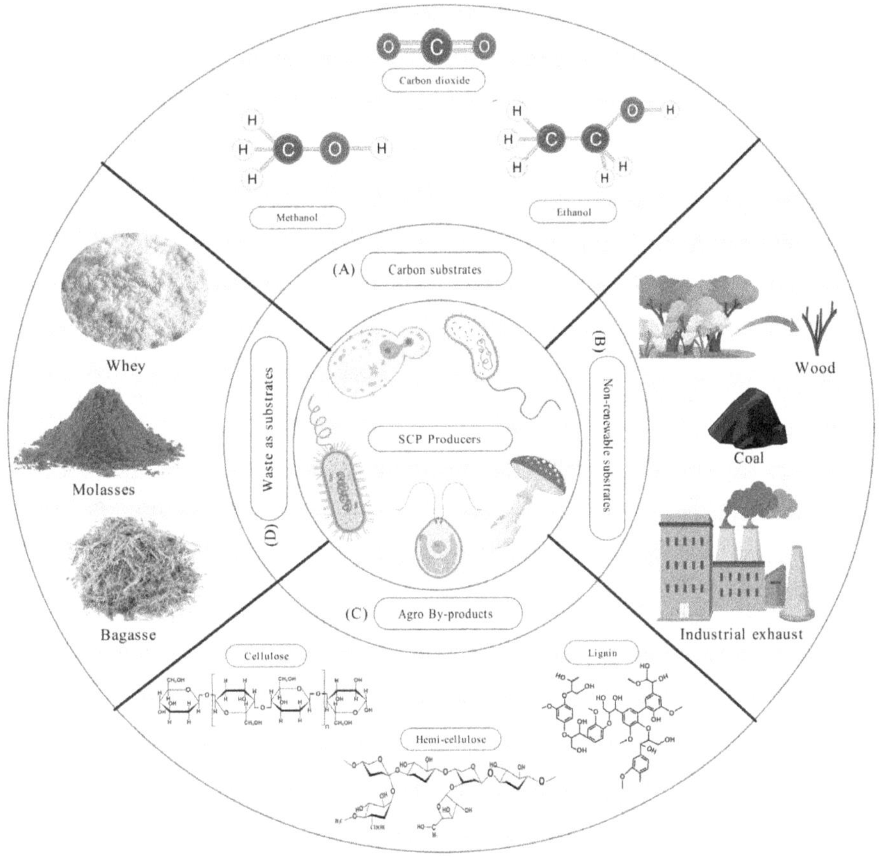

The illustration was created using GDP (Jiang et al., 2025; Pobiega et al., 2024; Sharif et al., 2021)

2. FUNGAL-BASED SINGLE CELL PROTEIN (FSCP) AND MYCOPROTEINS (MP): CHARACTERISTICS AND APPLICATIONS

The development of single-cell proteins (SCPs) from fungi and yeast is a significant advance in addressing protein needs in the food and feed industries (Jach et al., 2022). These proteins, synthesized during microbial growth, offer a nutrient-rich alternative to conventional sources like meat and soy. The growing global protein demand has intensified research in FSCP, particularly for its potential in food security and sustainable agriculture.

FSCPs have exceptional nutritional characteristics, with protein content over 50% and favorable amino acid profiles with high digestibility. Research has documented SCP production from various fungi, including *Fusarium*, *Aspergillus*, *Penicillium*, and yeast such as *Saccharomyces cerevisiae* (Ahmad et al., 2022). Mycoprotein (MP) is a specialized SCP developed for human consumption (Jach et al., 2022; Linder, 2024). Although often grouped with plant-based proteins, mycoprotein is distinct within the fungal kingdom, including mushrooms, yeasts, and filamentous fungi.

The relationship between mycoproteins and SCPs requires careful distinction. Mycoproteins are a refined SCP subset, denoting protein-rich products from fungal mycelium after specialized processing (Linder, 2024; Ravindra, 2020). This differentiation is more than terminology; mycoprotein production includes specific methods, like RNA reduction, for safety and palatability. Mushrooms and fermented foods with fungal mycelium, such as tempeh, do not qualify as mycoproteins despite their fungal origin (Linder, 2023; Lonchamp et al., 2024). Fungi have advantageous characteristics for SCP and mycoprotein production (Ali et al., 2017). Their rapid growth allows biomass to double in hours, facilitating efficient production (Ahmed et al., 2022). Filamentous species like *Fusarium venenatum* double in 2-3 hours, while yeast strains like *Candida utilis* have even shorter periods. This rapid growth reduces production time and resources, enabling continuous or semicontinuous systems. Fungi's metabolic versatility enhances their suitability for SCP production, utilizing diverse substrates, including agricultural residues and industrial waste (Kurcz et al., 2018, Kumar et al., 2024). For example, *Rhizopus oligosporus* generates SCP from cassava bagasse with protein yields of 40%, while *Yarrowia lipolytica* cultured on crude glycerol achieves protein content over 50% (Majumder et al., 2024). Methylotrophic yeasts, such as *Komagataella phaffi*i, demonstrate the capacity to metabolize methanol as the sole carbon source, contributing to carbon capture processes. The nutritional profiles of fungal proteins exhibit remarkable characteristics. Protein concentrations typically range from 30% to 50% of dry weight, with some strains achieving 60% under optimized conditions (Majumder et al., 2024). The amino acid composition is closely aligned with human nutritional requirements, as exemplified by *Fusarium venenatum* mycoprotein, which shows protein quality comparable to that of egg protein. Studies of Quorn™ mycoprotein reveal a protein digestibility corrected amino acid score (PDCAAS) of 0.996, indicating exceptional nutritional value (Lonchamp et al., 2024; Wiebe, 2002).

Beyond protein production, fungi generate valuable co-products that enhance process economics. These include industrial enzymes, organic acids, bioactive metabolites, and cell wall components with applications across food, pharmaceutical, and cosmetic industries (Majumder et al., 2024). This diversified product portfolio significantly improves the economic viability of fungal SCP production. The convergence of rapid growth rates, substrate flexibility, superior protein quality, and valuable co-product generation establishes fungi as promising candidates for sustainable SCP production. These attributes, combined with ongoing advancements in biotechnology and bioprocess engineering, position fungal SCPs as crucial contributors to addressing global protein demands while maintaining the principles of circular economy.

2.1. Overview of Fungal Species in SCP and Mycoprotein Production

The historical development of single cell protein (SCP) technology dates back to 1966 when the Massachusetts Institute of Technology (MIT) introduced the concept, recognizing the potential of unicellular organisms to utilize biomass for protein production. Carol Wilson's pioneering work in establishing terminology laid the foundation for subsequent research (Nalage et al., 2016). Fungi emerged as particularly promising candidates for SCP production due to their high lysine and threonine content, with typical protein concentrations ranging from 30-50% of biomass (Nasseri et al., 2011; Ravindra, 2000; Schulz & Oslage, 1976). Among yeasts, *Saccharomyces cerevisiae* is a prominent organism in SCP production, distinguished by its versatility for carbon source utilization (Table 1; i) and its established safety profile, as evidenced by its general recognition of safety (GRAS) status from the US Food and Drug Administration (Ferreira et al., 2010; Saravanan et al., 2023). Similarly, *Kluyveromyces marxianus*, a homothallic, thermotolerant, hemi-ascomycetous yeast, demonstrates remarkable metabolic diversity and the ability to convert agro-industrial waste into valuable protein products while producing β-galactosidase (Table 1; ii) (Nayeem et al., 2017; Pobiega et al., 2024). Conversion of waste materials into SCP represents a significant advance in sustainable protein production. *Candida utilis* exemplifies this capability through its efficient use of orange peel extracts, which contain substantial amounts of pectin (42.5% dry weight), hemicellulose (10.5% dry weight), and cellulose (9.21% dry weight) (Carranza-Méndez et al., 2022). Furthermore, this organism effectively processes potato waste into SCP, demonstrating the potential for industrial waste valorization (Table 1; iii). *Yarrowia lipolytica,* an aerobic ascomycete, achieves a protein content ranging from 30.5% to 56.4% in its biomass while producing valuable co-products, including vitamins B12 and E, enzymes, organic acids, melanins, and aromatic compounds. The GRAS status of the organism and its ability to utilize various substrates, including sugarcane molasses and cooking oils, enhance its industrial significance (Pobiega et al., 2024). *Komagataella phaffii* (formerly *Pichia pastoris*) presents unique capabilities in methanol-based biotransformation, although challenges exist regarding toxic intermediate formation and temperature control requirements during large-scale production (Gao et al., 2023). One carbon feedstock (C1), methanol (CH_3OH), has emerged as an attractive substrate for fermentation, due to its attributes such as availability; affordability and substantial reduction potential (Cereghino & Cregg, 2000). Among filamentous fungi, several species have shown remarkable potential for FSCP production. *Fusarium venenatum* achieves protein yields of 44-58% when cultivated on glucose and fiber sludge hydrolysate, respectively, showcasing its versatility in substrate utilization (Junaid et al., 2020). The success of this organism in commercial applications, particularly in Quorn™ production, validates its significance in the FSCP industry (Alriksson et al., 2014). *Aspergillus oryzae,* a highly studied filamentous fungus, exhibits exceptional versatility in both traditional fermentation processes and modern biotechnology applications. Its ability to convert okara (soymilk residue) into protein-rich biomass, achieving a protein content of 36.20% in fresh weight and 50.41% in dry weight, demonstrates its potential in waste valorization. Beyond protein production, *A. oryzae* contributes valuable cell wall constituents that enhance immune system function and provide essential minerals, including calcium, sodium, sulfur, potassium, and magnesium. Other notable filamentous fungi include *Pleurotus florida*, which achieves impressive protein yields of up to 63% when cultivated on wheat straw, and *Caetomium cellulolyticum*, which produces 37.9% protein content using delignified aspen wood as substrate (Table 1). *Paecilomyces variotii* demonstrates effective conversion of rayon pulp into SCP with 45% protein content, while various *Penicillium* species show promising results in utilizing diverse agricultural wastes, including orange, banana, and garlic waste media (Table 1). Filamentous fungi's ability to convert complex lignocellulosic materials into protein-rich biomass, along with their production of secondary metabolites and enzymes, makes them vital for sustainable protein production. Their diverse substrate use and high protein yields offer opportunities for industrial applications and waste valorization. Table 1 provides an overview of fungal species used in SCP production, showing their versatility in converting waste into protein-rich biomass, with yields from less than 1% to over 80%, indicating the need for bioprocess optimization and proper organism-substrate matching.

Table 1. Single cell protein (SCP) production by fungi

Yeast	Substrate	SCP Production (%)	Applications	References
(i) *Saccharomyces cerevisiae*	*Eichhornia crassipes* (water hyacinth)	1.725	(i) Reduce environmental hazards caused by *Eichhornia crassipes* (ii) Produce SCP with potent antioxidant activities.	(Saravanan et al., 2023)
(ii) *Kluyveromyces marxianus*	Whey	83.33	(i) Reduce the environmental degradation caused by dairy industry (ii) Produce a good amount of crude protein.	(Nayeem et al., 2017)
	Whey	43	SCP production using whey	(Yadav et al., 2014, 2016)
	Orange and potato pulp, molasses, whey, and digestible grains.	59	----	(Aggelopoulos et al., 2014)
(iii) *Candida utilis*	Potato wastewater and 5% glycerol.	1.22	Industrial waste efficiently harnessed for SCP production	(Kurcz et al., 2018)
(iv) *Candida tropicalis*	Sawdust N_2 Sources: Ammonium phosphate	48.22	Conversion of cheap sawdust resources to high-yielding protein-rich SCP	(Haider, 2021)
	Soy Molasses	56.42	----	(Gao et al., 2012)
(v) *Candida lipolytica*	Olive residues (supplemented with 0.4% peptone)	64.04	Agricultural waste could be efficiently converted into industrially important products.	(Rages et al., 2021)
(vi) *Candida sorboxylosa*	Coffee Wastewater (CWW)	64.4	Effective conversion of waste to economically valuable products	(Pillaca-Pullo et al., 2023)
(vii) *Yarrowia lipolytica*	Sugarcane Mollases	15.12	Utilization of low-grade carbon sources to produce SCP as well as lipases	(Yan et al., 2018)
	Waste Cooking Oil (WCC)	0.723	SCP produced using cheap food industry wastes.	(Spalvins et al., 2020)
(viii) *Pichia occidentalis* PG5	(i) Leftover food waste (LFWH) and (ii) Salad peel waste (SPWH) hydrolysate	(i) 4.2 (ii) 5.4	Protein production; waste management; and ensures global food security	(Sagar et al., 2024)
(ix) *Rhodotorula paludigena* CM33	YPD; Miao-2020; MMS, Liu lipid; Riberio-2019; and Liu seed medium	0.42	This strain shows potential in (β-carotene and SCP production.	(Thumkasem et al., 2023)

continued on following page

Table 1. Continued

Yeast	Substrate	SCP Production (%)	Applications	References
Filamentous Fungi	**Substrate used**	**SCP production (%)**	**Applications**	**References**
(x) *Pleurotus florida*	Wheat straw	63	----	(Ahmadi et al., 2010; Ritala et al., 2017)
(xi) *Fusarium venenatum*	Glucose	44	----	(Junaid et al., 2020)
	Fiber sludge hydrolysate	58	Residues from the forest industry serve as feed for the aquaculture industry.	(Alriksson et al., 2014)
(xii) *Aspergillus oryzae*	Okara (Soymilk residue) pH 5.0	36.20 (fresh weight) 50.41 (dry weight)	The insoluble Okara residue converted to SCP	(Yousufi, 2012)
(xiii) *Cephalosporium eichhorniae*	Cassava (i) 1% (ii) 2%	(i) 36 (ii) 34	SCP production	(Varavinit, et al., 1996)
(xiv) *Caetomium cellulolyticum*	Aspen Wood (Sodium Chlorite treated)	37.9	Delignification improves protein yield in the final product.	(Abdel-Azeem & Sheir, 2020)
(xv) *Paecilomyces variotii*	Rayon Pulp	45	An effective measure to reduce environmental pollution and convert waste products into SCP	(Bajpai & Bajpai, 1987)
(xvi) *Penicillium citrinum*	Orange Waste Media (OWM)	0.96	Waste Substances Utilized for SCP Production	(Ahmed et al., 2024)
(xvii) *P. crustosum*	Banana Waste Media (BWM)	0.94		
(xviii) *P. chermesinum*	Garlic Waste Media (GWM)	0.83		
(xix) *Trichoderma viridae*	Pineapple peel waste (fructose as a source of C)	18.35	Effective method of waste processing	(Anichebe et al., 2019)
	Banana peel waste (saccharose)	30.96		

3. GENERAL METHODOLOGIES TO PRODUCE SCP BY FUNGI

Industrial production of FSCP involves generalized steps of substrate selection based on availability, cost-effectiveness, and other parameters. This is followed by the utilization of substrate by fungi inside the fermentor. Fermentation can be performed through submerged, semi-solid, or solid-state means (Figure 2). Further steps in filtration, purification, and drying lead to the formation of single cell proteins (SCP) (Bratosin et al., 2021; Raziq et al., 2020). The fermentation process consists of both wet and dry processes. The wet preparation method is used for substrates with high moisture content, which commonly includes waste generated from fruit and vegetable peels. This entails a sequential step of washing with water and acid, followed by pulverization, filtration, and sterilization to obtain a contamination-free liquid medium. Similarly, for substrates with minimal moisture content, the dry method is commonly applied, which removes moisture content. The substrate then undergoes reduction and sifting generating a fine powder with a large particle size. This powder was then thoroughly mixed with water, filtered, and sterilized to obtain a moist solid medium (Figure 3) (Gorte et al., 2020; Li et al., 2024). The general procedure to produce SCP involves preliminary steps, such as preparation of the medium with a suitable carbon source, avoidance of contamination, and biomass synthesis followed by its processing (Suman et al., 2015).

Figure 2. The flow chart illustrates the generic steps of FSCP production. The procedural framework encompasses the application of a multitude of hazardous materials and waste products as substrates, followed by sterilization and microbial utilization. An array of fermentation methodologies are implemented, culminating in the production of FSCP and facilitating further downstream processing aimed at increasing its market value.

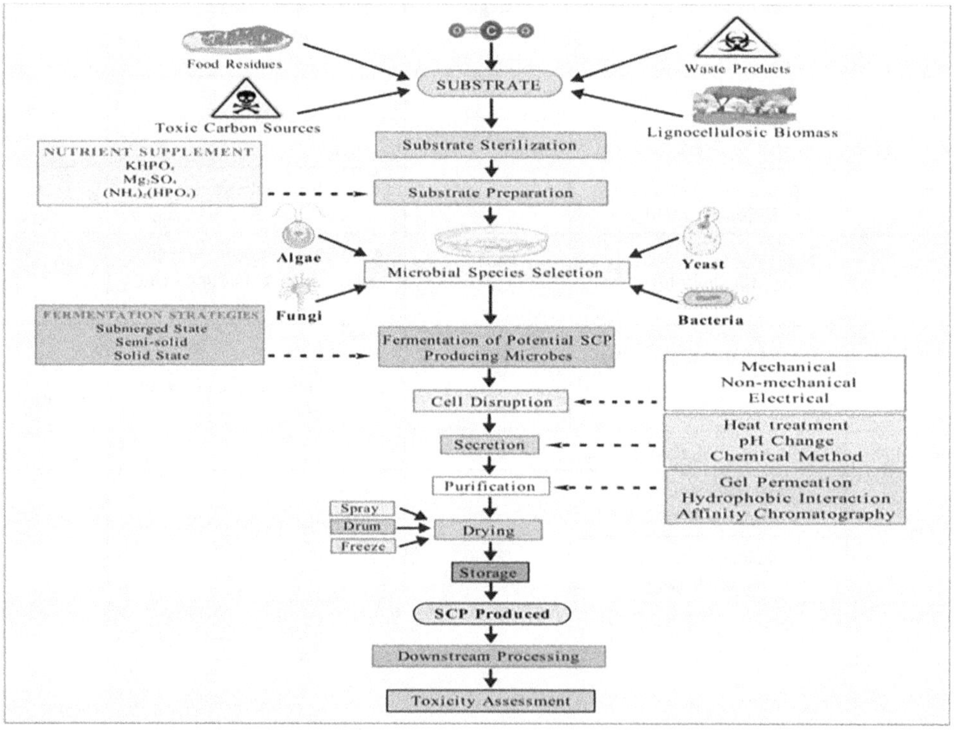

The illustration was created using GDP (Aidoo et al., 2023; Bratosin et al., 2021; Jiang et al., 2025; Li et al., 2024; Pobiega et al., 2024)

3.1 Selection of the suitable substrate

Substrates for FSCP production mainly involve plants, agrowastes, and industrial waste (Ritala et al., 2017; Spalvins et al., 2018). Molasses are inexpensive carbohydrate-rich substrates containing 40-60% fermentable carbohydrates, 20% fats and 10% mineral and nitrogen contents, usually derived from sugarcane, sugar beet, or soy proteins (Spalvins et al., 2018). Soy molasses is used to grow *Candida tropicalis*, containing protein content and sugars like sucrose, stachyose, and raffinose (Gao et al., 2012). Dairy industry by-products like whey contain high concentrations of lactose and protein suitable for SCP production, with cheese production contributing approximately 139 million tons of whey annually worldwide (Spalvins et al., 2018; Yadav et al., 2014). Agro-waste is also suitable, being non-toxic, regenerable, abundant and inexpensive (Dharumadurai et al., 2011). Important fruit wastes include banana (*Musa* spp.) and pineapple peel (*Ananas camosus*), containing high amounts of carbohydrates and proteins (Anichebe et al., 2019; Dharumadurai et al., 2011; Sankar et al., 2011). Substrates for SCP production are classified into carbohydrate-rich plant resources, high-energy-rich resources, and wastes from biological and industrial processes. Microbial strains undergo fermentation to produce single cell proteins (Raziq et al., 2020). Spent sulfite liquor is suitable for cultivating yeasts like *Candida utilis* and *Candida tropicalis*. Whey is utilized by *Kluyveromyces marxianus* for SCP production due to its lactose metabolism. Molasses is widely employed in SCP production, influenced by cost, availability, and composition. Fruit and vegetable waste offers carbohydrates and nutrients for yeast growth. Starchy waste from crops like cassava and potatoes is utilized after conversion into fermentable sugars. Coffee press waste, containing soluble carbohydrates and high chemical oxygen demand (COD), presents potential for FSCP production using *Trichoderma* species. Waste liquor has potential for SCP production. Other agricultural and industrial wastes, such as bagasse and soybean hulls, have been proposed for yeast-based SCP. The selection of waste substrates depends on factors like availability, cost, composition, safety, and necessary processing. Leveraging waste streams for SCP contributes to a circular economy and addresses protein necessity sustainably, mitigating waste and environmental consequences (Amara & El-Baky, 2023).

Figure 3. Preparation and potential applications of SCP-based substrates. Substrate preparation methodologies for SCP include (A) dry, (B) wet, and (C) direct approaches. The three strategies incorporate successive phases of pretreatment, pulverization, blending, filtration, and growth medium preparation. Subsequently, microbe inoculation, incubation, harvesting, drying, and SCP production processes are executed. The resultant SCP fulfills a multitude of roles across various domains.

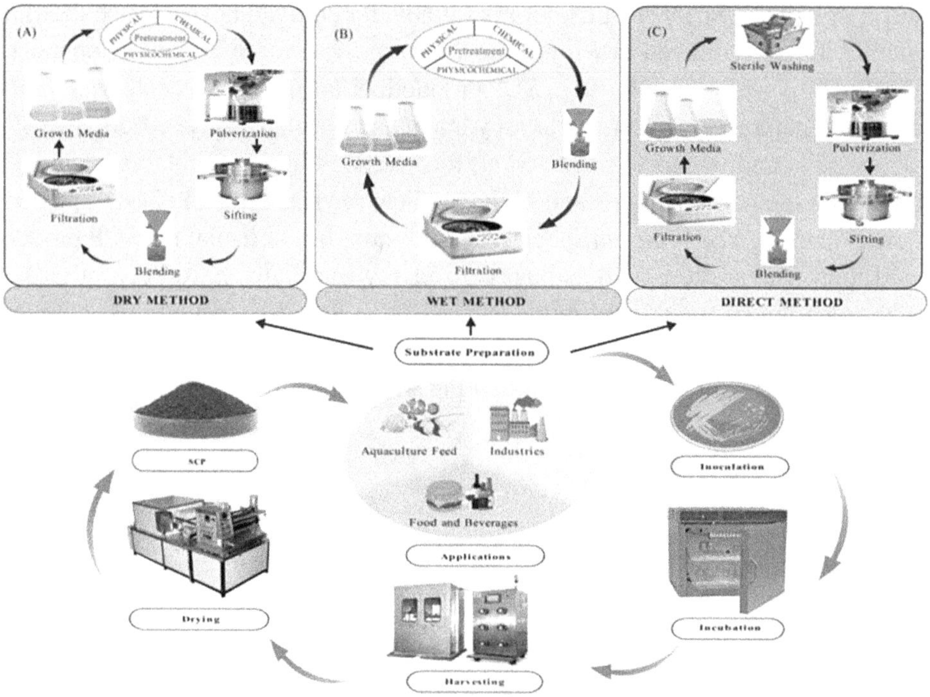

The illustration was created using GDP (Aidoo et al., 2023; Bratosin et al., 2021; Jiang et al., 2025; Koukoumaki et al., 2024; Li et al., 2024; Zhang et al., 2024)

3.2 Fermentation techniques

SCP production involves fermentation techniques such as submerged or liquid state fermentation (LSF), semisolid, and solid-state fermentation (SSF). In LSF, microbes utilize liquid substrates in an aerated environment, and biomass is obtained that requires constant harvesting (Reihani & Khosravi-Darani, 2019). The product is filtered, centrifuged, and dried. Heat is removed by a cooling device. Yeasts are removed by centrifugation and filamentous fungi by filtration. Water removal is achieved by drying. The process requires a liquid substrate with approximately 95% water content. It offers online monitoring and heat removal techniques, making it preferable to SSF. However, it is neither cost-effective nor environmentally friendly, generating substantial waste (Kargi et al., 1980; Varavinit et al., 1996; Zhang et al., 2024). The semisolid fermentation process uses a U-loop fermentor bioreactor to identify mass and energy transfer. It utilizes solid-state substrates with carbon sources like methanol, ethanol, or molasses. The culture process involves agitation, aeration, and heat removal (Raziq et al., 2020; Suman et al., 2015). SSF involves seeding microorganisms on solid substrates at room temperature for several days. In liquid culture, unicellular microorganisms grow in an aerobic environment with constant agitation. Parameters like temperature, oxygen supply, pH, and ionic strength are constantly monitored. Minimal additional water is required as the solid substrate contains sufficient moisture. Filamentous fungi and yeast species require a moisture content of 60-65% on the solid substrate. Fungi obtain nutrients through adsorption or penetration to produce protein biomass. SSF is environmentally and economically beneficial with minimal waste generation. However, limitations include hindrance in large-scale production, restrictions in online monitoring, and heat generation (Li et al., 2024; Suman et al., 2015).

3.3 Steps in Downstream Processing

The main aim of downstream processing is to eliminate unwanted substances from the product, SCP. Purification can be performed by centrifugation, where the desired supernatant is extracted. The desired protein is separated based on affinity or molecular weight. SCP production involves cell wall degradation and lowers nucleic acid content (Ali et al., 2017). Cell wall degradation is important because microbial cell walls could cause allergic reactions (Pobiega et al., 2024). Cell wall disruptions can be obtained through mechanical or non-mechanical methods. The mechanical method is non-selective and includes high-pressure cell homogenization (HPH), crushing, grinding, or crumbling. Nonmechanical methods employ physical, chemical, electrical, and enzyme-based techniques, which are selective and harmful to cells. The freeze-thaw technique involves gradual freezing of microbial cells, forming ice crystals within cells, causing disruption. Osmotic shock involves exposing cells to high osmotic pressure, leading to disruption. Chemical disintegration uses detergents, alkali, acid, organic solvents, or ionic liquids. Electrical methods employ pulsed electric field (PEF) and high voltage electrical discharge (HVED). Enzymatic degradation targets mannoprotein complexes and glucan backbone through autolysis or lytic enzymes, preserving cellular integrity (Pobiega et al., 2024; Ritala et al., 2017). After cell destruction, proteins are released into solvents and extracted through centrifugation. Proteins for the food sector require precipitation, involving heat treatment and pH alteration. The proteins extracted are further purified by chromatography or micro- and ultrafiltration techniques (Pobiega et al., 2024).

4. CASE STUDIES OF SUCCESSFUL SCP PRODUCED BY FUNGI

4.1 SCP production by *Pichia pastoris*

Pichia pastoris produces single cell proteins (SCP) using methanol but is intolerant to temperatures above 28-30°C. Thermotolerant strains can lower SCP production costs (Ritala et al., 2017). Meng et al. (2023) improved a strain to withstand 33°C and enhance methanol use. Inducing cell wall remodeling protects against temperature stress, while metabolic engineering boosts protein content through nitrogen metabolism, increasing SCP yield from methanol in batch cultures at pilot scale (Meng et al., 2023). Cell wall sensors help *P. pastoris* perceive methanol, effectively using it as a C1 source for SCP and other compounds (Schrader et al., 2009). Gao et al. (2023) found cell wall synthesis genes correlate with methanol tolerance: 3% methanol raised wall thickness by 22.8% over 0.5% methanol, showing cell wall sensors boost methylotrophy, enhanced by the PAS_0305 gene which maintains wall integrity (Gao et al., 2023).

4.2 SCP Production by *Candida utilis*

Candida utilis is used in various studies utilizing waste products as substrates, including tuber wastes (Ouedraogo et al., 2017), salad oil wastewater (Zheng et al., 2005), yellow wine production waste (Yao et al., 2018), and potato wastewater (Kurcz et al., 2018). Through submerged fermentation and using orange peel extracts, *C. utilis* produced 6.22% protein, showing potential for waste management (Carranza-Méndez et al., 2022). Lignocellulose is a renewable substrate with advantages like abundance and low cost, but its composition is challenging. It's composed of xylan and glucan in a lignin-carbohydrate complex, requiring pretreatments that generate inhibitors like furfurals, phenolics, and hydroxymethyl furfurals (HMFs), complicating the process. *Candida intermedia* (FL023) was shown to tolerate these inhibitors and produce single cell protein (Wu et al., 2018).

4.3 Mycoprotein of *Fusarium venenetum*

The purpose of SCP production is to synthesize a better alternative to the existing protein source, which would be both economically and abundant. In this sense, the best example that emerges is of 'mycoprotein'. It is a well-known safe alternative to meat. *F. venenetum* is known for its ability to produce mycoproteins and was officially sanctioned by the Ministry of Agriculture, Fisheries and Food in the United Kingdom (1984). Its market name is 'Quorn' (Ahmad et al., 2022; Hashempour-Baltork et al., 2020; Lee et al., 2024). The fermentation process yields a protein content of 65-76%. The lipid content of mycoprotein ranges from 2 to 3.5%, and the fatty acid content shows strong similarities with that of plants (Pobiega et al., 2024). The mycoprotein production process involves a pressure cycle reactor or a fermentor, to which glucose and ammonium are fed as carbon and nitrogen sources, respectively. For several weeks (approximately 6 weeks), the fermentation process is maintained at 28-30 °C and pH 6. Additionally, nucleic acid (especially RNA) is reduced by a heat shock at 64-65 °C, followed by centrifugation and chilling. This process generates a mycoprotein that undergoes several clinical trials and toxicological tests for human consumption (Wiebe, 2002).

4.4 Mycoprotein of the *Fusarium* strain *flavolapis*

Fy Protein™ is made from fermenting the *Fusarium strain flavolapis*, found in Yellowstone National Park, without genetic modification (Furey et al., 2022). Comprising mycelium, it is a complete protein with all essential amino acids (20–24% dry weight) and 40-50% total amino acids. It also includes fiber, fat, carbohydrates, vitamins, and minerals. Produced by Nature's Fynd, the process involves inoculating a sterilized medium with a pure culture of *F. str. flavolapis*. This medium contains carbon, nitrogen, and other nutrients, all high-quality and approved. The organism forms a semi-solid biomat, which is harvested and heat-treated. Nature's Fynd follows GMP and monitors safety, mycotoxins, and heavy metals. Recognized as safe by the FDA since 2021, Fy Protein™ showed no adverse effects in rat studies, with some cholesterol benefits. It has no mutagenic or allergenic risks, and mycotoxin levels are undetectable in controlled batches. The company claims Fy Protein™ uses 99% less land and water and emits 94% fewer greenhouse gases than beef, though these have not been independently verified (Furey et al., 2022).

4.5 Production of SCPs by *Penicillium citrinum*

The *Penicillium and Aspergillus* species have a high protein content and thus serve as potent SCP producers. *P. citrinum; P. crustosum and P. chermesinum* are known to use waste such as orange waste medium, banana waste medium and garlic waste medium, respectively, to produce SCP (Table 1, xvi; xvii and xviii) (Ahmed et al., 2024).

5. FSCP PRODUCTION AND PATENTING

Protein represents an essential component of human nutrition, with animal-derived products traditionally serving as primary sources of essential amino acids (Bratosin et al., 2021; Greene & McAdam, 1983). Global consumption of animal protein reached approximately 327,683 Kilotons during 2019-2021, with market projections indicating expansion to 7.3 trillion dollars by 2025 (Bohrer, 2019). However, the environmental impact of conventional animal protein production has necessitated exploration of alternative protein sources. Single cell proteins (SCPs) have emerged as a promising solution to these challenges, demonstrating significant advancement in microbial protein production methodologies over the past century. The commitment to SCP development is evidenced by substantial industrial investment, exemplified by Deep Branch Biotechnology's 2.5-million-euro allocation toward production platform advancement in 2020 (Ribeiro et al., 2023). Bibliometric analysis of alternative protein patents reveals significant research activity, with 360 documented articles comprising 271 research papers, 49 review articles, and 40 additional publications. China demonstrates prominence in this field, accounting for 70% of patent deposits and publications between 2001-2016. Patent concentrations span multiple sectors, including biochemistry, enzymology, beverages, microbiology, and genetic engineering, with 110 active patents primarily originating from industrial entities. (Ribeiro et al., 2023; Ritala et al., 2017). Single-cell proteins are frequently considered by-products capable of enhancing the economic viability of marginally profitable biorefinery processes and reducing downstream processing expenses linked to waste disposal. The utilization of residual biomass as animal feed is significantly advantageous compared to its application as fertilizer. This is well-documented in diverse publications and patents that address the transformation of specific waste products into SCP. Moreover, environmental considerations now prominently influence the development of new SCP products (Ritala et al., 2017). Additionally, economic factors propel the innovation of novel SCP advancements. In the production of fungal SCPs, 62% of total costs are ascribed to raw materials, whereas 19% pertain to the production process itself (Ugalde & Castrillo, 2002). Recent patents pertaining to fungi-based SCP production are enumerated in Table 2.

Table 2. Patents recently issued concerning SCP production in fungi. These patents were located and collected using a patent search tool available at https://www.lens.org/

Yeast	Patent Obtained	Patent No.	References
Pichia pastoris	The strain has high protein synthesis and multicarbon source, utilizing capacity. Additionally, the patent presents future directions on the ability of this strain to utilize cheap carbon sources for SCP production.	CN 116478838 A	(Xin & Le, 2023)
	Pichia pastoris HGD01 demonstrates efficiency in converting methanol into SCP	CN 103045494 A	(Guohou et al., 2013)
	P. pastoris simultaneously produces lipases and methanol protein	CN 105132301 A	(Wanxiao et al., 2015).
Debaromyces hansenii	For SCP production using biogas slurry	CN 113913309 A	(Dong et al., 2022)
Candida utilis	Application of the strain in SCP production	CN 114836331 A	(Zhanying et al., 2022)
Kluyveromyces marxiamus	Describing a novel fermentative process of SCP production; known as KMPROTEO 1 A protein content of 50-60% was obtained.	WO 2024/005625 A1	(Ángel, 2024)
Fungus	**Patent Obtained**	**Patent No.**	**References**
Thermophilic Fungi	For describing the process of SCP production utilizing thermophilic fungi	AU 2017/310017 B2	(Maria & Sebastian, 2023b)
Rhizomucor pusillus	Process of SCP production Growth of fungal strain above 45°C; pH < 3.8 The recovered protein content is 40% dry matter (w/v).	US 11707078 B2	(Maria & Sebastian, 2023a)

6. APPLICATIONS AND ADVANTAGES OF FUNGAL SINGLE CELL PROTEINS

The applications of fungal single cell proteins (FSCP) extend across multiple sectors, demonstrating significant potential in addressing global nutritional and environmental challenges. These applications can be systematically examined through their diverse implementation across various industries and their inherent advantages in sustainable protein production (Li et al., 2024).

In the realm of nutritional supplementation, FSCPs, particularly mycoproteins, represent a comprehensive nutrient source. These proteins deliver essential micronutrients, including vitamin B9 (folate), vitamin B12, and crucial minerals such as calcium, phosphorus, magnesium, and zinc (Li et al., 2023) The nutritional profile is further enhanced by significant fiber content, which promotes digestive health, and notable choline concentrations approximating 180 mg per 100g, comparable to conventional protein sources like salmon and pork (Gao et al., 2023). The integration of FSCPs into animal feed systems presents a strategic approach to sustainable livestock nutrition. This application addresses the growing concern over traditional feed sources' resource intensity while contributing to circular economy principles through waste material utilization. The regulatory framework governing these applications emphasizes thorough safety assessments of waste-derived substrates, ensuring compliance with animal consumption standards (Lapeña et al., 2020). The biotechnological applications of FSCP production systems extend beyond protein generation to include biofuel and biochemical synthesis. Yeast-based cell factories demonstrate promise in producing fatty acid-derived oleochemicals, offering sustainable alternatives to fossil fuels (Derbyshire & Delange., 2021). Advanced genetic engineering techniques have enhanced these processes, enabling efficient utilization of diverse carbon sources derived from agricultural waste streams. In food product development, FSCPs have established a significant presence, particularly in meat analog formulations. Products such as Quorn™ exemplify successful commercial implementation, utilizing fungal fermentation to achieve meat-like textures and flavors (Ma et al., 2023). This application addresses the expanding market for plant-based alternatives while considering crucial factors such as consumer acceptance and dietary preferences (Amara & El-Baky, 2023). The advantages of fungal and yeast-derived SCPs encompass multiple dimensions of sustainability and nutrition. Their nutritional profile matches or exceeds conventional protein sources, offering a complete spectrum of essential amino acids alongside significant micronutrient content (Jach et al., 2022). The environmental benefits are particularly noteworthy, as FSCP production systems demonstrate superior resource efficiency compared to traditional livestock farming, while effectively incorporating waste streams and potentially contributing to carbon dioxide sequestration (Sagar et al. 2024). The economic viability of FSCP production is enhanced through scalable continuous processing systems, which optimize operational costs across labor, energy, and raw material utilization. This economic efficiency extends to potential small-scale applications, suggesting adaptability across various production scales. Furthermore, research has demonstrated significant health benefits associated with mycoprotein consumption, including improved appetite regulation and glycemic control, positioning FSCPs as valuable contributors to contemporary dietary guidelines (Figure 4). This comprehensive analysis of FSCP applications and advantages underscores their potential in addressing global protein demands while advancing sustainable production practices. The convergence of nutritional value, environmental sustainability, economic efficiency, and health benefits establishes FSCPs as a crucial component in future food security strategies.

7. NUTRITIONAL ATTRIBUTES OF FUNGI-BASED SCP

The amino acid composition of fungal single cell proteins (FSCP) represents a crucial determinant of their nutritional value and potential applications in food and feed systems. Comprehensive analysis of amino acid profiles across various fungal species demonstrates their capacity to meet or exceed established nutritional standards (Ritala, et al., 2017).

Studies examining *Candida tropicalis* cultivated on soy molasses reveal exceptional essential amino acid content. The protein demonstrates elevated concentrations of critical amino acids: lysine (6.91%), threonine (4.35%), valine (4.58%), leucine (6.24%), and phenylalanine (3.71%), all surpassing Food and Agriculture Organization (FAO) reference standards (Gao et al., 2012). Similarly, *Pichia guillermondii*, when grown on n-alkane substrates, produces biomass with 50.1% crude protein content, featuring notably high lysine concentrations of 8.3% alongside substantial levels of arginine, glycine, histidine, isoleucine, leucine, and sulfur-containing amino acids (Schulz & Oslage, 1976). Recent investigations into waste valorization processes have yielded promising results. Research conducted in 2024 demonstrated that *Aspergillus* and *Penicillium* species effectively convert fruit processing wastes into protein-rich biomass. *Penicillium chermesinum* achieved remarkable amino acid yields, producing 1.56 g leucine and 1.02 g lysine per 100 g biomass, while *P. crustosum* generated 1.32 g threonine per 100 g (Ahmed et al., 2024). These findings underscore the potential for converting agricultural waste streams into nutritionally valuable protein sources. Particularly noteworthy is the amino acid profile of *Pichia pastoris* cultivated on methanol, achieving 41.78% of dry cell weight as protein, with essential amino acids constituting

21.74% of this fraction. The distribution includes methionine (1.25%), histidine (1.4%), valine (2.16%), arginine (2.31%), threonine (2.36%), isoleucine (2.385%), phenylalanine (2.89%), leucine (3.49%), and lysine (3.5%) (Meng et al., 2023). The presence of branched-chain amino acids (BCAAs), valine, isoleucine, and leucine, totaling 8.03%, is significant given their role in energy metabolism, glucose uptake regulation, and protein synthesis (Meng et al., 2023).

Mycoprotein's nutritional profile demonstrates comprehensive essential amino acid content, providing all nine essential amino acids required for human nutrition. These proteins support critical physiological functions: leucine facilitates protein synthesis and muscle repair, lysine enhances calcium absorption and immune function, while methionine serves as a precursor for various metabolic processes. Beyond amino acid content, mycoprotein offers additional nutritional advantages through its low saturated fat content, absence of cholesterol, and significant dietary fiber concentration (Derbyshire & Ayoob, 2019).

8. FUNCTIONAL ASPECTS AND HEALTH BENEFITS OF FUNGI-BASED SCP

Fungal single cell proteins demonstrate distinctive advantages over conventional plant-based protein alternatives, particularly in their nutritional completeness and bioavailability. While plant-based proteins often require fortification with essential nutrients such as zinc, selenium, iron, iodine, calcium, and vitamins (B2, B12, and D), microbial proteins naturally contain comprehensive vitamin profiles including riboflavin, choline, Biotin, thiamin, Niacin, folic acid, Pyridoxine, Pantothenic, and para-aminobenzoic acid (Figure 4) (Bakaloudi et al., 2021; Ng et al., 2024).

Figure 4. Dietary and functional traits of fungi-based SCP. FSCP serves as a significant reservoir of (A) essential vitamins and (E) indispensable amino acids. The most prevalent among these are the B complex vitamins, which encompass B1 (thiamine), B2 (riboflavin), B3 (Niacin), B5 (Pantothenic Acid), B6 (Pyridoxine), and B7 (Biotin); the critical amino acids consist of K (Lysine), M (Methionine), T (Threonine), C (Cysteine), I (Isoleucine), and L (Leucine). The functional characteristics comprise their (B) Capacity to regulate cholesterol concentrations, (C) Manage glycemic levels, and (D) Enhance overall gastrointestinal health.

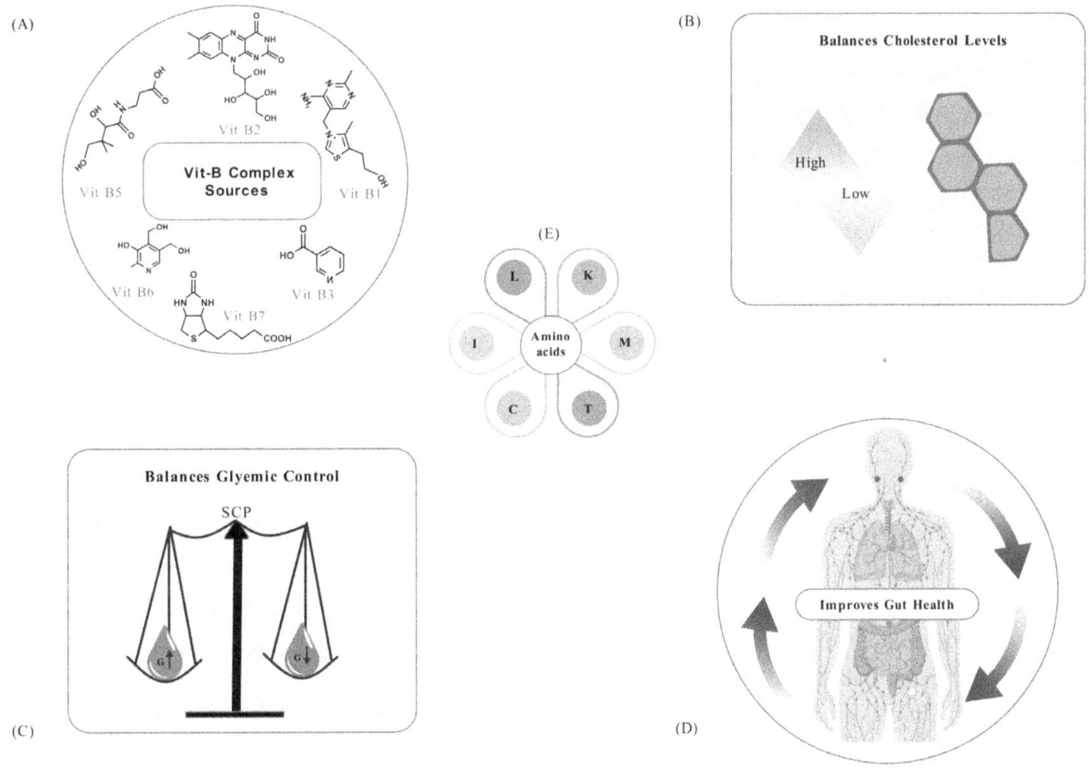

The illustration was created using GDP (Bakaloudi et al., 2021; Jiang et al., 2025; Pobiega et al., 2024; Schulz and Oslage, 1976; Udall et al., 1984)

The health benefits of fungal SCPs, particularly mycoproteins, extend across multiple physiological domains. Clinical research has demonstrated significant impacts on cardiovascular health, metabolic regulation, and digestive function (Majumder et al., 2024, Ng et al., 2024). Studies dating from the 1980s have consistently documented the cholesterol-lowering effects of *Fusarium venenatum* mycoprotein, with double-blind crossover studies confirming substantial reductions in serum cholesterol levels following sustained consumption (Turnbull et al., 1990). Similar effects have been observed in *Rhizopus oligosporus*, though the underlying mechanisms require further investigation (Lee et al., 2024).

Glycemic control represents another significant benefit of mycoprotein consumption. The high fiber content characteristic of these proteins modulates postprandial glucose and insulin responses by decreasing carbohydrate digestion and absorption rates (Cherta-Murillo et al., 2020). Mycoprotein, a high source of dietary fiber comprising 2/3 beta-glucan and 1/3 chitin, has demonstrated a positive influence on energy expenditure and appetite control through the regulation of metabolic hormones like ghrelin, GLP-1, leptin, and peptide tyrosine (PYY) (Majumder et al., 2024). This fiber content additionally contributes to enhanced satiety, supporting weight management through reduced caloric intake (Figure 4). The protein quality of fungal SCPs is particularly noteworthy. Their comprehensive essential amino acid profile and high protein digestibility-corrected amino acid score (PDCAAS) facilitate efficient muscle protein synthesis, positioning these proteins as viable alternatives to animal-based protein sources for muscle development and maintenance (Finnigan et al., 2024) (Figure 4). Yeast-based SCPs similarly support muscle growth and recovery, making them valuable for athletic performance and rehabilitation (Derbyshire & Ayoob 2019). Gastrointestinal health benefits emerge as a significant attribute of fungal proteins (Li et al., 2023). *Fusarium venenatum* mycoprotein's fiber content promotes beneficial gut microbiota populations, with clinical trials demonstrating reduced fecal genotoxicity and enhanced microbial diversity when substituted for red and processed meat (Li et al.,2023; Sirtori et al.,2023; Finnigan et al.,2024; Lee et al.,2024). *Aspergillus oryzae* contributes to gut health through its beta-glucan content, which exhibits prebiotic effects. Research has demonstrated improved iron bioavailability and gut microbiota composition in experimental models supplemented with iron-enriched *A. oryzae*. Recent investigations have revealed potential antitumor properties among certain fungal species. Studies of *Aspergillus luchuensis* powder demonstrate protective effects against carcinogenesis, while *A. niger* CJ6 extract exhibits pancreatic cancer cell growth suppression (Derbyshire & Delange, 2021; Lee et al., 2024). *Rhizopus* species have shown promise through their extracellular polysaccharides' anticancer potential, though these findings require further validation through human clinical trials (Derbyshire & Delange, 2021). The diverse health benefits of fungal SCPs extend beyond conventional nutritional parameters to include specific therapeutic applications. However, continued research is essential to elucidate precise mechanisms of action and validate preliminary findings through comprehensive human clinical trials. Safety assessments, particularly regarding mycotoxin production, remain crucial for establishing long-term consumption guidelines. The convergence of nutritional adequacy, therapeutic potential, and sustainability positions fungal SCPs as promising candidates for addressing both nutritional requirements and health promotion in contemporary food systems (Derbyshire & Delange, 2021). Yeast-derived SCPs further exemplify the nutritional potential of fungal proteins, containing a balanced amino acid profile that aligns with FAO recommendations (Ma et al., 2023). Their nutritional value extends beyond protein content to include significant concentrations of B-complex vitamins (thiamine, riboflavin, Pyridoxine, Biotin, and folic acid) and essential minerals including nickel, lithium, selenium, and chromium (Sharif et al., 2021). This comprehensive examination of amino acid profiles and nutritional attributes demonstrates the significant potential of fungal SCPs in addressing global nutritional needs while offering specific functional benefits for human health and metabolism. The ability to produce these nutrient-rich proteins from diverse substrates, including waste materials, further enhances their significance in sustainable food production systems.

9. CHALLENGES IN FUNGI-BASED SCP PRODUCTION

The development and implementation of fungal single cell protein (SCP) production systems encounter multifaceted challenges across technical, economic, and regulatory domains. These challenges necessitate systematic approaches for establishing viable and sustainable production systems (Aidoo et al., 2023).

9.1. Technical Challenges

Fermentation optimization represents a fundamental technical hurdle, requiring precise control of multiple parameters including temperature, pH, oxygen availability, and agitation rates. The complexity of maintaining optimal conditions is compounded by stringent sterility requirements, as contamination can compromise production integrity and introduce safety risks. Strain selection and improvement demand careful consideration of essential characteristics: protein content, growth kinetics, substrate utilization capabilities, and nucleic acid concentrations. Contemporary research focuses on screening programs and genetic engineering applications to develop strains optimized for specific substrates and industrial conditions. Scale-up processes present additional complexities, particularly in downstream operations. While continuous production systems demonstrate superior profitability, challenges in cell harvesting, drying, and nucleic acid reduction significantly impact nutritional value and process efficiency. These operations require careful optimization to maintain product quality while ensuring economic viability (Kumar et al., 2024; Li et al., 2023).

9.2. Economic Challenges

Raw material costs and production expenses significantly influence economic feasibility. Traditional substrates, particularly glucose derived from maize, constitute substantial production costs. Alternative carbon sources, including agricultural residues and food industry byproducts, offer potential cost reductions but often require additional preprocessing steps, necessitating careful economic analysis (Dalbanjan et al., 2024). Market acceptance presents complex challenges involving competition with established protein sources. Success requires demonstrating clear advantages in nutrition, sustainability, and cost-effectiveness while addressing consumer concerns regarding taste, texture, and safety through comprehensive education and marketing strategies (Sirtori et al., 2023).

9.3. Regulatory and Safety Considerations

Regulatory compliance demands rigorous adherence to food safety frameworks established by authorities such as European Food Safety Authority (EFSA) and the United States, Food and Drug Administration (FDA). These requirements encompass production processes, labeling standards, and contamination controls. Additionally, comprehensive allergenicity and toxicity evaluations are essential for establishing long-term safety profiles and securing consumer trust (Sirtori et al., 2023). The successful implementation of fungal SCP production systems requires integrated approaches addressing these multifaceted challenges while ensuring product safety and quality. Continued research and development efforts, coupled with effective regulatory compliance and market positioning strategies, remain crucial for establishing fungal SCPs as viable protein alternatives (Dalbanjan et al., 2024).

10. CONCLUSIONS

Fungal single cell proteins (FSCPs) represent a significant advancement in sustainable protein production technology, demonstrating remarkable potential in addressing global nutritional demands. The evolution from fundamental research to industrial application has yielded substantial progress in strain development, bioprocess engineering, and product formulation, though considerable potential remains unexplored. Key attributes of FSCPs include their exceptional nutritional profile, characterized by high protein content, optimized amino acid compositions, and enhanced digestibility comparable to animal proteins. Their biochemical complexity, including bioactive compounds such as β-glucans, further augments their nutritional value. Life cycle assessments consistently demonstrate their reduced environmental impact relative to conventional protein sources, particularly regarding resource utilization and greenhouse gas emissions. The versatility of FSCPs spans applications from meat analogues to animal feed, while their capacity to utilize diverse waste streams positions them advantageously within circular bioeconomy frameworks. Although current production costs exceed traditional protein sources, ongoing technological advances and scaling efforts steadily improve their economic viability.

Critical challenges requiring attention include optimization of scale-up processes, genetic enhancement through CRISPR-Cas9 technology, and bioprocess innovations. Additionally, regulatory harmonization and consumer acceptance remain crucial for widespread adoption. The integration of FSCP production with biorefinery systems offers enhanced resource efficiency, while novel applications in materials science and space nutrition present emerging research opportunities. The convergence of biotechnology, sustainability imperatives, and food security concerns positions FSCPs as integral components of future protein production systems. As research advances and production scales expand, FSCPs demonstrate increasing potential to contribute to sustainable global nutrition, marking a transformation from laboratory innovation to industrial implementation. This technological progression, coupled with growing environmental awareness and resource constraints, suggests FSCPs will play an increasingly significant role in reshaping protein production paradigms, promising continued innovation in addressing global nutritional needs while maintaining environmental stewardship.

REFERENCES

Abdel-Azeem, A. M., & Sheir, D. H. (2020). Bioconversion of Lignocellulosic Residues into Single-Cell Protein (SCP) by *Chaetomium*. In Abdel-Azeem, A. (Ed.), *Recent Developments on Genus Chaetomium. Fungal Biology*. Springer., DOI: 10.1007/978-3-030-31612-9_13

Aggelopoulos, T., Katsieris, K., Bekatorou, A., Pandey, A., Banat, I. M., & Koutinas, A. A. (2014). Solid state fermentation of food waste mixtures for single cell protein, aroma volatiles and fat production. *Food Chemistry*, *145*, 710–716. DOI: 10.1016/j.foodchem.2013.07.105

Ahmad, M. I., Farooq, S., Alhamoud, Y., Li, C., & Zhang, H. (2022). A review on mycoprotein: History, nutritional composition, production methods, and health benefits. *Trends in Food Science & Technology*, *121*, 14–29. DOI: 10.1016/j.tifs.2022.01.027

Ahmadi, A. R., Ghoorchian, H., Hajihosaini, R., & Khanifar, J. (2010). Determination of the amount of protein and amino acids extracted from the microbial protein (SCP) of lignocellulosic wastes. *Pakistan journal of biological sciences. Pakistan Journal of Biological Sciences*, *13*(8), 355–361. DOI: 10.3923/pjbs.2010.355.361

Ahmed, M. G., Gouda, S. A., Donia, S., & Hassanein, N. M. (2024). Production of single cell protein by fungi from different food wastes. *Biomass Conversion and Biorefinery*, •••, 1–16. DOI: 10.1007/s13399-024-05478-5

Aidoo, R., Kwofie, E. M., Adewale, P., Lam, E., & Ngadi, M. (2023). Overview of single cell protein: Production pathway, sustainability outlook, and digital twin potentials. *Trends in Food Science & Technology*, *138*, 577–598. Advance online publication. DOI: 10.1016/j.tifs.2023.07.003

Ali, S., Mushtaq, J., Nazir, F., & Sarfraz, H. (2017). Production and processing of single cell protein (SCP)-a review. *European Journal Pharmaceutical and Medical Research*, *4*(7), 86–94.

Alriksson, B., Hörnberg, A. N. D. R. E. A. S., Gudnason, A. E., Knobloch, S., Arnason, J., & Johannsson, R. (2014). Fish feed from wood. *Cellulose Chemistry and Technology*, *48*(9-10), 843–848.

Amara, A. A., & El-Baky, N. A. (2023). Fungi as a source of edible proteins and animal feed. *Journal of Fungi (Basel, Switzerland)*, *9*(1), 73. DOI: 10.3390/jof9010073

Ángel, N. L. M. (2024). *Production process and formulation of a single-cell protein by fermenting the yeast Kluyveromyces marxianus* (WO 2024/005625 A1). https://lens.org/060-340-418-255-658

Anichebe, C. O., Okoye, E. L., & Onochie, C. C. (2019). (Forthcoming). Comparative study on single cell protein (SCP) production by Trichoderma Viride from pineapple wastes and banana peels. *SSRN*. Advance online publication. https://papers.ssrn.com/sol3/papers.cfm?abstract_id=3448990. DOI: 10.2139/ssrn.3448990

Bajpai, P., & Bajpai, P. K. (1987). Single cell protein production from rayon pulp will waste by *Paecilomyces variotii.Journal of Fermentation Technology*, *65*(3), 349–351. DOI: 10.1016/0385-6380(87)90099-9

Bakaloudi, D. R., Halloran, A., Rippin, H. L., Oikonomidou, A. C., Dardavesis, T. I., Williams, J., Wickramasinghe, K., Breda, J., & Chourdakis, M. (2021). Intake and adequacy of the vegan diet. A systematic review of the evidence. *Clinical Nutrition (Edinburgh, Lothian)*, *40*(5), 3503–3521. DOI: 10.1016/j.clnu.2020.11.035

Bohrer, B. M. (2019). An investigation of the formulation and nutritional composition of modern meat analogue products. *Food Science and Human Wellness*, *8*(4), 320–329. DOI: 10.1016/j.fshw.2019.11.006

Bratosin, B. C., Darjan, S., & Vodnar, D. C. (2021). Single cell protein: A potential substitute in human and animal nutrition. *Sustainability (Basel)*, *13*(16), 9284. DOI: 10.3390/su13169284

Carranza-Méndez, R. C., Chávez-González, M. L., Sepúlveda-Torre, L., Aguilar, C. N., Govea-Salas, M., & Ramos-González, R. (2022). Production of single cell protein from orange peel residues by Candida utilis. *Biocatalysis and Agricultural Biotechnology*, *40*, 102298. DOI: 10.1016/j.bcab.2022.102298

Cereghino, J. L., & Cregg, J. M. (2000). Heterologous protein expression in the methylotrophic yeast Pichia pastoris. *FEMS Microbiology Reviews*, *24*(1), 45–66. DOI: 10.1111/j.1574-6976.2000.tb00532.x

Cherta-Murillo, A., Lett, A. M., Frampton, J. E., Chambers, E. S., Finnigan, T. J. A., & Frost, G. (2020). Effects of mycoprotein on glycaemic control and energy intake in humans: A systematic review. *British Journal of Nutrition*, *123*(12), 1321–1332. DOI: 10.1017/S0007114520000756

Cleland, J. (2013). World Population Growth; Past, Present and Future. *Environmental and Resource Economics*, *55*(4), 543–554. DOI: 10.1007/s10640-013-9675-6

Dalbanjan, N. P., Eelager, M. P., & Narasagoudr, S. S. (2024). Microbial protein sources: A comprehensive review on the potential usage of fungi and cyanobacteria in sustainable food systems. *Food and Humanity*, *3*, 100366. DOI: 10.1016/j.foohum.2024.100366

Derbyshire, E., & Ayoob, K.-T. (2019). Mycoprotein: Nutritional and health properties. *Nutrition Today*, *54*(1), 7–15. DOI: 10.1097/NT.0000000000000316

Derbyshire, E. J., & Delange, J. (2021). Fungal protein–what is it and what is the health evidence? A systematic review focusing on mycoprotein. *Frontiers in Sustainable Food Systems*, *5*, 581682. DOI: 10.3389/fsufs.2021.581682

Dharumadurai Dhanasekaran, D. D., Subramaniyan Lawanya, S. L., Subhasish Saha, S. S., Nooruddin Thajuddin, N. T., & Annamalai Panneerselvam, A. P. (2011). Production of single cell protein from pineapple waste using yeast. https://www.cabidigitallibrary.org/doi/full/10.5555/20113172515

Dong, L., Lu, Z., & Jiabao, L. (2022). *Alkali-resistant yeast and application thereof in production of single-cell protein by utilizing biogas slurry* (CN 113913309 A). https://lens.org/031-552-795-216-120

Ferreira, I. M. P. L. V. O., Pinho, O., Vieira, E., & Tavarela, J. G. (2010). Brewer's Saccharomyces yeast biomass: Characteristics and potential applications. *Trends in Food Science & Technology*, *21*(2), 77–84. DOI: 10.1016/j.tifs.2009.10.008

Finnigan, T., Mach, K., & Edlin, A. (2024). Mycoprotein: a healthy new protein with a low environmental impact. In *Sustainable protein sources* (pp. 539-566). Academic Press. https://doi.org/DOI: 10.1016/B978-0-323-91652-3.00011-3

Furey, B., Slingerland, K., Bauter, M. R., Dunn, C., Goodman, R. E., & Koo, S. (2022). Safety evaluation of Fy Protein™ (Nutritional Fungi Protein), a macroingredient for human consumption. *Food and Chemical Toxicology*, *166*, 113005. DOI: 10.1016/j.fct.2022.113005

Gao, L., Meng, J., Dai, W., Zhang, Z., Dong, H., Yuan, Q., Zhang, W., Liu, S., & Wu, X. (2023). Deciphering cell wall sensors enabling the construction of robust P. pastoris for single-cell protein production. *Biotechnology for Biofuels and Bioproducts*, *16*(1), 178. DOI: 10.1186/s13068-023-02428-7

Gao, Y., Li, D., & Liu, Y. (2012). Production of single cell protein from soy molasses using Candida tropicalis. *Annals of Microbiology*, *62*(3), 1165–1172. DOI: 10.1007/s13213-011-0356-9

Gorte, O., Hollenbach, R., Papachristou, I., Steinweg, C., Silve, A., Frey, W., Syldatk, C., & Ochsenreither, K. (2020). Evaluation of Downstream Processing, Extraction, and Quantification Strategies for Single Cell Oil Produced by the Oleaginous Yeasts *Saitozyma podzolica* DSM 27192 and *Apiotrichum porosum* DSM 27194. *Frontiers in Bioengineering and Biotechnology*, *8*, 355. DOI: 10.3389/fbioe.2020.00355

Greene, J. M., & McAdam, P. A. (1983). Nutritional Requirements and the Appropriate Use of Supplements. In *Nutrition in the middle and later years* (pp. 17–45). Butterworth-Heinemann., DOI: 10.1016/B978-0-7236-7046-9.50008-4

Guohou, Q., Lanfu, W., Wanxiao, G., Yuansen, H., Yin, Z., Min, C., Chong, F., Guojie, Z., Zonghao, X., Hongwei, W., Yapeng, T., Guangyou, Z., & Xia, W. (2013). *Pichia pastoris for efficiently converting methanol to produce single cell protein and application of Pichia pastoris* (CN 103045494 A). https://lens.org/042-289-199-791-43X

Haider, M. (2021). Bioconversion of saw dust powder acid hydrolysis to single cell protein by the yeast Candida tropicalis. *Journal of Life and Bio Sciences Research*, *2*(02), 31–35. DOI: 10.38094/jlbsr2241

Hashempour-Baltork, F., Hosseini, S. M., Assarehzadegan, M. A., Khosravi-Darani, K., & Hosseini, H. (2020). Safety assays and nutritional values of mycoprotein produced by Fusarium venenatum IR372C from date waste as substrate. *Journal of the Science of Food and Agriculture*, *100*(12), 4433–4441. DOI: 10.1002/jsfa.10483

Jach, M. E., Serefko, A., Ziaja, M., & Kieliszek, M. (2022). Yeast protein as an easily accessible food source. *Metabolites*, *12*(1), 63. DOI: 10.3390/metabo12010063

Jiang, S., Li, H., Zhang, L., Mu, W., Zhang, Y., Chen, T., Wu, J., Tang, H., Zheng, S., Liu, Y., Wu, Y., Luo, X., Xie, Y., & Ren, J. (2025). Generic Diagramming Platform (GDP): A comprehensive database of high-quality biomedical graphics. *Nucleic Acids Research*, *53*(D1), D1670–D1676. DOI: 10.1093/nar/gkae973

Junaid, F. K. L. A., Khawaja, L. A., & Ali, S. (2020). Single cell proteins as a potential meat substitute: A critical review. *World Journal of Pharmaceutical Research*, *9*, 141–161. DOI: 10.20959/wjpr20202-16686

Kargi, F., Shuler, M. L., Vashon, R., Seeley, H. W.Jr, Henry, A., & Austic, R. E. (1980). Continuous aerobic conversion of poultry waste into single-cell protein using a single reactor: Kinetic analysis and determination of optimal conditions. *Biotechnology and Bioengineering*, *22*(8), 1567–1600. DOI: 10.1002/bit.260220805

Kumar, R., Raj, T., Næss, G., Sørensen, M., & Dhawan, V. (2024). Opportunities and challenges in single-cell protein production using lignocellulosic material. *Biofuels, Bioproducts & Biorefining*, *18*(1), 310–321. DOI: 10.1002/bbb.2563

Kurcz, A., Błażejak, S., Kot, A. M., Bzducha-Wróbel, A., & Kieliszek, M. (2018). Application of industrial wastes for the production of microbial single-cell protein by fodder yeast Candida utilis. *Waste and Biomass Valorization*, *9*(1), 57–64. DOI: 10.1007/s12649-016-9782-z

Lapeña, D., Kosa, G., Hansen, L. D., Mydland, L. T., Passoth, V., Horn, S. J., & Eijsink, V. G. (2020). Production and characterization of yeasts grown on media composed of spruce-derived sugars and protein hydrolysates from chicken by-products. *Microbial Cell Factories*, *19*(1), 1–14. DOI: 10.1186/s12934-020-1287-6

Lee, D., Pan, J. H., Kim, D., Heo, W., Shin, E. C., Kim, Y. J., Shim, Y. Y., Reaney, M. J. T., Ko, S., Hong, S., Cho, H. T., Kim, T. G., Lee, K., & Kim, J. K. (2024). Mycoproteins and their health-promoting properties: Fusarium species and beyond. *Comprehensive Reviews in Food Science and Food Safety*, *23*(3), e13365. DOI: 10.1111/1541-4337.13365

Li, K., Qiao, K., Xiong, J., Guo, H., & Zhang, Y. (2023). Nutritional Values and Bio-Functional Properties of Fungal Proteins: Applications in Foods as a Sustainable Source. *Foods (Basel, Switzerland)*, *12*(24), 4388. DOI: 10.3390/molecules28114388

Li, Y. P., Ahmadi, F., Kariman, K., & Lackner, M. (2024). Recent advances and challenges in single cell protein (SCP) technologies for food and feed production. *npj Science of Food*, *8*(1), Article 66. https://doi.org/DOI: 10.1038/s41538-024-00299-2

Linder, T. (2023). Beyond Agriculture— How Microorganisms Can Revolutionize Global Food Production. *ACS Food Science & Technology*, *3*(7), 1144–1152. DOI: 10.1021/acsfoodscitech.3c00099

Linder, T. (2024). What next for mycoprotein? *Current Opinion in Food Science*, *101199*, 101199. Advance online publication. DOI: 10.1016/j.cofs.2024.101199

Lonchamp, J., Stewart, K., Munialo, C. D., Evans, L., Akintoye, M., Gordon, S., Clegg, P. S., Willoughby, N., & Euston, S. R. (2022). Mycoprotein as novel functional ingredient: Mapping of functionality, composition and structure throughout the Quorn fermentation process. *Food Chemistry*, *396*, 133736. DOI: 10.1016/j.foodchem.2022.133736

Ma, J., Sun, Y., Meng, D., Zhou, Z., Zhang, Y., & Yang, R. (2023). Yeast proteins: The novel and sustainable alternative protein in food applications. *Trends in Food Science & Technology*, *135*, 190–201. DOI: 10.1016/j.tifs.2023.04.003

Majumder, R., Miatur, S., Saha, A., & Hossain, S. (2024). Mycoprotein: Production and nutritional aspects: A review. *Sustainable Food Technology*, *2*(1), 81–91. DOI: 10.1039/D3FB00169E

Maria, D. L. W. T. A., & Sebastian, G. M. J. (2023). *Single cell protein from thermophilic fungi* (US 11707078 B2). U.S. Patent. https://lens.org/142-801-579-684-076

Meng, J., Liu, S., Gao, L., Hong, K., Liu, S., & Wu, X. (2023). Economical production of Pichia pastoris single cell protein from methanol at industrial pilot scale. *Microbial Cell Factories*, *22*(1), 198. DOI: 10.1186/s12934-023-02198-9

Nalage, D., Khedkar, G., Kalyankar, A., Sarkate, A., Ghodke, S., Bedre, V. B., & Khedkar, C. D. (2016). Single cell proteins. *Encyclopedia of food and health*, *4*, 790-794. http://dx.doi.org/DOI: 10.1016/B978-0-12-384947-2.00628-0

Nasseri, A. T., Rasoul-Amini, S., Morowvat, M. H., & Ghasemi, Y. (2011). Single cell protein: Production and process. *American Journal of Food Technology*, *6*(2), 103–116. DOI: 10.3923/ajft.2011.103.116

Nayeem, M., Chauhan, K., Khan, S., Rattu, G., Dhaka, R. K., & Sidduqui, H. (2017). Optimization of low-cost substrate for the production of single cell protein using Kluyveromyces marxianus. *Pharma Innov J*, *6*(8).

Ng, Z. Y., Kee, P. E., Abdullah, R., Lan, J. C. W., Ling, T. C., Jiang, J. J., Lim, J. W., & Khoo, K. S. (2024). Conversion of lignocellulosic biomass waste into mycoprotein: Current status and future directions for sustainable protein production. *Biomass Conversion and Biorefinery*, •••, 1–27. DOI: 10.1007/s13399-024-06130-y

Onyeaka, H., Anumudu, C. K., Okpe, C., Okafor, A., Ihenetu, F., Miri, T., ... & Anyogu, A. (2022). Single cell protein for foods and feeds: A review of trends. *The Open Microbiology Journal, 16*(1), e187428582206160. https://, 2022, 16, e187428582206160DOI: 10.2174/18742858-v16-e2206160

Ouedraogo, N., Savadogo, A., Somda, M. K., Zongo, C., & Traore, A. S. (2017). Effect of mineral salts and nitrogen source on yeast (Candida utilis NOY1) biomass production using tubers wastes. *African Journal of Biotechnology*, *16*(8), 359–365. DOI: 10.5897/AJB2016.15801

Pillaca-Pullo, O. S., Lopes, A. M., & Estela-Escalante, W. D. (2024). Reusing wastewater from Coffea arabica processing to produce single-cell protein using Candida sorboxylosa: Optimizing of culture conditions. *Biotechnology Progress*, *40*(1), e3393. DOI: 10.1002/btpr.3393

Pobiega, K., Sękul, J., Pakulska, A., Latoszewska, M., Michońska, A., Korzeniowska, Z., Macherzyńska, Z., Pląder, M., Duda, W., Szafraniuk, J., Kufel, A., Dominiak, Ł., Lis, Z., Kłusek, E., Kozicka, E., Wierzbicka, A., Trusińska, M., Rybak, K., Kot, A. M., & Nowacka, M. (2024). Fungal proteins: Sources, production and purification methods, industrial applications, and future perspectives. *Applied Sciences (Basel, Switzerland)*, *14*(14), 6259. DOI: 10.3390/app14146259

Rages, A. A., Haider, M. M., & Aydin, M. (2021). Alkaline hydrolysis of olive fruits wastes for the production of single cell protein by *Candida lipolytica.Biocatalysis and Agricultural Biotechnology*, *33*, 101999. DOI: 10.1016/j.bcab.2021.101999

Ravindra, P. (2000). Value-added food: Single cell protein. *Biotechnology Advances*, *18*(6), 459–479. DOI: 10.1016/S0734-9750(00)00045-8

Raziq, A., Lateef, M., Ullah, A., Ullah, H., & Khan, M. W. (2020). Single cell protein (SCP) production and potential substrates: A comprehensive review. [PAB]. *Pure and Applied Biology*, *9*(3), 1743–1754. DOI: 10.19045/bspab.2020.90185

Reihani, S. F. S., & Khosravi-Darani, K. (2019). Influencing factors on single-cell protein production by submerged fermentation: A review. *Electronic Journal of Biotechnology*, *37*, 34–40. DOI: 10.1016/j.ejbt.2018.11.005

Ribeiro, G. O., Rodrigues, L. D. A. P., Santos, T. B. S. D., Alves, J. P. S., Oliveira, R. S., Nery, T. B. R., Barbosa, J. D. V., & Soares, M. B. P. (2023). Innovations and developments in single cell protein: Bibliometric review and patents analysis. *Frontiers in Microbiology*, *13*, 1093464. DOI: 10.3389/fmicb.2022.1093464

Ritala, A., Häkkinen, S. T., Toivari, M., & Wiebe, M. G. (2017). Single cell protein—state-of-the-art, industrial landscape and patents 2001–2016. *Frontiers in microbiology, 8*, 2009. https://doi.org/DOI: 10.3389/fmicb.2017.02009

Sagar, P. R., Raol, G. G., Prajapati, D., Chaudhari, H. G., & Kiri, B. D. (2024). Statistical Evaluation for the formulation of low-cost Single cell protein production in Food waste medium by Pichia occidentalis: A sustainable valorization approach. *Waste and Biomass Valorization*, *15*(6), 3457–3475. DOI: 10.1007/s12649-023-02382-3

Sankar, N. R., Kumar, V. K., Shailaja, R., Saritha, K., & Naidu, N. V. (2011). Single cell protein production by *Trichoderma harzianum* using waste banana peel. http://idosi.org/ijmr/ijmr2(1)11/13a.pdf

Saravanan, K., Rajendran, M., & Kathirvel, P. (2023). Utilization of Water Hyacinth (Eichhornia crassipes) as a Feasible Substrate for the Production of Single Cell Protein Using Saccharomyces cerevisiae (Baker's Yeast). *Waste and Biomass Valorization*, *14*(12), 4231–4242. DOI: 10.1007/s12649-023-02138-z

Schrader, J., Schilling, M., Holtmann, D., Sell, D., Villela Filho, M., Marx, A., & Vorholt, J. A. (2009). Methanol-based industrial biotechnology: Current status and future perspectives of methylotrophic bacteria. *Trends in Biotechnology*, *27*(2), 107–115. DOI: 10.1016/j.tibtech.2008.10.009

Schulz, E., & Oslage, H. J. (1976). Composition and nutritive value of single-cell protein (SCP). *Animal Feed Science and Technology*, *1*(1), 9–24. DOI: 10.1016/0377-8401(76)90003-1

Sharif, M., Zafar, M. H., Aqib, A. I., Saeed, M., Farag, M. R., & Alagawany, M. (2021). Single cell protein: Sources, mechanism of production, nutritional value and its uses in aquaculture nutrition. *Aquaculture (Amsterdam, Netherlands)*, *531*, 735885. DOI: 10.1016/j.aquaculture.2020.735885

Sharma, R., Garg, P., Kumar, P., Bhatia, S. K., & Kulshrestha, S. (2020). Microbial fermentation and its role in quality improvement of fermented foods. *Fermentation (Basel, Switzerland)*, *6*(4), 106. DOI: 10.3390/fermentation6040106

Sirtori, C. R., & Lammi, C. (2023). Challenges and Opportunities for Mycoprotein-Based Food Products. *The American Journal of Clinical Nutrition*, *118*(1), 5–7. DOI: 10.1016/j.ajcnut.2023.05.019

Spalvins, K., Geiba, Z., Kusnere, Z., & Blumberga, D. (2020). Waste Cooking Oil as Substrate for Single Cell Protein Production by Yeast Yarrowia lipolytica. *Environmental and Climate Technologies*, *24*(3), 457–469. DOI: 10.2478/rtuect-2020-0116

Spalvins, K., Ivanovs, K., & Blumberga, D. (2018). Single cell protein production from waste biomass: Review of various agricultural by-products. https://doi.org/DOI: 10.15159/AR.18.129

Suman, G., Nupur, M., Anuradha, S., & Pradeep, B. (2015). Single cell protein production: A review. *International Journal of Current Microbiology and Applied Sciences*, *4*(9), 251–262.

Thumkasem, N., On-mee, T., Chittapun, S., Pornpukdeewattana, S., Ketudat-Cairns, M., Thongprajukaew, K., Antimanon, S., & Charoenrat, T. (2023). Rhodotorula paludigena CM33 cultivation process development for high β-carotene single cell protein production. *Biocatalysis and Agricultural Biotechnology*, *54*, 102926. DOI: 10.1016/j.bcab.2023.102926

Turnbull, W., Ar, L., & Edwards, G. (1990). Effect of mycoprotein on blood lipids. *The American Journal of Clinical Nutrition*, *52*(4), 646–650. DOI: 10.1093/ajcn/52.4.646

Ugalde, U. O., & Castrillo, J. I. (2002). Single cell proteins from fungi and yeasts. In *Applied mycology and biotechnology* (Vol. 2, pp. 123–149). Elsevier., https://www.sciencedirect.com/science/article/pii/S1874533402800089

United Nations, Department of Economic and Social Affairs, Population Division (2017). *World Population Prospects: The 2017 Revision, Key Findings and Advance Tables.* Working Paper No. ESA/P/WP/248.

Varavinit, S., Srithongkum, P., De-Eknamkul, C., Assavanig, A., & Charoensiri, K. (1996). Production of Single Cell Protein from Cassava Starch in Air-lift Fermenter by Cephalosporium eichhorniae. *Stärke*, *48*(10), 379–382. DOI: 10.1002/star.19960481007

Wanxiao, G., Gang, Y., Xiaohui, R., Yuansen, H., Hongwei, W., Yanchao, F., Guanzhang, B., Jun, X., Zhihong, L., Jingmei, H., Min, M., Qian, X., Yapei, D., Yanbing, C., & Congle, Y. (2015). Pichia pastoris for producing methanol protein and lipase at same time and application thereof (CN 105132301 A). https://lens.org/056-903-998-744-571

Wiebe, M. (2002). Myco-protein from Fusarium venenatum: A well-established product for human consumption. *Applied Microbiology and Biotechnology*, *58*(4), 421–427. DOI: 10.1007/s00253-002-0931-x

Wu, J., Hu, J., Zhao, S., He, M., Hu, G., Ge, X., & Peng, N. (2018). Single-cell Protein and Xylitol Production by a Novel Yeast Strain Candida intermedia FL023 from Lignocellulosic Hydrolysates and Xylose. *Applied Biochemistry and Biotechnology*, *185*(1), 163–178. DOI: 10.1007/s12010-017-2644-8

Xin, W., & Le, G. (2023). *Pichia pastoris with multi-carbon-source utilization and high-protein synthesis and application of Pichia pastoris* (CN 116478838 A). https://lens.org/122-059-031-651-763

Yadav, J. S. S., Bezawada, J., Ajila, C. M., Yan, S., Tyagi, R. D., & Surampalli, R. Y. (2014). Mixed culture of Kluyveromyces marxianus and Candida krusei for single-cell protein production and organic load removal from whey. *Bioresource Technology*, *164*, 119–127. DOI: 10.1016/j.biortech.2014.04.069

Yadav, J. S. S., Yan, S., Ajila, C. M., Bezawada, J., Tyagi, R. D., & Surampalli, R. Y. (2016). Food-grade single-cell protein production, characterization and ultrafiltration recovery of residual fermented whey proteins from whey. *Food and Bioproducts Processing*, *99*, 156–165. DOI: 10.1016/j.fbp.2016.04.012

Yan, J., Han, B., Gui, X., Wang, G., Xu, L., Yan, Y., Madzak, C., Pan, D., Wang, Y., Zha, G., & Jiao, L. (2018). Engineering Yarrowia lipolytica to simultaneously produce lipase and single cell protein from agro-industrial wastes for feed. *Scientific Reports*, *8*(1), 758. DOI: 10.1038/s41598-018-19238-9

Yao, K. Y., Zhang, T. Z., Wang, H. F., & Liu, J. X. (2018). Upgrading of by-product from beverage industry through solid-state fermentation with Candida utilis and Bacillus subtilis. *Letters in Applied Microbiology*, *67*(6), 557–563. DOI: 10.1111/lam.13078

Yousufi, M. K. (2012). Impact of pH on the single cell protein produced on okara-wheat grit substrates using Rhizopus oligosporus and Aspergillus oryzae. *IOSR J. Environ. Sci. Toxicol. Food Technol, 1*, 32-35. ISSN: 2319-2402, ISBN: 2319-2399. Volume 1, Issue 2 (Sep-Oct. 2012), PP 32-35 www.iosrjournals.org

Zhang, Z., Chen, X., & Gao, L. (2024). New strategy for the biosynthesis of alternative feed protein: Single-cell protein production from straw-based biomass. *Global Change Biology. Bioenergy*, *16*(2), e13120. DOI: 10.1111/gcbb.13120

Zhanying, L., Jing, Y., & Jinna, C. (2022). Candida utilis and application thereof (CN 114836331 A). https://lens.org/194-618-495-505-094

Zheng, S., Yang, M., & Yang, Z. (2005). Biomass production of yeast isolate from salad oil manufacturing wastewater. *Bioresource Technology*, *96*(10), 1183–1187. DOI: 10.1016/j.biortech.2004.09.022

KEY TERMS AND DEFINITIONS

SCP: Single Cell Protein is a nutrient-dense food supplement usually derived from micro-organisms.

Fungal SCP (FSCP): Fungal based SCPs are protein sources exclusively derived from yeast and fungi.

Mycoprotein (MP): It is a renowned, popular and safe meat substitute that offers plethora of functional attributes that also include controlling glycemic levels.

Sustainable Protein: These proteins are typically synthesized based on alternative eco-friendly production strategies.

Food security: It is the state of being where everyone has accessibility to safe, reliable and nutritious food sources.

Alternative protein: These protein sources critically meet the rising global protein demands, while minimizing strain on land and water resources.

Life Cycle Assessment (LCA): LCA is the process of rigorous environmental evaluation strategy. It examines ecological footprints of SCP production, further enhancing protein quality.

Functional foods: They are known to enhance additional health benefits by containing biologically active compounds in them, which are usually absent from basic nutrition.

Animal feed: It refers to food materials that meet the nutritional requirements and enhance growth as well as productivity.

Nutraceuticals: These are food sources that additionally provide health benefits by preventing health hazards caused by chronic diseases.

Chapter 9

Mycoproteins:

The Rise of Fungal-Based Meat Alternatives

Dipali Saxena
https://orcid.org/0000-0002-8177-319X
Shri Vaishnav Vidyapeeth Vishwavidyalaya, India

Neha Prasad
https://orcid.org/0009-0006-1857-0521
VLCC Ltd., India

ABSTRACT

The rise of mycoproteins is contextualized within the broader movement towards sustainable food systems, driven by the growing global population, climate change, and the urgent need to reduce the environmental footprint of food production. This chapter delves into the historical development, production processes, nutritional benefits, and environmental impacts of mycoproteins. Mycoproteins offer several advantages, including lower greenhouse gas emissions, reduced land and water usage, and the ability to produce protein efficiently and at scale. The nutritional profile of mycoproteins is highlighted, emphasizing their high protein content, essential amino acids, and fiber, alongside low fat and cholesterol levels. The chapter also addresses consumer acceptance and market trends, noting the increasing popularity of mycoprotein-based products in mainstream diets. Mycoproteins serves as a viable and innovative solution to some of the most pressing challenges in sustainable food production, underscoring their potential to transform the future of food.

1. INTRODUCTION

The increasing global demand for protein, driven by population growth and changing dietary habits, has highlighted the limitations and environmental impacts of traditional meat production. The demand for sustainable and ethical protein sources has driven intense innovation in food science, with mycoproteins, derived from fungi, emerging as one of the most promising solutions. Mycoprotein is a type of protein that derives from a fungus that is having similarity like mushrooms and can be used for human consumptions (Finnigan et al., 2019). In the recent era where most of the population is now going to

DOI: 10.4018/979-8-3693-8201-1.ch009

become totally vegan because of animal exploitation, here mycoprotein can be used as a best alternative for meat and meat-based products. The increased global population has resulted in increased need of foods, so the production of animal-based proteins includes limitations, including time, energy and cost as well as environmental consideration. Therefore, replacement of meat by alternative ingredients can be a useful approach and in the past few decades, plant-based protein has been explored as a viable source of red meat substitutes due to animal welfare, social, and environmental concerns (Apostolidis & Mcleay, 2019) and microbial protein, especially mycoproteins, can substitute partially or entirely animal-based protein foods such as meats. Use of agro-industrial wastes to produce mycoproteins is a multiple target, especially from environmental aspects.

2. STATUS OF GLOBAL DEMAND OF MEAT ALTERNATIVES

By 2050 the annual growth rate of 0.8%, the world's population will be close to 9.8 billion and ensuring the world's expanding population for the access of sufficient, sustainable, and nutrient-rich food is one of the biggest challenges (Nadathur et al.,2017). The data suggests that protein consumption will be drastically increased, which may lead to the prevalence of malnutrition across the globe. To counteract this situation for increased demand of protein, production of protein sources must be increased, however the difficulty is to increase the production sustainably. Therefore, looking for suitable alternatives to meat proteins which involve lower cost and require fewer raw materials is essential (Aiking 2014). Emerging foods, including insects, algae, and microbes, are highly demanded as sustainable and alternative protein sources. The meat industry won't be able to meet this increase in demand by using more resources because of the ecological significance of cattle and issues with animal welfare. Demand for economical and ecologically sustainable protein in the food system is increasing significantly (Fasolin et al.,2019). The growing Food-Based Dietary Guidelines (FBDG) are gradually including more plant-based dietary proteins. However, other popular dietary proteins, such as those derived from fungi, appear to have gained less study. Mycoproteins derived from fungi are gaining popularity due to their excellent nutritional value, low production costs, ecological advantages, and tolerance to environmental challenges, including drought and flooding (Zhang et al., 2023).

3. EXPLANATION OF MYCOPROTEINS

Mycoproteins were developed during the Green Revolution. Mycoprotein is a nutritious, fungal-based protein that is an alternative source to animal-based protein foods such as meat. Mycoprotein was derived from a fungus *Fusarium venenatum* that can be easily grown in soil and is not harmful for human consumption. Mycoprotein is protein rich & fibre rich, plant-based proteins as healthier than animal products. Food-grade mycoprotein contains proteins, fats, carbohydrates, fiber and essential minerals like Sodium, zinc, Iron (Hashempour, et al 2020). As the strain of *Fusarium venenatum* have thin fibres it can be easily imitate as the texture of meat (Derbyshire & Delange; Zhang et al.,2021)

As meat alternatives, mycoproteins have been developed into several types, each varying in texture, flavor, and formulation to meet diverse consumer preferences and culinary uses. Textured mycoproteins are processed to mimic the fibrous structure of meat, achieving a texture close to ground meats like beef and chicken. This type of mycoprotein undergoes alignment and pressing of fungal filaments to simulate muscle fibres, making it suitable for applications such as meatless burgers, sausages, and fillets. Mycoproteins are designed to meet demand for realistic meat substitutes and have shown high acceptance in markets such as Europe and North America (Grossmann & Weiss, 2021).

3.1 History of mycoprotein development

The discovery of mycoprotein was a result of "The Green Protein Revolution"-inspired British scientists looking for a sustainable alternative protein that could offset the global crisis fuelled by population growth, food shortage and greenhouse gas emissions. The history of mycoprotein as a meat alternative date back to research in the 1960s when the growing demand for protein sources, alongside concerns about global food security, led scientists to explore sustainable protein alternatives. Mycoprotein's development began in the United Kingdom, primarily due to the efforts of scientists at the British company Rank Hovis McDougall (RHM). They sought to cultivate a sustainable protein source from microorganisms, focusing on filamentous fungi due to their high protein content and efficient growth on low-cost feed stocks. In 1985, almost 20 years later, Marlow Foods became the first company to launch mycoprotein-based products on the market under the commercial name of Quorn (Finnigan et al., 2019).

Fusarium venenatum ATCC 2684 was developed in 1960 and is the prolific strain used to cultivate and harvest mycoprotein. Mycoprotein gained popularity in the UK and several European countries, driven by rising consumer demand for healthier and more sustainable food options. Research on mycoprotein's benefits for weight management, heart health, and cholesterol reduction helped strengthen its market position (Derbyshire, 2020). More recently, mycoprotein production has been recognized for its low environmental footprint compared to traditional livestock farming, which has further fuelled interest in its potential as a sustainable protein source (Grossmann & Weiss, 2021).

3.2 Types of mycoproteins

As meat alternatives, mycoproteins have been developed into several types, each varying in texture, flavor, and formulation to meet diverse consumer preferences and culinary uses. Untextured mycoproteins retain a softer, less structured consistency, making them ideal for sauces, soups, and dishes where the primary focus is on nutritional value rather than replicating meat texture. These types are typically used in processed forms where blending with other ingredients is common (Elzerman et al., 2015). Their high protein and fibre content make untextured mycoproteins a versatile base ingredient in diverse plant-based foods. Some mycoproteins are pre-seasoned or marinated to enhance flavour and simplify cooking. These versions allow for quick, easy preparation and are often flavoured with spices or herbs to emulate the taste of traditional meat dishes. Pre-flavoured mycoproteins have gained popularity due to their convenience and improved sensory appeal, offering consumers a ready-to-use option in stir-fries, salads, and wraps. Blended mycoprotein products combine mycoprotein with other plant-based proteins, such as soy or pea protein, to enhance the overall protein content and nutritional profile (Kyriakopoulou et al., 2021). These blends aim to deliver complementary nutrients and functional properties, meeting specific dietary needs while adding to the versatility and flavor variety of mycoprotein-based products.

Whole or fermented mycoproteins, in which the mycoprotein is used with minimal processing, offer a high-fiber, whole-food option. This type is generally preferred by health-conscious consumers who seek less processed options. Fermentation preserves more natural nutrients and provides additional gut health benefits, making whole mycoproteins a popular choice in minimally processed plant-based products (Finnigan et al., 2019). Quorn and promin are well-known mycoprotein products found all over the world that contain glucose and nutrients produced by fungi that is basically providing nutritious, healthy, and vegan food to consumers as a meat substitute for maintaining cholesterol levels, healthy blood cholesterol levels, and controlling glucose levels (Souza et.al 2019)

3.3 Method of Mycoprotein development

In several studies it is found that Mycoprotein is produced by fermenting agro-industrial waste through microbial fermentation processes. Agro-industrial wastes like seaweed waste (Salgado et al,.2021), waste from date (Reihani et al,.2018), by product and waste of soy products, (Liang et al,. 2023) pineapple peel waste, and pea process byproducts (Umesh et al,.2023) can be used to produce mycoprotein biomass. There are two fermentation processes for the mass production of mycoprotein: solid-state fermentation and submerged fermentation; another method known as the surface culture method can produce mycoprotein on a laboratory scale. It is produced through the process of fermentation, a process that uses harmless bacteria to change the form of a food over time for the production of Mycoprotein.

The development of mycoprotein involves a specialized fermentation process that transforms filamentous fungi, typically *Fusarium venenatum*, into a high-protein, fiber-rich product. Here is an outline of the primary stages and methods used in mycoprotein development, supported by scientific references.

1. Strain Selection and Optimization: *Fusarium venenatum* was identified as a suitable mycoprotein-producing strain due to its high protein content, ease of cultivation, and favorable texture. Extensive screening and optimization processes were conducted to maximize the fungus's protein yield and nutrient profile (Finnigan et al., 2019). Additionally, strain selection aims to ensure that the final product meets food safety and quality standards.
2. Fermentation Process: The mycoprotein production process is based on continuous aerobic fermentation, where *Fusarium venenatum* is grown in large bioreactors. The fungus is cultivated on a glucose substrate in a controlled environment with regulated pH, oxygen, and nutrient levels, promoting rapid fungal growth and high biomass yield. This continuous fermentation process enables the efficient production of mycoprotein at an industrial scale, where fungal biomass can double every five hours (Edwards & Cummings, 2010).
3. Harvesting and Cell Rupture: Once the fungal biomass reaches a desired density, it is harvested and subjected to cell rupture, typically using mechanical or enzymatic methods to improve digestibility and nutritional accessibility. The cell rupture process ensures that the fungal cell walls are broken down, reducing the risk of digestive discomfort associated with chitin, a structural polysaccharide in fungal cell walls (Derbyshire, 2020).
4. Heat Treatment for Safety: The harvested mycoprotein is subjected to a heat treatment to inactivate any residual fungal spores and to reduce RNA levels. High RNA content can be problematic due to its association with uric acid formation in humans. The heating process reduces RNA levels, making mycoprotein safer for long-term human consumption (Finnigan et al., 2019).
5. Texturization and Flavor Enhancement: The mycoprotein undergoes a texturization process that aligns the fungal filaments, creating a texture that mimics the fibrous structure of meat. This step is particularly important for producing mycoprotein-based products that resemble chicken or beef. In some cases, flavorings, seasonings, or marinades are added to improve taste and make the product ready for direct culinary use (Asgar et al., 2010).

6. Final Processing and Packaging: The mycoprotein is shaped, cooked, and packaged, depending on the intended product type. Mycoprotein is often frozen or dried to extend shelf life and maintain quality during distribution (Grossmann & Weiss, 2021). Quality control tests are conducted to ensure the final product's safety, texture, and nutritional profile.

4. NUTRITIONAL STATUS OF MYCOPROTEINS

Mycoproteins are a healthy source of various vitamins, carbohydrates, amino acids, and high fibre content. It helps control blood cholesterol and blood sugar. Many people who are vegan, vegetarian, or have allergies avoid eating animal-based products. Mycoproteins help them get enough protein and fibre if they don't eat meat. Mycoproteins are described as "Generally Recognized as Safe" (FDA, 2023). In general, 100 gm (dry matter basis) of mycoproteins contains 13 g of fats, 45 g of proteins, 10 g of carbohydrates, 25 g of fibers and several vitamins and minerals (Finnigan et al., 2019).

Due to its relatively high protein value, increased fiber and reduced saturated fatty acid content, mycoprotein is essential for a healthy diet. Mycoprotein may provide several nutritional advantages, including improving satiety and controlling blood sugar and helpful in managing cholesterol levels too. Mycoprotein must be of high protein quality if it is to be used in foods that are the main course of a meal. All of the necessary amino acids are present in mycoprotein. The ideal protein bioavailability-assessed amino acid level is 0.996, indicating that the protein was produced using the best techniques and is of high quality (Edwards et al,.2010). Mycoprotein contains 6 g of fiber per 100 g, which implies that it is "rich in fiber," according to European Commission requirements. The natural dietary fiber of mycoprotein comprises 12% soluble and 88% insoluble fibres, with a minor amount of chitin and predominant glucan (creating a "Fibrous chitin–glucan matrix" in the small intestine region). Both linear and branched a-glucans from grains and yeast have been identified to actively participate in physiological processes related to fat metabolization and boost the immune system.32 The mycoprotein's cellular wall mainly consists of non-soluble fibres, with approximately 65% b-glucan and 35% chitins, which forms a chitin-glucan. b-glucan and chitins are the most important polysaccharides in fungal cell walls, influencing the morphology and characteristics of both vegetative and pathogenic hyphae (Bottin et al., 2016).

Mycoprotein's energy makeup is about one-third fat since it contains little readily available carbohydrates. Majorly, mycoprotein contains less than 1.5 g of saturated long and short chain fatty acid per 100 g of solid. This fungal protein is rich in mono- and polyunsaturated fatty acids.33Additionally, every 100 g of dried mycoprotein comprises 4.3 g of omega-6 fatty acids and 6.9 g of omega-3 fatty acids. Derbyshire et al., 2021 found that beef's PUFA to SFA ratio is approximately 0.1, which declines with increasing fat concentration in meat. On the other hand, the PUFA to SFA ratio in mycoprotein is 1.44. Mycoprotein contains several water-soluble B vitamins, including pyridoxine (0.1 mg), folate (114 mg), and cobalamin (0.72 mg).34 According to nutrient claims made by the European Commission, fungal protein is a potent source of various mineral components, including zinc, phosphorus, calcium, iron, potassium, and many more (Cherta-Murillo et al,. 2022).

Furthermore, mycoprotein is a viable source of vitamin B2, vitamin B6, and vitamin D. Vitamin B2 facilitates the conversion of carbohydrates into glucose for energy production (Molfetta et al., 2022), while vitamin B6 is crucial for overall health and bodily function. Vitamin D promotes calcium absorption, which is crucial for developing strong bones. It also prevents osteoporosis, a condition where bones become thin and weak, making them more prone to fractures (Desbarats, 2020).

Table 1. Nutritional profile of mycoprotein, based on 100 grams of the product, as commonly found in products like Quorn™. Values may vary slightly depending on the specific formulation and processing

Nutrient Amount per 100g Description & Health Benefits
Energy 85 kcal Low-calorie, suitable for weight management and low-calorie diets
Protein 11-15 g High-quality, complete protein with all essential amino acids (Wiebe, 2004)
Total Fat 2-3 g Low in fat, especially in saturated fats, supporting cardiovascular health
Saturated Fat 0.5 g Low levels help maintain heart health (Derbyshire, 2020)
Carbohydrates 2-6 g Includes fiber, mainly chitin and beta-glucans, beneficial for digestion
Dietary Fiber 6-7 g High in fiber, improves satiety, aids digestion, supports gut health (Elzerman et al., 2015)
Sugars 0 g Contains no added sugars, beneficial for blood sugar management
Iron 2 mg Helps prevent iron deficiency, especially in vegetarian/vegan diets
Phosphorus 250 mg Essential for bone health and energy metabolism
Mycoprotein nutritional benefits
1. High in protein and fiber content
2. It is low in saturated fat and good source of Short Chain Fatty acid
3. It contains water-soluble B vitamins, including pyridoxine, folate, and cobalamin
4. Rich in mono- and polyunsaturated fatty acids
5. May help maintain healthy blood cholesterol levels
6. May promote muscle synthesis as the richest and cleanest from of plant-based protein
7. May control glucose and insulin levels as having lower glycemic index
8. Helpful increase satiety and improving metabolic rate of the body

5. HEALTH BENEFITS OF MYCOPROTEINS

Mycoproteins provide several health advantages, making them a valuable alternative protein source with potential positive effects on various aspects of health. Derived from the fungus *Fusarium venenatum*, they are high in quality protein, essential amino acids, and dietary fiber, while low in saturated fats. These characteristics contribute to benefits like supporting metabolic and cardiovascular health, aiding in digestion, and promoting overall wellness. So, incorporating mycoprotein in diet have several health benefits and increases its nutritional profile (Hosseini et al.,2011) like for example, consumption of mycoprotein helps to reduce the blood glucose level and modifies the secretion of gastric juices and motility of intestinal cells (Murillo et al., 2020). Different studies prove that the consumption and intake of mycoprotein help regulate the blood cholesterol level. As a result, it decreases the chance of coronary and cardiovascular disease, (Coelho et al.,2021) enhances the growth of good muscles, and reduces individuals' energy intake (Bottin et al.,2016).

It is found that Mycoprotein meals were found to be significantly having lower blood insulin incremental area under the curve (iAUC) (the area under the curve is produced from the oral glucose tolerance test, which is frequently used to diagnose impaired glucose tolerance). The incremental area under the curve is developed because of variations in baseline blood glucose. It essentially provides information on the glycemic index of foods, which indicates how much blood glucose is raised (after consuming any meal) at the first 60 min after consumption compared to matched-chicken meals, indicating that mycoprotein decreased the insulinogenic index (the ratio of insulin concentration at 30 min minus fasting insulin to the difference of glucose simultaneously) and increased peripheral sensitivity to insulin (Ahmad et al,.2022). Even in other studies it was found that intake of mycoprotein products led to a gradual and prolonged increase in serum insulin levels within 30 min, in contrast to milk proteins (Monteyne et al. 2020)

In several studies it was found that mycoprotein in diet has been found to enhance lipid biomarkers among overweight and obese individuals (Bottin et al., 2016). In a recent study it was analysed that consuming mycoproteins in diet improves blood cholesterol levels in healthy adults following a one-week diet change with controlled fish- or meat-based protein and mycoprotein diets. In the study the group that was following the diet change, that have consumed mycoprotein exhibited a substantial reduction in blood cholesterol levels, such as low-density lipoprotein (LDL) (19.33%), intermediate-density lipoprotein (14.47%), and very low-density lipoprotein (VLDL) (8.36%) in comparison to the control group (Coelho et al. (2021). Moreover, a noticeable reduction was observed in the levels of docosahexaenoic acid (17.53%) and x-3 fatty acids (17.26%) compared to the control diets (Dunlop et al., 2017).

The consumption of animal-based proteins is linked to greater muscle protein synthesis compared to plant derived proteins. Recent research has emphasised the anabolic capacity of plant-derived protein sources like soybean and wheat proteins, which have lower levels of EAAs and leucine (Gorissen et al., 2016). Mycoprotein is a fungal-based protein that contains around 6% leucine. Additionally, it contains a high amount of EAAs, comprising 45% of the total proteins, and has a high PDCAAS content (Finnigan et al., 2017). The current research revealed that consuming mycoprotein enhances the availability of branched-chain amino acids (BCAAs) and promotes muscle protein synthesis in young males examined the blood circulatory BCAAs levels in a group of youngsters after consuming boluses containing varying quantities of mycoprotein and milk proteins (20, 40, 60, and 80 g) (Dunlop et al. 2017).

The study found that postprandial plasma BCAA concentrations varied based on mycoprotein consumption (40, 60, and 80 g) and a 20 g milk bolus. An 80 g mycoprotein bolus also increased BCAA levels more significantly than the 20 g milk-based bolus. In another study it was investigated that the effect of a mycoprotein drink enriched with smaller quantities of BCAAs (18.7 g protein) versus a larger mycoprotein bolus (31 g protein) on muscle protein production in resistance-trained young individuals following resistance workouts (Monteyne et al. 2020) The results showed that lower doses of mycoprotein resulted in reduced muscle protein synthesis (0.031–0.041% post-resting and 0.027–0.052% post-exercise) than the higher doses of mycoprotein (0.025–0.057% post-resting and from 0.024–0.072% post-exercise). A higher intake of mycoprotein led to elevated levels of phenylalanine and tyrosine in the plasma compared to a lower intake. Furthermore, consuming less mycoprotein resulted in a rapid and more noticeable increase in the availability of leucine, isoleucine, and valine in the blood after a meal, compared to consuming a larger amount of mycoproteins.

According to one study it was found that mycoproteins have a higher satiety effect compared to chickens (Munkvold et al. 2021). Mycoprotein products are suitable substitutes for school meals, as they effectively induce satiety and sustained energy levels among children. Multiple studies have demonstrated that mycoprotein is a viable substitute for individuals seeking to achieve weight loss. It extends the feeling of satiety, reducing the probability of superfluous snacking between main meals. The empirical research also provided some evidence supporting the notion that the inclusion of mycoproteins into dietary patterns benefits the regulation of satiety. The results indicated that mycoproteins exhibit a significantly reduced satiety compared to chicken meats. The synthesis of short-chain fatty acids (SCFAs) via the gastrointestinal breakdown of mycoprotein fibres has been linked to the modulation of appetite, as it triggers satiation messages from the colon to the brain (Ahmad et al., 2020). However, it is important to note that a conclusive causal mechanism has yet to be established, and further research is currently being conducted to explore this particular facet (Bottin et al., 2016).

Mycoprotein production through microbes produces various bioactive metabolites, including flavonoids, hexadecanes, and phenolic compounds, which are widely recognised for their antioxidant and free-radical removing properties (Chhikara et al., 2021; Venthodika et al., 2021). Furthermore, it was also found that mycoprotein produced through the fermentation of *M. pyrifera* and seaweed wastes using *P. salina* after eight days exhibited high antioxidant activity in the range of 46.3 to 49.7% (Salgado et al. 2021). The produced mycoprotein had high total phenolic contents ranging from 99.2 to 100.3 mg EAG g _1. Research indicated that consuming foods rich in flavonoids and phenolics may help prevent cardiovascular diseases and type 2 diabetes (Gorissen et al., 2016). It can be revealed that mycoprotein serves as an intrinsic source of antioxidant compounds. Additionally, it may confer benefits in managing high cholesterol levels and various cardiovascular diseases. Hence, consuming mycoprotein may have various positive health impacts, including appetite regulation, glucose metabolism, lipid profile, muscle growth, and antioxidant activity.

Table 2. Health benefits of mycoprotein with supporting studies

High-Quality Protein Source	**Mycoprotein contains all essential amino acids, making it a complete protein suitable for muscle maintenance and growth.**	**Finnigan et al. (2019); Edwards & Cummings (2010); Derbyshire (2020)**
Cholesterol Reduction	Regular consumption of mycoprotein has been linked to lower blood cholesterol levels due to its high fiber and low fat.	Turnbull et al. (1992) - Reduced serum cholesterol study; Derbyshire (2020)
Blood Sugar Control	The fiber in mycoprotein helps stabilize blood sugar, which may benefit individuals with diabetes or insulin resistance.	Bottin et al. (2016) - Impact of mycoprotein on glycemic control; Derbyshire (2020)
Gut Health	The high fiber content, including beta-glucans, promotes digestive health and may support beneficial gut bacteria.	Finnigan et al. (2019) - Gut health benefits; Elzerman et al. (2015) - Role of mycoprotein fiber in digestion
Weight Management	Mycoprotein's fiber and protein promote satiety, helping to reduce calorie intake and support weight management.	Bottin et al. (2016); Turnbull et al. (1992)
Sustainable Protein Source	Mycoprotein production has a lower environmental footprint compared to meat, aiding in sustainable dietary practices.	Grossmann & Weiss (2021); Smetana et al. (2015)
Immune Support	Rich in essential minerals like zinc and selenium, which support immune function and antioxidant defenses.	Derbyshire (2020); Wiebe (2004)

6. ENVIRONMENTAL IMPACT OF MYCOPROTEINS

Meat-based foods have significant environmental impacts that are lower for plant-based meat alternatives.

i. Reduced greenhouse gas emissions compared to animal meat production

Mycoprotein has been evaluated for its environmental impact and shows advantages over both plant-based meat alternatives and traditional meats, particularly in greenhouse gas emissions (GHGe). According to comparative life cycle assessments (LCAs), mycoprotein emits 0.73 $kgCO_2eq/kg$, which is notably lower than emissions for soy protein (1.21 kgCO_2eq/kg) and pea protein concentrate (1.91 kgCO_2eq/kg). Mycoprotein-based products, such as burgers, mince, and sausages, either match or outperform equivalent plant-based products in terms of GHGe and are significantly lower than those of traditional meat products.

The ingredient production stage, a major contributor to overall emissions, shows lower GHGe for mycoprotein products compared to their plant-based counterparts. However, further data on land use and water scarcity impacts are needed to fully distinguish mycoprotein's environmental profile from other alternative proteins. Overall, mycoprotein represents a lower-emission alternative to meat, though additional research is necessary for a complete comparison with plant-based alternatives.

ii. Less water and land usage

Extensive research has explored land use for mycoprotein production and evaluated its environmental impact. Expanding land use for food production adversely affects the planet's sustainable development capacity (Allan et al., 2015). Mycoprotein is a potential alternative to animal products for reducing land use (Sturtewagen et al., 2016). Population growth may require expanding land resources to meet the increasing demand for food production. Mycoprotein is a land-efficient alternative for food production. Different studies have yielded varying results on the annual land area needed for mycoprotein production, likely due to methodological differences (Temme et al., 2013). It was found that mycoprotein production requires only a quarter of the land area compared to animal-based foods. Thus, replacing animals with mycoprotein is a practical alternative due to lower land requirements and reduced load on sustainable thresholds.

iii. Decreased deforestation and habitat destruction

Mycoprotein offers environmental advantages over traditional meat sources in terms of greenhouse gas emissions and land use. Mycoprotein production emits fewer greenhouse gases and requires less land than beef production or traditional animal farming. Although there are measurement discrepancies, the overall trend indicates that mycoprotein is a sustainable food source. As global food demand increases, integrating sustainable sources like mycoprotein is essential for a more environmentally conscious food system. Generally, the environmental impact of single-cell proteins is dependent on the use of renewable energy. The greater the use of renewable energy in processes, the better the environmental performance. However, the time required to produce a certain amount of biomass and, eventually, proteins must be taken into account.

iv. Consumer Acceptance and Market Growth of Mycoproteins

Mycoprotein Market size is estimated to reach \$1.4 billion by 2030, growing at a CAGR of 7.5% during the forecast period 2023-2030. The rising number of vegans across the globe, increasing demand for high-quality protein diets, and innovation in product development are propelling the Mycoprotein Market growth. Additionally, Mycoprotein is gaining popularity among vegan consumers as a source of protein-based nutrients that can be included in a healthy diet. Mycoprotein is high in fiber, high in protein, low in saturated fat, and free of cholesterol, which has a positive effect on the consumption pattern and demand dynamics of the end products. Mycoprotein products are similar to other meat alternatives like tempeh or seitan in terms of flavor and texture. Mycoprotein products are expected to see increased demand because of their high mineral and vitamin content as well as the high proportion of various minerals. These factors positively influence the Mycoprotein industry outlook during the forecast period.

7. CHALLENGES AND FUTURE RESEARCH ON MYCOPROTEINS

Meat analogs, or substitutes, have evolved because of both rising consumer demand and advancements in food processing technology. Among various alternative proteins, meat analogs are the most developed, with significant progress in recreating the texture, taste, and appearance of conventional meat (Grossmann and Weiss, 2021). Although extensive research has been conducted, particularly on plant-based alternatives, there is a growing need for fundamental research on other protein sources, including microalgae, mycoproteins, single-cell proteins, and cultured meat. The environmental impact of these alternatives remains less documented, and research is needed to examine trade-offs and synergies across environmental, social, and economic aspects.

Challenges related to mycoprotein basically includes improvement in its texture and taste for having better resemblance to meat, overcoming potential consumer apprehension due to its fungal origin, optimizing large-scale production methods to reduce cost, addressing potential allergic reactions in a small subset of individuals, and further research into its long-term health benefits and environmental impact. Future studies should concentrate on creating new fungal strains with better nutritional profiles, investigating cutting-edge processing methods to improve texture and flavor, and carrying out in-depth analyses of the sustainability and health advantages of mycoprotein consumption in a range of demographics.

In addition to environmental impacts, holistic studies are necessary to explore the balance between the environmental benefits and nutritional properties of meat analogs. Since the nutritional and environmental impacts do not align linearly, they each affect human health both directly, by supplying nutrients, and indirectly, by influencing environmental quality. Future studies should focus on understanding the complexities and potential health implications of these substitutes and compare them with traditional animal-based products to fully assess their viability and benefits.

7.1 Challenges

While mycoproteins offer a promising alternative to traditional protein sources, several challenges remain. Regulatory hurdles represent a primary barrier to widespread adoption, as mycoprotein is classified as a "novel food" in several regions, requiring rigorous safety assessments and approvals before commercialization. Mycoprotein has received GRAS (Generally Recognized as Safe) status, yet concerns about allergenic potential, especially due to its fungal origins, remain a topic of discussion among regulatory bodies. These regulatory processes can be lengthy and vary significantly between countries, creating obstacles for companies aiming for international distribution.

Nonacceptance pose another significant challenge. Despite growing interest in alternative proteins, acceptance of fungal-based proteins can be limited by cultural attitudes, unfamiliarity with fungal ingredients, and concerns about safety and allergenicity. For instance, research shows that mycoprotein's distinct taste and texture may appeal more to European markets, where Quorn™ has established a strong foothold, compared to other regions where plant-based proteins are more widely recognized. As companies focus on branding and education to highlight mycoprotein's unique health benefits, they face the task of distinguishing these benefits from those of other well-known alternatives like soy and pea protein

In the Mycoprotein Market analysis, the Food Grade Mycoprotein segment is estimated to grow with the highest CAGR of 7.9% during the forecast period. This dominance can be attributed to the growing consumer interest in plant-based and alternative protein sources, driven by health and sustainability concerns. Globally, infectious diseases are increasing. Approximately 8% of the world's population has died, and 640 million cases of Coronavirus Disease. Food Grade Mycoprotein is increasingly used in meat substitutes, vegetarian and vegan products, and as a protein-rich ingredient in various food items. Its versatility and ability to mimic the texture and taste of meat make it a preferred choice among consumers seeking healthier and eco-friendly dietary options. As a result, it has gained traction across a wider range of food applications, positioning it as the dominant segment in the Mycoprotein market.

7.2 Future Directions

The future of mycoprotein continued research and technological innovation aimed at enhancing production efficiency, scalability, and nutritional quality. Genetic engineering and metabolic optimization hold potential for improving *Fusarium venenatum*'s protein yield, growth rate, and amino acid profile, which could make mycoprotein even more competitive with animal proteins. Additionally, exploring other fungal strains may open newies for taste, texture, and nutritional diversity, potentially expanding the appeal of fungal proteins to different consumer segments.

Another key direction is market expansion through partnerships education. As companies work to promote mycoprotein's environmental benefits, collaborations with governments, NGOs, and industry groups could aid in raising consumer awareness and establishing mycoprotein as a mainstream protein option. Targeted education campaigns emphasizing its health benefits and environmental advantages, such as reduced greenhouse gas emissions and low resource requirements, can help shift public perceptions and drive demand. This awareness can also support market entry in regions like Asia and North America, where demand for sustainable foods is growing rapidly.

8. CONCLUSION

In conclusion, mycoproteins are positioned as an innovative, sustainable alternative to conventional animal proteins, offering both environmental and nutritional advantages. Through an efficient fermentation process using *Fusarium venenatum*, mycoprotein production uses significantly less land, water, and energy than livestock farming, aligning with global goals for reducing agriculture's environmental impact. Nutritionally, mycoprotein is rich in essential amino acids, dietary fiber, and low in saturated fats, supporting health benefits like enhanced satiety, improved lipid profiles, and potential weight management benefits. These qualities make mycoprotein not only a functional meat substitute but also a health-promoting option for consumers seeking eco-friendly dietary alternatives.

However, mycoprotein's success depends on overcoming regulatory challenges, increasing consumer acceptance, and differentiating itself from other alternative proteins like plant-based and cell-cultured meats. Effective education on its health benefits, environmental advantages, and food safety will be essential to encourage wider adoption across diverse consumer demographics. With continued research, technological advancements, and public engagement, mycoproteins are well-positioned to play a key role in the future of sustainable food systems, contributing to healthier, more sustainable global diets.

REFERENCES

Ahmad, M. I., Farooq, S., Alhamoud, Y., Li, C., & Zhang, H. (2022). A Review on Mycoprotein: History, Nutritional Composition, Production Methods, and Health Benefits. *Trends in Food Science & Technology*, *2022*(121), 14–29. DOI: 10.1016/j.tifs.2022.01.027

Aiking, H. (2014). Protein production: Planet, profit, plus people? *The American Journal of Clinical Nutrition*, *100*(1), 483S–489S. DOI: 10.3945/ajcn.113.071209 PMID: 24871472

Allan, E., Manning, P., Alt, F., Binkenstein, J., Blaser, S., Blüthgen, N., Böhm, S., Grassein, F., Hölzel, N., Klaus, V. H., Kleinebecker, T., Morris, E. K., Oelmann, Y., Prati, D., Renner, S. C., Rillig, M. C., Schaefer, M., Schloter, M., Schmitt, B., & Fischer, M. (2015). Land use intensification alters ecosystem multifunctionality via loss of biodiversity and changes to functional composition. *Ecology Letters*, *18*(8), 834–843. DOI: 10.1111/ele.12469 PMID: 26096863

Apostolidis, C., & Mcleay, F. (2019). To meat or not to meat? Comparing empowered meat consumers' and anti-consumers' preferences for sustainability labels. *Food Quality and Preference*, *77*, 109–122. DOI: 10.1016/j.foodqual.2019.04.008

Asgar, M. A., Fazilah, A., Huda, N., Bhat, R., & Karim, A. A. (2010). Nonmeat protein alternatives as meat extenders and meat analogs. *Comprehensive Reviews in Food Science and Food Safety*, *9*(5), 513–529. DOI: 10.1111/j.1541-4337.2010.00124.x PMID: 33467834

Bakratsas, G., Polydera, A., Nilson, O., Chatzikonstantinou, A. V., Xiros, C., Katapodis, P., & Stamatis, H. (2023). Polydera, A Nilson, O. A. V. Chatzikonstantinou, Xiros, C. Katapodis P. and Stamatis, H. Mycoprotein Production by Submerged Fermentation of the Edible Mushroom Pleurotus ostreatus in a Batch Stirred Tank Bioreactor Using Agro-Industrial Hydrolysate. *Foods*, *12*(12), 2295. DOI: 10.3390/foods12122295 PMID: 37372506

Bottin, J., Swann, J., Cropp, E., Chambers, E., Ford, H., Ghatei, M., & Frost, G. (2016). Mycoprotein reduces energy intake and postprandial insulin release without altering glucagon-like peptide-1 and peptide tyrosine-tyrosine concentrations in healthy overweight and obese adults: A randomised-controlled trial. *British Journal of Nutrition*, *116*(2), 360–374. DOI: 10.1017/S0007114516001872 PMID: 27198187

Bottin, J. H., Swann, J. R., Cropp, E., Chambers, E. S., & Ford, H. E. (2016). Mycoprotein reduces energy intake and postprandial insulin levels in overweight adults: A randomised controlled trial. *British Journal of Nutrition*, *116*(2), 360–374. DOI: 10.1017/S0007114516001872 PMID: 27198187

Bottin, J. H., Swann, J. R., Cropp, E., Chambers, E. S., Ford, H. E., Ghatei, M. A., & Frost, G. S. (2016). Mycoprotein reduces energy intake and postprandial insulin release without altering glucagon-like peptide-1 and peptide tyrosine-tyrosine concentrations in healthy overweight and obese adults: A randomized controlled trial. *British Journal of Nutrition*, *116*(2), 360–374. DOI: 10.1017/S0007114516001872 PMID: 27198187

Cebri'an, M., & Ibarruri, J. (2023). Filamentous fungi processing by solid-state fermentation, in Current Developments in Biotechnology and Bioengineering, Elsevier, 2023, 251–292.

Cherta-Murillo, A., & Frost, G. S. (2022). The association of mycoprotein-based food consumption with diet quality, energy intake and non-communicable diseases' risk in the UK adult population using the National Diet and Nutrition Survey (NDNS) years 2008/2009–2016/2017: A cross-sectional study. British Journal of Nutrition, 127, 1685–1694. *Frontiers in Sustainable Food Systems*, *5*, 581682.

Cherta-Murillo, A., Lett, A., Frampton, J., Chambers, E., Finniga, T., & Frost, G. (2020). Effects of mycoprotein on glycaemic control and energy intake in humans: A systematic review. *British Journal of Nutrition*, *123*(12), 1321–1332. DOI: 10.1017/S0007114520000756 PMID: 32100651

Chhikara, N., Kaur, A., Mann, S., Garg, M., Sofi, S.A. & Panghal, A. (2021). Bioactive compounds, associated health benefits and safety considerations of Moringa oleifera L.: an updated review.

Coelho, M., Monteyne, A., Dirks, M., Finnigan, T., Stephens, F., & Wall, B. (2021). Daily mycoprotein consumption for 1 week does not affect insulin sensitivity or glycaemic control but modulates the plasma lipidome in healthy adults: A randomised controlled trial. *British Journal of Nutrition*, *125*(2), 147–160. DOI: 10.1017/S0007114520002524 PMID: 32660657

Derbyshire, E. J. (2020). Flexitarian diets and health: A review of the evidence-based literature. *Frontiers in Nutrition*, *7*, 597. PMID: 28111625

Derbyshire, E.J. & Delange, J. (2021). Fungal protein–what is it and what is the health evidence? A systematic review focusing on mycoprotein.

Derbyshire, E. J., & Delange, J. (2021). Fungal Protein–What Is It and What Is the Health Evidence? A Systematic Review Focusing on Mycoprotein. *Frontiers in Sustainable Food Systems*, *5*, 581682. DOI: 10.3389/fsufs.2021.581682

Desbarats, J. (2020). Pyridoxal 50-phosphate to mitigate immune dysregulation and coagulopathy in Covid-19, 2020050144.

Dunlop, M. V., Kilroe, S. P., Bowtell, J. L., Finnigan, T. J., Salmon, D. L., & Wall, B. T. (2017). Mycoprotein represents a bioavailable and insulinotropic non-animal-derived dietary protein source: A dose–response study. *British Journal of Nutrition*, *118*(9), 673–685. DOI: 10.1017/S0007114517002409 PMID: 29017627

Edwards, D. G., & Cummings, J. H. (2010). The protein quality of mycoprotein. Proceedings of the Nutrition Society, 69(OCE4), E331.

Edwards, D. G., & Cummings, J. H. (2010). Mycoprotein: A meat-free dietary protein with a low environmental footprint. *The Proceedings of the Nutrition Society*, *69*(3), 322–329.

Edwards, D. G., & Cummings, J. H. (2010). Mycoprotein: A meat-free dietary protein with a low environmental footprint. *The Proceedings of the Nutrition Society*, *69*(3), 322–329.

Fasolin, L. H., Pereira, R. N., Pinheiro, A. C., Martins, J. T., Andrade, C. C. P., Ramos, O. L., & Vicente, A. A. (2019). Emergent food proteins–towards sustainability, health and innovation. *Food Research International*, *2019*(125), 108586. DOI: 10.1016/j.foodres.2019.108586 PMID: 31554037

Finnigan, T., Needham, L., & Abbott, C. (2017). Mycoprotein: a healthy new protein with a low environmental impact. In Nadathur, S. R., Wanasundara, J. P. D., & Scanlin, L. (Eds.), *Sustainable Protein Sources* (pp. 305–325). Academic Press. DOI: 10.1016/B978-0-12-802778-3.00019-6

Finnigan, T., Needham, L., & Pickering, K. (2019). Mycoprotein: A healthy new protein with a low environmental impact. *Current Developments in Nutrition, 3*(6), nzz021. DOI: 10.1093/cdn/nzz021 PMID: 31187084

Gorissen, S. H., Horstman, A. M., Franssen, R., Crombag, J. J. R., Langer, H., Bierau, J., Respondek, F., & van Loon, L. J. C. (2016). Ingestion of wheat protein increases in vivo muscle protein synthesis rates in healthy older men in a randomized trial. *The Journal of Nutrition, 146*(9), 1651–1659. DOI: 10.3945/jn.116.231340 PMID: 27440260

Grossmann, L., & Weiss, J. (2021). Alternative protein sources as technofunctional food ingredients. *Annual Review of Food Science and Technology, 12*(1), 93–117. DOI: 10.1146/annurev-food-062520-093642 PMID: 33472014

Hashempour-Baltork, F.. (2020). "Mycoproteins as Safe Meat Substitutes." Journal of Cleaner Production, vol. 253, Elsevier BV, Apr. 119958. Crossref, Liang, J., Xu, N. Nedele, A. K., Rigling, M., Zhu, L., and Zhang, Y. (2023).Upcycling of Soy Whey with Ischnoderma benzoinum toward Production of Bioflavors and Mycoprotein. *Journal of Agricultural and Food Chemistry, 71*(23), 9070–9079.

Meyer, V., Basenko, E. Y., Benz, J. P., Braus, G. H., Caddick, M. X., Csukai, M., De Vries, R., Frisvad, D., & Gunde-Cimerman, N. (2020). Growing a Circular Economy with Fungal Biotechnology: A White Paper. *Fungal Biology and Biotechnology, 7*(1), 1–23. DOI: 10.1186/s40694-020-00095-z PMID: 32280481

Molfetta, M., Morais, E. G., Barreira, L., Bruno, G. L., Porcelli, F., Dugat-Bony, E., Bonnarme, P., & Minervini, F. (2022). Protein sources alternative to meat: State of the art and involvement of fermentation. *Food, 11*(14), 2065. DOI: 10.3390/foods11142065 PMID: 35885308

Monteyne, A. J., Coelho, M. O., Porter, C., Abdelrahman, D. R., Jameson, T. S. O., Jackman, S. R., Blackwell, J. R., Finnigan, T. J. A., Stephens, F. B., Dirks, M. L., & Wall, B. T. (2020). Mycoprotein ingestion stimulates protein synthesis rates to a greater extent than milk protein in rested and exercised skeletal muscle of healthy young men: A randomized controlled trial. *The American Journal of Clinical Nutrition, 112*(2), 318–333. DOI: 10.1093/ajcn/nqaa092 PMID: 32438401

Munkvold, G. P., Proctor, R. H., & Moretti, A. (2021). Mycotoxin production in Fusarium according to contemporary species concepts. *Annual Review of Phytopathology, 59*(1), 373–402. DOI: 10.1146/annurev-phyto-020620-102825 PMID: 34077240

Nadathur, S. R., Wanasundara, S. R., & Scanlin, L. (2017). Feeding the globe nutritious food in 2050: obligations and ethical choices. In *Sustainable Protein Sources* (pp. 409–422). Elsevier Inc., Academic Press. DOI: 10.1016/B978-0-12-802778-3.00025-1

Okeudo-Cogan, M. C., Murray, B. S., Ettelaie, R., Connell, S. D., Radford, S. J., Micklethwaite, S., & Sarkar, A. (2023). Understanding the microstructure of a functional meat analogue: Demystifying interactions between fungal hyphae and egg white protein. *Food Hydrocolloids, 140*, 108606. DOI: 10.1016/j.foodhyd.2023.108606

Reihani, S. F. S., & Khosravi-Darani, K. (2018). Mycoprotein production from date waste using Fusarium venenatum in a submerged culture. *Applied Food Biotechnology*, *5*(4), 243–352.

Salgado, C. L., Muñoz, R., Blanco, A., & Lienqueo, M. E. (2021). Valorization and upgrading of the nutritional value of seaweed and seaweed waste using the marine fungi Paradendryphiella salina to produce mycoprotein. *Algal Research*, *53*, 102135. DOI: 10.1016/j.algal.2020.102135

Smetana, S., Mathys, A., Knoch, A., & Heinz, V. (2015). Meat alternatives: Life cycle assessment of most known meat substitutes. *The International Journal of Life Cycle Assessment*, *20*(9), 1254–1267. DOI: 10.1007/s11367-015-0931-6

Smetana, S., Mathys, A., Knoch, A., & Heinz, V. (2015). Meat alternatives: Life cycle assessment of most known meat substitutes. *The International Journal of Life Cycle Assessment*, *20*(9), 1254–1267. DOI: 10.1007/s11367-015-0931-6

Sturtewagen, L., De Soete, W., Dewulf, J., Lachat, C., Lauryssen, S., Heirman, B., Rossi, F., & Schaubroeck, T. (2016). Resource use profile and nutritional value assessment of a typical Belgian meal catered or home-cooked, with pork or QuornTM as a protein source. *Journal of Cleaner Production*, *112*, 196–204. DOI: 10.1016/j.jclepro.2015.09.006

Turnbull, W. H., Leeds, A. R., & Edwards, G. D. (1992). Mycoprotein reduces blood lipids in free-living subjects. *The American Journal of Clinical Nutrition*, *55*(2), 415–419. DOI: 10.1093/ajcn/55.2.415 PMID: 1734679

Van Loo, E. J., Caputo, V., & Lusk, J. L. (2020). Consumer Preferences for Farm-raised Meat, Lab-grown Meat, and Plant-based Meat Alternatives: Does Information or Brand Matter? *Food Policy*, *95*, 101931. DOI: 10.1016/j.foodpol.2020.101931

Venthodika, A., Chhikara, N., Mann, S., Garg, M. K., Sofi, S. A., & Panghal, A. (2021). Bioactive compounds of Aegle marmelos L., medicinal values and its food applications: A critical review. *Phytotherapy Research*, *35*(4), 1887–1907. DOI: 10.1002/ptr.6934 PMID: 33159390

Whittaker, J. A., Johnson, R. I., Finnigan, T. J., Avery, S. V., & Dyer, P. S. (2020). The Biotechnology of Quorn Mycoprotein: Past, Present and Future Challenges. In *Grand challenges fungal biotechnology* (pp. 59–79). Springer. DOI: 10.1007/978-3-030-29541-7_3

Wiebe, M. G. (2002). Myco-protein from Fusarium venenatum: A well-established product for human consumption. *Applied Microbiology and Biotechnology*, *58*(4), 421–427. DOI: 10.1007/s00253-002-0931-x PMID: 11954786

Wiebe, M. G. (2004). Quorn™ mycoprotein–Overview of a successful fungal product. *The Mycologist*, *18*(1), 17–20. DOI: 10.1017/S0269915X04001089

Zhang, C., Guan, X., Yu, S., Zhou, J., & Chen, J. (2023). Production of Meat Alternatives Using Live Cells, Cultures and Plant Proteins. *Current Opinion in Food Science*, *43*, 43–52. DOI: 10.1016/j.cofs.2021.11.002

Zhang, T., Dou, W., Zhang, X., Zhao, Y., Zhang, Y., Jiang, L., & Sui, X. (2021). The development history and recent updates on soy protein-based meat alternatives. *Trends in Food Science & Technology*, *109*, 702–710. DOI: 10.1016/j.tifs.2021.01.060

Chapter 10
Arbuscular Mycorrhizal Fungi:
The Metagenomics and the Circadian Rhythm Governed by the Environment

Fatima-Zahra Akensous
https://orcid.org/0000-0001-9518-8146
Faculty of Science Semlalia, Cadi Ayyad University, Morocco

ABSTRACT

Arbuscular mycorrhizal fungi (AMFs) are behind the crucial symbiosis that occurs within over 80% of the root systems of vascular plants, reputed for their potentialities regarding the plant, soil, and conjecture with the existing microbiome. Researchers inclined toward studying AMFs in their occurrence in the environment are currently embracing investigation at the holobiont level within the sphere of the natural habitat. Hence, the present book division looks at how the metagenomics can provide important information in this respect. With a keen interest in studying microbial communities retrieved from their environment, relying on advanced sequencing techniques, the metagenomics can investigate the composition and diversity of AMFs from a close-up lens. In the fungal verse, the circadian rhythm, pioneeringly explored in the Sordariales (Neurospora), is slowly but steadily being investigated in the Glomerales (Rhizoglomus), providing genetic and molecular discernments worth spotlighting into the bargain.

INTRODUCTION

Arbuscular mycorrhizal fungi (AMFs) represent telluric symbionts, mostly obligate biotroph endomycorrhiza that can colonize the root systems (cortical root cells) of over 80% of vascular plants, ranging from terrestrial to aquatic (Fusconi & Mucciarelli, 2018; Vierheilig, 2004). Cladded under the Glomeromycota division, with more than 220 characterized species, AMFs have been gaining momentum amongst the scientific community, ever since the 1970s, for their layered beneficial effects, both proven and speculated, at the broad plant and soil levels, as well as toward sustainable horticulture (Akensous et al., 2023a, b; Akensous et al., 2024; Hart et al., 2003; Khan et al., 2022; Willing et al., 2024) (Figure 1).

DOI: 10.4018/979-8-3693-8201-1.ch010

Figure 1. Notable roles of Arbuscular Mycorrhizal Fungi (AMFs) on the soil-plant nexus. **AMF,** *Arbuscular Mycorrhizal Fungal;* **AQPs,** *Aquaporins;* **C,** *Carbon;* **Ca,** *Calcium;* **Fe,** *Iron;* **GRSP,** *Glomalin-Related Soil Protein;* **$H_2PO_4^-$,** *Dihydrogen Phosphate;* **K,** *Potassium;* **Mg,** *Magnesium;* **N,** *Nitrogen;* **NH_4^+,** *Ammonium;* **NO_3^-,** *Nitrate;* **P,** *Phosphorus;* **ROS,** *Reactive Oxygen Species;* **Zn,** *Zinc*

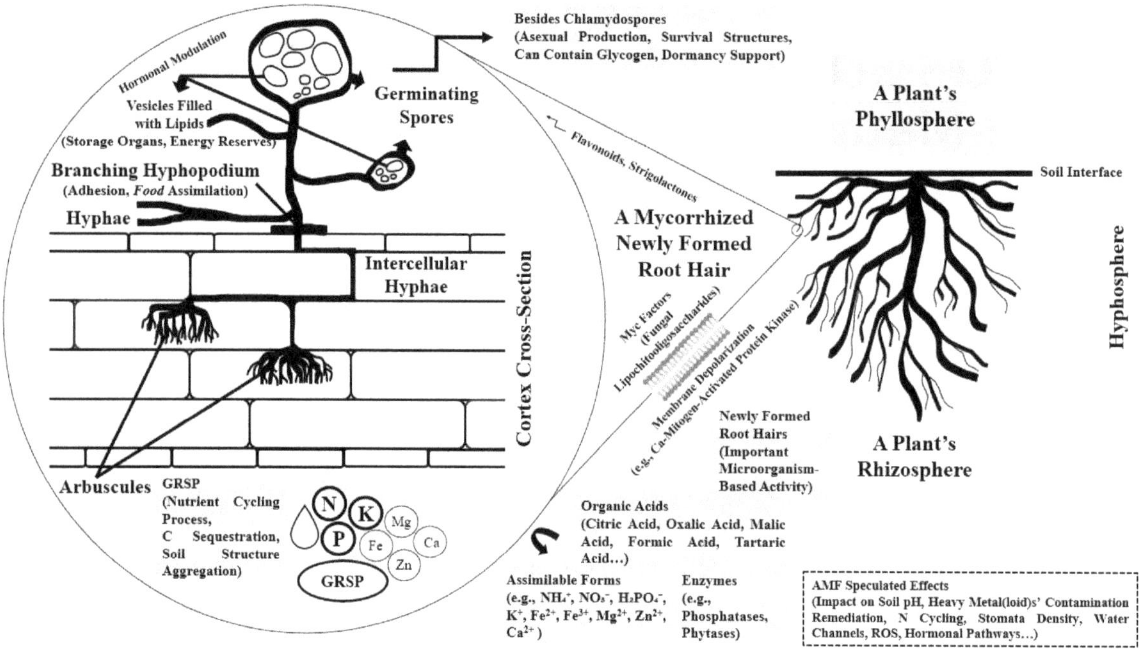

On the one hand, mycorrhized plants' root systems can explore important depths of the soil, allowing water *attraction* and uptake from remote soil areas, owing to their extended hyphal network that leads to augmenting the plants' root systems' absorbing surface (Kong et al., 2018). Besides, AMFs help with the assimilation of macronutrients like phosphorus (P) and nitrogen (N), as well as micronutrients, such as iron (Fe) and zinc (Zn), owing to the secretion of organic acids and key enzymes (e.g., phosphatases), as well as the presence of inorganic phosphate transport systems, for instance (Bhupenchandra et al., 2024). The ameliorated hydration and nutrition statuses help decrease the stress impact on plants that experience the same by promoting the production of osmolytes that help plants thrive under osmotic stress (Jogawat, 2019). On the other hand, AMFs produce considerable amounts of glomalin-related soil protein (GRSP), a waxy glycoprotein that contributes to gluing poorly-textured soil particles, thereby improving the soil's porosity profile, water retention ability, and organic matter pool, in addition to the sequestration role of up to third of the world's soil carbon (C), all the while preserving essential carbonaceous compounds against degradation (Wilson et al., 2009).

Molecularly, the increasing research inclination toward AMFs triggered curiosity to closely investigate the mycorrhizal symbiosis at the level of plants' root systems. Relying on advanced molecular techniques, carried out molecular investigation yielded genetically-demonstrating evidence. For reference, Wang and team investigated the developmental and conservational (functional) aspects of three genes, namely "*Doesn't Make Infections* 1" (DMI1), "*Doesn't Make Infection* 3" (DMI3), and "Interacting Protein of DMI3" (IPD3), necessary for mycorrhizal development within legume and rice (*Oryza sativa* L.) root systems and found that the genes were vertically inherited ever since the origin of land plants (Wang et al., 2010). Moreover, molecular advances were capitalized on to decipher the signaling pathways in AMFs and plants' root systems' symbiosis and the findings revealed that AMFs experienced physiological alterations due to the host plants' root system producing specific exudates, paving the way to the mycorrhizal colonization establishment (Hoang et al., 2022). The host plants' root systems' exudates initiate mycorrhizal genes necessary for spore-based compound respiration (e.g., activity in the mitochondria, production of enzymes), in a few hours (Tamasloukht et al., 2003). In addition, microbial as well as mycorrhizal ecologists developed a keen interest in AMFs in their occurring individually, communally, and within an ecosystem, beginning with the very spores, through multilocus genotyping, for instance (Martin & van der Heijden, 2024).

However, owing to the AMFs' diverse profiles, the non-occurrence of sporulation events within certain species, seasonal dependency and variability, uncultivability issues, the human-related error factor, multinucleate spores, and important genetical variation within clonal species make the conventional identification method of soil spore extraction and taxonomic morphological analysis (microscopic identification) convoluted (Marleau et al., 2011; Rosendahl, 2008). Consequently, it is highly possible that the research community misreported and/or misinterpreted the real mycorrhizal composition and communities occurring at a particular point of time, as well as the exact active or functioning AMFs during symbiosis, relying on these conventional methods (Brundrett & Tedersoo, 2019). Furthermore, AMFs present in the field soil can contain multiple associated mycorrhizal species, which makes targeting a particular species or a consortium of species complicated (Magkourilou et al., 2024). Hence, identifying closely *linked* AMFs requires the deployment of sophisticated molecular techniques.

To deal with the discrepancies and inaccuracy that occur relying on conventional microbial genome sequencing, in addition to the generalized insights, studying genetic material retrieved from samples through sequencing promises to reveal the exact intricate as well as undeciphered areas surrounding the diversity in the sphere of AMFs. The in-question molecular investigation falls under the designation "metagenomics" (Bhargava et al., 2019; Edwin et al., 2024). Metagenomics is composed of two parts: "meta" = many, and "genomics" = to get a genomic sequence. It refers to the study of genetic material retrieved from environmental samples (e.g., one gram of soil) through genetic material sequencing. This particular study takes an interest in investigating nucleic acids retrieved from the environment. Metagenomics differs from genomics in the sense the latter evaluates nucleic acids extracted from single organisms (Xu, 2006) (Figure 2).

Figure 2. Major differences between genomics and metagenomics and key historical lapses. **DNA,** *Deoxyribonucleic Acid;* **HTS,** *High-Throughput Sequencing;* **NGS,** *Next-Generation Sequencing*

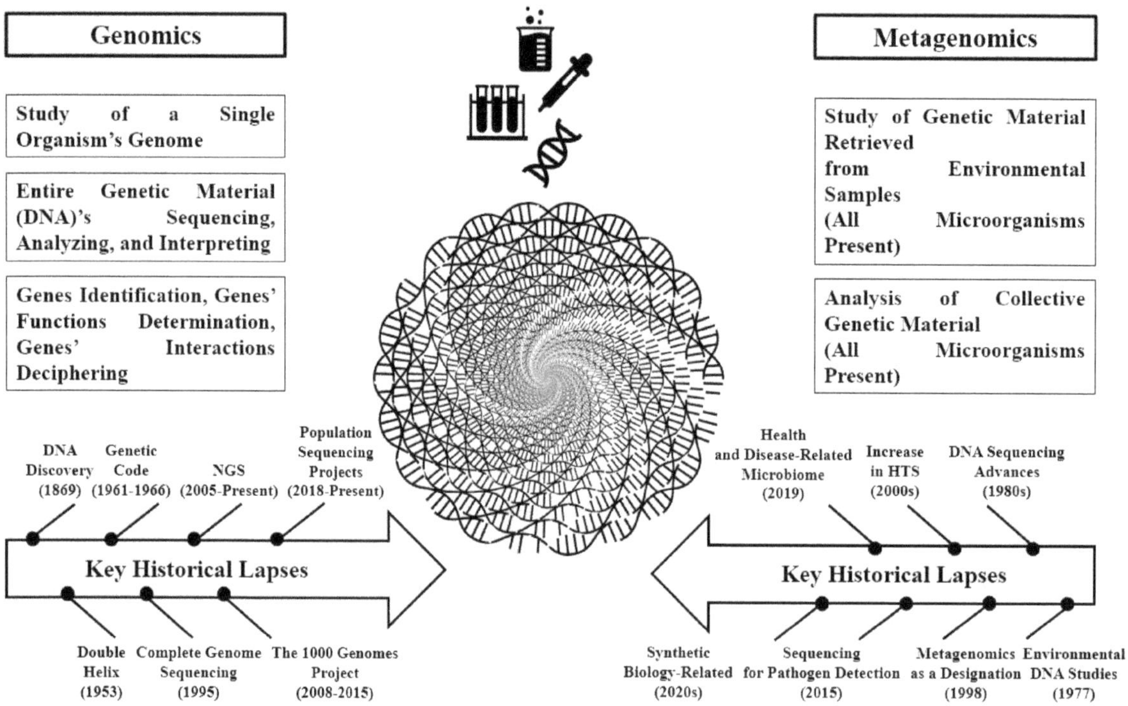

In light of the advances in molecular genetics, metagenomics allows analyzing microbial communities at the functional level with further confidence and in a more precise manner (Zhang et al., 2021). Metagenomics approaches genetic sequencing by targeting the identification of the microbial species present in a given sample, revealing their functionalities in their environment, and decoding the portion or ratio of each microbial species present (Sota et al., 2014). Indeed, it evaluates the entire genetic *reservoir* of the microorganisms present in a given sample retrieved from the environment, significantly contributing to uncovering unexplored areas in the composition of microbial communities, encoding functional diversity of the genomes, and discovering unique genes through Deoxyribonucleic Acid (DNA) pyrosequencing and metabarcoding, for instance (Bhargava et al., 2019).

The circadian rhythm designates the repetitive natural oscillation in a day (24 hours), within cells of a particular organism, in response to the environment (Pittendrigh, 1960). Probably, fungi represent the group of microorganisms that are highly linked to the higher plant and animal kingdoms, spotlighting curiosity to understand the circadian rhythm in fungi (Dunlap et al., 2007). What is precisely intriguing in the group of AMFs is that, as telluric microorganisms that are associated with plants' root systems, present within an environment that is almost deprived of light and temperature fluctuating events, AMFs carry a conserved circadian clock. The in-question circadian clock's Fungal Circadian Oscillator (*frq*) gene can be activated by blue light, similar to what occurs in *Neurospora crassa*, often studied as a model circadian fungus (Dunlap & Loros, 2006) (Figure 3). The presence of the *frq* gene was evidenced in *Rhizophagus irregularis*, as it was stimulated in response to brief lighting, paving the way to further explore the circadian clocks in AMFs (Lee et al., 2018).

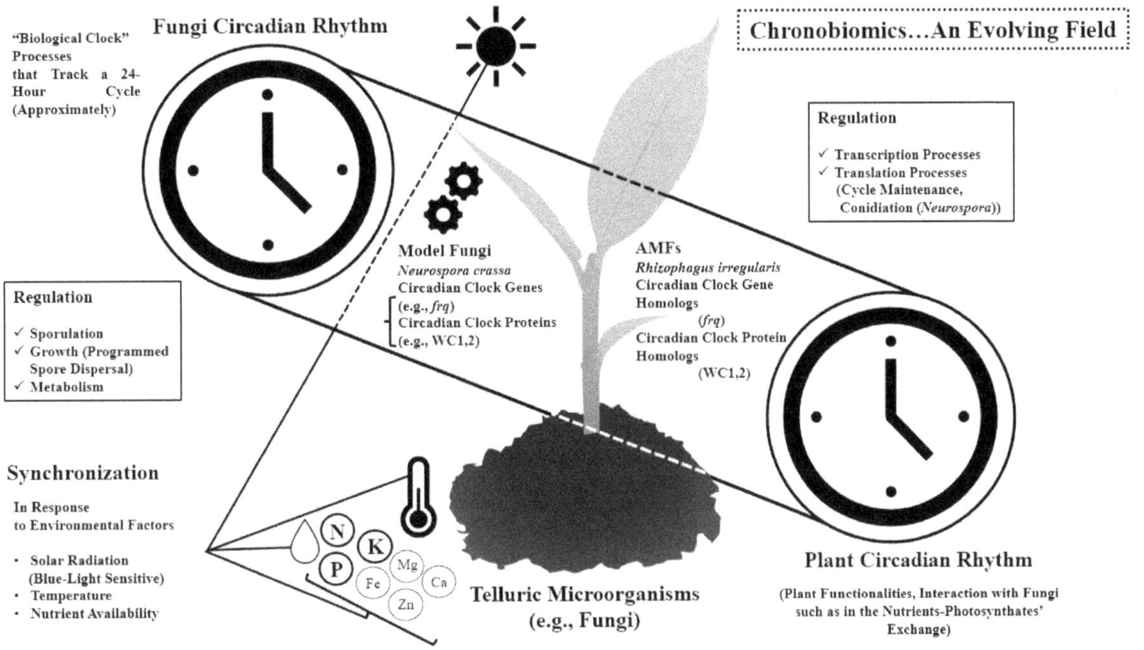

Figure 3. Circadian rhythm in fungi. **AMFs,** *Arbuscular Mycorrhizal Fungi;* **Ca,** *Calcium;* **Fe,** *Iron;* **frq,** *Fungal Circadian Oscillator gene;* **K,** *Potassium;* **Mg,** *Magnesium;* **N,** *Nitrogen;* **P,** *Phosphorus;* **WC1, 2,** *White Collar Proteins 1 and 2;* **Zn,** *Zinc*

The bridging aspect being the environmental factor, the present book division takes interest in the intrigued AMFs diversity and responses from the molecular lens through the sophisticated metagenomics broad approach. Besides, it singularly features advances in the *frq* gene and its role in the circadian rhythm in AMFs.

RESEARCH METHOD

To set up the present book division, the gathered analyzed data were collected through the secondary research method (Johnston, 2014) (Figure 4). The secondary research method implicates collecting, summarizing, and analyzing existing available data in the literature.

• *Customized Range, Search Terms and Phrases, and Boolean Search Method*

Data were retrieved from reputed databases, namely Google Scholar, Scopus, and Web of Science. Selected case studies from the last 14 years [2010-2024] were analyzed. Different combinations of terms and phrases were used to increase the possibilities of finding relevant material. The Boolean search method can lead to accurate retrieval of search results (Bello Aliyu, 2017). The Boolean operators that were used were "AND" and "OR", thereby refining and/or expanding the search scope.

• *Inclusion-Exclusion Criteria and The Range of the Analyzed Material*

The retrieved material exhibiting a strong correlation with the book chapter scope and research question was backed up, whereas material that showed minimum association was ruled out, based on Inclusion-Exclusion Criteria. The treated material ranged from original articles, review articles, and academic book chapters to proceedings, editorials, commentaries/opinions, and reports.

Figure 4. Research method

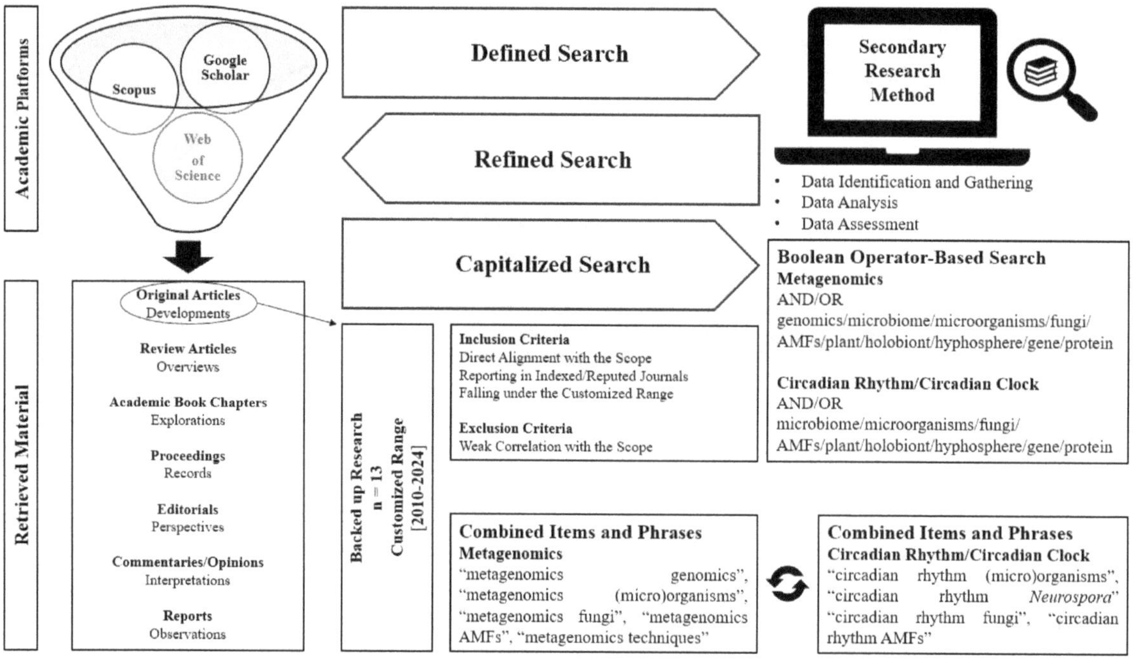

THE METAGENOMICS SEQUENCING TECHNIQUES

• Shotgun Sequencing

Shotgun sequencing implicates the random slicing of the genome into small DNA fragments (100-1000 base pairs (bp)). These DNA fragments are sequenced separately (Chen & Pachter, 2005). Historically, the cauliflower mosaic virus was initially sequenced in 1981 (Gardner et al., 1981). Through shotgun sequencing, DNA sequences are randomly fragmented into several small fragments, which are then resembled by figuring out areas of overlap, using the chain termination method (Anderson, 1981). Due to its speed and affordability, it is currently the most favored method for genome sequencing (Pinto & Bhatt, 2024). The conjugation of shotgun sequencing could bring additional metagenomics insights, such as regarding soil depolluting by AMFs and the mycorrhizal interaction with the existing telluric microbiome (Nuccio et al., 2022).

• Pyrosequencing

Pyrosequencing represents an alternate method for sequencing tiny DNA fragments. The pyrosequencing technique is a very flexible means for testing and experimental work in many different fields because of its interesting output, flexible analysis, and simplified processes (Ronaghi, 2001). Regarding the applicability in AMFs' genome studying, Boon et al. (2015) for instance investigated the genetic diversity and polymorphism at a genomic scale as well as putative single-copy loci. Relying on data retrieved from whole-genome pyrosequencing of four AMF isolates to describe genetic diversity, the authors compared AMF isolates to a panel of control fungal genomes and recorded significant variances in genome-wide diversity patterns, in addition to providing conservative estimations of *Rhizophagus* spp. genome sizes and even recording loci polymorphism.

• Polymerase Chain Reaction Sequencing

Polymerase chain reaction (PCR) sequencing technique is used to identify particular DNA strands to detect if a specific gene experienced a mutation event. Historically, Frederick Sanger is credited for the establishment of the technique in 1977, with the necessary components being nucleotide precursors, a primer, and a DNA polymerase, for which he won the Nobel Prize in 1980 (Kubista et al., 2006). Ever since the technique was operable and made available, analyzing fragments of DNA strands where it is difficult to culture axenically has become possible for microorganisms like AMFs (Redecker et al., 2003).

At the molecular identification level, research has focused on specific regions of the ribosomal genes. Few studies used other regions like the Random Amplified Polymorphic DNA (RAPD) fragment with an undetermined function (Lanfranco et al., 1995). The authors Simon et al. (1992) identified the first three 18S Small Subunits (SSU) sequences from fungal ribosomal DNA. Based on the obtained data, the authors created a particular PCR primer for AMFs, VANS1. Additional group-specific primers for AMF lineages were subsequently introduced (Simon et al., 1993). Further studies explored more SSU sequences and it was noted that the VANS1 priming site was not well-conserved in all AMFs (Redecker et al., 2003). The authors Clapp et al. (1999) examined an AMF field population and recorded a glaring discrepancy in some *Glomus* spp. between important root systems colonization and the absence of the respective spores. The 28S ribosomal Large Subunit (LSU) was used later. Specific primers aimed at the highly variable LSU's D2 region, each for a specific AMF species. A particular primer was used for a *Glomus* subgroup comprising *G. mossae* and *G. intraradices* for this region to investigate AMF field populations in symbiosis with the root systems of peas. A single primer (AM1) aimed at an SSU region to examine AMF communities colonizing root systems from horticultural sites and found that the AMF community almost corresponded to a particular AMF sequence type, that of *Glomus mossae* (Helgason et al., 1998; Kjøller & Rosendahl, 2001; Van Tuinen et al., 1998).

• Next-Generation Sequencing

In the domains of genetics and biochemistry chiefly, sequencing designates determining the primary structure of an unbranched biopolymer, resulting in a linear description, or a sequence, concisely summarizing a good portion of the atomic structure of the molecule subject to sequencing. Next-Generation Sequencing (NGS) of DNA/Ribonucleic Acid (RNA), also goes by the name Massive Parallel Sequencing and High-Throughput Sequencing, is a technique that emerged in the mid-90s (around 1998) and was made available in the early 2000s (2005), where DNA/RNA is sequenced and possible mutations are revealed (Behjati & Tarpey, 2013; L. Liu et al., 2012). The performance quality of NGS resides in the fact that it can lead to sequencing hundreds to thousands of genes or even whole a genome in a brief span. In this regard, the NGS technique provides both a precise and speedy performance (Pervez et al., 2022). Relying on this technique, the determination of the exact order of nucleotides that form a precise region of DNA/RNA, or whole a genome, is made possible. It is advantageous in the sense it allows the dealing with a wide range of settings and studying complicated biological systems, thereby addressing important questions in the metagenome of the studied organisms like AMFs (De Castro et al., 2018a).

In a study by De Castro et al. (2018), the NGS technique contributed to sequencing the rRNA's nuclear molecular markers SSU-Internal Transcribed Spacer (ITS)-LSU and ITS2, revealing the presence of AMF in symbiosis with *Pancratium maritimum* L., with a number of the obtained sequences matching, exhibiting similarity, or relating to the Glomeromycota division. The authors Kryukov et al. (2020) attempted to identify AMF strains from an available collection and could note eight out of nine isolates, with five being *Rhizophagus irregularis* and three being *R. invermaius*, *Paraglomus laccatum*, and *Claroideoglomus etunicatum*. The NGS technique is also often relied upon to identify particular extracted DNA regions encoding 16S rRNA as well as 18S rRNA based on similarities in a given obtainable gene sequence from a public database via amplification and sequencing (De Souza et al., 2004; Jovel et al., 2016). For reference, De Souza et al. (2004) studied the (inter-/intra)species variations in the 18S rRNA genes relating to the *Gigaspora* genus to evaluate the application of this marker for the *Gigaspora* isolate and Gigasporaceae population discrimination from environmental samples and concluded that the 18S rDNA heterogeneity's patterns within a specific region, V9 region, could be relied upon for the documented *Gigaspora* species identification.

THE METAGENOMICS AND ARBUSCULAR MYCORRHIZAL FUNGI

Metagenomics takes interest in sequencing nucleic acids retrieved from the environment, such as samples that contain (micro)organisms, subject of a particular study. Microorganisms like AMFs tend to occur in communities, either exhibiting synergistic or antagonistic behavior. The deployment of genetic material and sophisticated sequencing approaches continue to improve the scientific community's understanding in this regard. In the recent past, metagenomics was deployed to study AMFs' occurrence in a given environment, especially in terms of their genetic diversity and possible environmental factors influencing their telluric presence as well as symbiosis with plants' root systems (Table 1).

For instance, in a study carried out by Balestrini et al. (2010), the authors attempted to explore the composition of AMF species in two production vineyards with different edaphic properties (particularly from a loamy texture, pH, organic matter, and C:N ratio view angle), based on morphological (spores) and PCR sequencing approaches. The amplification of a segment (550 bp) of AM fungal SSU rDNA revealed different distributions of sequences within the two vineyard sites in the main glomeromycotan groups. One representative Operational Taxonomic Unit (OTU), out of OTUs corresponding to phylotypes with 98% similarity levels, was related to the *Glomus* spp. group within the two vineyard soils, out of a cohort of species, pointing out possible dominant influence by the type of the soil versus an impact to a less extent of the features relating to the host plant (e.g., vegetative stage) (Lumini et al., 2010b).

In another study conducted by Colombo et al. (2014), the focus was most especially on the (non-) agronomic practices and their effects on the AMFs community in studied soils from three horticultural exploitations, with previous wheat and soybean cultivation history, based on morphological (spores) and pyrosequencing approaches. What followed was the identification of 188 Molecular Operational Taxonomic Units (MOTUs) belonging to the Glomeromycota phylum, in addition to 29 morphological AMF species, with the most diversity recorded in the non-horticultural soils. The authors could classify 52.68% of the obtained MOTUs at the genus level, with *Glomus* showing the highest percentage values (24.9%), followed by *Rhizophagus* (11.29%), *Claroideoglomus* (8.49%), *Gigaspora* (2.85%), *Funneliformis* (2.74%), *Paraglomus* (0.98%), *Scutellospora* (0.70%), *Diversispora* (0.38%), *Ambispora* (0.17%), and *Archaeospora* (0.03%). The authors emphasized the fact that non-horticultural soils were substantially more biodiverse, corroborating the anthropogenic perturbance of the soils decreasing fungal species occurrence as well as richness (Lumini et al., 2010c).

Exploring AMFs as heterokaryotic microorganisms with coenocytic characteristics was made possible owing to the metagenomics progress. In this vein, Boon et al. (2015) explored the genetic diversity in AMFs, on the coenocytic level, bearing in mind that AMFs encompass thousands of nuclei and exhibit important levels of genetic variation for certain loci, and given the fact that the degree of as well as the physical polymorphism location, at the level of the mycorrhizal genome level, is undeciphered (Boon, 2012). Relying on data procured from whole-genome pyrosequencing of four AMF isolates (*Agrobacterium rhizogenens* transformed carrot roots), genetic diversity was described. When compared to a panel of control fungal genomes, AMF isolates displayed notable variations in genome-wide diversity patterns, with the obtaining of conservative estimations of the *Rhizophagus* spp. genome sizes. The authors further designed putative single-copy genomic markers and used large parallel amplicon sequencing to examine them for two isolates of *Rhizophagus irregularis* and one strain of *Rhizophagus* sp. With up to 103 alleles per marker, the majority of loci displayed significant polymorphism, indicating that the investigated *Rhizophagus* isolates may be heterokaryotic (Tisserant et al., 2013). The authors suggested that every *Rhizophagus* isolate harbored a population of differentiated genomes. Furthermore, they suggested that the observed extensive *Rhizophagus* single copy markers' genomic heterogeneity could be divided between and not within nuclei in the cytoplasm.

On a different note, Yuan et al. (2021) studied the temporal-spatial patterns that govern the soil fungal-bacterial occurrence in proximity to the wild oat grass' root system, grown in its naturalized soil in a greenhouse setting to investigate their potential cooperating or competing behavior/neighbors. Relying on amplicon sequences of the fungal ITS and bacterial 16S rRNA genes related to the rhizosphere and bulk soil samples retrieved at various plant growth stages, identifying fungal-bacterial interchange was initiated. Interestingly, fungal-bacterial networks were substantially different at the level of the rhizosphere than in the bulk soils. Notably, AMFs formed relatively important networks with bacteria at the level of bulk soils, with the indication of bacterial breakdown and fungal biomass consumption. Concerning AMFs, a total of 156 ITS OTUs were identified as Glomeromycota, which covered a wide range of the phylogeny of AMFs, with averages of 6.3% for the bulk soil and just 1.1% for the rhizosphere. The authors noted that the locations where sampling was performed had a substantial impact on the topology of the network, leading to distinct structured networks in the rhizosphere and bulk soils (Frey, 2007). The species' cooccurrence can follow from the associated effects of environmental filtering relating to niche sharing and biotic interactions (Bueno de Mesquita et al., 2016). The authors attributed the partition of distinct niches to the availability as well as forms of C substrates, noting that they may be the fundamental drivers of how microorganisms stemming from variable life histories and ecologies cooccur and potentially interact in complex soil environments (Bissett et al., 2010; Macalady et al., 2013).

In the same vein, Wang et al. (2023) attempted to explore the mechanisms by which AMFs interact with the hyphosphere's soil microbiome as well as impact the organic P mobilization, with the *Medicago-Rhizophagus irregularis* as the model system. Under inoculation with *R. irregularis* and P treatments, the authors studied the soil's P turnover processes, characterizing the microbiome functional profiles for P turnover. Interestingly, the hyphosphere soil that had *R. irregularis* exhibited inhabitation by a specific bacterial community, with their associated functional profiles for P turnover stimulated. The authors registered the recruitment of the genome binning (bin)2.39, harboring the quinoprotein glucose dehydrogenase (*gcd*) and alkaline phosphatase (*phoD*) encoding genes, and genome bin2.97, harboring the *phoD* gene, which explainingly characterized the shift in hyphosphere soil microbiome at the species level, synergistically driving non-soluble phytate mobilization (Xie et al., 2020). The authors suggested that AMFs may have recruited a specific hyphosphere soil microbiome and stimulated their P turnover functional profiles to increase phytate utilization (Qin et al., 2022). In addition, the authors attributed a substantial rise in alkaline phosphatase (ALP) activity with the presence of fungal hyphae to either an expanded bacteria population harboring the gene encoding phosphatase (e.g., *phoD*) or a noticeable shift in the bacterial community structure (Zheng et al., 2021).

Metagenomics was deployed to decipher the role of AMFs in remediating to Polycyclic Aromatic Hydrocarbons (PAHs) polluted soils, as depicted in the study by Li et al. (2024). In their study, the authors studied the effects of AMFs in the depolluting from PAHs as well as their impact on the composition, abundance, and function of the microbial community. In addition to their enhancing the removal of benzo[a]pyrene and enriching more PAH-degrading bacteria, the mycorrhizal input contributed to the enrichment of genes implicated in amino acid synthesis, aminoacyl-tRNA biosynthesis, and cysteine and methionine metabolism pathways. Besides, the authors mentioned that the inoculation with *Funneliformis mosseae* had a substantial impact on alpha- and beta-diversity of microbial genes. Regarding soil microbial biomass C and the organic matter pool, the mycorrhizal inoculation substantially increased these attributes, the authors stressed. The dissipation of PAHs was supported by the microbial community assembly as well as function *molded* by AM willow. The recorded results shed light on the complexity of the relationship between plants' root systems' exudates and indigenous microorganisms (Lenoir et al., 2017). The authors also noted that the application of environmentally exogenous AMFs could modify the microbiome abundance as well as the composition at the phylum and genus levels.

Table 1. Metagenomics breakthroughs in arbuscular mycorrhizal fungi (AMFs) in relation with their environment

Main Objective	Plant Species	Sequencing Approach	Molecular Features	Key Findings	Pivotal Conclusions	Reference
-Morphological and Molecular investigation of the diversity of two Piedmont vineyards with different soil characteristics -Comparison of the AMFs detected in the soil with AMFs colonizing the grapevine root system -Determination of the possible environmental factor influencing the symbiosis	Common grape vine (*Vitis vinifera* L.)	PCR sequencing	*-AMF Community composition* -A fragment (550 bp) of AM fungal SSU rDNA	-Different sequences' distribution in the main Glomeromycotan groups -One detected common OTU relating to *Glomus* spp. phenotype	It is possible that soil type has a greater impact on the overall structure and composition of AMF assemblages in vineyards than do host plant characteristics (e.g., age, vegetative stages), or management techniques	Balestrini et al. (2010)
-Establishment of the first and largest AMF soil-based sequence data set -Investigation of the diversity of AMFs in a Mediterranean ecosystem	Cork oak (*Quercus suber* L.) (Previously)	Pyrosequencing	Two 18S ribosomal RNA gene regions	-Production of 10924 sequences -Selection of 3189 and 1003 sequences to produce OTUs to assess the AMF community richness and similarity -Dominance of Glomerales and detection of Archaeosporales and Paraglomerales	The biodiversity of AMFs is shaped by human and ecological factors	Lumini et al. (2010)
-Assessment of agronomic practices' effects on soil AMF communities, with horticultural sites sharing similar horticultural history -Comparison between the inputs obtained through morphological and pyrosequencing approaches	Wheat (*Triticum* sp.)-Soybean (*Glycine max* L.) (Previously)	Pyrosequencing	Amplicons	-Identification of 188 molecular OTUs -Morphological identification of 29 AMF species	-Agronomic management affect the biodiversity of AMF communities in the investigated sites -Both morphological and pyrosequencing approaches complement each other	Colombo et al. (2014)

continued on following page

Table 1. Continued

Main Objective	Plant Species	Sequencing Approach	Molecular Features	Key Findings	Pivotal Conclusions	Reference
-Investigation of heterokaryons' presence in multicellular organisms -Measuring genetic diversity in AMF coenocytes and variation in loci -Examining polymorphism both on a genome scale and in putative single-copy loci	Carrot (*Daucus carota*) (*Agrobacterium rhizogenens* transformed roots)	Pyrosequencing	-Amplicons -Loci	-Diversity in AMF isolates genome patterns -Provision of conservative estimates on *Rhizophagus* spp. genomes sizes -Designation of novel putative single-copy genomic markers for two *Rhizophagus irregularis* and one *Rhizophagus* sp. isolates -Recording of high polymorphism (up to 103 alleles per marker) and speculation of heterokaryosis	The concept of the individual in *Rhizophagus* may need to be carefully considered in light of the levels of genetic diversity found in its cytoplasm, as well as the potential organization of the genetic diversity into differentiated nuclei	Boon et al. (2015)
-Understanding how cropping systems impact soil microbial communities -Characterization and comparison of soil microbial communities in restored prairies and three potential cellulosic biomass crops	Corn (*Zea mays* L.), switchgrass (*Panicum virgatum* L.), and mixed prairie grasses	Pyrosequencing	rRNA genes *nifH* genes	-Assessment of microbial community structure as well as composition -AMFs (*Paraglomus*)' biomass as the most important under perennial grasses and restored prairie, implying an important C pool	Several factors like location, soil type, history, landscape, cultivation duration, as well as climate determine the microbial communities present	Jesus et al. (2016)

continued on following page

Table 1. Continued

Main Objective	Plant Species	Sequencing Approach	Molecular Features	Key Findings	Pivotal Conclusions	Reference
Introduction of novel bioinformatics program, SeSaMe, intended for taxonomic classification of short sequences	-	NGS	3-codon DNA 9-mer encoding three consecutive amino acids located in a protein secondary structure	The mean percentages of correct predictions of the coding and non-coding sequences test sets at the genus level as 49% and 72% for AMF (*Rhizophagus irregularis*)	-SeSaMe as a bioinformatics means can estimate the gene reservoir of the reference taxonomic groups linked to AMF in addition to taxonomic diversity and abundance -SeSaMe can permit users to study the symbiotic functions of related microbes as well as be applied to soil metagenomes and other microbes	Eun Kang et al. (2020)
-Development of a simple procedure to improve the taxonomic assignment of AMF 18S sequences	*Tamarix* (*Tamarix aphylla* L.)	PCR sequencing	ASVs	-Identification of seven phylogenetic divergent clades of the *Glomus* genus, making for 27% of the ASVs -Attribution of 85% of the ASVs to the *Glomus* genus	Because the 18S gene is less variable than the ITS region and LSU gene, amplicon sequence variants—that is, sequences with 100% nucleotide identity—are probably the most suitable approach to avoid lumping closely related species	Stefani et al. (2020)
Exploring the temporal and spatial patterns of telluric fungal as well as bacterial co-occurrence in proximity to the wild oat grass root system, grown in its naturalized soil in a greenhouse setting at multiple plant growth stages	Wild oat grass (*Avena fatua* L.)	NGS	Amplicon sequences of the fungal internal transcribed spacer rRNA genes	-Identification of a total of 156 ITS OTUs as Glomeromycota enclosing a wide range of the AMF phylogeny -AMF OTUs being present at 6.3% in the bulk soil and 1.1% in the rhizosphere	Some of the fungal-bacterial connections found may be indicators of bacterial degradation and fungal biomass consumption	Yuan et al. (2021)

continued on following page

Table 1. Continued

Main Objective	Plant Species	Sequencing Approach	Molecular Features	Key Findings	Pivotal Conclusions	Reference
-To link the identity of wild microbes and their ecophysiological attributes and environmental functionalities -Testing of a designed semi-automated and high-throughput pipeline intended for supporting well-replicated and temporally resolved amplicon and metagenomics experiments for a soil microhabitat with substantial ecological importance (the hyphosphere zone adjacent to AMF hyphae)	-	Shotgun sequencing	^{13}C-AMF hyphosphere DNA	High enrichment of *Rhizophagus intraradices* despite the soils' overall low enrichment	-The semi-automated and high-throughput pipeline approach lowers operator time and ameliorates reproductivity by aiming at the most labor-intensive steps-fraction collection and cleanup -The phylogenetic composition as well as the gene content suggest predation, decomposition, and ammonia oxidation could be key factors in the cycling of hyphosphere nutrient cycling	Nuccio et al. (2022)
Study of the soil P turnover processes and characterization of the microbiome functional profiles for P turnover in the hyphosphere soil through metagenomic sequencing	Barrel clover (*Medicago truncatula* Gaertn.)	-PCR sequencing -Shotgun sequencing	*gcd* gene *phoD* gene	The recruitment of genome bin2.39, harboring both the *gcd* and *phoD* genes and genome bin2.97, harboring the phoD gene as indicative of the species-level change in the hyphosphere soil microbiome. These two genes probably working synergistically to drive the mobilization of nonsoluble phytate in the hyphosphere's soil	AMFs possibly recruited a specific hyphosphere soil microbiome and intrigued their functional profiles for P turnover to increase the phytate utilization	Wang et al. (2023)

continued on following page

Table 1. Continued

Main Objective	Plant Species	Sequencing Approach	Molecular Features	Key Findings	Pivotal Conclusions	Reference
-Investigation of the AMF inoculation effects on the growth, nutrient absorption as well heavy metals' uptake, and functional genes of the rhizosphere microbial community of Cr-stressed *Iris tectorum* under a greenhouse setting	Roof iris (*Iris tectorum* Maxim.)	PCR sequencing	Functional genes relating to nutrient cycling (N, P), heavy metal resistance (*chrA*, *arsB*), heavy metal transporter family (P-ATPase, MIT, CDF, ABC)	-Substantial alterations spanning the structure as well as the composition of the rhizosphere's microbial community -Upregulation of the functional genes associated with nutrient cycling and heavy metal resistance as well as transporter family -Improvement of the rhizosphere's microbial ecological network's complexity and stability and regulation of the synergy between microbial communities	Advancement of the understanding of the possible mechanism of plant remediation of Cr-contaminated soil under AMF symbiosis showed that AMF could control rhizosphere's microbial communities to enhance plant development and heavy metal stress tolerance	Zhao et al. (2023)
Assessment of the effects of AMFs in remediating the polycyclic aromatic hydrocarbons and the exploration of the microbial community composition, abundance, and function impact	The basket willow (*Salix viminalis* L.)	NGS	Amino acid synthesis, aminoacyl-tRNA biosynthesis, and cysteine and methionine metabolism pathway-related genes	-Improvement by AMF willow of the enrichment of the genes related to the metabolism pathways of cysteine and methionine, amino acid synthesis, and aminoacyl-tRNA production -Affecting by *Funneliformis mosseae* inoculation of alpha- and beta-diversity of microbial genes	The dissipation of polycyclic aromatic hydrocarbons was boosted by the microbial community assembly and function influenced by AM willow	Li et al. (2024)

ABC, ATP-binding Cassette; **AM,** Arbuscular Mycorrhizal; **AMF,** Arbuscular Mycorrhizal Fungus; **AMFs,** Arbuscular Mycorrhizal Fungi; ***arsB*,** Arsenic Efflux Pump; **ASVs,** Amplicon Sequence Variants; **bin,** binning; **bp,** base pair; **C,** Carbon; **CDF,** Cation Diffusion Facilitator; ***chrA***, Chromate Transport Protein; **Cr,** Chromium; ***gcd*,** Quinoprotein Glucose Dehydrogenase encoding gene; **ITS,** Internal Transcribed Spacer; **LSU,** Large Subunit; **MIT,** Metal Ion Transporter; **N,** Nitrogen; **NGS,** Next Generation Sequencing; ***nifH*,** Nitrogenase Iron Protein gene responsible for Dinitrogenase Reductase; **OTU,** Operational Taxonomic Unit; **P,** Phosphorus; ***phoD*,** alkaline phosphatase encoding gene; **P-ATPase,** P-type

Adenosine Triphosphatase; **RNA,** Ribonucleic Acid; **SeSaMe,** Spore-associated Symbiotic Microbes; **SSU rDNA,** Small Subunit ribosomal Deoxyribonucleic Acid; **t-RNA,** transfer Ribonucleic Acid

THE FUNGAL CIRCADIAN OSCILLATOR (*frq*) GENE AND ARBUSCULAR MYCORRHIZAL FUNGI

The daily oscillations, fluctuations in genes' expression profile as well as activity, observed in almost all groups of organisms are known as circadian rhythms (Bita-Nicolae & Dhyani, 2025; Pittendrigh, 1960). The circadian rhythms are propelled by an internal clock that incorporates environmental input to allow a given organism to align and synchronize its activities with the daily light (more of an entraining signal) and temperature (way more complex impact) cycles (Dunlap et al., 2004; Dunlap & Loros, 2004). In the group of fungi, a genus of the Ascomycetes, *Neurospora*, was probably pioneeringly studied in this regard (species: *Neurospora crassa*) (Lakin-Thomas et al., 1990). In several organisms, the clock core encompasses a negative feedback loop, where in *Neurospora*, this loop comprises a negative element, which is the *frq* gene, whose transcription is stimulated by the positive elements, which are the White Collar 1 (*wc-1*) and *wc-2* genes' products (Bell-Pedersen et al., 2005; Y. Liu, 2003). The FRQ protein can suppress its proper synthesis through interaction with and prompting the phosphorylation of the WC-1 and WC-2 proteins (Liu, 2003), generating a self-sustaining oscillation in the clock components (Colot et al., 2005). Nevertheless, FRQ can also positively prompt WC-1 and WC-2 levels. These positive feedback loops interlock with the negative feedback loop to foster the rhythm's robustness (Lee et al., 2000). Looking at AMFs, the mechanisms surrounding the circadian clock were rarely studied, though circadian growth in mycorrhizal hyphae in the field was reported (Lee et al., 2018).

Henceforward, advances in the homologs of *wc-1*, *wc-2*, and *frq*, as well as associated circadian clock genes were recorded in *Rhizoglomus irregulare* (synonym *Rhizophagus irregularis*) (Lee et al., 2018). The authors analyzed the expression of *wc-1*, *wc-2*, and *frq* genes through the Quantitative Reverse Transcription Polymerase Chain Reaction (RT-qPCR) on RNA retrieved from germinating spores and fungal material cultured *in vitro* with transformed carrot roots. The WC-1 and WC-2 proteins modulate fungal responses to blue light. Together with the *frq* gene, they form key putative of the internal circadian clock components. During the pre-mycorrhizal and post-mycorrhizal phases of *R. irregulare* development, it was discovered that all three core clock genes were expressed. Brief light exposure triggered the main circadian oscillator *frq* gene, just like in the model fungus *Neurospora crassa*. Interestingly, the authors recorded gene expression of the mentioned three core clock components, both at the pre- and post-mycorrhizal stages of *R. irregulare* growth. Both the observed WC-1 and WC-2 proteins present in *R. irregulare* were shown to be closely related to Mucoromycotina orthologs, orthologs that are related to a basal fungal subphylum proposed to be the closest neighbor of Glomeromycota in studies that deployed mitochondrial genomes and SSU rRNA sequences, the authors exemplified (Nadimi et al., 2012; Redecker et al., 2013). The authors also illustrated that, thus far, all the fungal species with conserved WC proteins and *frq* had a circadian clock (Liu & Bell-Pedersen, 2006; Lombardi & Brody, 2005; Salichos & Rokas, 2010).

Given the fact that circadian clocks are almost ubiquitously occurring timing mechanisms that shape the rhythmic behavior as well as gene expression in a wide range of organisms, it is worth looking at it from the holobiont perspective (Lee et al., 2018). However, the mechanisms governing holobionts' circadian clock development and interspecies temporal adaptation remain undeciphered. It is suggested that a better comprehension of the function and synchronization of circadian clocks in holobionts, in general, would be improved by examining the interconnected temporal organization of the AM symbiosis and identifying its underlying processes. In this sense, holobiont chronobiomics, which includes the chronobiology of the host and its associated microbiota, has recently emerged and acquired considerable interest due to the increased reliance on metagenomics' sequencing techniques (Hewitt & Shaikh, 2021).

KNOWLEDGE GAPS

The subfield of metagenomics falls under the field of molecular biology as a cornerstone in investigating AMFs in their assemblages, diversity, and genetic heterogeneity (Stefani et al., 2020). Hence, more advanced genetic sequencing-oriented investigation should be performed to ensure the analysis and comparison of AMF genetic patrimony with the available gene bank sequences worldwide and to better understand mycorrhizal-associated phylotypes (Balestrini et al., 2010). Whether the most commonly represented taxa are the most functionally active is one of the several unanswered complications regarding the functioning of the AMF communities. A distinct property that is worth the molecular investigation in this sense is the anastomosis role in the maintaining of mycorrhizal genetic variation (e.g., the *Rhizophagus* genus). The comparison of inter-isolate single-copy marker diversity lines could allow studying the governing linkage and recombination in mycorrhizal isolates (Boon et al., 2015). Comparative research on different regulator subpopulations may also provide enlightening insights into the creation of genetic markers that can allow the recording of alterations in the structures of microbial communities in response to environmental factors (Eun Kang et al., 2020).

In an ecological context, holobionts as a host and complex cohort of prokaryotic communities imply that every individual in this nested ecosystem will use their internal clock to govern their activity. Interestingly, the symbionts' and hosts' mutually dependent connection is also rhythmically synchronized. This connection is complicated, especially as there are probably several entrainment cues and the relative significance of each may vary depending on the host and symbiont. More research on the chronobiology of mycorrhizal symbioses is required, given the significance of plants and mycorrhizal symbiosis in terrestrial ecology and horticulture. By incorporating the time dimension, it would be easier to see how aboveground and belowground processes are related, and how subsurface metabolism is related to the periodic changes in day and night. A thorough investigation to separate the connections between host and symbiont-driven circadian modulation would be required (Hewitt & Shaikh, 2021). One intriguing question regarding the chronobiology of mycorrhizal symbiosis is whether photosynthesis will drive the symbiont rhythms during the daytime phase. The circadian clocks of crassulacean acid metabolism (CAM) plants also coordinate carbon dioxide (CO_2) fixation, malate buildup, and CAM plants' productivity during the nighttime, which is probably the case for C3 plants (Lee et al., 2019).

CONCLUDING REMARKS

The number of AMF fungal sequences that were found in the soil thus far is greatly outnumbered by the number of AMF sequences produced by a metagenomic technique applied to an AM fungus and based on a gene of taxonomic relevance, such as the ribosomal gene. Both ecological characteristics and human input most likely have an impact on AMF biodiversity. Metagenomics-based techniques may be able to decipher not only the species present but also the advantages that these microorganisms provide to their host plant. The compelled and analyzed case studies revealed that advanced sequencing techniques either consolidated or outperformed conventional spore-based characterizing methods related to the ecology of AMFs. The anthropogenic action (e.g., agronomic versus non-agronomic practices), soil properties, soil niches, surrounding climate, and plant features (perennial versus annual, terrestrial versus aquatic) as well as growth stages can more or less affect the distribution, occurrence, and composition of AMFs in their environment as well as their association with and their impact on existing microbial communities. Because an isolate may have a population of somewhat differentiated genomes rather than a single typical genome, the population of partially heterogeneous genomes in AMFs can be comparable to a pangenome (e.g., the *Rhizophagus* genus). In this respect, AMFs can exhibit important genetic diversity levels within their cytoplasm, and how this diversity can be subject to possible encompassment into differentiated nuclei, with a mindful consideration of the concept of the individual. The analyzed AMF medium, such as hyphosphere with important risk samples (the human error factor), can be made to be cautiously dealt with through reliance on metagenomics techniques, especially in terms of factors like limited soil volume and bulk soil enrichment. The conjugation of metagenomics techniques with sophisticated omics, such as metabolomics, can result in delivering important information on the microbial shifts, the exact microbiome interrelation, and the impact of AMFs on the existing telluric microbiome as well as their vital role in depolluting soils and enhancing its nutritional reservoir, making nutrients ready for the assimilation by plants.

Circadian clocks are almost universal timing systems that can control gene expression and rhythmic behavior in a variety of organisms. In fungal, animal, and plant model systems, clock mechanisms are getting more noticed. Yet, while studying the biology of these organisms, it is important to consider the rich and varied microbiota that surrounds them. The symbiotic associations between organisms that have coevolved with time to create a unit known as holobiont are of special interest. Coordinating changes in activity throughout the day for each partner appears to depend on the interrelated regulation of holobiont circadian rhythms. A very interesting observation that was recorded concerning *frq* is that, though it was previously suggested to have been acquired post the Ascomycota and Basidiomycota divergence from their common ancestor, a *frq* homolog was recorded in *Rhizoglomus irregulare*, since Glomeromycota phylum diverged before Dikarya, calling for a further phylogenic investigation in this regard.

REFERENCES

Akensous, F.-Z., Anli, M., & Meddich, A. (2024). Arbuscular Mycorrhizal Fungi as Solubilizers of Rock Phosphate and Compost Application Improve Date Palm (*Phoenix dactylifera* L.)'s Resilience to Drought. *Gesunde Pflanzen*, *76*, 161–179.

Akensous, F.-Z., El Mazouni, N., Mesnaoui, M., & Meddich, A. (2023a). Biostimulants for upgrading plant and soil health statuses: Testimonials from date palm (*Phoenix dactylifera* L.) and durum wheat (*Triticum durum* Desf.). In Agricultural Research Updates, *46*, 129–179.

Akensous, F.-Z., Sbbar, N., Ech-chatir, L., & Meddich, A. (2023b). Artificial Intelligence, Internet of Things, and Machine-Learning: To Smart Irrigation and Precision Agriculture. In Artificial Intelligence Applications in Water Treatment and Water Resource Management, 113–145.

Anderson, S. (1981). Shotgun DNA sequencing using cloned DNase I-generated fragments. *Nucleic Acids Research*, *9*(13), 3015–3027. DOI: 10.1093/nar/9.13.3015 PMID: 6269069

Balestrini, R., Magurno, F., Walker, C., Lumini, E., & Bianciotto, V. (2010). Cohorts of arbuscular mycorrhizal fungi (AMF) in *Vitis vinifera*, a typical Mediterranean fruit crop. *Environmental Microbiology Reports*, *2*(4), 594–604. DOI: 10.1111/j.1758-2229.2010.00160.x PMID: 23766230

Behjati, S., & Tarpey, P. S. (2013). What is next generation sequencing? *Archives of Disease in Childhood - Education and Practice*, *98*(6), 236–238. DOI: 10.1136/archdischild-2013-304340 PMID: 23986538

Bell-Pedersen, D., Cassone, V. M., Earnest, D. J., Golden, S. S., Hardin, P. E., Thomas, T. L., & Zoran, M. J. (2005). Circadian rhythms from multiple oscillators: Lessons from diverse organisms. *Nature Reviews. Genetics*, *6*(7), 544–556. DOI: 10.1038/nrg1633 PMID: 15951747

Bello Aliyu, M. (2017). Efficiency of Boolean Search strings for Information Retrieval. [AJER]. *American Journal of Engineering Research*, *6*, 216–222.

Bhargava, P., Vats, S., & Gupta, N. (2019). Metagenomics as a Tool to Explore Mycorrhizal Fungal Communities. In *Mycorrhizosphere and Pedogenesis*, 207–219. DOI: 10.1007/978-981-13-6480-8_13

Bhupenchandra, I., Chongtham, S. K., Devi, A. G., Dutta, P., Sahoo, M. R., Mohanty, S., Kumar, S., Choudhary, A. K., Devi, E. L., Sinyorita, S., Devi, S. H., Mahanta, M., Kumari, A., Devi, H. L., Josmee, R. K., Pusparani, A., Pathaw, N., Gupta, S., Meena, M., & Swapnil, P. (2024). Unlocking the Potential of Arbuscular Mycorrhizal Fungi: Exploring Role in Plant Growth Promotion, Nutrient Uptake Mechanisms, Biotic Stress Alleviation, and Sustaining Agricultural Production Systems. *Journal of Plant Growth Regulation*. Advance online publication. DOI: 10.1007/s00344-024-11467-9

Bissett, A., Richardson, A. E., Baker, G., Wakelin, S., & Thrall, P. H. (2010). Life history determines biogeographical patterns of soil bacterial communities over multiple spatial scales. *Molecular Ecology*, *19*(19), 4315–4327. DOI: 10.1111/j.1365-294X.2010.04804.x PMID: 25241408

Biţă-Nicolae, C., & Dhyani, P. (2025). Editorial: Plant diversity: the key to ecosystem resilience in a changing world. Frontiers in Plant Science. *Frontiers in Plant Science*, *15*, 1–3. DOI: 10.3389/fpls.2024.1534119

Boon, E. (2012). The evolution of inter-genomic variation in arbuscular mycorrhizal fungi. In *Biological sciences*.

Boon, E., Halary, S., Bapteste, E., & Hijri, M. (2015). Studying genome heterogeneity within the arbuscular mycorrhizal fungal cytoplasm. *Genome Biology and Evolution*, *7*(2), 505–521. DOI: 10.1093/gbe/evv002 PMID: 25573960

Brundrett, M., & Tedersoo, L. (2019). Misdiagnosis of mycorrhizas and inappropriate recycling of data can lead to false conclusions. *The New Phytologist*, *221*(1), 18–24. DOI: 10.1111/nph.15440 PMID: 30191568

Bueno de Mesquita, C. P., King, A. J., Schmidt, S. K., Farrer, E. C., & Suding, K. N. (2016). Incorporating biotic factors in species distribution modeling: Are interactions with soil microbes important? *Ecography*, *39*(10), 970–980. DOI: 10.1111/ecog.01797

Chen, K., & Pachter, L. (2005). Bioinformatics for Whole-Genome Shotgun Sequencing of Microbial Communities. *PLoS Computational Biology*, *1*(2), e24. DOI: 10.1371/journal.pcbi.0010024 PMID: 16110337

Clapp, J. P., Fitter, A. H., & Young, J. P. W. (1999). Ribosomal small subunit sequence variation within spores of an arbuscular mycorrhizal fungus, *Scutellospora* sp. *Molecular Ecology*, *8*(6), 915–921. DOI: 10.1046/j.1365-294x.1999.00642.x PMID: 10434413

Colombo, R. P., Fernández Bidondo, L., Silvani, V. A., Carbonetto, M. B., Rascovan, N., Bompadre, M. J., Pérgola, M., Cuenca, G., & Godeas, A. M. (2014). Diversity of arbuscular mycorrhizal fungi in soil from the pampa ondulada, Argentina, Assessed by pyrosequencing and morphological techniques. *Canadian Journal of Microbiology*, *60*(12), 819–827. DOI: 10.1139/cjm-2014-0364 PMID: 25408171

Colot, H. V., Loros, J. J., & Dunlap, J. C. (2005). Temperature-modulated Alternative Splicing and Promoter Use in the Circadian Clock Gene frequency. *Molecular Biology of the Cell*, *16*(12), 5563–5571. DOI: 10.1091/mbc.e05-08-0756 PMID: 16195340

De Castro, O., Avino, M., Di Maio, A., Menale, B., & Guida, M. (2018). Sanger and next generation sequencing in the characterisation of arbuscular mycorrhizal fungi (AMF) in *Pancratium maritimum* L. (Amaryllidaceae), a representative plant species of Mediterranean sand dunes. *Planta*, *248*(6), 1443–1453. DOI: 10.1007/s00425-018-2981-z PMID: 30128603

De Souza, F. A., Kowalchuk, G. A., Leeflang, P., Van Veen, J. A., & Smit, E. (2004). PCR-Denaturing Gradient Gel Electrophoresis Profiling of Inter- and Intraspecies 18S rRNA Gene Sequence Heterogeneity Is an Accurate and Sensitive Method to Assess Species Diversity of Arbuscular Mycorrhizal Fungi of the Genus *Gigaspora.Applied and Environmental Microbiology*, *70*(3), 1413–1424. DOI: 10.1128/AEM.70.3.1413-1424.2004 PMID: 15006761

Dunlap, J. C., & Loros, J. J. (2004). The *Neurospora* circadian system. *Journal of Biological Rhythms*, *19*(5), 414–424. DOI: 10.1177/0748730404269116 PMID: 15534321

Dunlap, J. C., & Loros, J. J. (2006). How fungi keep time: Circadian system in *Neurospora* and other fungi. *Current Opinion in Microbiology*, *9*(6), 579–587. DOI: 10.1016/j.mib.2006.10.008 PMID: 17064954

Dunlap, J. C., Loros, J. J., Colot, H. V., Mehra, A., Belden, W. J., Shi, M., Hong, C. I., Larrondo, L. F., Baker, C. L., Chen, C. H., Schwerdtfeger, C., Collopy, P. D., Gamsby, J. J., & Lambreghts, R. (2007). A circadian clock in *Neurospora*: How genes and proteins cooperate to produce a sustained, entrainable, and compensated biological oscillator with a period of about a day. *Cold Spring Harbor Symposia on Quantitative Biology*, *72*(1), 57–68. DOI: 10.1101/sqb.2007.72.072 PMID: 18522516

Dunlap, J. C., Loros, J. J., & DeCoursey, P. J. (2004). *Chronology: Biological timekeeping*. American Psychological Association.

Edwin, N. R., Fitzpatrick, A. H., Brennan, F., Abram, F., & O'Sullivan, O. (2024). An in-depth evaluation of metagenomic classifiers for soil microbiomes. *Environmental Microbiome*, *19*(1), 1–17. DOI: 10.1186/s40793-024-00561-w PMID: 38549112

Eun Kang, J., Ciampi, A., & Hijri, M. (2020). SeSaMe: Metagenome Sequence Classification of Arbuscular Mycorrhizal Fungi-associated Microorganisms. *Genomics, Proteomics & Bioinformatics*, *18*(5), 601–612. DOI: 10.1016/j.gpb.2018.07.010 PMID: 33346086

Frey, S. D. (2007). Spatial Distribution of Soil Organisms. In *Soil Microbiology* (pp. 283–300). Ecology and Biochemistry.

Fusconi, A., & Mucciarelli, M. (2018). How important is arbuscular mycorrhizal colonization in wetland and aquatic habitats? *Environmental and Experimental Botany*, *155*, 128–141. DOI: 10.1016/j.envexpbot.2018.06.016

Gardner, R. C., Howarth, A. J., Hahn, P., Brown-luedi, M., Shepherd, R. J., & Messing, J. (1981). The complete nucleotide sequence of an infectious clone of cauliflower mosaic virus by M13mp7 shotgun sequencing. *Nucleic Acids Research*, *9*(12), 2871–2888. DOI: 10.1093/nar/9.12.2871 PMID: 6269062

Hart, M. M., Reader, R. J., & Klironomos, J. N. (2003). Plant coexistence mediated by arbuscular mycorrhizal fungi. *Trends in Ecology & Evolution*, *18*(8), 418–423. DOI: 10.1016/S0169-5347(03)00127-7

Helgason, T., Daniell, T. J., Husband, R., Fitter, A. H., & Young, J. P. W. (1998). Ploughing up the wood-wide web? *Nature*, *394*(6692), 431–431. DOI: 10.1038/28764 PMID: 9697763

Hewitt, O. H., & Shaikh, H. M. (2021). The Rhythm of Many: Biological Rhythms in the Marine Environment, From Macro-Scale Planktonic Ecosystems to Micro-Scale Holobionts. *Frontiers in Marine Science*, *8*, 1–21. DOI: 10.3389/fmars.2021.744169

Hoang, D. T. T., Rashtbari, M., Anh, L. T., Wang, S., Tu, D. T., Hiep, N. V., & Razavi, B. S. (2022). Mutualistic interaction between arbuscular mycorrhiza fungi and soybean roots enhances drought resistant through regulating glucose exudation and rhizosphere expansion. *Soil Biology & Biochemistry*, *171*, 108728. DOI: 10.1016/j.soilbio.2022.108728

Jesus, E. da C., Liang, C., Quensen, J. F., Susilawati, E., Jackson, R. D., Balser, T. C., & Tiedje, J. M. (2016). Influence of corn, switchgrass, and prairie cropping systems on soil microbial communities in the upper Midwest of the United States. *Global Change Biology. Bioenergy*, *8*(2), 481–494. DOI: 10.1111/gcbb.12289

Jogawat, A. (2019). Osmolytes and their Role in Abiotic Stress Tolerance in Plants. In *Molecular Plant Abiotic Stress* (pp. 91–104). Biology and Biotechnology. DOI: 10.1002/9781119463665.ch5

Johnston, M. P. (2014). Secondary Data Analysis : A Method of which the Time Has Come. [QQML]. *Qualitatve and Quantative Methods in Libraryes*, *3*, 619–626.

Jovel, J., Patterson, J., Wang, W., Hotte, N., O'Keefe, S., Mitchel, T., Perry, T., Kao, D., Mason, A. L., Madsen, K. L., & Wong, G. K. S. (2016). Characterization of the Gut Microbiome Using 16S or Shotgun Metagenomics. *Frontiers in Microbiology*, *7*, 1–17. DOI: 10.3389/fmicb.2016.00459 PMID: 27148170

Khan, Y., Shah, S., & Hui, T. (2022). The Roles of Arbuscular Mycorrhizal Fungi in Influencing Plant Nutrients, Photosynthesis, and Metabolites of Cereal Crops—A Review. *Agronomy (Basel)*, *12*(9), 1–19. DOI: 10.3390/agronomy12092191

Kjøller, R., & Rosendahl, S. (2001). Molecular diversity of glomalean (arbuscular mycorrhizal) fungi determined as distinct *Glomus* specific DNA sequences from roots of field grown peas fungi. *Mycological Research*, *105*(9), 1027–1032. DOI: 10.1016/S0953-7562(08)61964-4

Kong, X., Jia, Y., Song, F., Tian, K., Lin, H., Bei, Z., Jia, X., Yao, B., Guo, P., & Tian, X. (2018). Insight into litter decomposition driven by nutrient demands of symbiosis system through the hypha bridge of arbuscular mycorrhizal fungi. *Environmental Science and Pollution Research International*, *25*(6), 5369–5378. DOI: 10.1007/s11356-017-0877-2 PMID: 29209973

Kryukov, A. A., Gorbunova, A. O., Machs, E. M., Mikhaylova, Y. V., Rodionov, A. V., Zhurbenko, P. M., & Yurkov, A. P. (2020). Perspectives of using Illumina MiSeq for identification of arbuscular mycorrhizal fungi. *Vavilovskii Zhurnal Genetiki i Selektsii*, *24*(2), 158–167. DOI: 10.18699/VJ19.38-o PMID: 33659795

Kubista, M., Andrade, J. M., Bengtsson, M., Forootan, A., Jonák, J., Lind, K., Sindelka, R., Sjöback, R., Sjögreen, B., Strömbom, L., Ståhlberg, A., & Zoric, N. (2006). The real-time polymerase chain reaction. *Molecular Aspects of Medicine*, *27*(2-3), 95–125. DOI: 10.1016/j.mam.2005.12.007 PMID: 16460794

Lakin-Thomas, P. L., Coté, G. G., & Brody, S. (1990). Circadian Rhythms in *Neurospora crassa*: Biochemistry and Genetics. *Critical Reviews in Microbiology*, *17*(5), 365–416. DOI: 10.3109/10408419009114762 PMID: 2147375

Lanfranco, L., Wyss, P., Marzachi, C., & Bonfante, P. (1995). Generation of RAPD-PCR primers for the identification of isolates of *Glomus mosseae*, an arbuscular mycorrhizal fungus. *Molecular Ecology*, *4*(1), 61–68. DOI: 10.1111/j.1365-294X.1995.tb00192.x PMID: 7711955

Lee, K., Loros, J. J., & Dunlap, J. C. (2000). Interconnected Feedback Loops in the *Neurospora* Circadian System. *Science*, *289*(5476), 107–110. DOI: 10.1126/science.289.5476.107 PMID: 10884222

Lee, S.-J., Kong, M., Morse, D., & Hijri, M. (2018). Expression of putative circadian clock components in the arbuscular mycorrhizal fungus *Rhizoglomus irregulare*.*Mycorrhiza*, *28*(5-6), 523–534. DOI: 10.1007/s00572-018-0843-y PMID: 29931403

Lee, S. J., Morse, D., & Hijri, M. (2019). Holobiont chronobiology: Mycorrhiza may be a key to linking aboveground and underground rhythms. *Mycorrhiza, 29*(5), 403–412. DOI: 10.1007/s00572-019-00903-4 PMID: 31190278

Lenoir, I., Fontaine, J., Tisserant, B., Laruelle, F., & Lounès-Hadj Sahraoui, A. (2017). Beneficial contribution of the arbuscular mycorrhizal fungus, *Rhizophagus irregularis*, in the protection of *Medicago truncatula* roots against benzo[a]pyrene toxicity. *Mycorrhiza, 27*(5), 465–476. DOI: 10.1007/s00572-017-0764-1 PMID: 28197735

Li, X., Song, C., Kang, X., Chen, F., Li, A., Wang, Y., Zou, J., Yin, J., Li, Y., Sun, Z., Ma, X., & Liu, J. (2024). Assembly and functional profile of rhizosphere microbial community during the *Salix viminalis*-AMF remediation of polycyclic aromatic hydrocarbon polluted soils. *Journal of Environmental Management, 370*, 122503. DOI: 10.1016/j.jenvman.2024.122503 PMID: 39299104

Liu, L., Li, Y., Li, S., Hu, N., He, Y., Pong, R., Lin, D., Lu, L., & Law, M. (2012). Comparison of Next-Generation Sequencing Systems. *Journal of Biomedicine & Biotechnology, 2012*, 1–11. PMID: 22829749

Liu, Y. (2003). Molecular Mechanisms of Entrainment in the *Neurospora* Circadian Clock. *Journal of Biological Rhythms, 18*(3), 195–205. DOI: 10.1177/0748730403018003002 PMID: 12828277

Liu, Y., & Bell-Pedersen, D. (2006). Circadian Rhythms in *Neurospora crassa* and Other Filamentous Fungi. *Eukaryotic Cell, 5*(8), 1184–1193. DOI: 10.1128/EC.00133-06 PMID: 16896204

Lombardi, L. M., & Brody, S. (2005). Circadian rhythms in *Neurospora crassa*: Clock gene homologues in fungi. *Fungal Genetics and Biology, 42*(11), 887–892. DOI: 10.1016/j.fgb.2005.06.005 PMID: 16154782

Lumini, E., Orgiazzi, A., Borriello, R., Bonfante, P., & Bianciotto, V. (2010). Disclosing arbuscular mycorrhizal fungal biodiversity in soil through a land-use gradient using a pyrosequencing approach. *Environmental Microbiology, 12*(8), 2165–2179. DOI: 10.1111/j.1462-2920.2009.02099.x PMID: 21966911

Macalady, J. L., Hamilton, T. L., Grettenberger, C. L., Jones, D. S., Tsao, L. E., & Burgos, W. D. (2013). Energy, ecology and the distribution of microbial life. *Philosophical Transactions of the Royal Society of London. Series B, Biological Sciences, 368*(1622), 20120383. DOI: 10.1098/rstb.2012.0383 PMID: 23754819

Magkourilou, E., Bell, C. A., Daniell, T. J., & Field, K. J. (2024). The functionality of arbuscular mycorrhizal networks across scales of experimental complexity and ecological relevance. *Functional Ecology*, •••, 1–16. DOI: 10.1111/1365-2435.14618

Marleau, J., Dalpé, Y., St-Arnaud, M., & Hijri, M. (2011). Spore development and nuclear inheritance in arbuscular mycorrhizal fungi. *BMC Evolutionary Biology, 11*(1), 51. DOI: 10.1186/1471-2148-11-51 PMID: 21349193

Martin, F. M., & van der Heijden, M. G. A. (2024). The mycorrhizal symbiosis: Research frontiers in genomics, ecology, and agricultural application. *The New Phytologist, 242*(4), 1486–1506. DOI: 10.1111/nph.19541 PMID: 38297461

Nadimi, M., Beaudet, D., Forget, L., Hijri, M., & Lang, B. F. (2012). Group I Intron–Mediated Trans-splicing in Mitochondria of *Gigaspora rosea* and a Robust Phylogenetic Affiliation of Arbuscular Mycorrhizal Fungi with Mortierellales. *Molecular Biology and Evolution, 29*(9), 2199–2210. DOI: 10.1093/molbev/mss088 PMID: 22411852

Nuccio, E. E., Blazewicz, S. J., Lafler, M., Campbell, A. N., Kakouridis, A., Kimbrel, J. A., Wollard, J., Vyshenska, D., Riley, R., Tomatsu, A., Hestrin, R., Malmstrom, R. R., Firestone, M., & Pett-Ridge, J. (2022). HT-SIP: A semi-automated stable isotope probing pipeline identifies cross-kingdom interactions in the hyphosphere of arbuscular mycorrhizal fungi. *Microbiome, 10*(1), 1–20. DOI: 10.1186/s40168-022-01391-z PMID: 36434737

Pervez, M. T., Hasnain, M. J. U., Abbas, S. H., Moustafa, M. F., Aslam, N., & Shah, S. S. M. (2022). A Comprehensive Review of Performance of Next-Generation Sequencing Platforms. *BioMed Research International, 2022*(1), 1–12. DOI: 10.1155/2022/3457806 PMID: 36212714

Pinto, Y., & Bhatt, A. S. (2024). Sequencing-based analysis of microbiomes. *Nature Reviews. Genetics, 25*(12), 829–845. DOI: 10.1038/s41576-024-00746-6 PMID: 38918544

Pittendrigh, C. S. (1960). Circadian Rhythms and the Circadian Organization of Living Systems. *Cold Spring Harbor Symposia on Quantitative Biology, 25*(0), 159–184. DOI: 10.1101/SQB.1960.025.01.015 PMID: 13736116

Qin, Y., Zhang, W., Feng, Z., Feng, G., Zhu, H., & Yao, Q. (2022). Arbuscular mycorrhizal fungus differentially regulates P mobilizing bacterial community and abundance in rhizosphere and hyphosphere. *Applied Soil Ecology, 170*, 104294. DOI: 10.1016/j.apsoil.2021.104294

Redecker, D., Hijri, I., & Wiemken, A. (2003). Molecular Identification of Arbuscular Mycorrhizal Fungi in Roots: Perspectives and Problems. *Folia Geobotanica, 38*(2), 113–124. DOI: 10.1007/BF02803144

Redecker, D., Schüßler, A., Stockinger, H., Stürmer, S. L., Morton, J. B., & Walker, C. (2013). An evidence-based consensus for the classification of arbuscular mycorrhizal fungi (Glomeromycota). *Mycorrhiza, 23*(7), 515–531. DOI: 10.1007/s00572-013-0486-y PMID: 23558516

Ronaghi, M. (2001). Pyrosequencing Sheds Light on DNA Sequencing. *Genome Research, 11*(1), 3–11. DOI: 10.1101/gr.150601 PMID: 11156611

Rosendahl, S. (2008). Communities, populations and individuals of arbuscular mycorrhizal fungi. *The New Phytologist, 178*(2), 253–266. DOI: 10.1111/j.1469-8137.2008.02378.x PMID: 18248587

Salichos, L., & Rokas, A. (2010). The diversity and evolution of circadian clock proteins in fungi. *Mycologia, 102*(2), 269–278. DOI: 10.3852/09-073 PMID: 20361495

Simon, L., Lalonde, M., & Bruns, T. D. (1992). Specific amplification of 18S fungal ribosomal genes from vesicular-arbuscular endomycorrhizal fungi colonizing roots. *Applied and Environmental Microbiology, 58*(1), 291–295. DOI: 10.1128/aem.58.1.291-295.1992 PMID: 1339260

Simon, L., Lévesque, R. C., & Lalonde, M. (1993). Identification of endomycorrhizal fungi colonizing roots by fluorescent single-strand conformation polymorphism-polymerase chain reaction. *Applied and Environmental Microbiology*, *59*(12), 4211–4215. DOI: 10.1128/aem.59.12.4211-4215.1993 PMID: 8285712

Sota, T., Kagata, H., Ando, Y., Utsumi, S., & Osono, T. (2014). Metagenomic Approach Yields Insights into Fungal Diversity and Functioning. In *Species Diversity and Community Structure*, 1–23. DOI: 10.1007/978-4-431-54261-2_1

Stefani, F., Bencherif, K., Sabourin, S., Hadj-Sahraoui, A. L., Banchini, C., Séguin, S., & Dalpé, Y. (2020). Taxonomic assignment of arbuscular mycorrhizal fungi in an 18S metagenomic dataset: A case study with saltcedar (*Tamarix aphylla*). *Mycorrhiza*, *30*(2-3), 243–255. DOI: 10.1007/s00572-020-00946-y PMID: 32180012

Tamasloukht, M., Séjalon-Delmas, N., Kluever, A., Jauneau, A., Roux, C., Bécard, G., & Franken, P. (2003). Root Factors Induce Mitochondrial-Related Gene Expression and Fungal Respiration during the Developmental Switch from Asymbiosis to Presymbiosis in the Arbuscular Mycorrhizal Fungus *Gigaspora rosea.Plant Physiology*, *131*(3), 1468–1478. DOI: 10.1104/pp.012898 PMID: 12644696

Tisserant, E., Malbreil, M., Kuo, A., Kohler, A., Symeonidi, A., Balestrini, R., & Martin, F. (2013). Genome of an arbuscular mycorrhizal fungus provides insight into the oldest plant symbiosis. *Proceedings of the National Academy of Sciences of the United States of America*, *110*(50), 20117–20122.

Van Tuinen, D., Jacquot, E., Zhao, B., Gollotte, A., & Gianinazzi Pearson, V. (1998). Characterization of root colonization profiles by a microcosm community of arbuscular mycorrhizal fungi using 25S rDNA-targeted nested PCR. *Molecular Ecology*, *7*(7), 879–887. DOI: 10.1046/j.1365-294x.1998.00410.x PMID: 9691489

Vierheilig, H. (2004). Further root colonization by arbuscular mycorrhizal fungi in already mycorrhizal plants is suppressed after a critical level of root colonization. *Journal of Plant Physiology*, *161*(3), 339–341. DOI: 10.1078/0176-1617-01097 PMID: 15077632

Wang, B., Yeun, L. H., Xue, J., Liu, Y., Ané, J., & Qiu, Y. (2010). Presence of three mycorrhizal genes in the common ancestor of land plants suggests a key role of mycorrhizas in the colonization of land by plants. *The New Phytologist*, *186*(2), 514–525. DOI: 10.1111/j.1469-8137.2009.03137.x PMID: 20059702

Wang, G., Jin, Z., George, T. S., Feng, G., & Zhang, L. (2023). Arbuscular mycorrhizal fungi enhance plant phosphorus uptake through stimulating hyphosphere soil microbiome functional profiles for phosphorus turnover. *The New Phytologist*, *238*(6), 2578–2593. DOI: 10.1111/nph.18772 PMID: 36694293

Willing, C. E., Wan, J., Yeam, J. J., Cessna, A. M., & Peay, K. G. (2024). Arbuscular mycorrhizal fungi equalize differences in plant fitness and facilitate plant species coexistence through niche differentiation. *Nature Ecology & Evolution*, *8*(11), 2058–2071. DOI: 10.1038/s41559-024-02526-1 PMID: 39251818

Wilson, G. W. T., Rice, C. W., Rillig, M. C., Springer, A., & Hartnett, D. C. (2009). Soil aggregation and carbon sequestration are tightly correlated with the abundance of arbuscular mycorrhizal fungi: Results from long-term field experiments. *Ecology Letters*, *12*(5), 452–461. DOI: 10.1111/j.1461-0248.2009.01303.x PMID: 19320689

Xie, Y., Wang, F., Wang, K., Yue, H., & Lan, X. (2020). Responses of bacterial phoD gene abundance and diversity to crop rotation and feedbacks to phosphorus uptake in wheat. *Applied Soil Ecology*, *154*, 103604. DOI: 10.1016/j.apsoil.2020.103604

Xu, J. (2006). Microbial ecology in the age of genomics and metagenomics: Concepts, tools, and recent advances. *Molecular Ecology*, *15*(7), 1713–1731. DOI: 10.1111/j.1365-294X.2006.02882.x PMID: 16689892

Yuan, M. M., Kakouridis, A., Starr, E., Nguyen, N., Shi, S., Pett-Ridge, J., Nuccio, E., Zhou, J., & Firestone, M. (2021). Fungal-Bacterial Cooccurrence Patterns Differ between Arbuscular Mycorrhizal Fungi and Nonmycorrhizal Fungi across Soil Niches. *mBio*, *12*(2), 1–19. DOI: 10.1128/mBio.03509-20 PMID: 33879589

Zhang, L., Chen, F. X., Zeng, Z., Xu, M., Sun, F., Yang, L., Bi, X., Lin, Y., Gao, Y. J., Hao, H. X., Yi, W., Li, M., & Xie, Y. (2021). Advances in Metagenomics and Its Application in Environmental Microorganisms. *Frontiers in Microbiology*, *12*, 1–15. DOI: 10.3389/fmicb.2021.766364 PMID: 34975791

Zhao, W., Chen, Z., Yang, X., Sheng, L., Mao, H., & Zhu, S. (2023). Metagenomics reveal arbuscular mycorrhizal fungi altering functional gene expression of rhizosphere microbial community to enhance *Iris tectorum*'s resistance to Cr stress. *The Science of the Total Environment*, *895*, 164970. DOI: 10.1016/j.scitotenv.2023.164970 PMID: 37343864

Zheng, M. M., Wang, C., Li, W. X., Guo, L., Cai, Z. J., Wang, B. R., Chen, J., & Shen, R. F. (2021). Changes of acid and alkaline phosphatase activities in long-term chemical fertilization are driven by the similar soil properties and associated microbial community composition in acidic soil. *European Journal of Soil Biology*, *104*, 103312. DOI: 10.1016/j.ejsobi.2021.103312

KEY TERMS AND DEFINITIONS

Alkaline Phosphatase (*phoD*) encoding gene: a gene encoding for alkaline phosphatase enzyme, implicated in phosphate groups hydrolysis under alkaline conditions.

Anastomosis (mycology): the process where hyphae of the same or different species are merged.

Chronobiomics: a rising field that merges chronobiology (the study of biological rhythms and internal clocks) and microbiomics (the study of microbiome).

Hyphosphere (mycology): the presence of fungi through their branching filaments (hyphae) and in the soil environment (sphere).

Internal Transcribed Spacer (ITS): a DNA region located between 18S, 5.8S, and 28S ribosomal RNA (rRNA) genes within eukaryotic organisms. It is implicated in the generation of rRNA.

Large Subunit (LSU): the larger of the two principal subunits that form the ribosome, crucial for protein synthesis.

Molecular Operational Taxonomic Unit (MOTU): clusters of sequences obtained by assembling molecular features like a conserved gene's or gene fragment's DNA sequences.

Nitrogenase Iron Protein gene responsible for Dinitrogenase Reductase (*nifH*): a gene that encodes for the dinitrogenase reductase protein, implicated in N fixation.

Operational Taxonomic Unit (OTU): a group of intimately related individuals that are categorized together based on common characteristics, typically genetic similarity.

Pangenome: whole genetic material from all individuals of a given species.

Polycyclic Aromatic Hydrocarbons (PAHs): a group of organic compounds consisting of several fused aromatic rings.

Polymorphism (genetics): the occurrence of several forms of a gene within a given population.

Small Subunit (SSU): the smaller of the two principal subunits that form the ribosome, crucial for protein synthesis.

White Collar 1,2 (WC1, 2) related gene: a gene related to the fungus species *Neurospora crassa*, implicated in the modulation of the circadian rhythm, specifically involved in the photoreception pathway.

Quinoprotein Glucose Dehydrogenase (*gcd*) encoding gene: a gene that encodes for the enzyme glucose dehydrogenase, a quinoprotein (a protein that contains a quinone). It catalyzes the oxidation of glucose to gluconolactone.

Quantitative Reverse Transcription Polymerase Chain Reaction (RT-qPCR): a laboratory technique deployed to quantify the amount of specific RNA in a given sample, combining the principles of reverse transcription and quantitative PCR.

Chapter 11
Fungi for Ameliorating Soil and Water Pollution: Mycoremediation

Khirabdhi Tanaya
https://orcid.org/0009-0005-6875-2035
Department of Botany, Mahatma Gandhi Central University, India

Durgeshwer Singh
https://orcid.org/0000-0003-3029-847X
Department of Botany, Mahatma Gandhi Central University, India

ABSTRACT

Pollution negatively impacts the environment due to human activities such as urbanization and industrialization. Pollutants that cause pollution can disrupt pH levels, nutrient balance, and microbial activity, resulting in lower crop yields and deteriorated soil and water quality. The removal of these pollutants is essential to ensure the sustainability of the environment. Many physicochemical treatment techniques are utilized. However, these methods often prove insufficient. Mycoremediation emerges as a promising alternative, harnessing fungi's natural abilities to degrade pollutants and sequester heavy metals. This process relies on fungal enzymatic activity and adsorption mechanisms, offering a more eco-friendly and cost-effective approach to environmental cleanup. This chapter delves into the various methods of mycoremediation, the specific enzymes responsible for breaking down pollutants, and the ways in which fungi accumulate heavy metals. The goal of this chapter is to develop more efficient, economically feasible, and eco-friendly pollution treatment strategies.

INTRODUCTION

Rapid industrialisation and usage of chemical compounds such as pesticides, heavy metals, dyes, polycyclic aromatic hydrocarbons (PAHs), and xenobiotics have elevated environmental pollution. The improper disposal of these toxic wastes can cause hazardous effects to water, soil, and air. They can cause potential health risks by entering into the food chain. Synthetic pollutants are non-biodegradable and thus, pile up in the water and soil. The accumulated waste in water can affect aquatic organisms.

DOI: 10.4018/979-8-3693-8201-1.ch011

Pollutants when accumulated in soil, degrade the quality and have a malevolent influence on agricultural products (Sakshi et al., 2019). The remediation of these contaminants has become very necessary. The treatment of pollutants can be performed via various methods. Physico-chemical methods of treatment can be very expensive, have high energy consumption, produce secondary pollutants, and are laborious. Thus, the utilisation of sustainable remediation methods becomes essential. The biological method of treatment or bioremediation becomes an excellent candidate for the degradation of such pollutants (Tanaya et al., 2024). Thus, bioremediation can be used to remove pollutants from plants, fungi, and bacteria. Among many, the Fungi's vigorous growth, extensive hyphal growth, production of extracellular enzymes, adaptability, and production of metal-binding protein make them an excellent contender for bioremediation (Akhtar & Mannan, 2020). The employment of fungi for remediating the environment and degradation of pollutants can be referred to as Mycoremediation. Keeping cost-effectiveness and sustainability in mind, it can be considered an economical, eco-friendly method that converts toxic pollutants into harmless products. The objective of this book chapter is to study the mechanisms involved in the removal or treatment of water and soil pollutants using fungi to obtain sustainable, green, cheap, and environmentally sound solutions. This book chapter will elucidate the types of mycoremediation, enzymes involved in the biodegradation of pollutants found in soil and water ecosystems, and how this technique provides a clean and environment-friendly method for removing toxic pollutants.

POLLUTION

Pollution can be defined as a situation when the pollutants create adverse impacts on nature. Pollution is categorised into soil, air, noise, light, water, and radiation pollution. Air pollution is the presence of harmful particulate matter or gases in the air that negatively impact its quality. The particulate matter mixes with air to form aerosols, fine dust, gases, and fumes. Air Pollution is the result of the emanation of various hazardous gases such as Nitrogen Dioxide, Sulphur Dioxide, Carbon Monoxide, Carbon Dioxide, etc. (Mia et al., 2019). Water Pollution is caused when the waste or the effluents from various industries get mixed with water bodies. They affect the Biological Oxygen Demand (BOD), Chemical Oxygen Demand (COD), pH, Total Dissolved Solute (TDS), and Total Suspended Solids (TSS). The imbalance of these parameters disbalances the condition of the water and disrupts the aquatic life (Tanaya et al., 2024). Soil Pollution is caused by accumulating contaminants such as PAHs, dyes, heavy metals, etc which affects the soil balance. Noise Pollution can be explained as excessive sound that can adversely impact human health. The World Health Organisation (WHO) describes noise to be very painful after reaching 120 decibels. It can be caused by traffic noises, animals, construction sites, etc. (Paudel & States, 2023). Light Pollution can be explained as the modification made to the natural light by man which disrupts the lifestyle of migratory birds, pollinators, sea turtles, etc. They also impact the circadian rhythm, production of hormones and in some cause breast cancer (Chepesiuk, 2009). Radioactive Pollution can be defined as the unconditioned release of radioactive elements. Radioactive elements such as ^{238}U, ^{232}Th, ^{222}Ra, and ^{87}Rb are found in rocks, sediments, soils, nuclear weapons, and power plants. Excessive exposure to such radionuclides can have carcinogenic and mutagenic effects (Hussain & Keçili, 2020).

Pollution has become a lingering global concern and its increase at an alarming rate has caused mayhem. Industrialization and reforms in agriculture have put a risk on the endurance of the environment by downgrading the quality of air, soil, and water. As reported earlier by the World Health Organisation

(WHO), a large population (2.2 billion) was not getting potable water sources and 144 million people were utilizing non-potable and contaminated water (Akhtar & Mannan, 2020). Reports from the European Environment Agency in 2015 suggested an increase in contaminated sites by 50% and an urgent need for the remediation of 3,40,000 sites by 2025 (Yadav et al., 2021). Pollution reports have provided information regarding the loss of Gross Domestic Product (GDP) by 5% in developing countries pushing the development of such countries further. Pollution also accounts for 16% of global deaths due to various effects caused by them (Akhtar & Mannan, 2020).

Sources of pollutants can be classified into anthropogenic and natural activities. The natural activities may include forest fires, exudates from trees, and volcanic eruptions. The anthropogenous activities may consist of the burning of fossil fuels, spillage of coal tar, crude oil, or petroleum, burning of solid biomass fuel including crop residues, dung cakes and wood, industrial effluents, petroleum refinery effluents, automotive emissions, etc. When released directly or indirectly into the environment, the pollutants contaminate air, water, and soil. Dyes, PAHs, and heavy metals contaminate the aquatic environment. Those organic pollutants change the chemical and physical nature of the water bodies declining the quality of aquatic life (Tanaya et al., 2024). When contaminated with organic and inorganic pollutants, soil also poses a high risk to the environment as they can be absorbed by various organisms and thus, enter the food chain. They remain resistant to degradation and their accumulation can affect human health via direct dermal contact. Air pollutants are harmful and can cause various health risks. They adversely affect nature as well as humans by posing various serious diseases such as respiratory diseases, cardiovascular diseases, and cancer (Saleem et al., 2022). Treatment of pollution has become imperative. Thus, the utilization of optimal remediation methods for such polluted sites has become very necessary.

DIFFERENT TYPES OF POLLUTANTS

Pollution can mainly affect air, soil or water. Soil Pollution is when the earth's crust is affected by organic and inorganic pollutants. Soil health is very important as it keeps a balance between chemical, biological, and physical factors. It includes the composition of organic matter and a diverse mix of microflora. Any divergence in the balance can cause pollution. Natural disasters such as landslides, earthquakes, hurricanes, and floods are responsible for soil pollution via natural causes. In contrast, industrial waste, and agricultural products such as fertilisers, pesticides, weedicides, and herbicides are responsible for anthropogenic pollution causes. Earth's surface contains approximately 71% water, but only 2.5% is covered by freshwater. Rapid industrialisation has caused a greater chance of pollution, and wastewater from various industries pollutes the rivers, lakes, ponds, etc., and devours water bodies of potable water. The main causes of water and soil pollution are textile industries, cosmetic industries, agricultural industries, leather industries, oil refineries, etc.

Various types of pollutants that are the reason for soil and water pollution are heavy metals, dyes, and polycyclic aromatic hydrocarbons. Table 1 shows the various pollutants and their effects on plants and animals. Heavy metals have a greater impact on terrestrial and aquatic ecosystems. Though heavy metals are found naturally, overuse and human interference increase the deposit and persistence of heavy metals in the ecosystem. Heavy metals can be sourced naturally as well as anthropogenically which include mining, smelting, cosmetic industry, agricultural products, Industrial wastes, etc. They could have many toxic effects (Neurotoxic, Carcinogenic, Mutagenic, Teratogenic). They cause many chronic health issues in humans and accumulate in plant tissues leading to toxicity, affecting homeostasis, nutrient uptake, and inhibiting root-shoot growth. Heavy metal accumulation in soil results in decreased species diversity of soil microorganisms and soil quality . Due to negligence in the correct processing method in the textile industries, dye molecules do not attach properly to fabrics. The residue flows down the waterbodies and soil causing pollution. Dyes cause lower penetration of light, lowering the rate of photosynthesis in aquatic flora. In Animals, dye molecules cause mortality, and interfere with enzymatic activities. The dye molecules decrease the soil health and quality of crops grown in those soils. In humans, they cause dermatitis, asthma, conjunctivitis, etc. (Tanaya et al., 2024). Petroleum is a non-renewable energy source containing alkanes, cycloalkanes, polycyclic aromatic hydrocarbons, asphaltenes, etc. However, excessive usage of them causes soil and water pollution which threatens the human health and well-being of other organisms. They cause toxic effects on the circulatory, endocrine, immune, and nervous systems and cause hormonal and metabolic disorders. It brings physicochemical changes to the soil which affects the microflora. They also cause toxic effects on plants by affecting seed germination and inhibiting water and mineral absorption (Ahmed et al., 2024). Plastics degrade and decompose at a very slow rate and take several decades. 60-70% of garbage consists of plastics such as Polyvinyl chloride, Polystyrene, Polypropylene, Polyethylene, etc. which persist and cause soil and water pollution. They interfere with various biochemical processes such as lipid peroxidation, carbohydrate metabolism, DNA damage, and oxidative stress and inhibit regulatory genes (Zolotova et al., 2022).

Table 1. Major environmental pollutants and their toxic effects

S. No	*Major Pollutants*	*Sources*	*Toxic Effects on Animals*	*Toxic Effects on Plants*	*References*
1.	**Heavy metals**				
i.	Mercury	Textile Industries, Petroleum Industries, Cosmetic Industries, Agricultural Industries	Neurodevelopmental impairment, Parkinson-type syndrome, Behavioural disorders, Reduction of intelligence quotient	----	(Münzel et al., 2023)
ii.	Lead	Textile Industries, Petroleum Industries, Cosmetic Industries, Agricultural Industries	Affects the Lymph nodes by altering immune responses, reduction in response to vaccine in children	----	(Münzel et al., 2023)
iii.	Cadmium	Textile Industries, Petroleum Industries, Cosmetic Industries, Agricultural Industries	Alters metabolism, reduce production of thyroid hormones, alters production of reproductive hormones, osteomalasia, bone impairment, disruption of circadian rhythms	----	(Münzel et al., 2023)(Li et al., 2020)

continued on following page

Table 1. Continued

S. No	*Major Pollutants*	*Sources*	*Toxic Effects on Animals*	*Toxic Effects on Plants*	*References*
iv.	Arsenic	Textile Industries, Petroleum Industries, Cosmetic Industries, Agricultural Industries	Skin irritation, inflammation, hyperpigmentation, hypopigmentation, cancer in urinary bladder, nausea, vomiting, asthma, and lung cancer	----	(Münzel et al., 2023)
v.	Chromium	Textile Industries, Petroleum Industries, Cosmetic Industries, Agricultural Industries	Leads to increase in cholesterol level, skin problems, liver cancer	----	(Münzel et al., 2023)
vi.	Antimony	Textile Industries, Petroleum Industries, Cosmetic Industries, Agricultural Industries	Causes Testicular Atrophy, reduction in testosterone, endometriosis, hormonal cancer,	----	(Münzel et al., 2023)
vii.	Copper	Agricultural Industries, household waste	Anemia and various other toxicity	----	(Raina et al., 2020)
viii.	Nickel	Industrial and Household waste	Lung fibrosis, skin allergies	----	(Raina et al., 2020)
2.	**Dyes**	Textile Industry, Leather Industries, Food Industries	Skin problems, asthma, conjunctivitis, rhinitis, causes problems in spermatogenesis and ovulation, are carcinogenic and mutagenic	Reduces rate of photosynthesis, Damages roots, fronds etc., caused necrosis and chlorosis in plants	(Tanaya et al., 2024)
3.	**Polycyclic aromatic hydrocarbons (petroleum, coal tar, gasoline)**	Petroleum Industry, Oil Refineries	Caused damages in circulatory, nervous, immune systems and various hormonal and metabolic disorders	Have toxic effect on seed germination, inhibits absorption of minerals and water, impacts plant growth, reduces root length and leaf area	(Ahmed et al., 2024)
4.	**Plastics**	Industrial, Agricultural, Medical, Household Wastes	Oxidative stress, DNA Damage, inhibits regulatory genes, inflammation, increased mortality, infertility,	Alters germination and root-shoot growth	(Hu & Palić, 2020; Zolotova et al., 2022; Mészáros et al., 2023)
5.	**Pesticides and herbicides**	Agricultural Industry	Pesticide poisoning, diabetes, cancer, asthma, nausea, skin irritation, hormonal disruption, reproductive abnormalities	Degrades the seed quality, makes plants more susceptible to disease, and reduces the yield of plants	(Kim et al., 2017; Aktar et al., 2009)

TREATMENT METHODS

Many environmentalists suggest pollution should be treated to stop the mayhem and restore the environment's health. Choice of treatment methods depends on factors such as the cost of the method, sustainability, compatibility with the environment, compliance with the treatment method, etc. The treatment of toxic pollutants is achieved via physical, chemical, and biological methods.

Physical methods of treatment utilize the physical phenomenon and techniques for the remediation process (Saravanan et al., 2021). They include filtration, adsorption, irradiation, sedimentation, solvent extraction, air sparging, vitrification, electrokinetic treatments, degasification and microwave heating (Sakshi et al., 2019; Tanaya et al., 2024) Physical treatments are not suitable when it comes to the sustainability of the environment as they cause the formation of secondary pollutants. They are costlier and not flexible as compared to other treatment methods. Chemical methods utilise various chemicals for the remediation of the environment. They include coagulation, flocculation, ozonation, oxidation, Fenton's reagent treatment, chemical precipitation, etc. Chemical methods cause high sludge production, high cost of operation, can be toxic when misused, and produce secondary pollutants (Saravanan et al., 2021). Physical and chemical treatments are non-ecofriendly, costly, and produce secondary pollutants. Bioremediation can be considered as a green alternative to physico-chemical methods of treatment.

BIOLOGICAL TREATMENT

Bioremediation converts toxic pollutants into non-toxic or less toxic molecules. Bioremediation is the process by which contaminants are broken down using various living organisms, including plants, fungi, and microbes (Saravanan et al., 2021). The appropriateness of using biological remediation is influenced by various factors, including the quantity and identity of the pollutants, the condition of the site to be remediated, and the population of native microflora. Bioremediation can be categorized into Phytoremediation, which involves the use of plants, and Microbial Bioremediation, which involves the use of microorganisms. Microbial bioremediation can further divide based on microbes used into Bacterial Bioremediation and Mycoremediation.

Phytoremediation

The bioremediation of pollutants by using proper agronomic techniques, soil amendments, and plants can be termed phytoremediation. Phytoremediation can help recover the polluted soil, keep soil health in mind, and cause no harm to soil diversity and fertility. Plants can decontaminate soil either through degradation or accumulation. The advantage of using phytoremediation is that it does not require any special labor or sophisticated equipment, does not cause any secondary pollutants, and can be used for a variety of pollutants. Phytoremediation involves various methods, including phytoaccumulation (phytoextraction), phytovolatilization, phytodegradation, and phytostabilization. Phytoaccumulation is the process where pollutants are absorbed from the soil and stored in the tissues as plants have a greater tolerance for those pollutants. Later, chelators are used which increases the accumulating power of the plant by many folds. Phytodegradation takes place inside and outside of plants that can produce degrading enzymes which degrade organic pollutants. Phytovolatilization happens by absorbing the contaminants from the soil and returning the contaminant to the environment in a less toxic form. Phytostabilisation occurs in plants that have an advanced root system which stops the flowing back of contaminants into soil and water bodies. Pollutants are absorbed and stabilized in the tissues of plants, where they are solubilized within the rhizosphere (Saravanan et al., 2021). Phytoremediation has many disadvantages. Pollutants get bioaccumulated in the edible parts, enter the food chain, and get biomagnified. Repetitive planting and harvesting of the phytoremediators are required for decontamination because of the production of low biomass. This method is considered to be very slow and also depends on different seasons for growth (Farraji et al., 2016).

Bioremediation

Bioremediation is an environment-friendly, economical method that produces no residue and where bacteria can accumulate or biodegrade organic pollutants. Bacterial bioremediation can be performed using two mechanisms: either by biosorption on biomass or enzymatic degradation of organic pollutants. The organic pollutants get converted into bioaccumulated biomass or atmospheric gases. They depend on biotic and abiotic factors such as oxygen, temperature, pH, availability of pollutants, and the ability of isolated bacterial strains to degrade (Tanaya et al., 2024). Bacterial bioremediation is distinguished into *in-situ* and *ex-situ* applications. *Ex-situ* bioremediation is when the contaminated water or soil is extracted from its natural environment to be treated in a different location. It includes biopiles, bioreactors, landfarming, and composting. Biopiles involve piling the contaminated material *ex-situ* above ground to enhance aeration, nutrients, treatment beds, and irrigation to enhance bacterial activities. Bioreactors are vessels where the pollutants are converted into less toxic forms by biological activities (Azubuike et al., 2016). *Ex-situ* landfarming is when extracting pollutants and degrading them via natural processes by controlling physical factors. Composting is the method for degrading solid organic pollutants and converting them into nutrient-rich compost, carbon dioxide, and water (Lukić et al., 2017). *In-situ* bioremediation applications can be classified as either bioaugmentation or biostimulation. Biostimulation can be explained as applying or adjusting physical factors such as pH, nutrients, and temperature to stimulate the bacteria present on-site to degrade the pollutants. Bioaugmentation is the process of adding remediation-promoting bacterial strains or consortia on-site to increase biodegradation. Bioaugmentation is carried out in soil that has the requisite nutrients but lacks the microflora (Tanaya et al., 2024). However, Bioremediation has many disadvantages. Bacteria have a lower adaptability to extreme climates. Lower pH decreases the rate of degradation, lower temperature interferes with the growth of the bacteria, and enzymatic activity needed for biodegradation gets lost at high salinity (Ahmed et al., 2024). Bioremediation needs skilled labor and optimum conditions for growth. They also take more time to degrade various ranges of pollutants. Bacterial interaction inconsistency may also cause problems (Tanaya et al., 2024).

Mycoremediation

Fungi are considered the second most diverse group on Earth and are important for the sustenance of life on Earth. The members of the Kingdom Fungi can be unicellular or multicellular, mycelial, or filamentous, and they can be either uninucleated or multinucleated. They have a saprophytic, parasitic, or mutualistic mode of life and can reproduce sexually or asexually. They have an Osmo-heterotrophic mode of nutrition. They obtain their food by absorbing organic material via the cell surface. Mostly, they form a network of hyphae called mycelium which penetrates through substrates and derives nutrition from within. These mycelia also help them branch out in the substratum and use organic substances. Fungi have diverse roles from leavening bread, and fermenting alcohol, to extracting the antibiotic "Penicillin". Enzymes extracted from Fungi are used in many industries as catalysts to enhance the chemical processes. Fungi are also key decomposers of dead and decaying organic substances and have a major role in recycling nutrients. Fungi have several ecological roles in various ecosystems which makes them co-exist and still function in distinctive niches. Fungi have hyphae entwined to rocks, roots, and soil particles which penetrate and break down organic substances. They also have adaptation and tolerance to extremities such as temperature, pH, nutrition, and accumulated pollutants. This makes them an efficient candidate for biodegradation of toxic contaminants (Akpasi et al., 2023). Mycoremediation is the method where the pollutants are broken down from toxic forms into less toxic or non-toxic forms by using fungi. Fungi are considered to be better than bacteria because of their extended lifecycle, fungal biomass, and hyphal network. Ecologically and biochemically, Fungi can break down organic substances and alleviate the risk caused by radionuclides, metalloids, and metals by impacting their availability. Fungi do not rely on toxic compounds as a growth substrate and have non-specific catabolic enzymes that break down many toxic pollutants. They can break down complex carbohydrates without hydrolysis by extracellular enzymes (Akpasi et al., 2023). The presence of mycelium helps with the penetration and mechanical support which enables extracellular enzymes to perform mycoremediation. They attach to the substratum and a larger surface area provides larger mechanical as well as enzymatic exposure to the pollutants present in the environment. Mycoremediation only happens on the surface or in aerobic conditions where hyphae can grow. The presence of degradative enzymes of extracellular nature allows fungi to have more endurance to the toxic material than when brought inside of the cell. Fungi have a greater role as decomposers in the ecosystem and thus, are efficient in breaking down pollutants via mycelium or the extracellularly secreted enzymes. To accomplish efficient mycoremediation, the efficacy of the fungal species has to be screened for biodegradation of toxic compounds, the physical factors such as temperature, and pH have to be optimised, and the availability of pollutants (the site) has to be evaluated. Various factors such as duration, types of substrates, water content, oxygen levels, nutrients, fungal lifecycle, temperature, and pH influence the efficiency of mycoremediation. Temperature has an impact on the availability of pollutants and fungal culture. Higher temperatures seemed to have a positive effect by increasing the degradation of pollutants and increased availability of contaminants. The optimal temperature is considered to be 25°C-30°C. The pH also has been proven to have shown direct effect on the bioavailability of pollutants (Akpasi et al., 2023).

Mycoremediation is carried out via two approaches either *in-situ* or *ex-situ* approaches. The *in-situ* approach describes the method to be done on-site where the toxic pollutant is present. The *ex-situ* approach describes the method where the toxic compounds are taken out physically from the site and the remediation method is done elsewhere. Decontaminating via the *in-situ* approach is cheaper than the *ex-situ* approach. The pollutant, in the *ex-situ* approach, is transferred to a different location rather than completely eradicating the pollutant, which in turn causes secondary pollutants. The uttermost objective of mycoremediation is to mineralise those contaminants and convert the hazardous compounds into simpler compounds such as Water, Nitrogen, and Carbon dioxide. Hydrochloric Acid etc. Inorganic pollutants (Radioactive materials and heavy metals) do not get decomposed rather they are converted into less toxic forms of pollutants and can be extracted from the ground by phytoaccumulation or myco-accumulation (Rhodes, 2014).

Mycoremediation should be the preferred way of bioremediation because of its unique ecological and biological capabilities. The Advantages of mycoremediation are adaptation to extremities, efficiency and diversity of enzymes present, versatility, and ability to create symbiotic relationships. Fungi can adapt and flourish in extremities such as a varied range of pH, temperature, and moisture content. Fungi have a greater diversity of enzymes such as peroxidases, laccases, and oxidoreductases which have the capability of breaking down complex organic molecules. Fungi when associated with other organisms such as Plants (mycorrhiza) facilitate greater uptake of toxic pollutants. Thus, Mycoremediation is an efficient and sustainable method for treating a diverse range of hazardous compounds (Dinakarkumar et al., 2024).

FUNGI INVOLVED IN MYCOREMEDIATION

Fungi are very efficient in degrading toxic compounds such as plastics, heavy metals, pesticides, herbicides, dye molecules, PAHs, and petroleum hydrocarbons (PHCs). Various Fungi have been helpful in the biodegradation of toxic pollutants which are listed in **Table 2.**El-Bondkly and El-Gendy (2022) investigated mycoremediation by using two fungal strains (*Penicillium* sp. and *Aspergillus* sp.) isolated from Petroleum refineries for their ability to biodegrade various heavy metals. They observed that Fe^{3+} and Co^{2+} were bioremoved by live biomass, which leads to efficient bioaccumulation and biosorption of the heavy metals. Gao et al. (2020) studied the decolorisation and detoxification of different triphenylmethane dyes (Cotton Blue, Malachite Green, Crystal Violet, and Methyl Violet) and observed approximately 90% decolorisation by using *Bjerkandera adusta* live cells for biosorption and enzymes like manganese peroxidase (MnP), lignin peroxidase (LiP). Bhattacharya et al. in (2014) observed 71.2% of biodegradation of Benzo[a]pyrene by using *Pleurotus ostreatus* enzymes (Laccase and MnP) in the presence of copper. In the presence of 1 mM 2,2′-azinobis-(3-ethylbenzothiazoline-6-sulfonate), the degradation was observed to be 74.2%, and the presence of 5mM vanillin showed degradation of 83.6%. Abdel-Fattah Mostafa et al. (2022) isolated fungi and tested these fungi against their capability of biodegradation of pesticides (Diazinon). Five strains of fungi *Aspergillus niger, Bactriola antennata, Fusarium graminearum, Rhizopus stolonifer,* and *Trichoderma viride* showed biodegradation capability to be 78.22%, 83.88%, 75.43%, 77.36% and 80.26% respectively.

Table 2. Fungi involved in degradation of major pollutants

S. No.	*Fungi Involved*	*Degradation Efficiency*	*Mechanism*	*References*
	Heavy Metal			
1.	*Aspergillus niger*	98% (Cadmium) 43% (Chromium)	Bioaccumulation	(Khan et al., 2019)
2.	*Curvularia coatesiae*	67.57% (Cadmium)	Biosorption	(Di et al., 2022)
3.	*Fusarium Sp.*	76.5-82.4% (Cadmium)	Biosorption and Bioaccumulation	(Saad et al., 2022)
4.	*Trichoderma lixii*	84.6% (Copper)	Bioaccumulation and Biosorption	(Kumar & Dwivedi, 2021)

continued on following page

Table 2. Continued

S. No.	Fungi Involved	Degradation Efficiency	Mechanism	References
	Heavy Metal			
5.	*Didymella glomerata*	97% (Mercury)	Biosorption	(Văcar et al., 2021)
6.	*Mortierella sp.*	84% (Copper)	Bioaccumulation	(Carvajal et al., 2023)
	Dyes			
1.	*Penicillium simplicissimum*	75.6-90.8% (Crystal Violet, Methyl Violet, Malachite Green, Cotton Blue)	Biosorption	(Chen et al., 2020)
2.	*Aspergillus niger*	100% (Direct Blue)	Biosorption and Extracellular Enzymes	(Ekanayake & Manage, 2022)
3.	*Aspergillus tamarii*	97.17% (Malachite green) 96.65% (Reactive Black)	Biosorption	(Rai & Vijayakumar, 2023)
4.	*Trichoderma lixii*	77.78% (Alizarin Red) 98.31% (Quinizarine Green)	Biosorption and Enzymatic Process	(Adnan et al., 2017)
5.	*Ganoderma lucidum*	77.8%	Biosorption	(Zahuri et al., 2024)
6.	*Bjerkandera adusta*	90% (Triphenylmethane dye)	Biosorption	(Gao et al., 2020)
7.	*Candida tropicalis*	97.2% (Azo Dye)	Enzymatic Processes	(Kolb et al., 2017)
	Polycyclic Aromatic Hydrocarbons			
1.	*Ganoderma lucidum*	99.65% (Phenanthrene) 99.58% (Pyrene)	Enzymatic Process	(Agrawal et al., 2018)
2.	*Trametes versicolor*	75.8% (Chrysene)	Enzymatic process	(Vipotnik et al., 2022)
3.	*Agaricus bisporus*	71.5% (PAHs)	-	(Mohammadi-Sichani et al., 2019)
4.	*Pleurotus ostreatus*	71.2% (Benzo[a]pyrene)	-	(Bhattacharya et al., 2014)
5.	*Aspergillus ochraceus*	87-94%	Enzymatic processes	(Bilen Ozyurek et al., 2021)
5.	*Mucor circinelloides*	78.1% (Crude Oil)	Enzymatic Processes	(Yehia, 2023)
6.	*Candida vishwanathii*	83% (Pyrene) 69% (Naphthalene)	Enzymatic Processes	(Priya et al., 2016)
7.	*Penicillium expansum*	81.76% (Petroleum Hydrocarbon)	-	(El Ayari et al., 2024)
	Pesticides and Herbicides			
1.	*Trametes versicolor*	64-73% (Methamidophos)	Enzymatic Processes	(Benites-Alfaro et al., 2023)
2.	*Pleurotus ostreatus*	26-64% (Hexachlorocyclohexane)	Biosorption	(Sadiq et al., 2018)
3.	*Trichoderma viride*	80.26% (Diazinon)	-	(Abdel-Fattah Mostafa et al., 2022)
4.	*Rhizopus stolonifer*	77.36% (Diazinon)	-	(Abdel-Fattah Mostafa et al., 2022)
5.	*Fusarium graminearum*	75.43% (Diazinon)	-	(Abdel-Fattah Mostafa et al., 2022)

Mechanisms of Mycoremediation

Mycoremediation is an efficacious, economical, and sustainable method of bioremediation that can be utilised for the bioremoval of various ranges of pollutants. Mycoremediation can be categorised into various types based on the mode of action. Different types of mycoremediation include Sequestration, Extrusion, Precipitation, Immobilisation, Vacuolization, and Surface sequestration. Sequestration is when the intracellular chelating compounds are synthesized to chelate the pollutants inside the cell to dilute the environment. The advantage is it can be used for a broad range of pollutants but this capability is found in very selective fungi. Extrusion is done by removing the pollutants and transporting them out of the biomass (Akpasi et al., 2023). Precipitation is the conversion of metal ions into insoluble forms. Organic Acids secreted by fungi cause reactions with these metal ions to form metal phosphates or oxalates. These phosphates and oxalates can further be extracted via filtration or sedimentation. (Dinakarkumar et al., 2024). The advantage of precipitation is that it can be used to remove large loads of pollutants and is useful for the separation of metal sulfide. In contrast, disadvantages are not useful to remove smaller loads of pollutants. Immobilisation is carried out by changing the chemical and physical, characteristics and immobilizing the pollutants. Fungal cells contain various functional groups such as hydroxyl, thiol, carbonyl, carboxyl, and amino groups which get bound to the toxic pollutants on the surface of the fungal cells. Stabilisation and solidification are carried out for immobilisation of the toxic pollutants. The advantage of immobilisation is it can be utilised in contaminated areas containing heavy metals. Vacuolisation is carried out in yeast and some filamentous fungi which contain vacuoles. The pollutants and their byproducts get into the cell and then these pollutants enter the vacuole and bioaccumulate inside the cell. The advantage of this method is it can extract pollutants from the contaminated site but the disadvantage is the pollutant is not removed completely, just from the site and remains inside the cell. Surface Sequestration involves the synthesis of extracellular enzymes that convert toxic pollutants into simpler molecules which are absorbed by the cell for further degradation. The Advantage of surface sequestration is it can be used to break down PAHs, pesticides, and insecticides. The disadvantage is only some fungi can carry out this process (Akpasi et al., 2023).

Mycoremediation can also be categorised into two types: The non-enzymatic and the enzymatic approach. The enzymatic approach includes biodegradation that happens due to the enzymatic behaviour of the fungi which can either be classified into intracellular enzyme reaction or extracellular enzyme reaction. Intracellular enzymatic reactions happen inside the cells where enzymes are secreted to break down the taken-up pollutant or byproduct of pollutants into smaller less toxic or non-toxic forms. Extracellular enzymatic reaction happens when the fungal cells secrete extracellular enzymes to break down the pollutants into smaller molecules so that they can be taken up by the cell for further processes. The non-enzymatic approach includes biomineralization, biosorption, bioaccumulation, and biotransformation (Dinakarkumar et al., 2024). Biosorption can be described as the binding of pollutants to live or dead biomass via adsorption, absorption, and covalent binding (Yadav et al., 2021). Bioaccumulation can be described as the accumulation and transportation of pollutants or their byproducts into the cell. Biomineralisation involves the complete mineralisation of pollutants by fungal cells into simpler compounds such as CO_2, H_2O, $NO_{3,}$ and inorganic compounds (Kumar, 2017). Biotransformation is the method of transforming recalcitrant pollutants by oxidizing, reducing, methylating, demethylating, and evaporating toxic pollutants into non-toxic or less toxic forms(Akpasi et al., 2023).

BIOSORPTION

Biosorption is an efficient method for degradation of toxic compounds from the site of contamination. Fungi have the capability of binding metals/toxic pollutants from the effluents. This phenomenon can be described as mycosorption. Fungi hold the biochemical and ecological capability of degrading organic pollutants. In Mycosorption, the pollutants interact with the fungal cell by ion exchange, adsorption, absorption, complexation, covalent binding, micro-precipitation, and chelation (Tanaya et al., 2024). Mycosorption involves physicochemical interactions between metal (pollutants) and functional groups that attach to metal (Yadav et al., 2021). Various functional groups such as hydroxyl, phosphate, amine, carboxyl, and sulfhydryl bind to the pollutants. The polar groups of proteins, structural carbohydrates (chitin, chitosan, glucans), lipids, and amino acids are said to be involved in binding of pollutants to the fungal cell (Kumar, 2017). The fungal cell has mycelium that contains hyphae which help with the absorption and adsorption of pollutants. Mycosorption can be achieved by both the live and dead biomass of fungi. Reports show dead biomass give better results than live biomass (Yadav et al., 2021).

Mycosorption can be classified based on the location of the pollutants into three parts: (1) Accumulation or precipitation extracellular to cell, (2) Precipitation or Sorption at the surface of the cell, (3) Accumulation inside the cell. Mycosorption can also be classified based on metabolism by the cell into Mycosorption, which is metabolism independent and Mycosorption which is metabolism dependent. The metabolic dependent mycosorption requires cell metabolic energy for the pollutant to bind to the biomass and enter the cell through the cell membrane whereas the metabolic-independent mycosorption is a comparatively quick method and does not require any metabolic energy for the process to happen (Yadav et al., 2021). According to the intra-particle diffusion model, mycosorption can also be seen to go through three steps: rapidly adsorbing at the surface, slowly shifting from external to internal, and then, moderately reaching the equilibrium. It depends on the type of contaminants, site of the functional group, and ionic form of the solution. The rate of mycosorption can be influenced by many factors, such as contact time, substrate concentration, temperature, pH, host cell wall composition, and biomass concentration. Advantages of Mycosorption include revival, mechanical strength, cost-effectiveness, reuse of the adsorbent, durability, and specificity to specific pollutants. The disadvantage is that they can be utilised on specific pollutants (Tanaya et al., 2024).

Many Fungal species have been used as biosorbents and have shown the capacity to adsorb or absorb various organic pollutants. Espinoza-Sánchez et al., (2019) investigated the fungal biomass of *Rhizopus* sp. and observed that they showed 99.5% removal of Cr (VI) via adsorption from aqueous solutions under optimum parameters such as temperature (20-40°C), pH (2.0), and agitation at 150rpm. Zaki et al. (2022) investigated two immobilised endophytic fungal strains for their ability of mycosorption and found *Chaetomium globosum* El26 mutant showed mycosorption of Lead (II) by 199.0 mg/g and *Alternaria alternata* S5 mutant showed biosorption of Cadmium (II) by 162.0 mg/g. They showed efficient results as nano-biosorbents. Ruscasso et al. (2021) investigated an Antarctic fungal strain, *Debaryomyces hansenii* F39A, for its biosorption capability against two textile dyes at optimum pH and temperature. They showed 90% decolourisation of Reactive Red 141, and after increasing the fungal biomass Reactive Blue 19 decolourisation has risen from 50% to 90%. Drumm et al. (2021) observed wastes extracted from *Agaricus bisporus* were used as a cheaper biosorbent for treating Acid red 97 and Crystal violet with an adsorption capacity of 165.9 mg/g. Al-Hawash et al. (2019) found that fungal strain *Aspergillus* sp. RFC-1 from contaminated sludge of Rumaila oil fields showed biosorption and biodegradation capability of 60.3% for Crude Oil, 97.4% for Naphthalene, 84.9% for Phenanthrene, and 90.7% for pyrene.

ENZYMATIC MECHANISM OF MYCOREMEDIATION

Fungi have a versatile ecological role as decomposers and can be used for the degradation of pollutants. Fungi are known to be used in bioreactors for treating wastewater and soil contaminated with PAHs, heavy metals, and dyes. For degradation, fungi use various mechanisms which have been explained earlier. However, the most important mechanism involves extracellular and intracellular enzymes. Fungi possess a diversity of enzymes such as laccase, oxidoreductases, and peroxidases, which aid in the degradation of various pollutants. These enzymes facilitate the biodegradation of persistent pollutants that are resistant to other degradation methods (Dinakarkumar et al., 2024). Extracellular enzymes such as laccases and various peroxidases are responsible for breaking down the pollutants into smaller parts, which are either used up as carbon sources or get into the cell. Intracellular enzymes catalyze these fragmented compounds by reduction or oxidation (Tanaya et al., 2024). Various enzymes that have a role in biodegradation are listed below:

Laccase

Laccases are multi-copper oxidases that typically range in size from 60 to 75 kDa. They are enzymes known for oxidizing various aromatic and inorganic compounds by reducing O^2 to water. Laccases can exist in monomeric, dimeric, or tetrameric forms and are glycoproteins that play a role in ligninolytic processes. In most cases, each monomer contains four copper atoms at the catalytic sites (Khan et al., 2012). These copper centers oxidise phenols, dyes, and other pollutants then convert them into reactive radicals which go through demethylation, quinone formation, dehalogenation, depolymerisation (Yadav et al., 2021). Laccases use copper ions and dioxygen as electron acceptors in oxidation reactions. Laccase is considered a potent enzyme because of its simple requirements, stability, and lack of inhibition (Khan et al., 2012). Bhattacharya et al. (2014) observed 83.6% biodegradation of Benzo[*a*]pyrene by Laccase in presence of mediators from *Pleurotus ostreatus*. Vipotnik et al. (2022) observed 75.8% degradation of chrysene, 99.2% of iron, 99.6% of aluminium was accomplished by laccase from *Trametes versicolor*. Rudakiya et al. (2020) investigated hardwood degrading fungus, *Tricholoma giganteum* AGDR1 and reported laccase could degrade 7% of profenofos, 72% of thiophanate methyl and 29% of chlorpyrifos and has an efficient use in pesticide degradation.

Manganese Peroxidases (MnPs)

They are peroxidases that can only degrade phenolic compounds. Manganese Peroxidases are recognised as a potential contender in lignin and lignin-like compound degradation because of their molecular structure. MnPs contains three acidic residues linked to Mn^{2+} catalytic binding sites that oxidise to Mn^{3+} and result in the chelation of toxic complexes. The productivity of MnPs depends on various factors such as pH (2.0-9.0), Temperature (25-70°C), growth phase of fungi, and media composition by affecting the expressivity of enzyme-producing genes (Yadav et al., 2021). Siddeeg et al. (2020) extracted Manganese Peroxidase from *Anthracophyllum discolor* and tested it for the removal of methylene blue (MB) and reactive orange 16 (RO 16). It was reported that 96% ± 2% of Methylene Blue and 98% ± 2% of reactive orange was removed by the extracted enzyme. Hoque et al. (2023) extracted novel MnP from *Trichoderma parestonica* which has various pathways for the catalysis of aromatic amines, phenols and wastewater treatment.

Lignin Peroxidases (LiP)

Lignin Peroxidases can also be known as 1, 2-bis (3,4 dimethoxyphenyl) propane-1,3-diol: hydrogen-peroxide oxidoreductase. LiP (38-47 KDa) is an enzyme that has a role in lignin oxidative breakdowns. Lignin Peroxidases have a stabilised structure due to the presence of four sites of disulphide bonds, two glycosylating sites, and two Ca^{2+} binding sites. These enzyme works against a variety of toxic pollutants as they are lignin like compounds (Tanaya et al., 2024). In contrast to MnP, LiP works against non-phenolic aromatic compounds. Lignin peroxidases help in the degradation of pollutants by performing oxidation of pollutants such as aromatic compounds by utilising O_2 and releasing water as a by-product (Yadav et al., 2021). Lignin or lignin like compounds are broken down by cleaving the β-O-4 linkage. The initial step in the process is using H_2O_2 and O_2 as electron sources and secondary metabolites as cofactors, breaking down the substrate, and completing the catalytic cycle. Bouacem et al. (2022) extracted lignin peroxidase enzyme (LiP PR40) from white-rot *Agaricomycetes* fungus identified as *Phlebia radiata* which showed better result than commercially available horseradish peroxidases in dye decolorisation. Guo et al. (2019) extracted and immobilised Lignin Peroxidases from *Pichia methanolica* and found that immobilised LiP showed efficacious results by degrading 100% of tetracycline, 100% of dibutyl phthalate, 100% of 5-chlorophenol, 100% of phenol, 79% of phenanthrene, 73% of fluoranthene and 65% of benzo(a)pyrene.

Lipases

Lipases are responsible for the modification of pollutants that contain oils and fats. These enzymes cause hydrolysis of ester links of triglycerides in the presence of water to form glycerol and fatty acids. They minimise the toxic effect of fat-containing pollutants and remediate the environment (Ortellado et al., 2022). Lipases can be distinguished into (1) Lipases which has enhanced enzymatic activity when present in emulsion form and (2) Lipases that have an active site which is full of protein. Lipases are versatile and has various attributes such as high selectivity of the substrate, increased specificity, stability and activity under several conditions (Ortellado et al., 2022). Lipases have a role in the reduction of various hydrocarbons. Lipase activity can be considered as the most common indication of hydrocarbon degradation. Lipase carries out various reactions such as hydrolysis, alcoholysis, aminolysis, esterification, etc (Gupta & Pathak, 2020). de Moura Dickel et al. (2023) investigated and found that *Aspergillus sclerotiorum* produces lipase. These lipase enzymes were responsible for 90% reduction of oil and grease from the wastewater sample. Amin et al. (2022) used lipase from *Aspergillus melleus* for degradation of poly (ε-caprolactone) and thus is an efficient candidate for mycoremediation of solid waste.

Catalase

Catalases are enzymes with tetrameric haem proteins (60-75 KDa) which make a dumbbell shape with four identical subunits of 500 amino acid sizes. Catalases are oxidoreductases that catalyse various substrates but are highly specific to H_2O_2 (Takio et al., 2021). Pollutants such as dyes, PAHs and heavy metals cause chemical reactions and influence the production of ROS. Reactive oxygen species (ROS) damage cellular macromolecules. The ROS are converted to H_2O_2 by the means of Superoxide Dismutases (SOD). The Hydrogen peroxide molecules are degraded by Catalase into water and oxygen. Fungi have the ability to produce monofunctional catalase and bifunctional catalase, which acts as a defence mechanism against produced ROS. Catalase has been proven to increase the tolerance of fungi for decontaminating pollutants. They are an effective tool that is utilised for Mycoremediation (Shanmugapriya et al., 2019). Shivakumar et al. (2014) observed increased catalase activity of fungal consortia (*Aspergillus niger, Rhizopus sp., Penicillium sp.*) due to elevated levels of heavy metals.

Cellulases

Cellulases are hydrolytic enzymes that have a role in hydrolyzing pollutants and breaking chemical bonds of hydrocarbons, insecticides, etc. Fungi are known to extracellularly produce membrane-bound cellulases. Cellulases can perform hydrolysis by three approaches. Cellulases can be of various types endoglucanases, endo-1,4-β-D-glucanase, carboxymethyl cellulose (CMCase), endo-1,4-β-glucanase, cellobiohydrolases (CBH), β-1,4-glucanase,1-4-endoglucan hydrolase, β-β-1,4-glucanase, and exogluconases, exocellulases, and β-glucosidases (Shanmugapriya et al., 2019). Endoglucanase acts upon the low crystalline region present on the cellulosic fibers and then breaks them into free chains. Exoglucanase reacts with cellulosic molecules and takes off cellobiose units by detaching them from free chain ends. β-glucosidase then hydrolyses these cellobiose units into glucose. These glucose molecules can be taken up by the cell as a carbon source (Gupta & Pathak, 2020). Al-Zaban et al. (2021) found cellulase enzymes from *Aspergillus flavus*, *Trichoderma harzianum*, and *Fusarium solani* could biodegrade Hydrocarbons.

Proteases

Proteases (18.2-75KDa) are hydrolytic enzymes that help with breaking down peptide bonds when present in aqueous phase. Proteases are responsible for breaking down various pollutants such as detergents, food, and pharmaceutical wastes (Gupta & Pathak, 2020). Metalloprotease is a type of protease enzyme that has catalytic mechanisms involving metal ions. They require Co^{2+}, Zn^{2+}, Ca^{2+}, and Mg^{2+} ions to elevate the productivity of the enzyme (Shanmugapriya et al., 2019).

Tyrosinases

Tyrosinases or Polyphenol oxidases are Cu^{2+} dependent enzymes that persuade hydroxylation of phenolic compounds which leads to an oxidation reaction by forming quinones such as o-benzoquinone and dopaquinone by reduction of oxygen to water. Enzymes' active site is influenced by different substrates in the reaction. The oxidation state of tyrosinase is interconverted from oxy form to met form to deoxy form and cycles back to oxy form. Oxy state of tyrosinase attaches to oxygen resulting in the cleavage of monophenols to diphenols in the presence of water and converting itself into met form. The most stable form of tyrosinase, the met form, changes to the deoxy form by converting diphenols into quinones. The deoxy form of tyrosinase converts themselves to oxy form by attaching themselves to oxygen (Tanaya et al., 2024). Some of the fungi that produce tyrosinase are *Agaricus bisporus, Amanita muscaria, Aspergillus oryzae, Lentinula edodes, Lentinula boryana, Neurospora crassa, and Pycnoporus sanguineus* (Singh et al., 2019).

Monooxygenases

Monooxygenase is an enzyme that metabolizes organic compounds by cleaving the aromatic ring, increasing their water solubility and reactivity. These enzymes are bound to the cell membrane. They work against a broad spectrum of pollutants which include aromatic and aliphatic compounds. Monooxygenases are responsible for dehalogenation, hydroxylation, ammonification, desulfurization, biotransformation, biodegradation, and denitrification of these compounds. Monooxygenases can be differentiated into two subclasses on the basis of the presence of cofactor: P450 monooxygenases and flavin-dependent monooxygenases. Based on the number of utilised oxygen for oxygenation, they can be classified as dioxygenases and monooxygenases (Singh et al., 2019). Ma et al. (2021) reported extraction of lytic polysaccharide monooxygenase TgAA11 from Fungal strain *Trichoderma guizhouense* NJAU 4742 which helped in biodegradation of chitin.

CONCLUSION AND FUTURE PROSPECTIVE

The proliferation of synthetic compounds and chemicals in developing industries has led to significant environmental contamination, particularly in the form of soil, water, and air pollution. In order to address these environmental challenges, it is crucial to implement effective remediation strategies. While physicochemical treatment methods have traditionally been used, bioremediation approaches have gained popularity due to their eco-friendly nature, sustainability, and economic viability. Mycoremediation, a method for using fungi to biodegrade pollutants, has shown particular promise among these methods. Mycoremediation uses various mechanisms for degradation such as sequestration, extrusion, precipitation, immobilisation, vacuolization, and surface sequestration. Fungi carry out mycosorption and enzymatic degradation of the pollutants. Enzymes such as laccase, lignin peroxidases, manganese peroxidases, lipase, catalase, protease, tyrosinases, and monooxygenases have shown the capability of biodegradation of various toxic compounds. However, further research is necessary to identify and optimize fungal species with enhanced biodegradation and biosorption capacities. Enzymatic degradation plays a crucial role in mycoremediation processes. The isolation, characterization, and utilization of fungal enzymes with biodegradative properties present a potential avenue for improving remediation efficiency. Recent advances in genetic engineering techniques provide opportunities to modify enzymes, potentially improving their specificity, stability, and catalytic activity under different environmental conditions. Furthermore, understanding the genetic basis of these enzymes could help in creating customized biocatalysts through protein engineering methods. While mycoremediation has demonstrated efficacy in laboratory conditions, its industrial-scale implementation remains limited. Challenges associated with fungal growth control and process optimization at larger scales need to be addressed. Genetic engineering strategies could be employed to develop fungi with regulated growth characteristics, ensuring controlled proliferation in remediation applications. To realize the full potential of mycoremediation, several key research areas warrant further investigation. These includes high-throughput screening of fungal species for enhanced biodegradative capabilities, elucidation of metabolic pathways and enzymatic mechanisms involved in pollutant degradation, development of robust, genetically engineered fungi with improved remediation performance and controlled growth, optimization of bioreactor designs and process parameters for industrial-scale mycoremediation, assessment of the ecological impacts and long-term sustainability of fungal-based remediation strategies, integration of mycoremediation with other remediation technologies for synergistic effects. By addressing these research priorities, mycoremediation has the potential to emerge as a powerful tool in environmental restoration efforts, contributing to the mitigation of anthropogenic pollution and the preservation of ecosystem integrity.

ACKNOWLEDGEMENT

The authors are grateful to the Vice-Chancellor, Mahatma Gandhi Central University, Dean School of Life Sciences, and Head, Department of Botany, Mahatma Gandhi Central University, Bihar, Motihari, India, for the encouragement and support.

REFERENCES

Abdel-Fattah Mostafa, A., Yassin, M. T., Dawoud, T. M., Al-Otibi, F. O., & Sayed, S. R. M. (2022). Mycodegradation of diazinon pesticide utilizing fungal strains isolated from polluted soil. *Environmental Research*, *212*, 113421. DOI: 10.1016/j.envres.2022.113421 PMID: 35568233

Adnan, L. A., Sathishkumar, P., Yusoff, A. R. M., Hadibarata, T., & Ameen, F. (2017). Rapid bioremediation of Alizarin Red S and Quinizarine Green SS dyes using Trichoderma lixii F21 mediated by biosorption and enzymatic processes. *Bioprocess and Biosystems Engineering*, *40*(1), 85–97. DOI: 10.1007/s00449-016-1677-7 PMID: 27663440

Agrawal, N., Verma, P., & Shahi, S. K. (2018). Degradation of polycyclic aromatic hydrocarbons (phenanthrene and pyrene) by the ligninolytic fungi Ganoderma lucidum isolated from the hardwood stump. *Bioresources and Bioprocessing*, *5*(1), 11. DOI: 10.1186/s40643-018-0197-5

Ahmed, S., Kumari, K., & Singh, D. (2024). Different strategies and bio-removal mechanisms of petroleum hydrocarbons from contaminated sites. *Arab Gulf Journal of Scientific Research*, *42*(2), 342–358. DOI: 10.1108/AGJSR-09-2022-0172

Akhtar, N., & Mannan, M. A. (2020). Mycoremediation: Expunging environmental pollutants. *Biotechnology Reports (Amsterdam, Netherlands)*, *26*, e00452. https://doi.org/https://doi.org/10.1016/j.btre.2020.e00452. DOI: 10.1016/j.btre.2020.e00452 PMID: 32617263

Akpasi, S. O., Anekwe, I. M., Tetteh, E. K., Amune, U. O., Shoyiga, H. O., Mahlangu, T. P., & Kiambi, S. L. (2023). Mycoremediation as a Potentially Promising Technology: Current Status and Prospects—A Review. In *Applied Sciences* (Vol. 13, Issue 8). DOI: 10.3390/app13084978

Aktar, W., Sengupta, D., & Chowdhury, A. (2009). Impact of pesticides use in agriculture: Their benefits and hazards. *Interdisciplinary Toxicology*, *2*(1), 1–12. https://doi.org/doi:10.2478/v10102-009-0001-7. DOI: 10.2478/v10102-009-0001-7 PMID: 21217838

Al-Hawash, A. B., Zhang, X., & Ma, F. (2019). Removal and biodegradation of different petroleum hydrocarbons using the filamentous fungus Aspergillus sp. RFC-1. *MicrobiologyOpen*, *8*(1), e00619. https://doi.org/https://doi.org/10.1002/mbo3.619. DOI: 10.1002/mbo3.619 PMID: 29577679

Al-Zaban, M. I., AlHarbi, M. A., & Mahmoud, M. A. (2021). Hydrocarbon biodegradation and transcriptome responses of cellulase, peroxidase, and laccase encoding genes inhabiting rhizospheric fungal isolates. *Saudi Journal of Biological Sciences*, *28*(4), 2083–2090. https://doi.org/https://doi.org/10.1016/j.sjbs.2021.01.009. DOI: 10.1016/j.sjbs.2021.01.009 PMID: 33935563

Amin, M., Bhatti, H. N., & Bilal, M. (2022). Kinetic and thermodynamic characterization of lipase from Aspergillus melleus and its biocatalytic performance for degradation of poly(ε-caprolactone). *Journal of Chemical Technology and Biotechnology*, *97*(2), 446–454. https://doi.org/https://doi.org/10.1002/jctb.6649. DOI: 10.1002/jctb.6649

Azubuike, C. C., Chikere, C. B., & Okpokwasili, G. C. (2016). Bioremediation techniques-classification based on site of application: Principles, advantages, limitations and prospects. *World Journal of Microbiology & Biotechnology*, *32*(11), 180. DOI: 10.1007/s11274-016-2137-x PMID: 27638318

Benites-Alfaro, E., Silva Carhuatocto, L. M., Guanilo Inigo, A. P., Jave Nakayo, J., Castaneda-Olivera, C. A., Lizarzaburu Aguinaga, D. A., & Cabrera Carranza, C. F. (2023). Pleurotus Ostreatus and Trametes Versicolor Fungi to Decontaminate Soils Containing Organophosphates Methamidophos and Cadmium. *Chemical Engineering Transactions, 100*, 97-102 SE-Research Articles. DOI: 10.3303/CET23100017

Bhattacharya, S., Das, A., Prashanthi, K., Palaniswamy, M., & Angayarkanni, J. (2014). Mycoremediation of Benzo[a]pyrene by Pleurotus ostreatus in the presence of heavy metals and mediators. *3 Biotech, 4*(2), 205–211. DOI: 10.1007/s13205-013-0148-y

Bilen Ozyurek, S., Avcioglu, N. H., & Seyis Bilkay, I. (2021). Mycoremediation potential of Aspergillus ochraceus NRRL 3174. *Archives of Microbiology*, *203*(10), 5937–5950. DOI: 10.1007/s00203-021-02490-5 PMID: 34599404

Bouacem, K., Allala, F., Zaraî Jaouadi, N., Hamdi, S., Mechri, S., Ighilahriz, K., Rekik, H., Hacene, H., Bouanane-Darenfed, A., & Jaouadi, B. (2022). A novel peroxidase from white-rot Agaricomycetes fungus Phlebia radiata strain KB-DZ15: Its purification, characterisation, and potential application for dye-decolorisation and lignin-biodegradation. *Biocatalysis and Biotransformation*, *40*(5), 365–377. DOI: 10.1080/10242422.2021.1939315

Carvajal, M., Jeldres, P., Vergara, A., Lobaina, E., Olivares, M., Meza, D., Velásquez, A., Dorta, F., Jorquera, F., Seeger, M., Cereceda-Balic, F., & Fadic, X. (2023). Bioremoval of copper by filamentous fungi isolated from contaminated soils of Puchuncaví-Ventanas Central Chile. *Environmental Geochemistry and Health*, *45*(7), 4275–4293. DOI: 10.1007/s10653-023-01493-z PMID: 36729229

Chen, S. H., Cheow, Y. L., Ng, S. L., & Ting, A. S. Y. (2020). Removal of triphenylmethane dyes in single-dye and dye-metal mixtures by live and dead cells of metal-tolerant Penicillium simplicissimum. *Separation Science and Technology*, *55*(13), 2410–2420. DOI: 10.1080/01496395.2019.1626422

Chepesiuk, R. (2009). Missing the dark: Health effects of light pollution. *Environmental Health Perspectives*, *117*(1), A20–A27. DOI: 10.1289/ehp.117-a20 PMID: 19165374

de Moura Dickel, J. D., Carvalho, J. K., Silveira, M. A. D., Menegotto dos Santos, P., Rodrigues, M. L. F., Fagundes-Klen, M. R., Rosa, C. A., Johann, S., Buzanello, C. V., da Silva de Lucca, R. A., de Oliveira Santos, A. R., & da Rosa, M. F. (2023). Aspergillus sclerotiorum lipolytic activity and its application in bioremediation of high-fat dairy wastewater environments. *Environmental Science and Pollution Research International*, *30*(13), 35517–35527. DOI: 10.1007/s11356-022-24669-z PMID: 36529799

Di, Z., Chaoyang, L., Mengxi, Z., Yunlin, Z., Zhenggang, X., & Guiyan, Y. (2022). Curvularia coatesiae XK8, a Potential Bioadsorbent Material for Adsorbing Cd(II) and Sb(III) Compound Pollution: Characteristics and Effects. *Frontiers in Microbiology*, *12*(January), 1–13. DOI: 10.3389/fmicb.2021.816312 PMID: 35154041

Dinakarkumar, Y., Ramakrishnan, G., Gujjula, K. R., Vasu, V., Balamurugan, P., & Murali, G. (2024). Fungal bioremediation: An overview of the mechanisms, applications and future perspectives. *Environmental Chemistry and Ecotoxicology, 6*, 293–302. https://doi.org/https://doi.org/10.1016/j.enceco.2024.07.002

Drumm, F. C., Franco, D. S. P., Georgin, J., Grassi, P., Jahn, S. L., & Dotto, G. L. (2021). Macro-fungal (Agaricus bisporus) wastes as an adsorbent in the removal of the acid red 97 and crystal violet dyes from ideal colored effluents. *Environmental Science and Pollution Research International*, *28*(1), 405–415. DOI: 10.1007/s11356-020-10521-9 PMID: 32812157

Ekanayake, M. S., & Manage, P. (2022). Mycoremediation Potential of Synthetic Textile Dyes by Aspergillus niger via Biosorption and Enzymatic Degradation: . *Environment and Natural Resources Journal, 20*(3), 234–245.DOI: 10.32526/ennrj/20/202100171

El Ayari, T., Bouhdida, R., Ouzari, H. I., & El Menif, N. T. (2024). Bioremediation of petroleum refinery wastewater by fungal stains isolated from the fishing harbour of Bizerte (Mediterranean Sea). *Biodegradation*, *35*(5), 755–767. DOI: 10.1007/s10532-024-10083-4 PMID: 38687419

El-Bondkly, A. M. A., & El-Gendy, M. M. A. A. (2022). Bioremoval of some heavy metals from aqueous solutions by two different indigenous fungi *Aspergillus sp.* AHM69 and *Penicillium sp.* AHM96 isolated from petroleum refining wastewater. *Heliyon*, *8*(7), e09854. Advance online publication. DOI: 10.1016/j.heliyon.2022.e09854 PMID: 35815132

Espinoza-Sánchez, M. A., Arévalo-Niño, K., Quintero-Zapata, I., Castro-González, I., & Almaguer-Cantú, V. (2019). Cr(VI) adsorption from aqueous solution by fungal bioremediation based using Rhizopus sp. *Journal of Environmental Management*, *251*, 109595. https://doi.org/https://doi.org/10.1016/j.jenvman.2019.109595. DOI: 10.1016/j.jenvman.2019.109595 PMID: 31561145

Farraji, H., Zaman, N. Q., Tajuddin, R., & Faraji, H. (2016). Advantages and disadvantages of phytoremediation: A concise review. *Int J Env Tech Sci*, *2*, 69–75.

Gao, T., Qin, D., Zuo, S., Peng, Y., Xu, J., Yu, B., Song, H., & Dong, J. (2020). Decolorization and detoxification of triphenylmethane dyes by isolated endophytic fungus, Bjerkandera adusta SWUSI4 under non-nutritive conditions. *Bioresources and Bioprocessing*, *7*(1), 53. DOI: 10.1186/s40643-020-00340-8

Guo, J., Liu, X., Zhang, X., Wu, J., Chai, C., Ma, D., Chen, Q., Xiang, D., & Ge, W. (2019). Immobilized lignin peroxidase on Fe3O4@SiO2@polydopamine nanoparticles for degradation of organic pollutants. *International Journal of Biological Macromolecules*, *138*, 433–440. https://doi.org/https://doi.org/10.1016/j.ijbiomac.2019.07.105. DOI: 10.1016/j.ijbiomac.2019.07.105 PMID: 31325503

Gupta, S., & Pathak, B. (2020). *Chapter 6 - Mycoremediation of polycyclic aromatic hydrocarbons* (P. Singh, A. Kumar, & A. B. T.-A. of E. P. Borthakur (eds.); pp. 127–149). Elsevier. https://doi.org/https://doi.org/10.1016/B978-0-12-818095-2.00006-0

Hoque, R. A., Yadav, M., Yadav, H. S., & Boruah, R. (2023). Purification and characterization of novel manganese peroxidase from Trichoderma parestonica and its bio-conversion study of toxic arylamine. *Analytical Chemistry Letters*, *13*(6), 641–659. DOI: 10.1080/22297928.2023.2299257

Hu, M., & Palić, D. (2020). Micro- and nano-plastics activation of oxidative and inflammatory adverse outcome pathways. *Redox Biology*, *37*, 101620. DOI: 10.1016/j.redox.2020.101620 PMID: 32863185

Hussain, C. M., & Keçili, R. (2020). *Chapter 1 - Environmental pollution and environmental analysis* (C. M. Hussain & R. B. T.-M. E. A. T. for P. Keçili (eds.); pp. 1–36). Elsevier. https://doi.org/https://doi.org/10.1016/B978-0-12-816934-6.00001-1

Khan, I., Aftab, M., Shakir, S., Ali, M., Qayyum, S., Rehman, M. U., Haleem, K. S., & Touseef, I. (2019). Mycoremediation of heavy metal (Cd and Cr)-polluted soil through indigenous metallotolerant fungal isolates. *Environmental Monitoring and Assessment*, *191*(9), 585. DOI: 10.1007/s10661-019-7769-5 PMID: 31440913

Khan, R., Pathak, B., & Fulekar, M. (2012). Microbial decolorization and degradation of synthetic dyes: A review. *Reviews in Environmental Science and Biotechnology*, *12*(1), 75–97. Advance online publication. DOI: 10.1007/s11157-012-9287-6

Kim, K.-H., Kabir, E., & Jahan, S. A. (2017). Exposure to pesticides and the associated human health effects. *The Science of the Total Environment*, *575*, 525–535. DOI: 10.1016/j.scitotenv.2016.09.009 PMID: 27614863

Kolb, M., Bahadir, M., & Teichgräber, B. (2017). Determination of chemical oxygen demand (COD) using an alternative wet chemical method free of mercury and dichromate. *Water Research*, *122*, 645–654. DOI: 10.1016/j.watres.2017.06.034 PMID: 28649029

Kumar, V. (2017). *Mycoremediation: A Step Toward Cleaner Environment.*, DOI: 10.1007/978-3-319-68957-9_10

Kumar, V., & Dwivedi, S. K. (2021). Bioremediation mechanism and potential of copper by actively growing fungus Trichoderma lixii CR700 isolated from electroplating wastewater. *Journal of Environmental Management*, *277*, 111370. DOI: 10.1016/j.jenvman.2020.111370 PMID: 32979751

Li, H., Kilgallen, A. B., Münzel, T., Wolf, E., Lecour, S., Schulz, R., Daiber, A., & Van Laake, L. W. (2020). Influence of mental stress and environmental toxins on circadian clocks: Implications for redox regulation of the heart and cardioprotection. *British Journal of Pharmacology*, *177*(23), 5393–5412. DOI: 10.1111/bph.14949 PMID: 31833063

Lukić, B., Panico, A., Huguenot, D., Fabbricino, M., van Hullebusch, E. D., & Esposito, G. (2017). A review on the efficiency of landfarming integrated with composting as a soil remediation treatment. *Environmental Technology Reviews*, *6*(1), 94–116. DOI: 10.1080/21622515.2017.1310310

Ma, L., Liu, Z., Kong, Z., Wang, M., Li, T., Zhu, H., Wan, Q., Liu, D., & Shen, Q. (2021). Functional characterization of a novel copper-dependent lytic polysaccharide monooxygenase TgAA11 from Trichoderma guizhouense NJAU 4742 in the oxidative degradation of chitin. *Carbohydrate Polymers*, *258*, 117708. https://doi.org/https://doi.org/10.1016/j.carbpol.2021.117708. DOI: 10.1016/j.carbpol.2021.117708 PMID: 33593577

Mészáros, E., Bodor, A., Kovács, E., Papp, S., Kovács, K., Perei, K., & Feigl, G. (2023). Impacts of Plastics on Plant Development: Recent Advances and Future Research Directions. *Plants*, *12*(18), 3282. Advance online publication. DOI: 10.3390/plants12183282 PMID: 37765446

Mia, R., Selim, M., Shamim, A. M., Chowdhury, M., Sultana, S., Armin, M., Hossain, M., Akter, R., Dey, S., & Naznin, H. (2019). Review on various types of pollution problem in textile dyeing & printing industries of Bangladesh and recommandation for mitigation. *Journal of Textile Engineering & Fashion Technology*, *5*(4). Advance online publication. DOI: 10.15406/jteft.2019.05.00205

Mohammadi-Sichani, M., Mazaheri Assadi, M., Farazmand, A., Kianirad, M., Ahadi, A. M., & Hadian-Ghahderijani, H. (2019). Ability of Agaricus bisporus, Pleurotus ostreatus and Ganoderma lucidum compost in biodegradation of petroleum hydrocarbon-contaminated soil. *International Journal of Environmental Science and Technology*, *16*(5), 2313–2320. DOI: 10.1007/s13762-017-1636-0

Münzel, T., Hahad, O., Daiber, A., & Landrigan, P. J. (2023). Soil and water pollution and human health: What should cardiologists worry about? *Cardiovascular Research*, *119*(2), 440–449. DOI: 10.1093/cvr/cvac082 PMID: 35772469

Ortellado, L. E., Villalba, L. L., Zapata, P. D., & Fonseca, M. I. (2022). In Udayanga, D., Bhatt, P., Manamgoda, D., & Saez, J. M. (Eds.), *Mycoremediation of Wastewater by Fungal Lipases BT - Mycoremediation Protocols* (pp. 213–219). Springer US., DOI: 10.1007/978-1-0716-2006-9_18

Paudel, S., & States, S. L. (2023). Urban green spaces and sustainability: Exploring the ecosystem services and disservices of grassy lawns versus floral meadows. *Urban Forestry & Urban Greening*, *84*, 127932. https://doi.org/https://doi.org/10.1016/j.ufug.2023.127932. DOI: 10.1016/j.ufug.2023.127932

Priya, A., Manab Sarma, P., & Lal, B. (2016). Isolation and characterization of Candida vishwanathii strain TERI MS1 for degradation of petroleum hydrocarbons in marine environment. *Desalination and Water Treatment*, *57*(46), 22099–22106. DOI: 10.1080/19443994.2015.1134351

Rai, R., & Vijayakumar, B. S. (2023). Myco-Remediation of Textile Dyes Via Biosorption by Aspergillus tamarii Isolated from Domestic Wastewater. *Water, Air, and Soil Pollution*, *234*(8), 542. DOI: 10.1007/s11270-023-06535-x

Raina, S. A., Yahmed, N. Ben, Bhat, R. A., & Dervash, M. A. (2020). Mycoremediation: a sustainable tool for abating environmental pollution. *Bioremediation and Biotechnology: Sustainable Approaches to Pollution Degradation*, 269–291.

Rhodes, C. J. (2014). Mycoremediation (bioremediation with fungi) – growing mushrooms to clean the earth. *Chemical Speciation and Bioavailability*, *26*(3), 196–198. DOI: 10.3184/095422914X14047407349335

Rudakiya, D. M., Patel, D. H., & Gupte, A. (2020). Exploiting the potential of metal and solvent tolerant laccase from Tricholoma giganteum AGDR1 for the removal of pesticides. *International Journal of Biological Macromolecules*, *144*, 586–595. https://doi.org/https://doi.org/10.1016/j.ijbiomac.2019.12.068. DOI: 10.1016/j.ijbiomac.2019.12.068 PMID: 31830449

Ruscasso, F., Bezus, B., Garmendia, G., Vero, S., Curutchet, G., Cavello, I., & Cavalitto, S. (2021). Debaryomyces hansenii F39A as biosorbent for textile dye removal. *Revista Argentina de Microbiologia*, *53*(3), 257–265. https://doi.org/https://doi.org/10.1016/j.ram.2020.10.004. DOI: 10.1016/j.ram.2020.10.004 PMID: 33454152

Saad, M. M. G., Saad, M. A., Saad, B. S., Zakaria, F. A., Husain, A.-R. A., & Abdelgaleil, S. A. M. (2022). Bioremediation and microbial-assisted phytoremediation of heavy metals by endophytic Fusarium species isolated from Convolvulus arvensis. *Bioremediation Journal*, *28*(2), 202–212. DOI: 10.1080/10889868.2022.2138256

Sadiq, S., Mahmood-ul-Hassan, M., Ahad, K., & Nazir, S. (2018). Bioremediation of hexachlorocyclohexane (HCH) in soil using spent mushroom compost of Pleurotus ostreatus. *Bioremediation Journal*, *22*(3–4), 126–135. DOI: 10.1080/10889868.2018.1516615

Sakshi, S., Singh, S. K., & Haritash, A. K. (2019). Polycyclic aromatic hydrocarbons: Soil pollution and remediation. *International Journal of Environmental Science and Technology*, *16*(10), 6489–6512. DOI: 10.1007/s13762-019-02414-3

Saleem, H., Zaidi, S. J., Ismail, A. F., & Goh, P. S. (2022). Advances of nanomaterials for air pollution remediation and their impacts on the environment. *Chemosphere*, *287*, 132083. https://doi.org/https://doi.org/10.1016/j.chemosphere.2021.132083. DOI: 10.1016/j.chemosphere.2021.132083 PMID: 34488054

Saravanan, A., Senthil Kumar, P., Jeevanantham, S., Karishma, S., Tajsabreen, B., Yaashikaa, P. R., & Reshma, B. (2021). Effective water/wastewater treatment methodologies for toxic pollutants removal: Processes and applications towards sustainable development. *Chemosphere*, *280*, 130595. https://doi.org/https://doi.org/10.1016/j.chemosphere.2021.130595. DOI: 10.1016/j.chemosphere.2021.130595 PMID: 33940449

Shanmugapriya, S., Manivannan, G., Selvakumar, G., & Sivakumar, N. (2019). *Extracellular Fungal Peroxidases and Laccases for Waste Treatment: Recent Improvement BT - Recent Advancement in White Biotechnology Through Fungi: Volume 3: Perspective for Sustainable Environments* (A. N. Yadav, S. Singh, S. Mishra, & A. Gupta (eds.); pp. 153–187). Springer International Publishing. DOI: 10.1007/978-3-030-25506-0_6

Shivakumar, C. K., Thippeswamy, B., & Krishnappa, M. (2014). Studies on Heavy Metals Detoxification Biomarkers in Fungal Consortia. *Caribbean Journal of Sciences and Technology, 2*(01 SE-Research Article), 496–502. https://www.caribjscitech.com/index.php/cjst/article/view/62

Siddeeg, S. M., Tahoon, M. A., Mnif, W., & Ben Rebah, F. (2020). Iron Oxide/Chitosan Magnetic Nanocomposite Immobilized Manganese Peroxidase for Decolorization of Textile Wastewater. In *Processes* (Vol. 8, Issue 1). DOI: 10.3390/pr8010005

Singh, N., Kumar, A., & Sharma, B. (2019). *Role of Fungal Enzymes for Bioremediation of Hazardous Chemicals BT - Recent Advancement in White Biotechnology Through Fungi: Volume 3: Perspective for Sustainable Environments* (A. N. Yadav, S. Singh, S. Mishra, & A. Gupta (eds.); pp. 237–256). Springer International Publishing. DOI: 10.1007/978-3-030-25506-0_9

Takio, N., Yadav, M., & Yadav, H. S. (2021). Catalase-mediated remediation of environmental pollutants and potential application – a review. *Biocatalysis and Biotransformation, 39*(6), 389–407. DOI: 10.1080/10242422.2021.1932838

Tanaya, K., Kumari, A., Singh, A. K., & Singh, D. (2024). Bioremediation: An Economical Approach for Treatment of Textile Dye Effluents. *Water, Air, and Soil Pollution, 235*(8), 516. DOI: 10.1007/s11270-024-07287-y

Văcar, C. L., Covaci, E., Chakraborty, S., Li, B., Weindorf, D. C., Fren iu, T., Pârvu, M., & Podar, D. (2021). Heavy Metal-Resistant Filamentous Fungi as Potential Mercury Bioremediators. *Journal of Fungi (Basel, Switzerland), 7*(5), 386. Advance online publication. DOI: 10.3390/jof7050386 PMID: 34069296

Vipotnik, Z., Michelin, M., & Tavares, T. (2022). Biodegradation of chrysene and benzo[a]pyrene and removal of metals from naturally contaminated soil by isolated Trametes versicolor strain and laccase produced thereof. *Environmental Technology & Innovation, 28*, 102737. https://doi.org/https://doi.org/10.1016/j.eti.2022.102737. DOI: 10.1016/j.eti.2022.102737

Yadav, P., Rai, S., Singh, M., & Mishra, V. (2021). Mycoremediation of environmental pollutants: A review with special emphasis on mushrooms. *Environmental Sustainability, 4*(4), 605–618. Advance online publication. DOI: 10.1007/s42398-021-00197-3

Yehia, R. S. (2023). Highlighting the potential for crude oil bioremediation of locally isolated Cunninghamella echinulata and Mucor circinelloides. *Brazilian Journal of Microbiology : Publication of the Brazilian Society for Microbiology, 54*(3), 1969–1981. DOI: 10.1007/s42770-023-01008-z PMID: 37249816

Zahuri, A. A., Wan Mohtar, W. H. M., Hanafiah, Z. M., Abdul Patah, M. F., Show, P.-L., Gafforov, Y., & Wan-Mohtar, W. A. A. Q. I. (2024). Mycoremediation of Industrial Textile Wastewater Using Ganoderma lucidum Pellets and Activated Dolomite in Batch Bioreactor. *Molecular Biotechnology*. Advance online publication. DOI: 10.1007/s12033-023-01035-z PMID: 38286973

Zaki, A. G., Hasanien, Y. A., & Abdel-Razek, A. S. (2022). Biosorption optimization of lead(II) and cadmium(II) ions by two novel nanosilica-immobilized fungal mutants. *Journal of Applied Microbiology, 133*(2), 987–1000. DOI: 10.1111/jam.15624 PMID: 35578549

Zolotova, N., Kosyreva, A., Dzhalilova, D., Fokichev, N., & Makarova, O. (2022). Harmful effects of the microplastic pollution on animal health: A literature review. *PeerJ, 10*, e13503. DOI: 10.7717/peerj.13503 PMID: 35722253

Chapter 12
Fungal Biodegradation:
Breaking Down Agricultural Wastes

P. Selvakumar
https://orcid.org/0000-0002-3650-4548
Nehru Institute of Technology, India

ABSTRACT

Fungal biodegradation represents a promising approach in addressing environmental challenges posed by recalcitrant pollutants. With increasing global concern over pollution and its ramifications for ecosystems and human health, the search for sustainable solutions has intensified. Fungi, with their remarkable metabolic versatility and adaptive capabilities, offer a unique toolkit for bioremediation. This introductory exploration delves into the mechanisms underlying fungal biodegradation, highlighting its applications, benefits, and current research trends.Fungi have evolved sophisticated enzymatic systems capable of degrading a wide range of organic compounds, including complex pollutants such as pesticides, petroleum hydrocarbons, dyes, and pharmaceuticals. Unlike bacteria, fungi possess ligninolytic enzymes like lignin peroxidase, manganese peroxidase, and laccase, enabling them to break down recalcitrant substances found in industrial waste and contaminated environments.

INTRODUCTION TO FUNGAL BIODEGRADATION: A SUSTAINABLE SOLUTION

Fungal biodegradation represents a promising approach in addressing environmental challenges posed by recalcitrant pollutants. With increasing global concern over pollution and its ramifications for ecosystems and human health, the search for sustainable solutions has intensified. Fungi, with their remarkable metabolic versatility and adaptive capabilities, offer a unique toolkit for bioremediation. This introductory exploration delves into the mechanisms underlying fungal biodegradation, highlighting its applications, benefits, and current research trends.Fungi have evolved sophisticated enzymatic systems capable of degrading a wide range of organic compounds, including complex pollutants such as pesticides, petroleum hydrocarbons, dyes, and pharmaceuticals. Unlike bacteria, fungi possess ligninolytic enzymes like lignin peroxidase, manganese peroxidase, and laccase, enabling them to break down recalcitrant substances found in industrial waste and contaminated environments. These enzymes act synergistically to oxidize and transform pollutants into simpler, less toxic metabolites or assimilate them as carbon and energy sources.The effectiveness of fungal biodegradation lies in the ability of fungi to thrive in diverse

DOI: 10.4018/979-8-3693-8201-1.ch012

environmental conditions, including extreme pH, temperature, and nutrient limitation. This adaptability makes fungi invaluable in bioremediation efforts across terrestrial and aquatic ecosystems, from contaminated soil and wastewater to polluted air and marine environments.

In addition to their enzymatic prowess, fungi exhibit other beneficial traits for biodegradation. They produce extracellular polymeric substances (EPS) that aid in biofilm formation, enhancing attachment to pollutant surfaces and facilitating degradation processes. Moreover, fungal spores and hyphae can penetrate soil matrices and densely packed pollutants, accessing otherwise inaccessible contaminants. These attributes contribute to the efficiency and versatility of fungal biodegradation systems, making them increasingly favored in environmental engineering and biotechnology.The application of fungal biodegradation extends beyond remediation to include biopulping, bioleaching, and the production of enzymes and bioactive compounds. Biopulping utilizes fungi to degrade lignin in wood, reducing chemical pulping requirements and environmental impact. Similarly, bioleaching employs fungi to extract metals from ores through oxidative processes, offering a greener alternative to traditional mining practices (Jorgensen H, et al., 2006). Fungal enzymes find applications in various industries, including textile, food, and pharmaceutical sectors, for their ability to catalyze specific biochemical reactions efficiently. Recent advances in molecular biology and genetic engineering have accelerated the development of fungal biodegradation technologies. Researchers manipulate fungal genomes to enhance enzyme production, substrate specificity, and environmental tolerance, paving the way for tailored bioremediation solutions. Metagenomics and transcriptomics reveal insights into fungal community dynamics and gene expression patterns in polluted environments, informing strategies for optimizing fungal consortia in biodegradation applications.

Despite these advancements, challenges persist in the practical implementation of fungal biodegradation. Factors such as fungal competition with native microbial communities, substrate availability, and operational conditions influence the efficacy of bioremediation processes. Moreover, regulatory frameworks and public acceptance of biotechnological approaches play critical roles in the widespread adoption of fungal biodegradation strategies (Kausar H, et al., 2010). Advances in synthetic biology offer opportunities to design synthetic fungal consortia optimized for specific pollutants or environmental conditions, further expanding the scope and efficiency of fungal biodegradation. As awareness of environmental sustainability grows, fungal biodegradation stands poised to emerge as a cornerstone of eco-friendly solutions to global pollution challenges. In conclusion, fungal biodegradation represents a sustainable and versatile approach to mitigating environmental pollution. Fungi's innate ability to degrade diverse pollutants, coupled with their adaptability and synergistic interactions, positions them as powerful agents in bioremediation efforts worldwide.

AGRICULTURAL WASTES: CHARACTERISTICS AND ENVIRONMENTAL IMPACTS

Agricultural wastes are diverse by-products generated from farming activities, encompassing a wide range of materials and substances that vary in composition, characteristics, and environmental impacts (Kaur A, et al., 2004). These wastes originate from various stages of agricultural processes, including cultivation, harvesting, processing, and post-harvest handling. The composition of agricultural wastes depends largely on the type of crop cultivated, farming practices employed, and the region's agricultural policies and practices. The type and quantity of crop residues generated also depend on factors such as crop yield, farming techniques (e.g., tillage practices), and climatic conditions. Another significant category of agricultural waste is animal manure, which comprises animal excreta along with bedding materials and uneaten feed. Animal manure is rich in organic matter, nitrogen, phosphorus, and potassium, making it a valuable resource for soil fertility improvement and crop productivity when managed properly. However, improper management of animal manure can lead to environmental issues such as nutrient leaching into water bodies, contributing to eutrophication and harmful algal blooms.

In addition to crop residues and animal manure, agricultural activities also generate processing wastes from food and beverage industries (Kanabkaew T, et al., 2011). These include pomace from fruit and vegetable processing, spent grains from brewing, and other by-products from food processing operations. These wastes can contain valuable nutrients and organic matter, but their disposal and management present challenges due to their perishable nature and potential for environmental pollution if not properly handled. The environmental impacts of agricultural wastes are multifaceted and can vary depending on waste type, management practices, and local environmental conditions. Nitrous oxide is released through microbial processes in soils amended with nitrogen-rich agricultural residues and fertilizers. Furthermore, agricultural wastes can impact water quality through nutrient runoff and leaching. Excess nutrients such as nitrogen and phosphorus from fertilizers and organic wastes can enter which adversely affect aquatic ecosystems and drinking water sources (Kasana RC, et al., 2008). Moreover, the presence of pathogens in animal manure poses risks to human health if not effectively managed during storage, handling, and land application. The disposal of agricultural wastes also poses challenges related to soil degradation and contamination. Improper disposal practices, such as burning crop residues or dumping untreated waste into water bodies, can degrade soil quality, reduce soil fertility, and contribute to soil erosion. Additionally, the accumulation of agricultural residues in water bodies can obstruct water flow, exacerbating flooding risks and disrupting aquatic habitats.

To address these environmental challenges, various strategies and technologies have been developed to manage agricultural wastes more sustainably. One approach is the promotion of integrated nutrient management practices that optimize the use of organic wastes, mineral fertilizers, and cover crops to improve soil health and reduce nutrient losses. Conservation tillage and crop residue management techniques can also help minimize soil erosion, improve water retention, and enhance carbon sequestration in agricultural soils (Milagres AMF, et al., 1993). In conclusion, agricultural wastes represent a significant challenge and opportunity for sustainable development. Effective management of these wastes is essential to mitigate their environmental impacts, conserve natural resources, and promote circular economy principles in agriculture. By adopting integrated waste management strategies, promoting technological innovations, and enhancing policy frameworks, stakeholders can work together to harness the potential of agricultural wastes as valuable resources for sustainable agriculture and environmental stewardship.

FUNGAL DIVERSITY AND ENZYMATIC CAPABILITIES

Fungal diversity encompasses a vast array of organisms found in almost every ecosystem on Earth, playing crucial roles in nutrient cycling, decomposition, symbiosis, and pathogenesis. These diverse fungi possess an extensive repertoire of enzymatic capabilities that are pivotal for their ecological functions and have significant implications for various industrial applications and biotechnological advancements (Mahadev Madgule, et al., 2023). Fungi belong to a separate kingdom, distinct from plants, animals, and bacteria, with estimates suggesting there may be anywhere from 2.2 to 3.8 million fungal species on the planet, although only about 120,000 species have been formally described. Fungi exhibit a remarkable diversity in morphology, physiology, and ecological niches, ranging from microscopic single-celled yeasts to complex multicellular molds and mushrooms. Their ability to thrive in diverse environments, including extreme habitats such as deep-sea vents, polar regions, and acidic soils, underscores their adaptability and evolutionary success. One of the defining characteristics of fungi is their capacity to secrete a wide range of enzymes, which are essential for breaking down complex organic matter into simpler compounds that can be assimilated as nutrients. These enzymatic capabilities enable fungi to degrade a broad spectrum of substrates, including cellulose, lignin, chitin, proteins, lipids, and various complex polysaccharides. This enzymatic activity is crucial not only for nutrient cycling in ecosystems but also for applications in biofuel production, bioremediation of environmental pollutants, and the pulp and paper industry (Maza M, et al., 2014). Furthermore, fungi are known for their production of hydrolytic enzymes such as proteases, lipases, amylases, and pectinases, which catalyze the hydrolysis of proteins, lipids, starches, and pectins, respectively. These enzymes play essential roles in nutrient acquisition from organic substrates and are also exploited in various industrial processes, including food and beverage production, textile processing, detergent formulation, and pharmaceutical manufacturing. In addition to their role in organic matter degradation, fungal enzymes have attracted considerable interest in the biotechnological sector due to their potential applications in enzyme technology and biocatalysis. Fungal enzymes are highly versatile and can operate under a wide range of environmental conditions, including extremes of pH, temperature, and salinity, which makes them valuable for industrial processes requiring robust and efficient biocatalysts.

The discovery and characterization of novel fungal enzymes with unique properties have been facilitated by advancements in genomic sequencing, metagenomics, and bioinformatics. These technologies have enabled researchers to explore the genetic diversity of fungi and identify genes encoding enzymes with desired characteristics, such as enhanced activity, stability, and substrate specificity. Metagenomic studies of fungal communities in diverse habitats have also revealed the presence of genes encoding novel enzymes that hold promise for biotechnological applications. Moreover, the application of genetic engineering and protein engineering techniques has further expanded the repertoire of fungal enzymes available for industrial use. Researchers can modify enzymes to improve their performance, optimize their biochemical properties, or engineer entirely new functionalities that are tailored to specific industrial applications (Lundell TK, et al., 2010). This approach has led to the development of enzyme cocktails for more efficient biomass conversion, biocatalysts for green chemistry processes, and enzymes with enhanced stability for use in harsh industrial conditions. Beyond their industrial significance, fungal enzymes also hold promise in medical and pharmaceutical applications. For example, fungal proteases and lipases are used in the production of therapeutic proteins and pharmaceutical formulations. Fungal-derived enzymes are also being investigated for their potential to degrade biofilms associated with human infections, offering new avenues for the treatment of antibiotic-resistant pathogens and chronic infections.

CELLULOSE DEGRADATION: UNLOCKING PLANT CELL WALL COMPONENTS

Cellulose degradation represents a critical process in the cycling of carbon and nutrients within ecosystems, facilitated primarily by microbial enzymes that hydrolyze cellulose, a major component of plant cell walls. Cellulose is a linear polysaccharide composed of repeating glucose units linked together by β-1,4 glycosidic bonds. This structural arrangement confers rigidity and strength to plant cell walls, making cellulose one of the most abundant biopolymers on Earth (Ogbonna AI, et al., 2015). Efficient degradation of cellulose is essential for carbon recycling in terrestrial and aquatic environments, influencing soil fertility, nutrient availability, and the global carbon cycle. Microorganisms are the primary agents responsible for cellulose degradation in natural ecosystems. Bacteria, fungi, and certain protists produce a variety of enzymes collectively known as cellulases, which work synergistically to hydrolyze cellulose into soluble sugars. These fungi produce a diverse array of cellulases tailored to degrade different forms of cellulose found in plant tissues. For example, crystalline cellulose, which is tightly packed and highly ordered, requires cellulases with strong cellulolytic activity to disrupt its crystalline structure and initiate degradation. In contrast, amorphous cellulose, which is less ordered and more accessible, may be degraded by cellulases with different enzymatic properties.

The enzymatic machinery of cellulose-degrading fungi is often regulated in response to environmental cues, such as substrate availability, nutrient status, and physical conditions (e.g., pH, temperature) (Liao H, et al., 2012). Induction of cellulase production occurs when fungi encounter cellulose or related polysaccharides, triggering the expression of cellulase genes and synthesis of the corresponding enzymes. This adaptive response allows fungi to efficiently depolymerize cellulose and utilize the released sugars for growth and metabolism.Beyond their ecological role, cellulose-degrading enzymes have significant implications for various industrial applications, particularly in biofuel production and biorefineries. The enzymatic hydrolysis of cellulose is a key step in the production of bioethanol. Efficient cellulose degradation is essential to release fermentable sugars that can be fermented by microorganisms into ethanol, providing a renewable alternative to fossil fuels (Paramasivam Selvakumar, et al., 2024). In addition to bioethanol production, cellulose-degrading enzymes are utilized in the production of other bio-based chemicals and materials, including platform chemicals (e.g., lactate, succinate), biodegradable plastics, and biochemicals used in pharmaceutical and cosmetic industries. The versatility of cellulases in catalyzing the hydrolysis of cellulose into simple sugars has spurred research efforts to optimize enzyme performance, enhance enzyme stability, and reduce production costs through enzyme engineering and bioprocess optimization.

Recent advancements in biotechnology and synthetic biology have enabled the development of engineered microbial strains and enzyme cocktails tailored for specific biomass feedstocks and industrial applications. Strategies such as metabolic engineering, directed evolution, and rational design have been employed to enhance the efficiency and specificity of cellulose-degrading enzymes, overcoming challenges related to substrate complexity, enzyme inhibition, and product inhibition in industrial processes. Moreover, the study of cellulose degradation has contributed to our understanding of microbial diversity, community dynamics, and ecological interactions in diverse environments. Microbial consortia involved in cellulose degradation often exhibit synergistic interactions, where different organisms cooperate to degrade complex substrates more efficiently than individual species alone. These interactions are influenced by factors such as substrate availability, environmental conditions, and microbial community composition, highlighting the complexity of cellulose degradation processes in natural ecosystems (Jin R, et al., 2012). In summary, cellulose degradation represents a fundamental process mediated by microbial enzymes that play crucial roles in carbon cycling, nutrient recycling, and bioenergy production. The diverse enzymatic capabilities of cellulolytic microorganisms, particularly fungi, enable efficient degradation of cellulose under various environmental conditions. Harnessing these enzymes for industrial applications holds promise for advancing sustainable biotechnologies and reducing reliance on fossil fuels through the production of renewable biofuels and biochemicals. As research continues to unravel the complexities of cellulose degradation and enzyme function, new opportunities emerge for innovation in biotechnology, environmental sustainability, and global carbon management.

LIGNIN DEGRADATION: BREAKING DOWN THE MOST RECALCITRANT POLYMER

Lignin degradation represents a pivotal process in the global carbon cycle and holds significant promise for industrial applications due to lignin's recalcitrant nature and complex structure. This network forms cross-links with cellulose and hemicellulose, contributing to the rigidity and impermeability of plant cell walls. Lignin's recalcitrance stems from its complex structure, which resists degradation by most organisms and enzymes (Sandhu DK, et al., 1985). Microbial degradation of lignin involves an array of enzymes collectively known as ligninases or ligninolytic enzymes, which are primarily produced by fungi but also found in some bacteria. These enzymes include lignin peroxidases (LiPs), manganese peroxidases (MnPs), versatile peroxidases (VPs), and laccases, each with distinct catalytic properties and roles in lignin degradation. Laccases, although not directly involved in lignin depolymerization, catalyze the oxidation of phenolic and non-phenolic lignin units, contributing to lignin degradation in conjunction with other enzymes. These fungi produce a diverse array of ligninolytic enzymes tailored to degrade different forms of lignin found in plant tissues. For example, white rot fungi are capable of degrading both lignin and cellulose components of wood, enabling efficient decomposition and nutrient cycling in forest ecosystems. The production of ligninolytic enzymes is often regulated in response to environmental cues, such as the presence of lignin-derived aromatic compounds or the availability of alternative carbon sources.

Beyond their ecological role, lignin-degrading enzymes have significant potential for industrial applications, particularly in the biofuel and biorefinery industries. Lignin valorization—conversion of lignin into high-value products such as biofuels, chemicals, and materials—is a key area of research aimed at enhancing the economic viability and sustainability of biomass utilization processes (Ramakrishnan T.,et al., 2022). Enzymatic depolymerization of lignin represents a green and sustainable approach compared to traditional chemical methods, which are often energy-intensive and generate environmentally harmful by-products. Directed evolution and rational design approaches have been employed to improve the efficiency, stability, and specificity of lignin-degrading enzymes, making them more suitable for industrial-scale biorefining applications. Moreover, the study of lignin degradation has broader implications for understanding microbial diversity, community dynamics, and biogeochemical cycling in natural ecosystems. Microbial consortia involved in lignin degradation often exhibit synergistic interactions, where different organisms cooperate to degrade lignin more efficiently than individual species alone. These interactions are influenced by factors such as substrate availability, environmental conditions, and microbial community composition, highlighting the complexity of lignin degradation processes in natural environments. In summary, lignin degradation represents a fundamental process mediated by microbial enzymes that play crucial roles in carbon cycling, nutrient recycling, and ecosystem functioning (Paramasivam Selvakumar, et al., 2024). The enzymatic capabilities of lignin-degrading fungi, particularly their ability to produce diverse ligninolytic enzymes, enable efficient depolymerization of lignin under various environmental conditions. Harnessing these enzymes for industrial applications holds promise for advancing sustainable biotechnologies and reducing reliance on fossil fuels through the production of renewable biofuels, biochemicals, and materials. As research continues to unravel the complexities of lignin degradation and enzyme function, new opportunities emerge for innovation in biotechnology, environmental sustainability, and global carbon management.

HEMICELLULOSE DEGRADATION: CONVERTING PENTOSE SUGARS

Hemicellulose degradation is a vital biochemical process that involves the breakdown of complex heteropolysaccharides found in plant cell walls, particularly alongside cellulose and lignin. Unlike cellulose, which consists of linear chains of glucose units, hemicelluloses are a diverse group of branched polysaccharides composed of various monosaccharides such as xylose, arabinose, mannose, galactose, and glucuronic acid. The composition of hemicelluloses varies depending on the plant species and tissue type, reflecting their roles in providing structural support, water retention, and as a carbon source for microbial and fungal communities involved in degradation processes (Reese ET, et al., 1963). Microorganisms play a central role in the degradation of hemicelluloses through the production of a range of enzymes known as hemicellulases. The degradation of other hemicellulose components, such as arabinose, galactose, mannose, and glucuronic acid, involves specific hemicellulases tailored to their respective structures. α-arabinofuranosidases and α-galactosidases hydrolyze arabinose and galactose residues from arabinoxylans and galactomannans, respectively, while α-glucuronidases target glucuronic acid side chains attached to xylan and other hemicelluloses. These enzymes work synergistically to depolymerize complex hemicellulose structures into simpler sugars and oligosaccharides that can be further metabolized by microorganisms. These fungi secrete a diverse array of hemicellulases tailored to degrade specific hemicellulosic substrates encountered in their respective habitats. The production of hemicellulases is often regulated by the availability of hemicellulosic substrates, triggering the expression of corresponding genes and enzyme synthesis to facilitate efficient utilization of hemicellulose-derived sugars.

In addition to their ecological roles, hemicellulose-degrading enzymes have significant applications in various industrial sectors, particularly in biofuel production and biorefineries. Hemicelluloses represent a significant fraction of lignocellulosic biomass—alongside cellulose and lignin—and their efficient depolymerization is crucial for maximizing the yield of fermentable sugars for bioethanol production (Sugumaran K. M, D., et al., 2024). Enzymatic hydrolysis of hemicelluloses into fermentable sugars, such as xylose and arabinose, complements the enzymatic conversion of cellulose into glucose, enabling more comprehensive utilization of biomass feedstocks for biofuel production. Moreover, hemicellulose-derived sugars can be converted into other bio-based products, including platform chemicals (e.g., xylitol, furfural, and levulinic acid) used in the chemical, pharmaceutical, and cosmetic industries. The versatility of hemicellulase enzymes in catalyzing the hydrolysis of hemicelluloses has driven research efforts to optimize enzyme performance, enhance enzyme stability, and reduce production costs through enzyme engineering and bioprocess optimization. Advances in metagenomic approaches and genetic engineering have facilitated the discovery and characterization of novel hemicellulases with improved catalytic properties and substrate specificities, expanding the toolkit for sustainable biomass conversion technologies (Sharma A, et al., 2014). Furthermore, the study of hemicellulose degradation provides insights into microbial diversity, community dynamics, and ecosystem functioning in diverse environments. Microbial consortia involved in hemicellulose degradation often exhibit complex interactions, where different organisms collaborate to degrade hemicelluloses more efficiently than individual species alone. These interactions are influenced by environmental factors such as substrate availability, nutrient status, and microbial community composition, highlighting the intricate biochemical processes involved in biomass decomposition and carbon cycling in natural ecosystems (Terrasan CR, et al., 2010).

PECTIN DEGRADATION: RELEASING FRUIT AND VEGETABLE WASTES

Pectin degradation is a crucial biochemical process that involves the breakdown of pectin, a complex polysaccharide found predominantly in the cell walls of fruits and vegetables. Pectin serves structural functions in plant tissues, contributing to cell adhesion and providing strength and stability to cell walls. Chemically, pectin is composed of a backbone. Microorganisms produce a range of enzymes collectively known as pectinases or pectinolytic enzymes, which catalyze the hydrolysis of pectin into smaller, soluble oligosaccharides and monosaccharides. The main types of pectinases include pectin methylesterases (PMEs), polygalacturonases (PGs), pectate lyases (PLs), and pectin lyases (PLys), each targeting different components of the pectin molecule (Vijayakumar G, et al., 2024). PMEs remove methyl ester groups from galacturonic acid residues, modulating the degree of esterification and influencing the physicochemical properties of pectin. Polygalacturonases cleave the α-1,4 glycosidic bonds between galacturonic acid residues, resulting in the depolymerization of pectin chains. Pectate lyases and pectin lyases cleave the α-1,4 glycosidic bonds in the backbone of pectin, particularly in regions of low esterification, releasing unsaturated oligogalacturonates and generating double bonds in the resulting fragments. These microorganisms possess the genetic machinery to synthesize and secrete pectinases in response to the presence of pectin-rich substrates. The enzymatic degradation of pectin by these microorganisms plays a crucial role in the decomposition of plant materials, particularly fruits and vegetables, during natural decay processes.

In addition to their ecological roles, pectinolytic enzymes have significant industrial applications, particularly in the food and beverage industries, where they are used for fruit juice extraction, clarification, and flavor enhancement (Tuomela M, et al., 2000). Pectinases facilitate the release of juices from fruits by degrading the pectin matrix that encapsulates juice vesicles, thereby improving juice yield and quality. In fruit processing, pectinases are also employed to break down pectin in fruit mash, facilitating the extraction of flavors, colors, and bioactive compounds for use in food products, beverages, and pharmaceutical formulations. Furthermore, pectinases play a crucial role in the production of pectin-derived products and additives, such as low-methoxy pectins used as gelling agents in jams, jellies, and confectionery products. Controlled enzymatic modification of pectin using PMEs allows for precise modulation of its gel-forming properties, enabling the production of gels with specific textures and functionalities. Pectin-derived oligosaccharides generated through enzymatic hydrolysis have also garnered interest for their potential health benefits, including prebiotic effects that promote gastrointestinal health and immune function. Beyond the food industry, pectinases are utilized in the textile industry for the degumming of natural fibers such as flax, hemp, and cotton. Pectinolytic enzymes efficiently remove pectinaceous impurities from fibers, improving their processing properties and enhancing fiber quality. In addition, pectinases have applications in the pulp and paper industry, where they aid in the bleaching and refining of paper pulp by degrading pectin impurities and improving pulp drainage and paper strength.

The biotechnological potential of pectinolytic enzymes extends to environmental applications, including the management and valorization of fruit and vegetable wastes generated from food processing industries. Enzymatic degradation of pectin-rich waste materials such as citrus peel, apple pomace, and grape marc can release fermentable sugars and bioactive compounds, which can be converted into value-added products such as biofuels, biochemicals, and dietary supplements (Zeng GM, et al., 2006). This approach not only reduces the environmental impact of organic waste disposal but also contributes to the development of sustainable biorefinery processes that maximize resource efficiency and minimize waste generation. In summary, pectin degradation mediated by microbial pectinases represents a fundamental biochemical process with diverse ecological, industrial, and biotechnological implications. The enzymatic capabilities of pectinolytic microorganisms, particularly their ability to depolymerize complex pectin structures into useful products and components, underscore their importance in natural ecosystems and industrial processes alike (Selvakumar, P. et al., 2024). Harnessing pectinolytic enzymes for sustainable biotechnologies offers opportunities to advance waste valorization, improve resource efficiency, and promote the transition towards a bio-based economy based on renewable biomass resources. As research continues to uncover the enzymatic diversity and functional properties of pectinolytic enzymes, new avenues emerge for innovation in biotechnology, food processing, environmental sustainability, and global waste management strategies.

PROTEIN DEGRADATION: UTILIZING NITROGEN-RICH AGRICULTURAL WASTES

Protein degradation is a fundamental biological process essential for nutrient recycling and energy generation in ecosystems. Proteins, composed of amino acids linked by peptide bonds, are ubiquitous in living organisms and play crucial roles in cellular structure, function, and metabolism. The turnover of proteins involves their continuous synthesis and degradation, maintaining cellular homeostasis and adapting to environmental changes. In natural environments, protein degradation is predominantly carried out by microorganisms through the secretion of proteolytic enzymes, which catalyze the hydrolysis of peptide bonds and the subsequent breakdown of proteins into peptides and amino acids.Microbial proteases are the primary enzymes responsible for protein degradation in various ecosystems, including soil, water, and the gastrointestinal tracts of animals (Ward OP, et al., 2005). These proteases exhibit diverse catalytic activities and specificities, targeting different peptide bonds within protein molecules. Endopeptidases cleave peptide bonds within the interior of protein chains, generating peptides of various lengths, while exopeptidases trim amino acids from the ends of peptides and proteins. Aminopeptidases specifically hydrolyze amino-terminal peptide bonds, releasing individual amino acids that can be assimilated by microorganisms for growth and metabolism. These microorganisms secrete a diverse array of proteases tailored to degrade proteins encountered in their respective habitats. The production of proteases is often regulated in response to environmental cues, such as nutrient availability, pH, temperature, and substrate complexity. Induction of protease synthesis occurs when microorganisms encounter nitrogen-rich proteinaceous substrates, triggering the expression of protease genes and the secretion of enzymes to facilitate nutrient acquisition and energy metabolism.

Beyond their ecological roles, proteolytic enzymes have significant industrial applications, particularly in the food, pharmaceutical, and detergent industries. In food processing, proteases are used for meat tenderization, dairy processing (e.g., cheese production), and protein hydrolysis to enhance flavor, texture, and nutritional properties. Enzymatic hydrolysis of proteins into peptides and amino acids generates bio-active peptides with potential health benefits, including antioxidant, antimicrobial, and antihypertensive activities. Proteases are also employed in pharmaceutical formulations for the production of therapeutic proteins, peptide-based drugs, and digestive enzyme supplementsMoreover, proteases play a crucial role in bioremediation and waste management by facilitating the degradation of organic matter, including nitrogen-rich agricultural wastes such as crop residues, animal manure, and food processing by-products. Enzymatic degradation of proteinaceous wastes releases nitrogenous compounds, peptides, and amino acids that serve as nutrients for soil microorganisms, promoting nutrient cycling and improving soil fertility (Yin ZW, et al., 2011). This process is particularly important in sustainable agriculture practices, where the recycling of organic wastes enhances soil health, reduces greenhouse gas emissions, and minimizes reliance on synthetic fertilizers.In the biotechnological sector, proteolytic enzymes are employed in the production of biofuels and biochemicals through microbial fermentation processes.

Recent advancements in enzyme engineering and bioprocess optimization have enhanced the efficiency, specificity, and stability of proteolytic enzymes for industrial applications. Techniques such as directed evolution, rational design, and metagenomics enable the discovery and development of novel proteases with tailored properties and enhanced performance under diverse environmental conditions. These innovations drive the sustainable use of proteolytic enzymes in biomass conversion, waste valorization, and the development of biobased products that support a circular economy model. In summary, protein degradation mediated by microbial proteases represents a fundamental biochemical process with diverse ecological, industrial, and biotechnological implications (Selvakumar, P., 2024). The enzymatic capabilities of proteolytic microorganisms, particularly their ability to degrade complex proteinaceous substrates into bioavailable nutrients, underscore their importance in natural ecosystems and industrial processes alike. Harnessing proteolytic enzymes for sustainable biotechnologies offers opportunities to advance waste valorization, improve resource efficiency, and promote the transition towards a bio-based economy based on renewable biomass resources. As research continues to uncover the enzymatic diversity and functional properties of proteases, new avenues emerge for innovation in biotechnology, food and agriculture, environmental sustainability, and global nutrient management strategies.

FUNGAL BIODEGRADATION PROCESSES: SOLID-STATE FERMENTATION AND COMPOSTING

Fungal biodegradation processes, particularly solid-state fermentation (SSF) and composting, play crucial roles in environmental sustainability and waste management. These processes harness the metabolic activities of fungi to degrade organic materials, transforming them into simpler compounds and reducing waste volumes. Solid-state fermentation involves the growth of fungi on solid substrates with limited moisture, mimicking natural conditions where fungi thrive on various organic matter like agricultural residues, food wastes, and lignocellulosic biomass. This method not only facilitates the breakdown of complex polymers such as cellulose and lignin but also generates valuable secondary metabolites like enzymes, organic acids, and bioactive compounds (Jahromi MF, et al., 2011). Composting, on the other hand, is a controlled aerobic degradation process that utilizes fungi along with bacteria and other microorganisms to convert organic waste into humus-like material known as compost. Fungi contribute significantly to the decomposition process by breaking down tougher compounds like lignin, thus accelerating the overall degradation rate compared to bacterial-dominated composting. The process is influenced by various factors such as substrate composition, moisture content, pH, temperature, and aeration, which must be carefully controlled to optimize fungal growth and enzyme production. SSF is particularly advantageous for industries seeking sustainable solutions to manage agricultural residues, food processing wastes, and lignocellulosic biomass (Inderjit S, et al., 2004). By converting these wastes into valuable products like enzymes and biofuels, SSF not only reduces waste disposal costs but also provides economic opportunities through the production of high-value bioproducts. Furthermore, the inherent tolerance of many fungal species to extreme environmental conditions and their ability to utilize a wide range of substrates make SSF a versatile biotechnological tool applicable in diverse industrial sectors.

Composting, on the other hand, operates under aerobic conditions that promote the growth of thermophilic fungi alongside bacteria, resulting in rapid degradation of organic matter. Fungal species such as Thermomyces and Humicola thrive in the elevated temperatures typical of thermophilic composting, contributing to the breakdown of lignocellulosic materials and the synthesis of stable humic substances. By diverting organic wastes from landfills, these processes minimize methane production and leachate contamination, thereby mitigating environmental pollution and contributing to climate change mitigation efforts. Moreover, the end products of fungal biodegradation processes, such as compost and bioactive compounds, can be utilized in agriculture, horticulture, and bioremediation, further promoting sustainable practices and circular economy principles. In conclusion, fungal biodegradation processes like solid-state fermentation and composting represent effective and sustainable approaches to managing organic wastes and promoting environmental stewardship (Selvakumar, P.,2024). These processes harness the enzymatic capabilities of fungi to break down complex organic compounds, generate valuable products, and enhance soil fertility without compromising environmental integrity. As the global community continues to prioritize sustainability and waste reduction, these biotechnological tools will play an increasingly vital role in achieving a circular economy and mitigating the impacts of organic waste disposal on the environment.

OPTIMIZING FUNGAL BIODEGRADATION: PH, TEMPERATURE, AND MOISTURE

Optimizing fungal biodegradation processes, such as solid-state fermentation (SSF) and composting, involves careful control and manipulation of key environmental factors like pH, temperature, and moisture content. These factors significantly influence fungal growth, enzyme production, and the efficiency of organic matter degradation, thereby impacting the overall effectiveness and economic viability of biotechnological applications. pH plays a critical role in fungal metabolism and enzyme activity. Different fungal species exhibit varying pH optima for growth and enzyme production. For instance, filamentous fungi like Aspergillus and Trichoderma typically thrive in slightly acidic to neutral conditions (pH 5.0-7.0), while yeast species such as Saccharomyces prefer slightly acidic to neutral pH ranges as well. Maintaining optimal pH levels ensures maximum enzymatic activity, as enzymes involved in cellulose, hemicellulose, and lignin degradation often exhibit pH-dependent activities. Adjusting pH levels in SSF and composting processes can be achieved through the addition of buffering agents or by selecting fungal strains adapted to specific pH ranges, thereby enhancing substrate utilization efficiency and overall biodegradation rates (Gardes M, et al., 1993). Temperature regulation is another crucial factor influencing fungal biodegradation processes. Fungi exhibit diverse temperature requirements for growth and enzyme production, ranging from mesophilic (20-40°C) to thermophilic (45-70°C) conditions. Mesophilic fungi like Aspergillus niger and Penicillium spp. are commonly used in SSF for their ability to produce a broad spectrum of hydrolytic enzymes under moderate temperature conditions. Thermophilic fungi, such as species within the genera Thermomyces and Humicola, are preferred in composting due to their optimal growth and enzymatic activity at elevated temperatures, which accelerate the decomposition of lignocellulosic materials and organic wastes. Proper temperature control during SSF and composting processes ensures fungal viability, enzyme stability, and efficient substrate degradation, thereby maximizing the production of valuable bioproducts like enzymes, biofuels, and compost.

Moisture content is a critical determinant of fungal activity and substrate accessibility in biodegradation processes. Fungi require adequate moisture levels for enzymatic activities, nutrient uptake, and metabolic processes. In SSF, maintaining optimal moisture content (typically 50-70% based on dry weight) is essential to support fungal growth and ensure uniform substrate colonization (Jain N, et al., 2014). Insufficient moisture can limit enzymatic hydrolysis of complex polymers, while excessive moisture may lead to anaerobic conditions, inhibiting fungal growth and causing substrate compaction. Composting also requires careful moisture management to facilitate microbial respiration and maintain aerobic conditions throughout the decomposition process. Monitoring and adjusting moisture levels through periodic watering or aeration are essential strategies to optimize fungal biodegradation efficiency, ensure process stability, and prevent microbial activity. Integration of advanced biotechnological techniques, such as process modeling and optimization strategies, enhances the efficiency and scalability of fungal biodegradation processes. Furthermore, the selection and genetic engineering of fungal strains with enhanced capabilities for substrate utilization, enzyme secretion, and stress tolerance represent promising avenues for advancing fungal biodegradation technologies. Engineered fungal species with tailored metabolic pathways and enhanced enzymatic activities offer opportunities to overcome substrate recalcitrance, improve process efficiency, and expand the applicability of fungal biodegradation in diverse industrial sectors, including bioenergy production, waste management, and environmental remediation. In conclusion, optimizing pH, temperature, and moisture content is essential for maximizing the efficiency, sustainability, and economic feasibility of fungal biodegradation processes like SSF and composting. By fine-tuning these environmental factors through advanced monitoring, control strategies, and biotechnological innovations, researchers and practitioners can harness the metabolic capabilities of fungi to transform organic wastes into valuable bioproducts, mitigate environmental impacts, and promote a circular economy approach to resource management

FUNGAL STRAINS FOR BIODEGRADATION: SELECTION AND IMPROVEMENT

Selecting and improving fungal strains for biodegradation involves a strategic approach aimed at enhancing their capabilities to efficiently degrade complex organic substrates and produce valuable bioproducts. Fungi possess diverse enzymatic systems and metabolic pathways that enable them to degrade a wide range of organic materials, including lignocellulosic biomass, agricultural residues, food wastes, and pollutants (Chavez R, et al., 2006). The selection process typically begins with identifying fungal species or strains with desirable characteristics such as high enzyme activity, substrate specificity, tolerance to environmental stresses, and compatibility with intended biotechnological applications. Several criteria guide the selection of fungal strains for biodegradation purposes. Firstly, the ability to produce a spectrum of hydrolytic enzymes like cellulases, hemicellulases, ligninases, and proteases is crucial for efficient degradation of complex polymers present in organic substrates. Fungal species known for their robust enzymatic activities include Aspergillus spp., Trichoderma spp., Penicillium spp., and certain yeast species like Saccharomyces cerevisiae and Candida spp. These fungi produce enzymes that act synergistically to break down cellulose, hemicellulose, lignin, and other recalcitrant compounds into simpler sugars, which can then be metabolized for energy or converted into valuable bioproducts. Secondly, fungal strains must exhibit adaptability to environmental conditions prevalent in biodegradation processes such as solid-state fermentation (SSF) and composting.

Moreover, the genetic diversity within fungal populations offers opportunities for strain improvement through selective breeding, mutagenesis, and genetic engineering. These approaches aim to enhance fungal traits such as enzyme productivity, substrate specificity, stress tolerance, and metabolic efficiency. Directed evolution strategies involve subjecting fungal populations to selective pressures under controlled conditions, thereby promoting the emergence of variants with improved biodegradation capabilities (Chang AJ, et al., 2012). Mutagenesis techniques, including chemical mutagenesis and radiation-induced mutagenesis, induce genetic mutations that may enhance enzyme activities or substrate utilization efficiencies in fungal strains. In recent years, genetic engineering has emerged as a powerful tool for targeted modification of fungal genomes to achieve specific biotechnological objectives. Techniques such as gene knockout, gene overexpression, and pathway engineering enable researchers to manipulate metabolic pathways involved in enzyme production, substrate utilization, and stress response mechanisms. For example, engineering fungal strains to overexpress key cellulolytic enzymes like cellobiohydrolases and endoglucanases can significantly enhance their ability to degrade cellulose-rich substrates in SSF processes. Furthermore, synthetic biology approaches facilitate the design and construction of novel fungal strains with customized metabolic pathways for the production of biofuels, bioplastics, and other high-value bioproducts from renewable feedstocks. By integrating synthetic biology principles with fungal biodegradation processes, researchers aim to develop sustainable biotechnological solutions that address global challenges related to waste management, environmental pollution, and resource scarcity.

The selection and improvement of fungal strains for biodegradation applications also involve considerations of regulatory compliance, safety, and environmental impact assessment. Ensuring the safe deployment of genetically modified fungi in industrial and environmental settings requires rigorous risk assessment protocols to evaluate potential ecological risks, unintended consequences, and biosafety measures. Regulatory frameworks governing the use of genetically modified organisms (GMOs) in biotechnological applications vary globally and must be adhered to during strain development, field trials, and commercialization phases. In conclusion, the selection and improvement of fungal strains for biodegradation represent a multidisciplinary endeavor encompassing microbiology, genetics, biotechnology, and environmental science (Abhilasha A., et al., 2022). By leveraging the natural capabilities of fungi and advancing biotechnological innovations, researchers and practitioners can optimize biodegradation processes, reduce dependency on fossil-based resources, and promote sustainable practices in waste management and bio-based industries

AGRICULTURAL WASTE STREAMS: CROP RESIDUES, MANURES, AND EFFLUENTS

Agricultural waste streams encompass a diverse array of organic materials generated during farming and livestock production, including crop residues, manures, and effluents. These waste streams are integral components of agricultural ecosystems, presenting both challenges and opportunities for sustainable resource management. Crop residues consist of non-edible parts of plants left in fields after harvest, such as stalks, leaves, and husks. These residues are rich in organic matter and nutrients like nitrogen, phosphorus, and potassium, essential for soil fertility and crop productivity. However, their disposal and decomposition can impact soil health, greenhouse gas emissions, and nutrient cycling. Techniques like mulching, composting, and bioenergy production from crop residues can mitigate these challenges, converting waste into valuable products like organic fertilizers or renewable energy sources. Manures, another significant agricultural waste stream, comprise animal excreta and bedding materials rich in organic nutrients. Implementing best management practices (BMPs) such as runoff controls, wetland buffers, and precision agriculture technologies can minimize these impacts, safeguarding water quality and ecosystem health. The sustainable management of agricultural waste streams requires integrated approaches that consider environmental, economic, and social dimensions. Promoting circular economy principles through waste valorization, resource recovery, and renewable energy production can enhance the resilience and sustainability of agricultural systems. Additionally, stakeholder engagement, policy support, and technological innovations play pivotal roles in advancing sustainable agricultural waste management practices. By optimizing the utilization of crop residues, manures, and effluents, agriculture can contribute to climate change mitigation, soil health improvement, and sustainable food production, thereby fostering a more resilient and environmentally conscious agricultural sector

VALUE-ADDED PRODUCTS FROM FUNGAL BIODEGRADATION: BIOFUELS, BIOPLASTICS, AND FERTILIZERS

Value-added products derived from fungal biodegradation processes encompass a diverse range of biotechnological innovations that harness the metabolic capabilities of fungi to convert organic wastes into economically valuable bioproducts. These products include biofuels, bioplastics, and fertilizers, each contributing to sustainable resource utilization, waste management, and environmental stewardship. Biofuels represent one of the prominent outputs from fungal biodegradation, addressing the global demand for renewable energy sources while mitigating carbon emissions. Fungi such as Aspergillus, Trichoderma, and Neurospora are utilized in solid-state fermentation (SSF) and submerged fermentation (SmF) processes to produce enzymes like cellulases and hemicellulases. These enzymes hydrolyze lignocellulosic biomass—such as agricultural residues, forestry waste, and dedicated energy crops—into fermentable sugars. Subsequently, fungal fermentation converts these sugars into biofuels such as ethanol and butanol. Ethanol, in particular, is widely used as a biofuel additive to gasoline, offering a renewable alternative that reduces reliance on fossil fuels and lowers greenhouse gas emissions. Bioplastics represent another innovative application of fungal biodegradation in the realm of sustainable materials production. These biopolymers exhibit properties similar to conventional plastics but are biodegradable and derived from renewable resources. By harnessing fungal metabolic pathways, biotechnologists can optimize the production of bioplastics with tailored properties suitable for various applications, including packaging, agriculture, and biomedical devices. Moreover, fungal-derived bioplastics offer the potential to reduce plastic pollution in marine and terrestrial environments, contributing to a circular economy approach to plastic waste management.

Fertilizers represent a third category of value-added products derived from fungal biodegradation processes. Fungi play critical roles in nutrient cycling and soil fertility enhancement through the production of enzymes and organic acids that mobilize essential nutrients from organic residues. Fungal-based composting and SSF technologies convert agricultural wastes, food residues, and manures into nutrient-rich composts and biofertilizers.. The development and commercialization of value-added products from fungal biodegradation require interdisciplinary collaborations across microbiology, biotechnology, engineering, and environmental sciences. Advances in genomic technologies, metabolic engineering, and fermentation process optimization enable researchers to tailor fungal strains for enhanced productivity, substrate specificity, and stress tolerance. Moreover, bioprocess engineering techniques such as bioreactor design, downstream processing, and life cycle assessment (LCA) facilitate the scalable production and market competitiveness of fungal-derived bioproducts. Furthermore, policy support, financial incentives, and regulatory frameworks play crucial roles in accelerating the adoption of fungal biodegradation technologies in industrial and agricultural sectors. Governments and industry stakeholders are increasingly recognizing the importance of sustainable biotechnologies in achieving climate change mitigation, resource conservation, and circular economy objectives. By promoting research and development initiatives that leverage fungal biodegradation for biofuels, bioplastics, and fertilizers, societies can foster a transition towards more sustainable and resilient bio-based economies. In conclusion, fungal biodegradation processes offer promising avenues for transforming organic wastes into value-added products that contribute to sustainable development goals. From renewable biofuels and biodegradable bioplastics to nutrient-rich fertilizers, fungi-based biotechnologies exemplify innovative solutions to global challenges related to energy security, waste management, and environmental sustainability.

SCALING UP FUNGAL BIODEGRADATION: INDUSTRIAL APPLICATIONS AND CHALLENGES

Scaling up fungal biodegradation from laboratory-scale processes to industrial applications involves overcoming various challenges while leveraging the potential of fungi to address pressing environmental and economic needs. Fungal biodegradation technologies, including solid-state fermentation (SSF), submerged fermentation (SmF), and composting, offer versatile solutions for converting organic wastes into valuable bioproducts such as enzymes, biofuels, bioplastics, and fertilizers. However, transitioning these technologies from pilot studies to large-scale production requires addressing key challenges related to process optimization, substrate variability, bioreactor design, downstream processing, economic feasibility, and regulatory compliance. Process optimization is critical for maximizing product yields and operational efficiency in industrial-scale fungal biodegradation. Optimizing these parameters through experimental design, statistical modeling, and process control strategies is essential for achieving consistent performance and scalability across different feedstocks and environmental conditions. Moreover, integrating advanced monitoring and automation technologies enhances process reliability, reduces variability, and facilitates real-time adjustments to optimize resource utilization and minimize waste.

Bioreactor design plays a crucial role in scaling up fungal biodegradation processes to meet industrial production demands. Bioreactors must accommodate high-volume throughput, ensure efficient mass transfer of nutrients and gases, provide optimal conditions for fungal growth and enzyme secretion, and facilitate downstream processing of bioproducts. Continuous advancements in bioreactor engineering, including novel materials, mixing mechanisms, and scalable production platforms, enhance the scalability, productivity, and cost-effectiveness of fungal biodegradation technologies. Downstream processing represents a critical step in industrial-scale fungal biodegradation for recovering and purifying target bioproducts from complex fermentation broths. Techniques such as filtration, centrifugation, precipitation, chromatography, and drying are employed to isolate enzymes, biofuels, bioplastics, and fertilizers from fermentation residues and by-products. Streamlining downstream processing operations through integrated process design, novel separation technologies, and cost-effective recovery strategies is essential for maximizing product recovery yields, minimizing energy consumption, and meeting stringent quality standards for commercial applications. Economic feasibility and competitiveness are key considerations in scaling up fungal biodegradation technologies for industrial deployment. Assessing production costs, market demand, product pricing, and return on investment (ROI) informs decision-making regarding process scalability, technology adoption, and business sustainability.

Regulatory compliance and environmental sustainability are paramount in the industrial deployment of fungal biodegradation processes. Ensuring compliance with local, national, and international regulations governing waste management, biosecurity, genetically modified organisms (GMOs), and product safety is essential for securing permits, approvals, and public acceptance. Implementing robust environmental impact assessments (EIAs), lifecycle assessments (LCAs), and sustainability certifications demonstrates the ecological benefits, resource efficiency, and social responsibility of fungal biodegradation technologies. Engaging with regulatory agencies, industry associations, and community stakeholders fosters transparency, trust, and regulatory alignment in advancing sustainable biotechnological solutions. In conclusion, scaling up fungal biodegradation from laboratory-scale research to industrial applications requires addressing technical, economic, regulatory, and environmental challenges while capitalizing on the versatility and sustainability of fungal biotechnologies. By optimizing process parameters, adapting to substrate variability, innovating bioreactor designs, streamlining downstream processing, ensuring economic feasibility, and complying with regulatory requirements, stakeholders can unlock the full potential of fungal biodegradation for addressing global challenges in waste management, resource conservation, and sustainable development. Collaborative efforts across academia, industry, government, and civil society are essential in advancing scalable and commercially viable fungal biodegradation solutions that contribute to a circular economy and resilient bio-based industries

FUTURE DIRECTIONS: INTEGRATING FUNGAL BIODEGRADATION INTO CIRCULAR AGRICULTURE

Future directions in integrating fungal biodegradation into circular agriculture hold significant promise for advancing sustainable practices in food production, waste management, and environmental stewardship. Fungal biodegradation technologies, including solid-state fermentation (SSF), submerged fermentation (SmF), and fungal-assisted composting, offer innovative solutions to mitigate agricultural waste, enhance soil health, and promote resource efficiency within agricultural systems. One key direction involves leveraging fungi to enhance the valorization of agricultural residues and by-products into high-value bioproducts. complex polymers that fungi can efficiently degrade through enzymatic hydrolysis. By employing fungi like Aspergillus, Trichoderma, and Pleurotus species, researchers can optimize enzyme production pathways and metabolic activities to convert lignocellulosic biomass into biofuels (e.g., ethanol), bioplastics, and bio-based chemicals.

Another promising direction involves exploring the potential of fungi in bioremediation and environmental remediation applications within agricultural contexts. Fungi possess unique metabolic capabilities to degrade organic pollutants, detoxify contaminated soils, and sequester heavy metals through mechanisms like biosorption and enzymatic transformation. Fungal species such as white-rot fungi (e.g., Phanerochaete chrysosporium) are particularly effective in degrading recalcitrant pollutants like polycyclic aromatic hydrocarbons (PAHs) and pesticides. Integrating fungal bioremediation technologies with agroecosystem management strategies can remediate polluted soils, restore ecosystem services, and safeguard human health while promoting sustainable land use practices. Moreover, advancing fungal biodegradation technologies requires interdisciplinary collaborations across microbiology, biotechnology, agronomy, and environmental sciences. Research efforts focused on elucidating fungal metabolic pathways, optimizing fermentation conditions, enhancing enzyme activities, and developing genetically modified fungal strains are critical for advancing the scalability, efficiency, and economic viability of fungal biodegradation processes. Additionally, promoting knowledge exchange, capacity building, and technology transfer among stakeholders—including researchers, farmers, industry partners, and policymakers—facilitates the adoption of sustainable agricultural practices that integrate fungal biotechnologies into circular agriculture frameworks. Addressing challenges such as regulatory compliance, public acceptance, and market access for fungal biodegradation technologies is essential for their widespread adoption and commercialization. Engaging with regulatory agencies to establish clear guidelines, standards, and safety protocols ensures the responsible deployment of genetically modified fungi and bioproducts derived from fungal biodegradation.

CONCLUSION

In conclusion, fungal biodegradation emerges as a pivotal technology in tackling the environmental challenges posed by agricultural wastes, offering innovative solutions that align with principles of sustainability, circular economy, and environmental stewardship. Through processes such as solid-state fermentation (SSF), submerged fermentation (SmF), and fungal-assisted composting, fungi demonstrate remarkable capabilities to degrade complex organic materials like crop residues, manures, and effluents into valuable bioproducts. These bioproducts include biofuels, bioplastics, fertilizers, and enzymes, each contributing to reducing waste volumes, minimizing environmental impacts, and promoting resource efficiency in agricultural systems. The integration of fungal biodegradation into agricultural practices not only addresses the pressing issue of waste management but also supports soil health and fertility enhancement. Fungi play integral roles in nutrient cycling, organic matter decomposition, and soil structure improvement, fostering conditions that support sustainable crop production and resilience to climate change. By harnessing fungal metabolic activities, agricultural stakeholders can optimize the utilization of agricultural residues and by-products, reducing dependency on external inputs such as synthetic fertilizers and fossil fuels while mitigating greenhouse gas emissions associated with organic waste decomposition. Future directions in fungal biodegradation research and application emphasize enhancing process efficiencies, scaling up production capacities, and exploring novel fungal species or genetically modified strains with tailored metabolic pathways. Advances in bioreactor design, bioprocess engineering, and fermentation optimization will enable the commercialization and widespread adoption of fungal biotechnologies in diverse agricultural contexts globally. Challenges such as substrate variability, economic viability, regulatory compliance, and public acceptance underscore the need for continued innovation and strategic investments in research and development. Overcoming these challenges requires concerted efforts to optimize fungal biodegradation processes, develop cost-effective production methods, navigate regulatory landscapes, and engage stakeholders in promoting the benefits of sustainable agricultural practices enabled by fungal biotechnologies. In conclusion, fungal biodegradation represents a transformative approach to addressing agricultural waste management challenges, unlocking opportunities for sustainable resource utilization, environmental conservation, and agricultural productivity enhancement. By harnessing the natural capabilities of fungi to break down organic wastes and convert them into valuable bioproducts, societies can move towards more resilient, regenerative, and environmentally responsible agricultural systems. Embracing fungal biodegradation as a cornerstone of circular agriculture holds promise for fostering sustainable development, achieving food security, and safeguarding natural resources for future generations

REFERENCES

Abhilasha, A., Sreenivasulu, A., & Manimozhi, T. Satheesh kumar P., Selvakumar P. and Singh P.,(2022). "The Model of Smart Sensing Device For Sensitive Nanoclusters Modification in Sensing Properties," 2022 2nd International Conference on Advance Computing and Innovative Technologies in Engineering (ICACITE), Greater Noida, India, 2022, pp. 1043-1047, DOI: 10.1109/ICACITE53722.2022.9823846

Chang, A. J., Fan, J., & Wen, X. (2012). Screening of fungi capable of highly selective degradation of lignin in rice straw. *International Biodeterioration & Biodegradation*, *72*, 26–30. DOI: 10.1016/j.ibiod.2012.04.013

Chavez, R., Bull, P., & Eyzaguirre, J. (2006). The xylanolytic enzyme system from the genus Penicillium. *Journal of Biotechnology*, *123*(4), 413–433. DOI: 10.1016/j.jbiotec.2005.12.036

Gardes, M., & Bruns, T. D. (1993). ITS primers with enhanced specificity for Basidiomycetes—Application to the identification of mycorrhizae and rusts. *Molecular Ecology*, *2*(2), 113–118. DOI: 10.1111/j.1365-294X.1993.tb00005.x

Inderjit, S., Rawat, D., & Foy, C. L. (2004). Multifaceted approach to determine rice straw phytotoxicity. *Canadian Journal of Botany*, *82*(2), 168–176. DOI: 10.1139/b03-137

Jahromi, M. F., Liang, J. B., Rosfarizan, M., Goh, Y. M., Shokryazdan, P., & Ho, Y. W. (2011). Efficiency of rice straw lignocelluloses degradability by Aspergillus terreus ATCC 74135 in solid state fermentation. *African Journal of Biotechnology*, *10*, 4428–4435.

Jain, N., Bhatia, A., & Pathak, H. (2014). Emission of air pollutants from crop residue burning in India. *Aerosol and Air Quality Research*, *14*(1), 422–430. DOI: 10.4209/aaqr.2013.01.0031

Jin, R., Liao, H., Xuanming, L., Zheng, M., Xiong, X., Liu, X., Zhang, L., & Zhu, Y. (2012). Identification and characterization of a fungal strain with lignin and cellulose hydrolysis activities. *African Journal of Microbiological Research*, *6*(36), 6545–6550. DOI: 10.5897/AJMR12.476

Jorgensen, H., & Olsson, L. (2006). Production of cellulases by Penicillium brasilianum IBT 20888: Effect of substrate on hydrolytic performance. *Enzyme and Microbial Technology*, *38*(3-4), 381–390. DOI: 10.1016/j.enzmictec.2005.06.018

Kanabkaew, T., & Oanh, N. T. K. (2011). Development of spatial and temporal emission inventory for crop residue field burning. *Environmental Modeling and Assessment*, *16*(5), 453–464. DOI: 10.1007/s10666-010-9244-0

Kanipriya, M., Sugumaran, D., Pandiyarajan, V., Yogananda, B. S., Selvakumar, P., & Mohan, S. (2024, January). Analysis of Electromyography Signals for Control of Mechanical Prosthesis using Machine Learning Techniques. In 2024 2nd International Conference on Intelligent Data Communication Technologies and Internet of Things (IDCIoT) (pp. 1059-1067). IEEE.

Kasana, R. C., Richa, S., Dhar, H., Dutt, S., & Gulati, A. (2008). A rapid and easy method for the detection of microbial cellulases on agar plates using Gram's Iodine. *Current Microbiology*, *57*(5), 503–507. DOI: 10.1007/s00284-008-9276-8

Kaur, A., & Sahota, P. P. (2004). Studies on the role of fungal inocula in composting of wheat straw. *Asian J Microbiol Biotechnol Environ Exp Sci*, *6*, 671–673.

Kausar, H., Sariah, M., Saud, H. M., Alam, M. Z., & Ismail, M. R. (2010). Development of compatible lignocellulolytic fungal consortium for rapid composting of rice straw. *International Biodeterioration & Biodegradation*, *64*(7), 594–600. DOI: 10.1016/j.ibiod.2010.06.012

Liao, H., Xu, C., Tan, S., Wei, Z., Ling, N., Yu, G., Raza, W., Zhang, R., Shen, Q., & Xu, Y. (2012). Production and characterization of acidophilic xylanolytic enzymes from Penicillium oxalicum GZ-2. *Bioresource Technology*, *123*, 117–124. DOI: 10.1016/j.biortech.2012.07.051

Lundell, T. K., Makela, M. R., & Hilden, K. (2010). Lignin-modifying enzymes in filamentous basidiomycetes—Ecological, functional and phylogenetic review. *Journal of Basic Microbiology*, *50*(1), 5–20. DOI: 10.1002/jobm.200900338

Mahadev Madgule, N. (2023). Vinayaka, Yeshwant M. Sonkhaskar, Dhiren Ramanbhai Patel, R. Karthikeyan, P. Selvakumar,(2023). Mechanical properties and microstructure of activated TIG welded similar joints of Inconel alloys by desirability approaches. *Materials Today: Proceedings*, *77*, 528–533. DOI: 10.1016/j.matpr.2022.12.250

Maza, M., Pajot, H. F., Amoroso, M. J., & Yasem, M. G. (2014). Post-harvest sugarcane residue degradation by autochthonous fungi. *International Biodeterioration & Biodegradation*, *87*, 18–25. DOI: 10.1016/j.ibiod.2013.10.020

Milagres, A. M. F., Lacis, L. S., & Prade, R. A. (1993). Characterization of xylanase production by a local isolate of Penicillium janthinellum. *Enzyme and Microbial Technology*, *15*(3), 248–253. DOI: 10.1016/0141-0229(93)90145-R

Ogbonna, A. I., Nwadiaro, P. O., Chuku, A., Ogbonna, C. I. C., & Onwuliri, F. C. (2015). Cellulase induction in three Aspergillus species isolated from Artemisia annua L. plantation soil using different cellulose substrates. *J Appl Biotechnol*, *3*(1), 63–74. DOI: 10.5296/jab.v3i1.6738

Ramakrishnan, T., Mohan Gift, M. D., Chitradevi, S., Jegan, R., Hency Jose, P. S., Nagaraja, H. N., & Hailegiorgis, S. M. (2022). Study of numerous resins used in polymer matrix composite materials. *Advances in Materials Science and Engineering*, *2022*(1), 1088926.

Reese, E. T., & Mandels, M. (1963). Enzymatic hydrolysis of cellulose and its derivatives. In Whistler, R. L. (Ed.), *Methods carbohydrate chemistry* (3rd ed., pp. 139–143). Academic Press.

Sandhu, D. K., & Arora, D. S. (1985). Laccase production by Polyporus sanguineus under different nutritional and environmental conditions. *Experientia*, *41*(3), 355–356. DOI: 10.1007/BF02004501

Selvakumar, P. (2024). Climate Change. In Sofia Idris, I. G. I. (Ed.), *Global Challenges for the Environment and Climate Change* (pp. 100–110). Global., DOI: 10.4018/979-8-3693-2845-3.ch006

Selvakumar, P. (2024). Reasons for Biodiversity Loss. In Ashok Kumar Rathoure, I. G. I. (Ed.), *Impact of Societal Development and Infrastructure on Biodiversity Decline* (pp. 37–49). Global., DOI: 10.4018/979-8-3693-6950-0.ch004

Selvakumar, P.. (2024). The Role of Corrosion Inhibitors. In Thakur, A., & Kumar, A. (Eds.), *Innovations in Nanomaterials-Based Corrosion Inhibitors* (pp. 130–141). IGI Global., DOI: 10.4018/979-8-3693-3088-3.ch006

Selvakumar, P., Mariappan, P., Shunmugam, S., Paul, R., & Jabez, S. (2024). Removal of chemicals from effluent using photobioreactor technology to improve environmental and health impacts. *Zeitschrift für Physikalische Chemie*, *2024*(7), 1295–1302. Advance online publication. DOI: 10.1515/zpch-2023-0450

Selvakumar, P., Palanisamy, S. K., Cinthaikinian, S., Palanisamy, V., Mariappan, R., & Selvakumar, P. K. (2024). Biosensors and its diverse applications in healthcare systems. *Zeitschrift für Physikalische Chemie*, *2024*. Advance online publication. DOI: 10.1515/zpch-2023-0406

Sharma, A., Sharma, R., Arora, A., Shah, R., Singh, A., Pranaw, K., & Nain, L. (2014). Insights into rapid composting of paddy straw augmented with efficient microorganism consortium. *International Journal of Recycling of Organic Waste in Agriculture*, *3*(2), 54. DOI: 10.1007/s40093-014-0054-2

Terrasan, C. R., Temer, B., Duarte, M. C., & Carmona, E. C. (2010). Production of xylanolytic enzymes by Penicillium janczewskii. *Bioresource Technology*, *101*(11), 4139–4143. DOI: 10.1016/j.biortech.2010.01.011

Tuomela, M., Vikman, M., Hatakka, A., & Itavaava, M. (2000). Biodegradation of lignin in a compost environment: A review. *Bioresource Technology*, *72*(2), 169–183. DOI: 10.1016/S0960-8524(99)00104-2

Vijayakumar, G., Rajkumar, M., Rajiv Chandar, N., Selvakumar, P., & Duraisamy, R. (2024). Muniyandi Rajkumar, Rajiv Chandar N, Selvakumar P and Ramesh Duraisamy, (2024). Environmentally friendly TDS removal from waste water by electrochemical ion exchange batch-type recirculation (EIR) technique. *Environmental Science. Water Research & Technology*, *10*(4), 826–835. DOI: 10.1039/D3EW00793F

Ward, O. P., Qin, W. M., Dhanjoon, J., Ye, J., & Singh, A. (2005). Physiology and biotechnology of Aspergillus. *Advances in Applied Microbiology*, *58*, 1–75. DOI: 10.1016/S0065-2164(05)58001-8

Yin, Z. W., Fan, B. Q., & Ren, P. (2011). Isolation and identification of a cellulose degrading fungus Y5 and its capability of degrading wheat straw. *Huan Jing Ke Xue*, *32*, 247–252.

Zeng, G. M., Yu, H. Y., Huang, H. L., Chen, Y. N., & Huang, G. H. (2006). Laccase activities of a soil fungus Penicilluim simplicissimum in relation to lignin degradation. *World Journal of Microbiology & Biotechnology*, *22*(4), 317–324. DOI: 10.1007/s11274-005-9025-0

Zhu, N. (2007). Effect of low initial C/N ratio on aerobic composting of swine manure with rice straw. *Bioresource Technology*, *98*(1), 9–13. DOI: 10.1016/j.biortech.2005.12.003

Chapter 13
Mycoremediation:
Leveraging Fungi for Ameliorating Soil and Water Pollution

Licon Kumar Acharya
https://orcid.org/0000-0003-3908-737X
National Research Centre for Integrated Pest Management, India

A. U. Akash
https://orcid.org/0000-0001-6694-1047
ICAR-Indian Agricultural Research Institute, India

Suryakant Manik
https://orcid.org/0009-0000-4415-0412
ICAR-Indian Institute of Agricultural Biotechnology, India

Amjada S. Khan
https://orcid.org/0009-0001-8401-9117
ICAR-Indian Agricultural Research Institute, India

Neha Singh
https://orcid.org/0009-0002-3052-9119
ICAR-Indian Agricultural Research Institute, India

ABSTRACT

The burgeoning population and increased demands for resources across the world have led to the industrial and agricultural revolution. As a result of anthropogenic activities, environmental pollutants are exacerbating day by day in nature leading to soil and water contamination. Mycoremediation, which uses fungi or their derivatives to ameliorate the pollutants has shown immense promise in the recent past. In addition, it is advantageous over other prevalent traditional methods due to its comparative cost-effectiveness, eco-friendliness, high efficiency, and its innate potential to degrade a wide range of contaminants across diverse environmental conditions. In this chapter, we explore the potential of fungi as effective agents for ameliorating soil and water pollution. We focus on the major classes of fungi that can be employed, the mechanisms involved, their applications, and real-world examples. Additionally, we address the major bottlenecks associated with mycoremediation and future directions to enhance

DOI: 10.4018/979-8-3693-8201-1.ch013

this strategy.

1. INTRODUCTION

Surge in population and incremental developmental activities propelled by modern technologies have altered the natural flow of materials thereby leading to the accumulation of hazardous contaminants in nature. Pollutants like pesticides and other agrochemical wastes, heavy metals (Cd, Cr, Zn, Ag, etc.), synthetic dyes, plastics, chlorinated compounds, pharmaceutical compounds, and several other organic pollutants enter the natural ecosystem leading to soil and water pollution (Kumar *et al.*, 2022). Pollutants like heavy metals are usually not destroyed easily and undergo several chemical transformations to remain in soil for thousands of years (Dinakarkumar *et al.*, 2024). Lack of pragmatic waste management solutions and the non-existence of strict guidelines have aggravated the problems leading to various human health hazards. The global scientific community is toiling hard to come up with an eco-friendly yet practical technology to tackle the menace.

Thereby, bioremediation has become a buzzworthy topic in the near past. It involves the use of biological systems to treat environmental contaminants and has evolved significantly since its formal application in the 1970s (Chapelle, 1999; Shukla *et al.*, 2010). In this process, microorganisms are exploited by various research groups for the decontamination of pollutants from soil and water due to their inherent ability to absorb contaminants and degrade or transform them into less toxic forms. Initially focused on petroleum hydrocarbon-contaminated groundwater, early techniques involved adding oxidants and nutrients to enhance microbial degradation (Chapelle, 1999). While bioremediation has proven efficient and cost-effective for certain pollutants, challenges remain due to substrate and environmental variability, limited biodegradation potential, and microbial viability (Garzón *et al.*, 2017).

Bioremediation can be classified into two main types: *in situ* and *ex situ* (Rajput *et al.*, 2025). *In situ* bioremediation treats contamination at the site, while *ex situ* involves removing contaminated material for treatment elsewhere. Various bioremediation methods include phytoremediation, bacterial remediation, rhizoremediation, phycoremediation, and mycoremediation (Gothalwal, 2020). While phytoremediation is effective for heavy metal sequestration, it is a slow process and limited to plant-root interactions. Bacterial bioremediation, though useful, requires specific environmental conditions and is less effective in degrading persistent organic pollutants. In contrast, mycoremediation leverages fungi's extensive hyphal networks, enabling deeper soil penetration and enzymatic breakdown of pollutants beyond bacterial capacities.

Mycoremediation is a subset of bioremediation where fungi are used to degrade or neutralize toxic pollutants in the environment to non-toxic or less toxic forms through biochemical reactions (Akpasi *et al.*, 2023). The idea of "mycoremediation" comes from its literal meaning *myco* (fungus) and *remediation* (to clean, reconstruct, or resolve) which combines the principles of both. Various fungi like white and brown rot fungi, mushrooms, endophytes, mycorrhiza, etc. are being used for ameliorating the pollutants from nature. The chemoheterotrophic mode of nutrition in fungi enables them to target a wide range of organic and inorganic chemicals and the harmful compounds are removed either by converting them as cellular building blocks or as a source of energy without releasing any toxic byproducts (Zaman *et al.*, 2023). Hence, it is a sustainable approach to detoxifying contaminated soil and water with reduced chemical, resource, and time inputs (Dinakarkumar *et al.*, 2024).

Figure 1. Mycoremediation: Leveraging fungi for ameliorating soil and water pollution.

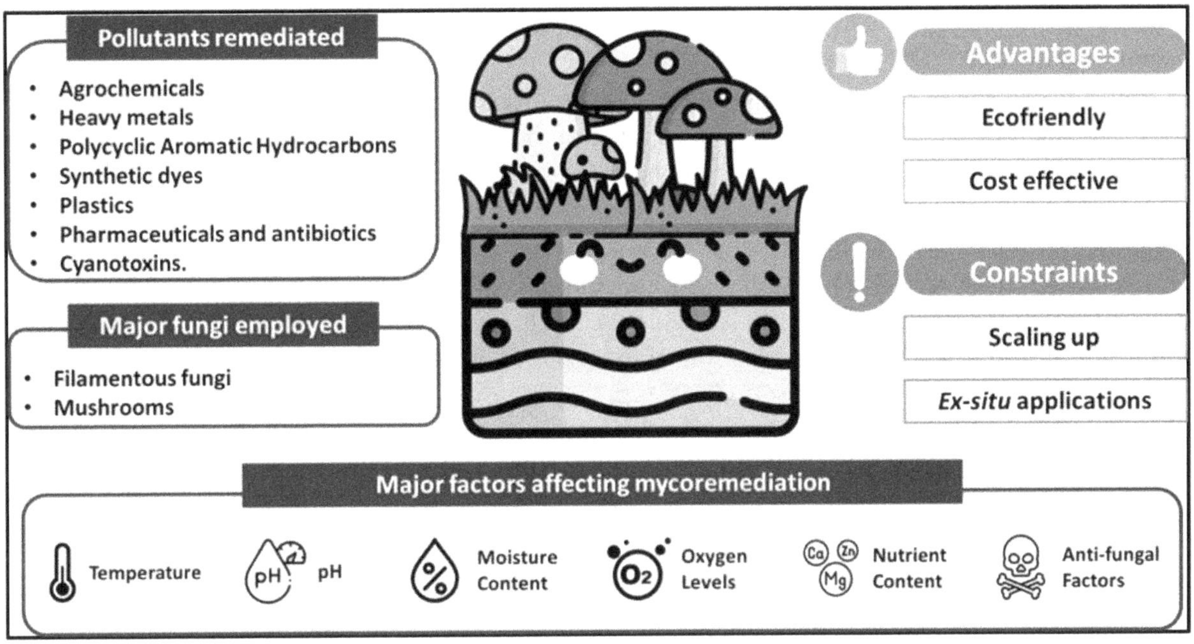

2. ENVIRONMENTAL POLLUTION: CURRENT STATUS

Environmental pollution arising from human interventions, is a critical concern, impacting the well-being of the global population and ecosystem with greater implications. It is recognized as one of the triple planetary crises faced by the United Nations Framework Convention on Climate Change (UNEP, 2023). Industrial and agricultural activities are key drivers of land and water resource contamination and degradation by hazardous chemicals. (Mir *et al.*, 2023; Ukaogo *et al.*, 2020; Sharma *et al.*, 2023). These pollutants encompass a range of forms and structures, such as heavy metals, petroleum hydrocarbons, industrial dyes, pharmaceutical compounds, persistent organic pollutants, plastics, radioactive substances, and numerous other hazardous chemicals (Singh, 2024; Egwim *et al.*, 2023).

The improper disposal of effluents and various wastes directly into water and soil has led to an increase in the concentration of many harmful metals, chemicals, and gases in the environment. Industries such as fertilizer production, metallurgy, electroplating, and electronics frequently release effluents containing heavy metals, including Lead (Pb), Chromium (Cr), Mercury (Hg), Cadmium (Cd), Copper (Cu), Arsenic (As), and Selenium (Se), which contribute to pollution (Tiwari *et al.*, 2023). The effluent water from paper-making industries not only releases heavy metals but also toxic volatile organic compounds. Most of these heavy metals, even in low concentrations, are toxic and fatal to living beings (Gupta *et al.*, 2023). Persistent organic pollutants such as dioxins, insecticides, and polychlorinated biphenyls (PCBs) are another set of pollutants that can endure natural degradation and remain in the environment for extended periods (Devendrapandi *et al.*, 2024). Polycyclic aromatic hydrocarbons (PAHs), produced mainly from incomplete organic combustion, are widespread environmental pollutants known for their carcinogenic, genotoxic, mutagenic, and cytotoxic properties. Common PAHs such as naphthalene, phenanthrene, anthracene, and benzo(a)pyrene are getting exposed to the environment by incomplete combustion of motor fuels, forest fires, volcanic eruptions, oil spills, and petroleum storage tank leaks (Samanta *et al.*, 2002; Thacharodi *et al.*, 2023).

The unmanaged discharge of plastics in the forms of capsules, microbeads, fibers, or microplastics is a major cause of soil and water pollution. Because of their non-degradability, persistence, and recalcitrant nature plastic pollution is giving severe health and ecological pressures to the different life forms (Anani and Adetunji, 2021). The release of synthetic dyes, detergents, dissolved solids, and agricultural chemicals from industries and fields is responsible for hazardous levels of water pollution. Globally, the centralized development of industry and agriculture has led to severe pollution of groundwater. Contaminants infiltrate through landfill sites, sewage discharge, waste accumulation, fertilizers, organic pollutants, and oil leaks, posing a serious threat to groundwater resources (Stuart *et al.*, 2012).

3. MYCOREMEDIATION: AN EFFICIENT FUNGAL-BASED SUSTAINABLE BIOREMEDIATION APPROACH

The chemoheterotrophic mode of nutrition in fungi enables them to target a wide range of organic and inorganic chemicals (Kulshreshtha *et al.*, 2014; Akpasi *et al.*, 2023, Zaman *et al.*, 2023). These harmful compounds are removed either by converting them as cellular building blocks or as a source of energy without releasing any toxic byproducts. Thus, mycoremediation can be recommended as a sustainable approach to detoxifying contaminated soil and environment with reduced chemical, resource, and time inputs (Dinakarkumar *et al.*, 2024).

There is an enormous diversity of fungal species at the global level but only a few species play an important role in bioremediation. Wood rot fungi (Brown-rot fungi or White-rot fungi) are the most dominant group among them, that plays a central role in mycoremediation. The first reported fungus capable of bioremediation was the white rot fungi, *Phanerochaete chrysosporium,* and is currently the model system for studying mycoremediation (Bumpus and Aust, 1987). Of all the literature on mycoremediation, approximately 30% is concerned with white rot fungi (Singh, 2006). Their simultaneous oxidative and reductive mechanisms allow fungi to be used in various situations, depending on the type and extent of pollution, oxygen availability, and the nature of the site itself (Kulshreshtha *et al.*, 2014; Rhodes, 2014). Several other fungi such as *Pleurotus, Bjerkandera, Trametes, Irpex, Lentinula, Agaricus, Cryptococcus, Trichoderma, Rhizopus, Curvularia, Aspergillus, Fusarium,* etc are also found to be promising to break down persistent xenobiotic chemicals (Singh, 2006; Adenipekun and Lawal, 2012; Rhodes, 2014; Akpasi *et al.*, 2023).

Endophytes are the microbes, predominantly bacteria and fungi that colonize the plant tissues without harming the host plant. More recently both bacterial and fungal endophytes have been exploited for bioremediation of pesticides, heavy metals, organic pollutants, and other deadly pollutants (Mello *et al.*, 2020). Fungal endophytes like *Pestalotiopsis microspore*, *Aspergillus niger,* and *Geomyces pannorum* were found effectively degrading some of the polyurethane (Russell *et al.*, 2011). *Drechslera* sp. a fungal endophyte isolated from a grass i.e., *Neurachne alopecuroidea* has shown promise for the removal of methyl tertiary-butyl ether contamination from soil (d'Errico *et al.*, 2021).

Soil-dwelling fungi such as *Trichoderma, Rhizopus, Penicillium, Paecilomyces, Aspergillus niger, Alternaria,* etc. are the potent mycoremediation agents of a variety of soil and water pollutants such as persistent pesticides, polycyclic aromatic hydrocarbons (PAHs), polychlorinated biphenyls (PCBs) and plastic pollutants like polyethylene, polystyrene, polypropylene, polyvinyl chloride etc. (Srikanth *et al.*, 2022; Khatua *et al.*, 2024).

Mycorrhizal fungi are used as an eco-friendly alternative for amelioration of various metal toxicity found in soil. Heavy metals are absorbed and retained by mycorrhizal structures like mycelium and vacuoles, thereby preventing their mobilization to higher plant parts. *Funneliformis geosporum*, a mycorrhiza fungus associated with the wheat crop was instrumental in increasing the growth and yield of the wheat crop in zinc contaminated soil (Abu-Elsaoud *et al.*, 2017).

Bioremediation mediated by fungi is advantageous over bacteria which require prior exposure to specific pollutants to induce the necessary enzymes for degradation. Additionally, bacterial enzyme expression is limited by pollutant concentration thresholds, restricting the applicability of bacterial bioremediation (Rhodes, 2014).

4. MECHANISM OF MYCOREMEDIATION

Fungi, with their saprophytic characteristics and strong enzymatic systems have immense potential to break down environmental contaminants and assist in the rehabilitation of ecosystems (Deshmukh *et al.*, 2016). Several fungi, mainly Ascomycota species like *Alternaria*, *Aspergillus*, *Cladosporium*, *Fusarium,* and *Penicillium* are also involved in degrading plastic pollutants such as polyethylene, polystyrene, polypropylene, and various biodegradable plastics (Srikanth *et al.*, 2022; Khatua *et al.*, 2024). Their intricate cell wall architecture aids in metal binding, and they secrete oxidoreductase enzymes to break down harmful substances (Dutta *et al.*, 2023; Mudaliar *et al.*, 2023). Fungal mycelia can also function as a mycofilter, aiming at specific organic and inorganic pollutants and microbial contaminants (Mnkandla and Otomo, 2021). Mycoremediation has been successfully used to address oil spills in soil and water, promoting enzymatic activity of microorganisms (Mudaliar *et al.*, 2023). Fungal bioremediation utilizes various strategies to alleviate contaminants, and their flexibility allows them to thrive in extreme environmental conditions (Picciotti *et al.*, 2023). They can neutralize or withstand heavy metals through biosorption, biotransformation, immobilization, and mobilization, making them powerful agents for environmental cleanup (Anjitha et al., 2021; Kulshreshtha *et al.*, 2014; Dinakarkumar *et al.*, 2024). The detailed mechanism of mycoremediation of environmental pollutants is given below.

4.1. Avoidance

Fungi have shown great potential in the adsorption of heavy metals from contaminated environments and avoid their further contamination by precipitating these pollutants within their cell wall. One notable example is the use of the fungus *Aspergillus niger* for the biosorption of heavy metals such as cadmium, lead, and zinc. This fungus has been found to effectively remove these metals from aqueous solutions due to its high metal-binding capacity (Abdel-Wareth, 2023).

4.2. Chelation and precipitation

Phosphate-solubilizing fungi, such as *Aspergillus niger*, play a critical role in the bioremediation of heavy metals by releasing organic acids like citric and oxalic acids, which are essential for the chelation and precipitation of these contaminants. Citric acid, in particular, is highly effective in forming stable complexes with metals such as lead, cadmium, and aluminum, thus reducing their bioavailability and toxicity in the environment (Khan *et al.*, 2019). *Aspergillus niger* is capable of producing up to 15 g/L of citric acid, which efficiently chelates metals like cadmium, facilitating their immobilization and removal from contaminated sites. This process not only reduces the harmful effects of heavy metals but also aids in their safe sequestration, preventing further environmental contamination.

In addition to citric acid, the secretion of oxalic acid by fungi further enhances their ability to remediate polluted environments. Oxalic acid promotes the formation of insoluble metal oxalates, which can be easily extracted from soils and water through sedimentation or filtration. This precipitation process transforms soluble metal ions into solid metal oxalates, making them more amenable to removal. Together, the production of citric and oxalic acids by phosphate-solubilizing fungi represents an effective dual strategy for both immobilizing and removing heavy metals from contaminated environments, further advancing the potential of mycoremediation in environmental cleanup efforts.

4.3. Biosorption and bioaccumulation

Biosorption is a key mechanism employed by fungi in the bioremediation of heavy metals, playing a crucial role in the removal of pollutants from contaminated environments. Fungal cell walls, rich in functional groups such as carboxyl, hydroxyl, and amine groups, facilitate the binding of metal ions through processes like ionic exchange, complexation, and chelation. This enables fungi to effectively capture and immobilize toxic metals. For instance, *Agaricus bisporus* and *Lactarius piperatus* are known for their ability to remove cadmium (Cd II) ions from contaminated environments (Nagy *et al.,* 2014). Furthermore, *Pleurotus sajor-caju* is known for its capacity to absorb various heavy metals (Jibran and Mol, 2011). By adsorbing metals onto their cell surfaces, fungi significantly reduce the mobility and bioavailability of these metals, making them less harmful to the surrounding ecosystem.

In addition to biosorption, fungi also employ bioaccumulation, a process where metals are taken up and stored internally within the fungal cells. This involves the sequestration of metals in vacuoles or their binding to intracellular proteins, further reducing their bioavailability and toxicity in the environment. Together, biosorption and bioaccumulation represent complementary strategies that fungi utilize to mitigate heavy metal pollution, reducing the environmental impact of these toxic substances.

4.4. Biodegradation

Biodegradation is the natural process through which organic substances in the environment are broken down by living organisms, recycling waste materials into essential nutrients without harming the ecosystem. During this process, complex compounds are mineralized into simpler, non-toxic forms such as CO_2, H_2O, NO_3, and other inorganic substances. Macrofungi, such as mushrooms, play a key role in this process by releasing a variety of extracellular enzymes that aid in the breakdown of complex organic pollutants (Nyanhongo *et al.*, 2007). These fungi are particularly effective in degrading non-polymeric, resistant pollutants, including both organic and synthetic dyes, nitrotoluenes, polycyclic aromatic hydrocarbons (PAHs), and pentachlorophenol (Haritash and Kaushik, 2009). Some of the examples of degradation by using enzymes are given below:

1. **Lignin Peroxidases**, produced by fungi such as *Phanerochaete chrysosporium*, *Pleurotus pulmonarius,* and *Trametes versicolor*, act on compounds including pyrene, bentazon (herbicide), aminodinitrotoluenes and polyaromatic hydrocarbons (PAHs). These enzymes catalyze single-electron oxidation reactions on aromatic compounds, driven by H_2O_2 (Arora and Sharma, 2010).
2. **Lipase**, found in various fungi including *Trichoderma*, *Curvularia*, *Drechslera,* and *Aspergillus* species, degrades PAHs and other organic pollutants. It works via single-electron oxidation reactions in the presence of H_2O_2 (Solanki *et al.*, 2019).
3. **Catalase**, derived from fungi such as *Curvularia* and *Aspergillus*, targets heavy metals through similar H_2O_2-driven single-electron oxidation mechanisms (Bosco and Mollea, 2019).
4. **Laccase**, produced by fungi like *Bjerkandera adusta*, *Phanerochaete chrysosporium*, and *Pleurotus pulmonarius*, acts on pesticides and other pollutants. It utilizes oxygen to oxidize organic compounds (Mouhamadou *et al.*, 2013).
5. **Peroxidase**, found in fungi such as *Geotrichum candidum*, *Aspergillus niger,* and *Trichoderma sp.*, is involved in the decolorization of textile dyes. Its activity is driven by H_2O_2, leading to single-electron oxidation reactions that degrade pollutants (Jakovljević and Vrvić, 2018).

4.5. Bioconversion/Biotransformation

Bioconversion transforms lignocellulosic materials, such as agricultural and forestry residues, into valuable bioproducts. These lignocellulosic by-products, which include materials like woody biomass, grass, and municipal solid waste, offer substantial potential for bioconversion due to their abundance and chemical composition. Various fungal species play a key role in this process by breaking down complex lignocellulosic structures. For instance, *Pleurotus ostreatus* var. *florida* has been shown to thrive on solid sludges and effluents from industries such as cardboard and handmade paper production, demonstrating its ability to bio-convert industrial waste into valuable bioproducts (Kulshreshtha *et al.*, 2014). Similarly, *Lentinula edodes* has been reported to grow on a variety of agricultural by-products, including oakwood sawdust, wheat straw, and corn cobs, further highlighting the role of fungi in utilizing lignocellulosic waste (Israilides and Philippoussis, 2003).

Additionally, several other fungal species, including *Macrocybe titans*, *Pleurotus ostreatus*, *Grifola frondosa,* and *Ganoderma lucidum*, have demonstrated the ability to convert passion fruit waste into β-glucosides, showcasing their capacity for biotransformation and the potential for utilizing fruit waste in industrial applications (Zilly *et al.*, 2012). These examples illustrate the promising applications of fungal bioconversion in both waste management and the production of high-value bioproducts from lignocellulosic materials.

It involves three processes:

1. Pretreatment – This process improves hydrolysis, boosts product yield, and makes cellulose more accessible to cellulolytic enzymes.:
   (a) chemical (acid/alkali);
   (b) physical/physicochemical (physicochemical steam explosion); and
   (c) biological pretreatment (living microorganisms).

2. Hydrolysis – This process breaks down cellulose and hemicellulose into monomeric soluble sugars, including hexoses and pentoses.
3. Fermentation – In fermentation, hydrolytic products are converted into valuable products like ethanol. *P. ostreatus*, *Agaricus blazei,* and *Flammulina velutipes* are used in wine production and show strong potential to create wine and related products. These mushrooms are also believed to offer protective or therapeutic benefits against cancer and thrombosis (Okamura *et al.*, 2001).

Figure 2. Different mechanisms involved in fungal mycoremediation

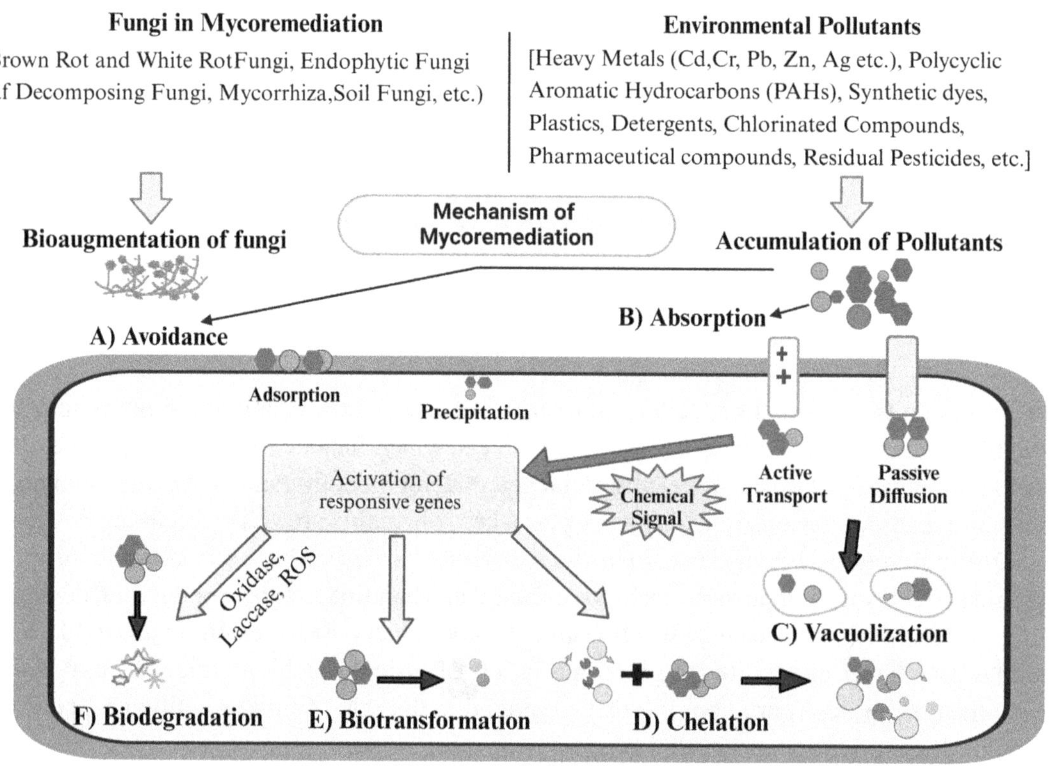

5. APPLICATIONS IN AMELIORATING SOIL AND WATER POLLUTION

Soil and water face serious degradation threats from various pollutants, including pesticides, heavy metals, dyes, polycyclic aromatic hydrocarbons, pharmaceuticals, and antibiotics, originating from diverse sources, particularly industries and agricultural fields. Soil and water pollution pose severe risks to both human health and ecosystems, with pollution-linked illnesses estimated to have caused 9 million premature deaths in 2015 - over three times the combined fatalities from malaria, AIDS, and tuberculosis (Landrigan *et al.*, 2018). Therefore, combating soil and water pollution has become an inevitable necessity to sustain the environment.

Various approaches are employed for the remediation of potentially toxic elements (PTE), including physical, chemical, and physico-chemical techniques. Bioremediation, which employs biologically mediated processes to transform, degrade, sequester, or completely remove PTE from the ecosystem, has proven to be promising. Bioremediation is either plant-mediated (phytoremediation), employing plants or their products, or microbe-mediated, employing microorganisms or their derivatives. Fungi-mediated mycoremediation emerges as an economical, eco-friendly, and effective strategy to combat soil and water pollution (Vaksmaa *et al.*, 2023). Fungi are an ideal candidate for bioremediation of various pollutants owing to their high growth rate, extensive hyphal network, vast arsenal of extracellular enzymes, high surface-area-to-volume ratio, heavy metal accumulation capacity, and ability to adapt to fluctuating and extreme environmental variables like pH and temperature (Akhtar and Mannan, 2020; Dinakarkumar *et al.*, 2024).

A diverse range of fungi, including biocontrol agents, mushrooms, and marine fungi, have been reported to possess remediation properties for various soil and water pollutants (Table 1). These pollutants include pesticides, heavy metals, dyes, Polycyclic aromatic hydrocarbons, Pharmaceuticals and antibiotics, and others. Fungi can be directly applied in bioremediation processes, where their natural growth and metabolic activities help degrade or sequester pollutants in soil and water, whether through *in-situ* or *ex-situ* remediation (Vaksmaa *et al.*, 2023; Dinakarkumar *et al.*, 2024). Apart from bioaugmentation approaches, advances in molecular biology, biotechnology, and enzymology have driven the development of engineered fungi and enzymes specifically for mycoremediation. A common strategy involves genetically enhancing native fungal strains by introducing new metabolic pathways to boost enzyme production and bioremediation potential (Yang *et al.*, 2024).

In addition to using whole fungi, fungal metabolites—such as organic acids, extracellular enzymes, and biosurfactants—can also be extracted and applied to target specific contaminants (Singh *et al.*, 2021). Further, fungi can be integrated into other techniques to improve the efficacy of the remediation process like biocomposites (Nagy *et al.*, 2014), biosorbent surface (Alam *et al.*, 2018), microbial fuel cells (Gorin *et al.*, 2024), among others. Another approach involves combining nanotechnology with mycoremediation to create fungi-based nanoremediation. Mycogenic nanoparticles, synthesized by fungi, enhance mycoremediation by facilitating the degradation of organic pollutants, immobilizing heavy metals, reducing toxicity and bioavailability of contaminants, increasing fungal resilience against environmental stresses, and working synergistically with fungal enzymes to improve overall remediation efficiency (Rai *et al.*, 2018; Manjula *et al.*, 2023).

Table 1. Fungal species employed for bioremediation of some major pollutants of soil and water.

Pollutant type/ source	Compound/class	Fungi	Reference
Agrochemicals/ Agriculture: soils and water	Heptachlor/ organochlorine	*Phlebia tremellosa, Phlebia brevispora, Phlebia acanthocystis*	Xiao *et al.*, 2011
		Pleurotus ostreatus	Purnomo *et al.*, 2014; Sadiq *et al.*, 2015
	Lindane/ organochlorine	*Phlebia brevispora, Phlebia lindtneri*	Xiao and Kondo, 2020
	Dicofol/ organochlorine	*Trametes versicolor*	Hu *et al.*, 2020
	Endosulfan/ organochlorine	*Pleurotus eryngii, Coprinus comatus*	Wang *et al.*, 2018
		Aspergillus niger, Aspergillus flavus, Penicillium chrysogenum	Ahmad, 2020
	DDT/ Organochlorine	*Trichoderma hamatum, Rhizopus arrhizus*	Russo *et al.*, 2019
		Aspergillus niger	Ebsa *et al.*, 2024
	Chlorpyriphos/ organophosphate	*Fusarium oxysporum, Teleromyces verruculosus, Albifimbria terrestris, Aspergillus terreus, Purpureocillium lilacinum*	Hasan *et al.*, 2024
		Trametes versicolor	Hu *et al.*, 2020
		Aspergillus sydowii	Soares *et al.*, 2021
	Methyl parathion/ organophosphate	*Aspergillus sydowii*	Soares *et al.*, 2021
	Profenofos/ organophosphate	*Aspergillus sydowii*	Soares *et al.*, 2021
		Trichoderma harzianum, Metarhizium anisopliae	Abd El-Ghany and Masmali, 2016
	Malathion/ organophosphate	*Trichoderma harzianum, Metarhizium anisopliae*	Abd El-Ghany and Masmali, 2016
	Diazinon/ organophosphate	*Trichoderma harzianum, Metarhizium anisopliae*	Abd El-Ghany and Masmali, 2016
	Cypermethrin/ pyrethroids	*Trametes versicolor*	Hu *et al.*, 2020
	Lambda-cyhalothrin/ Pyrethroids	*Aspergillus* sp. *Acremonium* sp. *Microsphaeropsis* sp. *Westerdykella* sp.	Birolli *et al.*, 2018
	Allethrin/ pyrethroids	*Fusarium proliferatum*	Bhatt *et al.*, 2020
	Atrazine/ Herbicide	*Aspergillus fumigatus, Fusarium verticillioides, Penicillium citrinum*	Bravim *et al.*, 2021
	Glyphosate/ herbicide	*Aspergillus fumigatus, Fusarium verticillioides, Penicillium citrinum*	Bravim *et al.*, 2021
	Pendimethalin/ herbicide	*Aspergillus fumigatus, Fusarium verticillioides, Penicillium citrinum*	Bravim *et al.*, 2021

continued on following page

Table 1. Continued

Pollutant type/ source	Compound/class	Fungi	Reference
Heavy metals/ Industrial effluents	Arsenic	*Aspergillus clavatus, Aspergillus niger, Trichoderma viride, Penicillium glabrum*	Urík *et al.*, 2007
		Trichoderma lixii	Kumar *et al.*, 2022
		Aspergillus ustus	Kamal *et al.*, 2023
	Cadmium	*Aspergillus niger, Aspergillus fumigatus*	Khan *et al.*, 2019
		Pleurotus ostreatus, Pleurotus cornucopiae	Xu *et al.*, 2021
		Trichoderma asperellum	Bandurska *et al.*, 2021
		Pleurotus eryngii	Goligar *et al.*, 2023
	Chromium	*Aspergillus tamarii*	Kamal *et al.*, 2023
		Aspergillus niger, Aspergillus fumigatus	Khan *et al.*, 2019
	Lead	*Trichoderma asperellum*	Bandurska *et al.*, 2021
		Pleurotus eryngii	Goligar *et al.*, 2023
	Nickel	*Trichoderma lixii*	Kumar *et al.*, 2022
	Copper	*Aspergillus hiratsukae, Aspergillus terreus*	Palanivel *et al.*, 2023
	Zinc	*Funneliformis geosporum*	Abu-Elsaoud *et al.*, 2017
	Mercury	*Cladosporium sp., Didymella glomerata, Fusarium oxysporum, Phoma costaricensis, Sarocladium kiliense*	Văcar *et al.*, 2021
Polycyclic aromatic hydrocarbons/ combustion, industrial sources, oil spills, etc.	Fluoranthene, anthracene, Phenanthrene	*Bjerkandera adusta, Schizophyllum commune, Pleurotus ostreatus, Trametes versicolor, Pleurotus ostreatus, Stropharia rugosoannulata, Agaricus bisporus*	Pozdnyakova *et al.*, 2018
	Benzo[α]pyrene	*Lasiodiplodia theobromae*	Cao *et al.*, 2020
	Naphthalene	*Fusarium proliferatum*	Gupta *et al.*, 2024
Plastics/ manufacturing, municipal waste, etc.	Poly(ethylene terephthalate) (PET)	*Microsphaeropsis arundinis*	Malafatti-Picca *et al.*, 2019
	Low Density Polyethylene (LDPE)	*Aspergillus oryzae*	Muhonja *et al.*, 2018

continued on following page

Table 1. Continued

Pollutant type/ source	Compound/class	Fungi	Reference
Dyes/ Textile, paper and pulp and leather industries, etc.	Reactive dyes (red 120, blue 4, orange 16, and black 5), Acid dyes (red 114, blue 62, orange 7, and black 172)	*Bjerkandera adusta*	Sodaneath *et al.*, 2017
	Scarlet RR	*Peyronellaea prosopidis*	Bankole *et al.*, 2018
	Reactive Black, Congo Red, Malachite Green, Methylene Blue	*Aspergillus niger, Penicillium sp.*	Alam *et al.*, 2018
	Malachite Green	*Pleurotus ostreatus*	Zhuo *et al.*, 2019
	Remazol Brilliant Blue R	*Aspergillus iizukae*	Noman *et al.*, 2020
	Azo, indigo, and anthraquinone dyes	*Emmia latemarginata Mucor circinelloides*	Juárez-Hernández *et al.*, 2021
	Azo dyes (Mb 17, AR 14, AR 18, FR 17, DR 28, AO 52)	*Cyberlindnera fabianii*	Danouche *et al.*, 2021
Pharmaceuticals and antibiotics/ Manufacturing units, hospital waste, etc.	Clomipramine, fluoxetine, mianserin, mycophenolic acid (MPA), paroxetine, sertraline	*Pleurotus ostreatus*	Sośnicka *et al.*, 2022
	Bleomycin, Vincristine	*Fomes fomentarius, Hypholoma fasciculare, Phyllotopsis nidulans, Pleurotus ostreatus, Trametes versicolor*	Jureczko *et al.* 2021
	Sulfonamides	*Pleurotus eryngii*	Baran *et al.*, 2021
	Sulfonamides	*Pleurotus ostreatus*	Mayans *et al.*, 2021
	Diclofenac	*Pleurotus ostreatus, Aspergillus niger, Penicillium roquefortii*	Blanco-Ortaa *et al.*, 2023
	Acetaminophen, Paraaminophenol Sulfanilamide	*Trichoderma harzianum*	Gorin *et al.*, 2024
Cyanotoxins and algal blooms/ Eutrophication	Microcystin-LR	*Mucor hiemalis*	Esterhuizen-Londt *et al.*, 2017
	Microcystin-LR, cylindrospermopsin	*Rhodotorula diobovata, Cladosporium cladosporioides, Dydimella sp., Trichoderma sp., Furcasterigmium furcatum, Penicillium brevicompactum*	Álvarez *et al.*, 2024

6. FACTORS AFFECTING MYCOREMEDIATION

Mycoremediation invariably involves fungi, which are living organisms and therefore affected by various factors, especially temperature, pH, moisture content, nutrient availability, and oxygen levels, among others. These factors not only influence fungal growth and development but also critically impact the efficiency of the remediation process. Also, certain fungal species that thrive well in a wide range of conditions perform mycoremediation efficiently only under specific optimum conditions (Purchase *et al.*, 2009).

Temperature plays a key role in affecting both the bioavailability of pollutants and fungal development. Mycoremediation is generally more efficient at higher temperatures, ideally between 25–30°C (Pathak *et al.*, 2015), as elevated temperatures accelerate the degradation of organic compounds, thereby enhancing the bioavailability of contaminants. Additionally, temperature affects enzyme activity, which is crucial for biodegradation since enzyme-mediated degradation is a primary mechanism in mycoremediation (Akpasi *et al.*, 2023). Another important factor affecting the mycoremediation process is pH, as it influences both the bioavailability of contaminants and the efficiency of their bio removal (Król *et al.*, 2020). Furthermore, pH also affects the diversity of fungal species; for instance, Penicillium species dominate in acidic soils, while Aspergillus populations are more abundant in alkaline soils (Akpasi *et al.*, 2023).

In addition to temperature and pH, other abiotic factors—such as moisture content, humidity, nutrient availability, oxygen levels, and the presence of antifungal agents—and biotic factors, including synergistic or antagonistic microbial populations, impact the efficiency of the mycoremediation process. For example, 70% relative humidity has been reported as optimal for mycoremediation using mushrooms like *Pleurotus ostreatus* (Aguilar-Rivera *et al.*, 2012). Nutrient requirements can be met through inorganic and organic manures; thus, combining fungi with nutrient sources like compost can significantly improve remediation effectiveness (Pathak *et al.*, 2015). Additionally, antifungal agents, soil geochemistry, surfactants, and chelating agents each influence mycoremediation by affecting fungal growth, pollutant accessibility, and contaminant degradation While antifungal agents may inhibit fungi, soil chemistry, surfactants, and chelators may enhance pollutant breakdown by improving contaminant bioavailability and reducing toxicity (Akpasi *et al.*, 2023). Microbial competition with other organisms, such as bacteria, can limit nutrient availability and inhibit fungal colonization and degradation, though some fungi form beneficial interactions with other microbes, enhancing pollutant degradation through synergistic actions.

7. CHALLENGES AND PROSPECTS

Mycoremediation holds substantial promise as an innovative and sustainable solution for pollution management. It contributes to the Sustainable Development Goals (SDGs) by improving water quality (SDG 6), addressing climate-induced pollution (SDG 13), and restoring contaminated soils (SDG 15) (Tiwari and Park, 2024). However, its practical application faces significant challenges. The precise mechanisms by which fungi degrade or transform pollutants remain elusive, highlighting the need for in-depth research to better understand the biochemical and metabolic pathways involved.

The effectiveness of mycoremediation varies depending on environmental conditions, the types of contaminants, and the fungal species employed. These factors influence the efficiency of fungal-based remediation in real-world settings, making it difficult to standardize or optimize for broader applications. Scaling up mycoremediation for large-scale environmental restoration requires further research to enhance its reliability and cost-effectiveness. Innovations are particularly needed to improve the resilience and adaptability of fungal strains in diverse and often unpredictable environments.

One promising approach is the integration of fungi with vegetation, as certain plants can enhance fungal remediation capacity, particularly in soils with mixed pollutants. This synergistic strategy could improve remediation effectiveness in areas where heavy metals coexist with organic pollutants. With targeted research and innovation, mycoremediation has the potential to address pressing pollution issues across both terrestrial and aquatic ecosystems.

By advancing our understanding and application of mycoremediation, this approach could significantly reduce environmental contaminants, promote healthier ecosystems, and offer a sustainable solution to the pollution crisis. Enhanced research and development in this field could pave the way for cleaner, more resilient environments, supporting the long-term restoration of polluted sites worldwide.

8. CONCLUSIONS

Mycoremediation holds vast potential as a sustainable solution for soil and water pollution, particularly amid escalating urbanization, industrialization, and environmental degradation. This approach, which uses fungi or fungal derivatives to degrade pollutants, leverages fungi's unique characteristics: an extensive mycelial network, enzymatic versatility, and resilience in diverse conditions. These properties enable fungi to effectively break down a wide range of contaminants, including agrochemicals, heavy metals, polycyclic aromatic hydrocarbons, synthetic dyes, plastics, pharmaceuticals and antibiotics, and cyanotoxins, which are prevalent in polluted environments. Compared to conventional remediation methods, mycoremediation offers several advantages. For start, it is cost-effective and eco-friendly, with the ability to target complex pollutants without generating secondary toxic by-products. Fungi's enzymatic system, notably involving oxidoreductases and other extracellular enzymes, allows them to metabolize contaminants without requiring energy-intensive processes or additional chemicals. Additionally, fungi can adapt to and thrive under various environmental conditions, enabling their application across diverse ecosystems and pollution types.

However, while mycoremediation is promising, challenges remain. Scaling up from laboratory to field applications presents logistical and efficiency obstacles, including ensuring consistent results in natural settings and overcoming variables like soil composition, moisture, and temperature. Additionally, optimizing fungal species selection and enhancing their pollutant-degrading capacities are areas requiring further research. Advances in fungal biotechnology and genetic engineering hold promise for improving the effectiveness of mycoremediation and addressing these limitations. In essence, by integrating fungal-based remediation into environmental strategies, we could significantly reduce soil and water contaminants, fostering healthier ecosystems. With continued research, mycoremediation could become a cornerstone of bioremediation, delivering long-term solutions to pressing pollution challenges in an environmentally responsible and economically viable manner.

REFERENCES

Abd El-Ghany, T. M., & Masmali, I. A. (2016). Fungal biodegradation of organophosphorus insecticides and their impact on soil microbial population. *Journal of Plant Pathology & Microbiology*, *7*(5), 1000349.

Abdel-Wareth, M. T. A. (2023). Sequestration and Detoxification of Heavy Metals by Fungi. In *Sustainable Industrial Wastewater Treatment and Pollution Control* (pp. 185-209). Singapore: Springer Nature Singapore.Abdel-Wareth, M. T. A. (2023). Sequestration and Detoxification of Heavy Metals by Fungi. In *Sustainable Industrial Wastewater Treatment and Pollution Control* (pp. 185-209). Singapore: Springer Nature Singapore. DOI: 10.1007/978-981-99-2560-5_10

Abu-Elsaoud, A. M., Nafady, N. A., & Abdel-Azeem, A. M. (2017). Arbuscular mycorrhizal strategy for zinc mycoremediation and diminished translocation to shoots and grains in wheat. *PLoS One*, *12*(11), e0188220. DOI: 10.1371/journal.pone.0188220 PMID: 29145471

Adenipekun, C. O., & Lawal, R. (2012). Uses of mushrooms in bioremediation: A review. *Biotechnology and Molecular Biology Reviews*, *7*(3), 62–68.

Aguilar-Rivera, N., Moran, A. C., Lagunes, D. A. R., & Gonzalez, J. M. (2012). Production of *Pleurotus ostreatus* (oyster mushroom) grown on sugar cane biomass (trash, bagasse and pith). *no, 1*, 149-152.

Ahmad, K. S. (2020). Remedial potential of bacterial and fungal strains (*Bacillus subtilis, Aspergillus niger, Aspergillus flavus* and *Penicillium chrysogenum*) against organochlorine insecticide Endosulfan. *Folia Microbiologica*, *65*(5), 801–810. DOI: 10.1007/s12223-020-00792-7 PMID: 32383069

Akhtar, N., & Mannan, M. A. U. (2020). Mycoremediation: Expunging environmental pollutants. *Biotechnology Reports (Amsterdam, Netherlands)*, *26*, e00452. DOI: 10.1016/j.btre.2020.e00452 PMID: 32617263

Akpasi, S. O., Anekwe, I. M. S., Tetteh, E. K., Amune, U. O., Shoyiga, H. O., Mahlangu, T. P., & Kiambi, S. L. (2023). Mycoremediation as a potentially promising technology: Current status and prospects—a review. *Applied Sciences (Basel, Switzerland)*, *13*(8), 4978. DOI: 10.3390/app13084978

Alam, M. Z., Khan, M. J. H., Kabbashi, N. A., & Sayem, S. A. (2018). Development of an effective biosorbent by fungal immobilization technique for removal of dyes. *Waste and Biomass Valorization*, *9*(4), 681–690. DOI: 10.1007/s12649-016-9821-9

Álvarez, Á. G. (2024). Fungal removal of cyanotoxins in constructed wetlands: The forgotten degraders. *The Science of the Total Environment*, *929*, 172590. DOI: 10.1016/j.scitotenv.2024.172590 PMID: 38642746

Anani, O. A., & Adetunji, C. O. (2021). Bioremediation of polythene and plastics using beneficial microorganisms. *Microbial Rejuvenation of Polluted Environment*, *3*, 281–302. DOI: 10.1007/978-981-15-7459-7_13

Anjitha, K. S., Sameena, P. P., & Puthur, J. T. (2021). Functional aspects of plant secondary metabolites in metal stress tolerance and their importance in pharmacology. *Plant Stress*, *2*, 100038. DOI: 10.1016/j.stress.2021.100038

Arora, D. S., & Sharma, R. K. (2010). Ligninolytic fungal laccases and their biotechnological applications. *Applied Biochemistry and Biotechnology*, *160*(6), 1760–1788. DOI: 10.1007/s12010-009-8676-y PMID: 19513857

Bandurska, K., Krupa, P., Berdowska, A., Jatulewicz, I., & Zawierucha, I. (2021). Mycoremediation of Soil Contaminated with Cadmium and Lead by *Trichoderma* sp. *Ecological Chemistry and Engineering. S*, *28*(2), 277–286. DOI: 10.2478/eces-2021-0020

Bankole, P. O., Adekunle, A. A., Obidi, O. F., Chandanshive, V. V., & Govindwar, S. P. (2018). Biodegradation and detoxification of Scarlet RR dye by a newly isolated filamentous fungus, *Peyronellaea prosopidis.Sustainable Environment Research*, *28*(5), 214–222. DOI: 10.1016/j.serj.2018.03.001

Baran, W., Adamek, E., Włodarczyk, A., Lazur, J., Opoka, W., & Muszyńska, B. (2021). The remediation of sulfonamides from the environment by *Pleurotus eryngii* mycelium. Efficiency, products and mechanisms of mycodegradation. *Chemosphere*, *262*, 128026. DOI: 10.1016/j.chemosphere.2020.128026 PMID: 33182090

Bhatt, P., Zhang, W., Lin, Z., Pang, S., Huang, Y., & Chen, S. (2020). Biodegradation of allethrin by a novel fungus *Fusarium proliferatum* strain CF2, isolated from contaminated soils. *Microorganisms*, *8*(4), 593. DOI: 10.3390/microorganisms8040593 PMID: 32325934

Birolli, W. G., Vacondio, B., Alvarenga, N., Seleghim, M. H., & Porto, A. L. (2018). Enantioselective biodegradation of the pyrethroid (±)-lambda-cyhalothrin by marine-derived fungi. *Chemosphere*, *197*, 651–660. DOI: 10.1016/j.chemosphere.2018.01.054 PMID: 29407829

Blanco-Orta, M. F., García-de la Cruz, R. F., Paz-Maldonado, L. M. T., Pedraza-González, D. A., Morales-Avila, M. M., Balderas-Hernández, V. E., González-Ortega, O., & Pérez-Martínez, A. S. (2023). Assessing three industrially produced fungi for the bioremediation of diclofenac. *Journal of Environmental Science and Health. Part A, Toxic/Hazardous Substances & Environmental Engineering*, *58*(7), 661–670. DOI: 10.1080/10934529.2023.2206353 PMID: 37128145

Bosco, F., & Mollea, C. (2019). Mycoremediation in soil. *Environmental chemistry and recent pollution control approaches*, 173.

Bravim, N. P. B., Alves, A. F., Orlanda, J. F. F., & Silva, P. B. R. (2021). Selection of filamentous fungi that are resistant to the herbicides atrazine, glyphosate and pendimethalin. *Acta Scientiarum. Agronomy*, *43*, e51656. DOI: 10.4025/actasciagron.v43i1.51656

Bumpus, J. A., & Aust, S. D. (1987). Biodegradation of DDT [1, 1, 1-trichloro-2, 2-bis (4-chlorophenyl) ethane] by the white rot fungus *Phanerochaete chrysosporium.Applied and Environmental Microbiology*, *53*(9), 2001–2008. DOI: 10.1128/aem.53.9.2001-2008.1987 PMID: 3674869

Cao, H., Wang, C., Liu, H., Jia, W., & Sun, H. (2020). Enzyme activities during Benzo [a] pyrene degradation by the fungus *Lasiodiplodia theobromae* isolated from a polluted soil. *Scientific Reports*, *10*(1), 865. DOI: 10.1038/s41598-020-57692-6 PMID: 31964981

Chapelle, F. H. (1999). Bioremediation of petroleum hydrocarbon-contaminated ground water: The perspectives of history and hydrology. *Groundwater*, *37*(1), 122–132. DOI: 10.1111/j.1745-6584.1999.tb00965.x d'Errico, G., Aloj, V., Flematti, G. R., Sivasithamparam, K., Worth, C. M., Lombardi, N., Ritieni, A., Marra, R., Lorito, M., & Vinale, F. (2021). Metabolites of a Drechslera sp. endophyte with potential as biocontrol and bioremediation agent. *Natural Product Research, 35*(22), 4508–4516. DOI: 10.1080/14786419.2020.1737058 PMID: 32159387

Danouche, M., Ferioun, M., Bahafid, W., & El Ghachtouli, N. (2021). Mycoremediation of azo dyes using *Cyberlindnera fabianii* yeast strain: Application of designs of experiments for decolorization optimization. *Water Environment Research, 93*(8), 1402–1416. DOI: 10.1002/wer.1499 PMID: 33331006

Deshmukh, R., Khardenavis, A. A., & Purohit, H. J. (2016). Diverse metabolic capacities of fungi for bioremediation. *Indian Journal of Microbiology, 56*(3), 247–264. DOI: 10.1007/s12088-016-0584-6 PMID: 27407289

Devendrapandi, G., Liu, X., Balu, R., Ayyamperumal, R., Arasu, M. V., Lavanya, M., & Karthika, P. C. (2024). Innovative remediation strategies for persistent organic pollutants in soil and water: A comprehensive review. *Environmental Research, 249*, 118404. DOI: 10.1016/j.envres.2024.118404 PMID: 38341071

Dinakarkumar, Y., Gnanasekaran, R., Reddy, G. K., Vasu, V., Balamurugan, P., & Murali, G. (2024). *Fungal bioremediation: An overview of the mechanisms, applications and future perspectives*. Environmental Chemistry and Ecotoxicology.

Dutta, U., Jyote, S., Ghosh, K., Pal, S., Sengupta, S., & Laha, A. (2023). Bioremediation of contaminated environments through mycology: a review of current advancements and future prospects: a short review. *Journal of Advanced Zoology, 44*(S-5), 2611–2616. DOI: 10.53555/jaz.v44iS5.3483

Ebsa, G., Gizaw, B., Admassie, M., Desalegn, A., & Alemu, T. (2024). Screening, characterization and optimization of potential dichlorodiphenyl trichloroethane (DDT) degrading fungi. *Heliyon, 10*(12), e33289. DOI: 10.1016/j.heliyon.2024.e33289 PMID: 39022069

Egwim, E. C., Oyewole, O. A., & Yakubu, J. G. J. (2023). *Fungal bioremediation of pollutant*. Bentham., DOI: 10.2174/9789815123494123010009

Esterhuizen-Londt, M., Hertel, S., & Pflugmacher, S. (2017). Uptake and biotransformation of pure commercial microcystin-LR versus microcystin-LR from a natural cyanobacterial bloom extract in the aquatic fungus *Mucor hiemalis.Biotechnology Letters, 39*(10), 1537–1545. DOI: 10.1007/s10529-017-2378-2 PMID: 28597371

Garzón, J. M., Rodríguez-Miranda, J. P., & Hernández-Gómez, C. (2017). Aporte de la biorremediación para solucionar problemas de contaminación y su relación con el desarrollo sostenible. *Universidad y salud, 19*(2), 309-318.

Goligar, N., Saadatmand, S., & Khavarinejad, R. A. (2023). Mycoremediation of lead and cadmium by lignocellulosic enzymes of *Pleurotus eryngii.AMB Express, 13*(1), 127. DOI: 10.1186/s13568-023-01626-8 PMID: 37964138

Gorin, M., Shabani, M., Votat, S., Lebrun, L., Mbokou, S. F., & Pontié, M. (2024). Application of fungal-based microbial fuel cells for biodegradation of pharmaceuticals: Comparative study of individual vs. mixed contaminant solutions. *Chemosphere*, *363*, 142849. DOI: 10.1016/j.chemosphere.2024.142849 PMID: 39009093

Gothalwal, R. (2020). Bioremediation. In Managing Biological and Ecological Systems (pp. 359-378). CRC Press.

Gupta, P. (2023). *Bioremediation of Heavy Metal in Paper Mill Effluent*. Bentham., DOI: 10.2174/9789815123494123010005

Gupta, S., Pathak, B., & Ravi, R. K. (2024). Biodegradation of naphthalene using biosurfactant producing *Fusarium proliferatum* WC416 isolated from refinery effluent. *Applied Biochemistry and Biotechnology*, *196*(5), 2549–2565. DOI: 10.1007/s12010-023-04364-6 PMID: 36689160

Haritash, A. K., & Kaushik, C. P. (2009). Biodegradation aspects of polycyclic aromatic hydrocarbons (PAHs): A review. *Journal of Hazardous Materials*, *169*(1-3), 1–15. DOI: 10.1016/j.jhazmat.2009.03.137 PMID: 19442441

Hasan, N. F. N., Wan Hanafi, W. N., Safian, M. F., & Zainal Ariffin, Z. (2024). Diversity of chlorpyrifos-degrading fungi isolated from chlorpyrifos-contaminated agricultural soil. [ScL]. *Science Letters*, *18*(1), 10–20.

Hu, K., Peris, A., Torán, J., Eljarrat, E., Sarrà, M., Blánquez, P., & Caminal, G. (2020). Exploring the degradation capability of *Trametes versicolor* on selected hydrophobic pesticides through setting sights simultaneously on culture broth and biological matrix. *Chemosphere*, *250*, 126293. DOI: 10.1016/j.chemosphere.2020.126293 PMID: 32234621

Israilides, C., & Philippoussis, A. (2003). Bio-technologies of recycling agro-industrial wastes for the production of commercially important fungal polysaccharides and mushrooms. *Biotechnology & Genetic Engineering Reviews*, *20*(1), 247–260. DOI: 10.1080/02648725.2003.10648045 PMID: 14997854

Jakovljević, V. D., & Vrvić, M. M. (2018). Potential of pure and mixed cultures of *Cladosporium cladosporioides* and *Geotrichum candidum* for application in bioremediation and detergent industry. *Saudi Journal of Biological Sciences*, *25*(3), 529–536. DOI: 10.1016/j.sjbs.2016.01.020 PMID: 29686515

Jibran, A., & Mol, J. (2011). *Pleurotus sajor-caju* protein: A potential biosorptive agent. *Advanced Biotechnology*, *11*, 25–27.

Juárez-Hernández, J., Castillo-Hernández, D., Pérez-Parada, C., Nava-Galicia, S., Cuervo-Parra, J. A., Surian-Cruz, E., Díaz-Godínez, G., Sánchez, C., & Bibbins-Martínez, M. (2021). Isolation of fungi from a textile industry effluent and the screening of their potential to degrade industrial dyes. *Journal of Fungi (Basel, Switzerland)*, *7*(10), 805. DOI: 10.3390/jof7100805 PMID: 34682227

Jureczko, M., Przystaś, W., Krawczyk, T., Gonciarz, W., & Rudnicka, K. (2021). White-rot fungi-mediated biodegradation of cytostatic drugs-bleomycin and vincristine. *Journal of Hazardous Materials*, *407*, 124632. DOI: 10.1016/j.jhazmat.2020.124632 PMID: 33359974

Kamal, N., Parshad, J., Saharan, B. S., Kayasth, M., Mudgal, V., Duhan, J. S., Mandal, B. S., & Sadh, P. K. (2023). Ecosystem protection through myco-remediation of chromium and arsenic. *Journal of Xenobiotics*, *13*(1), 159–171. DOI: 10.3390/jox13010013 PMID: 36976162

Khan, I., Aftab, M., Shakir, S., Ali, M., Qayyum, S., Rehman, M. U., Haleem, K. S., & Touseef, I. (2019). Mycoremediation of heavy metal (Cd and Cr)–polluted soil through indigenous metallotolerant fungal isolates. *Environmental Monitoring and Assessment*, *191*(9), 1–11. DOI: 10.1007/s10661-019-7769-5 PMID: 31440913

Khatua, S., Simal-Gandara, J., & Acharya, K. (2024). Myco-remediation of plastic pollution: Current knowledge and future prospects. *Biodegradation*, *35*(3), 249–279. DOI: 10.1007/s10532-023-10053-2 PMID: 37665521

Król, A., Mizerna, K., & Bożym, M. (2020). An assessment of pH-dependent release and mobility of heavy metals from metallurgical slag. *Journal of Hazardous Materials*, *384*, 121502. DOI: 10.1016/j.jhazmat.2019.121502 PMID: 31732354

Kulshreshtha, S., Mathur, N., & Bhatnagar, P. (2014). Mushroom as a product and their role in mycoremediation. *AMB Express*, *4*(1), 1–7. DOI: 10.1186/s13568-014-0029-8 PMID: 24949264

Kumar, V., Singh, G., & Dwivedi, S. K. (2022). Arsenic and Nickel Stress Response, Their Bioremediation Potential and Mechanism of *Trichoderma lixii* Isolated from Electroplating Wastewater. *Geomicrobiology Journal*, *39*(3-5), 186–198. DOI: 10.1080/01490451.2021.1956648

Landrigan, P. J., Fuller, R., Acosta, N. J., Adeyi, O., Arnold, R., Baldé, A. B., & Zhong, M. (2018). The Lancet Commission on pollution and health. *Lancet*, *391*(10119), 462–512. DOI: 10.1016/S0140-6736(17)32345-0 PMID: 29056410

Malafatti-Picca, L., de Barros Chaves, M. R., de Castro, A. M., Valoni, É., de Oliveira, V. M., Marsaioli, A. J., de Franceschi de Angelis, D., & Attili-Angelis, D. (2019). Hydrocarbon-associated substrates reveal promising fungi for poly (ethylene terephthalate)(PET) depolymerization. *Brazilian Journal of Microbiology : Publication of the Brazilian Society for Microbiology*, *50*(3), 633–648. DOI: 10.1007/s42770-019-00093-3 PMID: 31175657

Manjula, N. G. Tajunnisa, Mamani, V., Meghana, C. A., & Mayegowda, S. B. (2023). Fungal-Based Synthesis to Generate Nanoparticles for Nanobioremediation. In *Green Nanoremediation: Sustainable Management of Environmental Pollution* (pp. 83-108). Cham: Springer International Publishing.

Mayans, B., Camacho-Arévalo, R., García-Delgado, C., Antón-Herrero, R., Escolástico, C., Segura, M. L., & Eymar, E. (2021). An assessment of *Pleurotus ostreatus* to remove sulfonamides, and its role as a biofilter based on its own spent mushroom substrate. *Environmental Science and Pollution Research International*, *28*(6), 7032–7042. DOI: 10.1007/s11356-020-11078-3 PMID: 33025433

Mello, I. S., Targanski, S., Pietro-Souza, W., Stachack, F. F. F., Terezo, A. J., & Soares, M. A. (2020). Endophytic bacteria stimulate mercury phytoremediation by modulating its bioaccumulation and volatilization. *Ecotoxicology and Environmental Safety*, *202*, 110818. DOI: 10.1016/j.ecoenv.2020.110818 PMID: 32590206

Mir, R. A., Mantoo, A. G., Sofi, Z. A., Bhat, D. A., Bashir, A., & Bashir, S. (2023). Types of Environmental Pollution and Its Effects on the Environment and Society. In *Geospatial Analytics for Environmental Pollution Modeling: Analysis, Control and Management* (pp. 1–31). Springer Nature Switzerland. DOI: 10.1007/978-3-031-45300-7_1

Mnkandla, S. M., & Otomo, P. V. (2021). Effectiveness of mycofiltration for removal of contaminants from water: A systematic review protocol. *Environmental Evidence*, *10*(1), 1–8. DOI: 10.1186/s13750-021-00232-0

Mouhamadou, B., Faure, M., Sage, L., Marçais, J., Souard, F., & Geremia, R. A. (2013). Potential of autochthonous fungal strains isolated from contaminated soils for degradation of polychlorinated biphenyls. *Fungal Biology*, *117*(4), 268–274. DOI: 10.1016/j.funbio.2013.02.004 PMID: 23622721

Mudaliar, S., Agrawal, K., & Verma, P. (2023). Fungi: a veritable tool for refractory pollutant remediation. *Clean Technologies Toward a Sustainable Future*, 283.

Muhonja, C. N., Makonde, H., Magoma, G., & Imbuga, M. (2018). Biodegradability of polyethylene by bacteria and fungi from Dandora dumpsite Nairobi-Kenya. *PLoS One*, *13*(7), e0198446. DOI: 10.1371/journal.pone.0198446 PMID: 29979708

Nagy, B., Măicăneanu, A., Indolean, C., Mânzatu, C., Silaghi-Dumitrescu, L., & Majdik, C. (2014). Comparative study of Cd (II) biosorption on cultivated *Agaricus bisporus* and wild *Lactarius piperatus* based biocomposites. Linear and nonlinear equilibrium modelling and kinetics. *Journal of the Taiwan Institute of Chemical Engineers*, *45*(3), 921–929. DOI: 10.1016/j.jtice.2013.08.013

Noman, E., Al-Gheethi, A., Talip, B. A., Mohamed, R., & Kassim, A. H. (2021). Decolourization of dye wastewater by A Malaysian isolate of *Aspergillus iizukae* 605EAN strain: A biokinetic, mechanism and microstructure study. *International Journal of Environmental Analytical Chemistry*, *101*(11), 1592–1615. DOI: 10.1080/03067319.2019.1686146

Nyanhongo, G. S., Gűbitz, G., Sukyai, P., Leitner, C., Haltrich, D., & Ludwig, R. (2007). Oxidoreductases from *Trametes* spp. in biotechnology: A wealth of catalytic activity. *Food Technology and Biotechnology*, *45*(3), 250–268.

Okamura, T., Ogata, T., Minamoto, N., Takeno, T., Noda, H., Fukuda, S., & Ohsugi, M. (2001). Characteristics of wine produced by mushroom fermentation. *Bioscience, Biotechnology, and Biochemistry*, *65*(7), 1596–1600. DOI: 10.1271/bbb.65.1596 PMID: 11515544

Palanivel, T. M., Pracejus, B., & Novo, L. A. (2023). Bioremediation of copper using indigenous fungi *Aspergillus* species isolated from an abandoned copper mine soil. *Chemosphere*, *314*, 137688. DOI: 10.1016/j.chemosphere.2022.137688 PMID: 36584825

Pathak, A. P., Rathod, M., & Gudda, R. (2015). Chapter Cultivation by Optimizing Substrate Dependent Growth of Selected Oyster Mushroom Species under Environmentally Controlled Conditions: A Review. *Biotechnology of Mushroom*, 24.

Picciotti, U., Araujo Dalbon, V., Ciancio, A., Colagiero, M., Cozzi, G., De Bellis, L., Finetti-Sialer, M. M., Greco, D., Ippolito, A., Lahbib, N., Logrieco, A. F., López-Llorca, L. V., Lopez-Moya, F., Luvisi, A., Mincuzzi, A., Molina-Acevedo, J. P., Pazzani, C., Scortichini, M., Scrascia, M., & Porcelli, F. (2023). "Ectomosphere": Insects and microorganism interactions. *Microorganisms*, *11*(2), 440. DOI: 10.3390/microorganisms11020440 PMID: 36838405

Pozdnyakova, N. N., Balandina, S. A., Dubrovskaya, E. V., Golubev, C. N., & Turkovskaya, O. V. (2018, January). Ligninolytic basidiomycetes as promising organisms for the mycoremediation of PAH-contaminated Environments. [). IOP Publishing.]. *IOP Conference Series. Earth and Environmental Science*, *107*, 012071. DOI: 10.1088/1755-1315/107/1/012071

Purchase, D., Scholes, L. N. L., Revitt, D. M., & Shutes, R. B. E. (2009). Effects of temperature on metal tolerance and the accumulation of Zn and Pb by metal-tolerant fungi isolated from urban runoff treatment wetlands. *Journal of Applied Microbiology*, *106*(4), 1163–1174. DOI: 10.1111/j.1365-2672.2008.04082.x PMID: 19187150

Purnomo, A. S., Putra, S. R., Shimizu, K., & Kondo, R. (2014). Biodegradation of heptachlor and heptachlor epoxide-contaminated soils by white-rot fungal inocula. *Environmental Science and Pollution Research International*, *21*(19), 11305–11312. DOI: 10.1007/s11356-014-3026-1 PMID: 24840358

Rai, M., Ingle, A., Gaikwad, S., Gupta, I., Yadav, A., Gade, A., & Duran, N. (2018). Fungi: myconanofactory, mycoremediation and medicine. In *Fungi* (pp. 201-219). CRC Press. DOI: 10.1201/9781315369471-10

Rajput, P., Benjwal, S., & Pandey, R. (2025). A Comprehensive Review on the Role of Bioremediation in Heavy Metal Contamination. *Nature Environment & Pollution Technology, 24*.

Rhodes, C. J. (2014). Mycoremediation (bioremediation with fungi)–growing mushrooms to clean the earth. *Chemical Speciation and Bioavailability*, *26*(3), 196–198. DOI: 10.3184/095422914X14047407349335

Russell, J. R., Huang, J., Anand, P., Kucera, K., Sandoval, A. G., Dantzler, K. W., Hickman, D. S., Jee, J., Kimovec, F. M., Koppstein, D., Marks, D. H., Mittermiller, P. A., Núñez, S. J., Santiago, M., Townes, M. A., Vishnevetsky, M., Williams, N. E., Vargas, M. P. N., Boulanger, L.-A., & Strobel, S. A. (2011). Biodegradation of polyester polyurethane by endophytic fungi. *Applied and Environmental Microbiology*, *77*(17), 6076–6084. DOI: 10.1128/AEM.00521-11 PMID: 21764951

Russo, F., Ceci, A., Pinzari, F., Siciliano, A., Guida, M., Malusà, E., Tartanus, M., Miszczak, A., Maggi, O., & Persiani, A. M. (2019). Bioremediation of dichlorodiphenyltrichloroethane (DDT)-contaminated agricultural soils: Potential of two autochthonous saprotrophic fungal strains. *Applied and Environmental Microbiology*, *85*(21), e01720–e19. DOI: 10.1128/AEM.01720-19 PMID: 31444208

Sadiq, S., Inam, H. M., Ahmad, I., Ahad, K., & Rashid, A. (2015). Bioremediation potential of white rot fungi, *Pleurotus* spp against organochlorines. *Journal of Bioremediation & Biodegradation*, *6*(308), 2.

Samanta, S. K., Singh, O. V., & Jain, R. K. (2002). Polycyclic aromatic hydrocarbons: Environmental pollution and bioremediation. *Trends in Biotechnology*, *20*(6), 243–248. DOI: 10.1016/S0167-7799(02)01943-1 PMID: 12007492

Sharma, A. K., Sharma, M., Sharma, A. K., Sharma, M., & Sharma, M. (2023). Mapping the impact of environmental pollutants on human health and environment: A systematic review and meta-analysis. *Journal of Geochemical Exploration*, *255*, 107325. DOI: 10.1016/j.gexplo.2023.107325

Shukla, K. P., Singh, N. K., & Sharma, S. (2010). Bioremediation: Developments, current practices and perspectives. *Genetic Engineering and Biotechnology Journal*, *3*, 1–20.

Singh, A., Kumari, R., & Yadav, A. N. (2021). Fungal secondary metabolites for bioremediation of hazardous heavy metals. *Recent trends in mycological research: volume 2: environmental and industrial perspective*, 65-98.

Singh, H. (2006). *Mycoremediation: fungal bioremediation*. John Wiley & Sons. DOI: 10.1002/0470050594

Singh, V. (2024). Soil Pollution. In *Textbook of Environment and Ecology* (pp. 267–274). Springer Nature Singapore. DOI: 10.1007/978-981-99-8846-4_18

Soares, P. R. S., Birolli, W. G., Ferreira, I. M., & Porto, A. L. M. (2021). Biodegradation pathway of the organophosphate pesticides chlorpyrifos, methyl parathion and profenofos by the marine-derived fungus *Aspergillus sydowii* CBMAI 935 and its potential for methylation reactions of phenolic compounds. *Marine Pollution Bulletin*, *166*, 112185. DOI: 10.1016/j.marpolbul.2021.112185 PMID: 33640600

Sodaneath, H., Lee, J. I., Yang, S. O., Jung, H., Ryu, H. W., & Cho, K. S. (2017). Decolorization of textile dyes in an air-lift bioreactor inoculated with *Bjerkandera adusta* OBR105. *Journal of Environmental Science and Health. Part A, Toxic/Hazardous Substances & Environmental Engineering*, *52*(11), 1099–1111. DOI: 10.1080/10934529.2017.1340753 PMID: 28763254

Solanki, M. K., Kashyap, B. K., Solanki, A. C., Malviya, M. K., & Surapathrudu, K. (2019). *Helpful Linkages of Trichoderma s in the process of Mycoremediation and Mycorestoration. Plant Health Under Biotic Stress* (Vol. 2). Microbial Interactions.

Sośnicka, A., Kózka, B., Makarova, K., Giebułtowicz, J., Klimaszewska, M., & Turło, J. (2022). Optimization of white-rot fungi mycelial culture components for bioremediation of pharmaceutical-derived pollutants. *Water (Basel)*, *14*(9), 1374. DOI: 10.3390/w14091374

Srikanth, M., Sandeep, T. S. R. S., Sucharitha, K., & Godi, S. (2022). Biodegradation of plastic polymers by fungi: A brief review. *Bioresources and Bioprocessing*, *9*(1), 42. DOI: 10.1186/s40643-022-00532-4 PMID: 38647755

Stuart, M., Lapworth, D., Crane, E., & Hart, A. (2012). Review of risk from potential emerging contaminants in UK groundwater. *The Science of the Total Environment*, *416*, 1–21. DOI: 10.1016/j.scitotenv.2011.11.072 PMID: 22209399

Thacharodi, A., Hassan, S., Singh, T., Mandal, R., Chinnadurai, J., Khan, H. A., Hussain, M. A., Brindhadevi, K., & Pugazhendhi, A. (2023). Bioremediation of polycyclic aromatic hydrocarbons: An updated microbiological review. *Chemosphere*, *328*, 138498. DOI: 10.1016/j.chemosphere.2023.138498 PMID: 36996919

Tiwari, D., Hussain, A., Tiwari, S. K., Ahmed, S., Sultan, M. W., & Ahamed, M. I. (2023). *Removal of Heavy Metals using Microbial Bioremediation*. Bentham., DOI: 10.2174/9789815123494123010004

Tiwari, P., & Park, K. I. (2024). Advanced Fungal Biotechnologies in Accomplishing Sustainable Development Goals (SDGs): What Do We Know and What Comes Next? *Journal of Fungi (Basel, Switzerland), 10*(7), 506. DOI: 10.3390/jof10070506 PMID: 39057391

Ukaogo, P. O., Ewuzie, U., & Onwuka, C. V. (2020). Environmental pollution: causes, effects, and the remedies. In *Microorganisms for sustainable environment and health* (pp. 419–429). Elsevier. DOI: 10.1016/B978-0-12-819001-2.00021-8

United Nations Environment Programme (UNEP). (2023). Annual Report 2023. https://www.unep.org/resources/annual-report-2023

Urík, M., Čerňanský, S., Ševc, J., Šimonovičová, A., & Littera, P. (2007). Biovolatilization of arsenic by different fungal strains. *Water, Air, and Soil Pollution, 186*(1-4), 337–342. DOI: 10.1007/s11270-007-9489-7

Văcar, C. L., Covaci, E., Chakraborty, S., Li, B., Weindorf, D. C., Fren iu, T., Pârvu, M., & Podar, D. (2021). Heavy metal-resistant filamentous fungi as potential mercury bioremediators. *Journal of Fungi (Basel, Switzerland), 7*(5), 386. DOI: 10.3390/jof7050386 PMID: 34069296

Vaksmaa, A., Guerrero-Cruz, S., Ghosh, P., Zeghal, E., Hernando-Morales, V., & Niemann, H. (2023). Role of fungi in bioremediation of emerging pollutants. *Frontiers in Marine Science, 10*, 1070905. DOI: 10.3389/fmars.2023.1070905

Wang, Y., Zhang, B., Chen, N., Wang, C., Feng, S., & Xu, H. (2018). Combined bioremediation of soil co-contaminated with cadmium and endosulfan by *Pleurotus eryngii* and *Coprinus comatus.Journal of Soils and Sediments, 18*(6), 2136–2147. DOI: 10.1007/s11368-017-1762-9

Wu, J., & Yu, H. Q. (2007). Biosorption of 2, 4-dichlorophenol by immobilized white-rot fungus Phanerochaete chrysosporium from aqueous solutions. *Bioresource Technology, 98*(2), 253–259. DOI: 10.1016/j.biortech.2006.01.018 PMID: 16517157

Xiao, P., & Kondo, R. (2020). Potency of Phlebia species of white rot fungi for the aerobic degradation, transformation and mineralization of lindane. *journal of microbiology, 58*, 395-404.

Xiao, P., Mori, T., Kamei, I., & Kondo, R. (2011). Metabolism of organochlorine pesticide heptachlor and its metabolite heptachlor epoxide by white rot fungi, belonging to genus *Phlebia.FEMS Microbiology Letters, 314*(2), 140–146. DOI: 10.1111/j.1574-6968.2010.02152.x PMID: 21087297

Xu, F., Chen, P., Li, H., Qiao, S., Wang, J., Wang, Y., Wang, X., Wu, B., Liu, H., Wang, C., & Xu, H. (2021). Comparative transcriptome analysis reveals the differential response to cadmium stress of two *Pleurotus* fungi: *Pleurotus cornucopiae* and *Pleurotus ostreatus.Journal of Hazardous Materials, 416*, 125814. DOI: 10.1016/j.jhazmat.2021.125814 PMID: 33866290

Yang, J., Yue, H. R., Pan, L. Y., Feng, J. X., Zhao, S., Suwannarangsee, S., Champreda, V., Liu, C.-G., & Zhao, X. Q. (2023). Fungal strain improvement for efficient cellulase production and lignocellulosic biorefinery: Current status and future prospects. *Bioresource Technology, 385*, 129449. DOI: 10.1016/j.biortech.2023.129449 PMID: 37406833

Zaman, M. U., Pandit, B. A., Kumar, R., & Sofi, T. A. (2023). Mycoremediation: An Innovative and Sustainable Approach for Environmental Restoration. *International Journal for Research in Applied Science and Engineering Technology*, *11*(6), 2407–2413. DOI: 10.22214/ijraset.2023.53736

Zhuo, R., Zhang, J., Yu, H., Ma, F., & Zhang, X. (2019). The roles of *Pleurotus ostreatus* HAUCC 162 laccase isoenzymes in decolorization of synthetic dyes and the transformation pathways. *Chemosphere*, *234*, 733–745. DOI: 10.1016/j.chemosphere.2019.06.113 PMID: 31234090

Zilly, A., dos Santos Bazanella, G. C., Helm, C. V., Araújo, C. A. V., de Souza, C. G. M., Bracht, A., & Peralta, R. M. (2012). Solid-state bioconversion of passion fruit waste by white-rot fungi for production of oxidative and hydrolytic enzymes. *Food and Bioprocess Technology*, *5*(5), 1573–1580. DOI: 10.1007/s11947-011-0532-8

Chapter 14

Antifungal Properties of Essential Oils From Rosmarinus Officinalis, Citrus Aurantium, and Origanum Majorana: In Vitro Evaluation for Biological Control Agents

Chaymae El-Hajjaji
Ibn Tofail University, Morocco

Soukaina Msairi
Mohammed V University, Morocco

Oumaima Ninich
https://orcid.org/0000-0003-1402-0590
Mohammed V University, Morocco

Yassine Mouniane
https://orcid.org/0000-0002-6037-5301
Ibn Tofail University, Morocco

Soumaya Choukri
Ibn Tofail University, Morocco

Rihab Hedhiri
Centre of Research and Water Technologies, Tunisia

Taoufiq Benali
https://orcid.org/0000-0001-5732-2818
Cadi Ayyad University, Morocco

Amir Souissi
https://orcid.org/0000-0002-4418-5807
Swift Current Research and Development Center, Canada

Amira Hachana
https://orcid.org/0000-0002-7153-8006
National Institute for Agricultural Research of Tunisia, Tunisia & Carthage University, Tunisia

Jamila Dahmani
Ibn Tofail University, Morocco

ABSTRACT

With the prohibition of many chemical pesticides due to their impact on the health of living resources, as well as on the quality of food, soil and water, it has become essential to look for alternatives of natural origin that promote sustainability. This study focuses on the antifungal activity of Rosmarinus

DOI: 10.4018/979-8-3693-8201-1.ch014

officinalis, Origanum majorana and Citrus aurantium essential oils against Phytophthora palmivora and Alternaria alternata. The antifungal activity of six essential oil concentrations was tested in vitro by the direct contact method. After a seven-day follow-up, the results revealed a significant antifungal effect of Origanum majorana essential oil, with total inhibition of mycelial growth at a concentration of 1.5 µl/ml and 3.33 µl/ml on Phytophthora palmivora and Alternaria alternata respectively. While Rosmarinus officinalis and Citrus aurantium showed less antifungal activity, with MICs ranging from 6.66 µL/mL to 12.5 µL/mL. These promising results pave the way for in vivo efficacy tests to confirm its use in more sustainable agricultural practices.

1. INTRODUCTION

Fungal diseases represent a major challenge for agriculture, affecting the health and productivity of crops worldwide (Ristaino et al., 2021). They cause substantial economic losses every year (Nizamani et al., 2024), costing the global economy some $220 billion, with a loss of 20 to 40% of agricultural productivity (FAO, 2019). These pathologies, due to fungal pathogens, affect various parts of the plant, such as leaves, stems, flowers and fruit. Their damage can be from moderate (reduction in the aesthetic appeal of crops) to extreme (major crop losses) (Alegbeleye et al., 2022). Preventing such damage could make it possible to feed up to 600 million people a year (Nizamani et al., 2024). Among these pathogens, *Phytophthora palmivora* and *Alternaria alternata* are of particular concern. *P. palmivora* is an oomycete responsible for root rot and deterioration of tropical crops, notably cocoa and citrus (Bose et al., 2023). A. *alternata*, meanwhile, is an opportunistic fungal pathogen causing leaf diseases, fruit rot and mycotoxigenic toxins affecting crop quality (Philip et al., 2024). The pathogens studied, *P. palmivora* and *A. alternata*, pose serious threats to global agricultural production. *P. palmivora* causes significant losses in tropical crops, leading to a decline in cocoa and citrus production. For its part, *A. alternata* is responsible for post-harvest diseases affecting fruit and vegetable preservation, directly impacting the food supply chain (Manathunga et al., 2024).

To improve crop production and control pests, large quantities of chemical inputs, mainly synthetic pesticides, are used (Schreinemachers and Tipraqsa, 2012). Pesticides are widely recognized for their vital role in protecting crops against potential losses that could affect global food production (Lengai et al., 2020). They are used either to control crop diseases, or to kill or retard the growth of insects, rodents, fungi and weeds that hinder their growth (Tudi et al., 2021). As a result, pesticide residues pollute the environment and persist in agricultural crops, as well as having negative effects on the aquatic environment (Cui et al., 2023). Studies have also shown that pesticides have adverse effects on human health (Wei et al., 2023). On the other hand, the repeated and intensive use of pesticides exposes target organisms to intense selective pressure, which develops resistance in pests. This makes chemicals less effective over time and requires higher doses or stronger products to achieve the same results (Hawkins et al., 2019).

It is therefore essential to promote sustainable agricultural practices such as integrated pest management, the use of organic methods and the adoption of agricultural techniques that reduce dependence on chemicals (Baker et al., 2020). One such alternative is biopesticides. Unlike traditional chemical pesticides, biopesticides are natural or naturally derived substances, such as plant extracts, microorganisms, or pheromones (Akutse et al., 2020). They are also effective against pests and most importantly, do not interfere with the well-being of ecosystems (Souto et al., 2021). Indeed, the use of essential oils, with their bioactive components and biological properties such as antifungal, antibiotic and antiviral activity, has been seen as a green strategy and a promising ecological alternative to chemical pesticides for protecting agricultural production (Hou et al., 2022; Liang et al., 2024).

Rosmarinus officinalis, *Origanum majorana*, and *Citrus aurantium*, have significant economic value due to their essential oils, which are widely used in high-value industries such as pharmaceuticals, cosmetics, and food preservation. *Rosmarinus officinalis* is rich in bioactive compounds known for their antimicrobial and antioxidant properties, making it valuable in food preservation and medicine. *Origanum majorana* contains thymol and carvacrol, potent antifungal and antibacterial agents used in natural health products and agro-industry. *Citrus aurantium* is a key source of limonene, widely applied in pharmaceuticals and aromatherapy for its antimicrobial and therapeutic effects. These essential oils represent sustainable alternatives to synthetic chemicals in multiple industries (Pergola et al., 2024; Seyyedi-Mansour et al., 2024; Qiu et al., 2024; Nurzyńska-Wierdak & Walasek-Janusz, 2025).

In this study, three plants were selected for extraction of their essential oils: *Rosmarinus officinalis*, *Origanum majorana* and *Citrus aurantium*. They were chosen for their active compounds known from literature for their economic interest and potential antifungal activity (Ksouri et al., 2017; Metoui et al., 2015; Bouyahya et al., 2021). The aim of this research is to evaluate the antifungal activity of these essential oils *in vitro*, on the radial mycelial growth of *Phytophtora palmivora* and *Alternaria alternata*. By exploring these natural alternatives, we aim to contribute to the development of effective and sustainable control strategies against fungal plant diseases, while limiting the use of chemical fungicides that are potentially harmful to the ecosystem and human health and promoting the importance of these alternatives within the framework of sustainability and environmentally friendly agricultural practices.

2. MATERIALS AND METHODS

2.1. Plant material

Rosmarinus officinalis, *Origanum majorana* and *Citrus aurantium* leaves were collected from the botanical garden of Agency for Medicinal and Aromatic Plants in Taounate, northern Morocco, between March and April 2022.

2.2. Fungal material

The fungal material consists of two fungal species known for their pathogenicity to crops. Among common pathogens, *Phytophtora palmivora* and *Alternaria alternata* stand out as responsible for significant damage in many crops. While the former is associated with root and crown rot, the latter causes leaf diseases such as Alternaria leaf spot. Our strains were isolated at the Phytobiotechnology laboratory of the National Agency of Medicinal and Aromatic Plants from olive trees in Fès region showing symptoms of decline. The pathogens were subcultured on PDA medium for 7 days at 25°C. The resulting samples were then exposed to different concentrations of essential oils to assess their antifungal properties (Freire et al., 2017).

2.3. Essential oil extraction

Essential oils (EO) were extracted from 300g of plant material with one liter of distilled water by hydrodistillation in a Clevenger apparatus, a method recognised as highly effective (Katekar et al., 2022). This process involves immersing the plant material in a water-filled flask and heating it to boiling for 3 to 4 hours. The extracted essential oil is then collected in opaque glass bottles to protect it from oxidation and stored in a refrigerator at 4°C for optimum presevation (Lainez-Cerón et al., 2021).

2.3.1. Essential oil yield

Essential oil yield is defined as the ratio between the mass of oil extracted and the weight of the plant used, expressed as a percentage. It is calculated based on dry plant matter according to the following equation:

$$\text{Essential oil yield } (\%) = \frac{m}{M} \times 100$$

Where: m: mass of the extracted oil (g);
M: mass of the raw plant material (g).

2.3.2. Organoleptic properties of essential oils

The organoleptic properties of plant essential oils, including color, odor, appearance, and viscosity, are observed and evaluated using sensory methods based on vision and smell (Mariod, 2016).

2.3.3. Refractive index of essential oils

The refractive index is measured using a refractometer, by placing two (2) to three (3) drops of essential oil on the refractometer prism. Look through the eyepiece and adjust knob until the dividing line between the light and dark zones is at the center of the reticle (Ospina et al., 2016).

$$n20 = nT + 0{,}00045\ (T - 20)$$

With T: temperature at which reading is taken.

2.3.4. Rotary power

This assessment was carried out based on the French standard NF V 75-113, in accordance with the requirements of the international standard (AFNOR, 2000).

2.4. Antifungal activity

2.4.1. Direct contact method

The direct contact technique involves evaluating the efficacy of essential oils on fungal strains tested *in vitro* (Fandohan et al., 2004). In this method, the essential oil is brought into direct contact with the fungus by adding different concentrations of essential oil (12.5; 6.66; 3.33; 1.5; 0.5; 0.25 µl/ml) to the PDA culture medium.

After the culture medium solidifies, a 6 mm diameter mycelial disk, taken from the growth zone of 7-day-old cultures of *Phytophtora palmivora* and *Alternaria alternata*, is transplanted and placed in the center of Petri dishes containing the culture medium and varying concentrations of EO. The dishes were incubated at 25°C and monitored for 7 days. Three replicates were performed for each concentration to minimize experimental error, along with a control. The incubation was carried out in an oven at 25°C.

2.4.2. Determination of inhibition rate

According to (Sharma and Tripathi, 2008), the percentage inhibition of radial mycelial growth is calculated using the following equation:

$$\text{Mycelial growth inhibition (\%)} = \frac{dc - dt}{dc} \times 100$$

dc: diametral growth of the control;
dt: diametral growth of the fungus in the presence of a concentration (c) of fungicide.

2.4.3. Determination of mycelial growth rate

The mycelial growth rate was assessed according to (Qi et al., 2021).

2.5. Statistical analysis

The results are presented as mean ± standard deviation and the statistical analysis was conducted using IBM SPSS version 23 software. Values are considered significantly different if the p-value is less than 5%.

3. RESULTS

3.1. Essential oil yield

Analysis of the essential oil yields of *Rosmarinus officinalis*, *Origanum majorana* and *Citrus aurantium* reveals significant differences between the plants. *Origanum majorana* (3,02%) has the highest yield, followed by *Rosmarinus officinalis* (0,87%), while *Citrus aurantium* (0,126%) has the lowest (Figure 1).

Figure 1. Essential oil yields of Rosmarinus officinalis, Origanum majorana and Citrus aurantium (Values sharing different letters are significance at the 5% level)

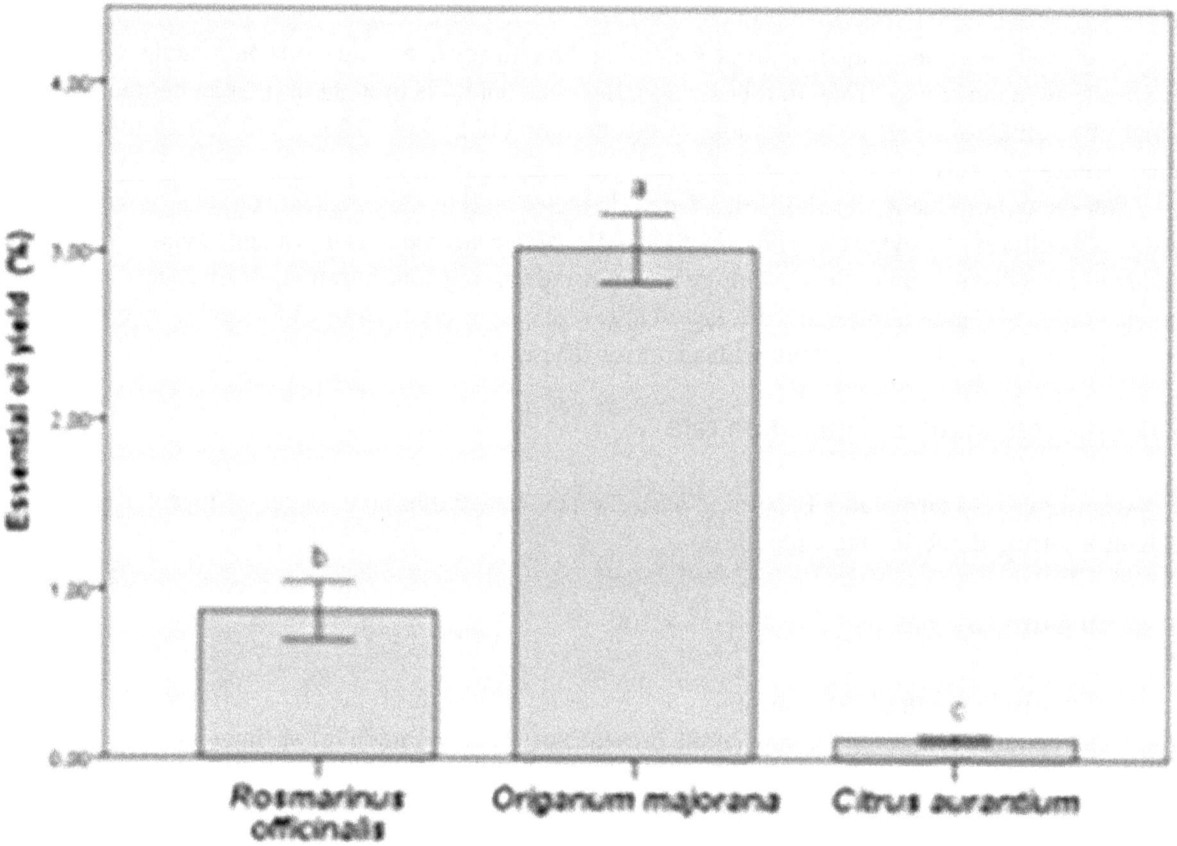

3.2. Organoleptic properties

The essential oils of *Citrus aurantium*, *Rosmarinus officinalis* and *Origanum majorana* are characterized by distinct variations in color, aspect and odor. Details of the results are shown in Table 1.

Table 1. Organoleptic properties of essential oils extracted from the leaves of Citrus aurantium, Rosmarinus officinalis, and Origanum majorana

Plant	Part of the plant	Organoleptic characteristics		
		Color	Aspect	Odor
Citrus aurantium	Leaves	Dark yellow	Mobile liquid and oily in nature	Characteristic odor fresh, lemony
Rosmarinus officinalis	Leaves	Dark yellow	clear liquid	Strong odor (pleasant)
Origanum majorana	Leaves	Light yellow	clear liquid	Cooler, lighter, more pleasant, characteristic

3.3. Refractive index

The results show a significative variation of the refractive index of the essential oils of the plants studied, varying between 1.4742 and 1.4651 and respecting the standard norms.

Table 2. Refractive index of Rosmarinus officinalis, Citrus aurantium and Origanum majorana

Plant	IR	T°(C)	ISO standard
Citrus aurantium	1.4695 ± 0,00003 [b]	20	1.464-1.473
Origanum majorana	1.4651 ± 0,00002 [c]	20	1.470-1.4780
Rosmarinus officinalis	1.4742 ± 0,00002 [a]	20	-

IR: Refractive index; T: Temperature in degrees (C)

3.4. Rotatory power

The results of the rotatory powers of the essential oils differ. In fact, *Rosmarinus officinalis* essential oil has the highest rotatory power at 3.5°, while *Origanum majorana* essential oil has a moderate rotatory power of 1.2°, and finally *Citrus aurantium* essential oil has the lowest rotatory power at 0.25°.

Table 3. Rotatory power of Rosmarinus officinalis, Citrus aurantium and Origanum majorana

Plant	Part used	Rotary power (at 20°C)
Citrus aurantium	Leaves	0.25°
Origanum majorana	Leaves	1.2°
Rosmarinus officinalis	Leaves	3.5°

3.5. Antifungal activity of essential oils

The results of the antifungal activity of *Rosmarinus officinalis*, *Citrus aurantium* and *Origanum majorana* essential oils are presented in Table 2. Regarding the antifungal activity of essential oils, promising results were observed, particularly for *Origanum majorana* essential oil. The latter showed significant antifungal power against the two fungal strains tested, effectively inhibiting their growth at relatively low concentrations. *Rosmarinus officinalis* and *Citrus aurantium* essential oils also showed inhibitory effects on fungal growth, albeit less pronounced than those observed with *Origanum majorana* essential oil.

Table 4. Antifungal activity of Rosmarinus officinalis, Citrus aurantium and Origanum majorana essential oils on Phytophtora palmivora et Alternaria alternata

essential oil	Concentration µL/mL	Fungi	
		Alternaria alternata	*Phytophtora palmivora*
Rosmarinus officinalis	12.5	+	-
	5	+	-
	2.5	+ + +	+
	0.5	+ + +	+
	0.25	+ + +	+ +
	0.08	+ + +	+ + +
Citrus aurantium	6.66	+	+
	3.33	+ +	+ +
	1.5	+ +	+ + +
	0.5	+ + +	+ + +
	0.25	+ + +	+ + +
Origanum majorana	6.66	-	-
	3.33	-	-
	1.5	+	-
	0.5	+ +	+
	0.25	+ + +	+ +

+++ Strong growth; ++ Medium growth; + Weak growth; - Inhibition

3.6. Mycelial growth rate

The results of the average growth rate of the two fungi *Phytophtora palmivora* and *Alternaria alternata* using different concentrations of the essential oils of *Rosmarinus officinalis, Origanum majorana, and Citrus aurantium* are shown in the following Figure 2. The mycelial growth rate decreases progressively with increasing essential oil concentration. Initially, the mycelial growth rate is highest in control, in other words, without the addition of essential oils. This rate decreases as essential oil concentrations increase.

More specifically, the decrease in growth rate is more marked as the concentration of *Origanum majorana, Rosmarinus officinalis* and *Citrus aurantium* essential oil increases. This decrease reaches a point where the growth rate becomes zero at a concentration of 6.66 µl for these three oils. This threshold represents the so-called Minimum Inhibitory Concentration (MIC) of the oil tested, i.e. the minimum concentration required to completely inhibit mycelial growth of the fungus studied.

Figure 2. Average growth rate of the two fungi Phytophtora palmivora and Alternaria alternata using different concentrations of essential oils of A: Rosmarinus officinalis; B: Origanum majorana; C: Citrus aurantium

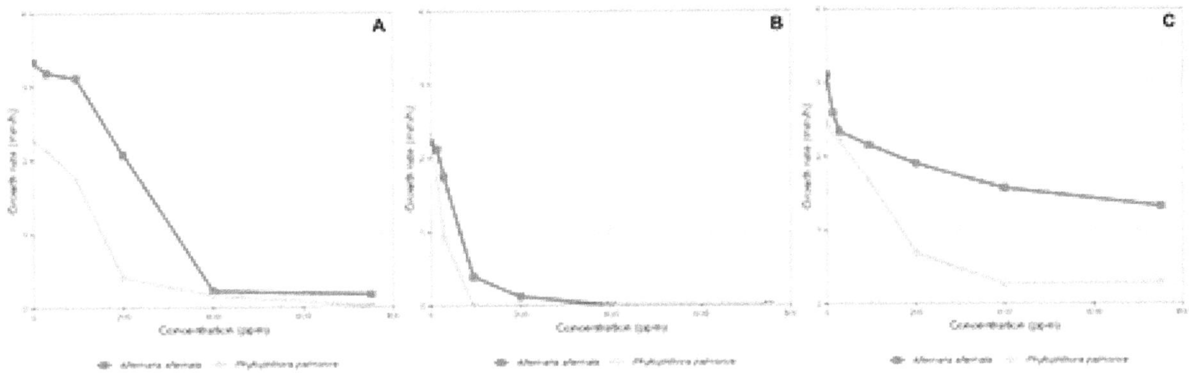

4. DISCUSSION

4.1. Essential oil yield

The results of the essential oil (EO) yield showed that the highest yield was for *Origanum majorana* (3.02%). Comparing these results to previous studies, we found that for *O. majorana*, the EO obtained through hydrodistillation from the aerial parts of the dry plant yielded 1.06% (Makrane et al., 2019). Another study conducted by Ben Salha et al. (2017) reported a yield of 1.7% (Ben Salha et al., 2017), while in Hungary, the EO yield of *O. majorana* was 1.85% (Hajlaoui et al., 2016). In Egypt, the EO yield was 1.20 ± 0.18% (Busatta et al., 2008). Based on these findings, our results were superior to those reported in the literature.

The second highest EO yield was for *Rosmarinus officinalis* with 0.87%. In a study conducted in Mexico, three techniques were used to extract rosemary EO: supercritical CO_2 extraction (SCE), hydrodistillation (HYDRO), and steam distillation (SD). For SCE, the yield ranged between 1.41% and 2.53% (w/w) from 100 g of dry rosemary. The study also showed that the best extraction results were obtained at higher pressures and lower temperatures. For the HYDRO method, a yield of 0.35% was achieved when extracting 25 g, while the highest yield (2.35%) was obtained using the SD method with 50 g of whole rosemary. The highest yield overall was achieved with the SCE method (Conde-Hernández et al., 2017).

In Morocco, using non-conventional hydrodistillation for *R. officinalis* EO yields ranged from 1.0% to 1.3%. While the yields of *R. officinalis* EO obtained through conventional hydrodistillation ranged from 0.58% to 2.81% (El Kharraf et al., 2021). Our results fall slightly above the lower range for both hydrodistillation techniques and below the highest reported yield (2.81%) for the conventional hydrodistillation.

Citron aurantium had the lowest yield among the species studied, with a yield of 0.126%. Compared to a study by Draoui et al. (2023), where essential oils were extracted from *C. aurantium* peels, the EO yield was 1.22%, significantly higher than the yield found in this study (Draoui et al., 2023).

The variation in yield percentage may be influenced by several factors, including climate conditions, the time of plant collection, the specific plant part used, and whether the plants were wild or cultivated (El Kharraf et al., 2021), the extraction technique employed and its parameters, as well as the duration of the extraction process (Conde-Hernández et al., 2017). Therefore, further research should investigate the effects of these factors on EO yield.

4.2. Organoleptic properties

The visual and olfactory characteristics of the essential oils were studied. In this research, *Citrus aurantium* essential oil (EO) was characterized by its dark yellow color, clear liquid appearance and characteristic fresh lemony odor. A study from China reported that the high percentage of two molecules (geranyl acetate and nerol acetate) were responsible for its scent and aroma. They also stated that the flower had a rich aroma and mellow taste, which were attributed to linalool. Additionally, other components like limonene, β-pinene, geraniol, linalyl acetate, α-terpineol, and nerolidol contributed to its profile. Alcohol was mainly responsible for the strong fragrance in the essential oil (Xie et al., 2024). In Egypt, the essential oil from the leaf was described as a very characteristic and pleasant odor (Farag et al., 2020). The sensory characteristics of *Rosmarinus officinalis* were also studied. It was found to have a dark yellow color, a clear appearance, and a very strong odor. Comparing these results to the literature, the species was referred to in ancient Greece as 'libanotis' because of its incense-like smell. Its fragrance was described as quite pleasant, which can be attributed to its richness in volatile oils. The most important flavor compounds were monoterpene hydrocarbons, oxygenated monoterpenes, sesquiterpene hydrocarbons, oxygenated sesquiterpenes, esters, ketones, phenols, alcohols, and others (Ribeiro-Santos et al., 2015) Lastly, *Origanum majorana* was characterized by its light-yellow color, clear liquid aspect, and very pleasant odor. *Origanum majorana* is recognized as a highly aromatic plant, containing many vital phytoconstituents with pharmacological and medicinal value (Ribeiro-Santos et al., 2015). In Nepal, the essential oil of *Origanum majorana* was described as a slightly viscous liquid with a colorless to pale yellow color, and a strong, sweet, and spicy odor (Narayan Paudel et al., 2022). In general, the sensory characteristics of essential oils are the result of the presence of specific compounds, which can be subjected to chemical alterations due to factors like high humidity, drying temperature, and processing methods. These factors can affect the shape of the flavor compounds and the oil's color (Xie et al., 2024). Our results were based on visual and sensory observations, suggesting the need for advanced technologies to obtain more accurate results and to analyze the chemical composition in greater detail.

4.3. Refractive index

The refractive index was measured in this study and compared to other studies. We found that the results were sometimes different and sometimes the same as our findings. Starting with *Rosmarinus officinalis* in Algeria, the refractive index of essential oil extracted using the innovative Microwave Hydrodiffusion and Gravity method showed an index of 1.468, which was close to our results. The second refractive index was 1.470 using traditional hydrodistillation, which was slightly higher than our results. Overall, all the results did not show a significant difference between the refractive index values (Bousbia et al., 2009). In a study conducted in Brazil, where purchased *Rosmarinus officinalis* L. essential oils were extracted by steam distillation, the refractive index at 20°C was 1.51, which is higher than our results. In the same study from Brazil, the second species, *Origanum vulgare* L., showed a refractive index at 20°C of 1.49 (de Medeiros Barbosa et al., 2020), which was a little higher than what we found in our study (1.4742 ± 0.00002). In Tunisia, where the essential oils were supplied from Italy, Citrus had a refractive index of 1.472–1.476 (Aloui et al., 2014), which was higher than both what we found in our results (1.4651 ± 0.00002) and the standard range (1.470–1.478). In general, essential oils can be differentiated by their characteristic olfactory properties (Rasheed et al., 2024).

4.4. Rotatory Power

The rotatory power of essential oils (EOs), which measures their ability to rotate the plane of polarized light, is an important physical property often used for quality control and characterization. Although essential oils are primarily hydrophobic and insoluble in water, they tend to dissolve in organic solvents and other oil-based substances such as vegetable oils and waxes. This property, along with their high refractive index and specific rotatory power, plays a crucial role in their analysis (Nurhaslina et al., 2022). Rotatory power is tied to the chiral nature of essential oil compounds, meaning that different enantiomers, or mirror-image molecules, within the oils can cause varying degrees of optical rotation. The intensity and direction of this rotation are influenced by the chemical composition of the essential oils (Afrokh et al., 2024), the wavelength of the light used, and the enantiomeric excess of the chiral compounds present (Lafhal et al., 2020). In Algeria, EOs from *Citrus sinensis* and *Citrus limonum* have been found to deviate polarized light to the right, classifying them as dextrorotatory. Where *Citrus limonum* exhibited a higher rotatory power indicating its richness in chiral compounds. The rotatory power values can vary significantly among essential oils due to their complex chemical composition and differing proportions of monoterpenes and oxygenated compounds. For instance, in our study *Rosmarinus officinalis* essential oil exhibited the highest rotatory power at 3.5°, reflecting its potent ability to rotate polarized light, possibly due to its high concentration of bioactive terpenes. In comparison, *Origanum majorana* essential oil showed a moderate rotatory power of 1.2°, while *Citrus aurantium* had the lowest value at 0.25°. These differences in rotatory power are significant as they provide insights into the chemical complexity of each oil. Essential oils that display higher rotatory power may contain a greater proportion of specific chiral compounds, which can be essential for both their aromatic properties and biological activity (Sharmeen et al., 2021).

4.5. Antifungal Activity

The antifungal activity of *Rosmarinus officinalis*, *Citrus aurantium*, and *Origanum majorana* essential oils demonstrated promising results (Table 4). *Origanum majorana* essential oil exhibited the most potent antifungal effects for both fungal strains (*Phytophthora palmivora* and *Alternaria alternata*). These results may be related to its rich terpenoid content, known for its antimicrobial properties (Liang et al., 2024). When it comes to *Rosmarinus officinalis*, it completely inhibited *Phytophthora palmivora* at the two highest concentrations tested, while *Citrus aurantium* leaf essential oils showed the lowest activity. The results of the mycelial growth rate (Figure 2) further support the antifungal efficacy of these oils. As the concentration of each essential oil increased, the mycelial growth rate of both fungi significantly. the growth rate became negligible, the more the concentration get higher. Essential oils are known to possess complex chemical profiles, with terpenes and terpenoids being the primary bioactive compounds (Liang et al., 2024). These compounds disrupt key cellular processes in fungi, including the integrity of the cell wall and the function of cell membranes. Moreover, studies have highlighted that specific component, such as 1,8-cineole, camphor, and eucalyptol in *Rosmarinus officinalis*, and limonene in *Citrus aurantium*, are largely responsible for their antifungal activity (Masyita et al., 2022; de Jesus Oliveira et al., 2024). The difference in antifungal potency observed among the three oils can be attributed to the differences in their chemical composition, which may be influenced by factors such as geographical origin, extraction method, and plant part used (Ribeiro-Santos et al., 2015). Furthermore, the findings align with previous studies that have suggested essential oils as effective alternatives to synthetic fungicides in both food preservation and agriculture (Elumalai et al., 2022).

5. CONCLUSION

In conclusion, this study highlights the potential of essential oils extracted from *Rosmarinus officinalis*, *Citrus aurantium* and *Origanum majorana* as promising new control strategies against plant diseases caused by antifungal agents *Phytophthora palmivora* and *Alternaria alternata*. It also highlight the importance of considering variations in the yields and biological activities of these essential oils as based on growing conditions and extraction methods.

Evaluation of the *in vitro* antifungal activity of *Rosmarinus officinalis*, *Citrus aurantium* and *Origanum majorana* essential oils on the mycelial growth of *Phytophtora palmivora* and *Alternaria alternata* reveals promising results. The essential oils of these plants demonstrated a significant ability to inhibit the fungal growth of the two pathogens tested. In particular, *Origanum majorana* essential oil showed strong antifungal activity, followed by *Rosmarinus officinalis* and *Citrus aurantium*.

These results suggest that the essential oils of *Rosmarinus officinalis*, *Citrus aurantium* and *Origanum majorana* could be considered as potential agents in the control of plant fungal diseases caused by *Phytophtora palmivora* and *Alternaria alternata*. This research paves the way for the development of new biological control strategies against pathogens, offering sustainable alternatives to conventional pesticides. However, further studies will be needed to assess the efficacy of these essential oils under field conditions, and their safety for the environment and human health.

REFERENCES

AFNOR. (2000). Recueil de normes: Les huiles essentielles. In La Défense, P. (Ed.), *Monographies Relatives aux Huiles Essentielles; AFNOR* (pp. 661–663).

Afrokh, M., El Mehrach, K., Chatoui, K., Ait Bihi, M., Sadki, H., Zarrouk, A., Tabyaoui, M., & Tahrouch, S. (2024). Quality criteria, chemical composition and antimicrobial activity of the essential oil of *Mentha suaveolens* Ehrh. *Heliyon, 10*(7), e28125. DOI: 10.1016/j.heliyon.2024.e28125 PMID: 38560258

Akutse, K. S., Subramanian, S., Maniania, N. K., Dubois, T., & Ekesi, S. (2020). Biopesticide Research and Product Development in Africa for Sustainable Agriculture and Food Security—Experiences From the International Centre of Insect Physiology and Ecology (Icipe). *Frontiers in Sustainable Food Systems, 4*, 563016. Advance online publication. DOI: 10.3389/fsufs.2020.563016

Alegbeleye, O., Odeyemi, O. A., Strateva, M., & Stratev, D. (2022). Microbial spoilage of vegetables, fruits and cereals. *Applied Food Research, 2*(1), 100122. DOI: 10.1016/j.afres.2022.100122

Baker, B. P., Green, T. A., & Loker, A. J. (2020). Biological control and integrated pest management in organic and conventional systems. *Biological Control, 140*, 104095. DOI: 10.1016/j.biocontrol.2019.104095

Ben Salha, G., Herrera Díaz, R., Labidi, J., & Abderrabba, M. (2017). Deterpenation of *Origanum majorana* L. essential oil by reduced pressure steam distillation. *Industrial Crops and Products, 109*, 116–122. DOI: 10.1016/j.indcrop.2017.08.016

Bose, T., Spies, C. F. J., Hammerbacher, A., & Coutinho, T. A. (2023). *Phytophthora*: An underestimated threat to agriculture, forestry, and natural ecosystems in sub-Saharan Africa. *Mycological Progress, 22*(11), 78. DOI: 10.1007/s11557-023-01926-0

Bousbia, N., Abert Vian, M., Ferhat, M. A., Petitcolas, E., Meklati, B. Y., & Chemat, F. (2009). Comparison of two isolation methods for essential oil from rosemary leaves: Hydrodistillation and microwave hydrodiffusion and gravity. *Food Chemistry, 114*(1), 355–362. DOI: 10.1016/j.foodchem.2008.09.106

Bouyahya, A., Chamkhi, I., Benali, T., Guaouguaou, F. E., Balahbib, A., El Omari, N., Taha, D., Belmehdi, O., Ghokhan, Z., & El Menyiy, N. (2021). Traditional use, phytochemistry, toxicology, and pharmacology of *Origanum majorana* L. *Journal of Ethnopharmacology, 265*, 113318. DOI: 10.1016/j.jep.2020.113318 PMID: 32882360

Busatta, C., Vidal, R. S., Popiolski, A. S., Mossi, A. J., Dariva, C., Rodrigues, M. R. A., Corazza, F. C., Corazza, M. L., Vladimir Oliveira, J., & Cansian, R. L. (2008). Application of *Origanum majorana* L. essential oil as an antimicrobial agent in sausage. *Food Microbiology, 25*(1), 207–211. DOI: 10.1016/j.fm.2007.07.003 PMID: 17993397

Conde-Hernández, L. A., Espinosa-Victoria, J. R., Trejo, A., & Guerrero-Beltrán, J. (2017). CO2-supercritical extraction, hydrodistillation and steam distillation of essential oil of rosemary (*Rosmarinus officinalis*). *Journal of Food Engineering, 200*, 81–86. DOI: 10.1016/j.jfoodeng.2016.12.022

Cui, J., Tian, S., Gu, Y., Wu, X., Wang, L., Wang, J., Chen, X., & Meng, Z. (2023). Toxicity effects of pesticides based on zebrafish (*Danio rerio*) models: Advances and perspectives. *Chemosphere, 340*, 139825. DOI: 10.1016/j.chemosphere.2023.139825 PMID: 37586498 de Jesus Oliveira, J., dos Passos, E. M., Alves, S. M., Sarmento, V. H. V., Bjerk, T. R., Cardoso, J. C., Blanco-Llamero, C., Souto, E. B., Severino, P., & da Costa Mendonça, M. (2024). Microemulsion of essential oil of *Citrus aurantium* var. *dulcis* for control of *Aleurocanthus woglumi* and evaluation of selectivity against *Aschersonia aleyrodis* and *Ceraeochrysa cornuta.Crop Protection (Guildford, Surrey), 178*, 106586. DOI: 10.1016/j.cropro.2024.106586

de Medeiros Barbosa, I., da Cruz Almeida, É. T., Gomes, A. C. A., & de Souza, E. L. (2020). Evidence on the induction of viable but non-culturable state in Listeria monocytogenes by *Origanum vulgare* L. and *Rosmarinus officinalis* L. essential oils in a meat-based broth. *Innovative Food Science & Emerging Technologies*, *62*, 102351. DOI: 10.1016/j.ifset.2020.102351

Draoui, A., Ouinten, M., Gourine, N., & Yousfi, M. (2023). Acute anti-inflammatory activity of Artemisia campestris and *Citrus aurantium*: Effects of essential oil combinations. *Biocatalysis and Agricultural Biotechnology*, *53*, 102860. DOI: 10.1016/j.bcab.2023.102860

El Kharraf, S., El-Guendouz, S., Farah, A., Bennani, B., Mateus, M. C., El Hadrami, E. M., & Miguel, M. G. (2021). Hydrodistillation and simultaneous hydrodistillation-steam distillation of *Rosmarinus officinalis* and *Origanum compactum*: Antioxidant, anti-inflammatory, and antibacterial effect of the essential oils. *Industrial Crops and Products*, *168*, 113591. DOI: 10.1016/j.indcrop.2021.113591

Elumalai, K., Krishnappa, K., Pandiyan, J., Alharbi, N. S., Kadaikunnan, S., Khaled, J. M., Barnard, D. R., Vijayakumar, N., & Govindarajan, M. (2022). Characterization of secondary metabolites from Lamiaceae plant leaf essential oil: A novel perspective to combat medical and agricultural pests. *Physiological and Molecular Plant Pathology*, *117*, 101752. DOI: 10.1016/j.pmpp.2021.101752

Fandohan, P., Gbenou, J. D., Gnonlonfin, B., Hell, K., Marasas, W. F. O., & Wingfield, M. J. (2004). Effect of essential oils on the growth of *Fusarium verticillioides* and fumonisin contamination in corn. *Journal of Agricultural and Food Chemistry*, *52*(22), 6824–6829. DOI: 10.1021/jf040043p PMID: 15506822

FAO. (2019). New standards to curb the global spread of plant pests and diseases.

Farag, M. A., Abib, B., Ayad, L., & Khattab, A. R. (2020). Sweet and bitter oranges: An updated comparative review of their bioactives, nutrition, food quality, therapeutic merits and biowaste valorization practices. *Food Chemistry*, *331*, 127306. DOI: 10.1016/j.foodchem.2020.127306 PMID: 32593794

Freire, J.C.P., Júnior, J.K.O., Silva, D.F., Sousa, J.P., Guerra, F.Q.S., de Oliveira Lima, E. (2017). Antifungal Activity of Essential Oils against *Candida albicans* Strains Isolated from Users of Dental Prostheses. *Evidence-Based Complementary and Alternative Medicine*. Article ID 7158756, 9 pages. https://doi.org/DOI: 10.1155/2017/7158756

Hajlaoui, H., Mighri, H., Aouni, M., Gharsallah, N., & Kadri, A. (2016). Chemical composition and *in vitro* evaluation of antioxidant, antimicrobial, cytotoxicity and anti-acetylcholinesterase properties of Tunisian *Origanum majorana* L. essential oil. *Microbial Pathogenesis*, *95*, 86–94. DOI: 10.1016/j.micpath.2016.03.003 PMID: 26997648

Hawkins, N. J., Bass, C., Dixon, A., & Neve, P. (2019). The Evolutionary Origins of Pesticide Resistance. *Biological Reviews of the Cambridge Philosophical Society*, *94*(1), 135–155. DOI: 10.1111/brv.12440 PMID: 29971903

Hou, T., Sana, S. S., Li, H., Xing, Y., Nanda, A., Netala, V. R., & Zhang, Z. (2022). Essential oils and its antibacterial, antifungal and anti-oxidant activity applications: A review. *Food Bioscience*, *47*, 101716. DOI: 10.1016/j.fbio.2022.101716

Katekar, V. P., Rao, A. B., & Sardeshpande, V. R. (2022). Review of the rose essential oil extraction by hydrodistillation: An investigation for the optimum operating condition for maximum yield. *Sustainable Chemistry and Pharmacy*, *29*, 100783. DOI: 10.1016/j.scp.2022.100783

Ksouri, S., Djebir, S., Bentorki, A. A., Gouri, A., Hadef, Y., & Benakhla, A. (2017). Antifungal activity of essential oils extract from *Origanum floribundum* Munby, *Rosmarinus officinalis* L. and *Thymus ciliatus* Desf. against *Candida albicans* isolated from bovine clinical mastitis. *Journal de Mycologie Médicale*, *27*(2), 245–249. DOI: 10.1016/j.mycmed.2017.03.004 PMID: 28454927

Lafhal, S., Bombarda, I., Dupuy, N., Jean, M., Ruiz, K., Vanloot, P., & Vanthuyne, N. (1610). 2020. Chiroptical fingerprints to characterize lavender and lavandin essential oils. *Journal of Chromatography. A*, *460568*. Advance online publication. DOI: 10.1016/j.chroma.2019.460568 PMID: 31586515

Lainez-Cerón, E., Jiménez-Munguía, M. T., López-Malo, A., & Ramírez-Corona, N. (2021). Effect of process variables on heating profiles and extraction mechanisms during hydrodistillation of eucalyptus essential oil. *Heliyon*, *7*(10), e08234. DOI: 10.1016/j.heliyon.2021.e08234 PMID: 34754975

Lengai, G. M. W., Muthomi, J. W., & Mbega, E. R. (2020). Phytochemical activity and role of botanical pesticides in pest management for sustainable agricultural crop production. *Scientific African*, *7*, e00239. DOI: 10.1016/j.sciaf.2019.e00239

Liang, H., Yang, M., Li, Q., Zhang, L., & Zhao, X. (2024). A comprehensive review of the main components of plant essential oils and the mechanisms responsible for the inhibitory effects on fungal growth and aflatoxin synthesis. *Innovative Food Science & Emerging Technologies*, *96*, 103747. DOI: 10.1016/j.ifset.2024.103747

Makrane, H., Aziz, M., Mekhfi, H., Ziyyat, A., Legssyer, A., Melhaoui, A., Berrabah, M., Bnouham, M., Alem, C., Elombo, F. K., Gressier, B., Desjeux, J. F., & Eto, B. (2019). Myorelaxant Activity of essential oil from *Origanum majorana* L. on rat and rabbit. *Journal of Ethnopharmacology*, *228*, 40–49. DOI: 10.1016/j.jep.2018.08.036 PMID: 30205180

Manathunga, K. K., Gunasekara, N. W., Meegahakumbura, M. K., Ratnaweera, P. B., Faraj, T. K., & Wanasinghe, D. N. (2024). Exploring Endophytic Fungi as Natural Antagonists against Fungal Pathogens of Food Crops. *Journal of Fungi (Basel, Switzerland)*, *10*(9), 606. DOI: 10.3390/jof10090606 PMID: 39330366

Mariod, A.A. (2016). Effect of Essential Oils on Organoleptic (Smell, Taste, and Texture) Properties of Food. *Essential Oils in Food Preservation, Flavor and Safety*. Elsevier. 131-137. https://doi.org/DOI: 10.1016/B978-0-12-416641-7.00013-4

Masyita, A., Mustika Sari, R., Dwi Astuti, A., Yasir, B., Rahma Rumata, N., Emran, T. B., Nainu, F., & Simal-Gandara, J. (2022). Terpenes and terpenoids as main bioactive compounds of essential oils, their roles in human health and potential application as natural food preservatives. *Food Chemistry: X*, *13*, 100217. DOI: 10.1016/j.fochx.2022.100217 PMID: 35498985

Metoui, N., Gargouri, S., Amri, I., Fezzani, T., Jamoussi, B., & Hamrouni, L. (2015). Activity antifungal of the essential oils; aqueous and ethanol extracts from *Citrus aurantium* L. *Natural Product Research, 29*(23), 2238–2241. DOI: 10.1080/14786419.2015.1007136 PMID: 26207731

Narayan Paudel, P., Satyal, P., Satyal, R., Setzer, W. N., Gyawali, R. 2022. molecules Chemical Composition, Enantiomeric Distribution, Antimicrobial and Antioxidant Activities of *Origanum majorana* L. Essential Oil from Nepal. https://doi.org/DOI: 10.3390/molecules27186136

Nizamani, M. M., Hughes, A. C., Zhang, H. L., & Wang, Y. (2024). Revolutionizing agriculture with nanotechnology: Innovative approaches in fungal disease management and plant health monitoring. *The Science of the Total Environment, 928*, 172473. DOI: 10.1016/j.scitotenv.2024.172473 PMID: 38615773

Nurhaslina, C. R., Andi Bacho, S., & Mustapa, A. N. (2022). Review on drying methods for herbal plants. *Materials Today: Proceedings, 63*, S122–S139. DOI: 10.1016/j.matpr.2022.02.052

Nurzyńska-Wierdak, R., & Walasek-Janusz, M. (2025). Chemical Composition, Biological Activity, and Potential Uses of Oregano (*Origanum vulgare* L.) and Oregano Essential Oil. *Pharmaceuticals (Basel, Switzerland), 18*(2), 267. DOI: 10.3390/ph18020267 PMID: 40006079

Ospina, J. D., Tovar, C. D. G., Flores, J. C. M., & Orozco, M. S. S. (2016). Relationship between refractive index and thymol concentration in essential oils of Lippia origanoides Kunth. *Chilean Journal of Agricultural & Animal Science, 32*(2), 127–133. DOI: 10.4067/S0719-38902016000200006

Pergola, M., De Falco, E., Belliggiano, A., & Ievoli, C. (2024). The Most Relevant Socio-Economic Aspects of Medicinal and Aromatic Plants through a Literature Review. *Agriculture, 14*(3), 405. DOI: 10.3390/agriculture14030405

Philip, B., Behiry, S. I., Salem, M. Z. M., Amer, M. A., El-Samra, I. A., Abdelkhalek, A., & Heflish, A. (2024). *Trichoderma afroharzianum* TRI07 metabolites inhibit *Alternaria alternata* growth and induce tomato defense-related enzymes. *Scientific Reports, 14*(1), 1874. DOI: 10.1038/s41598-024-52301-2 PMID: 38253713

Qi, Y., Zhang, R., Zhang, M., Wen, Q., & Shen, J. (2021). Effects of exogenous ascorbic acid on the mycelia growth and primordia formation of *Pleurotus ostreatus.Journal of Basic Microbiology, 61*(8), 736–744. DOI: 10.1002/jobm.202100143 PMID: 34252217

Qiu, K., Wang, S., Duan, F., Sang, Z., Wei, S., Liu, H., & Tan, H. (2024). Rosemary: Unrevealing an old aromatic crop as a new source of promising functional food additive—A review. *Comprehensive Reviews in Food Science and Food Safety, 23*(1), e13273. DOI: 10.1111/1541-4337.13273 PMID: 38284599

Rasheed, H. A., Rehman, A., Li, C., Bai, M., Karim, A., Dai, J., Cui, H., & Lin, L. (2024). Fabrication of *Citrus bergamia* essential oil-loaded sodium caseinate/peach gum nanocomplexes: Physicochemical, spectral, and structural characterization. *International Journal of Biological Macromolecules, 260*, 129475. DOI: 10.1016/j.ijbiomac.2024.129475 PMID: 38262830

Ribeiro-Santos, R., Carvalho-Costa, D., Cavaleiro, C., Costa, H. S., Albuquerque, T. G., Castilho, M. C., Ramos, F., Melo, N. R., & Sanches-Silva, A. (2015). A novel insight on an ancient aromatic plant: The rosemary (*Rosmarinus officinalis* L.). *Trends in Food Science & Technology, 45*(2), 355–368. DOI: 10.1016/j.tifs.2015.07.015

Ristaino, J. B., Anderson, P. K., Bebber, D. P., Brauman, K. A., Cunniffe, N. J., Fedoroff, N. V., & Wei, Q. (2021). The persistent threat of emerging plant disease pandemics to global food security. *Proceedings of the National Academy of Sciences of the United States of America, 118*(23), e2022239118. DOI: 10.1073/pnas.2022239118 PMID: 34021073

Schreinemachers, P., & Tipraqsa, P. (2012). Agricultural pesticides and land use intensification in high, middle and low income countries. *Food Policy, 37*(6), 616–626. DOI: 10.1016/j.foodpol.2012.06.003

Seyyedi-Mansour, S., Carpena, M., Donn, P., Barciela, P., Perez-Vazquez, A., Echave, J., Pereira, A. G., & Prieto, M. A. (2024). Citrus Seed Waste and Circular Bioeconomy: Insights on Nutritional Profile, Health Benefits, and Application as Food Ingredient. *Applied Sciences (Basel, Switzerland), 14*(20), 9463. DOI: 10.3390/app14209463

Sharma, N., & Tripathi, A. (2008). Effects of *Citrus sinensis* (L) Obseck epecarp essential oil on growth and morphogenesis of *Aspergillus niger* (L)." Van Tieghem. *Microbiological Research, 163*(3), 337–344. DOI: 10.1016/j.micres.2006.06.009 PMID: 16870411

Sharmeen, J. B., Mahomoodally, F. M., Zengin, G., & Maggi, F. (2021). Essential oils as natural sources of fragrance compounds for cosmetics and cosmeceuticals. *Molecules (Basel, Switzerland), 26*(3), 666. DOI: 10.3390/molecules26030666 PMID: 33514008

Souto, A. L., Sylvestre, M., Tölke, E. D., Tavares, J. F., Barbosa-Filho, J. M., & Cebrián-Torrejón, G. (2021). Plant-Derived Pesticides as an Alternative to Pest Management and Sustainable Agricultural Production: Prospects, Applications and Challenges. *Molecules (Basel, Switzerland), 26*(16), 4835. DOI: 10.3390/molecules26164835 PMID: 34443421

Tudi, M., Ruan, H. D., Wang, L., Lyu, J., Sadler, R., Connell, D., & Chu, C. (2021). Agriculture Development, Pesticide Application and Its Impact on the Environment. *International Journal of Environmental Research and Public Health, 18*(3), 1112. DOI: 10.3390/ijerph18031112 PMID: 33513796

Wei, Y., Wang, L., & Liu, J. (2023). The diabetogenic effects of pesticides: Evidence based on epidemiological and toxicological studies. *Environmental Pollution, 331*, 121927. DOI: 10.1016/j.envpol.2023.121927 PMID: 37268216

Xie, D., Zhou, Y., Wang, Y., Xu, L., Yang, C., Zhang, G., Zhang, H., Du, Q., & Jin, P. (2024). Insight into sensory characteristics and volatile differences of freeze-dried *Citrus aurantium* L. var. amara Engl. flower essential oil-casein phospholipid nanocomposites sustained-release scented tea by combining GC-E-nose and SBSE-GC-MS. *Food Control, 166*, 110714. DOI: 10.1016/j.foodcont.2024.110714

Chapter 15
Fungal Bioprospecting:
Discovering Novel Bioactive Compounds

P. Selvakumar
https://orcid.org/0000-0002-3650-4548
Nehru Institute of Technology, India

ABSTRACT

Introduction to fungal bioprospecting opens doors to exploring the vast diversity of fungi for discovering novel bioactive compounds, enzymes, and biotechnological applications. Fungi represent a diverse group of eukaryotic microorganisms found in various ecological niches, from terrestrial soils and plant surfaces to extreme environments like deep-sea vents and polar regions. Their metabolic diversity and evolutionary adaptations enable fungi to produce an array of secondary metabolites, enzymes, and biocatalysts with unique chemical structures and biological activities. The field of fungal bioprospecting focuses on systematically screening fungal species and isolates for their potential to produce valuable compounds and bioproducts beneficial to medicine, agriculture, biotechnology, and environmental management. Fungal enzymes, particularly lignocellulolytic enzymes like cellulases, hemicellulases, and ligninases, are essential in bioremediation, biofuel production, and biomass conversion industries.

INTRODUCTION TO FUNGAL BIOPROSPECTING: UNLOCKING THE SECRETS OF FUNGAL DIVERSITY

Introduction to fungal bioprospecting opens doors to exploring the vast diversity of fungi for discovering novel bioactive compounds, enzymes, and biotechnological applications. Fungi represent a diverse group of eukaryotic microorganisms found in various ecological niches, from terrestrial soils and plant surfaces to extreme environments like deep-sea vents and polar regions. Their metabolic diversity and evolutionary adaptations enable fungi to produce an array of secondary metabolites, enzymes, and biocatalysts with unique chemical structures and biological activities. The field of fungal bioprospecting focuses on systematically screening fungal species and isolates for their potential to produce valuable compounds and bioproducts beneficial to medicine, agriculture, biotechnology, and environmental management. Historically, fungi have been a prolific source of bioactive natural products, including antibiotics (e.g., penicillin from Penicillium spp.), immunosuppressants (e.g., cyclosporine from Tolypocladium inflatum), and anticancer agents (e.g., taxol from Taxomyces andreanae). These discoveries have revo-

DOI: 10.4018/979-8-3693-8201-1.ch015

lutionized medical treatments and inspired biotechnological innovations.Fungal enzymes, particularly lignocellulolytic enzymes like cellulases, hemicellulases, and ligninases, are essential in bioremediation, biofuel production, and biomass conversion industries. Fungi play critical roles in organic matter decomposition and nutrient cycling in ecosystems, contributing to soil fertility, plant health, and carbon sequestration. Moreover, fungal bioprospecting extends to exploring fungal biodiversity for sustainable solutions in agriculture, including biocontrol agents against plant pathogens and symbiotic relationships that enhance nutrient uptake in crops.

Advancements in genomic sequencing, metagenomics, and bioinformatics have accelerated the discovery and characterization of fungal diversity and their genetic potential. High-throughput screening techniques and molecular biology tools enable researchers to identify novel fungal strains, isolate key genes responsible for biosynthetic pathways, and optimize production processes for bioactive compounds and enzymes. These technological innovations facilitate the translation of fungal biodiversity into valuable biotechnological applications, driving innovation in pharmaceuticals, agriculture, industrial biotechnology, and environmental bioremediation.Challenges in fungal bioprospecting include accessing diverse fungal habitats, culturing previously uncultivable species, and scaling up production of bioactive compounds economically. Preservation of fungal biodiversity, ethical considerations in bioprospecting, and equitable sharing of benefits from fungal genetic resources are also critical issues. International frameworks such as the Nagoya Protocol provide guidelines for fair and equitable access and benefit-sharing (ABS) related to genetic resources and traditional knowledge associated with fungi. The future of fungal bioprospecting lies in interdisciplinary collaborations across microbiology, genetics, chemistry, ecology, and biotechnology. By harnessing the vast reservoir of fungal diversity and leveraging cutting-edge technologies, researchers aim to discover new drugs, enzymes, biofuels, and agricultural solutions that address global challenges in health, food security, energy sustainability, and environmental conservation. Fungal bioprospecting not only contributes to scientific knowledge but also supports sustainable development goals by promoting biodiversity conservation, ecosystem resilience, and innovation-driven economic growth.

FUNGAL SECONDARY METABOLITES: A TREASURE TROVE OF BIOACTIVE COMPOUNDS

Fungal secondary metabolites constitute a vast and diverse reservoir of bioactive compounds that have captivated scientific interest for their pharmaceutical, agricultural, industrial, and biotechnological applications. These compounds are synthesized by fungi as secondary products of metabolism, distinct from primary metabolites essential for growth and basic cellular functions. Fungal secondary metabolites exhibit an impressive array of chemical structures and biological activities, ranging from antibiotics and anticancer agents to immunosuppressants, antifungals, and enzyme inhibitors (Chun J, et al., 2010). Their complex chemical diversity and potent bioactivities make them valuable sources for drug discovery, crop protection, food preservation, and biocatalysis. Historically, fungi have yielded numerous groundbreaking pharmaceuticals that have revolutionized medicine. Perhaps the most famous example is penicillin, discovered by Alexander Fleming in 1928 from the fungus Penicillium notatum, which marked the beginning of the antibiotic era. Since then, fungi have continued to be prolific sources of antibiotics such as cephalosporins (from Cephalosporium spp.) and macrolides (e.g., erythromycin from Saccharopolyspora erythraea). These antibiotics are vital in combating bacterial infections and saving countless lives worldwide. In addition to antibiotics, fungi produce a plethora of other bioactive compounds with diverse pharmacological activities. For instance, the fungus Tolypocladium inflatum produces cyclosporine, an immunosuppressant used in organ transplantation to prevent rejection. Various fungi also produce anticancer agents such as taxol (from Taxomyces andreanae), which is used to treat ovarian, breast, and lung cancers. Moreover, fungal metabolites like statins (e.g., lovastatin from Aspergillus terreus) are used to lower cholesterol levels and prevent cardiovascular diseases, illustrating the broad therapeutic potential of fungal secondary metabolites.

Beyond pharmaceuticals, fungal secondary metabolites play crucial roles in agriculture as biocontrol agents against plant pathogens. Fungi produce compounds like trichothecenes (e.g., deoxynivalenol) and polyketides (e.g., aflatoxins) that exhibit antifungal, antiviral, and insecticidal properties, protecting crops from pests and diseases. Mycotoxins produced by fungi can also pose risks to human and animal health when contaminating food and feed, highlighting the dual nature of fungal secondary metabolites as both beneficial and detrimental compounds in agricultural settings (Jobstvogt N, et al., 2014).Industrial applications of fungal secondary metabolites extend to enzyme production and biocatalysis. Fungi produce a wide range of enzymes such as cellulases, hemicellulases, ligninases, and proteases, which are crucial for biomass degradation, biofuel production, and bioremediation processes. Enzymes derived from fungi are employed in industries ranging from pulp and paper to textile manufacturing, where they enable efficient bioprocessing and waste valorization. Moreover, fungal enzymes play pivotal roles in biocatalytic transformations for pharmaceutical synthesis, chemical manufacturing, and environmental remediation, offering sustainable alternatives to traditional chemical catalysts (Keating G, et al., 2003). Advancements in fungal genomics, metabolomics, and bioinformatics have revolutionized the discovery and exploitation of fungal secondary metabolites. High-throughput screening techniques, coupled with genome sequencing and bioinformatics analyses, enable researchers to identify gene clusters responsible for biosynthesis and predict chemical structures of novel metabolites. Genetic engineering approaches facilitate the manipulation and optimization of fungal metabolic pathways to enhance production yields and diversify metabolite profiles. These technological innovations accelerate the discovery and development of next-generation bioactive compounds with enhanced therapeutic efficacy, environmental sustainability, and industrial applicability.

Challenges in harnessing fungal secondary metabolites include the identification of novel compounds, optimization of production processes, and overcoming regulatory hurdles for commercialization. Accessing fungal biodiversity, culturing previously uncultivable species, and scaling up production under controlled conditions are ongoing challenges in fungal biotechnology. Moreover, ensuring sustainable practices in fungal metabolite production, addressing environmental impacts, and navigating intellectual property rights and regulatory frameworks are critical considerations in advancing fungal secondary metabolites from discovery to market. In conclusion, fungal secondary metabolites represent a treasure trove of bioactive compounds with vast potential for addressing global challenges in human health, agriculture, industry, and environmental sustainability (Gomes AR, et al., 2014). By exploring and exploiting the chemical diversity of fungi, researchers and biotechnologists continue to unlock new therapeutic agents, biopesticides, industrial enzymes, and biocatalysts that contribute to innovation-driven economic growth and sustainable development. Embracing interdisciplinary approaches, collaboration across sectors, and ethical stewardship of fungal biodiversity are essential in realizing the full potential of fungal secondary metabolites for a healthier, more resilient future

EXPLORING FUNGAL BIODIVERSITY: FROM SOIL TO MARINE ECOSYSTEMS

Exploring fungal biodiversity spans a vast spectrum of ecosystems, from terrestrial soils to marine environments, unveiling the remarkable diversity, ecological roles, and biotechnological potential of fungi across diverse habitats. Fungi, as eukaryotic microorganisms, exhibit a wide array of morphologies, lifestyles, and metabolic strategies that enable them to thrive in virtually every ecological niche on Earth. This diversity is not only crucial for ecosystem functioning but also holds immense promise for discovering novel bioactive compounds, enzymes, and biotechnological applications essential for human health, agriculture, industry, and environmental management. Terrestrial soils represent one of the richest reservoirs of fungal biodiversity, harboring an estimated 2.2 to 3.8 million fungal species, although only a fraction have been characterized. Soil fungi play pivotal roles in organic matter decomposition, nutrient cycling, and plant symbioses, influencing soil fertility, carbon sequestration, and ecosystem resilience (Gloer JB., 2007). Saprotrophic fungi, such as species from the genera Aspergillus, Penicillium, and Trichoderma, degrade complex organic materials like cellulose, lignin, and chitin through enzymatic activities, contributing to soil organic matter turnover and nutrient availability for plants. Additionally, mycorrhizal fungi form symbiotic associations with plant roots, enhancing nutrient uptake—particularly phosphorus—and promoting plant growth and stress tolerance in diverse terrestrial ecosystems.Forests, as complex ecosystems, host diverse fungal communities that interact intimately with trees, understory plants, and soil habitats. Wood-decaying fungi, including white-rot and brown-rot species, play critical roles in forest ecology by decomposing fallen trees and woody debris, thereby recycling nutrients and shaping forest nutrient cycles (Krause J, et al., 2013). Some fungi exhibit mutualistic relationships with plants, such as ectomycorrhizal fungi (e.g., Amanita, Boletus) that form sheaths around tree roots, facilitating nutrient exchange and enhancing tree resilience to environmental stresses. Forest fungi also produce a plethora of secondary metabolites, including antibiotics, antifungals, and enzymes, with potential applications in pharmaceuticals, biotechnology, and bioremediation.

Grasslands, wetlands, and agricultural soils harbor diverse fungal communities adapted to specific environmental conditions and land use practices. Agricultural fungi play dual roles as plant pathogens and beneficial symbionts, influencing crop productivity, disease resistance, and soil health. Plant pathogens such as Fusarium spp., Botrytis spp., and Alternaria spp. cause significant yield losses in crops worldwide, highlighting the economic and ecological impacts of fungal diseases in agriculture. Conversely, beneficial fungi like mycorrhizal species and plant growth-promoting fungi (e.g., Trichoderma) enhance nutrient uptake, improve soil structure, and suppress pathogen proliferation, promoting sustainable agricultural practices and reducing dependency on chemical inputs (Blunt JW, et al., 2015). Beyond terrestrial ecosystems, fungi exhibit remarkable adaptability to aquatic environments, including freshwater habitats, estuaries, and marine ecosystems. Aquatic fungi, predominantly belonging to the phylum Chytridiomycota and the class Oomycetes, play crucial roles in nutrient cycling, organic matter decomposition, and microbial community dynamics in aquatic ecosystems. These fungi decompose submerged plant material, algae, and animal remains, contributing to carbon and nutrient fluxes in freshwater and marine food webs (Selvakumar, P., 2024). Some marine fungi, such as species from the genus Corollospora, exhibit unique adaptations to marine conditions and produce bioactive metabolites with potential pharmaceutical and industrial applications. Exploration of marine fungal biodiversity holds promise for discovering novel enzymes, antibiotics, and antifungals adapted to saline environments, expanding the biotechnological toolkit for marine bioprospecting and sustainable resource utilization.

Technological advancements in molecular biology, genomics, and metagenomics have revolutionized the study of fungal biodiversity by enabling high-throughput sequencing, genome mining, and bioinformatics analyses of fungal communities in diverse habitats. Metagenomic approaches facilitate the discovery of novel fungal taxa and biosynthetic gene clusters responsible for producing bioactive compounds. Additionally, genome sequencing of individual fungal species provides insights into their metabolic potential, ecological roles, and evolutionary adaptations across different environments. These tools empower researchers to uncover hidden fungal diversity, decipher functional traits, and harness biotechnological applications from diverse fungal habitats (Ramakrishnan T.,et al., 2022).Challenges in exploring fungal biodiversity include accessing remote and extreme environments, culturing previously uncultivable species, and integrating multidisciplinary approaches to characterize complex fungal communities. Conservation of fungal biodiversity, ethical considerations in bioprospecting, and equitable sharing of benefits from genetic resources are critical for sustainable management and stewardship of fungal ecosystems. International collaborations, biodiversity conservation initiatives, and policy frameworks such as the Nagoya Protocol facilitate responsible research and development practices in fungal biotechnology, ensuring biodiversity conservation and fair benefit-sharing among stakeholders. In conclusion, exploring fungal biodiversity from soil to marine ecosystems illuminates the vast potential of fungi as sources of bioactive compounds, enzymes, and biotechnological innovations essential for sustainable development. By unraveling the ecological roles, evolutionary adaptations, and biotechnological applications of fungi in diverse habitats, researchers advance knowledge, foster innovation, and promote stewardship of fungal biodiversity for societal benefits. Embracing interdisciplinary approaches, technological innovations, and global partnerships enhances our understanding and utilization of fungal diversity, paving the way for transformative advancements in medicine, agriculture, industry, and environmental sustainability

FUNGAL ISOLATION AND CULTIVATION: TECHNIQUES AND STRATEGIES

Fungal isolation and cultivation techniques are fundamental to studying fungal biodiversity, exploring their metabolic capabilities, and harnessing their biotechnological potential across various applications (Sugumaran K. M, D., et al., 2024). Fungi, as diverse eukaryotic microorganisms, inhabit virtually every ecosystem on Earth, from terrestrial soils and plant surfaces to extreme environments like polar regions and deep-sea habitats. Their metabolic versatility and ecological roles make fungi invaluable sources of bioactive compounds, enzymes, and biocatalysts that contribute to pharmaceuticals, agriculture, industrial processes, and environmental management. The isolation of fungi from natural environments begins with sample collection, where diverse substrates such as soil, plant tissues, water, air, and decaying organic matter are sampled to capture fungal diversity. Sampling strategies consider factors like habitat type, microbial community composition, and seasonal variations to maximize fungal recovery and diversity. Techniques such as soil dilution plating, baiting methods (using selective substrates or host organisms), and molecular approaches (metagenomics, amplicon sequencing) facilitate the isolation and identification of fungal taxa based on morphological characteristics, physiological traits, and molecular markers. Once isolated, fungi are cultivated under controlled laboratory conditions to establish pure cultures for detailed characterization and biotechnological applications. Fungal cultivation techniques aim to replicate natural growth conditions while providing optimal nutrient sources, temperature, pH, and humidity levels conducive to fungal growth and sporulation (Selvakumar, P. et al., 2024). Common culture media for fungal isolation and maintenance include agar-based formulations supplemented with carbon sources (e.g., glucose, malt extract), nitrogen sources (e.g., peptones, yeast extract), vitamins, and trace minerals. Selective media containing antibiotics or inhibitors may also be used to suppress bacterial contaminants and promote fungal growth.

Molecular identification tools, such as DNA barcoding and metagenomic analyses, provide insights into fungal taxonomy, phylogeny, and functional diversity within microbial communities. These approaches aid in distinguishing between closely related species, identifying novel taxa, and elucidating fungal metabolic pathways associated with bioactive compound production (Proksch P, et al., 2002). Cultivation strategies for fungi extend beyond traditional agar plates to include liquid (submerged) fermentation and solid-state fermentation (SSF) methods. Submerged fermentation involves growing fungi in liquid media within bioreactors or shake flasks, where oxygen and nutrients are efficiently supplied, facilitating higher biomass production and metabolite yields. SSF, on the other hand, mimics natural environments by cultivating fungi on solid substrates (e.g., agricultural residues, sawdust) with limited moisture content, promoting the secretion of enzymes and secondary metabolites relevant to biotechnological applications such as enzyme production, biofuel synthesis, and bioremediation.

Optimizing fungal cultivation conditions—such as temperature, pH, agitation, aeration, substrate composition, and fermentation duration—is critical for maximizing biomass productivity and metabolite yields in biotechnological processes. Process optimization strategies include factorial experimental designs, response surface methodologies, and computational modeling to identify optimal growth parameters and production conditions for specific fungal strains and target metabolites. Moreover, scaling up fungal cultivation from laboratory-scale to industrial production requires engineering bioreactor systems that ensure efficient mass transfer, process control, and economic feasibility for large-scale bioprocessing applications (Paramasivam Selvakumar, et al., 2024).Challenges in fungal isolation and cultivation include accessing rare or uncultivable fungal species, overcoming interspecies interactions in microbial consortia, and mimicking complex environmental conditions in laboratory settings. Innovations in cultivation technologies, genomic approaches, and bioinformatics tools continue to advance the frontiers of fungal biotechnology by enhancing our understanding of fungal diversity, metabolic capabilities, and ecological functions. These advancements pave the way for discovering novel bioactive compounds, enzymes, and sustainable biotechnological solutions that address global challenges in health, food security, energy sustainability, and environmental conservation .In conclusion, fungal isolation and cultivation techniques are pivotal in unlocking the biotechnological potential of fungi, from biodiversity exploration to industrial applications. By integrating diverse isolation strategies, advanced cultivation methods, and molecular characterization tools, researchers and biotechnologists expand the repertoire of fungal resources available for developing innovative products and processes (Vijayakumar G, et al., 2024). Embracing interdisciplinary approaches, technological innovations, and collaborative efforts across academia, industry, and regulatory sectors ensures sustainable utilization and stewardship of fungal biodiversity for societal benefits and environmental sustainability.

SCREENING FOR BIOACTIVITY: ASSAYS AND METHODS FOR DETECTING NOVEL COMPOUNDS

Screening for bioactivity is a critical step in drug discovery and development, essential for identifying novel compounds with therapeutic potential. This process involves a variety of assays and methods designed to detect specific biological activities that indicate a compound's ability to interact with a biological target or pathway of interest. One of the most common assays used is the enzymatic assay, which measures the activity of enzymes involved in crucial biochemical pathways. This type of assay is valuable for identifying inhibitors or activators of enzyme function, which can lead to the development of drugs targeting diseases such as cancer or metabolic disorders. Cell-based assays represent another vital category, utilizing cultured cells to assess how compounds affect cellular processes like proliferation, apoptosis, or signal transduction (Selvakumar, P., 2024). These assays provide more complex and physiologically relevant information compared to enzymatic assays, offering insights into how compounds might behave in living organisms. Moreover, high-throughput screening (HTS) techniques have revolutionized bioactivity screening by allowing researchers to test thousands to millions of compounds rapidly. HTS systems automate the process of testing large compound libraries against biological targets, increasing efficiency and throughput in drug discovery programs.In addition to these methods, receptor-binding assays play a crucial role in identifying compounds that interact with specific receptors on cell membranes or within cells. By measuring the binding affinity and specificity of compounds to receptors, researchers can identify potential candidates for modulating receptor activity, important in designing drugs that target diseases involving hormone regulation, neurotransmission, or immune response. Furthermore, bioactivity screening encompasses biophysical methods such as surface plasmon resonance (SPR) and nuclear magnetic resonance (NMR) spectroscopy, which directly measure interactions between compounds and biological molecules. These techniques provide detailed information on binding kinetics, affinity, and structural changes induced by compound binding, aiding in the optimization of drug candidates.

Moreover, advances in computational methods have significantly enhanced bioactivity screening by predicting how compounds might interact with biological targets based on structural data. Molecular docking simulations, for example, predict the binding modes of compounds to target proteins, guiding experimental efforts towards the most promising candidates (Saslis-Lagoudakis CH, et al., 2012). Virtual screening approaches complement experimental methods by screening vast chemical libraries against three-dimensional models of target proteins, identifying compounds likely to exhibit desired bioactivities. Furthermore, bioactivity screening extends beyond traditional drug discovery to include the identification of compounds with potential agricultural, environmental, or industrial applications. In agriculture, for instance, bioactivity assays help identify herbicides, pesticides, or growth regulators that target specific pathways in plants or pests. Similarly, environmental screening focuses on compounds that affect ecosystems, such as pollutants or remediation agents. Industrial applications involve screening for compounds that improve manufacturing processes, enhance product performance, or have novel material properties (Paramasivam Selvakumar, et al., 2024). The integration of diverse bioactivity screening approaches facilitates the discovery of compounds with therapeutic, agricultural, environmental, or industrial potential. By combining experimental assays, computational modeling, and biophysical techniques, researchers can identify lead compounds with desired bioactivities and optimize their chemical structures for improved efficacy, safety, and specificity. Ultimately, bioactivity screening represents a cornerstone of modern drug discovery and other scientific disciplines, driving innovation and advancing our understanding of biological systems.

NATURAL PRODUCTS FROM FUNGI: ANTIBIOTICS, ANTICANCER AGENTS, AND MORE

Natural products derived from fungi have garnered significant attention due to their diverse array of bioactive properties, including antibiotics, anticancer agents, immunosuppressants, and more. Fungi are prolific producers of secondary metabolites, chemical compounds not directly involved in growth or reproduction but crucial for their interactions with the environment (Molinski TF, et al., 2009). These metabolites often possess unique chemical structures and biological activities, making fungi a valuable source for drug discovery and development. One of the most renowned groups of natural products from fungi is antibiotics. Penicillin, the first antibiotic discovered by Alexander Fleming from the fungus Penicillium notatum, revolutionized medicine by treating bacterial infections. Since then, numerous antibiotics have been isolated from fungi, such as cephalosporins from Cephalosporium spp. and macrolides from Streptomyces spp., which continue to play vital roles in combating bacterial infections worldwide. Fungal antibiotics exhibit diverse mechanisms of action, targeting bacterial cell walls, protein synthesis, or nucleic acid synthesis, thereby offering therapeutic options against a broad spectrum of pathogens. In addition to antibiotics, fungi produce a wide range of anticancer agents. Compounds like paclitaxel (taxol), isolated from Taxus brevifolia, were initially sourced from plants but are now also produced by fungal endophytes of yew trees. Paclitaxel interferes with microtubule function, inhibiting cell division and thereby showing efficacy against various cancers. Fungal-derived anticancer agents include vinca alkaloids (vincristine and vinblastine) from Catharanthus roseus, which are used to treat leukemia and lymphomas, as well as camptothecin derivatives from Camptotheca acuminata, exhibiting potent activity against solid tumors.

Moreover, fungi produce immunosuppressants that find applications in organ transplantation and autoimmune diseases (Zhang D, et al., 2012). The discovery and development of natural products from fungi involve various methods, including traditional bioassay-guided fractionation and modern techniques like genome mining and metagenomics. Bioassay-guided fractionation entails isolating and purifying bioactive compounds based on their activity in biological assays, while genome mining involves scanning fungal genomes for biosynthetic gene clusters responsible for producing secondary metabolites. Metagenomics explores the genetic material recovered directly from environmental samples to discover novel biosynthetic pathways and bioactive compounds (Mayer AMS, et al., 2010).In conclusion, natural products from fungi represent a vast and valuable resource for drug discovery and development, offering compounds with diverse biological activities including antibiotics, anticancer agents, immunosuppressants, and more. Their complex chemical structures and unique mechanisms of action continue to inspire researchers in the search for new therapies against human diseases. As scientific techniques advance, unlocking the potential of fungal biodiversity holds promise for addressing current and emerging health challenges globally.

FUNGAL ENZYMES: INDUSTRIAL APPLICATIONS AND BIOTECHNOLOGICAL POTENTIAL

Fungal enzymes are pivotal in various industrial applications and hold immense biotechnological potential due to their versatility, efficiency, and eco-friendly nature. Fungi produce an extensive array of enzymes that catalyze diverse biochemical reactions, offering solutions across several industries including food and beverages, biofuels, textiles, pulp and paper, pharmaceuticals, and environmental remediation. In the food and beverage industry, fungal enzymes play critical roles in food processing and production (Matz C, et al., 2008). For instance, enzymes like amylases, proteases, lipases, and cellulases derived from fungi are used to improve the texture, flavor, and nutritional properties of food products. Amylases break down starch into simpler sugars, proteases facilitate protein hydrolysis for meat tenderization or cheese ripening, lipases enhance flavor in cheese and dairy products, and cellulases aid in fruit juice extraction or enhance clarity in wine production. These enzymes not only improve food quality but also increase process efficiency and reduce production costs. In the biofuels sector, fungal enzymes are integral to the production of bioethanol and other renewable fuels. Cellulases and hemicellulases from fungi are employed in biomass degradation, breaking down complex plant materials (cellulose and hemicellulose) into fermentable sugars (Martins A, et al., 2014). This enzymatic hydrolysis step is crucial in bioethanol production from agricultural residues, forestry wastes, or energy crops. Fungal enzymes are also explored for lignin degradation, a challenging but essential process for the complete utilization of lignocellulosic biomass in biofuel production.

Textile industries utilize fungal enzymes for fabric processing and finishing. Enzymes like cellulases are employed to achieve stone-washed denim effects by selectively hydrolyzing cellulose fibers on denim fabric, creating a distressed appearance. Moreover, fungal enzymes are pivotal in biobleaching of textiles, replacing harsh chemical bleaching agents with eco-friendly enzymatic processes that reduce environmental impact and operational costs. In the pulp and paper industry, fungal enzymes contribute to the reduction of chlorine-based chemicals used in bleaching processes. Laccases and peroxidases derived from fungi are effective in delignification and bleaching of pulp, enhancing paper quality while minimizing environmental pollution. These enzymes catalyze the degradation of lignin, facilitating pulp bleaching with reduced energy consumption and chemical usage.

Environmental applications of fungal enzymes include bioremediation and waste management. Fungi produce ligninolytic enzymes such as lignin peroxidases, manganese peroxidases, and versatile peroxidases, which degrade recalcitrant pollutants like polycyclic aromatic hydrocarbons (PAHs) and chlorinated compounds. These enzymes play a vital role in the cleanup of contaminated soils, wastewater treatment, and detoxification of industrial effluents, contributing to environmental sustainability (Mahadev Madgule, et al., 2023). The biotechnological potential of fungal enzymes is further enhanced by advances in enzyme engineering, protein engineering, and metagenomics. Directed evolution and rational design approaches are used to enhance enzyme properties such as substrate specificity, stability, and activity under various industrial conditions. Metagenomics allows for the discovery of novel fungal enzymes from environmental samples, expanding the enzymatic toolkit available for biotechnological applications. In conclusion, fungal enzymes are indispensable in industrial processes across multiple sectors, offering sustainable and efficient solutions for diverse applications. Their ability to catalyze specific reactions under mild conditions makes them preferable over traditional chemical catalysts, aligning with the growing demand for green and sustainable technologies. As research continues to unveil new fungal enzymes and improve existing ones through biotechnological approaches, their role in shaping a sustainable bioeconomy becomes increasingly significant.

FUNGAL VOLATILES: UNLOCKING THE SECRETS OF FUNGAL COMMUNICATION

Fungal volatiles, often overlooked compared to other forms of fungal communication like chemical signaling or mycelial networks, play a crucial role in mediating interactions within fungal communities and with their environment. These volatile organic compounds (VOCs) are small molecules released by fungi into the surrounding air, where they serve diverse functions ranging from defense mechanisms to influencing growth and development of neighboring organisms. The study of fungal volatiles has gained attention in recent years due to their potential applications in agriculture, medicine, industry, and environmental management. In natural ecosystems, fungal volatiles contribute significantly to intra- and interspecies communication. Fungi release volatiles as part of their chemical language, enabling communication between individuals of the same species and even across different species. This communication is essential for coordinating behaviors such as spore germination, hyphal growth, mating, and defense against predators or competitors. For example, some fungi release volatiles that attract or repel specific organisms, influencing their spatial distribution and ecological interactions within microbial communities (Tarman K, et al., 2011).Moreover, fungal volatiles play crucial roles in plant-fungal interactions. Many fungi establish symbiotic relationships with plants, forming mycorrhizal associations or acting as endophytes. In these interactions, fungal volatiles can function as signaling molecules that modulate plant growth, enhance nutrient uptake, or induce systemic resistance against pathogens. Conversely, some fungal volatiles act as virulence factors that promote pathogenicity by suppressing plant defenses or facilitating host colonization. Understanding these interactions is crucial for developing sustainable agricultural practices and enhancing crop productivity.

In addition to ecological roles, fungal volatiles have promising applications in various industries and technologies. For instance, certain fungal volatiles possess antimicrobial properties, making them potential candidates for developing natural preservatives in food and cosmetics. These compounds inhibit the growth of spoilage microbes and foodborne pathogens, offering safer alternatives to synthetic preservatives (Mora C, et al., 2011). Similarly, fungal volatiles with insecticidal properties have implications in pest management strategies, potentially reducing reliance on chemical pesticides. Medical research has also begun exploring the therapeutic potential of fungal volatiles. Some volatile compounds produced by fungi exhibit antimicrobial activities against human pathogens, including bacteria, fungi, and even multidrug-resistant strains. These compounds may serve as leads for developing new antimicrobial agents or adjunct therapies in treating infectious diseases. Furthermore, fungal volatiles are being investigated for their role in modulating immune responses and inflammatory processes, suggesting potential applications in immune-modulatory therapies or treatments for inflammatory disorders.

Environmental applications of fungal volatiles include their role in bioremediation and waste management. Certain fungi produce volatiles that facilitate the degradation of organic pollutants in contaminated environments. These compounds promote the growth of pollutant-degrading microorganisms or enhance enzymatic activities involved in biodegradation processes. Harnessing fungal volatiles in bioremediation strategies holds promise for addressing environmental challenges associated with industrial pollution and agricultural runoff (Wildman HG., 1999).Advances in analytical techniques such as gas chromatography-mass spectrometry (GC-MS) and molecular biology tools have facilitated the identification and characterization of fungal volatiles. These methods enable researchers to profile complex volatile mixtures produced by fungi under different growth conditions or in interactions with other organisms. Furthermore, genetic and genomic approaches provide insights into the biosynthetic pathways responsible for volatile production, paving the way for manipulating fungal volatiles for desired applications through metabolic engineering or synthetic biology. In conclusion, fungal volatiles represent a fascinating frontier in fungal biology, offering insights into the intricate mechanisms of fungal communication and their ecological significance. Beyond ecological roles, these compounds hold promise for diverse applications in agriculture, medicine, industry, and environmental management. Continued research into fungal volatiles promises to uncover new bioactive compounds, improve our understanding of fungal interactions with their environment, and inspire innovative solutions to global challenges in agriculture, healthcare, and sustainability.

GENOMIC APPROACHES TO FUNGAL BIOPROSPECTING: MINING FUNGAL GENOMES FOR NOVEL COMPOUNDS

Genomic approaches have revolutionized fungal bioprospecting by facilitating the exploration and exploitation of fungal genomes for the discovery of novel bioactive compounds with potential applications in medicine, agriculture, industry, and environmental management. Fungi are prolific producers of diverse secondary metabolites, many of which exhibit valuable biological activities such as antimicrobial, anticancer, immunosuppressive, and enzyme-modulating properties (Leal MC, et al., 2012). Traditional methods of discovering these compounds relied on cultivation-based approaches, which often overlooked the vast majority of fungi that do not readily grow under laboratory conditions. Genomic tools have overcome these limitations by enabling the study of fungal genetic diversity, metabolic potential, and biosynthetic pathways, thereby unlocking the vast reservoir of untapped fungal biodiversity. One of the primary genomic approaches in fungal bioprospecting is genome sequencing. Advances in high-throughput sequencing technologies, coupled with bioinformatics tools for genome assembly and annotation, have facilitated the rapid and cost-effective sequencing of fungal genomes. This has led to the generation of large databases containing genomic information from diverse fungal species, including both known and previously uncultured fungi from environmental samples. Comparative genomics analyses provide insights into the genetic basis of fungal diversity, evolution, and adaptation to various ecological niches. Importantly, genome sequencing allows researchers to predict and prioritize potential biosynthetic gene clusters (BGCs) responsible for the production of secondary metabolites, including novel compounds of biotechnological interest.

Functional genomics approaches complement genome sequencing by elucidating the biological functions and regulatory mechanisms of fungal genes involved in secondary metabolism. Techniques such as transcriptomics, proteomics, and metabolomics provide insights into gene expression patterns, protein synthesis, and metabolite profiles under different growth conditions or in response to environmental stimuli. This approach has led to the discovery of novel biosynthetic pathways and bioactive compounds from previously inaccessible microbial resources (Demunshi Y, et al., 2010). Metagenomic libraries enable functional screening of environmental DNA for specific enzymatic activities or bioactivities, facilitating the identification of novel enzymes and natural products with potential applications in biotechnology and drug discovery. Moreover, synthetic biology and metabolic engineering approaches harness genomic insights to engineer fungal hosts for enhanced production of bioactive compounds. By manipulating biosynthetic pathways and regulatory networks, researchers can optimize metabolic flux towards desired metabolites, improve production yields, and create derivatives with enhanced pharmacological or industrial properties. Synthetic biology tools enable the construction of genetic constructs for heterologous expression of fungal biosynthetic pathways in well-characterized host organisms, providing scalable platforms for sustainable production of natural products and their analogs. In conclusion, genomic approaches have transformed fungal bioprospecting into a data-driven and systematic endeavor, accelerating the discovery and development of novel bioactive compounds from fungi. These approaches leverage genomic sequencing, bioinformatics, functional genomics, metagenomics, and synthetic biology to uncover the metabolic potential of fungi and translate genomic information into biotechnological applications (Zhang Y, et al., 2011). By expanding our understanding of fungal biodiversity and metabolic diversity, genomic bioprospecting holds promise for addressing global challenges in healthcare, agriculture, environmental sustainability, and industrial innovation through the sustainable utilization of fungal resources.

FUNGAL-HOST INTERACTIONS: UNCOVERING THE SECRETS OF SYMBIOTIC RELATIONSHIPS

Fungal-host interactions represent a complex and diverse array of symbiotic relationships that play pivotal roles in ecosystems, agriculture, and human health. These interactions encompass a spectrum from mutualistic to pathogenic associations, where fungi interact with various hosts including plants, animals, and other fungi. Understanding the mechanisms and dynamics of these interactions is crucial for harnessing their benefits in agriculture, biotechnology, and medicine while mitigating their detrimental effects on crops and human health. In agriculture, fungal symbioses are predominantly beneficial, particularly in the form of mycorrhizal associations between fungi and plant roots (Liu X, et al., 2010). Arbuscular mycorrhizal fungi (AMF), for instance, form symbiotic relationships with the majority of land plants, enhancing nutrient uptake, water absorption, and overall plant health. These fungi extend their hyphae into the soil, effectively increasing the root surface area and facilitating the exchange of nutrients such as phosphorus and nitrogen between the fungus and the plant. This mutualistic interaction is essential for sustainable agriculture, improving crop productivity, nutrient efficiency, and resilience to environmental stresses such as drought and soil salinity. Conversely, some fungal-host interactions in agriculture can be detrimental, leading to plant diseases and yield losses. Pathogenic fungi cause various plant diseases by colonizing plant tissues, secreting toxins, or disrupting plant physiological processes. For example, fungal pathogens like Fusarium spp., Botrytis cinerea, and Phytophthora infestans cause devastating diseases in crops such as wheat, grapes, and potatoes, respectively. Understanding the molecular mechanisms underlying these pathogenic interactions is crucial for developing effective strategies for disease management, including genetic resistance in crops and environmentally sustainable fungicide treatments.

In the context of human health, fungal-host interactions also encompass beneficial and harmful relationships. Fungi colonize the human body as part of the microbiota, contributing to host health through metabolic activities, immune modulation, and protection against pathogens. For example, Candida albicans and Malassezia spp (Bhatnagar I, et al., 2010). are commensal fungi found on human skin and mucosal surfaces, where they play roles in maintaining microbial balance and barrier function. However, under certain conditions such as immunosuppression or disrupted microbiota, these fungi can cause opportunistic infections ranging from superficial skin infections to life-threatening systemic diseases (Bowler C, et al., 2009). Furthermore, fungal pathogens are responsible for a significant burden of human diseases worldwide, including superficial infections (e.g., dermatophytosis), respiratory infections (e.g., aspergillosis), and systemic infections (e.g., candidiasis and cryptococcosis). These pathogens exploit host vulnerabilities to establish infections, evade immune responses, and cause tissue damage through mechanisms such as adhesion, invasion, toxin production, and immune evasion strategies. Understanding the virulence factors and host-pathogen interactions of these fungi is crucial for developing novel antifungal therapies, vaccines, and diagnostic tools to combat fungal infections and improve patient outcomes. Beyond agriculture and human health, fungal-fungal interactions also play critical roles in shaping microbial communities and ecosystem dynamics. Fungi interact with each other through competition for resources, cooperation in nutrient acquisition, and chemical signaling. These interactions influence fungal community structure, biodiversity, and ecosystem functions such as decomposition, nutrient cycling, and soil fertility. For example, saprotrophic fungi decompose organic matter, releasing nutrients that support plant growth and ecosystem productivity. Additionally, fungal interactions can lead to the production of secondary metabolites with ecological roles, such as antibiotics and toxins that mediate competition or defense against predators.

Research into fungal-host interactions leverages multidisciplinary approaches including genomics, transcriptomics, proteomics, metabolomics, and advanced imaging techniques to unravel the molecular mechanisms underlying symbiotic relationships and pathogenicity. Genomic studies reveal the genetic basis of fungal adaptations to host environments, including the evolution of symbiotic traits and virulence factors. Transcriptomic analyses provide insights into gene expression dynamics during host colonization and response to environmental cues. Proteomics and metabolomics elucidate the biochemical pathways and metabolite profiles involved in fungal-host interactions, offering targets for therapeutic intervention and biotechnological applications (Bose U, et al., 2014).In conclusion, fungal-host interactions encompass a diverse range of symbiotic relationships that profoundly impact ecosystems, agriculture, and human health. These interactions span from mutualistic associations that enhance nutrient cycling and plant productivity to pathogenic interactions that threaten crops and human well-being. Understanding the mechanisms underlying these interactions is essential for harnessing their benefits in sustainable agriculture, biotechnological innovation, and disease management, while developing strategies to mitigate their negative impacts. Advances in genomic and systems biology approaches continue to deepen our understanding of fungal-host interactions, paving the way for novel solutions to global challenges in food security, environmental sustainability, and human health.

FUNGAL BIOPROSPECTING FOR SUSTAINABLE AGRICULTURE: PESTICIDES, FERTILIZERS, AND MORE

Fungal bioprospecting holds significant promise for advancing sustainable agriculture through the development of environmentally friendly pesticides, biofertilizers, and other innovative solutions. Fungi possess a remarkable diversity of biochemical pathways and metabolic capabilities that can be harnessed to address challenges such as pest control, nutrient management, soil health improvement, and sustainable crop production practices. One of the key areas where fungal bioprospecting shows promise is in the development of biopesticides. Biopesticides derived from fungi offer alternatives to synthetic chemical pesticides, which can have adverse effects on human health, non-target organisms, and the environment. Fungi produce a variety of bioactive compounds with pesticidal properties, including secondary metabolites such as polyketides, peptides, and terpenoids (Beattie AJ, et al., 2011). For example, fungi like Beauveria bassiana and Metarhizium anisopliae are well-known for their insecticidal properties and are used as biocontrol agents against agricultural pests such as aphids, whiteflies, and beetles. These fungi infect and kill pests through mechanisms such as cuticle penetration and secretion of toxins, providing effective and sustainable pest management solutions. In addition to biopesticides, fungi contribute to the development of biofertilizers that enhance soil fertility and plant nutrient uptake. Mycorrhizal fungi form symbiotic associations with plant roots, improving nutrient acquisition by extending hyphal networks into the soil and facilitating the uptake of phosphorus, nitrogen, and micronutrients. Arbuscular mycorrhizal fungi (AMF) are particularly beneficial in agricultural systems, enhancing crop yields, water use efficiency, and tolerance to environmental stresses such as drought and salinity. Moreover, some fungi have the ability to solubilize phosphate and fix atmospheric nitrogen, further enhancing soil fertility and reducing the dependence on chemical fertilizers.

Furthermore, fungi contribute to the development of biostimulants that promote plant growth and enhance resilience to biotic and abiotic stresses. Fungal-derived biostimulants contain compounds such as phytohormones, amino acids, vitamins, and polysaccharides that stimulate plant growth, improve nutrient uptake, and activate defense mechanisms against pathogens and environmental stressors (Abida H, et al., 2013). For example, extracts from fungi like Trichoderma spp. and Mycelium can stimulate plant growth, induce systemic resistance, and enhance crop productivity under challenging conditions. These biostimulants offer sustainable alternatives to chemical growth promoters and help optimize agricultural practices for improved resource use efficiency. Moreover, fungi contribute to soil health and ecosystem resilience through their roles in organic matter decomposition and nutrient cycling. Saprotrophic fungi play crucial roles in breaking down complex organic materials such as crop residues and leaf litter, releasing nutrients that are essential for plant growth and microbial activity. Fungal decomposition processes contribute to soil structure improvement, water retention capacity, and carbon sequestration, thereby enhancing soil fertility and mitigating climate change impacts. Additionally, fungal communities influence soil biodiversity and ecosystem stability through interactions with other microorganisms and plants, further supporting sustainable agricultural practices. Advances in biotechnological and genomic tools have accelerated fungal bioprospecting efforts by enabling the discovery and optimization of beneficial fungal traits for agricultural applications. Genomic sequencing, bioinformatics, and metagenomics facilitate the identification and characterization of fungal genes involved in the synthesis of bioactive compounds, nutrient cycling, and symbiotic interactions with plants. Functional genomics approaches elucidate the molecular mechanisms underlying fungal-host interactions, stress tolerance, and biocontrol activities, providing insights into how fungi can be manipulated or engineered for enhanced agricultural performance.

FUNGAL BIOPROSPECTING FOR HUMAN HEALTH: NOVEL THERAPEUTICS AND DIAGNOSTICS

Fungal bioprospecting for human health represents a burgeoning field with immense potential to discover novel therapeutics, diagnostics, and biomedical applications. Fungi produce a diverse array of bioactive compounds, ranging from antibiotics and anticancer agents to immunomodulators and enzymes, which have been harnessed for their medicinal properties. These compounds offer promising leads for drug discovery, addressing emerging infectious diseases, antibiotic resistance, cancer therapy, and various other health challenges. One of the most significant contributions of fungal bioprospecting to human health is in the discovery of antibiotics. These antibiotics target diverse bacterial pathogens by inhibiting cell wall synthesis, protein synthesis, or nucleic acid replication, offering critical therapies against infections that are increasingly resistant to conventional antibiotics (Abhilasha A., et al., 2022). In addition to antibiotics, fungi produce a wide range of bioactive compounds with potential applications in cancer therapy. Examples include taxanes such as paclitaxel (Taxol) from Taxus brevifolia and camptothecin derivatives from Camptotheca acuminata, which interfere with microtubule dynamics and topoisomerase activity, respectively, leading to inhibition of cancer cell proliferation and induction of apoptosis. Fungal-derived anticancer agents like these have been developed into chemotherapeutic drugs for treating various types of cancer, including breast cancer, ovarian cancer, and lung cancer, illustrating their clinical relevance and therapeutic efficacy.

Moreover, fungi contribute to the development of immunomodulatory agents that regulate immune responses and inflammatory processes. For example, fungal-derived beta-glucans stimulate immune cells such as macrophages and dendritic cells, enhancing their phagocytic activity and cytokine production. Beta-glucans also modulate adaptive immune responses by promoting antibody production and T-cell activation, which can be beneficial in boosting immune function against infections or as adjuvants in vaccines. Additionally, fungal metabolites like cyclosporin A from Tolypocladium inflatum act as potent immunosuppressants, inhibiting T-cell activation and cytokine production, and are used to prevent organ rejection in transplant patients and treat autoimmune diseases. Furthermore, fungal enzymes play pivotal roles in biotechnological and diagnostic applications. Enzymes such as lipases, proteases, and amylases derived from fungi are used in industrial processes for food processing, pharmaceutical production, and environmental remediation. Fungal enzymes are also employed in diagnostic assays for detecting biomarkers of diseases, pathogens, or metabolic disorders.

Moreover, fungal bioprospecting extends beyond natural product discovery to include synthetic biology approaches for designing novel molecules and biologics. Synthetic biology techniques enable the engineering of fungal hosts for heterologous expression of biosynthetic pathways, production of complex molecules, and development of next-generation therapeutics such as antibody-drug conjugates or biopharmaceuticals. This interdisciplinary approach integrates principles of molecular biology, genetic engineering, and bioinformatics to accelerate innovation in drug discovery and biomedicine, addressing unmet medical needs and advancing personalized medicine strategies. In conclusion, fungal bioprospecting for human health holds immense promise for discovering novel therapeutics, diagnostics, and biotechnological applications. From antibiotics and anticancer agents to immunomodulators and diagnostic enzymes, fungi offer a rich source of bioactive compounds that can address global health challenges, including infectious diseases, cancer, autoimmune disorders, and metabolic conditions. Continued research and innovation in fungal biotechnology are essential for translating these discoveries into clinical applications, improving patient outcomes, and advancing healthcare solutions in an increasingly complex and interconnected world.

FUNGAL BIOPROSPECTING FOR ENVIRONMENTAL REMEDIATION: MYCOREMEDIATION AND MORE

Fungal bioprospecting for environmental remediation, particularly through mycoremediation, represents a promising approach to addressing pollution and environmental degradation caused by contaminants such as heavy metals, pesticides, petroleum hydrocarbons, and emerging pollutants. Fungi possess unique metabolic capabilities and enzymatic machinery that enable them to degrade, transform, or sequester pollutants in soil, water, and air environments, offering sustainable and eco-friendly solutions to mitigate environmental impacts. One of the key applications of fungal bioprospecting in environmental remediation is mycoremediation, which utilizes fungi to degrade or detoxify pollutants in contaminated ecosystems. Moreover, fungi contribute to the bioremediation of heavy metal-contaminated soils and sediments through processes such as biosorption and bioaccumulation. Certain fungi possess metal-binding proteins and chelating compounds, such as phytochelatins and metallothioneins, which sequester heavy metals and reduce their bioavailability to plants and organisms. For instance, fungi like Aspergillus niger and Saccharomyces cerevisiae have been engineered or utilized in bioremediation strategies to immobilize or detoxify metals such as cadmium, lead, mercury, and arsenic in contaminated soils and industrial wastewaters.

Furthermore, fungi play roles in the degradation of agricultural pollutants and agrochemical residues, including pesticides and herbicides. Fungi possess enzymatic systems capable of metabolizing pesticide molecules, breaking down their chemical structures, and detoxifying their residues in soil and water environments. For example, species of Trichoderma and Fusarium have been studied for their ability to degrade pesticides such as atrazine, chlorpyrifos, and glyphosate through enzymatic pathways involving hydrolases, oxidases, and esterases. Fungal bioremediation strategies offer sustainable alternatives to chemical methods for managing pesticide contamination in agricultural soils, protecting biodiversity, and safeguarding water quality. Additionally, fungi contribute to the biodegradation of petroleum hydrocarbons and hydrocarbon-based pollutants in contaminated environments. Fungi like species of the genus Aspergillus, Penicillium, and Fusarium produce enzymes such as lipases, oxidases, and peroxidases that degrade hydrocarbons by breaking down their carbon-carbon bonds and converting them into simpler compounds. These fungi participate in natural attenuation processes in oil-contaminated soils, wetlands, and marine environments, contributing to the restoration of ecosystems affected by oil spills and industrial discharges. Fungal bioremediation approaches complement traditional remediation methods by enhancing the degradation efficiency and reducing the environmental footprint associated with hydrocarbon contamination. Moreover, fungi contribute to the remediation of emerging pollutants and contaminants of emerging concern (CECs), including pharmaceuticals, personal care products, and endocrine-disrupting chemicals. Fungi possess diverse enzymatic activities and metabolic pathways that can transform or degrade these complex organic molecules into non-toxic or less harmful forms. For example, fungi like Pleurotus ostreatus and Bjerkandera adusta have demonstrated the ability to degrade pharmaceutical compounds such as antibiotics and non-steroidal anti-inflammatory drugs (NSAIDs) through enzymatic oxidation and conjugation processes. Fungal bioprospecting offers innovative solutions for mitigating the environmental impacts of emerging pollutants, reducing their persistence and ecological risks in aquatic and terrestrial ecosystems.

Advances in biotechnological and genomic tools have enhanced fungal bioprospecting efforts for environmental remediation by enabling the discovery and optimization of fungal strains and enzymes with specific biodegradative capabilities. Genomic sequencing, transcriptomics, and metagenomics facilitate the identification and characterization of fungal genes encoding enzymes involved in pollutant degradation pathways. Functional genomics approaches elucidate the molecular mechanisms underlying fungal metabolism, enzymatic activities, and adaptive responses to environmental contaminants. These insights inform the selection, manipulation, and optimization of fungal candidates for bioremediation applications, enhancing their efficiency and applicability across diverse contaminated environments. Furthermore, synthetic biology techniques enable the engineering of fungal strains for enhanced biodegradation capabilities, metabolic pathway optimization, and production of tailored enzymes for specific pollutant classes. By harnessing genetic and metabolic engineering approaches, researchers can design fungal biocatalysts with improved tolerance to pollutants, enhanced enzyme activities, and expanded substrate specificities, thereby accelerating the development of sustainable bioremediation solutions.

CHALLENGES AND OPPORTUNITIES IN FUNGAL BIOPROSPECTING: FUTURE DIRECTIONS

Fungal bioprospecting holds significant promise and faces several challenges as it moves forward into the future, offering opportunities to address global challenges in agriculture, medicine, environmental remediation, and biotechnology. One of the primary challenges in fungal bioprospecting is the vast untapped diversity of fungi in natural environments that remain unculturable or poorly characterized under laboratory conditions. Many fungi have complex nutritional and environmental requirements that are difficult to replicate in artificial culture settings, limiting our ability to access their potential for producing novel bioactive compounds and enzymes. Overcoming this challenge requires innovative approaches in microbial ecology, genomics, and biotechnology to cultivate and exploit these fungi for biotechnological applications. Furthermore, the discovery of novel bioactive compounds from fungi often involves lengthy and labor-intensive screening processes, which can be resource-intensive and time-consuming. Screening for bioactivity typically relies on traditional bioassays and chemical analysis techniques, which may miss subtle or novel activities that are not targeted by existing assays. Addressing this challenge requires the development of high-throughput screening platforms, advanced analytical techniques, and computational approaches to expedite the identification and characterization of bioactive compounds from fungal extracts. Integrating omics technologies such as genomics, metabolomics, and proteomics can provide comprehensive insights into fungal metabolic pathways and facilitate targeted screening for specific bioactivities of interest. Moreover, scaling up production of bioactive compounds from fungi for commercial applications poses significant challenges related to fermentation optimization, yield enhancement, and cost-effectiveness. Many fungal metabolites are produced in low quantities or are difficult to extract and purify from fermentation broths, limiting their feasibility for large-scale industrial production. Improving fermentation processes through strain engineering, fermentation optimization, and bioprocess engineering is essential for increasing yields, reducing production costs, and ensuring consistency in product quality for commercialization. Additionally, exploring alternative production platforms such as synthetic biology approaches for heterologous expression of biosynthetic pathways in well-characterized hosts offers potential solutions to overcome scalability limitations in fungal bioprospecting.

However, challenges such as regulatory hurdles, public perception of biotechnology, and market acceptance of bio-based products can hinder the adoption of fungal bioproducts in agricultural practices. Educating stakeholders about the benefits of fungal biotechnology, demonstrating the efficacy and safety of biopesticides and biostimulants, and navigating regulatory frameworks for product registration are essential for fostering innovation and commercialization in agricultural bioprospecting. Collaborative efforts between researchers, industry partners, policymakers, and farmers are crucial for translating scientific advancements into practical applications that address global food security challenges and promote sustainable agricultural practices. Furthermore, fungal bioprospecting holds promise for environmental applications, including the bioremediation of contaminated soils, water bodies, and industrial waste streams. Challenges such as the complexity and persistence of environmental pollutants, site-specific remediation requirements, and scalability of bioremediation technologies necessitate interdisciplinary approaches and tailored solutions for different contamination scenarios. Integrating fungal bioremediation strategies with traditional remediation techniques, optimizing fungal consortia for synergistic pollutant degradation, and monitoring long-term environmental impacts are essential for achieving effective and sustainable remediation outcomes.

CASE STUDIES IN FUNGAL BIOPROSPECTING: SUCCESS STORIES AND LESSONS LEARNED

Case studies in fungal bioprospecting illustrate successful applications of fungi in various sectors, offering valuable insights and lessons for advancing biotechnological innovation. One prominent success story is the discovery and development of the antibiotic penicillin from the fungus Penicillium notatum by Alexander Fleming in 1928. This breakthrough marked the beginning of the antibiotic era and revolutionized medicine by providing a potent treatment for bacterial infections. Penicillin and subsequent antibiotics derived from fungi, such as cephalosporins and macrolides, have saved countless lives worldwide and continue to be essential therapies in modern healthcare. The success of penicillin highlights the potential of fungi as prolific producers of bioactive compounds with therapeutic value, inspiring ongoing efforts in fungal bioprospecting for novel antibiotics and antimicrobial agents. Another compelling case study in fungal bioprospecting is the development of immunosuppressant drugs such as cyclosporin A from the fungus Tolypocladium inflatum. Cyclosporin A revolutionized organ transplantation by preventing rejection of transplanted organs and tissues, thereby improving patient outcomes and extending survival rates. This fungal-derived compound selectively inhibits T-cell activation and cytokine production, offering targeted immunosuppression with reduced systemic side effects compared to conventional therapies. The discovery of cyclosporin.

Moreover, fungal bioprospecting has contributed to environmental remediation efforts through mycoremediation, utilizing fungi to degrade or detoxify pollutants in contaminated ecosystems. White rot fungi like Phanerochaete chrysosporium and Pleurotus ostreatus are capable of breaking down complex pollutants such as polycyclic aromatic hydrocarbons (PAHs) and chlorinated compounds through enzymatic oxidation processes. These fungi produce extracellular enzymes like lignin peroxidases and laccases that catalyze the degradation of recalcitrant pollutants into simpler, less toxic compounds, facilitating their removal from soil and water environments. Mycoremediation exemplifies the potential of fungi as versatile biocatalysts for environmental sustainability, offering cost-effective and eco-friendly solutions to address industrial contamination and restore polluted ecosystems. Furthermore, fungal bioprospecting has yielded innovations in industrial biotechnology, particularly in enzyme production and biocatalysis. Fungal enzymes such as lipases, proteases, and amylases are widely used in various industrial processes, including food production, pharmaceutical manufacturing, and biofuel production. For instance, lipases from fungi like Candida antarctica are employed in biodiesel production to catalyze the transesterification of vegetable oils into biodiesel fuel, enhancing process efficiency and reducing environmental impact.

CONCLUSION

Fungal bioprospecting represents a promising frontier in the search for novel bioactive compounds with diverse applications. Through extensive exploration of fungal biodiversity, researchers have unearthed a wealth of compounds that exhibit pharmaceutical, agricultural, and industrial potential. These bioactive compounds often possess unique chemical structures and biological activities, making them valuable candidates for drug discovery, crop protection, and industrial processes. The methodologies employed in fungal bioprospecting have evolved significantly, leveraging advancements in genomics, metabolomics, and bioinformatics to streamline the discovery process. Despite these advancements, challenges such as regulatory hurdles, scaling production, and sustainable sourcing remain significant barriers to the commercialization of fungal-derived bioactives. Collaborative efforts among researchers, industry stakeholders, and policymakers will be crucial in overcoming these challenges and harnessing the full potential of fungal bioprospecting. Looking ahead, continued investment in research and development, coupled with a deeper understanding of fungal ecology and metabolism, promises to unveil even more novel bioactive compounds that can address global challenges in health, agriculture, and sustainable development. Fungal bioprospecting stands poised not only to expand our pharmacopeia but also to contribute significantly to the development of sustainable and innovative solutions for a rapidly changing world. Fungal enzymes also play crucial roles in bioremediation, diagnostic assays, and biopharmaceutical production, underscoring their versatility and economic value in biotechnological applications. However, despite these successes, fungal bioprospecting faces several challenges that impact the translation of research findings into practical applications. Challenges include the difficulty in cultivating and manipulating non-model fungi with complex growth requirements, variability in bioactive compound production, and scalability of bioprocesses for industrial production. Addressing these challenges requires interdisciplinary approaches integrating microbiology, genetics, bioinformatics, and engineering to optimize fungal strains, enhance product yields, and overcome barriers to commercialization.

REFERENCES

Abhilasha, A., Sreenivasulu, A., & Manimozhi, T. KUMAR, P. S., Selvakumar, P., & Singh, P. (2022, April). The Model of Smart Sensing Device For Sensitive Nanoclusters Modification in Sensing Properties. In 2022 2nd International Conference on Advance Computing and Innovative Technologies in Engineering (ICACITE) (pp. 1043-1047). IEEE.

Abida, H., Ruchaud, S., Rios, L., Humeau, A., Probert, I., De Vargas, C., Bach, S., & Bowler, C. (2013). Bioprospecting marine plankton. *Marine Drugs*, *11*(11), 4594–4611. DOI: 10.3390/md11114594 PMID: 24240981

Beattie, A. J., Hay, M., Magnusson, B., de Nys, R., Smeathers, J., & Vincent, J. F. V. (2011). Ecology and bioprospecting. *Austral Ecology*, *36*(3), 341–356. DOI: 10.1111/j.1442-9993.2010.02170.x PMID: 22737038

Bhatnagar, I., & Kim, S.-K. (2010). Immense essence of excellence: Marine microbial bioactive compounds. *Marine Drugs*, *8*(10), 2673–2701. DOI: 10.3390/md8102673 PMID: 21116414

Blunt, J. W., Copp, B. R., Keyzers, R. A., Munro, M. H. G., & Prinsep, M. R. (2015). Marine natural products. *Natural Product Reports*, *32*(2), 116–211. DOI: 10.1039/C4NP00144C PMID: 25620233

Bose U, Hodson MP, Shaw PN, Fuerst JA, Hewavitharana AK (2014) Bacterial production of the fungus-derived cholesterol-lowering agent mevinolin. Biomed chromatog 28(9):1163–1166

Bowler, C., Karl, D. M., & Colwell, R. R. (2009). Microbial oceanography in a sea of opportunity. *Nature*, *459*(7244), 180–184. DOI: 10.1038/nature08056 PMID: 19444203

Chun, J., & Hartung, H.-P. (2010). Mechanism of action of oral fingolimod (FTY720) in multiple sclerosis. *Clinical Neuropharmacology*, *33*(2), 91–101. DOI: 10.1097/WNF.0b013e3181cbf825 PMID: 20061941

Demunshi, Y., & Chugh, A. (2010). Role of traditional knowledge in marine bioprospecting. *Biodiversity and Conservation*, *19*(11), 3015–3033. DOI: 10.1007/s10531-010-9879-9

Gloer, J. B. (2007). Applications of fungal ecology in the search for new bioactive natural products. In Kubicek, C. P., & Druzhinina, I. S. (Eds.), *Environmental and microbial relationships* (pp. 257–283). Springer.

Gomes, A. R., Freitas, A. C., Rocha-Santos, T. A. P., & Duarte, A. C. (2014). Bioactive compounds derived from echinoderms. *RSC Advances*, *4*(56), 29365–29382. DOI: 10.1039/C4RA03352C

Jobstvogt, N., Hanley, N., Hynes, S., Kenter, J., & Witte, U. (2014). Twenty thousand sterling under the sea: Estimating the value of protecting deep-sea biodiversity. *Ecological Economics*, *97*, 10–19. DOI: 10.1016/j.ecolecon.2013.10.019

Kanipriya, M., Sugumaran, D., Pandiyarajan, V., Yogananda, B. S., Selvakumar, P., & Mohan, S. (2024, January). Analysis of Electromyography Signals for Control of Mechanical Prosthesis using Machine Learning Techniques. In 2024 2nd International Conference on Intelligent Data Communication Technologies and Internet of Things (IDCIoT) (pp. 1059-1067). IEEE.

Keating, G., & Figgitt, D. (2003). Caspofungin: A review of its use in oesophageal candidiasis, invasive candidiasis and invasive aspergillosis. *Drugs*, *63*(20), 2235–2263. DOI: 10.2165/00003495-200363200-00008 PMID: 14498760

Krause, J., & Tobin, G. (2013). Discovery, development, and regulation of natural products. In Marianna, K. (Ed.), *Using old solutions to new problems – natural drug discovery in the 21st century* (pp. 3–36). InTech. DOI: 10.5772/56424

Leal, M. C., Puga, J., Serôdio, J., Gomes, N. C. M., & Calado, R. (2012). Trends in the discovery of new marine natural products from invertebrates over the last two decades – where and what are we bioprospecting? *PLoS One*, *7*(1), e30580. DOI: 10.1371/journal.pone.0030580 PMID: 22276216

Liu, X., Ashforth, E., Ren, B., Song, F., Dai, H., Liu, M., Wang, J., Xie, Q., & Zhang, L. (2010). Bioprospecting microbial natural product libraries from the marine environment for drug discovery. *The Journal of Antibiotics*, *63*(8), 415–422. DOI: 10.1038/ja.2010.56 PMID: 20606699

Mahadev Madgule, N. (2023). Mechanical properties and microstructure of activated TIG welded similar joints of Inconel alloys by desirability approaches. *Materials Today: Proceedings*, *77*, 528–533. DOI: 10.1016/j.matpr.2022.12.250

Martins, A., Vieira, H., Gaspar, H., & Santos, S. (2014). Marketed marine natural products in the pharmaceutical and cosmeceutical industries: Tips for success. *Marine Drugs*, *12*(2), 1066–1101. DOI: 10.3390/md12021066 PMID: 24549205

Matz, C., Webb, J. S., Schupp, P. J., Phang, S. Y., Penesyan, A., Egan, S., Steinberg, P., & Kjelleberg, S. (2008). Marine biofilm bacteria evade eukaryotic predation by targeted chemical defense. *PLoS One*, *3*(7), e2744. DOI: 10.1371/journal.pone.0002744 PMID: 18648491

Mayer, A. M. S., Glaser, K. B., Cuevas, C., Jacobs, R. S., Kem, W., Little, R. D., McIntosh, J. M., Newman, D. J., Potts, B. C., & Shuster, D. E. (2010). The odyssey of marine pharmaceuticals: A current pipeline perspective. *Trends in Pharmacological Sciences*, *31*(6), 255–265. DOI: 10.1016/j.tips.2010.02.005 PMID: 20363514

Molinski, T. F., Dalisay, D. S., Lievens, S. L., & Saludes, J. P. (2009). Drug development from marine natural products. *National Review*, *8*(1), 69–85. PMID: 19096380

Mora, C., Tittensor, D. P., Adl, S., Simpson, A. G. B., & Worm, B. (2011). How many species are there on Earth and in the ocean? *PLoS Biology*, *9*(8), 1–8. DOI: 10.1371/journal.pbio.1001127 PMID: 21886479

Proksch, P., Edrada, R. A., & Ebel, R. (2002). Drugs from the seas – current status and microbiological implications. *Applied Microbiology and Biotechnology*, *59*(2–3), 125–134. PMID: 12111137

Ramakrishnan, T., Mohan Gift, M. D., Chitradevi, S., Jegan, R., Hency Jose, P. S., Nagaraja, H. N., & Hailegiorgis, S. M. (2022). Study of numerous resins used in polymer matrix composite materials. *Advances in Materials Science and Engineering*, *2022*(1), 1088926.

Saslis-Lagoudakis, C. H., Savolainen, V., Williamson, E. M., Forest, F., Wagstaff, S. J., Baral, S. R., Watson, M. F., Pendry, C. A., & Hawkins, J. A. (2012). Phylogenies reveal predictive power of traditional medicine in bioprospecting. *Proceedings of the National Academy of Sciences of the United States of America, 109*(39), 15835–15840. DOI: 10.1073/pnas.1202242109 PMID: 22984175

Selvakumar, P. (2024). Climate Change. In Sofia Idris, I. G. I. (Ed.), *Global Challenges for the Environment and Climate Change* (pp. 100–110). Global., DOI: 10.4018/979-8-3693-2845-3.ch006

Selvakumar, P. (2024). Reasons for Biodiversity Loss. In Ashok Kumar Rathoure, I. G. I. (Ed.), *Impact of Societal Development and Infrastructure on Biodiversity Decline* (pp. 37–49). Global., DOI: 10.4018/979-8-3693-6950-0.ch004

Selvakumar, P.. (2024). The Role of Corrosion Inhibitors. In Thakur, A., & Kumar, A. (Eds.), *Innovations in Nanomaterials-Based Corrosion Inhibitors* (pp. 130–141). IGI Global., DOI: 10.4018/979-8-3693-3088-3.ch006

Selvakumar, P., Mariappan, P., Shunmugam, S., Paul, R., & Jabez, S. (2024). Removal of chemicals from effluent using photobioreactor technology to improve environmental and health impacts. *Zeitschrift für Physikalische Chemie, 2024*(7), 1295–1302. Advance online publication. DOI: 10.1515/zpch-2023-0450

Selvakumar, P., Palanisamy, S. K., Cinthaikinian, S., Palanisamy, V., Mariappan, R., & Selvakumar, P. K. (2024). Biosensors and its diverse applications in healthcare systems. *Zeitschrift für Physikalische Chemie, 2024*. Advance online publication. DOI: 10.1515/zpch-2023-0406

Tarman, K., Lindequist, U., Wende, K., Porzel, A., Arnold, N., & Wessjohann, L. A. (2011). Isolation of a new natural product and cytotoxic and antimicrobial activities of extracts from fungi of indonesian marine habitats. *Marine Drugs, 9*(3), 294–306. DOI: 10.3390/md9030294 PMID: 21556160

Vijayakumar, G., Rajkumar, M., Rajiv Chandar, N., Selvakumar, P., & Duraisamy, R. (2024). Muniyandi Rajkumar, Rajiv Chandar N, Selvakumar P and Ramesh Duraisamy, (2024). Environmentally friendly TDS removal from waste water by electrochemical ion exchange batch-type recirculation (EIR) technique. *Environmental Science. Water Research & Technology, 10*(4), 826–835. DOI: 10.1039/D3EW00793F

Wildman, H. G. (1999). *Pharmaceutical bioprospecting and its relationship to the conservation and utilization of bioresources*. IUPAC.

Zhang, D., Satake, M., Fukuzawa, S., Sugahara, K., Niitsu, A., Shirai, T., & Tachibana, K. (2012). Two new indole alkaloids, 2-(3,3-dimethylprop-1-ene)-costaclavine and 2-(3,3-dimethylprop-1-ene)-epicostaclavine, from the marine-derived fungus Aspergillus fumigatus. *Journal of Natural Medicines, 66*(1), 222–226. DOI: 10.1007/s11418-011-0565-3 PMID: 21792727

Zhang, Y., Arends, J. B. A., Van de Wiele, T., & Boon, N. (2011). Bioreactor technology in marine microbiology: From design to future application. *Biotechnology Advances, 29*(3), 312–321. DOI: 10.1016/j.biotechadv.2011.01.004 PMID: 21251973

Chapter 16
Edible Mushrooms: A Sustainable Superfood for the Future

Vishal Shridhar Moger
https://orcid.org/0009-0003-5175-7433
St. Aloysius University, India

Harshitha Gujaran
https://orcid.org/0009-0002-6246-2967
St. Aloysius University, India

Shreya Gowda
St. Aloysius University, India

Pallavi Baliga
https://orcid.org/0000-0002-5485-95370
St. Aloysius University, India

ABSTRACT

Edible mushrooms are termed as superfoods owing to their exceptional nutritional profile. This chapter highlights the dietary benefits of edible mushrooms derived from their high content of protein, fiber, complex carbohydrates, vitamins, minerals, and antioxidants. It details the health benefits of bioactive compounds in mushrooms, which include immune modulation, roles in chronic disease management, and reducing inflammation. In addition, this chapter explores various ways in which mushrooms might be exploited in food processing because of their culinary versatility and functional properties. Focus is laid on the environmentally sustainable aspects of mushroom farming, which requires little water and land resources and promotes waste recycling. This chapter discusses the challenges associated with mushroom cultivation and future research trends to improve cultivation and nutritional enhancement. Overall, this chapter provides a comprehensive overview of edible mushrooms, highlighting their nutritional and health benefits, ecofriendly nature, and future directions in this field.

DOI: 10.4018/979-8-3693-8201-1.ch016

1. INTRODUCTION

The current era is marked by a shift towards the intake of ultra-processed foods and sedentary lifestyles, resulting in a surge in heart diseases, cancer, and obesity (Popkin et al. 2022). Due to this, there is an increase in interest in 'Superfoods', nutrient-dense foods that promote good health and prevent illness (Cobos & Díaz 2023; Jagdale et al. 2021). They provide the best nutrition for minimal calories (Jagdale et al. 2021). Superfoods offer several health benefits, such as boosting the immune system, regulating hormones, and promoting overall well-being (Fernández-Ríos et al. 2022). Among most superfoods, mushrooms stand out due to their exceptional functional and nutritional properties. Mushrooms have been regarded as a gourmet component in many cuisines worldwide due to their distinct flavor and are valued as culinary treasures (Valverde et al. 2015). The cultural and historical significance of mushrooms as food and medicine spans across regions (Bhambri et al. 2022). In ancient China, mushrooms were regarded as the "elixir of life," while the Romans considered them "food of the gods" (Valverde, et al. 2015). Indigenous cultures in North America and Africa incorporated mushrooms into traditional healing practices. This historical appreciation for mushrooms continues in modern times, with global mushroom production on the rise. China currently leads in production, accounting for over 75% of the world's total (Dimopoulou et al. 2022). Other countries, like India, have also witnessed significant growth in mushroom cultivation, increasing 13.8-fold between 1990 and 2020 (Okuda et al. 2022). This increased production reflects the growing recognition of the nutritional and economic value of mushrooms, particularly for rural communities (Raut et al. 2019).

Mushrooms are a good source of proteins, vitamins, minerals, fiber, and unique bioactive compounds like ergothioneine (Valverde et al. 2015) and prebiotics that promote gut health (Bell et al. 2022). This rich nutritional profile, coupled with growing consumer awareness of its nutritional value, is driving a major and ongoing boom in the global mushroom market. The market, currently worth tens of billions of US dollars, is experiencing robust growth, with a compound annual growth rate (CAGR) of 7% to over 10% through 2030 and beyond (https://essfeed.com/global-mushroom-market-size/). Asia-Pacific currently leads the world in both production and consumption. North America and Europe are witnessing significant growth driven by health-conscious consumers and the growing demand for specialty mushrooms like Shiitake and Oyster.

Beyond health benefits, mushrooms also offer environmental advantages. Their cultivation requires less water and land than animal farming, and they can be grown on lignocellulosic wastes such as sawdust and straw (Grimm et al. 2018), contributing to a circular economy. Besides being a nutritious food source, mushrooms also find applications in bioremediation. By exploiting the specific enzymatic properties of fungi to break down toxins and improve ecosystem health, mycoremediation holds significant potential for mitigating the consequences of climate change (Dinakarkumar et al. 2024).This includes developments in efficient remediation of contaminated soils and water by using particular mushroom species to decompose heavy metals, persistent organic contaminants, and even plastics (Dinakarkumar et al. 2024). Additionally, the spent mushroom substrate is a valuable resource for soil amendment, improving soil structure and fertility, supporting carbon capture, and reducing reliance on synthetic fertilizers. The extensive mycelial network of mushrooms also contributes to carbon sequestration by storing it in the soil (Shin et al. 2025). The present chapter aims to explain the numerous health and environmental benefits of mushrooms and reiterates their significance as an ecologically sustainable superfood rich in nutrition.

2. NUTRITIONAL COMPOSITION OF EDIBLE MUSHROOMS

The nutritional content of mushrooms is influenced by several factors, including species, stage of development, and environmental conditions. This section will detail the nutritional composition of various edible mushrooms, highlighting the variations between species.

2.1 Overall Composition

Mushrooms typically consist of 85–95% moisture, 35–70% carbohydrates, 15–34.7% protein, 10–20% fat, 6–10.9% minerals, and 3–8% nucleic acids. Additionally, it has a high concentration of vitamins, including thiamine (1.4–2.2 mg/%), folic acid (1.2–1.4 mg/100 g on dry weight basis), pantothenic acid (21.1–33.3 mg/%), niacin (60.6–73.3 mg/%), ascorbic acid (92–144 mg/%), and riboflavin (6.7–9.0 mg/%). While not a primary source of fat, mushrooms do contain essential fatty acids like linoleic, oleic, and linolenic acid (Sande et al. 2019). The minerals in mushrooms are calcium, iron, zinc, magnesium, manganese, and selenium. Because their fruit bodies contain significant levels of minerals like potassium, iron, copper, zinc, and manganese, they are regarded as nutritious food (Assemie & Abaya, 2022). Table 1 presents the nutritional profiles of four important mushrooms: common, enoki, shiitake, and oyster. These mushrooms are known for their nutritional value, culinary uses, as well as their diverse nutritional profiles.

Table 1. Nutritional values of common, enoki, shiitake, and oyster mushrooms (Li et al. 2023)

Nutrient (g/100g)	Common mushroom	Enoki mushroom	Shiitake mushroom	Oyster mushroom
Fat	0.34	0.29	0.49	0.41
Ash	0.85	0.91	0.73	1.01
Carbohydrate	3.26	7.81	6.79	6.09
Dietary fiber	1.0	2.7	2.5	2.3
Protein	3.09	2.66	2.24	3.31
Moisture	92.45	88.34	89.74	89.18

As shown in the table, Oyster mushrooms are the richest in protein and minerals, while Enoki mushrooms excel in dietary fiber and carbohydrates. Common mushrooms have the highest moisture content but comparatively lower amounts of other nutrients. Shiitake mushrooms, while generally balanced in nutrients, have the highest fat content among the four.

2.2 Protein

The dry matter of mushrooms contains a significant amount of protein (Wani et al. 2010). On a dry weight basis, their protein content ranges from 14% to 39%. Mushroom proteins are a valuable source of several essential amino acids (Dimopoulou et al. 2022). They offer a complete amino acid profile, including histidine, crucial for newborns, and all eight essential amino acids required by humans (Wang & Zhao, 2023). *Agaricus bisporus* (button mushroom) and *Pleurotus ostreatus* (oyster mushroom) are particularly good sources of protein (Table 2).

Table 2. Protein contents of some common mushrooms (Wang & Zhao, 2023)

Species	Protein weight (g/kg, dry weight)	Protein weight (g/kg)	Protein content (%)
A.aegerita	248	231(DW)	n.d.
A.bisporus	83.7	387	14.1–27.1
A.auricula	122.0	121 (DW)	n.d.
F.velutipes	113.3	240 (FW)	18.6–28.0
L. edodes	83.9	200 (DW)	17.1–24.6
P. eryngii	144.8	17	n.d.
P. ostreatus	194.3	19 (DW)	21.1–34.5

Note: DW indicates dry weight, FW indicates fresh weight, and n.d. indicates not determined.

2.3 Dietary Fiber

Edible mushrooms are regarded as a novel source of nutritional fiber, particularly insoluble dietary fibers (Cheung, 2013). Mushrooms have between 4% and 9% soluble fiber and between 22% and 30% insoluble fiber (Assemie & Abaya, 2022). Their carbohydrates consist of non-digestible oligosaccharides like trehalose and cell wall polysaccharides like chitin, mannans, and β-glucans. Mushroom sclerotium includes a high concentration of β-glucans (>80% DM), which display favorable biopharmacological effects for humans (Cheung, 2013).

2.4 Complex Carbohydrates

Edible mushrooms are rich in complex carbohydrates, including mannitol, glucose, and glycogen, while containing low levels of simple sugars (Assemie & Abaya, 2022). This complex carbohydrate profile contributes to slower absorption and digestion, potentially beneficial for blood sugar management in diabetic patients (Arora et al. 2018).

2.5 Vitamins and Minerals

Mushrooms are a good source of B vitamins, including thiamine (B1), riboflavin (B2), niacin (B3), pantothenic acid (B5), and pyridoxine (B6), with varying levels across species (Table 3; Wani et al. 2010). They also contain minerals like calcium, iron, zinc, magnesium, manganese, selenium, and potassium (Assemie & Abaya, 2022). Oyster mushrooms are particularly rich in iron, magnesium, potassium, and phosphorus, while shiitake mushrooms are a good source of manganese and zinc (Table 4). Due to their high potassium-to-sodium ratio, mushrooms are beneficial for those who have hypertension and high blood pressure. It is also well known that mushrooms contain selenium, which has demonstrated immunomodulating qualities. It strengthens teeth, hair, and nails and increases bone strength, enhancing bone health. Over 90% of the iron content in mushrooms can be absorbed by the body, making them a good source of iron that supports RBC formation (Arora et al. 2018). Mushrooms can also be a source of vitamin D when exposed to UV light, converting ergosterol to vitamin D2 (Org et al. 2023; Malik et al. 2022).

Table 3. Vitamin content of some of the mushrooms (Li et al. 2023)

Vitamins (mg/100g)	Common mushroom	Oyster mushroom	Enoki mushroom	Shiitake mushroom
Thiamine (B1)	0.081	0.125	0.225	0.015
Riboflavin (B2)	0.40	0.35	0.20	0.22
Niacin (B3)	3.61	4.96	7.03	3.88
Pantothenic acid (B5)	1.50	1.29	1.35	1.50
Pyridoxine (B6)	0.10	0.11	0.10	0.29

Table 4. Mineral contents of some of the mushrooms (Li et al. 2023)

Minerals (mg/100g)	Common mushroom	Oyster mushroom	Enoki mushroom	Shiitake mushroom
Fe	0.5	1.33	1.15	0.41
Mg	9	18	16	20
Mn	0.05	0.11	0.08	0.23
K	318	420	359	304
Zn	0.52	0.77	0.65	1.03
P	86	120	105	112
Na	5	18	3	9
Cu	0.32	0.24	0.11	0.14
Ca	3	3	0	2

2.6 Antioxidants

Mushrooms contain a variety of bioactive compounds with antioxidant properties, including phenolics, indoles, carotenoids, vitamins, polysaccharides, ergothioneine, tocopherols, glycosides, and ascorbic acid (Chun et al. 2021; Podkowa et al. 2021). These compounds can help protect against oxidative stress and contribute to various health benefits (Kozarski et al. 2015; Mwangi et al. 2022). Multiple spectrophotometric assays, nuclear magnetic resonance (NMR), Fourier transform infrared (FT-IR), UV-VIS spectroscopy, and high-performance liquid chromatography (HPLC) and gas chromatography (GC) have been used to identify and quantify these antioxidant chemicals from mushrooms (Chun et al. 2021). For example, studies have shown that methanolic extracts of mushrooms exhibit significant antioxidant activity, with variations in phenolic and flavonoid content depending on the species and extraction method (Sudha et al. 2012).

2.7 Nutritional Benefits vs. Other Aspects

The cultivation of mushrooms holds great potential, but it also faces challenges (Sangeeta et al. 2024). These include the need for economic models that address market volatility and high initial investment costs (Bijla & Sharma, 2023), efficient supply chain logistics to manage the perishable nature of mushrooms (Onwude et al. 2020), and continued technological advancements in cultivation methods and pest control (Kumar et al. 2024). Addressing these challenges through targeted investment, improved infrastructure, and knowledge dissemination will be crucial for unlocking the full potential of edible mushrooms (Kumar et al. 2024).

3. HEALTH BENEFITS OF EDIBLE MUSHROOMS

Edible mushrooms offer a wealth of health benefits (Figure 1), acting as functional foods that can help prevent certain diseases. They are packed with bioactive compounds often absent in typical plant and animal diets (Łysakowska et al. 2023). Mushrooms have been used in medicine for nearly 2,000 years (Reis et al. 2022). Their popularity as a source of medicine and nutraceuticals stems from their antioxidant, anticancer, and antibacterial properties. Furthermore, their high protein and low fat/energy content make them a valuable dietary addition (Kumar et al. 2015). Numerous edible species such as *Cordyceps militaris*, *Pleurotus ostreatus*, *Ganoderma lucidum*, *Lentinus edodes*, *Grifola frondosa*, *Agaricus blazei*, and *Hericium erinaceous*, have been studied for their medicinal potential (Lakhanpal et al. 2005).

Mushrooms contain a variety of active ingredients, including dietary fibers especially (1,3) (1,6)-β-d-glucans), oligosaccharides, polysaccharides, triterpenoids, peptides, proteins, alcohols, phenols (Das et al. 2021), chitin, hemicellulose, mannans, xylans, and galactans (Jayachandran et al. 2017), as well as minerals (zinc, copper, iodine, selenium, iron), vitamins, and amino acids. These components contribute to a range of health benefits, including immune system support, hepatoprotective and anti-hypercholesterolemia effects, antiviral and anti-HIV activity, and mitigation of the harmful effects of radiation and chemotherapy (Das et al. 2021). The non-digestible polysaccharides act as prebiotics, promoting beneficial gut bacteria and inhibiting pathogens (Sousa et al. 2011).

3.1 Immunomodulation

Mushrooms are potent immunomodulators, containing compounds like terpenes, terpenoids, lectins, fungal immunomodulatory proteins (FIPs), and especially polysaccharides (β-d-glucans, polysaccharide proteins, and polysaccharopeptides) (Zhao et al. 2020).

3.1.1 Polysaccharides

Polysaccharides, particularly β-d-glucans, are among the most studied immunomodulatory compounds in mushrooms (Chakraborty et al. 2019). These complex carbohydrates exhibit diverse structures and linkages, influencing their biological activity. Several mushroom-derived polysaccharides, such as lentinan, schizophyllan, and polysaccharide K, are used in various therapeutic applications, including cancer immunotherapy (Lindequist et al. 2005). For example, lentinan, derived from *Lentinus edodes* (shiitake), and schizophyllan, from *Schizophyllum commune*, have demonstrated anticancer and immunomodulatory effects, often attributed to their β-1,3-d-glucan structure with β-1,6 branching (Ina et al. 2013). These polysaccharides can activate various immune cells, including NK cells, macrophages, and dendritic cells, and induce the production of cytokines like IL-8, IFN-γ, and TNF (Firenzuoli et al. 2008). Other mushroom species, such as *Agaricus brasiliensis* and *Naematelia aurantialba*, also contain polysaccharides with immunomodulatory properties (Du et al. 2010; Hobb et al. 2005).

3.1.2 Terpenoids

Mushrooms are a source of bioactive terpenoids, including triterpenoids like Ganoderic acid C, which have various effects, including cholesterol reduction and antioxidant activity (Rathee et al. 2012). Other terpenoid derivatives in mushrooms also exhibit immunomodulatory effects.

Figure 1. This diagram illustrates the beneficial roles of various bioactive compounds in medicinal mushrooms that enhance immunity. These compounds regulate blood sugar, improve mental health, and inhibit viral replication. They also help mitigate allergies, inflammation, and display anticancer properties. Overall, these bioactive compounds promote immunity, health and well-being.

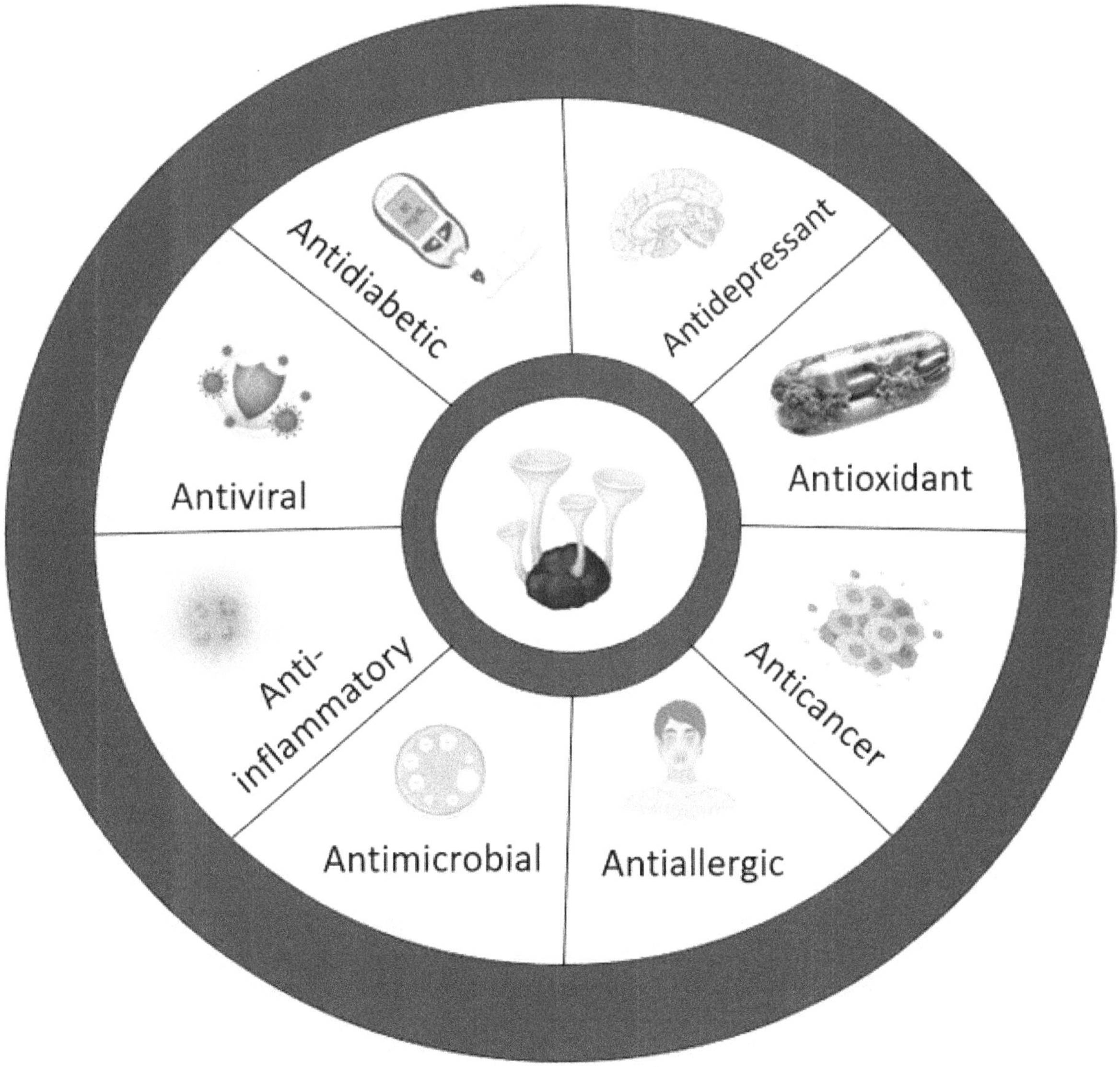

3.2 Role of Mushroom in Managing Chronic Diseases

Mushrooms play a role in managing chronic diseases like diabetes and cancer due to the abundance of bioactive compounds (Table 5).

Table 5. Bioactive compounds in edible mushrooms and their health benefits (Kumar et al. 2021)

Name of the Mushroom (Common Name)	Bioactive Compounds	Health Benefits
Lentinula edodes (Shiitake)	Mannoglucan, glucan, lentin (protein), catechin flavonoids, eritadenine, Lentinan, glucan, fucomannogalactan	Antioxidant, hypolipidemic activity anti-tumor, anti-inflammatory, antioxidant, antibacterial, antifungal, anti-fungal
Volvariella volvacea (Paddy straw mushroom)	Fip-vvo	Immunomodulatory
Pleurotus ostreatus (Oyster mushroom)	Proteoglycans pleuran (β-1, 3-glucan with galactose, and mannose), glucans, proteoglycan, laccase, pleurostrin (peptide), Functional proteins (ubiquinone-9, ubiquitin-like peptide, nebrodeolysin, and glycoprotein)	Anti-fungal, anti-tumor, antioxidant, anti-viral, Immunomodulatory, hyperglycemia
Agaricus bisporus (White mushroom)	Hydroxybenzoic acid derivatives, Pyrogallol, flavonoids, lectins	Enhanced insulin secretion, anti-aging properties, Anti-inflammatory

3.2.1 Role in the Management of Diabetes Mellitus

Mushroom polysaccharides can regulate glucose homeostasis, enhance lipid metabolism, and improve insulin resistance (Friedman et al. 2016; Das et al. 2022). They can modulate genes involved in glucose regulation and improve insulin sensitivity (Das et al. 2022). Shiitake mushrooms, for example, may boost insulin production and lower blood sugar (De Silva et al. 2012).

3.2.2 Role in Cancer Treatment

Mushrooms have emerged as a promising source of natural anticancer agents due to their rich content of bioactive compounds. Compounds such as polysaccharides, proteins, fats, phenolics, alkaloids, ergosterol, selenium, folate, enzymes, and organic acids, exhibit a wide range of anticancer mechanisms. Some of the key mechanisms include inducing apoptosis, inhibiting cell proliferation, modulating the immune system, anti-angiogenic activity, and synergistic effects with chemotherapy (Nandi et al. 2024; Park et al. 2022). Several mushroom species have demonstrated significant anticancer activity in preclinical and clinical studies (Joseph et al. 2018). Some notable examples include *Phellinus linteus*, *Pleurotus ostreatus*, *Agaricus blazei*, *Ganoderma lucidum*, and *Trametes versicolor*. Polysaccharides like lentinan, schizophyllan, AHCC, Maitake D-fraction, polysaccharide-K, and polysaccharide-P have been studied in human cancers (Daba et al. 2003). Numerous clinical trials are underway to evaluate the potential of mushroom extracts and compounds in cancer treatment. While some mushroom-derived drugs such as Lentinan from *Lentinula edodes*; schizophyllan from *Schizophyllum commune*, and Krestin from *Trametes versicolor* have been approved as prescription anticancer drugs in Japan and other countries, further research is needed to fully understand the anticancer mechanisms of mushrooms and develop novel therapeutic strategies (Park et al. 2022). *Agaricus blazei* Murill (ABM) has shown potential for cancer prevention and adjuvant treatment (Firenzuoli et al. 2008). Overall, mushrooms represent a valuable source of natural anticancer agents with diverse mechanisms of action. Their potential use as adjuncts in cancer therapy or as standalone anticancer agents warrants further investigation.

3.3 Anti-inflammatory property

Mushrooms contain compounds that can act as anti-inflammatory agents by inhibiting the production of inflammatory mediators (Fryer et al. 2024). Terpenoids, phenolic and indole compounds, and polysaccharides in mushrooms can block inflammatory signaling pathways and enzymes (Jedinak et al. 2011; Yahfoufi et al. 2018). Oyster mushroom concentrate, for example, has been shown to reduce the release of inflammatory cytokines from macrophages (Jedinak et al. 2011).

3.4 Antioxidant Property

Mushrooms contain various antioxidant phytochemicals, including phenols, flavonoids, polysaccharides, vitamins, carotenoids, and ergothioneine (Liuzzi et al. 2023). They exhibit significant antioxidant activity. For example, the ORAC (Oxygen Radical Absorbance Capacity) value of cultivated mushrooms, such as Portabella and Criminis (*A. bisporus*), was 9.7 and 9.5 μmoles trolox/g, respectively. According to the ORAC test, the value of carrots and green beans was 5, and red pepper was 10 (Zhou et al. 2006). Due to their increased ability to scavenge free radicals, oyster mushrooms are more commonly used in dietary supplements to help prevent disease (Bjelakovic et al. 2007). Antioxidants found in mushrooms show their protective effects in several ways and at different points in the oxidation process (Brewer et al. 2011). Mushrooms contain two types of antioxidants: primary antioxidants, which break the chain by scavenging free radicals, and secondary or defense antioxidants (Kozarski et al. 2015). Due to the action of primary antioxidants lipid hydroperoxides are blocked or degraded and singlet oxygen species are quenched or regenerated (Kozarski et al. 2011).

3.4.1 Ergothioneine (ERT)

This potent antioxidant is found in high concentrations in some mushroom species such as *Ganoderma neo-japonicum*, *G. applanatum*, and *Paecilomyces tenuipes* (Lee et al. 2009; Cheah et al. 2012). It has been studied for its protective role in conditions like rheumatoid arthritis and Crohn's disease, which involve chronic inflammation (Ey et al. 2007).

3.4.2 Vitamin D

Mushrooms are a source of vitamin D, particularly D2, which is important for various physiological processes such as calcium and phosphorus metabolism, neuromuscular and skeletal homeostasis, and resistance to several diseases (Strong et al. 2022; Keegan et al. 2013). Consuming vitamin D-enhanced mushrooms can help address vitamin D deficiency (Cardwell et al. 2018; Urbain et al. 2011).

4. SUSTAINABILITY, SOCIO-ECONOMIC IMPACT, AND POLICY RECOMMENDATIONS FOR MUSHROOM CULTIVATION

Achieving the sustainable development goals (SDGs) by 2030 is undoubtedly a monumental undertaking. As a nutrient-dense, reasonably priced food supply that addresses malnutrition and promotes food security, mushrooms are directly related to SDG 2 (Zero Hunger) (Hettiarachchi et al. 2020). By minimizing waste, establishing a circular economy, and using agricultural waste as a substrate, their creation also supports SDG 12 (Responsible Consumption and Creation) (Jasińska et al. 2018). Additionally, compared to typical agriculture, mushroom farming reduces greenhouse gas emissions and sequesters carbon, which helps achieve SDG 13 (Climate Action) (Shin et al. 2025). Mushroom farming indirectly supports SDG 6 (Clean Water and Sanitation) because it requires less water than other crops (Robinson et al. 2019); SDG 11 (Sustainable Cities and Communities) because it is integrated into urban farming initiatives; SDG 15 (Life on Land) because it promotes biodiversity and soil health (Jayaraman et al. 2024); and SDG 17 (Partnerships for the Goals) by encouraging cross-sector collaboration. Figure 2 presents the importance of mushroom cultivation in achieving the Sustainable Development Goals (SDGs).

Figure 2. The image illustrates how mushroom farming promotes several sustainable development goals (SDGs) by offering affordable, nutrient-dense food (SDG 2) and using less water and agricultural waste (SDG 6). Integrating mushroom production into urban farming contributes to SDG 11 while turning waste into valuable products aids SDG 12. Decomposing organic matter and enhancing soil health, contributes to SDG 13 by lowering greenhouse gas emissions, enhancing biodiversity, and repairing degraded land, which aligns with SDG 15. Achieving these goals fosters SDG 17 by encouraging collaboration across sectors for sustainable agricultural development

As an example, *A. bisporus* is grown with casing soil and compost, and 90% of its weight is made up of water (Caitano et al. 2023). In contrast to many other agricultural practices, mushroom farming has a substantially smaller water impact (Robinson et al. 2019). Spent mushroom substrate (SMS) handling is a crucial component of sustainability. SMS is not a waste product but a useful resource for soil conditioning, fertilization, and even bioremediation, which breaks down contaminants and heavy metals (Wang et al. 2016). To generate sustainable energy, SMS can also be utilized to produce biogas (Pérez-Chávez et al. 2019). The various uses of SMS, as seen in Figure 3, demonstrate how it may be used to minimize environmental effects and establish a closed-loop system. Additionally, by increasing nutrient content and encouraging beneficial microbial activity, mushroom cultivation improves soil health (Hu et al. 2021; Hu et al. 2022). However, by optimizing the composting process and maintaining proper

C/N ratios, it is imperative to address potential environmental challenges associated with composting, such as the release of greenhouse gases like methane and nitrous oxide (Onwosi et al. 2017; Peng et al. 2023). Although the cultivation of shiitake mushrooms has a smaller carbon footprint than that of many animal protein sources, more work is necessary to reduce CO_2 emissions through sustainable procedures (Tongpool et al. 2013).

Figure 3. A diagram showcasing various applications of spent mushroom substrate (SMS). The central element, SMS, can be utilized in multiple sectors such as organic farming, bioremediation, animal feeding, wastewater treatment, biofertilizer production, soil health improvement, recultivation of mushrooms, and biogas production. These diverse uses highlight SMS as a valuable by-product of the mushroom industry

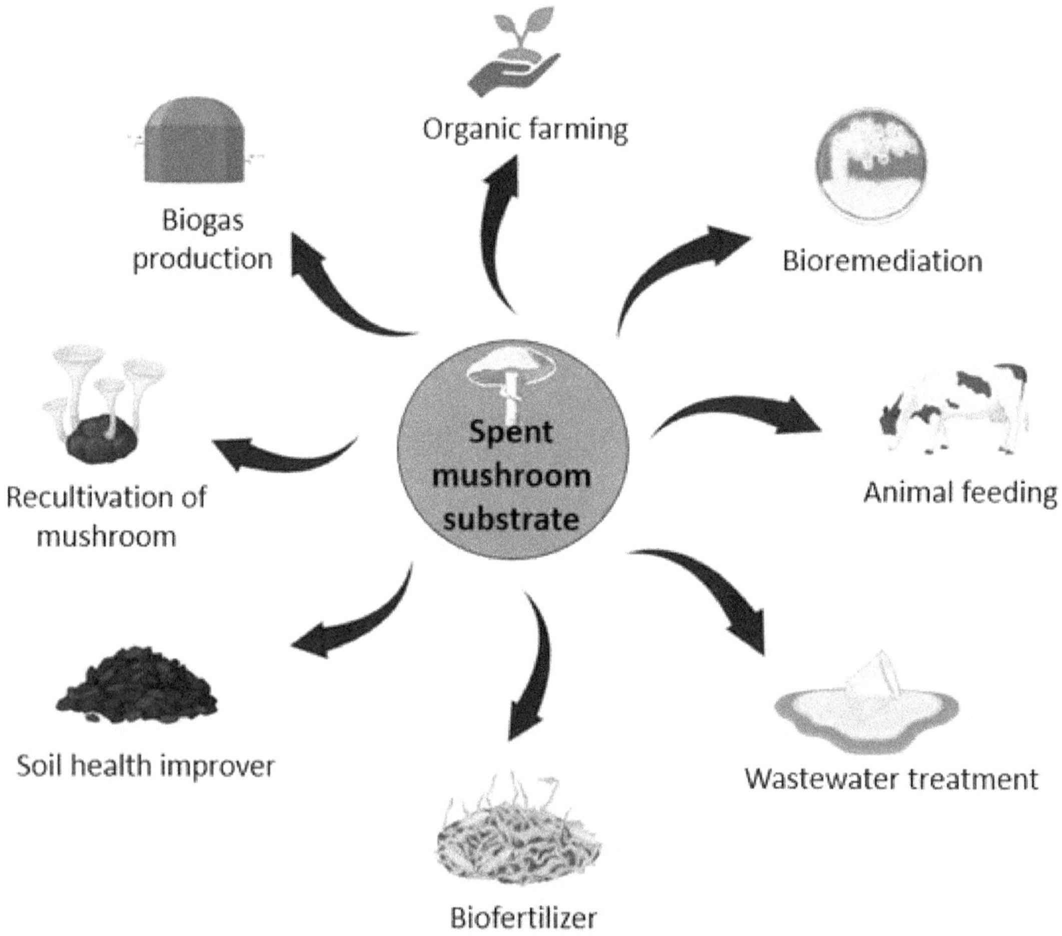

In addition to having positive effects on the environment, mushroom cultivation has enormous socio-economic potential, especially for rural areas. Because specialty mushrooms have a high market value, quicker returns, and lower input costs than traditional agriculture. Cost-benefit evaluations show that it is economically viable (Jayaraman et al. 2024; Xu et al. 2024). The scalability of mushroom farming from backyard operations to massive commercial facilities offers a variety of revenue-generating opportunities (Rana et al. 2024). Many successful projects demonstrate the transforming influence of mushroom farming on livelihoods, especially for women and underprivileged populations (Dey et al. 2020; Prajapati et al. 2023). Examples include the "Mushroom Village" in Bhutan and cooperative models in Thailand and India. These instances highlight how crucial community involvement, specialized interventions, and resource accessibility are necessary for their effective execution.

Policies that encourage mushroom growth are essential to maximizing its potential. Financial incentives, technical training, market access facilitation, and support for research and development are all ways that governments can play a critical role (Prajapati et al. 2023). It is also crucial to promote sustainable techniques and incorporate mushroom cultivation into national agricultural strategies (Khan et al. 2024). Additionally, knowledge exchange and innovation depend on government agencies, non-governmental organizations, and research institutions working together (Shirur et al. 2022). To maximize profitability and market access, farmers should prioritize high-quality mushroom production, use locally sourced substrates, adopt integrated farming approaches, prioritize sustainable cultivation practices, and investigate value-added products (Shirur et al. 2022). Policymakers and farmers may establish a win-win ecosystem that guarantees the long-term viability and success of the mushroom industry by putting these combined tactics into practice. This will support socio-economic growth and environmental sustainability.

5. ECONOMIC FEASIBILITY & SCALABILITY

According to Yusuf and Peker (2009), cost-benefit evaluations show that mushroom cultivation offers a strong economic alternative to conventional agricultural methods. Mushroom farming frequently uses less area, water, and time to produce a harvest than conventional crops, which lowers input costs and speeds up returns (Jayaraman et al. 2024). Additionally, using agricultural waste as a substrate lowers input costs considerably and encourages a circular economy (Xu et al. 2024). Profitability is further increased by the high market value of specialty mushrooms and the rising demand for therapeutic kinds. The potential for high yields and premium pricing eventually outweighs the initial setup expenditures, which can be substantial for cultivation in a controlled environment (Singh et al. 2011). Since mushroom farming may be scaled to different sizes, from backyard operations to large-scale commercial facilities, scalability is a significant benefit (Rana et al. 2024).

The growth of the mushroom industry is greatly aided by government and non-governmental organization assistance systems. These programs include improving market access, offering technical training and extension services, and supplying financial support in the form of grants and subsidies. Numerous countries are realising how growing mushrooms may improve food security, provide jobs in rural areas, and advance sustainable agriculture (Shirur et al. 2022). NGOs frequently work to empower underserved communities by giving them the tools and knowledge they need to grow mushrooms on a modest scale, which helps them make more money and enhance their standard of living. To create and share best practices, foster innovation, and guarantee the mushroom industry's long-term viability, cooperation between governmental organizations, non-governmental organizations, and research institutes is crucial (Shirur et al. 2022).

6. CHALLENGES AND FUTURE DIRECTIONS

The global mushroom market is experiencing significant growth, driven by increasing demand for elite varieties, functional foods, and plant-based diets. However, to realize the full potential of the mushroom industry, one needs to address several interconnected challenges. These challenges span from production and market access to consumer perception and environmental sustainability.

6.1 Production and Market Access Challenges

It is essential to maintain consistency in production for commercial success in specific market outlets. This starts with selecting appropriate mushroom species and reliable spawn suppliers (Raut et al. 2019). Other key issues include securing resources for growing systems, meeting market quality standards, and navigating the complexities of post-harvest handling. The short shelf life of mushrooms presents significant challenges related to packaging, storage, and transportation. Furthermore, rising raw substrate costs, price volatility, inadequate marketing channels, inconsistent spawn quality, limited access to financial assistance, and a lack of scientific research on mushroom farming all contribute to the difficulties faced by producers (Jayaraman et al. 2024). These challenges are illustrated in Figure 4. Inconsistent market availability and lack of adequate regulations, outdated technology, quality control, and certification issues also hinder the growth of the mushroom industry.

Figure 4. This figure represents various challenges faced by the mushroom industry, which include inconsistent market availability, low-quality spawn production, inadequate scientific research, financial uncertainty from unstable prices, poor post-harvest management, a lack of adequate laws, outdated technology, quality and certification issues. These factors collectively hinder the growth and development of industries.

6.2 Consumer Perception and Market Dynamics

Consumer perception also plays a critical role. Cultural associations with toxic mushrooms and a lack of awareness regarding the nutritional benefits of edible varieties can limit demand. Focused education and culturally sensitive marketing are essential to overcome these barriers. The absence of cold chain logistics in some regions further restricts shelf life and availability, requiring customized solutions to address diverse market needs.

6.3 Environmental Sustainability

The mushroom industry also faces environmental challenges. Sustainable waste management practices are crucial, as is mitigating potential greenhouse gas emissions during composting. However, mushrooms also offer a solution in this regard. Agricultural waste, a globally abundant resource, can be used for mushroom cultivation (Jayaraman et al. 2024). Despite the lignocellulosic content of some agricultural waste, like rice straw, mushrooms can be successfully cultivated on it. The rice-wheat cropping system (RWCS), vital for global food security (Banjara et al. 2022), generates substantial crop waste. Managing rice straw, with its higher silica content, is particularly challenging. Burning this residue has become a major environmental concern, contributing to air pollution, and accelerating global warming (Somasundaram et al. 2020). Black carbon emissions from biomass burning are a significant factor in global warming (Tubiello, 2014). The scale of this issue is substantial, with millions of tons of rice and wheat residue burned annually, particularly in regions like India. Cultivating mushrooms on paddy straw offers a promising solution, preventing agricultural waste burning and providing a valuable product (Jayaraman et al. 2024). Several mushroom species, including *A. bisporus*, *V. volvacea*, and *Pleurotus* spp., can be effectively grown on this substrate.

6.3.1. Strategies for Sustainable Development

To unlock the full potential of the mushroom industry and ensure its sustainable development, a multi-pronged approach is required. This includes:

Promoting research and development: Continued research is crucial for improving cultivation techniques, developing new varieties, and optimizing post-harvest processing.

Investing in infrastructure: Investment in cold chain logistics, transportation, and other infrastructure is essential to enhance product availability and shelf life.

Enhancing knowledge sharing: Providing farmers with access to information on best practices, financial assistance, and market opportunities is vital.

Developing supportive policies: Governments should implement policies that encourage sustainable cultivation, promote market access, and support research.

Fostering collaboration: Collaboration among stakeholders, including government agencies, NGOs, research institutions, and farmers, is crucial for addressing the complex challenges of the industry.

By proactively addressing these challenges and implementing the recommended strategies, the mushroom industry can achieve sustainable growth, contributing significantly to global food security, economic development, and environmental sustainability.

7. CULINARY VERSATILITY OF EDIBLE MUSHROOMS

Mushrooms are increasingly recognized not only for their nutritional and health benefits but also for their culinary versatility. They offer a unique combination of flavor, texture, and adaptability, making them a valuable ingredient in diverse cuisines. Mushrooms are considered to have all three food functions: flavor, nutrition, and physiological functions. They possess a savory, slightly sweet taste known as umami, derived from naturally occurring amino acids like glutamic and aspartic acid, as well as 5′-nucleotides (Zhang et al. 2013). This umami flavor is further enhanced by specific peptides, such as those found in dried shiitake mushrooms, which can interact with other volatile compounds to create complex and nuanced flavor profiles (Kong et al. 2019).

7.1 Nutritional Profile and Culinary Uses

Mushrooms are a nutrient-dense food, low in calories, sodium, and unhealthy fats while being rich in essential vitamins and minerals (Hossain et al. 2007). They are also a valuable source of protein, containing all nine essential amino acids, making them an excellent meat alternative, especially in vegetarian diets (Aida et al. 2009; Ganeshpurkar et al. 2010). Because of this nutritional profile, mushrooms are not only used directly as food but also as raw materials for functional foods and nutraceuticals (Sabaratnam et al. 2013).

Mushrooms are incredibly versatile in the kitchen, lending themselves to a wide range of culinary applications (Figure 5). Some of the most popular edible mushroom species include:

- ***Agaricus bisporus* (Button/Portobello):** The most cultivated mushroom, known for its mild flavor. Button mushrooms are excellent in sauces, soups, and pizzas (Figure 6a). Portobello mushrooms, a mature form of *A. bisporus*, can be grilled, stuffed, roasted, or added to soups, stews, and sauces (Andi et al. 2024).
- ***Hypsizygus tessellatus* (Shimeji):** These mushrooms have a crunchy texture and a savory, nutty, umami flavor, like parmesan cheese or anchovies (Michael et al. 2023). They are a great addition to omelets, stews, sauces, rice, noodles, and tempura dishes.
- ***Pleurotus ostreatus* (Oyster):** With a mild flavor and meaty texture, oyster mushrooms are popular in Chinese and Japanese cuisine (Figure 6c). They can be marinated and grilled, or incorporated into stir-fries, sauces, and soups (Andi et al. 2024).

Other notable edible mushrooms include *Lentinus*, *Auricularia*, *Hericium*, *Grifola*, *Flammulina*, and *Tremella fuciformis*, some of which also possess medicinal and functional properties (Ganeshpurkar et al. 2010). Certain species, due to their texture and flavor, may be better suited for specific preparations or even primarily recognized for their therapeutic qualities.

Figure 5. Common culinary uses of mushrooms include their incorporation into sauces, stir-fries, soups, and pizza fillings

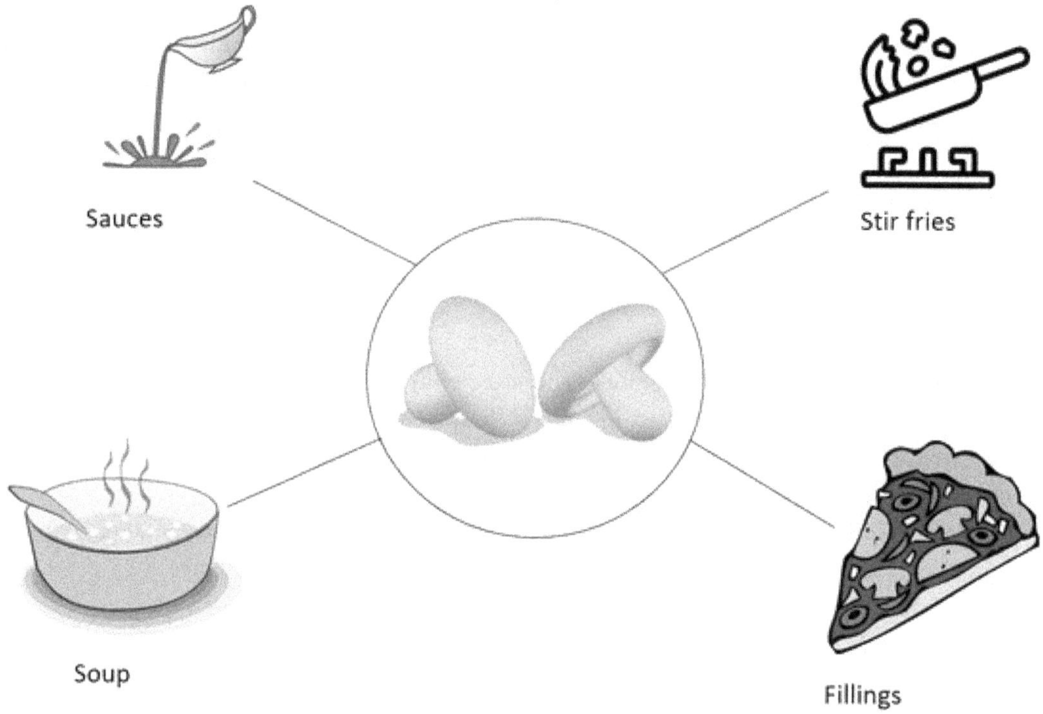

Figure 6. Commercially available edible mushrooms

a) *Agaricus bisporus*

b) *Hypsizygus tessulatus*

c) *Pleurotus ostreatus*

8. CONCLUSION

Mushrooms deserve to be called superfoods due to their remarkable nutritional profile and associated health benefits. The mushroom bioactive serves as an immunomodulatory agent, anticancer, anti-inflammatory agent, and antioxidant conferring protection from chronic ailments. This makes them invaluable for modern diets. In the long run, mushroom farming promises to provide food security by being an environmentally sustainable practice. The challenges associated with optimizing production and market accessibility must be overcome to successfully scale up mushroom production. With continued research and innovation, mushrooms will undoubtedly play a pivotal role in future global food systems.

REFERENCES

Aida, F. M. N. A., Shuhaimi, M., Yazid, M., & Maaruf, A. G. (2009). Mushroom as a potential source of prebiotics: A review. *Trends in Food Science & Technology*, *20*(11–12), 567–575. DOI: 10.1016/j.tifs.2009.07.007

Arora, B., Kamal, S., Sethi, S., Joshi, A., & Sharma, V. P. (2018). Mushrooms: Reservoir of vital nutrients and bioactives. [). www.nap.edu]. *Mushroom Research*, *27*(1), •••.

Assemie, A., & Abaya, G. (2022). The Effect of Edible Mushroom on Health and Their Biochemistry. In *International Journal of Microbiology* (Vol. 2022). Hindawi Limited., DOI: 10.1155/2022/8744788

Banjara, T. R., Bohra, J. S., Kumar, S., Ram, A., & Pal, V. (2022). Diversification of Rice–Wheat Cropping System Improves Growth, Productivity and Energetics of Rice in the Indo-Gangetic Plains of India. *Agricultural Research*, *11*(1), 48–57. DOI: 10.1007/s40003-020-00533-9

Bell, V., Silva, C. R. P. G., Guina, J., & Fernandes, T. H. (2022). Mushrooms as future generation healthy foods. *Frontiers in Nutrition*, *9*, 1050099. DOI: 10.3389/fnut.2022.1050099 PMID: 36562045

Bhambri, A., Srivastava, M., Mahale, V. G., Mahale, S., & Karn, S. K. (2022). Mushrooms as Potential Sources of Active Metabolites and Medicines. In Frontiers in Microbiology (Vol. 13). Frontiers Media S.A. DOI: 10.3389/fmicb.2022.837266

Bijla, S., & Sharma, V. P. (2023). Status of mushroom production: Global and national scenario. *Mushroom Research*, *32*(2), 91–98. DOI: 10.36036/MR.32.2.2023.146647

Bjelakovic, G., Nikolova, D., Gluud, L. L., Simonetti, R. G., & Gluud, C. (2007). Mortality in randomized trials of antioxidant supplements for primary and secondary prevention: Systematic review and meta-analysis. *Journal of the American Medical Association*, *297*(8), 842–857. DOI: 10.1001/jama.297.8.842 PMID: 17327526

Brewer, M. S. (2011). Natural Antioxidants: Sources, Compounds, Mechanisms of Action, and Potential Applications. *Comprehensive Reviews in Food Science and Food Safety*, *10*(4), 221–247. DOI: 10.1111/j.1541-4337.2011.00156.x

Caitano, C. E. C., Wagner, G. V. J., Soares, D. M. M., Lucas da Silva, A., Bianca de Barros, N., Pardo-Giménez, A., Stevani, C. V., & Zied, D. C. (2023). Management of Water Supply in the Cultivation of Different Agaricus Bisporus Strains. *Agronomy (Basel)*, *13*(10), 2626. Advance online publication. DOI: 10.3390/agronomy13102626

Cardwell, G., Bornman, J. F., James, A. P., Black, L. J., & Au, A. P. J. (2018). A Review of Mushrooms as a Potential Source of Dietary Vitamin D. DOI: 10.3390/nu10101498

Chakraborty, I., Sen, I. K., Mondal, S., Rout, D., Bhanja, S. K., Maity, G. N., & Maity, P. (2019). Bioactive polysaccharides from natural sources: A review on the antitumor and immunomodulating activities. In *Biocatalysis and Agricultural Biotechnology* (Vol. 22). Elsevier Ltd., DOI: 10.1016/j.bcab.2019.101425

Cheah, I. K., & Halliwell, B. (2012). Ergothioneine; antioxidant potential, physiological function, and role in disease. In Biochimica et Biophysica Acta - Molecular Basis of Disease (Vol. 1822, Issue 5, pp. 784–793). DOI: 10.1016/j.bbadis.2011.09.017

Cheung, P. C. K. (2013). Mini-review on Edible Mushrooms as Source of Dietary Fiber: Preparation and Health Benefits. In Food Science and Human Wellness (Vol. 2, Issues 3–4, pp. 162–166). Elsevier B.V. DOI: 10.1016/j.fshw.2013.08.001

Chun, S., Gopal, J., & Muthu, M. (2021). Antioxidant activity of mushroom extracts/polysaccharides —their antiviral properties and plausible antiCOVID-19 properties. In Antioxidants (Vol. 10, Issue 12). MDPI. DOI: 10.3390/antiox10121899

Cobos, Á., & Díaz, O. (2023). "Superfoods": Reliability of the Information for Consumers Available on the Web. *Foods*, *12*(3), 546. Advance online publication. DOI: 10.3390/foods12030546 PMID: 36766074

Daba, A. S., & Ezeronye, O. U. (2003). Anti-cancer effect of polysaccharides isolated from higher basidiomycetes mushrooms. *African Journal of Biotechnology*, *2*(12), 672–678. https://www.academicjournals.org/AJB. DOI: 10.5897/AJB2003.000-1123

Das, A., Chen, C. M., Mu, S. C., Yang, S. H., Ju, Y. M., & Li, S. C. (2022). Medicinal Components in Edible Mushrooms on Diabetes Mellitus Treatment. *Pharmaceutics*, *14*(2), 436. Advance online publication. DOI: 10.3390/pharmaceutics14020436 PMID: 35214168

Das, A. K., Nanda, P. K., Dandapat, P., Bandyopadhyay, S., Gullón, P., Sivaraman, G. K., McClements, D. J., Gullón, B., & Lorenzo, J. M. (2021). Edible mushrooms as functional ingredients for development of healthier and more sustainable muscle foods: A flexitarian approach. *Molecules (Basel, Switzerland)*, *26*(9), 2463. Advance online publication. DOI: 10.3390/molecules26092463 PMID: 33922630

De Silva, D. D., Rapior, S., Hyde, K. D., & Bahkali, A. H. (2012). Medicinal mushrooms in prevention and control of diabetes mellitus. *Fungal Diversity*, *56*(1), 1–29. DOI: 10.1007/s13225-012-0187-4 PMID: 23097638

de Sousa, V. M. C., dos Santos, E. F., & Sgarbieri, V. C. (2011). The Importance of Prebiotics in Functional Foods and Clinical Practice. *Food and Nutrition Sciences*, *02*(02), 133–144. DOI: 10.4236/fns.2011.22019

Dey, A., Patel, S., & Rashid, M. M. (2020). Small-scale mushroom production unit for the upliftment of rural economy and women empowerment in India: A Review. *Current Journal of Applied Science and Technology*, *39*(2), 38–46.

Dimopoulou, M., Kolonas, A., Mourtakos, S., Androutsos, O., & Gortzi, O. (2022). Nutritional Composition and Biological Properties of Sixteen Edible Mushroom Species. *Applied Sciences (Basel, Switzerland)*, *12*(16), 8074. Advance online publication. DOI: 10.3390/app12168074

Dinakarkumar, Y., Ramakrishnan, G., Gujjula, K. R., Vasu, V., Balamurugan, P., & Murali, G. (2024). Fungal bioremediation: An overview of the mechanisms, applications and future perspectives. *Environmental Chemistry and Ecotoxicology*, *6*, 293–302. DOI: 10.1016/j.enceco.2024.07.002

Du, X. J., Zhang, J. S., Yang, Y., Tang, Q. J., Jia, W., & Pan, Y. J. (2010). Purification, chemical modification and immunostimulating activity of polysaccharides from Tremella aurantialba fruit bodies. *Journal of Zhejiang University. Science. B.*, *11*(6), 437–442. DOI: 10.1631/jzus.B0900402 PMID: 20506575

Ey, J., Schömig, E., & Taubert, D. (2007). Dietary sources and antioxidant effects of ergothioneine. *Journal of Agricultural and Food Chemistry*, *55*(16), 6466–6474. DOI: 10.1021/jf071328f PMID: 17616140

Fernández-Ríos, A., Laso, J., Hoehn, D., Amo-Setién, F. J., Abajas-Bustillo, R., Ortego, C., Fullana-i-Palmer, P., Bala, A., Batlle-Bayer, L., Balcells, M., Puig, R., Aldaco, R., & Margallo, M. (2022). A critical review of superfoods from a holistic nutritional and environmental approach. *Journal of Cleaner Production*, *379*, 134491. DOI: 10.1016/j.jclepro.2022.134491

Firenzuoli, F., Gori, L., & Lombardo, G. (2008). The medicinal mushroom Agaricus blazei murrill: Review of literature and pharmaco-toxicological problems. In Evidence-based Complementary and Alternative Medicine (Vol. 5, Issue 1, pp. 3–15). DOI: 10.1093/ecam/nem007

Friedman, M. (2016). Mushroom Polysaccharides: Chemistry and Antiobesity, Antidiabetes, Anticancer, and Antibiotic Properties in Cells, Rodents, and Humans. *Foods*, *5*(4), 1–40. DOI: 10.3390/foods5040080 PMID: 28231175

Fryer, C. (2024). Mushrooms for Inflammation: Understanding their anti-inflammatory benefits. https://maoi.org/health/mushrooms-for-inflammation/

Ganeshpurkar, A., Rai, G., & Jain, A. (2010). Medicinal Mushrooms: Towards a New Horizon. *Pharmacognosy Reviews*, *4*(8), 127–135. DOI: 10.4103/0973-7847.70904 PMID: 22228952

Grimm, D., & Wösten, H. A. B. (2018). Mushroom cultivation in the circular economy. *Applied Microbiology and Biotechnology*, *102*(18), 7795–7803. DOI: 10.1007/s00253-018-9226-8 PMID: 30027491

Hettiarachchi, H., Caucci, S., & Schwärzel, K. (2020). Organic Waste Composting through Nexus Thinking.

Hossain, S., & Rahman, M. A. (2007). Essential fatty acid contents of Pleurotus ostreatus, Ganoderma lucidum and Agaricus bisporus. https://www.researchgate.net/publication/258050987

Hu, Y., Bandara, A. R., Xu, J., Kakumyan, P., Hyde, K. D., & Mortimer, P. E. (2022). 'The use of agaricus subrufescens for rehabilitation of agricultural soils'. *Agronomy (Basel)*, *12*(9), 2034. Advance online publication. DOI: 10.3390/agronomy12092034

Hu, Y., Mortimer, P. E., Hyde, K. D., Kakumyan, P., & Thongklang, N. (2021). 'Mushroom cultivation for soil amendment and bioremediation'. *Circular Agricultural Systems*, *1*(1), 1–14. DOI: 10.48130/CAS-2021-0011

Ina, K., Kataoka, T., & Ando, T. (2013). The use of lentinan for treating gastric cancer. *Anti-cancer Agents in Medicinal Chemistry*, *13*(5), 681–688. DOI: 10.2174/1871520611313050002 PMID: 23092289

Jagdale, Y. D., Mahale, S. V., Zohra, B., Nayik, G. A., Dar, A. H., Khan, K. A., Abdi, G., & Karabagias, I. K. (2021). Nutritional Profile and Potential Health Benefits of Super Foods: A Review. *Sustainability (Basel)*, *13*(16), 9240. Advance online publication. DOI: 10.3390/su13169240

Jasińska, A. (2018). Spent mushroom compost (SMC)–retrieved added value product closing loop in agricultural production.

Jayachandran, M., Xiao, J., & Xu, B. (2017). A critical review on health promoting benefits of edible mushrooms through gut microbiota. In International Journal of Molecular Sciences (Vol. 18, Issue 9). MDPI AG. DOI: 10.3390/ijms18091934

Jayaraman, S., Yadav, B., Dalal, R. C., Naorem, A., Sinha, N. K., & Rao, C. S. (2024). 'Mushroom Farming: A Review Focusing on Soil Health, Nutritional Security and Environmental Sustainability'. *Farming System*, *2*(3), 100098. DOI: 10.1016/j.farsys.2024.100098

Jedinak, A., Dudhgaonkar, S., Wu, Q. L., Simon, J., & Sliva, D. (2011). Anti-inflammatory activity of edible oyster mushroom is mediated through the inhibition of NF-κB and AP-1 signaling. *Nutrition Journal*, *10*(1), 1–10. DOI: 10.1186/1475-2891-10-52 PMID: 21575254

Joseph, M. (2023). *Types Of Mushrooms From A to Z (With Photos) | Live Eat*. https://www.liveeatlearn.com/types-of-mushrooms-how-to-use-them/

Joseph, T. P., Chanda, W., Padhiar, A. A., & Batool, S. LiQun, S., Zhong, M. T., & Huang, M. (2018). A Preclinical Evaluation of the Antitumor Activities of Edible and Medicinal Mushrooms: A Molecular Insight. In Integrative Cancer Therapies (Vol. 17, Issue 2, pp. 200–209). SAGE Publications Inc. DOI: 10.1177/1534735417736861

Keegan, R.-J. H., Lu, Z., Bogusz, J. M., Williams, J. E., & Holick, M. F. (2013). Photobiology of vitamin D in mushrooms and its bioavailability in humans. *Dermato-Endocrinology*, *5*(1), 165–176. DOI: 10.4161/derm.23321 PMID: 24494050

Khan, A., Murad, W., Salahuddin, , Ali, S., Shah, S. S., Halim, S. A., Khalid, A., Kashtoh, H., Khan, A., & Al-Harrasi, A. (2024). Contribution of mushroom farming to mitigating food scarcity: Current status, challenges and potential future prospects in Pakistan. *Heliyon*, *10*(23), e40362. DOI: 10.1016/j.heliyon.2024.e40362 PMID: 39660206

Kong, Y., Zhang, L. L., Zhao, J., Zhang, Y. Y., Sun, B. G., & Chen, H. T. (2019). Isolation and identification of the umami peptides from shiitake mushroom by consecutive chromatography and LC-Q-TOF-MS. *Food Research International*, *121*, 463–470. DOI: 10.1016/j.foodres.2018.11.060 PMID: 31108770

Kozarski, M., Klaus, A., Jakovljevic, D., Todorovic, N., Vunduk, J., Petrović, P., Niksic, M., Vrvic, M. M., & Van Griensven, L. (2015). Antioxidants of edible mushrooms. In Molecules (Vol. 20, Issue 10, pp. 19489–19525). MDPI AG. DOI: 10.3390/molecules201019489

Kozarski, M., Klaus, A., Niksic, M., Jakovljevic, D., Helsper, J. P. F. G., & Van Griensven, L. J. L. D. (2011). Antioxidative and immunomodulating activities of polysaccharide extracts of the medicinal mushrooms Agaricus bisporus, Agaricus brasiliensis, Ganoderma lucidum and Phellinus linteus. *Food Chemistry*, *129*(4), 1667–1675. DOI: 10.1016/j.foodchem.2011.06.029

Kumar, K. (2015). Role of edible mushroom as functional foods: A review. South Asian Journal of Food Technology and Environment, 01(3and4), 211–218. DOI: 10.46370/sajfte.2015.v01i03and04.02

Kumar, K., Mehra, R., Guiné, R. P. F., Lima, M. J., Kumar, N., Kaushik, R., Ahmed, N., Yadav, A. N., & Kumar, H. (2021). Edible Mushrooms: A Comprehensive Review on Bioactive Compounds with Health Benefits and Processing Aspects. Foods 2021, Vol. 10, Page 2996, 10(12), 2996. https://doi.org/ DOI: 10.3390/foods10122996

Kumar, V., Sharma, K., Kedam, N., Patel, A., Kate, T., & Rathnayake, U. (2024). A Comprehensive Review on Smart and Sustainable Agriculture Using IoT Technologies. *Smart Agricultural Technology*, *8*, 100487. Advance online publication. DOI: 10.1016/j.atech.2024.100487

Lakhanpal, T. N., & Rana, M. (2005). Medicinal and nutraceutical genetic resources of mushrooms. *Plant Genetic Resources*, *3*(2), 288–303. DOI: 10.1079/PGR200581

Lee, W. Y., Park, E.-J., Ahn, J. K., & Ka, K.-H. (2009). Ergothioneine Contents in Fruiting Bodies and Their Enhancement in Mycelial Cultures by the Addition of Methionine. *Microbiology*, *37*(1), 43. DOI: 10.1099/00221287-124-1-43 PMID: 23983506

Li, Z., Forester, S., Jennings-Dobbs, E., & Heber, D. (2023). Perspective: A Comprehensive Evaluation of Data Quality in Nutrient Databases. *Advances in Nutrition*, *14*(3), 379–391. DOI: 10.1016/j.advnut.2023.02.005 PMID: 36849083

Lindequist, U., Niedermeyer, T. H. J., & Jülich, W. D. (2005). The pharmacological potential of mushrooms. In Evidence-based Complementary and Alternative Medicine (Vol. 2, Issue 3, pp. 285–299). https://doi.org/DOI: 10.1093/ecam/neh107

Liuzzi, G. M., Petraglia, T., Latronico, T., Crescenzi, A., & Rossano, R. (2023). Antioxidant Compounds from Edible Mushrooms as Potential Candidates for Treating Age-Related Neurodegenerative Diseases. *Nutrients*, *15*(8), 1913. DOI: 10.3390/nu15081913 PMID: 37111131

Łysakowska, P., Sobota, A., & Wirkijowska, A. (2023). Medicinal Mushrooms: Their Bioactive Components, Nutritional Value and Application in Functional Food Production—A Review. In Molecules (Vol. 28, Issue 14). Multidisciplinary Digital Publishing Institute (MDPI). DOI: 10.3390/molecules28145393

Malik, M. A., Jan, Y., Al-Keridis, L. A., Haq, A., Ahmad, J., Adnan, M., Alshammari, N., Ashraf, S. A., & Panda, B. P. (2022). Effect of Vitamin-D-Enriched Edible Mushrooms on Vitamin D Status, Bone Health and Expression of CYP2R1, CYP27B1 and VDR Gene in Wistar Rats. *Journal of Fungi (Basel, Switzerland)*, *8*(8), 864. Advance online publication. DOI: 10.3390/jof8080864 PMID: 36012852

Morales, D., Rutckeviski, R., Villalva, M., Abreu, H., Soler-Rivas, C., Santoyo, S., Iacomini, M., & Smiderle, F. R. (2020). Isolation and comparison of α- and β-D-glucans from shiitake mushrooms (Lentinula edodes) with different biological activities. *Carbohydrate Polymers*, *229*, 115521. DOI: 10.1016/j.carbpol.2019.115521 PMID: 31826486

Mustafa, F., Chopra, H., Baig, A. A., Avula, S. K., Kumari, S., Mohanta, T. K., Saravanan, M., Mishra, A. K., Sharma, N., & Mohanta, Y. K. (2022). Edible Mushrooms as Novel Myco-Therapeutics: Effects on Lipid Level, Obesity, and BMI. In *Journal of Fungi* (Vol. 8, Issue 2). MDPI. https://doi.org/DOI: 10.3390/jof8020211

Mwangi, R. W., Macharia, J. M., Wagara, I. N., & Bence, R. L. (2022). The antioxidant potential of different edible and medicinal mushrooms. *Biomedicine and Pharmacotherapy*, *147*, 112621. DOI: 10.1016/j.biopha.2022.112621 PMID: 35026489

Nandi, S., Sikder, R., Rapior, S., Arnould, S., Simal-Gandara, J., & Acharya, K. (2024). A review for cancer treatment with mushroom metabolites through targeting mitochondrial signaling pathway: In vitro and in vivo evaluations, clinical studies and future prospects for mycomedicine. *Fitoterapia*, *172*, 105681. DOI: 10.1016/j.fitote.2023.105681 PMID: 37743029

Okuda, Y. (2022). Sustainability perspectives for future continuity of mushroom production: The bright and dark sides. *Frontiers in Sustainable Food Systems*, *6*, 1026508. Advance online publication. DOI: 10.3389/fsufs.2022.1026508

Onwosi, C. O., Igbokwe, V. C., Odimba, J. N., Eke, I. E., Nwankwoala, M. O., Iroh, I. N., & Ezeogu, L. I. (2017a). 'Composting Technology in Waste Stabilization: On the Methods, Challenges and Future Prospects'. *Journal of Environmental Management*, *190*, 140–157. DOI: 10.1016/j.jenvman.2016.12.051 PMID: 28040590

Onwude, D. I., Chen, G., Eke-emezie, N., Kabutey, A., Khaled, A. Y., & Sturm, B. (2020). Recent Advances in Reducing Food Losses in the Supply Chain of Fresh Agricultural Produce. *Processes (Basel, Switzerland)*, *8*(11), 1431. DOI: 10.3390/pr8111431

Org, I., Elkhateeb, W. A., Daba, G. M., & Elkhateeb, W. A. (2023). Vitamin D Enriched Edible Mushrooms: A. [Review]. *Journal of Pharmacy and Drug Development*, *2*(1). Advance online publication. DOI: 10.58489/2836-2322/013

Park, H. J. (2022). Current Uses of Mushrooms in Cancer Treatment and Their Anticancer Mechanisms. *International Journal of Molecular Sciences*, *23*(18), 10502. DOI: 10.3390/ijms231810502 PMID: 36142412

Peng, L., Tang, R., Wang, G., Ma, R., Li, Y., Li, G., & Yuan, J. (2023). 'Effect of Aeration Rate, Aeration Pattern, and Turning Frequency on Maturity and Gaseous Emissions during Kitchen Waste Composting'. *Environmental Technology & Innovation*, *29*, 102997. DOI: 10.1016/j.eti.2022.102997

Pérez-Chávez, A. M., Mayer, L., & Albertó, E. (2019). 'Mushroom Cultivation and Biogas Production: A Sustainable Reuse of Organic Resources'. *Energy for Sustainable Development*, *50*, 50–60. DOI: 10.1016/j.esd.2019.03.002

Podkowa, A., Kryczyk-Poprawa, A., Opoka, W., & Muszyńska, B. (2021). Culinary–medicinal mushrooms: a review of organic compounds and bioelements with antioxidant activity. In European Food Research and Technology (Vol. 247, Issue 3, pp. 513–533). Springer Science and Business Media Deutschland GmbH. DOI: 10.1007/s00217-020-03646-1

Popkin, B. M., & Ng, S. W. (2022). The nutrition transition to a stage of high obesity and noncommunicable disease prevalence dominated by ultra-processed foods is not inevitable. *Obesity Reviews*, *23*(1), e13366. https://doi.org/https://doi.org/10.1111/obr.13366. DOI: 10.1111/obr.13366 PMID: 34632692

Prajapati, S. K., Rawat, V., Singh, D., Saroj, S., Ravikant, D., & Verma, S. (2023). Mushroom Cultivation: A Sustainable Approach to Future Agriculture to Ensure Quality Food and Nutritional Security of the Current Population in India. *International Journal of Multidisciplinary Research and Growth Evaluation., 4*, 697–705.

Rana, R. K., Monga, S., & Kaur, T. (2024). Scaling up Mushroom Production for Sustainable Entrepreneurship.

Raut, J. K. (2019). Current Status, Challenges, and Prospects of Mushroom Industry in Nepal. *Journal of Agricultural Economics*, *4*(4), 154–160. DOI: 10.11648/j.ijae.20190404.13

Reis, A. (2024). 18 Different Types Of Edible Mushrooms And How To Use Them. https://cookingchew.com/different-types-of-edible-mushrooms-and-how-to-use-them.html

Reis, F. S., Hu, Y., & Fernandes, T. H. (2022). Mushrooms as future generation healthy foods.

Robinson, B., Winans, K., Kendall, A., Dlott, J., & Dlott, F. (2019). 'A Life Cycle Assessment of Agaricus Bisporus Mushroom Production in the USA'. *The International Journal of Life Cycle Assessment*, *24*(3), 456–467. DOI: 10.1007/s11367-018-1456-6

Sabaratnam, V., Kah-Hui, W., Naidu, M., & David, P. (2013). Neuronal health - can culinary and medicinal mushrooms help? *Journal of Traditional and Complementary Medicine*, *3*(1), 62–68. DOI: 10.4103/2225-4110.106549 PMID: 24716157

Sande, D., Oliveira, G. P., Moura, M. A. F. E., Martins, B. A., Lima, M. T. N. S., & Takahashi, J. A. (2019). Edible mushrooms as a ubiquitous source of essential fatty acids. *Food Research International*, *125*, 108524. DOI: 10.1016/j.foodres.2019.108524 PMID: 31554069

Sangeeta, S. D., Ramniwas, S., Mugabi, R., Uddin, J., & Nayik, G. A. (2024, August 25). Revolutionizing Mushroom processing: Innovative techniques and technologies. *Food Chemistry: X*, *23*, 101774. DOI: 10.1016/j.fochx.2024.101774 PMID: 39280230

Santos, A., Bustamante, M. A., Tortosa, G., Moral, R., & Bernal, M. P. (2016). 'Gaseous Emissions and Process Development during Composting of Pig Slurry: The Influence of the Proportion of Cotton Gin Waste'. *Journal of Cleaner Production*, *112*, 81–90. DOI: 10.1016/j.jclepro.2015.08.084

Shin, H.-J., Ro, H.-S., Kawauchi, M., & Honda, Y. (2025). Review on mushroom mycelium-based products and their production process: From upstream to downstream. *Bioresources and Bioprocessing*, *12*(1), 3. DOI: 10.1186/s40643-024-00836-7 PMID: 39794674

Shirur, M. (2022). Leveraging government policies and schemes for promoting mushroom entrepreneurship in India. *Mushroom Research*, *30*(2).

Singh, M., Vijay, B., Kamal, S., & Wakchaure, G. C. (2011). Mushrooms Cultivation, Marketing, and Consumption Editors.

Singh, Y., & Sidhu, H. S. (2014). Management of Cereal Crop Residues for Sustainable Rice-Wheat Production System in the Indo-Gangetic Plains of India. *Proceedings of the Indian National Science Academy. Part A, Physical Sciences*, *80*(1), 95–114. DOI: 10.16943/ptinsa/2014/v80i1/55089

Somasundaram, J., Sinha, N. K., & Ram, C. (2020). 'No-Till Farming and Conservation Agriculture in South Asia–Issues, Challenges, Prospects, and Benefits'. *Critical Reviews in Plant Sciences*, *39*(3), 236–279. DOI: 10.1080/07352689.2020.1782069

Strong, P. J., Self, R., Allikian, K., Szewczyk, E., Speight, R., O'Hara, I., & Harrison, M. D. (2022). Filamentous fungi for future functional food and feed. *Current Opinion in Biotechnology*, *76*, 102729. DOI: 10.1016/j.copbio.2022.102729 PMID: 35525176

Sudha, G., Vadivukkarasi, S., Shree, R. B. I., & Lakshmanan, P. (2012). Antioxidant activity of various extracts from an edible mushroom Pleurotus. *Food Science and Biotechnology*, *21*(3), 661–668. DOI: 10.1007/s10068-012-0086-1 PMID: 30263280

Tongpool, R., & Pongpat, P. (2013). Analysis of shiitake environmental performance via life cycle assessment. *International Journal of Environmental Sciences and Development*, *4*, 552–557. DOI: 10.7763/IJESD.2013.V4.412

Tubiello, F. N. 2014. 'Agriculture, Forestry, and Other Land Use Emissions by Sources and Removals by Sinks: 1990-2011 Analysis'. DOI: 10.13140/2.1.4143.4245

Uffelman, C. N., Chan, N. I., Davis, E. M., Wang, Y., McGowan, B. S., & Campbell, W. W. (2023). An Assessment of Mushroom Consumption on Cardiometabolic Disease Risk Factors and Morbidities in Humans: A Systematic Review. In Nutrients (Vol. 15, Issue 5). MDPI. DOI: 10.3390/nu15051079

Urbain, P., Singler, F., Ihorst, G., Biesalski, H. K., & Bertz, H. (2011). Bioavailability of vitamin D2 from UV-B-irradiated button mushrooms in healthy adults deficient in serum 25-hydroxyvitamin D: a randomized controlled trial. European Journal of Clinical Nutrition 2011 65:8, 65(8), 965–971. DOI: 10.1038/ejcn.2011.53

Valverde, M. E., Hernández-Pérez, T., & Paredes-López, O. (2015). Edible mushrooms: Improving human health and promoting quality life. *International Journal of Microbiology*, *376387*, 1–14. Advance online publication. DOI: 10.1155/2015/376387 PMID: 25685150

Wang, M., & Zhao, R. (2023). A review on nutritional advantages of edible mushrooms and its industrialization development situation in protein meat sanalogues. In Journal of Future Foods (Vol. 3, Issue 1, pp. 1–7). Beijing Academy of Food Sciences. DOI: 10.1016/j.jfutfo.2022.09.001

Wang, X., Selvam, A., & Wong, J. W. C. (2016). Influence of Lime on Struvite Formation and Nitrogen Conservation during Food Waste Composting. *Bioresource Technology*, *217*, 227–232. DOI: 10.1016/j.biortech.2016.02.117 PMID: 27020123

Wani, B. A., Bodha, R. H., & Wani, A. H. (2010). Nutritional and medicinal importance of mushrooms. In Journal of Medicinal Plants Research (Vol. 4, Issue 24, pp. 2598–2604). Academic Journals. DOI: 10.5897/JMPR09.565

Xu, Q., Zhang, T., Niu, Y., Mukherjee, S., Abou-Elwafa, S. F., Nguyen, N. S. H., al Aboud, N. M., Wang, Y., Pu, M., Zhang, Y., Tran, H. T., Almazroui, M., Hooda, P. S., Bolan, N. S., Rinklebe, J., & Shaheen, S. M. (2024). A comprehensive review of agricultural waste utilization through sustainable conversion techniques, with a focus on the additives effect on the fate of phosphorus and toxic elements during composting process. *The Science of the Total Environment*, *942*, 173567. DOI: 10.1016/j.scitotenv.2024.173567 PMID: 38848918

Yahfoufi, N., Alsadi, N., Jambi, M., & Matar, C. (2018). The Immunomodulatory and Anti-Inflammatory Role of Polyphenols. Nutrients 2018, Vol. 10, Page 1618, 10(11), 1618. DOI: 10.3390/nu10111618

Yusuf, Ç., & Peker, K. (2009). Benefit/cost analysis of mushroom production for diversification of income in developing countries. https://www.researchgate.net/publication/267956799

Zhang, Y., Venkitasamy, C., Pan, Z., & Wang, W. (2013). Recent developments on umami ingredients of edible mushrooms - A review. In Trends in Food Science and Technology (Vol. 33, Issue 2, pp. 78–92). DOI: 10.1016/j.tifs.2013.08.002

Zhao, S., Gao, Q., Rong, C., Wang, S., Zhao, Z., Liu, Y., & Xu, J. (2020). Immunomodulatory effects of edible and medicinal mushrooms and their bioactive immunoregulatory products. In Journal of Fungi (Vol. 6, Issue 4, pp. 1–37). MDPI AG. DOI: 10.3390/jof6040269

Zhou, K., & Yu, L. (2006). Total phenolic contents and antioxidant properties of commonly consumed vegetables grown in Colorado. *Lebensmittel-Wissenschaft + Technologie*, *39*(10), 1155–1162. DOI: 10.1016/j.lwt.2005.07.015

Chapter 17
Mycotoxins:
Challenges and Strategies for Mitigation

Tarun Kumar Patel
https://orcid.org/0009-0003-3002-8100
Sant Guru Ghasidas Government P.G. College, India

ABSTRACT

This chapter provides a comprehensive overview of mycotoxins, focusing on their occurrence, impact, detection, and mitigation strategies. It examines the major mycotoxins of concern, including aflatoxins, ochratoxins, fumonisins, trichothecenes, zearalenone, and patulin, highlighting their sources, health risks, and economic consequences. Key factors influencing mycotoxin contamination, such as environmental conditions and agricultural practices, are discussed. The chapter also explores various detection and quantification methods, including advanced technologies for accurate and efficient monitoring. A review of global regulatory frameworks, including those of the Codex Alimentarius Commission, the European Union, and the U.S. FDA, outlines current safety standards. Mitigation strategies, from pre-harvest practices to post-harvest and processing interventions, are detailed, alongside emerging technologies such as genetic engineering and biocontrol. The chapter concludes with future directions and policy recommendations to enhance global mycotoxin management efforts.

INTRODUCTION

Mycotoxins, a diverse group of toxic secondary metabolites produced by certain molds, have emerged as a critical issue in agricultural production, food safety, and public health. These toxic compounds, which commonly contaminate crops like maize, wheat, barley, and peanuts, pose significant risks to both human and animal health when consumed at unsafe levels. Mycotoxin-producing fungi thrive in specific environmental conditions, primarily in warm and humid climates, creating a persistent challenge for food production systems worldwide. Due to the global nature of food trade, mycotoxin contamination is not just a local concern; contaminated crops can affect food safety in regions far removed from the point of production (Awuchi et al. 2021).

Among the most well-known mycotoxins are aflatoxins, produced by *Aspergillus flavus* and *Aspergillus parasiticus*, ochratoxins from *Aspergillus* and *Penicillium* species, and fumonisins, typically produced by *Fusarium* species. Aflatoxins, in particular, are known for their high toxicity and carcinogenic po-

DOI: 10.4018/979-8-3693-8201-1.ch017

tential, especially aflatoxin B1, which has been strongly linked to liver cancer in humans. The economic implications of mycotoxin contamination are vast, affecting the profitability of the agricultural sector through crop losses, the cost of mitigation strategies, and potential barriers to international trade, where stringent regulations on mycotoxin levels in food and feed are in place (El-Sayed et al. 2022).

The challenge of mycotoxin contamination is exacerbated by climate change, which is reshaping global agricultural patterns and altering the geographical distribution of mycotoxin-producing fungi. Rising temperatures, shifts in rainfall patterns, and extreme weather events such as droughts and floods create environments that can increase the incidence and severity of mycotoxin outbreaks. Furthermore, the integration of global food markets means that crops grown in one region can affect food supplies far beyond their point of origin, potentially leading to widespread contamination in international markets. This global reach of the mycotoxin problem underscores the urgency for effective mitigation strategies that can be applied across diverse agricultural systems and climates (Kos et al. 2023).

Managing mycotoxin contamination requires a multi-faceted approach, including pre-harvest, post-harvest, and regulatory measures. Pre-harvest strategies focus on crop management techniques, such as the use of resistant crop varieties, biocontrol agents like *Aspergillus flavus* non-toxigenic strains (Patel et al. 2014), and integrated pest management (IPM) practices to reduce fungal infection in the field. Post-harvest strategies aim to minimize contamination during storage and processing by ensuring optimal storage conditions, utilizing appropriate drying techniques, and employing chemical or biological detoxification methods. Regulatory frameworks play an equally important role, setting maximum allowable limits for mycotoxins in food and feed products, and establishing guidelines for monitoring and control. However, these strategies face limitations, particularly in developing countries, where regulatory enforcement may be weaker, and resources for implementing advanced mitigation techniques may be limited (Habschied et al. 2021; Leslie et al. 2021).

In addition to these strategies, ongoing research into innovative technologies and methods offers hope for more effective mycotoxin mitigation in the future. Advances in molecular biology, such as genetic modification of crops for enhanced resistance to fungal infection, hold promise for long-term solutions to mycotoxin contamination. Likewise, the development of novel detection methods, including rapid on-site testing and the use of nanotechnology, could revolutionize the ability of producers and regulators to monitor and control mycotoxins in real time. The chapter will explore these challenges and opportunities in greater detail, offering a comprehensive overview of the current state of mycotoxin management and the future directions for reducing their impact on global food security and public health (Loi et al. 2023).

OCCURRENCE AND IMPACT OF MYCOTOXINS

Mycotoxins are toxic secondary metabolites produced by various fungi, particularly those belonging to the *Aspergillus*, *Penicillium*, and *Fusarium* genera, and they can contaminate a wide range of agricultural commodities such as grains, nuts, fruits, and animal feed. Their occurrence is closely linked to environmental factors like temperature, humidity, and improper storage conditions, which create an optimal environment for fungal growth and toxin production. Mycotoxin contamination is widespread, especially in tropical and subtropical regions, where hot and humid climates make crops particularly vulnerable. However, these toxins can also occur in cooler and temperate regions, often affecting crops like maize, wheat, barley, and peanuts (Pandey et al. 2023; Santos et al. 2022).

The impact of mycotoxins on human and animal health is severe. For instance, aflatoxins, which are common in maize, peanuts, and tree nuts, are known for their potent carcinogenic effects, particularly aflatoxin B1, which is closely linked to liver cancer. Ochratoxins, found in cereals, coffee, and dried fruits, can lead to kidney damage and other health complications, while fumonisins, typically contaminating maize, are associated with esophageal cancer in humans and neural tube defects in fetuses. Zearalenone, an estrogenic mycotoxin, can disrupt reproductive health, particularly in livestock, leading to issues like infertility and abortion, while trichothecenes such as deoxynivalenol (DON) can cause a range of symptoms, including nausea, immunosuppression, and growth inhibition in animals (Gurikar et al. 2023; Awuchi et al. 2021; Zain 2011).

The economic implications of mycotoxin contamination are vast, affecting the agricultural sector globally. Crop losses due to mycotoxin contamination are common, leading to reduced yields and lower quality produce. In severe cases, entire harvests can be deemed unfit for consumption, causing significant financial losses for farmers. Additionally, contaminated animal feed can reduce livestock productivity, leading to decreased milk production, weight loss, and fertility issues in animals, further impacting food supply and economic stability. The trade impact is equally critical, as many countries, particularly in regions like the European Union, have strict regulations governing mycotoxin levels in food and feed products. Contaminated crops can result in rejected exports, financial penalties, and long-term trade restrictions (Awuchi et al. 2021; Zain 2011).

Furthermore, mycotoxins are a major challenge for global food security. In regions where staple crops are frequently contaminated, food availability can be reduced, exacerbating hunger and malnutrition, particularly in vulnerable populations. Climate change is likely to intensify the problem, as rising temperatures and shifting rainfall patterns are expected to increase the prevalence of mycotoxin-producing fungi in many regions, further threatening food safety (Kos et al. 2023).

In conclusion, mycotoxins present a complex, multifaceted challenge, with far-reaching consequences for health, agriculture, and trade. The widespread occurrence of these toxins, combined with their toxicological impact on both humans and animals, underscores the urgent need for effective mitigation strategies to minimize contamination and protect food security globally.

MAJOR MYCOTOXINS OF CONCERN

Among the numerous mycotoxins identified, a few are of particular concern due to their high toxicity, prevalence, and economic impact. The most significant mycotoxins include aflatoxins, ochratoxins, fumonisins, trichothecenes, zearalenone, patulin, and a few others, which are regulated globally due to their widespread occurrence and potential to affect food safety (El-Sayed et al. 2022).

Aflatoxins

Aflatoxins are among the most studied and dangerous mycotoxins, primarily produced by *Aspergillus flavus* and *Aspergillus parasiticus*. A culture of *A. flavus* growing on Sabouraud Dextrose Agar (SDA) is shown in Figure 1. There are four main types of aflatoxins: B1, B2, G1, and G2, with aflatoxin B1 being the most potent and toxic. Aflatoxins are particularly problematic in warm and humid regions, where conditions favor fungal growth and toxin production, affecting crops such as maize, peanuts, tree nuts, and cottonseed.

Figure 1. Aspergillus flavus growing on SDA media

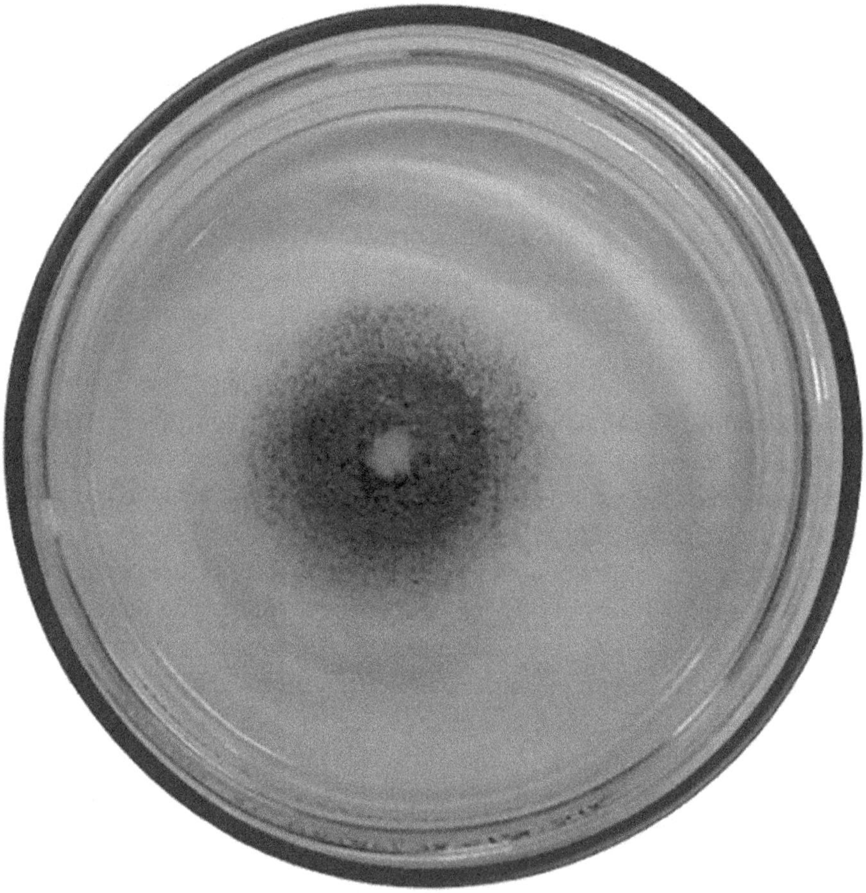

Health Impact: Aflatoxins are potent carcinogens, particularly affecting the liver. Chronic exposure to aflatoxin B1 has been strongly linked to liver cancer, especially in populations with high rates of hepatitis B infection, which synergistically increases the risk of cancer. Acute aflatoxicosis, which occurs from high levels of aflatoxin exposure, can cause severe liver damage, jaundice, vomiting, and even death. In animals, aflatoxins can cause reduced growth rates, immune suppression, and reproductive problems (Dhakal et al. 2023).

Economic and Agricultural Impact: Aflatoxin contamination leads to significant crop losses, both in terms of quantity and quality. In many cases, contaminated crops cannot be sold for human or animal consumption, resulting in financial losses for farmers and industries. Countries with stringent food safety regulations, such as those in the European Union, have strict limits on acceptable aflatoxin levels in food and feed, which can lead to trade restrictions or rejection of exports from contaminated regions (Ajmal et al. 2022; Kumar et al. 2021).

Table 1. Common mycotoxins, their sources, and associated health risks

Mycotoxin	Source (Fungi)	Commonly Affected Crops	Health Impact (Human/ Animal)	Maximum Allowable Limits (EU, FDA, Codex)
Aflatoxins	*Aspergillus flavus, A. parasiticus*	Maize, peanuts, cottonseed, tree nuts	Carcinogenic, hepatotoxic (liver damage)	EU: 4 μg/kg in food, FDA: 20 ppb in food, Codex: 15 μg/kg in food
Ochratoxin A	*Aspergillus ochraceus, Penicillium verrucosum*	Cereals (oats, barley), coffee, grapes	Nephrotoxic (kidney damage), carcinogenic	EU: 5 μg/kg in cereals, FDA: 5 ppb in cereals, Codex: 5 μg/kg
Zearalenone	*Fusarium graminearum, F. culmorum*	Maize, wheat, barley, oats, sorghum	Estrogenic effects, reproductive disorders in animals	EU/FDA/Codex: Varies by commodity
Deoxynivalenol (DON or Vomitoxin)	*Fusarium graminearum, F. culmorum*	Wheat, barley, maize, oats, rice	Gastrointestinal issues, immunosuppressive effects	EU: 1250 μg/kg in cereals, FDA: 1000-2000 ppb, Codex: Varies
Fumonisins	*Fusarium verticillioides*	Maize, other cereals	Hepatotoxic, carcinogenic (liver/kidney damage), neural tube defects	EU/FDA: Varies by commodity, Codex: Varies
Patulin	*Penicillium expansum*	Apples, apple products, pears	Gastrointestinal issues, immunotoxic, neurotoxic	EU: 50 μg/kg in fruit juice, FDA: 50 ppb in fruit juice
T-2 Toxin	*Fusarium sporotrichioides*	Barley, oats, wheat, maize	Immunosuppressive, hematotoxic, dermatotoxic	EU: Varies by commodity, FDA/Codex: No specific limit
Citrinin	*Penicillium citrinum, Aspergillus spp.*	Cereals, rice, wheat, maize	Nephrotoxic, associated with kidney damage	EU: No specific limit, FDA/Codex: No specific limit
Ergot Alkaloids	*Claviceps purpurea*	Rye, wheat, barley, oats	Neurotoxic, vasoconstriction, gangrene, convulsions	EU: 1000 μg/kg in cereals, FDA/Codex: No specific limit
Sterigmatocystin	*Aspergillus versicolor*	Cereals (wheat, maize), coffee	Carcinogenic, hepatotoxic, nephrotoxic	EU/FDA/Codex: No specific limit
Alternariol	*Alternaria alternata*	Tomatoes, wheat, sunflower seeds	Potential carcinogenic effects, DNA damage	EU/FDA/Codex: No specific limit

Ochratoxins

Ochratoxins are a group of mycotoxins produced by *Aspergillus* and *Penicillium* species. The most common and concerning member of this group is ochratoxin A (OTA), which can be found in a variety of food products, including cereals, coffee, wine, dried fruits, and spices. OTA contamination is prevalent in regions with temperate climates and is particularly problematic in improperly stored foods (Ben Miri et al. 2024).

Health Impact: Ochratoxin A is primarily nephrotoxic, meaning it can cause kidney damage in humans and animals. Chronic exposure to OTA has been linked to chronic kidney disease and urinary tract tumors (Khoi et al. 2021). In addition to its nephrotoxic effects, ochratoxin A is classified as a possible human carcinogen (Group 2B) by the International Agency for Research on Cancer (IARC). It is also immunosuppressive, potentially making individuals more susceptible to infections.

Economic and Agricultural Impact: Similar to aflatoxins, ochratoxin contamination affects crop quality and safety. Products like coffee and wine, which have significant economic value, are particularly vulnerable to OTA contamination. Regulatory limits on ochratoxin levels in food and beverages are strict, and exceeding these limits can result in the rejection of exports and financial losses for producers (Li et al. 2021).

Fumonisins

Fumonisins are primarily produced by *Fusarium verticillioides* and *Fusarium proliferatum*, and they predominantly contaminate maize and maize-based products. The most toxic and widely studied fumonisin is fumonisin B1. These toxins are prevalent in maize-growing regions worldwide, particularly in areas with warm climates (Kamle et al. 2019).

Health Impact: Fumonisin B1 has been linked to esophageal cancer in humans, especially in populations with high maize consumption. Additionally, fumonisins interfere with sphingolipid metabolism, which can lead to neural tube defects in developing fetuses. In animals, fumonisin consumption can cause a range of disorders, including equine leukoencephalomalacia in horses and porcine pulmonary edema in pigs, both of which can be fatal (Kamle et al. 2019).

Economic and Agricultural Impact: Fumonisin contamination can have significant economic consequences, particularly for maize-dependent economies. Crops contaminated with fumonisins may not meet regulatory limits for food or feed safety, leading to reduced marketability and export opportunities. In addition to direct crop losses, the cost of mitigation measures, such as improved drying and storage techniques, adds to the economic burden on producers (Kamle et al. 2019).

Trichothecenes

Trichothecenes are a diverse group of mycotoxins produced by *Fusarium* species, including deoxynivalenol (DON), T-2 toxin, and HT-2 toxin. These toxins primarily contaminate cereals like wheat, barley, oats, and maize, especially in regions with cool and wet climates, which favor the growth of *Fusarium* fungi (Polak-Śliwińska et al. 2021).

Health Impact: Deoxynivalenol (DON), also known as vomitoxin, is the most common trichothecene and can cause nausea, vomiting, diarrhea, and immune suppression in humans and animals. While DON is not considered highly carcinogenic, its acute toxicity can significantly impact animal feed intake and health. T-2 and HT-2 toxins are more potent than DON and can cause a range of adverse effects, including immune suppression, gastrointestinal problems, and hemorrhaging. In livestock, chronic exposure to trichothecenes can reduce feed efficiency, weight gain, and reproductive performance, leading to economic losses for farmers (Polak-Śliwińska et al. 2021).

Economic and Agricultural Impact: Trichothecene contamination is a major problem for cereal crops, especially in regions with frequent rainfall during the growing season. The presence of DON and other trichothecenes in animal feed can lead to reduced feed intake and growth performance, particularly in pigs, which are highly sensitive to these toxins. Regulatory limits for DON and other trichothecenes in food and feed products are enforced in many countries, and failure to meet these limits can result in trade restrictions and economic losses (Polak-Śliwińska et al. 2021).

Zearalenone

Zearalenone (ZEN) is another mycotoxin produced by *Fusarium* species, particularly *Fusarium graminearum* and *Fusarium culmorum*. It primarily contaminates maize, wheat, barley, and other cereals. Zearalenone is known for its estrogenic effects, which can mimic the hormone estrogen in animals and humans (Mahato et al. 2021a).

Health Impact: Due to its estrogenic properties, zearalenone can cause reproductive issues in livestock, particularly in pigs, which are highly sensitive to its effects. Symptoms of zearalenone exposure in animals include infertility, reduced litter size, abortions, and other reproductive disorders. In humans, although the risk is lower, chronic exposure to zearalenone could potentially disrupt endocrine functions and reproductive health (Mahato et al. 2021a).

Economic and Agricultural Impact: Zearalenone contamination is a significant problem in animal feed, particularly in regions where maize is a primary feed component. The reproductive issues caused by zearalenone exposure in livestock can lead to reduced productivity and financial losses for farmers. In addition, contamination of food products intended for human consumption can lead to economic losses due to regulatory restrictions on zearalenone levels in food (Mahato et al. 2021a).

Patulin

Patulin is a mycotoxin primarily produced by *Penicillium* and *Aspergillus* species, and it is commonly found in moldy fruits, particularly apples and apple products like juice and cider. It can also contaminate other fruits, such as pears and grapes, but it is most frequently associated with apples (Mahato et al. 2021b).

Health Impact: Patulin has been shown to cause gastrointestinal problems in humans, such as nausea, vomiting, and diarrhea. Although it is not considered a potent carcinogen, patulin can be toxic in large quantities, and its presence in fruit-based products intended for consumption by children is particularly concerning. Due to its association with apples, patulin contamination is closely monitored in the fruit juice industry (Mahato et al. 2021b).

Economic and Agricultural Impact: Patulin contamination can significantly affect the fruit industry, particularly apple growers and juice producers. Regulatory limits on patulin in fruit juices are strict, especially in products intended for infants and children. Contaminated fruit must be discarded or processed in a way that reduces patulin levels, leading to potential economic losses for producers (Zhong et al. 2018).

Other Mycotoxins of Concern

In addition to the major mycotoxins mentioned above, several other mycotoxins pose significant risks to food safety and agricultural productivity. These include citrinin, ergot alkaloids, and sterigmatocystin, each of which has unique sources and health impacts.

Citrinin: Citrinin is produced by *Penicillium* and *Aspergillus* species and is commonly found in stored grains, rice, and cheese. It is nephrotoxic, causing kidney damage in animals and potentially in humans. Citrinin contamination is particularly problematic in improperly stored grains, where conditions favor fungal growth (Kamle et al. 2022).

Ergot Alkaloids: Ergot alkaloids are produced by *Claviceps purpurea*, a fungus that infects cereal crops like rye, wheat, and barley. These toxins cause ergotism, a condition characterized by severe vasoconstriction, which can lead to gangrene and neurological symptoms such as hallucinations and convulsions. Ergot alkaloid contamination is less common today due to improved agricultural practices, but it remains a concern in some regions (Agriopoulou 2021).

Sterigmatocystin: Sterigmatocystin is produced by *Aspergillus* species and is considered a precursor to aflatoxin biosynthesis. It is commonly found in cereals and has been shown to cause liver and kidney damage in animals. Sterigmatocystin is also a possible human carcinogen, and its presence in food products is closely monitored (Viegas 2018).

The occurrence and impact of mycotoxins represent a significant challenge to global food safety, human and animal health, and agricultural productivity. Aflatoxins, ochratoxins, fumonisins, trichothecenes, zearalenone, patulin, and other mycotoxins can have devastating effects on both food security and economic stability. Mycotoxin contamination leads to crop losses, reduced livestock productivity, and trade restrictions, particularly in regions with inadequate food safety infrastructure. Efforts to mitigate mycotoxin contamination through improved agricultural practices, better storage conditions, and effective regulatory frameworks are essential to protect public health and ensure the safety of the global food supply.

FACTORS INFLUENCING MYCOTOXIN CONTAMINATION

Mycotoxin contamination is a complex issue influenced by a variety of factors that range from environmental conditions to agricultural practices and storage methods. Fungi that produce mycotoxins, such as those from the *Aspergillus*, *Penicillium*, and *Fusarium* genera, thrive under specific conditions that encourage their growth and toxin production. Understanding the factors that influence mycotoxin contamination is critical for developing strategies to mitigate its occurrence and impact on food safety and agricultural productivity.

Environmental Conditions

Environmental factors are some of the most significant determinants of mycotoxin contamination. The growth of mycotoxin-producing fungi is heavily influenced by temperature, humidity, and rainfall patterns. These conditions not only affect fungal proliferation during crop growth but also play a role in post-harvest contamination.

Temperature: Different fungi have specific temperature ranges in which they thrive. For instance, *Aspergillus* species, responsible for aflatoxin production, grow optimally in warm temperatures (25°C to 35°C), making aflatoxin contamination more prevalent in tropical and subtropical regions. Previous studies show that temperature has a significant impact on aflatoxin B1 production in *Aspergillus flavus* (Patel et al. 2014). *Fusarium* species, which produce fumonisins and trichothecenes, tend to flourish in cooler climates, typically between 15°C and 30°C. Variations in temperature during the growing season or improper storage at elevated temperatures can increase the risk of mycotoxin contamination (Daou et al. 2021).

Humidity and Moisture: Moisture is another critical factor that supports fungal growth. High humidity levels, particularly during flowering and ripening stages of crop growth, can increase the susceptibility of crops to fungal invasion. Post-harvest conditions, such as inadequate drying or high moisture content in storage environments, also promote mycotoxin production. Crops like maize and peanuts, which are often stored with moisture levels exceeding 14%, are particularly vulnerable to contamination by aflatoxins and other mycotoxins (Nji et al. 2022; Neme & Mohammed 2017).

Rainfall Patterns: Rainfall during pre-harvest, particularly excessive rain or drought conditions, can promote the colonization of crops by mycotoxin-producing fungi. Wet conditions during harvest can exacerbate fungal infection, while prolonged drought can weaken plants and make them more susceptible to fungal attack. In regions prone to erratic rainfall or prolonged wet seasons, the risk of mycotoxin contamination increases significantly (Mesterhazy et al. 2022).

Agricultural Practices

Agricultural practices, including crop selection, rotation, irrigation methods, and the use of fertilizers and pesticides, play an essential role in influencing mycotoxin contamination. These practices can either help control fungal growth or inadvertently increase the risk of contamination.

Crop Rotation and Monoculture: Continuous planting of the same crop (monoculture) increases the likelihood of mycotoxin contamination. For example, maize is particularly susceptible to aflatoxins, fumonisins, and zearalenone. When maize is continuously planted in the same field without crop rotation, the build-up of fungal spores in the soil can increase the chances of infection. Crop rotation, especially with non-host crops, can disrupt fungal life cycles and reduce the risk of contamination (Borràs-Vallverdú et al. 2022; Islam et al. 2022; Marete et al. 2020).

Planting and Harvest Timing: The timing of planting and harvesting also affects mycotoxin contamination. Late planting or delayed harvesting, especially in wet or humid conditions, increases the crop's exposure to favorable conditions for fungal growth. Early harvesting, combined with proper drying, can reduce the chances of fungal colonization and mycotoxin accumulation (Kochiieru et al. 2021).

Irrigation Practices: Over-irrigation or poor water management can lead to waterlogged soils, which create favorable conditions for fungal growth. On the other hand, drought stress weakens plants and makes them more susceptible to infection. Efficient irrigation practices, such as drip irrigation, can help regulate soil moisture levels and reduce the likelihood of mycotoxin-producing fungi infecting crops (Herrera et al. 2023).

Fertilization and Pesticide Use: The overuse or improper application of fertilizers, particularly nitrogen, can increase plant susceptibility to fungal infections by altering plant metabolism and making plants more attractive to pests, which act as vectors for fungi. Similarly, inadequate use of fungicides or pest control measures can lead to higher incidences of fungal contamination. Proper management of these inputs is essential for minimizing fungal infection and mycotoxin production (Singh & Sarkar 2021).

Post-Harvest Handling and Storage

Post-harvest practices are critical to preventing fungal growth and mycotoxin production after crops are harvested. Improper drying, inadequate storage conditions, and delayed processing are all factors that can contribute to mycotoxin contamination.

Drying Practices: One of the most effective methods of preventing mycotoxin contamination post-harvest is rapid drying to reduce moisture content to levels below which fungi can grow. For example, maize should be dried to a moisture content of 13% or less to prevent aflatoxin production. Improper or delayed drying can allow fungi to colonize and produce toxins during storage. In regions with high humidity, mechanical drying may be necessary to achieve adequate moisture reduction (Liu et al. 2016).

Storage Conditions: Poor storage conditions, such as high humidity, warm temperatures, and poor ventilation, create an ideal environment for fungal growth. Insects and rodents that damage stored grains can further increase the risk of fungal contamination by creating entry points for fungi. Grain silos and storage facilities must be well-ventilated, regularly inspected, and maintained at low humidity and temperature levels to prevent fungal growth and mycotoxin accumulation (Chulze 2010).

Packaging and Transportation: Mycotoxin contamination can also occur during transportation and storage due to poor handling practices. Packaging materials that do not protect crops from moisture, pests, or heat can increase the risk of contamination. During transportation, crops that are stored in unhygienic conditions or exposed to moisture may become contaminated with mycotoxin-producing fungi (Zhang et al. 2023).

Pest and Disease Pressure

Pests and plant diseases play a critical role in mycotoxin contamination, as they often act as vectors for mycotoxin-producing fungi or create wounds that make plants more susceptible to fungal infection.

Insect Damage: Insects such as corn borers and weevils can directly damage crops, creating wounds through which fungal spores can enter and colonize plant tissues. In addition, insect infestations during storage can introduce fungal spores and create conditions conducive to fungal growth. Proper pest control measures, both in the field and in storage facilities, are essential for minimizing mycotoxin contamination (Awuchi et al. 2021).

Fungal Diseases: Certain fungal diseases, such as ear rot in maize caused by *Fusarium* species, are closely associated with mycotoxin contamination. Crops infected by pathogenic fungi during the growing season are at a higher risk of mycotoxin production. The use of disease-resistant crop varieties and appropriate fungicides can help mitigate this risk (Bryła et al. 2022).

Genetic Susceptibility of Crops

The genetic makeup of crops influences their susceptibility to mycotoxin contamination. Some crop varieties are more resistant to fungal infection and mycotoxin production, while others are highly susceptible. Plant breeding programs aimed at developing mycotoxin-resistant varieties have been successful in reducing contamination levels in some crops (Fountain et al. 2015).

Climate Change

Climate change is emerging as a significant factor influencing mycotoxin contamination. Changes in temperature, rainfall patterns, and humidity levels are expected to alter the distribution and prevalence of mycotoxin-producing fungi. For instance, increasing global temperatures may expand the geographic range of aflatoxin-producing *Aspergillus* species, leading to higher contamination risks in regions that were previously less affected (Zingales et al. 2022).

Increased Fungal Growth: Warmer temperatures and increased humidity due to climate change may create more favorable conditions for fungal growth and mycotoxin production. This is especially concerning for crops grown in regions where climate conditions are becoming more variable and unpredictable (Perrone et al. 2020).

Shifts in Crop Susceptibility: Changes in environmental conditions may also affect the susceptibility of crops to fungal infection and mycotoxin contamination. Crops that are stressed by drought or heat may become more vulnerable to fungal attack, leading to increased mycotoxin levels (Zingales et al. 2022).

Mycotoxin contamination is influenced by a range of interrelated factors, including environmental conditions, agricultural practices, post-harvest handling, pest and disease pressure, genetic susceptibility of crops, and climate change. Understanding these factors is critical for developing effective strategies to prevent and mitigate mycotoxin contamination. Integrated approaches, such as improving crop management practices, optimizing post-harvest storage conditions, and developing resistant crop varieties, are essential to reducing the occurrence and impact of mycotoxins in the global food supply (Daou et al. 2021).

DETECTION AND QUANTIFICATION METHODS

The detection and quantification of mycotoxins in food and feed are essential for ensuring food safety, compliance with regulatory standards, and protecting human and animal health. Given the toxic nature of mycotoxins and their potential to cause severe health problems, there is a strong need for accurate, sensitive, and reliable methods to detect even trace amounts of these toxins in agricultural products. Several detection techniques have been developed, ranging from traditional chemical analysis to advanced biosensor technologies, each with its strengths and limitations. These methods can be classified into two broad categories: chromatographic methods and immunoassay-based methods, with newer techniques such as biosensors and molecular approaches emerging in recent years.

Chromatographic Methods

Chromatographic techniques are among the most reliable and widely used methods for the detection and quantification of mycotoxins. They offer high sensitivity, accuracy, and specificity, making them suitable for the analysis of complex food matrices.

High-Performance Liquid Chromatography (HPLC): High-Performance Liquid Chromatography (HPLC) is the most commonly used chromatographic method for mycotoxin analysis. HPLC separates the individual components of a sample based on their interactions with a stationary phase and a mobile phase. This technique is often coupled with various detectors such as UV-Vis, fluorescence, or mass spectrometry (MS) for sensitive and accurate quantification of mycotoxins (Keskin & Eyupoglu 2023).

HPLC is highly sensitive, capable of detecting very low concentrations of mycotoxins. It can simultaneously analyze multiple mycotoxins in a single run and is suitable for various food matrices, including cereals, nuts, fruits, and beverages. The main drawbacks of HPLC include the need for extensive sample preparation and relatively high operational costs. Additionally, specialized equipment and trained personnel are required to operate HPLC systems (De Girolamo et al. 2022).

Liquid Chromatography-Mass Spectrometry (LC-MS): Liquid Chromatography-Mass Spectrometry (LC-MS) combines the separation power of HPLC with the detection capabilities of mass spectrometry. LC-MS is highly effective in identifying and quantifying multiple mycotoxins in complex samples, including metabolites and modified mycotoxins (Keskin & Eyupoglu 2023).

LC-MS offers high sensitivity and specificity, making it ideal for the simultaneous detection of multiple mycotoxins, including those present in trace amounts. It can also differentiate between different forms of mycotoxins and their metabolites. LC-MS systems are expensive to acquire and maintain, and they require highly skilled personnel for operation and data interpretation. Additionally, complex sample matrices can sometimes interfere with the analysis (De Santis et al. 2017). In addition, direct tandem mass spectrometric detection method with an electrospray ionization interface in Multiple Reaction Monitoring mode was also developed (Patel et al. 2014). Mass spectrum of aflatoxin B1 standard is shown in Figure 2.

Figure 2. Mass spectrum of Aflatoxin B standard

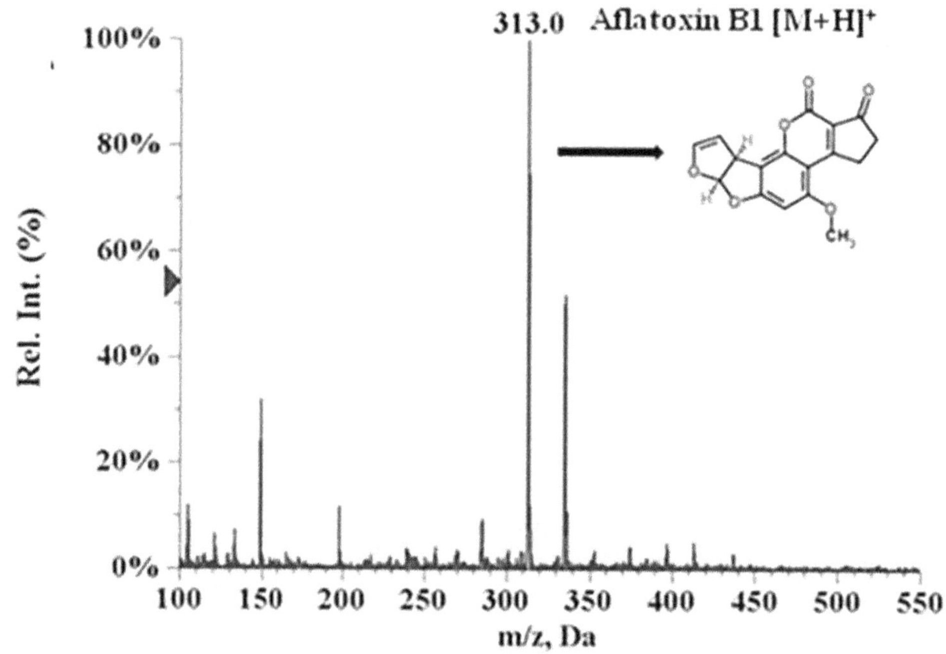

Gas Chromatography (GC): Gas Chromatography (GC) is another chromatographic technique used for mycotoxin analysis, particularly for volatile or semi-volatile compounds. GC is often coupled with mass spectrometry (GC-MS) to provide accurate and sensitive detection of mycotoxins. Gas Chromatography is highly effective for certain mycotoxins that can be made volatile through derivatization. It offers high resolution and sensitivity, making it suitable for low-level detection. GC requires derivatization of non-volatile mycotoxins before analysis, which can be time-consuming. Like HPLC and LC-MS, GC also involves complex sample preparation and requires expensive equipment (Agriopoulou et al. 2020).

Immunoassay-Based Methods

Immunoassays are widely used for the detection of mycotoxins due to their simplicity, rapid results, and relatively low cost. These methods rely on the interaction between specific antibodies and mycotoxins, allowing for the detection and quantification of mycotoxins in food and feed samples (Agriopoulou et al. 2020).

Enzyme-Linked Immunosorbent Assay (ELISA): The Enzyme-Linked Immunosorbent Assay (ELISA) is one of the most popular immunoassay techniques for mycotoxin detection. It is based on the use of antibodies that specifically bind to mycotoxins, and the detection is typically achieved through colorimetric, fluorescent, or chemiluminescent reactions. ELISA is a cost-effective, rapid, and user-friendly method for screening large numbers of samples. It does not require extensive sample preparation and can be easily performed in basic laboratory settings. ELISA kits are commercially available for a wide range of mycotoxins, including aflatoxins, ochratoxins, and fumonisins. While ELISA offers high sensitivity, it can sometimes suffer from cross-reactivity with structurally similar compounds, leading to false-positive or false-negative results. Additionally, ELISA provides semi-quantitative results, which may require confirmation by more specific methods such as HPLC or LC-MS (Agriopoulou et al. 2020).

Lateral Flow Immunoassay (LFIA): Lateral Flow Immunoassay (LFIA) is a rapid, on-site detection method similar to home pregnancy tests. LFIA uses antibodies immobilized on a test strip, where the presence of mycotoxins is indicated by the appearance of colored lines (Agriopoulou et al. 2020). LFIA is simple, portable, and does not require sophisticated laboratory equipment. It is particularly useful for field testing and rapid screening of mycotoxins in food and feed samples. The results can be obtained within minutes. LFIA is less sensitive and specific compared to ELISA or chromatographic methods. It is typically used for qualitative or semi-quantitative analysis, and positive results may need further confirmation through more accurate methods (Agriopoulou et al. 2020).

Biosensors: Biosensor technology has emerged as a promising alternative for mycotoxin detection. Biosensors combine a biological recognition element (such as an antibody or enzyme) with a transducer that converts the biological interaction into a measurable signal. Various types of biosensors, including optical, electrochemical, and piezoelectric sensors, have been developed for mycotoxin detection (Agriopoulou et al. 2020). Biosensors offer high sensitivity and specificity, rapid response times, and the potential for real-time, on-site detection of mycotoxins. They are portable and can be easily integrated into food monitoring systems. The development of biosensors for mycotoxins is still in the early stages, and their commercial availability is limited. Additionally, biosensors may suffer from stability issues, and their performance can be affected by complex sample matrices (Agriopoulou et al. 2020).

Molecular Techniques

Molecular techniques, such as polymerase chain reaction (PCR), are increasingly being used to detect mycotoxin-producing fungi by identifying specific fungal DNA or RNA sequences. While these methods do not directly detect mycotoxins, they can indicate the presence of mycotoxin-producing organisms in food or feed samples.

Polymerase Chain Reaction (PCR): PCR is a molecular biology technique used to amplify and detect specific DNA sequences. For mycotoxin detection, PCR is used to identify fungal species capable of producing mycotoxins, such as *Aspergillus* or *Fusarium* species (Agriopoulou et al. 2020). PCR is highly sensitive and specific, allowing for the detection of low levels of fungal contamination. It can be used to detect multiple mycotoxin-producing fungi simultaneously and can provide early warnings of potential mycotoxin contamination. PCR does not directly quantify mycotoxins but rather the presence of mycotoxin-producing fungi. Additionally, PCR requires specialized equipment and trained personnel, making it less suitable for routine on-site testing.

Quantitative PCR (qPCR): Quantitative PCR (qPCR) builds on traditional PCR by allowing for the quantification of fungal DNA in real time. This technique is particularly useful for monitoring the fungal load in food and feed samples and estimating the potential for mycotoxin production. qPCR provides both qualitative and quantitative data, offering a more comprehensive analysis of fungal contamination. It is highly sensitive and can detect small amounts of fungal DNA. Like conventional PCR, qPCR does not directly measure mycotoxin levels. Additionally, the accuracy of qPCR can be influenced by the quality of the DNA extracted from the sample (Agriopoulou et al. 2020).

The detection and quantification of mycotoxins are critical for ensuring the safety of food and feed products. A wide range of methods is available, each with its advantages and limitations. Chromatographic techniques such as HPLC and LC-MS provide high accuracy and sensitivity, making them the gold standard for mycotoxin analysis, while immunoassays like ELISA offer rapid and cost-effective screening options. Emerging technologies such as biosensors and molecular techniques are also showing promise for future applications in mycotoxin detection. In practice, the choice of method depends on factors such as the type of mycotoxin, the complexity of the sample matrix, the required sensitivity, and the available resources. By combining different detection methods, it is possible to achieve comprehensive and reliable mycotoxin monitoring across the food supply chain.

REGULATORY FRAMEWORK AND GUIDELINES

Mycotoxins, due to their severe health risks to humans and animals, have become a major concern for governments and international organizations that regulate food and feed safety. Regulatory frameworks and guidelines are essential to control the levels of mycotoxins in food and feed products, ensure public health, and facilitate international trade. Various countries and international bodies have developed maximum allowable limits for different mycotoxins in specific food products. These regulations are based on risk assessments, toxicological data, and scientific research, ensuring that the population is protected from the harmful effects of mycotoxins.

The major organizations involved in the establishment of guidelines and regulations for mycotoxins include the Codex Alimentarius Commission, the European Union (EU), the United States Food and Drug Administration (FDA), and other national regulatory bodies. These organizations work together to harmonize standards and ensure food safety at a global level.

Codex Alimentarius Commission

The Codex Alimentarius Commission, established by the Food and Agriculture Organization (FAO) and the World Health Organization (WHO), is the international body responsible for setting global food safety standards, including those for mycotoxins. The Codex standards serve as reference points for international trade and provide a framework for developing national food safety regulations. The Codex Alimentarius Commission has established maximum limits for several mycotoxins in food and feed. These limits are based on extensive risk assessments conducted by the Joint FAO/WHO Expert Committee on Food Additives (JECFA). The key mycotoxins regulated by Codex include:

Codex has set maximum limits for aflatoxins in various food commodities, particularly in peanuts, maize, and dried fruits. The permissible limit for aflatoxins in peanuts and maize destined for direct human consumption is 10 μg/kg, while the limit for aflatoxin B1 (the most toxic form) is set at 5 μg/kg. Codex has set a maximum limit of 5 μg/kg for ochratoxin A in cereals and cereal-based products, and 2 μg/kg in wine. For fumonisins, Codex has established a limit of 2-4 mg/kg in maize and maize-based products. Codex has set a maximum level of 50 μg/kg for patulin in apple juice and apple-based products (FAO 1995). The Codex standards for mycotoxins are critical for international trade, as they provide harmonized guidelines that are recognized by the World Trade Organization (WTO). This helps in resolving trade disputes related to food safety and ensures that food products meet international safety standards (Lee et al. 2021).

European Union (EU)

The European Union has one of the most stringent regulatory frameworks for mycotoxins, with detailed regulations governing the allowable limits of mycotoxins in various food and feed products. The EU regulations are enforced by the European Food Safety Authority (EFSA), which provides scientific assessments on the risks of mycotoxins in food. The EU has established specific maximum levels for several mycotoxins, including aflatoxins, ochratoxin A, fumonisins, zearalenone, deoxynivalenol (DON), and patulin, across different food and feed products. The limits are specified in **Commission Regulation (EC) No 1881/2006** and its amendments.

The EU has set stringent limits for aflatoxins in various foodstuffs. The maximum allowable limit for aflatoxin B1 in cereals is 2 μg/kg, and the total aflatoxins (B1, B2, G1, G2) are limited to 4 μg/kg in foodstuffs like nuts, dried fruits, and cereals. The EU regulation specifies a maximum limit of 3 μg/kg in cereals and 5 μg/kg in dried vine fruits. For maize-based products, the EU limit for fumonisins is set at 1,000 μg/kg (1 mg/kg) for fumonisins B1 and B2 combined. The EU limits zearalenone levels in cereals at 100 μg/kg, with lower limits for products intended for infants and young children. The maximum allowable limit for DON in cereals is 750 μg/kg. The EU has established a limit of 50 μg/kg for patulin in apple juice and 10 μg/kg in products intended for infants (European Commission 2006).

EU member states are required to monitor food and feed products for compliance with these regulations. The Rapid Alert System for Food and Feed (RASFF) is used to notify member states and the European Commission of any non-compliant products found in the market. The EU also imposes strict import regulations to prevent mycotoxin-contaminated products from entering the European market.

United States Food and Drug Administration (FDA)

The United States Food and Drug Administration (FDA) is the primary regulatory authority responsible for overseeing mycotoxin contamination in food and feed within the U.S. The FDA has established action levels for mycotoxins, which are the maximum allowable concentrations of mycotoxins in food and feed products that are considered safe for human or animal consumption. The FDA has set specific action levels for aflatoxins, fumonisins, and patulin in food products. The FDA has established an action level of 20 μg/kg for total aflatoxins (B1, B2, G1, G2) in food and feed. For milk, the action level for aflatoxin M1 is set at 0.5 μg/kg. The action level for fumonisins in maize and maize-based products is 2-4 mg/kg, depending on the intended use of the product.

The FDA has set a limit of 50 μg/kg for patulin in apple juice (FDA 2000). The FDA also provides guidance for the industry on how to manage and control mycotoxin contamination. This includes recommendations on good agricultural practices (GAPs), good manufacturing practices (GMPs), and proper storage conditions to prevent fungal growth and mycotoxin production (FDA 2000). The FDA enforces these action levels through inspections, sampling, and testing of food and feed products. Non-compliant products are subject to recall, seizure, or rejection, particularly for imported products that do not meet U.S. safety standards.

Other National Regulatory Frameworks

In addition to the Codex Alimentarius, the EU, and the FDA, many other countries have developed their own regulatory frameworks for mycotoxins, often based on international guidelines. China has established mycotoxin regulations under the National Food Safety Standards (GB standards). The country has specific limits for aflatoxins, ochratoxins, fumonisins, zearalenone, and deoxynivalenol in various food and feed products. For example, the limit for aflatoxins in peanuts and maize is 20 μg/kg, similar to the FDA's action level (USDA, FAS 2018). Japan's Ministry of Health, Labour and Welfare (MHLW) regulates mycotoxins, including aflatoxins and ochratoxins, in food products. The country has strict limits for aflatoxins in food, set at 10 μg/kg for total aflatoxins in products like nuts and grains (MHLW 2020). Australia and New Zealand regulate mycotoxins under the Food Standards Australia New Zealand (FSANZ) framework. FSANZ has set limits for aflatoxins, ochratoxins, and fumonisins in food products. The maximum allowable level for aflatoxins in nuts and grains is 15 μg/kg (FSANZ 2023).

International Harmonization Efforts

International harmonization of mycotoxin regulations is important for facilitating trade and ensuring global food safety. Organizations such as the World Trade Organization (WTO) and the International Food Safety Authorities Network (INFOSAN) work to promote harmonized standards based on Codex guidelines (FAO 2023). These efforts help reduce trade barriers and ensure that food products meet consistent safety standards worldwide.

The regulatory framework for mycotoxins is essential for protecting human and animal health and ensuring the safety of the global food supply. International bodies like the Codex Alimentarius Commission, along with regional and national authorities such as the European Union, the U.S. FDA, and others, have established strict guidelines and limits for mycotoxin levels in food and feed. These regulations are based on scientific risk assessments and are designed to mitigate the risks posed by mycotoxins. By adhering to these guidelines, producers, manufacturers, and regulatory bodies can work together to minimize mycotoxin contamination and its impact on public health and international trade.

STRATEGIES FOR MYCOTOXIN MITIGATION

Mitigating mycotoxin contamination in food and feed is crucial for ensuring the safety of agricultural products and minimizing health risks. Due to the toxic effects of mycotoxins on both humans and animals, industries and regulators have developed a variety of strategies aimed at reducing mycotoxin production and contamination across the entire food chain from pre-harvest through post-harvest and processing. Effective mycotoxin mitigation requires an integrated approach combining good agricultural practices, biological control, chemical treatments, and advanced food processing technologies.

The strategies can be classified into preventive measures and post-contamination control. Below are detailed approaches used to manage and mitigate mycotoxins at different stages of food production.

Pre-Harvest Strategies

Pre-harvest strategies are essential for preventing fungal growth and mycotoxin production during crop cultivation. Several factors influence the likelihood of mycotoxin contamination, including environmental conditions, plant stress, and agricultural practices. Implementing proper crop management techniques can significantly reduce the risk of mycotoxin formation.

Good Agricultural Practices (GAPs): Adhering to good agricultural practices is one of the most effective ways to reduce mycotoxin contamination in crops. GAPs involve the use of proper cultivation methods that minimize plant stress and reduce fungal infection (Dövényi-Nagy 2020). Key elements of GAPs include: Rotating crops can reduce the buildup of mycotoxin-producing fungi in the soil. For example, rotating maize with non-host crops like legumes can help control fungal pathogens such as *Aspergillus* and *Fusarium* (Dövényi-Nagy 2020). Planting crop varieties that are resistant to fungal infections can lower the risk of mycotoxin contamination. Breeding and genetically engineering crops for resistance to fungi and environmental stressors is an active area of research (Figlan & Mwadzingeni 2020). Proper irrigation techniques can help control moisture levels, which is critical in preventing fungal growth. Over-irrigation or drought stress can make crops more susceptible to fungal infection and mycotoxin production (Herrera et al. 2023). Maintaining healthy soil through proper fertilization and drainage helps reduce the risk of fungal contamination. Poor soil conditions can stress plants and increase susceptibility to fungi (FAO & WHO 2017). Too early or delaying harvest can expose crops to higher moisture levels and increase the risk of fungal growth. Timely harvesting at the optimal moisture content is crucial for reducing contamination (Mutungi et al. 2019; Kumar & Kalita 2017).

Biological Control: Biological control methods involve using naturally occurring organisms to inhibit the growth of mycotoxin-producing fungi. These methods are environmentally friendly and have shown promise in reducing mycotoxin levels in crops (Snini & Mathieu 2020). One of the most successful biological control strategies involves the use of non-toxigenic strains of *Aspergillus* and *Fusarium* to outcompete toxigenic strains. For example, non-toxigenic strains of *Aspergillus flavus* have been shown to reduce aflatoxin contamination in maize and peanuts by competitively excluding the toxigenic strains (Snini & Mathieu 2020). Some microorganisms, such as certain bacteria and yeast strains, can inhibit the growth of mycotoxin-producing fungi. These microbial antagonists are applied to crops as biopesticides or inoculants to reduce fungal infection (Zadravec et al. 2022). Endophytic fungi and bacteria that live inside plants can also provide protection against mycotoxin-producing fungi. These endophytes enhance the plant's resistance to stress and pathogens, reducing the risk of fungal contamination (Venkatesh & Keller 2019).

Chemical Fungicides: Chemical fungicides can be used to control fungal infections during crop growth, although their use must be carefully managed to avoid resistance development and environmental impacts. Fungicides are often applied as a preventive measure during critical stages of crop development, such as flowering, when the risk of fungal infection is highest. However, the overuse of fungicides can lead to the development of resistant fungal strains, which necessitates integrated pest management strategies.

Post-Harvest Strategies

Post-harvest handling and storage play a critical role in controlling mycotoxin contamination. Poor storage conditions, including high humidity and temperature, can promote fungal growth and mycotoxin production even after the crop has been harvested. Several techniques can be employed to reduce contamination risks during storage, transportation, and processing (Figure 3).

Figure 3. Post harvest strategies to reduce mycotoxin contamination

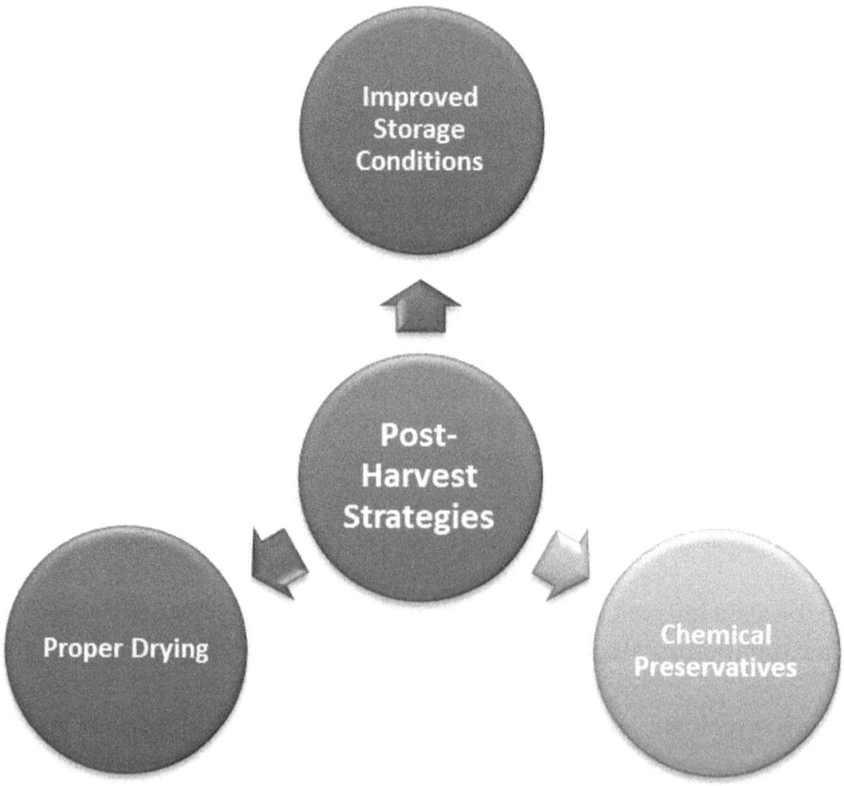

Proper Drying: One of the most important post-harvest strategies for mitigating mycotoxin contamination is reducing the moisture content of the harvested crops. High moisture levels promote fungal growth, so drying the crops to a safe moisture level (typically below 14% for cereals and nuts) is essential for preventing mycotoxin production (Kumar et al. 2022). Crops can be dried in the field or through solar drying methods, although this approach depends on favorable weather conditions. In humid climates or when quick drying is necessary, mechanical drying systems such as grain dryers are used. These systems are more reliable and provide better control over moisture levels (Chojnacka et al. 2021).

Improved Storage Conditions: Once crops are dried, proper storage conditions are critical for preventing fungal growth. Key factors in post-harvest storage include: Storing crops in cool, dry conditions helps inhibit fungal growth. Silos and storage facilities should be well-ventilated and equipped with moisture and temperature control systems to maintain optimal conditions (Bradford et al. 2018). Aeration systems in storage facilities can help prevent hot spots that lead to moisture accumulation and fungal contamination. Insects and rodents can damage stored crops, creating entry points for fungi. Implementing effective pest control measures is important for maintaining crop integrity during storage.

Chemical Preservatives: Chemical preservatives, such as organic acids (e.g., propionic acid) and fungistatic compounds, can be applied to stored grains and feed to prevent fungal growth. These preservatives inhibit the growth of mycotoxin-producing fungi, thereby reducing the risk of contamination. However, the use of chemical preservatives is regulated in many countries, and their application must comply with food safety standards (Awuchi et al. 2021; Leyva Salas et al. 2017).

Processing Techniques

Food processing methods can also reduce the levels of mycotoxins in contaminated products. While processing cannot completely eliminate mycotoxins, it can significantly lower their concentration and toxicity.

Sorting and Cleaning: Mechanical sorting and cleaning methods can remove contaminated grains, nuts, or seeds before processing. Techniques such as color sorting, density separation, and flotation can help identify and remove mycotoxin-contaminated particles from bulk shipments. Sorting is particularly effective for visible contamination, such as kernels affected by *Fusarium* or *Aspergillus* fungi (Karlovsky et al. 2016).

Thermal Processing: Certain mycotoxins, such as aflatoxins and fumonisins, are relatively heat-stable, but thermal processing methods like roasting, baking, and extrusion can reduce mycotoxin levels. High temperatures and long processing times can degrade some mycotoxins, making them less toxic. For example, extrusion cooking, which combines high heat, pressure, and moisture, has been shown to reduce fumonisin levels in maize-based products (Karlovsky 2016).

Fermentation: Fermentation processes, particularly lactic acid fermentation, can reduce mycotoxin levels in food products. The microbial activity during fermentation can degrade or detoxify certain mycotoxins. For example, lactic acid bacteria have been shown to degrade aflatoxins and ochratoxin A during the fermentation of products like yogurt and fermented maize (Guan et al. 2023; Luz et al. 2018).

Chemical Detoxification: Chemical treatments, such as the addition of ammonia or alkaline compounds, can detoxify certain mycotoxins in food and feed. Ammoniation, for instance, is used to detoxify aflatoxins in animal feed by breaking down the toxin's chemical structure. This process has been approved in several countries for use in animal feed but is less commonly used in human food products due to regulatory restrictions (Čolović et al. 2019).

Post-Processing Strategies

In cases where mycotoxins cannot be fully eliminated during processing, additional strategies are used to manage their presence in food and feed products.

Mycotoxin Binders: Mycotoxin binders, such as clay minerals (e.g., bentonite, zeolite) and activated charcoal, are commonly added to animal feed to adsorb mycotoxins and prevent their absorption in the gastrointestinal tract. These binders can significantly reduce the bioavailability of mycotoxins, minimizing their toxic effects on animals (Kihal et al. 2022).

Enzymatic Detoxification: Enzymatic detoxification involves the use of specific enzymes that can degrade or modify mycotoxins into non-toxic compounds. For example, certain enzymes can break down fumonisins and zearalenone in animal feed. This strategy is gaining attention as an environmentally friendly and targeted approach to mycotoxin mitigation (Liu et al. 2024).

Mycotoxin mitigation requires an integrated, multi-faceted approach that combines preventive measures during crop cultivation with effective post-harvest and processing strategies. Pre-harvest practices such as good agricultural practices, biological control, and proper crop management can reduce the risk of fungal infection and mycotoxin production. Post-harvest handling, including drying, storage, and sorting, further minimizes contamination risks, while advanced processing techniques and detoxification methods can lower mycotoxin levels in food and feed products. By implementing these strategies, food safety can be enhanced, and the impact of mycotoxins on human and animal health can be significantly reduced.

FUTURE DIRECTIONS AND RECOMMENDATIONS

As the global understanding of mycotoxins evolves, future research and policy development will be pivotal in addressing the challenges posed by these toxins. The following future directions and recommendations outline key areas for advancing mycotoxin management.

Advancing Detection Technologies

Future efforts should prioritize developing more robust and cost-effective detection methods for mycotoxins. While current techniques like chromatography and ELISA are effective, they often require specialized equipment and can be costly. Innovations in biosensors, incorporating nanomaterials and molecular recognition elements, offer the potential for enhanced sensitivity and portability, allowing for real-time, on-site detection. The integration of digital technologies, such as IoT and AI, with detection systems can further enhance monitoring capabilities, enabling continuous and automated risk assessment.

Exploring Genetic and Biotechnological Solutions

Developing crop varieties that are resistant to fungal infections is a long-term strategy for reducing mycotoxin contamination. Resistant varieties can prevent fungal colonization and reduce the need for chemical fungicides. For instance, maize varieties resistant to aflatoxin-producing fungi are being developed through both traditional breeding and genetic engineering approaches.

Genetic engineering and biotechnological interventions hold promise for reducing mycotoxin contamination. Research should focus on utilizing CRISPR-Cas9 and RNA interference technologies to develop crop varieties with enhanced resistance to mycotoxin-producing fungi and to inhibit fungal toxin production. Additionally, synthetic biology can be explored to engineer microorganisms or plants that detoxify or degrade mycotoxins. These advancements can provide targeted solutions for managing mycotoxin risks and improving crop resilience.

Enhancing Understanding of Molecular Mechanisms

A comprehensive understanding of the molecular mechanisms underlying mycotoxin production and resistance is crucial for developing effective mitigation strategies. Research should investigate the genetic and biochemical pathways involved in toxin biosynthesis and how plants and animals naturally resist mycotoxin effects. Insights into these mechanisms can lead to targeted interventions, such as developing resistant crop varieties or identifying compounds that can inhibit toxin production.

Strengthening International Collaboration

Addressing mycotoxin challenges requires global cooperation and knowledge sharing. Future efforts should focus on establishing international research networks to facilitate data sharing and collaborative research. Global initiatives can pool resources, standardize practices, and promote the development and implementation of innovative solutions. Joint funding programs and collaborative research projects can accelerate progress and address complex mycotoxin issues more effectively.

Improving Regulatory Frameworks and Policy

Strengthening regulatory frameworks and enforcement mechanisms is essential for effective mycotoxin management. Efforts should include harmonizing mycotoxin standards and regulations across countries to facilitate international trade and ensure consistent safety practices. Enhanced enforcement mechanisms, including improved inspection and monitoring systems, are necessary to ensure compliance with regulations. Regular updates to regulations based on new scientific knowledge and technological advancements will keep safety standards relevant and effective.

Promoting Public-Private Partnerships

Public-private partnerships can drive innovation and accelerate the development of new technologies for mycotoxin mitigation. Collaborative research projects between government agencies, research institutions, and private companies can lead to the development of practical solutions and the commercialization of new technologies. Industry involvement in developing standards and guidelines ensures that practical considerations are addressed and fosters the adoption of innovative solutions.

Enhancing Capacity Building and Training

Building the capacity of farmers and food processors is crucial for effective mycotoxin management. Future initiatives should focus on developing comprehensive training programs that educate stakeholders about mycotoxin risks and best practices. Extension services and public awareness campaigns can support the adoption of effective management practices and promote a culture of safety in agriculture and food processing.

CONCLUSIONS

Mycotoxin contamination poses significant risks to health, agriculture, and food security worldwide. To tackle these challenges effectively, a comprehensive approach is essential. Advancing detection technologies, such as enhanced biosensors and digital solutions, will improve our ability to monitor and manage mycotoxin levels in real-time. Understanding the molecular mechanisms behind mycotoxin production and resistance, along with exploring genetic and biotechnological innovations, will drive the development of more resilient crops and effective mitigation strategies.

Strengthening regulatory frameworks and enforcement mechanisms is crucial for ensuring consistent safety standards and compliance. Harmonizing regulations internationally and fostering public-private partnerships can accelerate technological advancements and promote best practices. Furthermore, enhancing capacity building and training for farmers and food processors is vital for the practical implementation of these strategies. International collaboration and knowledge sharing are key to addressing mycotoxin contamination on a global scale. By pooling resources, standardizing practices, and supporting joint research initiatives, we can develop and implement effective solutions more efficiently.

In conclusion, addressing the complex issue of mycotoxin contamination requires a multifaceted effort involving research, technology, policy, and collaboration. Through continued innovation and global cooperation, we can improve our ability to manage and mitigate the risks associated with mycotoxins, ensuring safer food and healthier environments.

REFERENCES

Agriopoulou, S. (2021). Ergot alkaloids mycotoxins in cereals and cereal-derived food products: Characteristics, toxicity, prevalence, and control strategies. *Agronomy (Basel)*, *11*(5), 931. DOI: 10.3390/agronomy11050931

Agriopoulou, S., Stamatelopoulou, E., & Varzakas, T. (2020). Advances in Analysis and Detection of Major Mycotoxins in Foods. *Foods*, *9*(4), 518. DOI: 10.3390/foods9040518 PMID: 32326063

Ajmal, M., Bedale, W., Akram, A., & Yu, J. H. (2022). Comprehensive Review of Aflatoxin Contamination, Impact on Health and Food Security, and Management Strategies in Pakistan. *Toxins*, *14*(12), 845. DOI: 10.3390/toxins14120845 PMID: 36548742

Awuchi, C. G., Ondari, E. N., Ogbonna, C. U., Upadhyay, A. K., Baran, K., Okpala, C. O. R., Korzeniowska, M., & Guiné, R. P. F. (2021). Mycotoxins affecting animals, foods, humans, and plants: Types, occurrence, toxicities, action mechanisms, prevention, and detoxification strategies-A revisit. *Foods*, *10*(6), 1279. DOI: 10.3390/foods10061279 PMID: 34205122

Ben Miri, Y., Benabdallah, A., Chentir, I., Djenane, D., Luvisi, A., & De Bellis, L. (2024). Comprehensive insights into ochratoxin A: Occurrence, analysis, and control strategies. *Foods*, *13*(8), 1184. DOI: 10.3390/foods13081184 PMID: 38672856

Borràs-Vallverdú, B., Ramos, A. J., Cantero-Martínez, C., Marín, S., Sanchis, V., & Fernández-Ortega, J. (2022). Influence of Agronomic Factors on Mycotoxin Contamination in Maize and Changes during a 10-Day Harvest-Till-Drying Simulation Period: A Different Perspective. *Toxins*, *14*(9), 620. DOI: 10.3390/toxins14090620 PMID: 36136558

Bradford, K. J., Dahal, P., Van Asbrouck, J., Kunusoth, K., Bello, P., Thompson, J., & Wu, F. (2018). The dry chain: Reducing postharvest losses and improving food safety in humid climates. *Trends in Food Science & Technology*, *71*, 84–93. DOI: 10.1016/j.tifs.2017.11.002

Bryła, M., Pierzgalski, A., Zapaśnik, A., Uwineza, P. A., Ksieniewicz-Woźniak, E., Modrzewska, M., & Waśkiewicz, A. (2022). Recent Research on *Fusarium* Mycotoxins in Maize-A Review. *Foods*, *11*(21), 3465. DOI: 10.3390/foods11213465 PMID: 36360078

Chojnacka, K., Mikula, K., Izydorczyk, G., Skrzypczak, D., Witek-Krowiak, A., Moustakas, K., Ludwig, W., & Kułażyński, M. (2021). Improvements in drying technologies - Efficient solutions for cleaner production with higher energy efficiency and reduced emission. *Journal of Cleaner Production*, *320*, 128706. DOI: 10.1016/j.jclepro.2021.128706

Chulze, S. N. (2010). Strategies to reduce mycotoxin levels in maize during storage: A review. *Food Additives & Contaminants. Part A, Chemistry, Analysis, Control, Exposure & Risk Assessment*, *27*(5), 651–657. DOI: 10.1080/19440040903573032 PMID: 20349375

Čolović, R., Puvača, N., Cheli, F., Avantaggiato, G., Greco, D., Đuragić, O., Kos, J., & Pinotti, L. (2019). Decontamination of Mycotoxin-Contaminated Feedstuffs and Compound Feed. *Toxins*, *11*(11), 617. DOI: 10.3390/toxins11110617 PMID: 31731462

Daou, R., Joubrane, K., Maroun, R. G., Rabbaa Khabbaz, L., Ismail, A., & El Khoury, A. (2021). Mycotoxins: Factors influencing production and control strategies. *AIMS Agriculture and Food*, *6*(1), 416–447. DOI: 10.3934/agrfood.2021025

De Girolamo, A., Lippolis, V., & Pascale, M. (2022). Overview of Recent Liquid Chromatography Mass Spectrometry-Based Methods for Natural Toxins Detection in Food Products. *Toxins*, *14*(5), 328. DOI: 10.3390/toxins14050328 PMID: 35622576

De Santis, B., Debegnach, F., Gregori, E., Russo, S., Marchegiani, F., Moracci, G., & Brera, C. (2017). Development of a LC-MS/MS Method for the Multi-Mycotoxin Determination in Composite Cereal-Based Samples. *Toxins*, *9*(5), 169. DOI: 10.3390/toxins9050169 PMID: 28524101

Dhakal, A., Hashmi, M. F., & Sbar, E. (2023). Aflatoxin Toxicity. In *StatPearls*. StatPearls Publishing.

Dövényi-Nagy, T., Rácz, C., Molnár, K., Bakó, K., Szláma, Z., Jóźwiak, Á., Farkas, Z., Pócsi, I., & Dobos, A. C. (2020). Pre-Harvest Modelling and Mitigation of Aflatoxins in Maize in a Changing Climatic Environment-A Review. *Toxins*, *12*(12), 768. DOI: 10.3390/toxins12120768 PMID: 33291729

El-Sayed, R. A., Jebur, A. B., Kang, W., & El-Demerdash, F. M. (2022). An overview on the major mycotoxins in food products: Characteristics, toxicity, and analysis. *Journal of Future Foods*, *2*(2), 91–102. DOI: 10.1016/j.jfutfo.2022.03.002

European Commission. (2006). *Commission Regulation (EC) No 1881/2006 of 19 December 2006 setting maximum levels for certain contaminants in foodstuffs*. EUR-Lex. https://eur-lex.europa.eu/legal-content/EN/TXT/PDF/?uri=CELEX:02006R1881-20230326

Figlan, S., & Mwadzingeni, L. (2022). Breeding Tools for Assessing and Improving Resistance and Limiting Mycotoxin Production by *Fusarium graminearum* in Wheat. *Plants*, *11*(15), 1933. DOI: 10.3390/plants11151933 PMID: 35893637

Food and Agriculture Organization of the United Nations. (1995). *General standard for contaminants and toxins in food and feed (CXS 193-1995)*. https://www.fao.org/fileadmin/user_upload/agns/pdf/CXS_193e.pdf

Food and Agriculture Organization of the United Nations. (2023). *International Food Safety Authorities Network (INFOSAN)*. https://www.fao.org/sustainable-development-goals-helpdesk/champion/article-detail/international-food-safety-authorities-network-(infosan)/en

Food and Agriculture Organization of the United Nations & World Health Organization. (2017). *Code of practice for the prevention and reduction of mycotoxin contamination in cereals (CXC 78-2017)*. https://www.fao.org/fao-who-codexalimentarius/sh-proxy/en/?lnk=1&url=https%253A%252F%252Fworkspace.fao.org%252Fsites%252Fcodex%252FStandards%252FCXC%2B78-2017%252FCXC_078e.pdf

Food Standards Australia New Zealand. (2023). *Peanut, pistachio, and aflatoxins*. https://www.foodstandards.gov.au/sites/default/files/2023-11/Peanut%20pistachio%20and%20aflatoxins.pdf

Fountain, J. C., Khera, P., Yang, L., Nayak, S. N., Scully, B. T., Lee, R. D., Chen, Z.-Y., Kemerait, R. C., Varshney, R. K., & Guo, B. (2015). Resistance to *Aspergillus flavus* in maize and peanut: Molecular biology, breeding, environmental stress, and future perspectives. *The Crop Journal*, *3*(3), 229–237. DOI: 10.1016/j.cj.2015.02.003

Guan, Y., Lv, H., Wu, G., Chen, J., Wang, M., Zhang, M., Pang, H., Duan, Y., Wang, L., & Tan, Z. (2023). Effects of Lactic Acid Bacteria Reducing the Content of Harmful Fungi and Mycotoxins on the Quality of Mixed Fermented Feed. *Toxins*, *15*(3), 226. DOI: 10.3390/toxins15030226 PMID: 36977117

Gurikar, C., Shivaprasad, D. P., Sabillón, L., Gowda, N. A. N., & Siliveru, K. (2023). Impact of mycotoxins and their metabolites associated with food grains. *Grain & Oil Science and Technology*, *6*(1), 1–9. DOI: 10.1016/j.gaost.2022.10.001

Habschied, K., Krstanović, V., Zdunić, Z., Babić, J., Mastanjević, K., & Šarić, G. K. (2021). Mycotoxins Biocontrol Methods for Healthier Crops and Stored Products. *Journal of Fungi (Basel, Switzerland)*, *7*(5), 348. DOI: 10.3390/jof7050348 PMID: 33946920

Herrera, M., Cavero, J., Franco-Luesma, S., Álvaro-Fuentes, J., Ariño, A., & Lorán, S. (2023). Mycotoxins and crop yield in maize as affected by irrigation management and tillage practices. *Agronomy (Basel)*, *13*(3), 798. DOI: 10.3390/agronomy13030798

Islam, M. N., Banik, M., Sura, S., Tucker, J. R., & Wang, X. (2022). Implications of Crop Rotation and Fungicide on *Fusarium* and Mycotoxin Spectra in Manitoba Barley, 2017-2019. *Toxins*, *14*(7), 463. DOI: 10.3390/toxins14070463 PMID: 35878201

Kamle, M., Mahato, D. K., Devi, S., Lee, K. E., Kang, S. G., & Kumar, P. (2019). Fumonisins: Impact on Agriculture, Food, and Human Health and their Management Strategies. *Toxins*, *11*(6), 328. DOI: 10.3390/toxins11060328 PMID: 31181628

Kamle, M., Mahato, D. K., Gupta, A., Pandhi, S., Sharma, N., Sharma, B., Mishra, S., Arora, S., Selvakumar, R., Saurabh, V., Dhakane-Lad, J., Kumar, M., Barua, S., Kumar, A., Gamlath, S., & Kumar, P. (2022). Citrinin Mycotoxin Contamination in Food and Feed: Impact on Agriculture, Human Health, and Detection and Management Strategies. *Toxins*, *14*(2), 85. DOI: 10.3390/toxins14020085 PMID: 35202113

Karlovsky, P., Suman, M., Berthiller, F., De Meester, J., Eisenbrand, G., Perrin, I., Oswald, I. P., Speijers, G., Chiodini, A., Recker, T., & Dussort, P. (2016). Impact of food processing and detoxification treatments on mycotoxin contamination. *Mycotoxin Research*, *32*(4), 179–205. DOI: 10.1007/s12550-016-0257-7 PMID: 27554261

Keskin, E., & Eyupoglu, O. E. (2023). Determination of mycotoxins by HPLC, LC-MS/MS and health risk assessment of the mycotoxins in bee products of Turkey. *Food Chemistry*, *400*, 134086. DOI: 10.1016/j.foodchem.2022.134086 PMID: 36075166

Khoi, C. S., Chen, J. H., Lin, T. Y., Chiang, C. K., & Hung, K. Y. (2021). Ochratoxin A-Induced Nephrotoxicity: Up-to-Date Evidence. *International Journal of Molecular Sciences*, *22*(20), 11237. DOI: 10.3390/ijms222011237 PMID: 34681895

Kihal, A., Rodríguez-Prado, M., & Calsamiglia, S. (2022). The efficacy of mycotoxin binders to control mycotoxins in feeds and the potential risk of interactions with nutrient: A review. *Journal of Animal Science*, *100*(11), skac328. Advance online publication. DOI: 10.1093/jas/skac328 PMID: 36208465

Kochiieru, Y., Mankevičienė, A., Cesevičienė, J., Semaškienė, R., Ramanauskienė, J., Gorash, A., Janavičienė, S., & Venslovas, E. (2021). The impact of harvesting time on *Fusarium* mycotoxins in spring wheat grain and their interaction with grain quality. *Agronomy (Basel)*, *11*(4), 642. DOI: 10.3390/agronomy11040642

Kos, J., Anić, M., Radić, B., Zadravec, M., Janić Hajnal, E., & Pleadin, J. (2023). Climate Change-A Global Threat Resulting in Increasing Mycotoxin Occurrence. *Foods*, *12*(14), 2704. DOI: 10.3390/foods12142704 PMID: 37509796

Kumar, A., Pathak, H., Bhadauria, S., & Sudan, J. (2021). Aflatoxin contamination in food crops: Causes, detection, and management: A review. *Food Production, Processing and Nutrition*, *3*(17), 17. Advance online publication. DOI: 10.1186/s43014-021-00064-y

Kumar, D., & Kalita, P. (2017). Reducing Postharvest Losses during Storage of Grain Crops to Strengthen Food Security in Developing Countries. *Foods*, *6*(1), 8. DOI: 10.3390/foods6010008 PMID: 28231087

Kumar, P., Gupta, A., Mahato, D. K., Pandhi, S., Pandey, A. K., Kargwal, R., Mishra, S., Suhag, R., Sharma, N., Saurabh, V., Paul, V., Kumar, M., Selvakumar, R., Gamlath, S., Kamle, M., Enshasy, H. A. E., Mokhtar, J. A., & Harakeh, S. (2022). Aflatoxins in Cereals and Cereal-Based Products: Occurrence, Toxicity, Impact on Human Health, and Their Detoxification and Management Strategies. *Toxins*, *14*(10), 687. DOI: 10.3390/toxins14100687 PMID: 36287956

Lee, J. G., Lee, Y., Kim, C. S., & Han, S. B. (2021). Codex Alimentarius commission on ensuring food safety and promoting fair trade: Harmonization of standards between Korea and codex. *Food Science and Biotechnology*, *30*(9), 1151–1170. DOI: 10.1007/s10068-021-00943-7 PMID: 34483698

Leslie, J. F., Moretti, A., Mesterházy, Á., Ameye, M., Audenaert, K., Singh, P. K., Richard-Forget, F., Chulze, S. N., Ponte, E. M. D., Chala, A., Battilani, P., & Logrieco, A. F. (2021). Key Global Actions for Mycotoxin Management in Wheat and Other Small Grains. *Toxins*, *13*(10), 725. DOI: 10.3390/toxins13100725 PMID: 34679018

Leyva Salas, M., Mounier, J., Valence, F., Coton, M., Thierry, A., & Coton, E. (2017). Antifungal Microbial Agents for Food Biopreservation-A Review. *Microorganisms*, *5*(3), 37. DOI: 10.3390/microorganisms5030037 PMID: 28698479

Li, X., Ma, W., Ma, Z., Zhang, Q., & Li, H. (2021). The Occurrence and Contamination Level of Ochratoxin A in Plant and Animal-Derived Food Commodities. *Molecules (Basel, Switzerland)*, *26*(22), 6928. DOI: 10.3390/molecules26226928 PMID: 34834020

Liu, M., Zhang, X., Luan, H., Zhang, Y., Xu, W., Feng, W., & Song, P. (2024). Bioenzymatic detoxification of mycotoxins. *Frontiers in Microbiology*, *15*, 1434987. DOI: 10.3389/fmicb.2024.1434987 PMID: 39091297

Liu, Z., Zhang, G., Zhang, Y., Jin, Q., Zhao, J., & Li, J. (2016). Factors controlling mycotoxin contamination in maize and food in the Hebei province, China. *Agronomy for Sustainable Development, 36*(39), 39. Advance online publication. DOI: 10.1007/s13593-016-0374-x

Loi, M., Logrieco, A. F., Pusztahelyi, T., Leiter, É., Hornok, L., & Pócsi, I. (2023). Advanced mycotoxin control and decontamination techniques in view of an increased aflatoxin risk in Europe due to climate change. *Frontiers in Microbiology, 13*, 1085891. DOI: 10.3389/fmicb.2022.1085891 PMID: 36762096

Luz, C., Ferrer, J., Mañes, J., & Meca, G. (2018). Toxicity reduction of ochratoxin A by lactic acid bacteria. *Food and Chemical Toxicology, 112*, 60–66. DOI: 10.1016/j.fct.2017.12.030 PMID: 29274433

Mahato, D. K., Devi, S., Pandhi, S., Sharma, B., Maurya, K. K., Mishra, S., Dhawan, K., Selvakumar, R., Kamle, M., Mishra, A. K., & Kumar, P. (2021a). Occurrence, Impact on Agriculture, Human Health, and Management Strategies of Zearalenone in Food and Feed: A Review. *Toxins, 13*(2), 92. DOI: 10.3390/toxins13020092 PMID: 33530606

Mahato, D. K., Kamle, M., Sharma, B., Pandhi, S., Devi, S., Dhawan, K., Selvakumar, R., Mishra, D., Kumar, A., Arora, S., Singh, N. A., & Kumar, P. (2021b). Patulin in food: A mycotoxin concern for human health and its management strategies. *Toxicon, 198*, 12–23. DOI: 10.1016/j.toxicon.2021.04.027 PMID: 33933519

Marete, G. N., Kanja, L. W., Mbaria, J. M., Okumu, M. O., Ateku, P. A., Korhonen, H., & Joutsjoki, V. (2020). Effects of the use of good agricultural practices on aflatoxin levels in maize grown in Nandi County, Kenya. *Sci, 2*(4), 85. DOI: 10.3390/sci2040085

Mesterhazy, A., Szieberth, D., Tóth, E. T., Nagy, Z., Szabó, B., Herczig, B., Bors, I., & Tóth, B. (2022). The role of preharvest natural infection and toxin contamination in food and feed safety in maize, South-East Hungary, 2014–2021. *Journal of Fungi (Basel, Switzerland), 8*(10), 1104. DOI: 10.3390/jof8101104 PMID: 36294669

Ministry of Health. Labour and Welfare, Japan. (2020). *Standards and criteria for food under the Food Sanitation Act (abstract)*. Retrieved from https://www.mhlw.go.jp/english/topics/importedfoods/19/appendix1.html

Mutungi, C., Muthoni, F., Bekunda, M., Gaspar, A., Kabula, E., & Abass, A. (2019). Physical quality of maize grain harvested and stored by smallholder farmers in the Northern highlands of Tanzania: Effects of harvesting and pre-storage handling practices in two marginally contrasting agro-locations. *Journal of Stored Products Research, 84*, 101517. DOI: 10.1016/j.jspr.2019.101517

Neme, K., & Mohammed, A. (2017). Mycotoxin occurrence in grains and the role of postharvest management as a mitigation strategy: A review. *Food Control, 78*, 412–425. DOI: 10.1016/j.foodcont.2017.03.012

Nji, Q. N., Babalola, O. O., Ekwomadu, T. I., Nleya, N., & Mwanza, M. (2022). Six Main Contributing Factors to High Levels of Mycotoxin Contamination in African Foods. *Toxins, 14*(5), 318. DOI: 10.3390/toxins14050318 PMID: 35622564

Pandey, A. K., Samota, M. K., Kumar, A., Silva, A. S., & Dubey, N. K. (2023). Fungal mycotoxins in food commodities: Present status and future concerns. *Frontiers in Sustainable Food Systems, 7*, 1162595. Advance online publication. DOI: 10.3389/fsufs.2023.1162595

Patel, T. K., Anand, R., Singh, A. P., Shankar, J., & Tiwary, B. N. (2014). Evaluation of aflatoxin B1 biosynthesis in *A. flavus* isolates from central India and identification of atoxigenic isolates. *Biotechnology and Bioprocess Engineering; BBE, 19*(6), 1105–1113. DOI: 10.1007/s12257-014-0464-z

Perrone, G., Ferrara, M., Medina, A., Pascale, M., & Magan, N. (2020). Toxigenic Fungi and Mycotoxins in a Climate Change Scenario: Ecology, Genomics, Distribution, Prediction and Prevention of the Risk. *Microorganisms, 8*(10), 1496. DOI: 10.3390/microorganisms8101496 PMID: 33003323

Polak-Śliwińska, M., & Paszczyk, B. (2021). Trichothecenes in Food and Feed, Relevance to Human and Animal Health and Methods of Detection: A Systematic Review. *Molecules (Basel, Switzerland), 26*(2), 454. DOI: 10.3390/molecules26020454 PMID: 33467103

Santos, A. R., Carreiró, F., Freitas, A., Barros, S., Brites, C., Ramos, F., & Sanches Silva, A. (2022). Mycotoxins Contamination in Rice: Analytical Methods, Occurrence and Detoxification Strategies. *Toxins, 14*(9), 647. DOI: 10.3390/toxins14090647 PMID: 36136585

Singh, A., & Sarkar, S. (2021). Effect of excess use of fertilizers on insect/pest infestation in *Oryza sativa*: A review. *Journal of Entomology and Zoology Studies, 9*(5), 246–252. DOI: 10.22271/j.ento.2021.v9.i5c.8834

Snini, S. P., & Mathieu, F. (2020). Biocontrol Agents and Natural Compounds Against Mycotoxinogenic Fungi. *Toxins, 12*(6), 353. DOI: 10.3390/toxins12060353 PMID: 32481521

U.S. Department of Agriculture, Foreign Agricultural Service. (2018). *China releases standard for maximum levels of mycotoxins in foods*. https://apps.fas.usda.gov/newgainapi/api/report/downloadreportbyfilename?filename=China%20Releases%20Standard%20for%20Maximum%20Levels%20of%20Mycotoxins%20in%20Foods%20_Beijing_China%20-%20Peoples%20Republic%20of_5-9-2018.pdf

U.S. Food and Drug Administration. (2000). *Guidance for industry: Action levels for poisonous or deleterious substances in human food and animal feed*. https://www.fda.gov/regulatory-information/search-fda-guidance-documents/guidance-industry-action-levels-poisonous-or-deleterious-substances-human-food-and-animal-feed

Venkatesh, N., & Keller, N. P. (2019). Mycotoxins in conversation with bacteria and fungi. *Frontiers in Microbiology, 10*, 403. DOI: 10.3389/fmicb.2019.00403 PMID: 30941105

Viegas, C., Nurme, J., Piecková, E., & Viegas, S. (2018). Sterigmatocystin in foodstuffs and feed: Aspects to consider. *Mycology, 11*(2), 91–104. DOI: 10.1080/21501203.2018.1492980 PMID: 32923018

Zadravec, M., Markov, K., Lešić, T., Frece, J., Petrović, D., & Pleadin, J. (2022). Biocontrol methods in avoidance and downsizing of mycotoxin contamination of food crops. *Processes (Basel, Switzerland), 10*(4), 655. DOI: 10.3390/pr10040655

Zain, M. E. (2011). Impact of mycotoxins on humans and animals. *Journal of Saudi Chemical Society, 15*(2), 129–144. DOI: 10.1016/j.jscs.2010.06.006

Zhang, C., Jiang, J., Li, J., Zhang, J., Zhang, X., & Wang, H. (2023). Long transportation duration affects nutrient composition, mycotoxins and microbial community in whole-plant corn silage. *Frontiers in Veterinary Science, 10*, 1189358. DOI: 10.3389/fvets.2023.1189358 PMID: 37275604

Zhong, L., Carere, J., Lu, Z., Lu, F., & Zhou, T. (2018). Patulin in Apples and Apple-Based Food Products: The Burdens and the Mitigation Strategies. *Toxins*, *10*(11), 475. DOI: 10.3390/toxins10110475 PMID: 30445713

Zingales, V., Taroncher, M., Martino, P. A., Ruiz, M. J., & Caloni, F. (2022). Climate Change and Effects on Molds and Mycotoxins. *Toxins*, *14*(7), 445. DOI: 10.3390/toxins14070445 PMID: 35878185

Chapter 18
Hazards of Mycotoxin in Agriculture and Effective Mitigation Strategies

Suryakant Manik
https://orcid.org/0009-0000-4415-0412
ICAR-Indian Institute of Agricultural Biotechnology, India

Sujit Kumar Bishi
https://orcid.org/0000-0002-4108-1826
ICAR-Indian Institute of Agricultural Biotechnology, India

S. Shreedevasena
Directorate of Medicinal and Aromatic Plant Research, India

Kuleshwar Prasad Sahu
College of Agriculture, Rani Lakshmi Bai Central Agricultural University, India

Asharani Patel
https://orcid.org/0009-0002-7064-1228
ICAR-Indian Agricultural Research Institute, India

Licon Kumar Acharya
https://orcid.org/0000-0003-3908-737X
Division of Plant Pathology, ICAR-Indian Agriculture Research Institute, India, ICAR-National Research Centre for Integrated Pest Management, India

ABSTRACT

Mycotoxins, toxic metabolites produced by fungi, represent a significant global health concern due to their prevalence and high toxicity. This chapter explores the major mycotoxins including aflatoxins, ochratoxin A, fumonisins, deoxynivalenol, T-2 toxins, zearalenone, etc. Both acute and chronic exposure to these contaminants can result in severe health outcomes demanding immediate and sustained attention. The importance of rapid and reliable detection of mycotoxin contamination in food and animal feed is emphasized as a cornerstone of health risk mitigation. The chapter highlights the role of advanced agricultural practices, encompassing both pre- and post-harvest stages, and the proper storage of agricultural commodities. Furthermore, it underscores the integration of routine monitoring with innovative technologies like biotechnological, biological and nanotechnological approaches as essential strategies to reduce mycotoxin risks. Together, these measures provide a comprehensive framework for addressing the complex challenges posed by mycotoxins to global health and food security.

DOI: 10.4018/979-8-3693-8201-1.ch018

1. INTRODUCTION

Among the various fundamental necessities for the sustenance of human beings, food constitutes the primary component. It fundamentally satisfies the nutritional requirements and provides energy, thus facilitating the maintenance of health and the promotion of physiological well-being. In addition to this, it plays a significant role in the economic advancement of a nation, as various agricultural or food-based industries serve as primary economic drivers while ensuring food security and sustainability on both national and global platforms. Foods and crops have been consistently infected and contaminated by a plethora of hazardous fungal pathogens throughout history, with documentation of these issues beginning in the 20th century. Mycotoxins are key contaminants in food that significantly compromise food quality and indirectly affect animal and human health they by posing a serious threat to global food security (CAST 2003). Currently, over 300 distinct mycotoxins have been identified, though only about 30 are considered critical for humans and animals. Notably, among these 30 mycotoxins, several are of particular concern, including aflatoxins (AFs), ochratoxins (OTs), fumonisins (FMNs), trichothecenes, patulin (PAT), zearalenone (ZEN), and alternariol (AOH) (Agriopoulou et al., 2020). Mycotoxins are secondary metabolites produced by specific fungi, including *Aspergillus, Fusarium, Penicillium,* and *Alternaria* and contaminate crops at multiple stages during preharvest stages like vegetative growth, flowering stage and fruiting stage as well as during harvesting stage and also in post-harvest stages like processing, storage and transportation particularly under conditions that favor fungal development (Pickova et al., 2021). This chapter deals with the various types of mycotoxins, their effects on different agricultural commodities and human health, the factors associated with their contamination, and recent advancements in their detection and management. It provides a comprehensive overview of the serious issues related to mycotoxins.

2. HISTORICAL SIGNIFICANCE AND PREVALENCE OF MYCOTOXINS IN FOOD AND FEED

One of the earliest and most notorious illnesses associated with mycotoxins was ergotism, which emerged from the ingestion of grains, particularly rye, contaminated with *Claviceps purpurea*, a fungus that synthesizes ergot alkaloids. Ergotism caused extensive epidemics throughout medieval Europe, resulting in severe clinical manifestations such as hallucinations, gangrene, convulsions, and ultimately mortality. In the 10th century, this affliction was designated as "St. Anthony's Fire," attributed to the monks of St. Anthony's order who were renowned for their treatment of those impacted (Dongen and Groot, 1995). The subsequent identification of aflatoxins was prompted by a notable outbreak termed Turkey X Disease in England in 1960, during which over 100,000 turkeys succumbed after ingesting contaminated peanut meal (CAST, 2003). This incident catalyzed the recognition of aflatoxins and instigated worldwide investigations into mycotoxins, culminating in the formulation of regulations aimed at monitoring and mitigating mycotoxin contamination in food and animal feed. Throughout World War II, there were occurrences of alimentary toxic aleukia (ATA), a condition attributed to mycotoxins produced by *Fusarium* species, including trichothecenes, within the former Soviet Union. A significant number of people perished during a famine as a consequence of consuming mold-infested grains, predominantly wheat and barley, which were contaminated with Fusarium mycotoxins. Symptoms of

ATA encompassed vomiting, fever, and suppression of bone marrow, ultimately leading to profound immunological compromise and fatality (Lutsky and Mor 1981).

The prevalence of these mycotoxins varies globally across regions like Northern Europe, Central Europe, Southern Europe, Eastern Europe, North America, Central America, South America, Middle East, Sub-Saharan Africa, South Africa, South Asia, Southeast Asia, East Asia, and Central Asia. Based on the survey conducted by different scientists and compiled by Gruber-Dorninger et al., 2019, it was observed that six mycotoxins Deoxynivalenol (DON), Zearalenone (ZEN), Fumonisin (FUM), Ochratoxin (OTA), T-2 and Aflatoxin B1 (AFB1) were predominantly contaminated in the food samples taken for observation in the above countries or continents. Among such mycotoxins, Deoxynivalenol (DON) was found to be the highest in the percentage of contamination to foods in 6 regions (Northern Europe, Central Europe, Eastern Europe, North America, East Asia, and Central Asia) and it ranged from 84.8% in East Asia to 23.1% in South Asia. The second most prevalent mycotoxin was found to be Fumonisin (FUM) and it was found highest in regions like (Southern Europe, Central America, South America, Middle East/North and Southeast Asia) and the percentage of contamination ranged from 22% in Northern Europe to 82% in Central America. The prevalence of Zearalenone (ZEN) was found to be third highest, followed by Aflatoxin B1 (AFB1), Ochratoxin (OTA) and the least one was T-2 toxin. Contamination of Aflatoxin B1, considered as highly carcinogenic mycotoxin, was found to be more in the South Asia (contamination rate ~83%) and followed by Sub-Saharan Africa (contamination rate ~76%). The prevalence of such mycotoxins and their quantitative amount (µg/ Kg of sample) in the samples tested for is detailed in the Table 1.

Table 1. Global prevalence of six mycotoxins contaminating different food products surveyed by different scientists and compiled by Gruber-Dorninger et al., 2019

Region	Prevalence of Mycotoxins in food sample
Northern Europe (n = 1958)	**DON**(74.2%; 504µg/kg) > **T-2**(30.3%; 34µg/kg) > **ZEN**(28.9%; 35µg/kg) > **FUM**(22.4%; 186 µg/kg) > **OTA**(8.1%; 19 µg/kg) > **ABF1**(5.9%; 3.1 µg/kg)
Central Europe (n = 21,036)	**DON**(69.8%; 428 µg/kg) > **ZEN**(45%; 40 µg/kg) > **FUM**(43.2%; 187µg/kg) > **T-2** (30.7%; 11µg/kg) > **AFB1**(12.7%; 1.6µg/kg) > **OTA**(11.9%; 2.8 µg/kg)
Southern Europe (n = 3527)	**FUM**(74.9%; 607µg/kg) > **DON**(52.9%; 324µg/kg) > **ZEN**(36.3%; 44µg/kg) > **AFB1**(28.9%; 2.1µg/kg) > **OTA**(21.2%; 2.6µg/kg) >**T-2**(11.7%; 25µg/kg)
Eastern Europe (n = 2382)	**DON**(59.9%; 153µg/kg) > **T-2**(48.2%; 21µg/kg) > **ZEN**(42.5%; 15µg/kg) > **OTA**(36.4%; 3.6µg/kg) > **FUM**(33.6%; 87µg/kg) > **AFB1**(17.0%; 3.4µg/kg)
North America (n = 5471)	**DON**(64.1%; 505µg/kg) > **FUM**(47.7%; 652µg/kg) > **ZEN**(31.7%; 102µg/kg) > **AFB1**(10.5%; 8.7µg/kg) > **OTA**(4.3%; 2.4µg/kg) > **T-2**(3.9%; 29µg/kg)
Central America (n = 367)	**FUM**(81.8%; 929µg/kg) > **DON**(70.0%; 316µg/kg) > **ZEN**(38.2%; 60µg/kg) > **AFB1**(8.6%; 3.9µg/kg) > **T-2**(4.1%; 3.1µg/kg) > **OTA**(3.8%; 2.5µg/kg)
South America (n = 17,332)	**FUM**(75.3%; 1390µg/kg) > **ZEN**(46.9%; 51µg/kg) > **DON**(26.9%; 344µg/kg) >**ABF1**(23.5%; 3.2µg/kg) > **T-2**(21.5%; 31µg/kg) > **OTA**(4.9%; 17µg/kg)
Middle East/North Africa (n =1075)	**FUM**(66.6%;347 µg/kg) > **DON**(47.0%;236 µg/kg) > **ZEN**(44.8%;31 µg/kg) > **ABF1**(22.2%; 2.4µg/kg) > **OTA** (20.3%; 3.1µg/kg) > **T-2**(0.5%; 14 µg/kg)
Sub-Saharan Africa (n = 208)	**ABF1**(76.0%; 23 µg/kg) > **FUM**(72.6%; 789µg/kg) > **ZEN** (52.2%; 38µg/kg) > **DON**(49.5%; 352µg/kg) > **OTA**(31.9%; 7.2µg/kg) > **T-2**(3.0%; 3.0µg/kg)
South Africa (n = 1077)	**ABF1**(9.0%; 2.2µg/kg) > **FUM**(62.6%; 266µg/kg) > **ZEN** (41.6%; 30µg/kg) >**DON**(63.2%; 36 µg/kg) > **OTA**(5.6%; 2.2µg/kg) > **T-2**(1.2%; 4.4µg/kg)

continued on following page

Table 1. Continued

Region	Prevalence of Mycotoxins in food sample
South Asia (n = 1136)	**ABF1**(82.2%; 20µg/kg) > **FUM**(69.0%; 288µg/kg) > **OTA**(50.4%; 4.6µg/kg) > **ZEN**(45.9%; 43µg/kg) > **DON**(23.1%; 96µg/kg) > **T-2**(0.9%;13µg/kg)
Southeast Asia (n = 4310)	**FUM**(69.6%; 573µg/kg) > **ZEN** (58.2%; 90µg/kg) **> ABF1**(57.4%; 10µg/kg) > **DON**(42.5%; 137µg/kg) > **OTA**(15.2%; 3.0µg/kg) > **T-2** (2.7%;26 µg/kg)
East Asia (n = 13,232)	**DON**(84.8%; 418µg/kg) > **FUM**(60.7%; 810µg/kg) > **ZEN**(23.1%; 1.5µg/kg) > **OTA**(14.1%; 29µg/kg) > **T-2**(11.0%;16µg/kg) > **ABF1**(10.4%; 7.0µg/kg)
Central Asia (n = 15)	**DON**(28.7%; 28µg/kg) > **FUM**(25.0%; 13µg/kg) > **ZEN** (23.1%; 22µg/kg) > **OTA**(13.3%; 22µg/kg) > **ABF1**(7.7%; 1.4µg/kg) > **T-2**(6.7%; 25µg/kg)

(n= number of samples taken into observation; Six mycotoxins include: DON-Deoxynivalenol, ZEN-Zearalenone, FUM-Fumonisin, OTA-Ochratoxin, T-2 and AFB1-Aflatoxin B1. The unit 'µg/kg' indicates the amount of mycotoxin identified per kg of sample taken.)

3. TYPES OF MYCOTOXINS

Mycotoxins, as predominant contaminants of food grains derived from cereals, pulses, spices, and horticultural crops, are systematically categorized according to their chemical composition, origin, and the specific agricultural commodities they affect, as well as their implications for animal health and related health concerns (Figure 1).

Figure 1. Fungal pathogen infections and mycotoxin production leading to contamination of food crops and mycotoxicosis in humans and animals upon consumption of such contaminated food and feed stuffs

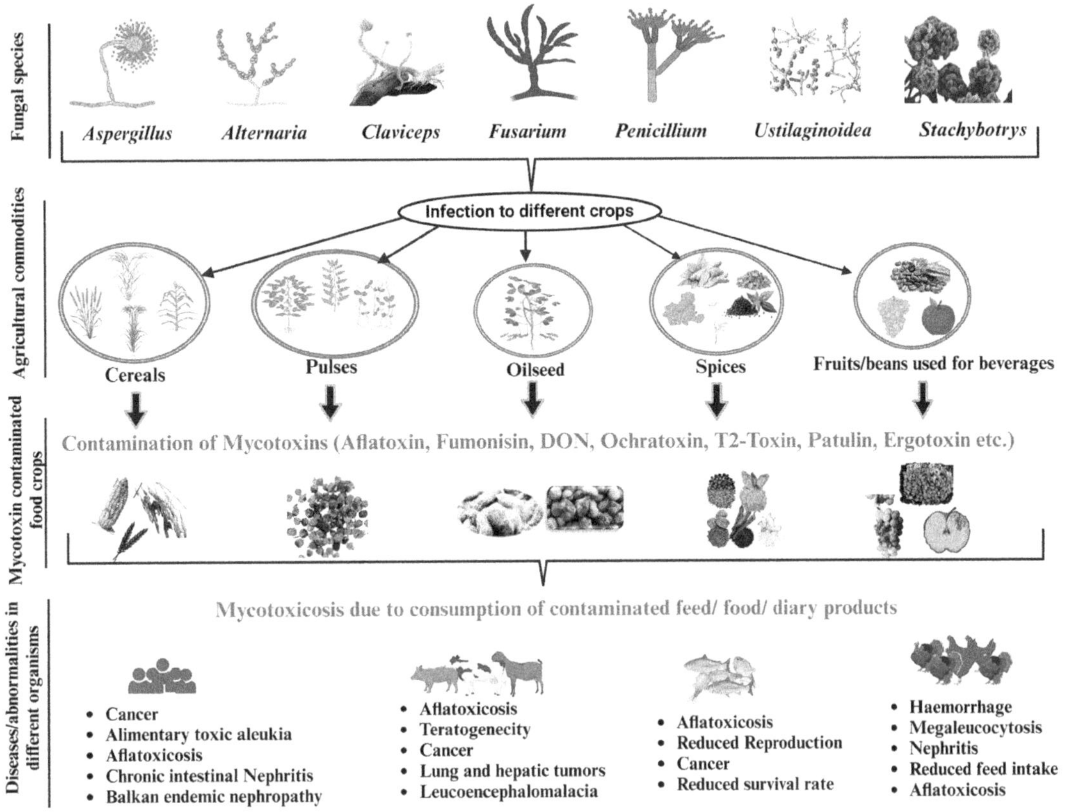

3.1. Based on their chemical nature

Mycotoxins, which are deleterious substances synthesized by specific fungal species, may be systematically categorized in accordance with their chemical characteristics into several main categories. Mycotoxins are of polyketide derivatives, carboxylic acid derivatives, estrogenic compounds, etc. So, on a broad aspect mycotoxins are classified in to different groups which are detailed in Table 2.

Aflatoxin: B1, B2, G1, G2 and M1 and M2

Fumonisins: B1, B2 and B3

Trichothecenes: Type A (T-2, HT-2, Diacetoxyscirpenol and related compounds), Type B (Deoxynivalenol/DON/Vmitoxin, Nivalenol and related compounds), C and D

Zearalenone: F-2 Toxin

Ochratoxins: Ochratoxin A, B and C

Other mycotoxins: Moniliformin, Fusaric acid, Fusaproliferin, Beauvericin, Cyclopiazonic acid, Sterigmatocystin, Citrinin, Penicillic acid, Ustilaginoidin A, Rubratoxin B

3.2. Based on their fungal source

A consortium of fungal species and the mycotoxins they synthesize are implicated in the adulteration of food grains and are accountable for inducing adverse effects upon the ingestion of such contaminated foodstuffs in our dietary practices. The major fungal pathogens associated with mycotoxin biosynthesis are *Aspergillus* spp. (*A. flavus, A. parasiticus, A. nomius, A. pseudotamarii, A. ochraceus, etc.*), *Alternaria* spp. (*A. alternate, A. solani,* etc.), *Fusarium* spp. (*F. graminearum, F. proliferatum, F. oxysporum, F. subglutinans, F. verticillioides*), *Penicillium* spp. (*P. chrysogenum, P. citrinum, P. expansum, P. viridicatum, etc.*), *Ustilagenoidea virens, Stachybotrys chartarum* and so on. Based on the source of fungal pathogens mycotoxins are classified as:

Aspergillus Toxin: Aflatoxin (B1, B2, G1, G2, M1 and M2), Ochratoxin, Cyclopiazonic Acid, Sterigmatocystin

Alternaria Toxin: Altertoxins, Citrinin

Claviceps: Ergotoxin

Fusarium Toxin: Aurofusarin, Moniliformin, Trichothecenes (Deoxyniva, Nivalen, HT-2, T-2, Diacetoxyscirpenol), Zearalenone, Fumonisins, Fusaproliferin, Fusaric acid

Penicillium Toxin: Citrinin, Penicillic Acid, Rubratoxins

Ustilagenoidea Toxin: Ustiloxins

Stachybotrys Toxin: Satratoxin

3.3. Based on their agricultural commodities affected

Various agricultural products, including cereals, pulses, oilseeds, spices, fruit crops, beverages, as well as milk and dairy derivatives, are subject to contamination by diverse mycotoxins synthesized by a specific group of fungal pathogens (Begum and Samajpati, 2000; Acuña-Gutiérrez et al., 2022; Demirhan and Demirhan, 2022; Quintela, 2020). This contamination results in the degradation of food quality and emerges as a primary contributor to health issues upon ingestion by both animals and humans. The agricultural commodities and their common mycotoxin contaminants are:

a) Cereal mycotoxins
   i) *Maize:* Aflatoxins, Fumonisins, Zearalenone, Deoxynivalenol, Citrinin, Penicillic Acid Rubratoxins, etc.
   ii) *Rice:* Aflatoxins, Ochratoxin A, Fumonisins (B1, B2, B3), Zearalenone, Ustiloxins, etc.
   iii) *Wheat:* Deoxynivalenol (DON), Zearalenone
   iv) *B*arley: Deoxynivalenol (DON), Ochratoxin
   v) *Sorghum:* Aflatoxins, Fumonisins, Ochratoxin A
   vi) *Rye:* Ergotoxin
b) Pulses mycotoxins
   i) *Soybean:* Ochratoxins, Zearalenone
   ii) *Chickpea:* Aflatoxins, Ochratoxin A
   iii) *Mung bean:* Ochratoxin A,
c) Oilseeds mycotoxins
   i) *Groundnut: Aflatoxin,* Ochratoxin, Patulin
d) Spices mycotoxins

i) *Black Pepper:* Aflatoxins, Ochratoxins, Zearalenone-ZEN
ii) *Turmeric:* Aflatoxins, Ochratoxins
iii) *Ginger:* Aflatoxins, Ochratoxins
iv) *Cumin:* Aflatoxins, Ochratoxins, Zearalenone-ZEN
v) *Cardamom*: Aflatoxins, Ochratoxins

e) Mycotoxins of Beverages
i) *Coffee bean:* Ochratoxins, Aflatoxin
ii) *Grape juice, Wine, Fruit wine:* Ochratoxin
iii) *Apple juice and other fruit jiuce:* Patulin

f) Mycotoxins of Milk and dairy derivatives
i) *Milk/Dairy Products:* Aflatoxin M1 and M2 (AFM1 and AFM2)

3.4. Based on their adverse impact on health of Animal and Human

Mycotoxins, which are synthesized by fungal organisms, represent a significant global health issue attributable to their ubiquitous presence and formidable toxicity. These toxic compounds remain in food products despite the application of processing techniques such as cooking and sterilization, due to their inherent chemical stability and resilience against degradation (Khan, 2024). The consumptions of mycotoxin contaminated food grains or feed cause acute and/or chronic health effect in animals and humans which is collectively called as mycotoxicoses. Based on the effect of mycotoxins on heath of animal and human, major mycotoxins are categories into following categories:

a) **Carcinogenic:** Aflatoxins (B1, B2, G1, G2, M1), Fumonisins (B1, B2), Ochratoxin A, B, C
b) **Mutagenic:** Aflatoxins B1, B2, G1, G2, M1, Trichothecenes (Type B: DON), Ochratoxin A, B, C
c) **Estrogenic:** Zearalenone
d) **Nephrotoxic:** Ochratoxin (A, B, C), Citrinin
e) **Neurotoxic:** Patulin, Trichothecenes (Type B: DON)
f) **Ergotism (gangrene):** Ergot alkaloids
g) **Hematotoxic:** T-2/HT-2 toxins, Nivalenol

Table 2. Types of mycotoxins based on their chemical structure

Name of Mycotoxin	Chemical group	Structure of the compound	Molecular formula
Aflatoxin B1	Difuran ring fused with a coumarin structure and a lactone ring	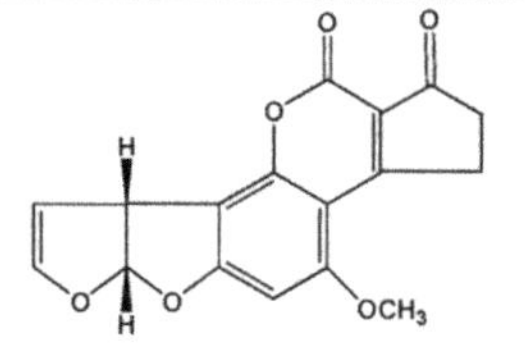	$C_{17}H_{12}O_6$
Aflatoxin B2	Similar to Aflatoxin B1 but with a saturated bond between C8 and C9	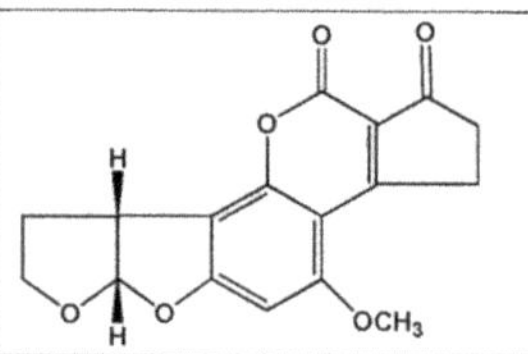	$C_{17}H_{14}O_6$
Aflatoxin G1	Difuran ring and a fused lactone ring with an additional oxygen	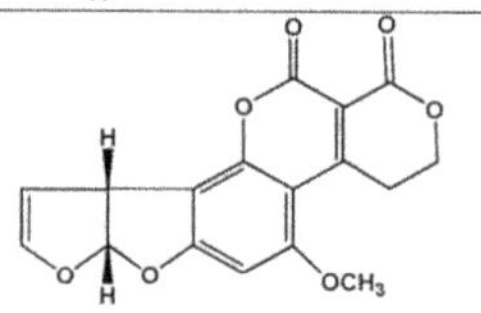	$C_{17}H_{12}O_7$
Aflatoxin G2	Similar to Aflatoxin G1 and with a saturated bond between C8 and C9	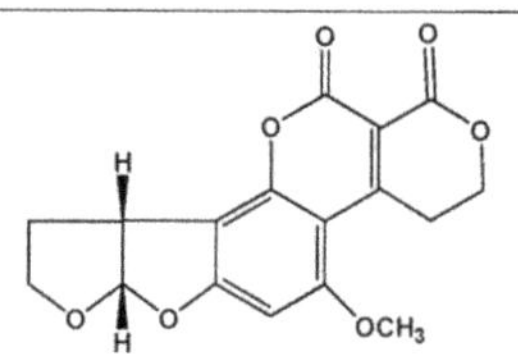	$C_{17}H_{14}O_7$
Fumonisins	1,2,3-Propanetricarboxylic acid	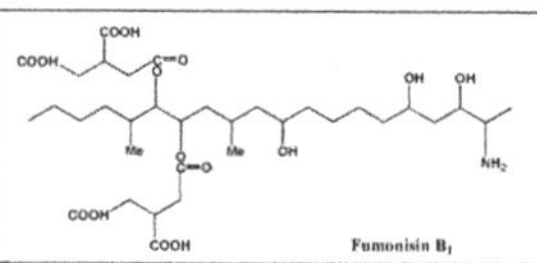	$C_{34}H_{59}NO_{15}$
Deoxynivalenol	3a, 7a,15-trihydroxy-12,13-epoxytrichothec-9-en-8-one	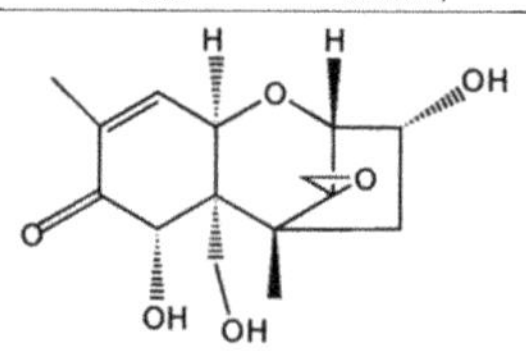	$C_{15}H_{20}O_6$

continued on following page

Table 2. Continued

T-2 toxin	3a-hydroxy-4,15-diacetoxy-8a-(3-methylbutyryloxy)-12,13-epoxytrichothec-9-ene	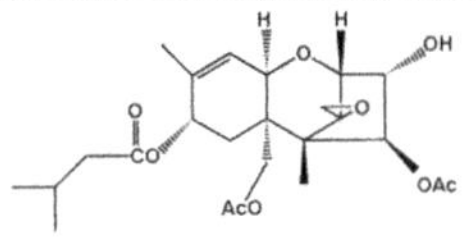	$C_{24}H_{34}O_9$
HT-2 toxin	3a,4b-dihydroxy-4,15-diacetoxy-8a-(3-methylbutyryloxy)-12,13 epoxytrichothec-9-ene	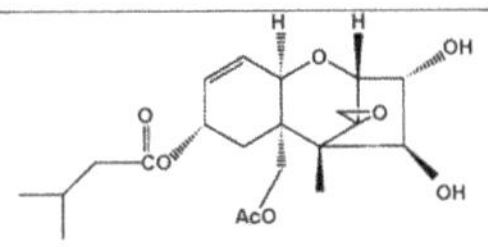	$C_{22}H_{23}O_8$
Diacetoxyscirpenol	3a-hydroxy-4,15-diacetoxy-12,13 epoxytrichothec-9-ene	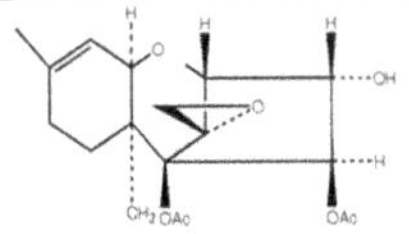	$C_{19}H_{26}O_7$
Nivalenol	3a,4b,7a,15-tetrahydroxy-12,13-epoxytrichothec 9-en-8-one	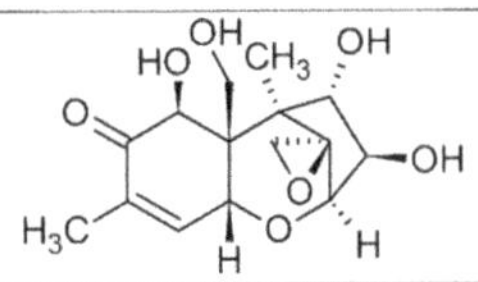	$C_{15}H_{20}O_7$
Zearalenone	6-(10-hydroxy-6-oxo-trans-1 undecenyl)-b-resorcylic acid lactone	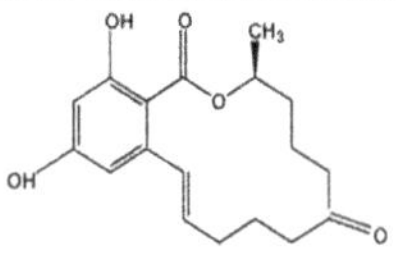	$C_{18}H_{22}O_5$
Moniliformin	potassium or sodium salt of 1-hydroxycyclobut-1-ene-3,4 dione	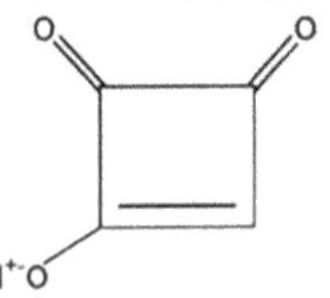	$C_4H_2O_3$
Fusaric acid	5-(butyl)-2-pyridinecarboxylic acid	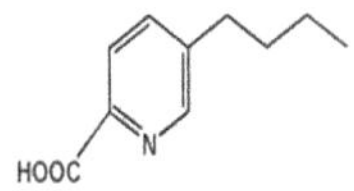	$C_{10}H_{13}NO_2$
Fusaproliferin	bicyclic sesterterpene	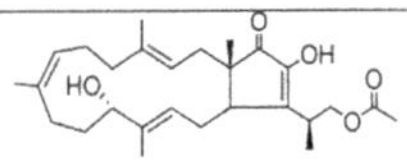	$C_{27}H_{40}O_5$

continued on following page

Table 2. Continued

Beauvericin	cyclic hexadepsipeptide	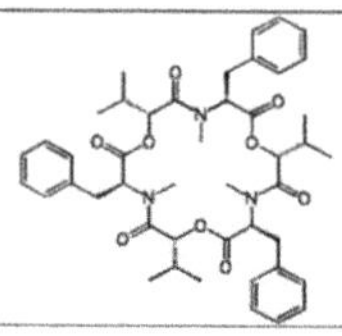	$C_{45}H_{57}N_3O_9$
Ochratoxin A	Isocoumarins with an amide linkage to L-phenyl-alanine.	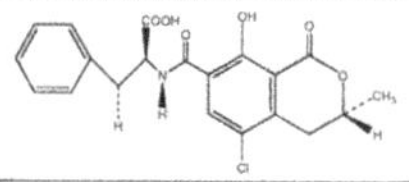	$C_{20}H_{18}ClNO_6$
Cyclopiazonic acid	Indole-tetramic acid	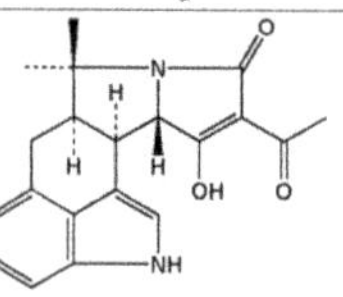	$C_{20}H_{20}N_2O_3$
Sterigmatocystin	Furofuran group of mycotoxins (3a,12C-dihydro-8-hydroxy-6-methoxyfuro [30 ,20 4,5]furo[3,2C]xanthene-7-one)	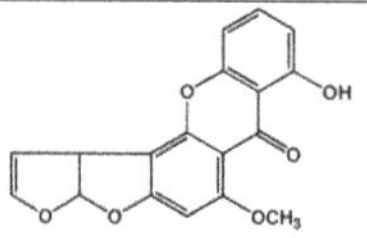	$C_{18}H_{12}O_6$
Citrinin	(3R-trans-)-4,6-dihyfdro-8-hydroxy-3,4,5-trimethyl-6-oxo-3H-2-benzopyran-7-carboxylic acid	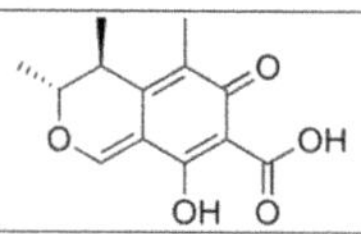	$C_{13}H_{14}O_5$
Penicillic acid	3-methoxy-5-methyl-4-oxo-2,5-hexadienoic acid	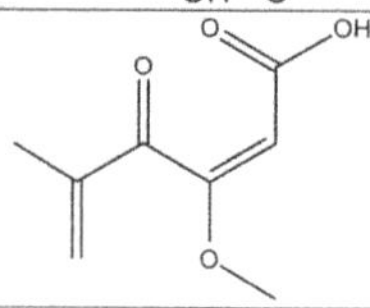	$C_8H_{10}O_4$
Ustilaginoidin A	Benzochromenone, a member of phenols, a naphtho-gamma-pyrone and a binaphthopyran.	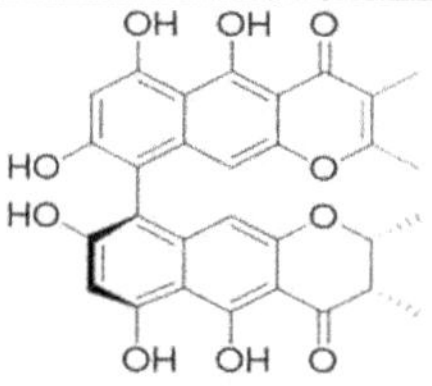	$C_{28}H_{18}O_{10}$
Rubratoxin B	carbonyl compound	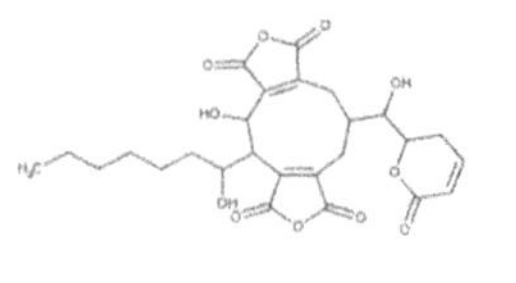	$C_{26}H_{30}O_{11}$

4. EFFECT OF MYCOTOXIN CONTAMINATION ON CROP YIELD AND QUALITY

Various mycotoxins, such as Aflatoxin B1, Ochratoxin A, Patulin, Fumonisin B1, Zearalenone, T-2 Toxin, Deoxynivalenol (DON), Nivalenol, Citrinin, and Sterigmatocystin, significantly reduce both the yield and quality of agricultural products, leading to substantial financial losses for farmers and producers. The effects of these mycotoxins go beyond economic damage in agriculture, as they also present serious health risks to both humans and animals consuming contaminated food or feed. Aflatoxin B1 (AFB1) is the most powerful and extensively studied aflatoxin, commonly found in crops like maize, peanuts, and cottonseed. Its presence not only decreases crop yield but also severely impacts quality, as it is a strong hepatocarcinogen classified as a Group 1 carcinogen by the International Agency for Research on Cancer (IARC, 2012). Affected crops often show signs of stunted growth and discoloration, which directly affects their marketability. Additionally, crops that may not show obvious signs of contamination are frequently considered unsuitable for sale due to food safety regulations, resulting in significant financial losses (Williams et al., 2004). Ochratoxin A (OTA) is mainly produced by *Aspergillus* and *Penicillium* species, causing contamination in cereals, dried fruits, coffee, and wine. Patulin is mainly found in decaying fruits, especially apples, with apple juice from contaminated or damaged apples being the most common source of patulin contamination. From an agricultural viewpoint, patulin contamination negatively impacts both yield and quality of fruits, especially during storage. Crops that appear damaged or rotten due to fungal growth are often discarded, leading to a reduction in marketable produce (Murphy et al., 2006). Fumonisin contamination in maize generally happens during the growing season, especially in plants under drought stress, and can greatly reduce crop yield. Infected maize shows discoloration, kernel rot, and poor development, adversely affecting both quality and market value. Contaminated crops are frequently excluded from food supply chains due to strict safety regulations, resulting in significant economic losses (Wu, 2015). The presence of ZEA in crops like cereals, including maize, wheat, and barley, can greatly affect quality, especially in livestock feed. Infected grains may show signs of discoloration or fungal growth, reducing their market appeal. T-2 toxin in crops negatively influences yield due to poor plant growth and increases the risk of post-harvest losses during storage. DON contamination often leads to insufficient grain development and lower kernel weight regarding crop yield. While infected crops may seem healthy, DON presence lowers their quality, resulting in exclusion from food and feed markets (Pestka, 2010). Nivalenol (NIV), another mycotoxin from Fusarium species, is closely related to DON. It affects crops like wheat, barley, and maize. In agriculture, NIV contamination reduces crop yield and quality, as infected grains often show less growth and are barred from food markets due to strict safety measures (Pestka & Smolinski, 2005). Citrinin mainly impacts grains like rice, maize, and wheat during post-harvest storage. Contaminated grains may show clear signs of fungal growth, leading to quality decline and financial losses from market rejection (Flajs & Peraica, 2009). Sterigmatocystin, a precursor to aflatoxin B1, impacts various crops, including grains, coffee beans, and spices during post-harvest storage, resulting in lower crop yield and quality. Infected crops may display visible fungal growth, leading to exclusion from food and feed markets due to strict safety standards (Peraica et al., 1999).

5. CHRONIC AND ACUTE HEALTH ISSUES IN ANIMAL AND HUMAN

Aflatoxins, produced by *Aspergillus* species, are highly carcinogenic, particularly to the liver, and are responsible for acute poisoning known as aflatoxicosis. Acute exposure to mycotoxins can cause symptoms such as nausea, vomiting, and diarrhea. Severe cases can lead to liver damage, immunosuppression, and even death (Bennett & Klich, 2003). Animals exposed to high levels of mycotoxins may exhibit symptoms such as feed refusal, reduced growth, and gastrointestinal distress (CAST, 2003). Chronic exposure to low levels of mycotoxins can result in long-term health problems such as liver cancer, kidney damage, and immune system suppression (Wild & Gong, 2010; Benkerroum, 2020). Chronic mycotoxin exposure in animals can lead to reduced productivity, reproductive issues, and increased susceptibility to diseases (Murphy et al., 2006). Chronic exposure to aflatoxins can lead to liver cancer, with evidence suggesting genotoxic effects that cause DNA damage in humans (IARC, 2012). Aflatoxins are commonly found in contaminated crops such as maize, peanuts, and tree nuts, as well as in the milk of animals consuming contaminated feed (Bennett & Klich, 2003). Ochratoxin A primarily affects the kidneys, leading to nephrotoxicity and potentially kidney cancer in animals, although evidence in humans is less clear (Pfohl-Leszkowicz & Manderville, 2007). This mycotoxin is commonly found in stored cereals, coffee beans, dried fruits, and wine. Fumonisins are primarily found in maize and maize-based products and are linked to esophageal cancer and neural tube defects in regions with high maize consumption (Harrison et al., 1990). Zearalenone, acts as an estrogenic compound, disrupting reproductive health and is commonly found in cereals such as maize, wheat, and barley (Pestka & Smolinski, 2005). The chronic exposure to these mycotoxins often leads to long-term health issues, including immune suppression, carcinogenesis, and reproductive disorders. Acute effects, meanwhile, are generally more severe and immediate, often causing liver damage, gastrointestinal distress, and even death. Yellow rice toxins, associated with the contamination of rice by molds such as *Penicillium and Aspergillus*, cause serious health issues, particularly under poor storage conditions in warm and humid environments (Benkerroum, 2020). Citreoviridin, one of the most concerning mycotoxins in this category, is known to cause beriberi-like symptoms, including peripheral neuropathy, cardiovascular problems, and gastrointestinal distress (Perrone et al., 2020). Chronic exposure to citreoviridin can lead to persistent neurological impairments, cardiovascular damage, and immune system suppression. Chromic and acute health issues with different teratogenic, mutagenic and carcinogenic effects of different mycotoxins are described in Table 3 and Table 4.

Table 3. Acute and chronic effects of major mycotoxins

Mycotoxin	Food crops infected	Acute Effects	Chronic Effects
Aflatoxins	Peanuts, cottonseeds, cereals, fruits, and spices	Aflatoxicosis: Severe liver damage, acute poisoning, potentially fatal	Liver cancer, immune suppression, genotoxicity
Ochratoxin A	Cereals, coffee beans, vegetables, fish, and meat	Kidney damage, gastrointestinal distress	Nephrotoxicity, kidney cancer, immune suppression
Patulin	Apples, peaches, plums, and apricots	Liver, spleen, and kidney damage;	Long-term liver and spleen damage, potential carcinogenic effects
Fumonisins	Corn, cereals	Pulmonary edema, liver and kidney toxicity	Esophageal cancer, neural tube defects, hepatotoxicity

Mycotoxin	Food crops infected	Acute Effects	Chronic Effects
Zearalenone	Corn, barley, and wheat	Reproductive disorders, hormonal imbalance	Endocrine disruption, infertility, estrogenic effects
Deoxynivalenol (DON)	Barley, wheat, oats	Gastrointestinal issues, diarrhea, Edema, pulmonary issues	Impaired immune response, reproductive toxicity, developmental abnormalities
Sterigmatocystin	Cereal grains, coffee, spices, nuts and beer.	Hepatotoxicity, liver inflammation	Liver cancer, pulmonary toxicity
Citrinin	Barley, rice, wheat, and rye	Nephrotoxicity, kidney pain	Chronic kidney disease, renal dysfunction
Ergot Alkaloids	Cereals, rye	Vasoconstriction, gangrene, Nausea, hallucinations	Chronic vasoconstriction, neurotoxicity, Cardiovascular issues, reproductive toxicity

Table 4. Carcinogenic, teratogenic, and mutagenic effects of mycotoxins

Mycotoxin	Carcinogenic Effects	Teratogenic Effects	Mutagenic Effects
Aflatoxin B1	Potent liver carcinogen, linked to hepatocellular carcinoma	Potential teratogenic effects, especially during pregnancy	Strong mutagenic effects, causes DNA adduct formation
Ochratoxin A	Associated with kidney and liver tumors	Teratogenic in rodents, leading to fetal malformations	Induces chromosomal aberrations, considered mutagenic
Patulin	Weak carcinogenic potential, unclear in humans	Limited evidence, but teratogenic in animal studies	Causes DNA strand breaks, mutagenic in various test systems
Fumonisin B1	Linked to esophageal cancer and liver toxicity	Causes neural tube defects in embryos, teratogenic in animals	Interferes with sphingolipid metabolism, indirect mutagenicity
Zearalenone	Potential carcinogen, linked to reproductive organ tumors	Disrupts fetal development, teratogenic in animal studies	Weak mutagenic activity, concerns mainly hormone disruption
T-2 Toxin	Immunosuppressive, potential carcinogen, linked to leukemia	Teratogenic effects, causing developmental delays in embryos	Causes DNA fragmentation, highly mutagenic
Deoxynivalenol (DON)	Not a potent carcinogen, but may contribute to tumor promotion	Limited teratogenic effects, but potential reproductive toxicity	Weak mutagenic effects, primarily causes DNA damage
Nivalenol	Not classified as carcinogenic but may have tumor-promoting properties	Potential teratogenic effects, limited studies available	Mutagenic in vitro, affects cell division
Citrinin	Linked to kidney cancer in animal studies	Teratogenic in some animal models, causes embryotoxicity	Causes DNA damage, considered mutagenic
Sterigmatocystin	Precursor to aflatoxin B1, strong liver carcinogen	Potential teratogenic effects, especially in developing embryos	Causes mutations, especially in the liver

6. DETECTION AND ANALYTICAL METHODS

Most mycotoxins are chemically stable and can withstand storage, processing, and high temperatures like those used to bake bread or cereal. This makes it crucial to prevent mycotoxin-forming environments, which is not always achievable or feasible. Mycotoxins are challenging to eliminate; thus, prevention is better (Bullerman et al., 1984). Approximately 90% of the inaccuracies in mycotoxin tests may be ascribed to the methodology used in the initial sample collection. Due to the uneven distribution of mycotoxins in grains and mixed feeds, collecting a representative sample for mycotoxin analysis is considered challenging (Turner, et al., 2009; Lauren et al., 2006). The toxicity of most mycotoxins at minimal concentrations necessitates the use of sensitive and accurate detection technologies. Owing to the diverse structures of these compounds, it is impossible to develop a single standard methodology for the detection of all mycotoxins, since each necessitates a distinct procedure.

6.1. Traditional quantitative methods

Thin-layer chromatography (TLC), high-performance liquid chromatography (HPLC) in conjunction with ultraviolet (UV), fluorescence (FD), diode array (DAD) or mass spectrometry (MS) detectors, gas chromatography (GC) in conjunction with flame ionization (FID), and electron capture (ECD) or mass spectrometry (MS) detectors are the most widely used chromatographic techniques for determining mycotoxin levels in cereals. Commercial immunometric tests, including membrane-based immunoassays or ELISAs, are often employed for selection purposes in addition to the previously described techniques.

6.1.1. Thin layer chromatography (TLC)

TLC has traditionally been the most widely used technique for mycotoxin analysis because it provides an affordable means of screening a large number of samples. Both quantitative and semi-quantitative applications of TLC analysis for mycotoxins remain common. This is because of its high sample throughput, cheap operating costs, and simplicity in identifying target molecules using UV spectral analysis. One of the main prerequisites for TLC, however, is the basic requirement for sample preparation; the kind of cleanup procedure used depends on the characteristics and kind of toxin under investigation.

6.1.2. High-performance liquid chromatography (HPLC)

HPLC is regarded as the most widely used chromatographic technique with a variety of detection techniques. HPLC is a key component of modern mycotoxin analysis, using a variety of adsorbents based on the mycotoxin's chemical and physical makeup. For HPLC, ultraviolet (UV) and fluorescence (FLD) detectors are often used. Analytes are identified using UV detectors, which measure how much light in different wavelengths is absorbed by the sample. Toxins were separated and purified based on their polarity using both normal and reversed-phase columns. Mycotoxin standards were prepared using large-scale preparative columns, whereas samples were pre-treated using small micro columns (Reinhard and Zimmerli, 1999). The majority of the procedures utilised for mycotoxin HPLC detection are essentially very similar.

6.1.3. Gas chromatography–Mass spectrometry (GC-MS)

The primary basis for GC is the differential partitioning of analytes between the two stages of sample analysis. Between the stationary phase and the mobile phase, the various chemical components in the sample will disperse. Using a mass spectrometer and an electron capture detector (ECD) or FID, the volatile compounds are detected after the separation process is complete. A GC-MS approach was developed by Rodríguez-Carrasco et al., 2014 to analyse mycotoxins (ZEA, PAT, and trichothecenes) in human urine and grain products. This method is very sensitive and specific for any mycotoxins that can be converted into a volatile enough product for gas chromatography. According to (Pettersson and Langseth, 2002), the primary issues with GC analysis for mycotoxin detections are column blockage, drifting responses, weak fluorescent and absorption groups, nonlinearity of calibration curves, reminiscence properties from previous samples, and the possibility of contamination in comparison to HPLC and LC methods. The presence of mycotoxins in food samples is often detected and measured by GC, and several procedures have been created for these substances. To identify the volatile compounds, the device is often connected to detection methods such as Fourier transform infrared spectroscopy (FTIR), flame ionisation detectors (FID), or mass spectrometry (MS) (Young et al., 1994; SUZUKI et al., 1981; Onji et al., 1998). Since the majority of mycotoxins are not volatile, they must be derivatised before being analysed by GC (Scott et al., 1995).

6.1.4. ELISA

Immunological tests, such ELISA, have been more common for screening for mycotoxins since the late 1970s. The ELISA test does not need cleaning steps and may analyse mycotoxin extract immediately. Even though the number of matrices evaluated is restricted and these immunoassays sometimes lack accuracy at low doses, they still provide quick and affordable results. However, conjugate and antibody binding may be hindered by structurally similar mycotoxins interference, making us susceptible to inaccuracies in quantitative mycotoxin readings. Using an immunoaffinity column sample prepared with the same antibody used by (Klarić, Cvetnić, Pepeljnjak, and Kosalec 2009; Tang et al., 2014) validated an indirect ELISA detection approach and discovered that it was very sensitive at 0.02 µg/L. The majority of researches focused on altering the standard ELISA methodology in order to restore high sensitivity. Because of its ease of use and very inexpensive cost, ELISA has lately gained a lot of popularity (Morgan et al., 1989; Goryacheva et al., 2007). Commercially available ELISA kits for mycotoxin detection are typically based on a competitive assay format that employs an enzyme-target conjugate or a primary antibody specific for the target molecule (Morgan et al., 1989; Stanker et al., 2008; PESTKA, J. and ABOUZIED M, 1995). After that, the complex will interact with a chromogenic substrate to produce a quantifiable outcome. They are easy to use, portable, quick, and very specific. These kits' drawback is that they are single-use only, which might raise the price of mass screening. ELISA based detection of mycotoxins like AFB1 (Gaur et al., 1981), DON, Zearalenone (Cavaliere et al., 2005), OTA (Matrella et al., 2006) has been conducted in different contaminated food grains and animal meet.

6.2. Advances in rapid detection methods

To address the legislation and ensure food safety, the development of analytical methods with high sensitivity and accuracy is of great demand. Current analytical methods include confirmatory methods and screening methods. TLC was the predominant method in early days. Although it is still used by some laboratories, it has almost been replaced by HPLC and GC. The instrumental methods are usually used as the gold standard. Nevertheless, despite their accurate and precise determination, sophisticated instrumental methods have some limitations related to high cost, long detection time, and the requirement of skilled operator (Castegnaro et al., 2006; Ringot et al., 2006). In response to these limitations, a couple of rapid methods with high sensitivity and specificity have been developed for the identification and quantification of mycotoxins (Baggian et al., 2015; Pichon et al., 2016; Huang et al., 2015; Wang et al., 2017; Xu et al., 2016). Furthermore, researchers are still working on developing novel methods with improved sensitivity, specificity, robustness, time-saving, and cost-efficiency.

6.2.1 Biosensors in Mycotoxins Detection

By offering important advantages such quick, simple, and affordable sample analysis, repeatability, stability, accuracy, and on-site sample testing, biosensors may help the food sector lower the amounts of mycotoxins present (Pirinçci et al., 2018; Slaughter, 2006). Recent developments in the use of biosensors for the identification of mycotoxins in food were reported by (Oliveira et al., 2019). The most common transducers for mycotoxin detection are electrochemical (impedimetric, potentiometric, and amperometric), optical (surface plasmon resonance, or SPR and fluorescence), and piezoelectric (quartz crystal microbalance, or QCM) (Tothill et al., 2011; Schulz et al., 2019). Peptides, enzymes, antibodies, cells, and nucleic acids are common recognition elements; however, aptamers, molecularly imprinted polymers, and recombinant antibodies are all possible materials (Oliveira et al., 2019). To increase the biosensor's sensitivity, metal nanoparticles, carbon nanotubes, and nanofibers have all been explored. These materials have unique physicochemical properties, such a high surface-to-volume ratio, and are biocompatible (Malekzad et al., 2017; Doria et al., 2012).

6.2.2 Lateral Flow Assays

Lateral flow assays are popular in food safety due to their cheap prices, user-friendliness, speed (<15 min), portability, high shelf life, and ease of use by non-specialists. All the chemicals needed for a measurable test should fit on a membrane, strip, or capillary. It should be feasible to deposit a volume of sample on the carrier or dip it into the liquid sample (e.g., sample extract) and detect the target analyte by colour depth or band length. The main components are a sample pad (for liquid addition), conjugate pad (with adsorbed, labelled target antibody probe, i.e., the primary antibody), nitrocellulose membrane with fixed test line (T-line, with adsorbed analyte protein conjugate) and control line (C-line, with adsorbed secondary antibody specific for the primary antibody; this line confirms that the test worked correctly), and absorption pad. After adding sample extract to the strip, mycotoxin will attach to the labelled antibody probe and move down the membrane via capillary forces with unbound antibody. Mycotoxin-protein conjugate on the T-line competes with sample extract free toxin for labelled antibody binding. A commercial strip reader can quantify the target based on T-line strength or T/C signal ratio. Fluorescent nanoparticles like quantum dots (Shandong Lvdu Bio-Sciences & Technology Co., LTD.,

Binzhou, China) and Lanthanide chelates (Beijing WDWK Biotechnology Co., LTD., Beijing, China) are employed to enhance test sensitivity.

7. FACTORS RESPONSIBLE FOR MYCOTOXIN CONTAMINATION

Mycotoxigenic fungi represent a prevalent category of pathogens encountered across all agricultural zones globally. These fungi possess the capability to infiltrate and proliferate on an extensive variety of crops, exhibiting considerable diversity that facilitates the synthesis of mycotoxins under a myriad of conditions, including environmental factors which are categories into following factors:

a) ***Host Factors***: In the host plant-pathogen interaction, host plays the vital role in the successful establishment of susceptible or resistance relationship in the development of disease. The types of host plant, variety, crop growth stages of crops are the main determinant of the fungal infection and corresponding mycotoxin contamination. In addition to that the physiology of plants like osmotic pressure, type of carbohydrate, nutritional status and pH of the plant modulate the growth, sporulation of fungi and also regulate biochemical pathways of mycotoxin production. In the presence of simple sugar, particularly those that are easily accessible for catabolism, the proliferation of fungi will be more prevalent in contrast to complex polysaccharides; fungal proliferation will occur at a diminished rate due to the necessity for extensive enzymatic digestion to convert them into readily assimilable carbon units. An elevation in the concentration of simple sugar such as glucose, maltose, and sucrose, has been shown to have a direct correlation with both the proliferation and mycotoxin synthesis, particularly aflatoxin B1, in the fungal species *A. flavus*, as documented by Uppala et al., 2013; Liu et al., 2016.
b) ***Fungal Factors:*** The characteristics and quality of mycotoxins are significantly affected by the particular fungal species responsible for their production. These characteristics fluctuate based on their genetic makeup and the environmental adaptability. Among the same species recognized for their toxigenic potential, not every strain demonstrates this property. The distribution of toxin-producing strains is contingent upon the fungal species, the geographic region, and the characteristics of the substrate. For instance, certain species of Aspergillus, such as *A. flavus*, *A. ochraceous*, *A. parasiticus*, and *A. nominus*, are known to produce significant mycotoxins, whereas some species do not synthesize mycotoxins; among those that do, some generate Aflatoxins (specifically *A. flavus* and *A. parasiticus*), while others produce different toxins, such as Ochratoxin (exemplified by *A. ochraceous*). Moreover, there exists variability in the types of aflatoxins produced by various species of *Aspergillus*. Generally, *A. flavus* predominantly synthesizes AFB1 and AFB2, while *A. parasiticus* is capable of producing AFB1, AFB2, AFG1, and AFG2. In addition to that the S strain (diameter of <400 μm sclerotia) of *A. flavus* can produce all the four types of aflatoxin, while the strain L (diameter of >400 μm sclerotia) able to produce only two types of aflatoxins i.e., AFB1 and AFB2.
c) *Climatic Factors*: Indeed, the conventional disease triangle, which encompasses the pathogen, host plant, and environment, is undergoing transformation due to the shifting significance and pressure of the environmental component. Climate factors like relative humidity, temperature, availability of moisture or water activity play key roles in determining not only fungal occurrence but also make host plants more succumb to fungal infection. In general, relative humidity > 70%, water activity >9.0 and temperature > 15 °C and < 40 °C promote fungal growth as well as mycotoxin production.

Drought condition makes crops more prone to *Aspergillus* infection and lead to production of more aflatoxin (Daou et al., 2021). Each species of fungi possesses a specific optimal range of temperature and water activity conducive to its growth, germination, and mycotoxin synthesis (Daou et al., 2021; Awuchi et al., 2021; Erazo et al., 2023) (Table 5). Consequently, it is impractical to establish a universally applicable range of temperature and water activity that can be deemed as stimulating fungal activity.

d) ***Other biological Factors:*** Insects such as weevils and mites, along with birds and rodent species, play an indirect role in the biosynthesis of mycotoxins by acting as vectors for fungal spores, thereby facilitating their entry into seeds via the lesions they inflict. Consequently, the presence of *A. flavus* and aflatoxins in peanuts, maize, and cottonseeds prior to harvest is frequently associated with infestations of these insect species (Mohapatra et al., 2022).

e) ***Anthropogenic Factors:*** Selection of crop, crop management, pre-harvest and post-harvest treatment of fungal infection in crop are set by human beings. Neglect in the management of pre-harvest crop infection by mycotoxigenic fungi may lead to progress of disease as well as accumulation of more mycotoxins. Post-harvest activities like transport and storage of the harvested crops may get contaminated by fungal pathogen if not carried out properly. Improper intercultural operations carried by human being may also lead to spread of fungal infection to a greater fold.

Table 5. Optimum climatic conditions responsible for mycotoxin biosynthesis by different fungi

Mycotoxins	Produced by Fungus	Optimum conditions for mycotoxin Production		
		Relative Humidity	Temperature Range	Water Activity (a_w)
Aflatoxin	*A. Flavus* and *A. parasiticus*	80 to 90%	28 °C - 35 °C	0.95 - 0.99
Ochratoxin	*A. ochraceous*	80 to 90%	25 °C - 30 °C	0.98
	P. verrucosum	78 to 85%	15 °C – 20 °C	0.86
Fumonisin	*F. moniliforme* and other species of *Fusarium spp*	> 75%	20 °C - 25 °C	0.99
	Aspergillus niger	> 75%	25 °C - 30 °C	0.985 - 0.97
Deoxynivalenol (DON)	*F. graminearum, F. culmorum, F. cerealis*	> 80%	20 °C - 30 °C	0.93 - 0.99
Zearalenone	*Fusarium* spp.	> 80%	15 °C - 30 °C	0.95 - 0.99
T-2	*Fusarium* spp.	> 80%	15 °C - 30 °C	0.95 - 0.99

(Date compiled based on report given by **Daou et al., 2021,Awuchi et al., 2021, Erazo et al., 2023**)

8. MANAGEMENT OF MYCOTOXIN CONTAMINATION

8.1. Traditional Approaches

The likelihood of mycotoxin contamination in agricultural produce is significantly influenced by the quality of inputs and raw materials utilized, the production methodologies employed, as well as the agricultural practices adhered to during the crop cultivation phase. According to a well-known adage, a small measure of preventive action is more valuable than a substantial remedy. Therefore, the implementation of a comprehensive management strategy encompassing pre-harvest, post-harvest, and processing stages is imperative for the interception, control, and elimination of deleterious mycotoxins from the food supply chain. Numerous methodologies have been developed to either mitigate mycotoxin prevalence, inhibit fungal proliferation, or effectively eradicate and degrade mycotoxins present in food products (Xing et al., 2017).

a) Preharvest management:

Pre-harvest management involves strategies implemented in the field to reduce contamination in crops destined for human and animal consumption. Typically, these approaches focus on preventive measures designed to prevent contamination and to modify factors that promote the production of mycotoxins.

Good agricultural practices (GAP): It refers to the best approaches, principles, and standards applied to on-site farm production to assure the safety of food and non-food agricultural products. It is a universally recommended practice to avoid and reduce mycotoxin production in different crops. GAP has been recommended by the FAO with uniform guidance provided for all countries to consider (FAO 2003), with various agricultural development initiatives tailoring local, stakeholder-informed solutions for smallholder production systems. Fungal infection starts at the pre-harvest stage in crops and hence GAP needs to be implemented from the production stage to minimize mycotoxin contamination (Mahuku et al., 2019). Sowing healthy seeds of the recommended resistant or tolerant varieties adapted to the given agroecological conditions, following optimal planting time, and maintaining appropriate plant density can evade preferred congenial conditions for mycotoxigenic fungi (Abbas et al., 2009). Seed treatment with bacterial biological control agents such as *Bacillus amyloliquefaciens, Enterobacter hormaechei* have been found effective in reducing mycotoxin contamination in crops (Pereira et al., 2010; Sartori et al., 2013).

b) Post-harvest strategies

Good agricultural practices can reduce mycotoxin contamination in crops, but not eliminate it. Whenever, external environmental conditions during storage, transportation, and processing of harvested crops fulfils the growth conditions for fungi, fungi will continue to flourish and produce mycotoxins. Standard post-harvest practices are, therefore, also an important aspect to avoid further mycotoxin contamination in different crops. Contamination of major mycotoxins in crops like AF and OTA can occur before and after harvest, whereas, Fusarium toxins such as FB, ZEN, and DON though mainly a pre-harvest problem, but congenial external environment during post-harvest stages leads to mycotoxin production.

Mycotoxin contamination during storage is mainly related to fungal load, moisture content, temperature, relative humidity, insect activities and packaging methods (Okori et al., 2022). For example, contamination with AFs can increase 10-fold in three days when harvested crops are stored under high moisture content (Hell et al., 2010). Maintaining optimum moisture during storage i.e., 12-14% at 30 °C (corresponding to water activity below 0.70), temperature (<20 °C) and relative humidity (<80%) and maintaining proper ventilation seizes the growth of fungi and thereby the production of mycotoxins (Channaiah and Maier 2014).

The main aim in the stage of processing is to detoxify or inactivate or remove mycotoxins and simultaneously maintaining nutritional value and deliciousness of foodstuff. There are several ways to reduce mycotoxin contamination from raw materials, which can be broadly classified into physical, chemical, and biological methods.

Physical Methods: Various physical methods can effectively mitigate mycotoxin contamination. These include sorting based on colour, density, or appearance, mechanical separation, removing fine particles, density segregation, applying thermal treatments, and using gamma, ultraviolet (UV), and visible light radiation.

Chemical methods: Chemical based degradation of mycotoxins encompasses a variety of methodologies, including ammoniation, the application of acidic or basic agents, the utilization of oxidizing agents (such as ozone), or reducing agents (such as sodium bisulfite), as well as enzymatic degradation techniques. The most prominent illustration of a chemical approach for mycotoxin remediation involves the employment of ammonia or ammonium hydroxide. Moreover, calcium or sodium hydroxide, along with other alkaline substances, has demonstrated efficacy in the decontamination of feed products contaminated with aflatoxins, T-2 toxin, diacetoxyscirpenol, or zearalenone. A range of commercial formulations is available for the biotransformation of mycotoxins, including but not limited to Mycofix, FUMzyme, Biomin BBSH 797, and Biomin MTV. Different enzymes like laccases and peroxidases derived from plants or microbes, have also been reported to be effective for aflatoxins decontamination (Loi et al., 2017).

Biological methods: Biological strategies for the management of fungal growth and mycotoxin production, which can be applied during the post-harvest phase, offer environmentally sustainable and chemical-free alternatives. Several strategies exist for mycotoxin decontamination in corn and other grain commodities, encompassing the inhibition of mold proliferation, mycotoxin degradation, and mycotoxin adsorption, all of which involve the utilization of various biological agents (Liu et al., 2021; Zadravec et al., 2022). Certain proteins and peptides possess the ability to impede the proliferation of microbes, therefore earning the designation of antifungal proteins (AFPs) and antimicrobial peptides (AMPs). Lactic acid bacteria and yeasts, which had a notable impact on mold growth and mycotoxin production are also used as biological agents to reduce the fungal infection to crops.

Biological enzymatic degradation includes the use of living cells and bioactive metabolites of bacteria and/or fungi which undergo different processes like acetylation, glycosylation, ring cleavage, hydrolysis, deamination, and decarboxylation and are capable of degrading or altering toxic form of mycotoxin to lesser toxin form. Various microorganisms, such as fungi (including *Rhizopus oryzae*, *Trichoderma reesei*, *Clonostachys rosea*, *Trichosporonmycotoxinivorans*), bacteria (such as *Flavobacteriumaurantiacum*, *Gliocladiumroseum*, *Eubacterium* BBSH 797, *Devosia*, *Eggerthella* spp., *Bacillus*, *Brevibacterium*, and more), and certain yeasts (like *Candidakrusei* AUMC 8161, *Pichiaanomala*, *Pichiaguilliermondii*, *Sac-*

charomycescerevisiae, *Candidautilis*, *Yarrowialipolytica*), are potential candidates for the degradation of various mycotoxins found in food grains or animal feed, including aflatoxins, zearalenone, fumonisin, and deoxynivalenol (Zadravec et al., 2022). Out of such micro-organisms, Lactic acid bacteria (LAB) are preferred because of their metabolites are able to eliminate mycotoxins from foods without damaging a living organism.

Similar to BCA, different botanical substances exhibit diverse mechanisms of action in mitigating fungal infection in crops and preventing contamination of various mycotoxins. These mechanisms encompass safeguarding the integrity of plant cells and macromolecules through the scavenging of free radicals as antioxidants, impeding the growth of mycotoxin-producing fungi by inducing cytotoxicity in these fungal cells, disrupting cellular permeability and function, inhibiting cytoplasmic and mitochondrial enzymes, and modifying osmotic and redox equilibrium (Makhuvele et al., 2020).

9. EMERGING TECHNOLOGIES AND INNOVATIONS FOR THE MANAGEMENT OF MYCOTOXIN CONTAMINATION

The management of mycotoxins encompasses methodologies that are implemented to pre-harvest, harvesting and drying processes, as well as in the postharvest phase. Cultivating resistant or tolerant crop varieties, alongside the adoption of appropriate agronomic practices (such as optimal planting dates, tillage methods, crop rotation, plant density, irrigation, and the application of crop protectants including fungicides and insecticides) constitutes prevalent strategies utilized in the pre-harvest management of fungal infections in crops. During the harvest and transportation phases, grains may sustain physical damage, which exacerbates the likelihood of fungal colonization and subsequent mycotoxin synthesis. The mechanical combines employed for the harvesting of the majority of cereal crops can result in superficial injuries or fragmentation of kernels. The quality of harvested grain can also be enhanced by increasing the fan speed of the combine, thus facilitating the expulsion of low-density, mold-infested kernels from the rear of the machinery. The reduction of grain moisture through artificial drying serves as a pivotal measure to inhibit fungal proliferation and mycotoxin synthesis. Various physical and chemical methodologies are employed to mitigate fungal infections and the ensuing production of mycotoxins. Nevertheless, in the context of the prevailing climate change scenario and the emergence of more virulent pathogens, these traditional management practices are insufficient to avert the contamination of food grains with mycotoxins. To address these challenges, a variety of contemporary technologies and strategies have been developed, which encompass: i) Biotechnological methodologies, ii) Advanced biological approaches iii) Nanoparticle-based approaches

9.1. Biotechnological approaches

Biotechnology serves a pivotal function in the reduction of mycotoxins and provides progressive and inventive strategies for the management of mycotoxins throughout various phases of the food supply chain. Biotechnology offers a complex and diverse methodology for the reduction of mycotoxins, encompassing the development of genetically resistant crop varieties or livestock breeds, genetically modified or transgenic crops, incorporation of gene editing techniques utilizing CRISPR/Cas9, development of vaccines, the synthesis of enzymes capable of detoxifying mycotoxins, and, crucially, the enhancement of biocontrol agents such as fungi, yeast, and bacteria.

CRISPR/Cas9 technology: The utilization of CRISPR/Cas9 technology to facilitate the precise gene deletion of *verA*, which is integral to the biosynthetic pathway of verrucosidin, a neurotoxic compound, in *Penicillium polonicum* has effectively rendered the fungal organism incapable of synthesizing this mycotoxin (Valente et al., 2021). CRISPR/cas system can also be used for the detection of genes responsible for the biosynthesis of mycotoxins in fungi (Wu, 2022).

RNA interference: The utilization of RNA interference approach has facilitated the advancement of transgenic maize (Masanga et al., 2015) and groundnut cultivars (Arias et al., 2015) aimed at the suppression of aflatoxin biosynthetic genes, demonstrating a notable diminution of aflatoxin levels upon exposure to an aflatoxigenic strain of *A. flavus*.

Enzymatic detoxification: Aflatoxin oxidase (AFO) is the first degrading enzyme identified for degrading aflatoxins (AFB1). Various enzymes sourced from microorganisms such as *Rhodococcus erythropolis*, *Trametes versicolor*, *Myxococcus fulvus* ANSM068, *Phoma glomerata* PG41, *Bacillus velezensis* DY3108, *B. subtilis*, and *Armillaria tabescens* have demonstrated a biotransformation efficacy ranging from 70% to 100% for the toxic moieties of mycotoxins, or the detoxification of such mycotoxins, across diverse *in vitro* and *in vivo* bioassays.

9.2. Advanced biological approaches

***Microbes as bioweapon*:** The use of microorganisms (bacteria, yeast and fungi) in the control of mycotoxins in foods and feeds have been widely reported. Various bacterial species or strains, including *Lactobacillus paracasei, Bacillus pumilus, Bacillus subtilis, Lactobacillus kefiri, Acetobacter syzygii, Pseudomonas putida, Rhodococcus erythropolis,* and *Flavobacterium aurantiacum*; alongside certain yeast species such as *Saccharomyces cerevisiae, S. bayanus, Yarrowia lipolytica, Pichia anomala*, as well as fungi like *Streptomyces lividans*, *Fusarium aurantiacum*, and *Trichoderma* spp, are recognized as prevalent biocontrol agents that have demonstrated efficacy in mitigating mycotoxin contamination through mechanisms such as the suppression of fungal proliferation and sporulation, utilization of mycotoxins by these microorganisms as carbon sources, and the adsorption or detoxification of mycotoxins present in food and animal feed (Taheur et al., 2019; Khan et al., 2024).

9.3. Nanoparticle-based approaches

Various magnetic materials, nanoparticles, and the conjugation of nanoparticles with enzymatic extracts from botanical sources or antimicrobial agents have demonstrated significant efficacy in addressing mycotoxin-related challenges in alimentary and feed products. These nanoparticles impede the germination of fungal spores, mycelial growth, sporulation and the biosynthesis of mycotoxins via the engagement of molecular pathways such as oxidative stress responses, interactions with nucleic acids and inflammatory mechanisms, which work to reduce microbial entities in both plant and animal systems. Interactions with cellular membranes may instigate apoptotic processes, produce reactive oxygen species, suppress mitochondrial functions, and promote either lipid peroxidation or autophagic pathways. Chitosan-encapsulated magnetic particles (Fe3O4), nano-cellulose, retinoic acid, surface-active maghemite nanoparticles (SAMNs), silver nanoparticles (AgNPs), copper nanoparticles (CuNPs), selenium nanoparticles (SeNPs), and zinc oxide nanoparticles (ZnO NPs) and nano-fungicides comprising phytochemicals such as tannins, phloretin, catechols, eugenol, and thymol have demonstrated efficacy in inhibiting the proliferation of *Aspergillus*, *Penicillium*, and *Fusarium* species, as well as in diminishing the synthesis of various mycotoxins including Aflatoxin, Patulin, Deoxynivalenol, Fumonisin, by as much as 80% (Khan et al., 2024).

10. REGULATORY FRAMEWORK FOR MITIGATING MYCOTOXIN ISSUES

Regulatory organizations play a crucial role in managing mycotoxin levels in corn and other agricultural products. The Codex Alimentarius Commission, established by the United Nations in 1963, formulates global food safety and hygiene principles. The World Health Organization (WHO) and the Food and Agriculture Organization (FAO) provide international guidance on mycotoxin safety standards, impacting corn trade (FAO, 2002). The U.S. Department of Agriculture (USDA) and European Food Safety Authority (EFSA) monitor mycotoxin levels, provide data to farmers, and conduct research on management practices. The Food and Drug Administration (FDA) enforces compliance with mycotoxin limits. In India, the Agricultural and Processed Food Products Export Development Authority (APEDA) and Food Safety and Standards Authority of India (FSSAI) implement measures to oversee safe levels of mycotoxins in food products (Table-6). In response to the constantly evolving trade standards, regulatory agencies in India, in collaboration with analytical scientists at the National Reference Laboratory of the Indian Council of Agricultural Research's (ICAR) National Research Centre for Grapes (NRL-NRCG) in Pune, have introduced cutting-edge testing techniques for quick detection of mycotoxins (Chatterjee et al., 2023).

Table 6. Maximum permissible limits of different mycotoxins in food products prescribed by FSSAI, India (Chatterjee et al., 2023)

Mycotoxin	**Food Product**	**FSSAI Regulatory Limit (μg/kg)**
AFs (AFB1, AFB2, AFG1, AFG2)	Cereal and Cereal Products	15
	Pulses	15
	Nuts and Nuts for further processing	15
	Ready to eat	10
	Dried figs	10
	Oilseeds or oil; Oilseeds for further processing	15
	Spices	30
	Betelnut/Arecanut	15
	Cereal and Cereal Products	15
AF M1	Milk	0.5
DON	Wheat	1000
OTA	Wheat, barley and rye	20
PAT	Apple juice and Apple juice ingredients in other beverages	50

11. CHALLENGES AND FUTURE PROSPECTS

Mycotoxins, toxic compounds produced by fungi which include AFs, OTA, FMNs, DON, T-2 toxins, ZEN, PAT, etc., present significant challenges to food safety and public health. Aflatoxins, particularly AFB1, present hepatotoxic, immunotoxic, and carcinogenic hazards, resulting in both acute and chronic health repercussions. Ochratoxin A primarily affects the renal system and provokes nephrotoxic, mutagenic, carcinogenic, and teratogenic ramifications, whereas its genotoxic properties continue to be a contentious topic. Fusarium mycotoxins are associated with hepatocellular carcinoma and developmental disorders in humans. T-2 and HT-2 toxins inhibit protein biosynthesis, yielding immunosuppressive effects, in addition to gastrointestinal and dermatological disorders. Zearalenone, classified as a non-carcinogenic estrogenic mycotoxin, impacts reproductive systems and various other organ systems, while Patulin induces oxidative stress alongside acute toxic effects. Consequently, it is essential to routinely monitor foodstuffs and animal feed, as well as to enact preventive strategies to effectively diminish the health threats posed by mycotoxins. There persists a necessity for ongoing inquiry into the toxicity, risk evaluation, and the development of more refined and precise detection methodologies for emergent mycotoxins and their degradation metabolites. In this regard, the advancement of enhanced recognition elements, analytical techniques, and diagnostic tools can serve a crucial function in safeguarding food safety in the face of the continually evolving challenges presented by mycotoxins. The reduction of fungal contamination in food products requires the application of a comprehensive strategy that encompasses physical, chemical, and microbiological detoxification methodologies. Furthermore, the advent of novel fungal strains has stimulated scientific exploration into innovative techniques, including biotechnological interventions, polyphenolic inhibitors, magnetic materials, and nanoparticles, among others. The efficacy of these methodologies may be affected by a range of internal and external variables, such as substrate type, contaminants, moisture content, chemical constituents and processing conditions. Further research is needed to understand the mechanisms and efficacy of advanced mycotoxin management strategies, especially concerning variations in mycotoxins, food matrices, and processing methods. While magnetic materials and nanoparticles show promise for mycotoxin adsorption, additional studies are necessary to ensure their safe use in food, agriculture, and livestock industries. Ongoing research is crucial for mitigating mycotoxin risks, ensuring food safety, and protecting consumer health.

REFERENCES

Abbas, H. K., Wilkinson, J. R., Zablotowicz, R. M., Accinelli, C. E. S. A. R. E., Abel, C. A., Bruns, H. A., & Weaver, M. A. (2009). Ecology of Aspergillus flavus, regulation of aflatoxin production, and management strategies to reduce aflatoxin contamination of corn. *Toxin Reviews*, *28*(2-3), 142–153. DOI: 10.1080/15569540903081590

Acuña-Gutiérrez, C., Jiménez, V. M., & Müller, J. (2022). Occurrence of mycotoxins in pulses. *Comprehensive Reviews in Food Science and Food Safety*, *21*(5), 4002–4017. DOI: 10.1111/1541-4337.13008 PMID: 35876644

Agriopoulou, S., Stamatelopoulou, E., & Varzakas, T. (2020). Advances in occurrence, importance, and mycotoxin control strategies: Prevention and detoxification in foods. *Foods*, *9*(2), 137. DOI: 10.3390/foods9020137 PMID: 32012820

Arias, R. S., Dang, P. M., & Sobolev, V. S. (2015). RNAi-mediated control of aflatoxins in peanut: Method to analyze mycotoxin production and transgene expression in the peanut/Aspergillus pathosystem. *Journal of Visualized Experiments*, •••, 106. PMID: 26709851

Awuchi, C. G., Ondari, E. N., Eseoghene, I. J., Twinomuhwezi, H., Amagwula, I. O., & Morya, S. (2021). Fungal growth and mycotoxins production: Types, toxicities, control strategies, and detoxification. *Fungal reproduction and growth, 100207.*

Baggiani, C., Giovannoli, C., & Anfossi, L. (2015). Man-made synthetic receptors for capture and analysis of ochratoxin A. *Toxins*, *7*(10), 4083–4098. DOI: 10.3390/toxins7104083 PMID: 26473924

Begum, F., & Samajpati, N. (2000). Mycotoxin production on rice, pulses and oilseeds. *Naturwissenschaften*, *87*(6), 275–277. DOI: 10.1007/s001140050720 PMID: 10929292

Benkerroum, N. (2020). Aflatoxins: Producing-molds, structure, health issues and incidence in Southeast Asian and Sub-Saharan African countries. *International Journal of Environmental Research and Public Health*, *17*(4), 1215. DOI: 10.3390/ijerph17041215 PMID: 32070028

Bennett, J. W., & Klich, M. (2003). Mycotoxins. *Clinical Microbiology Reviews*, *16*(3), 497–516. DOI: 10.1128/CMR.16.3.497-516.2003 PMID: 12857779

Bullerman, L. B., Schroeder, L. L., & Park, K. Y. (1984). Formation and control of mycotoxins in food. *Journal of Food Protection*, *47*(8), 637–646. DOI: 10.4315/0362-028X-47.8.637 PMID: 30934406

CAST. 2003, Mycotoxins: Risks in Plant, Animal, Human Systems, Task Force Report No. 38, Ames.

Castegnaro, M., Tozlovanu, M., Wild, C., Molinié, A., Sylla, A., & Pfohl-Leszkowicz, A. (2006). Advantages and drawbacks of immunoaffinity columns in analysis of mycotoxins in food. *Molecular Nutrition & Food Research*, *50*(6), 480–487. DOI: 10.1002/mnfr.200500264 PMID: 16671060

Cavaliere, C., D'Ascenzo, G., Foglia, P., Pastorini, E., Samperi, R., & Laganà, A. (2005). Determination of type B trichothecenes and macrocyclic lactone mycotoxins in field contaminated maize. *Food Chemistry*, *92*(3), 559–568. DOI: 10.1016/j.foodchem.2004.10.008

Channaiah, L. H., & Maier, D. E. (2014). Best stored maize management practices for the prevention of mycotoxin contamination. *Mycotoxin reduction in grain chains*, 78-88.

Chatterjee, S., Dhole, A., Krishnan, A. A., & Banerjee, K. (2023). Mycotoxin monitoring, regulation and analysis in India: A success story. *Foods*, *12*(4), 705. DOI: 10.3390/foods12040705 PMID: 36832780

Daou, R., Joubrane, K., Maroun, R. G., Khabbaz, L. R., Ismail, A., & El Khoury, A. (2021). Mycotoxins: Factors influencing production and control strategies. *AIMS Agriculture and Food*, *6*(1), 416–447. DOI: 10.3934/agrfood.2021025

Demirhan, B., & Demirhan, B. E. (2023). Analysis of multi-mycotoxins in commonly consumed spices using the LC-MS/MS method for assessing food safety risks. *Microorganisms*, *11*(7), 1786. DOI: 10.3390/microorganisms11071786 PMID: 37512958

Doria, G., Conde, J., Veigas, B., Giestas, L., Almeida, C., Assunção, M., Rosa, J., & Baptista, P. V. (2012). Noble metal nanoparticles for biosensing applications. *Sensors (Basel)*, *12*(2), 1657–1687. DOI: 10.3390/s120201657 PMID: 22438731

FAO. (2003). Development of a framework for good agricultural practices. Committee on Agriculture. https://www.fao.org/3/Y8704e/Y8704e.htm

Flajs, D., & Peraica, M. (2009). Toxicological properties of citrinin. *Arhiv za Higijenu Rada i Toksikologiju*, *60*(4), 457–464. DOI: 10.2478/10004-1254-60-2009-1992 PMID: 20061247

Gaur, P. K., Lau, H. P., Pestka, J. J., & Chu, F. S. (1981). Production and characterization of aflatoxin B2a antiserum. *Applied and Environmental Microbiology*, *41*(2), 478–482. DOI: 10.1128/aem.41.2.478-482.1981 PMID: 7235694

Goryacheva, I. Y., Saeger, S. D., Eremin, S. A., & Peteghem, C. V. (2007). Immunochemical methods for rapid mycotoxin detection: Evolution from single to multiple analyte screening: A review. *Food Additives and Contaminants*, *24*(10), 1169–1183. DOI: 10.1080/02652030701557179 PMID: 17886190

Gruber-Dorninger, C., Jenkins, T., & Schatzmayr, G. (2019). Global mycotoxin occurrence in feed: A ten-year survey. *Toxins*, *11*(7), 375. DOI: 10.3390/toxins11070375 PMID: 31252650

Harrison, L. R., Colvin, B. M., Greene, J. T., Newman, L. E., & Cole, J. R.Jr. (1990). Pulmonary edema and hydrothorax in swine produced by fumonisin B1, a toxic metabolite of *Fusarium moniliforme.Journal of Veterinary Diagnostic Investigation : Official Publication of the American Association of Veterinary Laboratory Diagnosticians, Inc*, *2*(3), 217–221. DOI: 10.1177/104063879000200312 PMID: 2094448

Hell, K., Mutegi, C., & Fandohan, P. (2010). Aflatoxin control and prevention strategies in maize for Sub-Saharan Africa. *Julius-Kühn-Archiv*, (425), 534.

Huang, Y., Chen, X., Duan, N., Wu, S., Wang, Z., Wei, X., & Wang, Y. (2015). Selection and characterization of DNA aptamers against Staphylococcus aureus enterotoxin C1. *Food Chemistry*, *166*, 623–629. DOI: 10.1016/j.foodchem.2014.06.039 PMID: 25053102

IARC. (2012). Aflatoxins. []. Lyon, France: International Agency for Research on Cancer.]. *IARC Monographs on the Evaluation of Carcinogenic Risks to Humans*, *100F*, 225–248.

Khan, R. (2024). Mycotoxins in food: Occurrence, health implications, and control strategies-A comprehensive review. *Toxicon*, *248*, 108038. DOI: 10.1016/j.toxicon.2024.108038 PMID: 39047955

Khan, R., Anwar, F., & Ghazali, F. M. (2024). A comprehensive review of mycotoxins: Toxicology, detection, and effective mitigation approaches. *Heliyon*, *10*(8), e28361. DOI: 10.1016/j.heliyon.2024.e28361 PMID: 38628751

Lauren, D. R., Jensen, D. J., & Smith, W. A. (2006). Mycotoxin contamination in graded fractions of maize (Zea mays) in New Zealand. *New Zealand Journal of Crop and Horticultural Science*, *34*(1), 63–72. DOI: 10.1080/01140671.2006.9514389

Liu, J., Sun, L., Zhang, N., Zhang, J., Guo, J., Li, C., & Qi, D. (2016). Effects of nutrients in substrates of different grains on aflatoxin B1 production by *Aspergillus flavus.BioMed Research International*, *2016*(1), 7232858. PMID: 27294129

Liu, N., Liu, C., Dudaš, T. N., Loc, M. Č., Bagi, F. F., & Van der Fels-Klerx, H. J. (2021). Improved aflatoxins and fumonisins forecasting models for maize (PREMA and PREFUM), using combined mechanistic and Bayesian Network modeling—Serbia as a case study. *Frontiers in Microbiology*, *12*, 643604. DOI: 10.3389/fmicb.2021.643604 PMID: 33967981

Loi, M., Fanelli, F., Liuzzi, V. C., Logrieco, A. F., & Mulè, G. (2017). Mycotoxin biotransformation by native and commercial enzymes: Present and future perspectives. *Toxins*, *9*(4), 111. DOI: 10.3390/toxins9040111 PMID: 28338601

Lutsky, I. I., & Mor, N. (1981). Alimentary toxic aleukia (septic angina, endemic panmyelotoxicosis, alimentary hemorrhagic aleukia): T-2 toxin-induced intoxication of cats. *American Journal of Pathology*, *104*(2), 189. PMID: 6973281

Mahuku, G., Nzioki, H. S., Mutegi, C., Kanampiu, F., Narrod, C., & Makumbi, D. (2019). Pre-harvest management is a critical practice for minimizing aflatoxin contamination of maize. *Food Control*, *96*, 219–226. DOI: 10.1016/j.foodcont.2018.08.032 PMID: 30713368

Makhuvele, R., Naidu, K., Gbashi, S., Thipe, V. C., Adebo, O. A., & Njobeh, P. B. (2020). The use of plant extracts and their phytochemicals for control of toxigenic fungi and mycotoxins. *Heliyon*, *6*(10), e05291. DOI: 10.1016/j.heliyon.2020.e05291 PMID: 33134582

Malekzad, H., Sahandi Zangabad, P., Mirshekari, H., Karimi, M., & Hamblin, M. R. (2017). Noble metal nanoparticles in biosensors: Recent studies and applications. *Nanotechnology Reviews*, *6*(3), 301–329. DOI: 10.1515/ntrev-2016-0014 PMID: 29335674

Masanga, J. O., Matheka, J. M., Omer, R. A., Ommeh, S. C., Monda, E. O., & Alakonya, A. E. (2015). Downregulation of transcription factor aflR in Aspergillus flavus confers reduction to aflatoxin accumulation in transgenic maize with alteration of host plant architecture. *Plant Cell Reports*, *34*(8), 1379–1387. DOI: 10.1007/s00299-015-1794-9 PMID: 25895735

Matrella, R., Monaci, L., Milillo, M. A., Palmisano, F., & Tantillo, M. G. (2006). Ochratoxin A determination in paired kidneys and muscle samples from swines slaughtered in southern Italy. *Food Control*, *17*(2), 114–117. DOI: 10.1016/j.foodcont.2004.08.008

Mohapatra, D., Rath, S. K., & Mohapatra, P. K. (2022). Soil fungi for bioremediation of pesticide toxicants: A perspective. *Geomicrobiology Journal*, *39*(3-5), 352–372. DOI: 10.1080/01490451.2021.2019855

Morgan, M. R. (1989). Mycotoxin immunoassays (with special reference to ELISAs). *Tetrahedron*, *45*(8), 2237–2249. DOI: 10.1016/S0040-4020(01)83429-3

Murphy, P. A., Hendrich, S., Landgren, C., & Bryant, C. M. (2006). Food mycotoxins: An update. *Journal of Food Science*, *71*(5), R51–R65. DOI: 10.1111/j.1750-3841.2006.00052.x

Okori, F., Cherotich, S., Baidhe, E., Komakech, A. J., & Banadda, N. (2022). Grain hermetic storage and post-harvest loss reduction in sub-saharan Africa: Effects on grain damage, weight loss, germination, insect infestation, and mold and mycotoxin contamination. *Journal of Biosystems Engineering*, *47*(1), 48–68. DOI: 10.1007/s42853-022-00130-4

Oliveira, I. S., da Silva, A. G.Junior, de Andrade, C. A. S., & Oliveira, M. D. L. (2019). Biosensors for early detection of fungi spoilage and toxigenic and mycotoxins in food. *Current Opinion in Food Science*, *29*, 64–79. DOI: 10.1016/j.cofs.2019.08.004

Onji, Y., Aoki, Y., Tani, N., Umebayashi, K., Kitada, Y., & Dohi, Y. (1998). Direct analysis of several Fusarium mycotoxins in cereals by capillary gas chromatography–mass spectrometry. *Journal of Chromatography. A*, *815*(1), 59–65. DOI: 10.1016/S0021-9673(98)00357-4 PMID: 9718707

Peraica, M., Radic, B., Lucic, A., & Pavlovic, M. (1999). Toxic effects of mycotoxins in humans. *Bulletin of the World Health Organization*, *77*(9), 754–766. PMID: 10534900

Pereira, P., Nesci, A., Castillo, C., & Etcheverry, M. (2010). Impact of bacterial biological control agents on fumonisin B1 content and Fusarium verticillioides infection of field-grown maize. *Biological Control*, *53*(3), 258–266. DOI: 10.1016/j.biocontrol.2010.02.001

Perrone, G., Ferrara, M., Medina, A., Pascale, M., & Magan, N. (2020). Toxigenic fungi and mycotoxins in a climate change scenario: Ecology, genomics, distribution, prediction, and prevention of the risk. *Microorganisms*, *8*(10), 1496. DOI: 10.3390/microorganisms8101496 PMID: 33003323

PESTKA, J., & ABOUZIED, M. (1995). Immunological assays for mycotoxin detection: Immunoassay applications to food analysis. *Food technology (Chicago)*, *49*(2), 120–128.

Pestka, J. J. (2010). Deoxynivalenol: Mechanisms of action, human exposure, and toxicological relevance. *Archives of Toxicology*, *84*(9), 663–679. DOI: 10.1007/s00204-010-0579-8 PMID: 20798930

Pestka, J. J., & Smolinski, A. T. (2005). Deoxynivalenol: Toxicology and potential effects on humans. *Journal of Toxicology and Environmental Health. Part B, Critical Reviews*, *8*(1), 39–69. DOI: 10.1080/10937400590889458 PMID: 15762554

Pfohl-Leszkowicz, A., & Manderville, R. A. (2007). Ochratoxin A: An overview on toxicity and carcinogenicity in animals and humans. *Molecular Nutrition & Food Research*, *51*(1), 61–99. DOI: 10.1002/mnfr.200600137 PMID: 17195275

Pichon, V., & Combès, A. (2016). Selective tools for the solid-phase extraction of Ochratoxin A from various complex samples: Immunosorbents, oligosorbents, and molecularly imprinted polymers. *Analytical and Bioanalytical Chemistry*, *408*(25), 6983–6999. DOI: 10.1007/s00216-016-9886-0 PMID: 27585915

Pickova, D., Ostry, V., & Malir, F. (2021). A recent overview of producers and important dietary sources of aflatoxins. *Toxins*, *13*(3), 186. DOI: 10.3390/toxins13030186 PMID: 33802572

Pirinçci, Ş. Ş., Ertekin, Ö., Laguna, D. E., Özen, F. Ş., Öztürk, Z. Z., & Öztürk, S. (2018). Label-free QCM immunosensor for the detection of ochratoxin A. *Sensors (Basel)*, *18*(4), 1161. DOI: 10.3390/s18041161 PMID: 29641432

Quintela, S. (2020). Mycotoxins in beverages: occurrence, regulation, economic impact and cost-effectiveness of preventive and removal methods. In *Safety issues in beverage production* (pp. 147–186). Academic Press. DOI: 10.1016/B978-0-12-816679-6.00005-X

Reinhard, H., & Zimmerli, B. (1999). Reversed-phase liquid chromatographic behavior of the mycotoxins citrinin and ochratoxin A. *Journal of Chromatography. A*, *862*(2), 147–159. DOI: 10.1016/S0021-9673(99)00929-2 PMID: 10596972

Ringot, D., Chango, A., Schneider, Y. J., & Larondelle, Y. (2006). Toxicokinetics and toxicodynamics of ochratoxin A, an update. *Chemico-Biological Interactions*, *159*(1), 18–46. DOI: 10.1016/j.cbi.2005.10.106 PMID: 16293235

Rodríguez-Carrasco, Y., Berrada, H., Font, G., & Mañes, J. (2012). Multi-mycotoxin analysis in wheat semolina using an acetonitrile-based extraction procedure and gas chromatography–tandem mass spectrometry. *Journal of Chromatography. A*, *1270*, 28–40. DOI: 10.1016/j.chroma.2012.10.061 PMID: 23182289

Sartori, M. V., Nesci, A. V., Castillo, C., & Etcheverry, M. G. (2013). Biological control of fumonisins production in maize at field level.

Schulz, K., Pöhlmann, C., Dietrich, R., Märtlbauer, E., & Elßner, T. (2019). Electrochemical biochip assays based on anti-idiotypic antibodies for rapid and automated on-site detection of low molecular weight toxins. *Frontiers in Chemistry*, *7*, 31. DOI: 10.3389/fchem.2019.00031 PMID: 30775361

Scott, P. M. (1995). Mycotoxin methodology. *Food Additives and Contaminants*, *12*(3), 395–403. DOI: 10.1080/02652039509374321 PMID: 7664934

Šegvić Klarić, M., Cvetnić, Z., Pepeljnjak, S., & Kosalec, I. (2009). Co-occurrence of aflatoxins, ochratoxin A, fumonisins, and zearalenone in cereals and feed, determined by competitive direct enzyme-linked immunosorbent assay and thin-layer chromatography. *Arhiv za Higijenu Rada i Toksikologiju*, *60*(4), 427–433. DOI: 10.2478/10004-1254-60-2009-1975 PMID: 20061243

Slaughter, G. (2006). Current advances in biosensor design and fabrication. *Encyclopedia of Analytical Chemistry: Applications, Theory and Instrumentation*, 1-25.

Stanker, L. H., Merrill, P., Scotcher, M. C., & Cheng, L. W. (2008). Development and partial characterization of high-affinity monoclonal antibodies for botulinum toxin type A and their use in analysis of milk by sandwich ELISA. *Journal of Immunological Methods*, *336*(1), 1–8. DOI: 10.1016/j.jim.2008.03.003 PMID: 18452945

Suzuki, T., Kurisu, M., Nose, N., & Watanabe, A. (1981). Determination of butenolide (Fusarium toxin) by gas chromatography with an electron-capture detector. Food Hygiene and Safety Science (Shokuhin Eiseigaku Zasshi), 22(3), 197-202_1.

Taheur, F. B., Kouidhi, B., Al Qurashi, Y. M. A., Salah-Abbès, J. B., & Chaieb, K. (2019). Biotechnology of mycotoxins detoxification using microorganisms and enzymes. *Toxicon*, *160*, 12–22. DOI: 10.1016/j.toxicon.2019.02.001 PMID: 30772433

Tang, X., Li, X., Li, P., Zhang, Q., Li, R., Zhang, W., Ding, X., Lei, J., & Zhang, Z. (2014). Development and application of an immunoaffinity column enzyme immunoassay for mycotoxin zearalenone in complicated samples. *PLoS One*, *9*(1), e85606. DOI: 10.1371/journal.pone.0085606 PMID: 24465616

Tothill, I. J. W. M. J. (2011). Biosensors and nanomaterials and their application for mycotoxin determination. *World Mycotoxin Journal*, *4*(4), 361–374. DOI: 10.3920/WMJ2011.1318

Turner, N. W., Subrahmanyam, S., & Piletsky, S. A. (2009). Analytical methods for determination of mycotoxins: A review. *Analytica Chimica Acta*, *632*(2), 168–180. DOI: 10.1016/j.aca.2008.11.010 PMID: 19110091

Uppala, S. S., Bowen, K. L., & Woods, F. M. (2013). Pre-harvest aflatoxin contamination and soluble sugars of peanut. *Peanut Science*, *40*(1), 40–51. DOI: 10.3146/PS12-9.1

Valente, S., Piombo, E., Schroeckh, V., Meloni, G. R., Heinekamp, T., Brakhage, A. A., & Spadaro, D. (2021). CRISPR-Cas9-based discovery of the verrucosidin biosynthesis gene cluster in Penicillium polonicum. *Frontiers in Microbiology*, *12*, 660871. DOI: 10.3389/fmicb.2021.660871 PMID: 34093475

Van Dongen, P. W., & de Groot, A. N. (1995). History of ergot alkaloids from ergotism to ergometrine. *European Journal of Obstetrics, Gynecology, and Reproductive Biology*, *60*(2), 109–116. DOI: 10.1016/0028-2243(95)02104-Z PMID: 7641960

Wang, F., Li, X., Li, Y., Han, J., Chen, Y., Zeng, J., Su, M., Zhuo, J., Ren, H., Liu, H., Hou, L., Fan, Y., Yan, X., Song, S., Zhao, J., Jin, D., Zhang, M., & Pei, Y. (2021). Arabidopsis P4 ATPase-mediated cell detoxification confers resistance to *Fusarium graminearum* and *Verticillium dahliae.Nature Communications*, *12*(1), 6426. DOI: 10.1038/s41467-021-26727-5 PMID: 34741039

Wang, J., Yu, J., Yang, Q., McDermott, J., Scott, A., Vukovich, M., Lagrois, R., Gong, Q., Greenleaf, W., Eisenstein, M., Ferguson, B. S., & Soh, H. T. (2017). Multiparameter particle display (MPPD): A quantitative screening method for the discovery of highly specific aptamers. *Angewandte Chemie*, *129*(3), 762–765. DOI: 10.1002/ange.201608880 PMID: 27933702

Wild, C. P., & Gong, Y. Y. (2010). Mycotoxins and human disease: A largely ignored global health issue. *Carcinogenesis*, *31*(1), 71–82. DOI: 10.1093/carcin/bgp264 PMID: 19875698

Williams, J. H., Phillips, T. D., Jolly, P. E., Stiles, J. K., Jolly, C. M., & Aggarwal, D. (2004). Human aflatoxicosis in developing countries: A review of toxicology, exposure, potential health consequences, and interventions. *The American Journal of Clinical Nutrition*, *80*(5), 1106–1122. DOI: 10.1093/ajcn/80.5.1106 PMID: 15531656

Wu, F. (2015). Global impacts of aflatoxin in maize: Trade and human health. *World Mycotoxin Journal*, *8*(2), 137–142. DOI: 10.3920/WMJ2014.1737

Wu, F. (2022). Mycotoxin risks are lower in biotech corn. *Current Opinion in Biotechnology*, *78*, 102792. DOI: 10.1016/j.copbio.2022.102792 PMID: 36088737

Xing, F., Liu, X., Wang, L., Selvaraj, J. N., Jin, N., Wang, Y., Zhao, Y., & Liu, Y. (2017). Distribution and variation of fungi and major mycotoxins in pre-and post-nature drying maize in North China Plain. *Food Control*, *80*, 244–251. DOI: 10.1016/j.foodcont.2017.03.055

Xu, L., Zhang, Z., Zhang, Q., & Li, P. (2016). Mycotoxin determination in foods using advanced sensors based on antibodies or aptamers. *Toxins*, *8*(8), 239. DOI: 10.3390/toxins8080239 PMID: 27529281

Young, J. C., & Games, D. E. (1994). Analysis of Fusarium mycotoxins by gas chromatography-Fourier transform infrared spectroscopy. *Journal of Chromatography. A*, *663*(2), 211–218. DOI: 10.1016/0021-9673(94)85247-2 PMID: 8173667

Zadravec, M., Markov, K., Lešić, T., Frece, J., Petrović, D., & Pleadin, J. (2022). Biocontrol methods in avoidance and downsizing of mycotoxin contamination of food crops. *Processes (Basel, Switzerland)*, *10*(4), 655. DOI: 10.3390/pr10040655

Compilation of References

Aamir, M., Rai, K. K., Zehra, A., Kumar, S., Yadav, M., Shukla, V., & Upadhyay, R. S. (2020). Fungal endophytes: Classification, diversity, ecological role, and their relevance in sustainable agriculture. In Microbial Endophytes (pp. 291–323). Elsevier., 0-12-818734-0.00012-7.DOI: 10.1016/B978-0-12-818734-0.00012-7

Abbas, H. K., Wilkinson, J. R., Zablotowicz, R. M., Accinelli, C. E. S. A. R. E., Abel, C. A., Bruns, H. A., & Weaver, M. A. (2009). Ecology of Aspergillus flavus, regulation of aflatoxin production, and management strategies to reduce aflatoxin contamination of corn. *Toxin Reviews*, *28*(2-3), 142–153. DOI: 10.1080/15569540903081590

Abd El-Ghany, T. M., & Masmali, I. A. (2016). Fungal biodegradation of organophosphorus insecticides and their impact on soil microbial population. *Journal of Plant Pathology & Microbiology*, *7*(5), 1000349.

Abdel-Azeem, A. M., & Sheir, D. H. (2020). Bioconversion of Lignocellulosic Residues into Single-Cell Protein (SCP) by *Chaetomium*. In Abdel-Azeem, A. (Ed.), *Recent Developments on Genus Chaetomium. Fungal Biology*. Springer., DOI: 10.1007/978-3-030-31612-9_13

Abdelaziz, A. M., Kalaba, M. H., Hashem, A. H., Sharaf, M. H., & Attia, M. S. (2022). Biostimulation of tomato growth and biocontrol of Fusarium wilt disease using certain endophytic fungi. *Botanical Studies*, *63*(1), 34. DOI: 10.1186/s40529-022-00364-7 PMID: 36484866

Abdel-Fattah Mostafa, A., Yassin, M. T., Dawoud, T. M., Al-Otibi, F. O., & Sayed, S. R. M. (2022). Mycodegradation of diazinon pesticide utilizing fungal strains isolated from polluted soil. *Environmental Research*, *212*, 113421. DOI: 10.1016/j.envres.2022.113421 PMID: 35568233

Abdel-Fattah, G. M., & Asrar, A.-W. A. (2012). Arbuscular mycorrhizal fungal application to improve growth and tolerance of wheat (Triticum aestivum L.) plants grown in saline soil. *Acta Physiologiae Plantarum*, *34*(1), 267–277. DOI: 10.1007/s11738-011-0825-6

Abdel-Wareth, M. T. A. (2023). Sequestration and Detoxification of Heavy Metals by Fungi. In *Sustainable Industrial Wastewater Treatment and Pollution Control* (pp. 185-209). Singapore: Springer Nature Singapore.Abdel-Wareth, M. T. A. (2023). Sequestration and Detoxification of Heavy Metals by Fungi. In *Sustainable Industrial Wastewater Treatment and Pollution Control* (pp. 185-209). Singapore: Springer Nature Singapore. DOI: 10.1007/978-981-99-2560-5_10

Abhilasha, A., Sreenivasulu, A., & Manimozhi, T. KUMAR, P. S., Selvakumar, P., & Singh, P. (2022, April). The Model of Smart Sensing Device For Sensitive Nanoclusters Modification in Sensing Properties. In 2022 2nd International Conference on Advance Computing and Innovative Technologies in Engineering (ICACITE) (pp. 1043-1047). IEEE.

Abhilasha, A., Sreenivasulu, A., & Manimozhi, T. Satheesh kumar P., Selvakumar P. and Singh P.,(2022). "The Model of Smart Sensing Device For Sensitive Nanoclusters Modification in Sensing Properties," 2022 2nd International Conference on Advance Computing and Innovative Technologies in Engineering (ICACITE), Greater Noida, India, 2022, pp. 1043-1047, DOI: 10.1109/ICACITE53722.2022.9823846

Abida, H., Ruchaud, S., Rios, L., Humeau, A., Probert, I., De Vargas, C., Bach, S., & Bowler, C. (2013). Bioprospecting marine plankton. *Marine Drugs*, *11*(11), 4594–4611. DOI: 10.3390/md11114594 PMID: 24240981

Abirami, S., Sree Gayathri, S., & Usha, C. (2022). Trichoderma as biostimulant - a plausible approach to alleviate abiotic stress for intensive production practices. In Singh, H. B., & Vaishnav, A. (Eds.), *New and Future Developments in Microbial Biotechnology and Bioengineering* (pp. 57–84). Elsevier., https://doi .org/https://doi.org/10.1016/B978-0-323-85577-8.00004-4 DOI: 10.1016/B978-0-323-85577-8.00004-4

Abu-Elsaoud, A. M., Nafady, N. A., & Abdel-Azeem, A. M. (2017). Arbuscular mycorrhizal strategy for zinc mycoremediation and diminished translocation to shoots and grains in wheat. *PLoS One*, *12*(11), e0188220. DOI: 10.1371/journal.pone.0188220 PMID: 29145471

Acuña-Gutiérrez, C., Jiménez, V. M., & Müller, J. (2022). Occurrence of mycotoxins in pulses. *Comprehensive Reviews in Food Science and Food Safety*, *21*(5), 4002–4017. DOI: 10.1111/1541-4337.13008 PMID: 35876644

Adenipekun, C. O., & Lawal, R. (2012). Uses of mushrooms in bioremediation: A review. *Biotechnology and Molecular Biology Reviews*, *7*(3), 62–68.

Adeyemi, N. O., Atayese, M. O., Sakariyawo, O. S., Azeez, J. O., Olubode, A. A., Ridwan, M., Adebiyi, A., Oni, O., & Ibrahim, I. (2021). Influence of different arbuscular mycorrhizal fungi isolates in enhancing growth, phosphorus uptake and grain yield of soybean in a phosphorus deficient soil under field conditions. *Communications in Soil Science and Plant Analysis*, *52*(10), 1171–1183. DOI: 10.1080/00103624.2021.1879117

Adnan, L. A., Sathishkumar, P., Yusoff, A. R. M., Hadibarata, T., & Ameen, F. (2017). Rapid bioremediation of Alizarin Red S and Quinizarine Green SS dyes using Trichoderma lixii F21 mediated by biosorption and enzymatic processes. *Bioprocess and Biosystems Engineering*, *40*(1), 85–97. DOI: 10.1007/s00449-016-1677-7 PMID: 27663440

AFNOR. (2000). Recueil de normes: Les huiles essentielles. In La Défense, P. (Ed.), *Monographies Relatives aux Huiles Essentielles; AFNOR* (pp. 661–663).

Afrokh, M., El Mehrach, K., Chatoui, K., Ait Bihi, M., Sadki, H., Zarrouk, A., Tabyaoui, M., & Tahrouch, S. (2024). Quality criteria, chemical composition and antimicrobial activity of the essential oil of *Mentha suaveolens* Ehrh. *Heliyon*, *10*(7), e28125. DOI: 10.1016/j.heliyon.2024.e28125 PMID: 38560258

Aggelopoulos, T., Katsieris, K., Bekatorou, A., Pandey, A., Banat, I. M., & Koutinas, A. A. (2014). Solid state fermentation of food waste mixtures for single cell protein, aroma volatiles and fat production. *Food Chemistry*, *145*, 710–716. DOI: 10.1016/j.foodchem.2013.07.105

Agrawal, P., Pant, K. K., Sonkar, M., & Chandra, V. (2023). Algal Biofuel. *Algae Refinery*, 187-208. Mehariya, S., & Verma, P. (eds.). (2023). Algae Refinery: Up- and Downstream Processes, CRC Press.

Agrawal, N., Verma, P., & Shahi, S. K. (2018). Degradation of polycyclic aromatic hydrocarbons (phenanthrene and pyrene) by the ligninolytic fungi Ganoderma lucidum isolated from the hardwood stump. *Bioresources and Bioprocessing*, *5*(1), 11. DOI: 10.1186/s40643-018-0197-5

Agriopoulou, S. (2021). Ergot alkaloids mycotoxins in cereals and cereal-derived food products: Characteristics, toxicity, prevalence, and control strategies. *Agronomy (Basel)*, *11*(5), 931. DOI: 10.3390/agronomy11050931

Agriopoulou, S., Stamatelopoulou, E., & Varzakas, T. (2020). Advances in Analysis and Detection of Major Mycotoxins in Foods. *Foods*, *9*(4), 518. DOI: 10.3390/foods9040518 PMID: 32326063

Agriopoulou, S., Stamatelopoulou, E., & Varzakas, T. (2020). Advances in occurrence, importance, and mycotoxin control strategies: Prevention and detoxification in foods. *Foods*, *9*(2), 137. DOI: 10.3390/foods9020137 PMID: 32012820

Aguilar-Rivera, N., Moran, A. C., Lagunes, D. A. R., & Gonzalez, J. M. (2012). Production of *Pleurotus ostreatus* (oyster mushroom) grown on sugar cane biomass (trash, bagasse and pith). *no, 1*, 149-152.

Ahlawat, O. P., Manikandan, K., & Singh, M. (2016). Proximate composition of different mushroom varieties and effect of UV light exposure on vitamin D content in Agaricus bisporus and Volvariella volvacea. *Mushroom Research*, *25*(1).

Ahmad, I., Jiménez-Gasco, M. D. M., Luthe, D. S., Shakeel, S. N., & Barbercheck, M. E. (2020). Endophytic Metarhizium robertsii promotes maize growth, suppresses insect growth, and alters plant defense gene expression. *Biological Control*, *144*, 104167. DOI: 10.1016/j.biocontrol.2019.104167

Ahmadi, A. R., Ghoorchian, H., Hajihosaini, R., & Khanifar, J. (2010). Determination of the amount of protein and amino acids extracted from the microbial protein (SCP) of lignocellulosic wastes. *Pakistan journal of biological sciences. Pakistan Journal of Biological Sciences*, *13*(8), 355–361. DOI: 10.3923/pjbs.2010.355.361

Ahmad, K. S. (2020). Remedial potential of bacterial and fungal strains (*Bacillus subtilis, Aspergillus niger, Aspergillus flavus* and *Penicillium chrysogenum*) against organochlorine insecticide Endosulfan. *Folia Microbiologica*, *65*(5), 801–810. DOI: 10.1007/s12223-020-00792-7 PMID: 32383069

Ahmad, M. I., Farooq, S., Alhamoud, Y., Li, C., & Zhang, H. (2022). A review on mycoprotein: History, nutritional composition, production methods, and health benefits. *Trends in Food Science & Technology*, *121*, 14–29. DOI: 10.1016/j.tifs.2022.01.027

Ahmed, M. G., Gouda, S. A., Donia, S., & Hassanein, N. M. (2024). Production of single cell protein by fungi from different food wastes. *Biomass Conversion and Biorefinery*, •••, 1–16. DOI: 10.1007/s13399-024-05478-5

Ahmed, S., Kumari, K., & Singh, D. (2024). Different strategies and bio-removal mechanisms of petroleum hydrocarbons from contaminated sites. *Arab Gulf Journal of Scientific Research*, *42*(2), 342–358. DOI: 10.1108/AGJSR-09-2022-0172

Aida, F. M. N. A., Shuhaimi, M., Yazid, M., & Maaruf, A. G. (2009). Mushroom as a potential source of prebiotics: A review. *Trends in Food Science & Technology*, *20*(11–12), 567–575. DOI: 10.1016/j.tifs.2009.07.007

Aidoo, R., Kwofie, E. M., Adewale, P., Lam, E., & Ngadi, M. (2023). Overview of single cell protein: Production pathway, sustainability outlook, and digital twin potentials. *Trends in Food Science & Technology*, *138*, 577–598. Advance online publication. DOI: 10.1016/j.tifs.2023.07.003

Aiking, H. (2014). Protein production: Planet, profit, plus people? *The American Journal of Clinical Nutrition*, *100*(1), 483S–489S. DOI: 10.3945/ajcn.113.071209 PMID: 24871472

Ajmal, M., Bedale, W., Akram, A., & Yu, J. H. (2022). Comprehensive Review of Aflatoxin Contamination, Impact on Health and Food Security, and Management Strategies in Pakistan. *Toxins*, *14*(12), 845. DOI: 10.3390/toxins14120845 PMID: 36548742

Akensous, F.-Z., El Mazouni, N., Mesnaoui, M., & Meddich, A. (2023a). Biostimulants for upgrading plant and soil health statuses: Testimonials from date palm (*Phoenix dactylifera* L.) and durum wheat (*Triticum durum* Desf.). In Agricultural Research Updates, *46*, 129–179.

Akensous, F.-Z., Sbbar, N., Ech-chatir, L., & Meddich, A. (2023b). Artificial Intelligence, Internet of Things, and Machine-Learning: To Smart Irrigation and Precision Agriculture. In Artificial Intelligence Applications in Water Treatment and Water Resource Management, 113–145.

Akensous, F.-Z., Anli, M., & Meddich, A. (2024). Arbuscular Mycorrhizal Fungi as Solubilizers of Rock Phosphate and Compost Application Improve Date Palm (*Phoenix dactylifera* L.)'s Resilience to Drought. *Gesunde Pflanzen*, *76*, 161–179.

Akhtar, M. S., & Abdullah, S. N. A. (2014). Mass production techniques of arbuscular mycorrhizal fungi: Major advantages and disadvantages: A review. In *Biosciences Biotechnology Research Asia* (Vol. 11, Issue 3, pp. 1199–1204). Oriental Scientific Publishing Company. DOI: 10.13005/bbra/1506

Akhtar, N., & Mannan, M. A. (2020). Mycoremediation: Expunging environmental pollutants. *Biotechnology Reports (Amsterdam, Netherlands)*, *26*, e00452. https://doi.org/https://doi.org/10.1016/j.btre.2020.e00452. DOI: 10.1016/j.btre.2020.e00452 PMID: 32617263

Akiyama, K., Matsuzaki, K., & Hayashi, H. (2005). Plant sesquiterpenes induce hyphal branching in arbuscular mycorrhizal fungi. *Nature*, *435*(7043), 824–827. DOI: 10.1038/nature03608 PMID: 15944706

Akpasi, S. O., Anekwe, I. M., Tetteh, E. K., Amune, U. O., Shoyiga, H. O., Mahlangu, T. P., & Kiambi, S. L. (2023). Mycoremediation as a Potentially Promising Technology: Current Status and Prospects—A Review. In *Applied Sciences* (Vol. 13, Issue 8). DOI: 10.3390/app13084978

Aktar, W., Sengupta, D., & Chowdhury, A. (2009). Impact of pesticides use in agriculture: Their benefits and hazards. *Interdisciplinary Toxicology*, *2*(1), 1–12. https://doi.org/doi:10.2478/v10102-009-0001-7. DOI: 10.2478/v10102-009-0001-7 PMID: 21217838

Akutse, K. S., Subramanian, S., Maniania, N. K., Dubois, T., & Ekesi, S. (2020). Biopesticide Research and Product Development in Africa for Sustainable Agriculture and Food Security—Experiences From the International Centre of Insect Physiology and Ecology (Icipe). *Frontiers in Sustainable Food Systems*, *4*, 563016. Advance online publication. DOI: 10.3389/fsufs.2020.563016

Akutse, K., Maniania, N., Fiaboe, K., Van den Berg, J., & Ekesi, S. (2013). Endophytic colonization of Vicia faba and Phaseolus vulgaris (Fabaceae) by fungal pathogens and their effects on the life-history parameters of Liriomyza huidobrensis (Diptera: Agromyzidae). *Fungal Ecology*, *6*(4), 293–301. DOI: 10.1016/j.funeco.2013.01.003

Al Mazra'awi, M. S., Shipp, J. L., Broadbent, A. B., & Kevan, P. G. (2006). Dissemination of Beauveria bassiana by honey bees (Hymenoptera: Apidae) for control of tarnished plant bug (Hemiptera: Miridae) on Canola. *Environmental Entomology*, *35*(6), 1569–1577. DOI: 10.1603/0046-225X(2006)35[1569:DOBB-BH]2.0.CO;2

Alam, B., L , J., Gě, Q., Khan, M. A., Gōng, J., Mehmood, S., Yuán, Y., & G ng, W. (2021). Endophytic Fungi: From Symbiosis to Secondary Metabolite Communications or Vice Versa? *Frontiers in Plant Science*, *12*, 791033. Advance online publication. DOI: 10.3389/fpls.2021.791033 PMID: 34975976

Alam, M. Z., Khan, M. J. H., Kabbashi, N. A., & Sayem, S. A. (2018). Development of an effective biosorbent by fungal immobilization technique for removal of dyes. *Waste and Biomass Valorization*, *9*(4), 681–690. DOI: 10.1007/s12649-016-9821-9

Albracht, C., Eisenhauer, N., Vogel, A., Wagg, C., Buscot, F., & Heintz-Buschart, A. (2023). Effects of recurrent summer droughts on arbuscular mycorrhizal and total fungal communities in experimental grasslands differing in plant diversity and community composition. *Frontiers in Soil Science*, *3*, 1129845. DOI: 10.3389/fsoil.2023.1129845

Albrecht, U. (n.d.). *Plant Biostimulants: Definiition and Overview of Categories and Effects.*

Aleahmad, P., & Ebrahimi, L. (2023). The Possible Applications of Endophytic Fungi. *Research Journal of Pharmacognosy*, *10*(4), 81–94. DOI: 10.22127/rjp.2023.179187

Alegbeleye, O., Odeyemi, O. A., Strateva, M., & Stratev, D. (2022). Microbial spoilage of vegetables, fruits and cereals. *Applied Food Research*, *2*(1), 100122. DOI: 10.1016/j.afres.2022.100122

Al-Hawash, A. B., Zhang, X., & Ma, F. (2019). Removal and biodegradation of different petroleum hydrocarbons using the filamentous fungus Aspergillus sp. RFC-1. *MicrobiologyOpen*, *8*(1), e00619. https://doi.org/https://doi.org/10.1002/mbo3.619. DOI: 10.1002/mbo3.619 PMID: 29577679

Ali, S., Mushtaq, J., Nazir, F., & Sarfraz, H. (2017). Production and processing of single cell protein (SCP)-a review. *European Journal Pharmaceutical and Medical Research*, *4*(7), 86–94.

Al-Karaki, G. N., & Al-Raddad, A. (2004). Effects of arbuscular mycorrhizal fungi and drought stress on growth and nutrient uptake of two wheat genotypes differing in drought resistance. *Mycorrhiza*, *14*(4), 263–269. DOI: 10.1007/s00572-003-0265-2 PMID: 12942358

Allan, E., Manning, P., Alt, F., Binkenstein, J., Blaser, S., Blüthgen, N., Böhm, S., Grassein, F., Hölzel, N., Klaus, V. H., Kleinebecker, T., Morris, E. K., Oelmann, Y., Prati, D., Renner, S. C., Rillig, M. C., Schaefer, M., Schloter, M., Schmitt, B., & Fischer, M. (2015). Land use intensification alters ecosystem multifunctionality via loss of biodiversity and changes to functional composition. *Ecology Letters*, *18*(8), 834–843. DOI: 10.1111/ele.12469 PMID: 26096863

Alriksson, B., Hörnberg, A. N. D. R. E. A. S., Gudnason, A. E., Knobloch, S., Arnason, J., & Johannsson, R. (2014). Fish feed from wood. *Cellulose Chemistry and Technology*, *48*(9-10), 843–848.

Álvarez, Á. G. (2024). Fungal removal of cyanotoxins in constructed wetlands: The forgotten degraders. *The Science of the Total Environment*, *929*, 172590. DOI: 10.1016/j.scitotenv.2024.172590 PMID: 38642746

Al-Zaban, M. I., AlHarbi, M. A., & Mahmoud, M. A. (2021). Hydrocarbon biodegradation and transcriptome responses of cellulase, peroxidase, and laccase encoding genes inhabiting rhizospheric fungal isolates. *Saudi Journal of Biological Sciences*, *28*(4), 2083–2090. https://doi.org/https://doi.org/10.1016/j.sjbs.2021.01.009. DOI: 10.1016/j.sjbs.2021.01.009 PMID: 33935563

Amara, A. A., & El-Baky, N. A. (2023). Fungi as a source of edible proteins and animal feed. *Journal of Fungi (Basel, Switzerland)*, *9*(1), 73. DOI: 10.3390/jof9010073

Amaranthus, M. P. (1998). *The importance and conservation of ectomycorrizal fungal diversity in forest ecosystems: lessons from Europe and the Pacific Northwest.* DOI: 10.2737/PNW-GTR-431

Amin, M., Bhatti, H. N., & Bilal, M. (2022). Kinetic and thermodynamic characterization of lipase from Aspergillus melleus and its biocatalytic performance for degradation of poly(ε-caprolactone). *Journal of Chemical Technology and Biotechnology*, *97*(2), 446–454. https://doi.org/https://doi.org/10.1002/jctb.6649. DOI: 10.1002/jctb.6649

Anani, O. A., & Adetunji, C. O. (2021). Bioremediation of polythene and plastics using beneficial microorganisms. *Microbial Rejuvenation of Polluted Environment*, *3*, 281–302. DOI: 10.1007/978-981-15-7459-7_13

Anderson, S. (1981). Shotgun DNA sequencing using cloned DNase I-generated fragments. *Nucleic Acids Research*, *9*(13), 3015–3027. DOI: 10.1093/nar/9.13.3015 PMID: 6269069

Andrino, A., Guggenberger, G., Kernchen, S., Mikutta, R., Sauheitl, L., & Boy, J. (2021). Production of organic acids by arbuscular mycorrhizal fungi and their contribution in the mobilization of phosphorus bound to iron oxides. *Frontiers in Plant Science*, *12*, 661842. DOI: 10.3389/fpls.2021.661842 PMID: 34335645

Anees, M., Tronsmo, A., Edel-Hermann, V., Hjeljord, L. G., Héraud, C., & Steinberg, C. (2010). Characterization of field isolates of Trichoderma antagonistic against Rhizoctonia solani. *Fungal Biology*, *114*(9), 691–701. DOI: 10.1016/j.funbio.2010.05.007 PMID: 20943179

Aneja, K.R., Kumar, V., Jiloha, P., Kaur, M., Sharma, C., Surain, P., Dhiman, R., & Aneja, A. (2013). Potential Bioherbicides: Indian perspectives. *Biotechnology: Prospects and Applications*, 197-215.

Ángel, N. L. M. (2024). *Production process and formulation of a single-cell protein by fermenting the yeast Kluyveromyces marxianus* (WO 2024/005625 A1). https://lens.org/060-340-418-255-658

Anichebe, C. O., Okoye, E. L., & Onochie, C. C. (2019). (Forthcoming). Comparative study on single cell protein (SCP) production by Trichoderma Viride from pineapple wastes and banana peels. *SSRN*. Advance online publication. https://papers.ssrn.com/sol3/papers.cfm?abstract_id=3448990. DOI: 10.2139/ssrn.3448990

An, J., Zeng, T., Ji, C., de Graaf, S., Zheng, Z., Xiao, T. T., Deng, X., Xiao, S., Bisseling, T., Limpens, E., & Pan, Z. (2019). A *Medicago truncatula* SWEET transporter implicated in arbuscule maintenance during arbuscular mycorrhizal symbiosis. *The New Phytologist*, *224*(1), 396–408. DOI: 10.1111/nph.15975 PMID: 31148173

Anjitha, K. S., Sameena, P. P., & Puthur, J. T. (2021). Functional aspects of plant secondary metabolites in metal stress tolerance and their importance in pharmacology. *Plant Stress*, *2*, 100038. DOI: 10.1016/j.stress.2021.100038

Anli, M., Baslam, M., Tahiri, A., Raklami, A., Symanczik, S., Boutasknit, A., Ait-El-Mokhtar, M., Ben-Laouane, R., Toubali, S., Ait Rahou, Y., Ait Chitt, M., Oufdou, K., Mitsui, T., Hafidi, M., & Meddich, A. (2020). Biofertilizers as Strategies to Improve Photosynthetic Apparatus, Growth, and Drought Stress Tolerance in the Date Palm. *Frontiers in Plant Science*, *11*, 516818. Advance online publication. DOI: 10.3389/fpls.2020.516818 PMID: 33193464

Ansari, M., Devi, B. M., Sarkar, A., Chattopadhyay, A., Satnami, L., Balu, P., Choudhary, M., Shahid, M. A., & Jailani, A. A. K. (2023). Microbial Exudates as Biostimulants: Role in Plant Growth Promotion and Stress Mitigation. *Journal of Xenobiotics*, *13*(4), 572–603. DOI: 10.3390/jox13040037 PMID: 37873814

Anthony, M. A., Tedersoo, L., De Vos, B., Croisé, L., Meesenburg, H., Wagner, M., Andreae, H., Jacob, F., Lech, P., Kowalska, A., Greve, M., Popova, G., Frey, B., Gessler, A., Schaub, M., Ferretti, M., Waldner, P., Calatayud, V., Canullo, R., & Averill, C. (2024). Fungal community composition predicts forest carbon storage at a continental scale. *Nature Communications*, *15*(1), 2385. DOI: 10.1038/s41467-024-46792-w PMID: 38493170

Apostolidis, C., & Mcleay, F. (2019). To meat or not to meat? Comparing empowered meat consumers' and anti-consumers' preferences for sustainability labels. *Food Quality and Preference*, *77*, 109–122. DOI: 10.1016/j.foodqual.2019.04.008

Ardakani, S., Heydari, A., Khorasani, N., Arjmandi, R., & Ehteshami, M. (2009). Preparation of new Biofungicides using antagonistic bacteria and mineral compounds for controlling cotton seedling damping-off disease. *Journal of Plant Protection Research*, *49*(1). Advance online publication. DOI: 10.2478/v10045-009-0007-3

Arias, R. S., Dang, P. M., & Sobolev, V. S. (2015). RNAi-mediated control of aflatoxins in peanut: Method to analyze mycotoxin production and transgene expression in the peanut/Aspergillus pathosystem. *Journal of Visualized Experiments*, ●●●, 106. PMID: 26709851

Arnold, A. E., & Lutzoni, F. (2007). Diversity and host range of foliar fungal endophytes: Are tropical leaves biodiversity hotspots? *Ecology*, *88*(3), 541–549. DOI: 10.1890/05-1459 PMID: 17503580

Arora, B., Kamal, S., Sethi, S., Joshi, A., & Sharma, V. P. (2018). Mushrooms: Reservoir of vital nutrients and bioactives. []. www.nap.edu]. *Mushroom Research*, *27*(1), •••.

Arora, D. S., & Sharma, R. K. (2010). Ligninolytic fungal laccases and their biotechnological applications. *Applied Biochemistry and Biotechnology*, *160*(6), 1760–1788. DOI: 10.1007/s12010-009-8676-y PMID: 19513857

Asensio D, Rapparini F, Peñuelas J (2012) AM fungi root colonization increases the production of essential isoprenoids vs. nonessential isoprenoids especially under drought stress conditions or after jasmonic acid application. Phytochemistry 77:149 161.DOI: 10.1016/j.phytochem.2011.12.012

Asgar, M. A., Fazilah, A., Huda, N., Bhat, R., & Karim, A. A. (2010). Nonmeat protein alternatives as meat extenders and meat analogs. *Comprehensive Reviews in Food Science and Food Safety*, *9*(5), 513–529. DOI: 10.1111/j.1541-4337.2010.00124.x PMID: 33467834

Asrar, A.-W. A., & Elhindi, K. M. (2011). Alleviation of drought stress of marigold (Tagetes erecta) plants by using arbuscular mycorrhizal fungi. *Saudi Journal of Biological Sciences*, *18*(1), 93–98. https://doi.org/https://doi.org/10.1016/j.sjbs.2010.06.007. DOI: 10.1016/j.sjbs.2010.06.007 PMID: 23961109

Asrar, A.-W., Abdel-Fattah, G., & Elhindi, K. M. (2012). Improved growth, flower yield, and water relations of snapdragon (Antirhinum majus L.) plants grown under well-watered and water-stress conditions using arbuscular mycorrhizal fungi. *Photosynthetica*, *50*(2), 305–316. DOI: 10.1007/s11099-012-0024-8

Assemie, A., & Abaya, G. (2022). The Effect of Edible Mushroom on Health and Their Biochemistry. In *International Journal of Microbiology* (Vol. 2022). Hindawi Limited., DOI: 10.1155/2022/8744788

Atanasova, L., Crom, S. L., Gruber, S., Coulpier, F., Seidl-Seiboth, V., Kubicek, C. P., & Druzhinina, I. S. (2013). Comparative transcriptomics reveals different strategies of Trichodermamycoparasitism. *BMC Genomics*, *14*(1), 121. DOI: 10.1186/1471-2164-14-121 PMID: 23432824

Atkins, S. D., Clark, I. M., Pande, S., Hirsch, P. R., & Kerry, B. R. (2005). The use of real-time PCR and species-specific primers for the identification and monitoring of Paecilomyces lilacinus. *FEMS Microbiology Ecology*, *51*(2), 257–264. DOI: 10.1016/j.femsec.2004.09.002 PMID: 16329874

Augé, R. M., Toler, H. D., & Saxton, A. M. (2015). Arbuscular mycorrhizal symbiosis alters stomatal conductance of host plants more under drought than under amply watered conditions: A meta-analysis. *Mycorrhiza*, *25*(1), 13–24. DOI: 10.1007/s00572-014-0585-4 PMID: 24831020

Augustyniuk-Kram, A., & Kram K.J. (2012). Entomopathogenic fungi as an important natural regulator of insect outbreaks in forests (Review). *Forest Ecosystems - More than Just Trees.*

Awuchi, C. G., Ondari, E. N., Eseoghene, I. J., Twinomuhwezi, H., Amagwula, I. O., & Morya, S. (2021). Fungal growth and mycotoxins production: Types, toxicities, control strategies, and detoxification. *Fungal reproduction and growth, 100207.*

Awuchi, C. G., Ondari, E. N., Ogbonna, C. U., Upadhyay, A. K., Baran, K., Okpala, C. O. R., Korzeniowska, M., & Guiné, R. P. F. (2021). Mycotoxins affecting animals, foods, humans, and plants: Types, occurrence, toxicities, action mechanisms, prevention, and detoxification strategies-A revisit. *Foods*, *10*(6), 1279. DOI: 10.3390/foods10061279 PMID: 34205122

Ayimbila, F., & Keawsompong, S. (2023). Nutritional Quality and Biological Application of Mushroom Protein as a Novel Protein Alternative. *Current Nutrition Reports*, *12*(2), 290–307. DOI: 10.1007/s13668-023-00468-x PMID: 37032416

Azubuike, C. C., Chikere, C. B., & Okpokwasili, G. C. (2016). Bioremediation techniques-classification based on site of application: Principles, advantages, limitations and prospects. *World Journal of Microbiology & Biotechnology*, *32*(11), 180. DOI: 10.1007/s11274-016-2137-x PMID: 27638318

Baeva, E., Bleha, R., Lavrova, E., Sushytskyi, L., Čopíková, J., Jablonsky, I., Klouček, P., & Synytsya, A. (2019). Polysaccharides from Basidiocarps of Cultivating Mushroom *Pleurotus ostreatus*: Isolation and Structural Characterization. *Molecules (Basel, Switzerland)*, *24*(15), 2740. DOI: 10.3390/molecules24152740 PMID: 31357717

Baggiani, C., Giovannoli, C., & Anfossi, L. (2015). Man-made synthetic receptors for capture and analysis of ochratoxin A. *Toxins*, *7*(10), 4083–4098. DOI: 10.3390/toxins7104083 PMID: 26473924

Bagyaraj, D., & Revanna, A. (2017). *Soil Biodiversity: Role in Sustainable Horticulture* (pp. 1 18).

Bailey, K.L. (2014). The Bioherbicide approach to weed control using plant pathogens. *Integrated Pest Management*, 245-266.

Bailey, K., Boyetchko, S., & Längle, T. (2010). Social and economic drivers shaping the future of biological control: A Canadian perspective on the factors affecting the development and use of microbial biopesticides. *Biological Control*, *52*(3), 221–229. DOI: 10.1016/j.biocontrol.2009.05.003

Bajpai, P., & Bajpai, P. K. (1987). Single cell protein production from rayon pulp will waste by *Paecilomyces variotii.Journal of Fermentation Technology*, *65*(3), 349–351. DOI: 10.1016/0385-6380(87)90099-9

Bakaloudi, D. R., Halloran, A., Rippin, H. L., Oikonomidou, A. C., Dardavesis, T. I., Williams, J., Wickramasinghe, K., Breda, J., & Chourdakis, M. (2021). Intake and adequacy of the vegan diet. A systematic review of the evidence. *Clinical Nutrition (Edinburgh, Lothian)*, *40*(5), 3503–3521. DOI: 10.1016/j.clnu.2020.11.035

Baker, B. P., Green, T. A., & Loker, A. J. (2020). Biological control and integrated pest management in organic and conventional systems. *Biological Control*, *140*, 104095. DOI: 10.1016/j.biocontrol.2019.104095

Baker, R. E., Mahmud, A. S., Miller, I. F., Rajeev, M., Rasambainarivo, F., Rice, B. L., Takahashi, S., Tatem, A. J., Wagner, C. E., Wang, L.-F., Wesolowski, A., & Metcalf, C. J. E. (2022). Infectious disease in an era of global change. *Nature Reviews. Microbiology*, *20*(4), 193–205. DOI: 10.1038/s41579-021-00639-z PMID: 34646006

Bakratsas, G., Polydera, A., Nilson, O., Chatzikonstantinou, A. V., Xiros, C., Katapodis, P., & Stamatis, H. (2023). Polydera, A Nilson, O. A. V. Chatzikonstantinou, Xiros, C. Katapodis P. and Stamatis, H. Mycoprotein Production by Submerged Fermentation of the Edible Mushroom Pleurotus ostreatus in a Batch Stirred Tank Bioreactor Using Agro-Industrial Hydrolysate. *Foods*, *12*(12), 2295. DOI: 10.3390/foods12122295 PMID: 37372506

Balasubramaniam, T., Shen, G., Esmaeili, N., & Zhang, H. (2023). Plants' Response Mechanisms to Salinity Stress. *Plants*, *12*(12), 2253. Advance online publication. DOI: 10.3390/plants12122253 PMID: 37375879

Balestrini, R., and Kottke, I. (2016). Structure and development of ectomycorrhizal roots. *Molecular Mycorrhizal Symbiosis*, 47–61. https://doi.org/DOI: 10.1002/9781118951446.ch4

Balestrini, R., Magurno, F., Walker, C., Lumini, E., & Bianciotto, V. (2010). Cohorts of arbuscular mycorrhizal fungi (AMF) in *Vitis vinifera*, a typical Mediterranean fruit crop. *Environmental Microbiology Reports*, *2*(4), 594–604. DOI: 10.1111/j.1758-2229.2010.00160.x PMID: 23766230

Baltazar, M., Correia, S., Guinan, K. J., Sujeeth, N., Bragança, R., & Gonçalves, B. (2021). Recent Advances in the Molecular Effects of Biostimulants in Plants: An Overview. *Biomolecules*, *11*(8), 1096. DOI: 10.3390/biom11081096 PMID: 34439763

Bamisile, B. S., Akutse, K. S., Siddiqui, J. A., & Xu, Y. (2021). Model Application of Entomopathogenic Fungi as Alternatives to Chemical Pesticides: Prospects, Challenges, and Insights for Next-Generation Sustainable Agriculture. *Frontiers in Plant Science*, *12*, 741804. Advance online publication. DOI: 10.3389/fpls.2021.741804 PMID: 34659310

Bandurska, K., Krupa, P., Berdowska, A., Jatulewicz, I., & Zawierucha, I. (2021). Mycoremediation of Soil Contaminated with Cadmium and Lead by *Trichoderma* sp. *Ecological Chemistry and Engineering. S*, *28*(2), 277–286. DOI: 10.2478/eces-2021-0020

Banjara, T. R., Bohra, J. S., Kumar, S., Ram, A., & Pal, V. (2022). Diversification of Rice–Wheat Cropping System Improves Growth, Productivity and Energetics of Rice in the Indo-Gangetic Plains of India. *Agricultural Research*, *11*(1), 48–57. DOI: 10.1007/s40003-020-00533-9

Bankole, P. O., Adekunle, A. A., Obidi, O. F., Chandanshive, V. V., & Govindwar, S. P. (2018). Biodegradation and detoxification of Scarlet RR dye by a newly isolated filamentous fungus, *Peyronellaea prosopidis.Sustainable Environment Research*, *28*(5), 214–222. DOI: 10.1016/j.serj.2018.03.001

Bapaume, L., Laukamm, S., Darbon, G., Monney, C., Meyenhofer, F., Feddermann, N., Chen, M., & Reinhardt, D. (2019). VAPYRIN marks an endosomal trafficking compartment involved in arbuscular mycorrhizal symbiosis. *Frontiers in Plant Science*, *10*, 666. DOI: 10.3389/fpls.2019.00666 PMID: 31231402

Baran, W., Adamek, E., Włodarczyk, A., Lazur, J., Opoka, W., & Muszyńska, B. (2021). The remediation of sulfonamides from the environment by *Pleurotus eryngii* mycelium. Efficiency, products and mechanisms of mycodegradation. *Chemosphere*, *262*, 128026. DOI: 10.1016/j.chemosphere.2020.128026 PMID: 33182090

Baron, N. C., Costa, N. T. A., Mochi, D. A., & Rigobelo, E. C. (2018). First report of Aspergillus sydowii and Aspergillus brasiliensis as phosphorus solubilizers in maize. *Annals of Microbiology*, *68*(12), 863–870. DOI: 10.1007/s13213-018-1392-5

Baron, N. C., & Rigobelo, E. C. (2022a). Endophytic fungi: A tool for plant growth promotion and sustainable agriculture. *Mycology*, *13*(1), 39–55. DOI: 10.1080/21501203.2021.1945699 PMID: 35186412

Baron, N. C., Rigobelo, E. C., & Zied, D. C. (2019). Filamentous fungi in biological control: Current status and future perspectives. *Chilean Journal of Agricultural Research*, *79*(2), 307–315. DOI: 10.4067/S0718-58392019000200307

Bartholdy, B. A., Berreck, M., & Haselwandter, K. (n.d.). *Hydroxamate siderophore synthesis by Phialocephala fortinii, a typical dark septate fungal root endophyte.*

Baum, C., El-Tohamy, W., and Gruda, N. (2015). Increasing the productivity and product quality of vegetable crops using arbuscular mycorrhizal fungi: a review. *Scientia Horticulturae, 187*, 131 141. https://doi.org/DOI: 10.1016/j.scienta.2015.03.002

Beattie, A. J., Hay, M., Magnusson, B., de Nys, R., Smeathers, J., & Vincent, J. F. V. (2011). Ecology and bioprospecting. *Austral Ecology*, *36*(3), 341–356. DOI: 10.1111/j.1442-9993.2010.02170.x PMID: 22737038

Begum, F., & Samajpati, N. (2000). Mycotoxin production on rice, pulses and oilseeds. *Naturwissenschaften*, *87*(6), 275–277. DOI: 10.1007/s001140050720 PMID: 10929292

Begum, N., Qin, C., Ahanger, M. A., Raza, S., Khan, M. I., Ashraf, M., Ahmed, N., & Zhang, L. (2019). Role of Arbuscular Mycorrhizal Fungi in Plant Growth Regulation: Implications in Abiotic Stress Tolerance. *Frontiers in Plant Science*, *10*, 1068. Advance online publication. DOI: 10.3389/fpls.2019.01068 PMID: 31608075

Behie, S. W., & Bidochka, M. J. (2014). Ubiquity of Insect-Derived Nitrogen Transfer to Plants by Endophytic Insect-Pathogenic Fungi: An Additional Branch of the Soil Nitrogen Cycle. *Applied and Environmental Microbiology*, *80*(5), 1553–1560. DOI: 10.1128/AEM.03338-13 PMID: 24334669

Behjati, S., & Tarpey, P. S. (2013). What is next generation sequencing? *Archives of Disease in Childhood - Education and Practice*, *98*(6), 236–238. DOI: 10.1136/archdischild-2013-304340 PMID: 23986538

Bello Aliyu, M. (2017). Efficiency of Boolean Search strings for Information Retrieval. [AJER]. *American Journal of Engineering Research*, *6*, 216–222.

Bell-Pedersen, D., Cassone, V. M., Earnest, D. J., Golden, S. S., Hardin, P. E., Thomas, T. L., & Zoran, M. J. (2005). Circadian rhythms from multiple oscillators: Lessons from diverse organisms. *Nature Reviews. Genetics*, *6*(7), 544–556. DOI: 10.1038/nrg1633 PMID: 15951747

Bell, V., Silva, C. R. P. G., Guina, J., & Fernandes, T. H. (2022). Mushrooms as future generation healthy foods. *Frontiers in Nutrition*, *9*, 1050099. DOI: 10.3389/fnut.2022.1050099 PMID: 36562045

Ben Miri, Y., Benabdallah, A., Chentir, I., Djenane, D., Luvisi, A., & De Bellis, L. (2024). Comprehensive insights into ochratoxin A: Occurrence, analysis, and control strategies. *Foods*, *13*(8), 1184. DOI: 10.3390/foods13081184 PMID: 38672856

Ben Salha, G., Herrera Díaz, R., Labidi, J., & Abderrabba, M. (2017). Deterpenation of *Origanum majorana* L. essential oil by reduced pressure steam distillation. *Industrial Crops and Products*, *109*, 116–122. DOI: 10.1016/j.indcrop.2017.08.016

Benites-Alfaro, E., Silva Carhuatocto, L. M., Guanilo Inigo, A. P., Jave Nakayo, J., Castaneda-Olivera, C. A., Lizarzaburu Aguinaga, D. A., & Cabrera Carranza, C. F. (2023). Pleurotus Ostreatus and Trametes Versicolor Fungi to Decontaminate Soils Containing Organophosphates Methamidophos and Cadmium. *Chemical Engineering Transactions, 100*, 97-102 SE-Research Articles. DOI: 10.3303/CET23100017

Benkerroum, N. (2020). Aflatoxins: Producing-molds, structure, health issues and incidence in Southeast Asian and Sub-Saharan African countries. *International Journal of Environmental Research and Public Health*, *17*(4), 1215. DOI: 10.3390/ijerph17041215 PMID: 32070028

Bennett, J. W., & Klich, M. (2003). Mycotoxins. *Clinical Microbiology Reviews*, *16*(3), 497–516. DOI: 10.1128/CMR.16.3.497-516.2003 PMID: 12857779

Besufekad, Y., Mekonnen, A., Girma, B., Daniel, R., Tassema, G., Melkamu, J., Asefa, M., Fikiru, T., & Denboba, L. (2020). Selection of appropriate substrate for production of oyster mushroom (Pleurotus ostreatus). *Journal of Yeast and Fungal Research*, *11*(1), 15–25. DOI: 10.5897/JYFR2019.0187

Bhambri, A., Srivastava, M., Mahale, V. G., Mahale, S., & Karn, S. K. (2022). Mushrooms as Potential Sources of Active Metabolites and Medicines. In Frontiers in Microbiology (Vol. 13). Frontiers Media S.A. DOI: 10.3389/fmicb.2022.837266

Bhardwaj, D., Ansari, M. W., Sahoo, R. K., & Tuteja, N. (2014). Biofertilizers function as key player in sustainable agriculture by improving soil fertility, plant tolerance and crop productivity. *Microbial Cell Factories*, *13*(1), 66. DOI: 10.1186/1475-2859-13-66 PMID: 24885352

Bhargava, P., Vats, S., & Gupta, N. (2019). Metagenomics as a Tool to Explore Mycorrhizal Fungal Communities. In *Mycorrhizosphere and Pedogenesis*, 207–219. DOI: 10.1007/978-981-13-6480-8_13

Bhatnagar, I., & Kim, S.-K. (2010). Immense essence of excellence: Marine microbial bioactive compounds. *Marine Drugs*, *8*(10), 2673–2701. DOI: 10.3390/md8102673 PMID: 21116414

Bhattacharya, S., Das, A., Prashanthi, K., Palaniswamy, M., & Angayarkanni, J. (2014). Mycoremediation of Benzo[a]pyrene by Pleurotus ostreatus in the presence of heavy metals and mediators. *3 Biotech*, *4*(2), 205–211. DOI: 10.1007/s13205-013-0148-y

Bhattacharyya, P. N., & Jha, D. K. (2012). Plant growth-promoting rhizobacteria (PGPR): Emergence in agriculture. *World Journal of Microbiology & Biotechnology*, *28*(4), 1327–1350. DOI: 10.1007/s11274-011-0979-9 PMID: 22805914

Bhatt, P., Zhang, W., Lin, Z., Pang, S., Huang, Y., & Chen, S. (2020). Biodegradation of allethrin by a novel fungus *Fusarium proliferatum* strain CF2, isolated from contaminated soils. *Microorganisms*, *8*(4), 593. DOI: 10.3390/microorganisms8040593 PMID: 32325934

Bhupenchandra, I., Chongtham, S. K., Devi, A. G., Dutta, P., Sahoo, M. R., Mohanty, S., Kumar, S., Choudhary, A. K., Devi, E. L., Sinyorita, S., Devi, S. H., Mahanta, M., Kumari, A., Devi, H. L., Josmee, R. K., Pusparani, A., Pathaw, N., Gupta, S., Meena, M., & Swapnil, P. (2024). Unlocking the Potential of Arbuscular Mycorrhizal Fungi: Exploring Role in Plant Growth Promotion, Nutrient Uptake Mechanisms, Biotic Stress Alleviation, and Sustaining Agricultural Production Systems. *Journal of Plant Growth Regulation*. Advance online publication. DOI: 10.1007/s00344-024-11467-9

Bhupenchandra, I., Chongtham, S. K., Devi, E. L., Choudhary, A. K., Salam, M. D., Sahoo, M. R., & Khaba, C. I. (2022). Role of biostimulants in mitigating the effects of climate change on crop performance. *Frontiers in Plant Science, 13*, 967665.

Bihal, R., Al-Khayri, J. M., Banu, A. N., Kudesia, N., Ahmed, F. K., Sarkar, R., Arora, A., & Abd-Elsalam, K. A. (2023). Entomopathogenic Fungi: An Eco-Friendly Synthesis of Sustainable Nanoparticles and Their Nanopesticide Properties. *Microorganisms, 11*(6), 1617. DOI: 10.3390/microorganisms11061617 PMID: 37375119

Bijla, S., & Sharma, V. P. (2023). Status of mushroom production: Global and national scenario. *Mushroom Research, 32*(2), 91–98. DOI: 10.36036/MR.32.2.2023.146647

Bilen Ozyurek, S., Avcioglu, N. H., & Seyis Bilkay, I. (2021). Mycoremediation potential of Aspergillus ochraceus NRRL 3174. *Archives of Microbiology, 203*(10), 5937–5950. DOI: 10.1007/s00203-021-02490-5 PMID: 34599404

Bilesky-Jose, N., Maruyama, C., Germano-Costa, T., Campos, E., Carvalho, L., Grillo, R., ... & De Lima, R. (2021). Biogenic α-Fe2O3 nanoparticles enhance the biological activity of Trichoderma against the plant pathogen Sclerotinia sclerotiorum.

Birolli, W. G., Vacondio, B., Alvarenga, N., Seleghim, M. H., & Porto, A. L. (2018). Enantioselective biodegradation of the pyrethroid (±)-lambda-cyhalothrin by marine-derived fungi. *Chemosphere, 197*, 651–660. DOI: 10.1016/j.chemosphere.2018.01.054 PMID: 29407829

Bisht, A., Lather, S., Goyal, P., & Garg, N. (2024). Molecular Determinants and Regulatory Mechanisms of Nutrient Exchange Between Plant and AMF. In *Arbuscular Mycorrhizal Fungi in Sustainable Agriculture: Nutrient and Crop Management* (pp. 77–97). Springer Nature Singapore. DOI: 10.1007/978-981-97-0300-5_3

Bisht, I. S., Rana, J. C., & Pal Ahlawat, S. (2020). The Future of Smallholder Farming in India: Some Sustainability Considerations. *Sustainability (Basel), 12*(9), 3751. DOI: 10.3390/su12093751

Bissett, A., Richardson, A. E., Baker, G., Wakelin, S., & Thrall, P. H. (2010). Life history determines biogeographical patterns of soil bacterial communities over multiple spatial scales. *Molecular Ecology, 19*(19), 4315–4327. DOI: 10.1111/j.1365-294X.2010.04804.x PMID: 25241408

Biţă-Nicolae, C., & Dhyani, P. (2025). Editorial: Plant diversity: the key to ecosystem resilience in a changing world. Frontiers in Plant Science. *Frontiers in Plant Science, 15*, 1–3. DOI: 10.3389/fpls.2024.1534119

Bjelakovic, G., Nikolova, D., Gluud, L. L., Simonetti, R. G., & Gluud, C. (2007). Mortality in randomized trials of antioxidant supplements for primary and secondary prevention: Systematic review and meta-analysis. *Journal of the American Medical Association, 297*(8), 842–857. DOI: 10.1001/jama.297.8.842 PMID: 17327526

Blanco-Orta, M. F., García-de la Cruz, R. F., Paz-Maldonado, L. M. T., Pedraza-González, D. A., Morales-Avila, M. M., Balderas-Hernández, V. E., González-Ortega, O., & Pérez-Martínez, A. S. (2023). Assessing three industrially produced fungi for the bioremediation of diclofenac. *Journal of Environmental Science and Health. Part A, Toxic/Hazardous Substances & Environmental Engineering, 58*(7), 661–670. DOI: 10.1080/10934529.2023.2206353 PMID: 37128145

Blunt, J. W., Copp, B. R., Keyzers, R. A., Munro, M. H. G., & Prinsep, M. R. (2015). Marine natural products. *Natural Product Reports, 32*(2), 116–211. DOI: 10.1039/C4NP00144C PMID: 25620233

Bogati, K., & Walczak, M. (2022). The impact of drought stress on soil microbial community, enzyme activities and plants. *Agronomy (Basel), 12*(1), 189. DOI: 10.3390/agronomy12010189

Bohrer, B. M. (2019). An investigation of the formulation and nutritional composition of modern meat analogue products. *Food Science and Human Wellness, 8*(4), 320–329. DOI: 10.1016/j.fshw.2019.11.006

Bonfante, P., & Genre, A. (2010). Mechanisms underlying beneficial plant–fungus interactions in mycorrhizal symbiosis. *Nature Communications, 1*(1), 48. DOI: 10.1038/ncomms1046 PMID: 20975705

Boon, E. (2012). The evolution of inter-genomic variation in arbuscular mycorrhizal fungi. In *Biological sciences.*

Boon, E., Halary, S., Bapteste, E., & Hijri, M. (2015). Studying genome heterogeneity within the arbuscular mycorrhizal fungal cytoplasm. *Genome Biology and Evolution, 7*(2), 505–521. DOI: 10.1093/gbe/evv002 PMID: 25573960

Borràs-Vallverdú, B., Ramos, A. J., Cantero-Martínez, C., Marín, S., Sanchis, V., & Fernández-Ortega, J. (2022). Influence of Agronomic Factors on Mycotoxin Contamination in Maize and Changes during a 10-Day Harvest-Till-Drying Simulation Period: A Different Perspective. *Toxins, 14*(9), 620. DOI: 10.3390/toxins14090620 PMID: 36136558

Bosco, F., & Mollea, C. (2019). Mycoremediation in soil. *Environmental chemistry and recent pollution control approaches*, 173.

Bose U, Hodson MP, Shaw PN, Fuerst JA, Hewavitharana AK (2014) Bacterial production of the fungus-derived cholesterol-lowering agent mevinolin. Biomed chromatog 28(9):1163–1166

Bose, T., Spies, C. F. J., Hammerbacher, A., & Coutinho, T. A. (2023). *Phytophthora*: An underestimated threat to agriculture, forestry, and natural ecosystems in sub-Saharan Africa. *Mycological Progress, 22*(11), 78. DOI: 10.1007/s11557-023-01926-0

Bottin, J., Swann, J., Cropp, E., Chambers, E., Ford, H., Ghatei, M., & Frost, G. (2016). Mycoprotein reduces energy intake and postprandial insulin release without altering glucagon-like peptide-1 and peptide tyrosine-tyrosine concentrations in healthy overweight and obese adults: A randomised-controlled trial. *British Journal of Nutrition, 116*(2), 360–374. DOI: 10.1017/S0007114516001872 PMID: 27198187

Bouacem, K., Allala, F., Zaraî Jaouadi, N., Hamdi, S., Mechri, S., Ighilahriz, K., Rekik, H., Hacene, H., Bouanane-Darenfed, A., & Jaouadi, B. (2022). A novel peroxidase from white-rot Agaricomycetes fungus Phlebia radiata strain KB-DZ15: Its purification, characterisation, and potential application for dye-decolorisation and lignin-biodegradation. *Biocatalysis and Biotransformation, 40*(5), 365–377. DOI: 10.1080/10242422.2021.1939315

Bousbia, N., Abert Vian, M., Ferhat, M. A., Petitcolas, E., Meklati, B. Y., & Chemat, F. (2009). Comparison of two isolation methods for essential oil from rosemary leaves: Hydrodistillation and microwave hydrodiffusion and gravity. *Food Chemistry, 114*(1), 355–362. DOI: 10.1016/j.foodchem.2008.09.106

Bouyahya, A., Chamkhi, I., Benali, T., Guaouguaou, F. E., Balahbib, A., El Omari, N., Taha, D., Belmehdi, O., Ghokhan, Z., & El Menyiy, N. (2021). Traditional use, phytochemistry, toxicology, and pharmacology of *Origanum majorana* L. *Journal of Ethnopharmacology*, *265*, 113318. DOI: 10.1016/j.jep.2020.113318 PMID: 32882360

Bowler, C., Karl, D. M., & Colwell, R. R. (2009). Microbial oceanography in a sea of opportunity. *Nature*, *459*(7244), 180–184. DOI: 10.1038/nature08056 PMID: 19444203

Boyette, C. D., Hoagland, R. E., & Stetina, K. C. (2019). Extending the host range of the bioherbicidal fungus *Colletotrichum gloeosporioides* f. sp. *aeschynomene.Biocontrol Science and Technology*, *29*(7), 720–726. DOI: 10.1080/09583157.2019.1581130

Boyno, G., & Demir, S. (2022). Plant-mycorrhiza communication and mycorrhizae in inter-plant communication. *Symbiosis*, *86*(2), 155–168. DOI: 10.1007/s13199-022-00837-0

Bozhinova, R. (2023). Yield and chemical composition of oriental tobacco as affected by biostimulant application. [). https://cabidigitallibrary.org]. *Bulgarian Journal of Agricultural Science*, *29*(1), •••.

Bradford, K. J., Dahal, P., Van Asbrouck, J., Kunusoth, K., Bello, P., Thompson, J., & Wu, F. (2018). The dry chain: Reducing postharvest losses and improving food safety in humid climates. *Trends in Food Science & Technology*, *71*, 84–93. DOI: 10.1016/j.tifs.2017.11.002

Brar, B., Bala, K., Saharan, B. S., Sadh, P. K., & Duhan, J. S. (2024). Bio-boosting agriculture: Harnessing the potential of fungi-bacteria-plant synergies for crop improvement. *Discover Plants*, *1*(1), 21. DOI: 10.1007/s44372-024-00023-0

Bratosin, B. C., Darjan, S., & Vodnar, D. C. (2021). Single cell protein: A potential substitute in human and animal nutrition. *Sustainability (Basel)*, *13*(16), 9284. DOI: 10.3390/su13169284

Bravim, N. P. B., Alves, A. F., Orlanda, J. F. F., & Silva, P. B. R. (2021). Selection of filamentous fungi that are resistant to the herbicides atrazine, glyphosate and pendimethalin. *Acta Scientiarum. Agronomy*, *43*, e51656. DOI: 10.4025/actasciagron.v43i1.51656

Brewer, M. S. (2011). Natural Antioxidants: Sources, Compounds, Mechanisms of Action, and Potential Applications. *Comprehensive Reviews in Food Science and Food Safety*, *10*(4), 221–247. DOI: 10.1111/j.1541-4337.2011.00156.x

Bristy, A. T., Islam, T., Ahmed, R., Hossain, J., Reza, H. M., & Jain, P. (2022). Evaluation of Total Phenolic Content, HPLC Analysis, and Antioxidant Potential of Three Local Varieties of Mushroom: A Comparative Study. *International Journal of Food Sciences*, *2022*, 3834936. DOI: 10.1155/2022/3834936 PMID: 36310853

Brotman, Y., Briff, E., Viterbo, A., & Chet, I. (2008). Role of Swollenin, an expansin-like protein from-*Trichoderma*, in plant root colonization. *Plant Physiology*, *147*(2), 779–789. DOI: 10.1104/pp.108.116293 PMID: 18400936

Brundrett, M., & Tedersoo, L. (2019). Misdiagnosis of mycorrhizas and inappropriate recycling of data can lead to false conclusions. *The New Phytologist*, *221*(1), 18–24. DOI: 10.1111/nph.15440 PMID: 30191568

Bryła, M., Pierzgalski, A., Zapaśnik, A., Uwineza, P. A., Ksieniewicz-Woźniak, E., Modrzewska, M., & Waśkiewicz, A. (2022). Recent Research on *Fusarium* Mycotoxins in Maize-A Review. *Foods, 11*(21), 3465. DOI: 10.3390/foods11213465 PMID: 36360078

Bücking, H., Liepold, E., & Ambilwade, P. (2012). *The Role of the Mycorrhizal Symbiosis in Nutrient Uptake of Plants and the Regulatory Mechanisms Underlying These Transport Processes* (N. K. Dhal and S. C. Sahu (eds.); p. Ch. 4). IntechOpen. DOI: 10.5772/52570

Budzyńska, S., Siwulski, M., Gąsecka, M., Magdziak, Z., Kalač, P., Niedzielski, P., & Mleczek, M. (2022). Biofortification of Three Cultivated Mushroom Species with Three Iron Salts—Potential for a New Iron-Rich Superfood. *Molecules (Basel, Switzerland), 27*(7), 2328. DOI: 10.3390/molecules27072328 PMID: 35408727

Bueno de Mesquita, C. P., King, A. J., Schmidt, S. K., Farrer, E. C., & Suding, K. N. (2016). Incorporating biotic factors in species distribution modeling: Are interactions with soil microbes important? *Ecography, 39*(10), 970–980. DOI: 10.1111/ecog.01797

Bullerman, L. B., Schroeder, L. L., & Park, K. Y. (1984). Formation and control of mycotoxins in food. *Journal of Food Protection, 47*(8), 637–646. DOI: 10.4315/0362-028X-47.8.637 PMID: 30934406

Bumpus, J. A., & Aust, S. D. (1987). Biodegradation of DDT [1, 1, 1-trichloro-2, 2-bis (4-chlorophenyl) ethane] by the white rot fungus *Phanerochaete chrysosporium.Applied and Environmental Microbiology, 53*(9), 2001–2008. DOI: 10.1128/aem.53.9.2001-2008.1987 PMID: 3674869

Busatta, C., Vidal, R. S., Popiolski, A. S., Mossi, A. J., Dariva, C., Rodrigues, M. R. A., Corazza, F. C., Corazza, M. L., Vladimir Oliveira, J., & Cansian, R. L. (2008). Application of *Origanum majorana* L. essential oil as an antimicrobial agent in sausage. *Food Microbiology, 25*(1), 207–211. DOI: 10.1016/j.fm.2007.07.003 PMID: 17993397

Caitano, C. E. C., Wagner, G. V. J., Soares, D. M. M., Lucas da Silva, A., Bianca de Barros, N., Pardo-Giménez, A., Stevani, C. V., & Zied, D. C. (2023). Management of Water Supply in the Cultivation of Different Agaricus Bisporus Strains. *Agronomy (Basel), 13*(10), 2626. Advance online publication. DOI: 10.3390/agronomy13102626

Calvet, C., Camprubí, A., Pérez-Hernández, A., & Lovato, P. (2013). Plant Growth Stimulation and Root Colonization Potential of In Vivo versus In Vitro Arbuscular Mycorrhizal Inocula. *HortScience: A Publication of the American Society for Horticultural Science, 48*, 897–901. DOI: 10.21273/HORTSCI.48.7.897

Calvo, P., Nelson, L., & Kloepper, J. W. (2014). Agricultural uses of plant biostimulants. *Plant and Soil, 383*(1), 3–41. DOI: 10.1007/s11104-014-2131-8

Canellas, L. P., Olivares, F. L., Aguiar, N. O., Jones, D. L., Nebbioso, A., Mazzei, P., & Piccolo, A. (2015). Humic and fulvic acids as biostimulants in horticulture. *Scientia Horticulturae, 196*, 15–27. DOI: 10.1016/j.scienta.2015.09.013

Cao, H., Wang, C., Liu, H., Jia, W., & Sun, H. (2020). Enzyme activities during Benzo [a] pyrene degradation by the fungus *Lasiodiplodia theobromae* isolated from a polluted soil. *Scientific Reports, 10*(1), 865. DOI: 10.1038/s41598-020-57692-6 PMID: 31964981

Cao, Y., Wu, X., Zhukova, A., Tang, Z., Weng, Y., Li, Z., & Yang, Y. (2020). Arbuscular mycorrhizal fungi (AMF) species and abundance exhibit different effects on saline-alkaline tolerance in Leymus chinensis. *Journal of Plant Interactions*, *15*(1), 266–279. DOI: 10.1080/17429145.2020.1802524

Cardwell, G., Bornman, J. F., James, A. P., & Black, L. J. (2018). A Review of Mushrooms as a Potential Source of Dietary Vitamin D. *Nutrients*, *10*(10), 1498. DOI: 10.3390/nu10101498 PMID: 30322118

Carranza-Méndez, R. C., Chávez-González, M. L., Sepúlveda-Torre, L., Aguilar, C. N., Govea-Salas, M., & Ramos-González, R. (2022). Production of single cell protein from orange peel residues by Candida utilis. *Biocatalysis and Agricultural Biotechnology*, *40*, 102298. DOI: 10.1016/j.bcab.2022.102298

Carvajal, M., Jeldres, P., Vergara, A., Lobaina, E., Olivares, M., Meza, D., Velásquez, A., Dorta, F., Jorquera, F., Seeger, M., Cereceda-Balic, F., & Fadic, X. (2023). Bioremoval of copper by filamentous fungi isolated from contaminated soils of Puchuncaví-Ventanas Central Chile. *Environmental Geochemistry and Health*, *45*(7), 4275–4293. DOI: 10.1007/s10653-023-01493-z PMID: 36729229

Cashman, K. D. (2022). Global differences in vitamin D status and dietary intake: A review of the data. *Endocrine Connections*, *11*(1), 11. DOI: 10.1530/EC-21-0282 PMID: 34860171

CAST. 2003, Mycotoxins: Risks in Plant, Animal, Human Systems, Task Force Report No. 38, Ames.

Castegnaro, M., Tozlovanu, M., Wild, C., Molinié, A., Sylla, A., & Pfohl-Leszkowicz, A. (2006). Advantages and drawbacks of immunoaffinity columns in analysis of mycotoxins in food. *Molecular Nutrition & Food Research*, *50*(6), 480–487. DOI: 10.1002/mnfr.200500264 PMID: 16671060

Castiglione, A. M., Mannino, G., Contartese, V., Bertea, C. M., & Ertani, A. (2021). Microbial Biostimulants as Response to Modern Agriculture Needs: Composition, Role and Application of These Innovative Products. *Plants*, *10*(8), 1533. DOI: 10.3390/plants10081533 PMID: 34451578

Cavaliere, C., D'Ascenzo, G., Foglia, P., Pastorini, E., Samperi, R., & Laganà, A. (2005). Determination of type B trichothecenes and macrocyclic lactone mycotoxins in field contaminated maize. *Food Chemistry*, *92*(3), 559–568. DOI: 10.1016/j.foodchem.2004.10.008

Cebri'an, M., & Ibarruri, J. (2023). Filamentous fungi processing by solid-state fermentation, in Current Developments in Biotechnology and Bioengineering, Elsevier, 2023, 251–292.

Cereghino, J. L., & Cregg, J. M. (2000). Heterologous protein expression in the methylotrophic yeast Pichia pastoris. *FEMS Microbiology Reviews*, *24*(1), 45–66. DOI: 10.1111/j.1574-6976.2000.tb00532.x

Cerletti, C., Esposito, S., & Iacoviello, L. (2021). Edible mushrooms and beta-glucans: Impact on human health. *Nutrients*, *13*(7), 2195. DOI: 10.3390/nu13072195 PMID: 34202377

Chakraborty, I., Sen, I. K., Mondal, S., Rout, D., Bhanja, S. K., Maity, G. N., & Maity, P. (2019). Bioactive polysaccharides from natural sources: A review on the antitumor and immunomodulating activities. In *Biocatalysis and Agricultural Biotechnology* (Vol. 22). Elsevier Ltd., DOI: 10.1016/j.bcab.2019.101425

Chandrasekaran, M. (2022). Arbuscular Mycorrhizal Fungi Mediated Enhanced Biomass, Root Morphological Traits and Nutrient Uptake under Drought Stress: A Meta-Analysis. In *Journal of Fungi* (Vol. 8, Issue 7). DOI: 10.3390/jof8070660

Chandra, V., Tiwari, A., Pant, K. K., & Bhatt, R. (2022). Animal Cell Culture: Basics and Applications. In Verma, P. (Ed.), *Industrial Microbiology and Biotechnology*. Springer., DOI: 10.1007/978-981-16-5214-1_24

Chang, A. J., Fan, J., & Wen, X. (2012). Screening of fungi capable of highly selective degradation of lignin in rice straw. *International Biodeterioration & Biodegradation*, *72*, 26–30. DOI: 10.1016/j.ibiod.2012.04.013

Channaiah, L. H., & Maier, D. E. (2014). Best stored maize management practices for the prevention of mycotoxin contamination. *Mycotoxin reduction in grain chains*, 78-88.

Chapelle, F. H. (1999). Bioremediation of petroleum hydrocarbon-contaminated ground water: The perspectives of history and hydrology. *Groundwater*, *37*(1), 122–132. DOI: 10.1111/j.1745-6584.1999.tb00965.x

Charudattan, R. (2005). Ecological, practical, and political inputs into selection of weed targets: What makes a good biological control target? *Biological Control*, *35*(3), 183–196. DOI: 10.1016/j.biocontrol.2005.07.009

Chatterjee, S., Dhole, A., Krishnan, A. A., & Banerjee, K. (2023). Mycotoxin monitoring, regulation and analysis in India: A success story. *Foods*, *12*(4), 705. DOI: 10.3390/foods12040705 PMID: 36832780

Chaurasia, P. K., & Bharati, S. L. (2021). Applicability of fungi in agriculture and environmental sustainability. In *Microbes in Land Use Change Management* (pp. 155–172). Elsevier., DOI: 10.1016/B978-0-12-824448-7.00010-3

Chavez, R., Bull, P., & Eyzaguirre, J. (2006). The xylanolytic enzyme system from the genus Penicillium. *Journal of Biotechnology*, *123*(4), 413–433. DOI: 10.1016/j.jbiotec.2005.12.036

Cheah, I. K., & Halliwell, B. (2012). Ergothioneine; antioxidant potential, physiological function, and role in disease. In Biochimica et Biophysica Acta - Molecular Basis of Disease (Vol. 1822, Issue 5, pp. 784–793). DOI: 10.1016/j.bbadis.2011.09.017

Chelladurai, G., Yadav, T. K., & Pathak, R. K. (2021). Chemical composition and nutritional value of paddy straw milky mushroom (Calocybe indica). *Nature Environment and Pollution Technology*, *20*(3), 1157–1164. DOI: 10.46488/NEPT.2021.v20i03.023

Chen, K., & Pachter, L. (2005). Bioinformatics for Whole-Genome Shotgun Sequencing of Microbial Communities. *PLoS Computational Biology*, *1*(2), e24. DOI: 10.1371/journal.pcbi.0010024 PMID: 16110337

Chen, M., Arato, M., Borghi, L., Nouri, E., & Reinhardt, D. (2018). Beneficial services of arbuscular mycorrhizal fungi–from ecology to application. *Frontiers in Plant Science*, *9*, 1270. DOI: 10.3389/fpls.2018.01270 PMID: 30233616

Chen, S. H., Cheow, Y. L., Ng, S. L., & Ting, A. S. Y. (2020). Removal of triphenylmethane dyes in single-dye and dye-metal mixtures by live and dead cells of metal-tolerant Penicillium simplicissimum. *Separation Science and Technology*, *55*(13), 2410–2420. DOI: 10.1080/01496395.2019.1626422

Chen, X. W., Wu, L., Luo, N., Mo, C. H., Wong, M. H., & Li, H. (2019). Arbuscular mycorrhizal fungi and the associated bacterial community influence the uptake of cadmium in rice. *Geoderma*, *337*, 749–757. DOI: 10.1016/j.geoderma.2018.10.029

Chepesiuk, R. (2009). Missing the dark: Health effects of light pollution. *Environmental Health Perspectives*, *117*(1), A20–A27. DOI: 10.1289/ehp.117-a20 PMID: 19165374

Cherta-Murillo, A., & Frost, G. S. (2022). The association of mycoprotein-based food consumption with diet quality, energy intake and non-communicable diseases' risk in the UK adult population using the National Diet and Nutrition Survey (NDNS) years 2008/2009–2016/2017: A cross-sectional study. British Journal of Nutrition, 127, 1685–1694. *Frontiers in Sustainable Food Systems*, *5*, 581682.

Cherta-Murillo, A., Lett, A. M., Frampton, J. E., Chambers, E. S., Finnigan, T. J. A., & Frost, G. (2020). Effects of mycoprotein on glycaemic control and energy intake in humans: A systematic review. *British Journal of Nutrition*, *123*(12), 1321–1332. DOI: 10.1017/S0007114520000756

Cheung, P. C. K. (2013). Mini-review on Edible Mushrooms as Source of Dietary Fiber: Preparation and Health Benefits. In Food Science and Human Wellness (Vol. 2, Issues 3–4, pp. 162–166). Elsevier B.V. DOI: 10.1016/j.fshw.2013.08.001

Chhikara, N., Kaur, A., Mann, S., Garg, M., Sofi, S.A. & Panghal, A. (2021). Bioactive compounds, associated health benefits and safety considerations of Moringa oleifera L.: an updated review.

Chiu, C. H., & Paszkowski, U. (2019). Mechanisms and impact of symbiotic phosphate acquisition. *Cold Spring Harbor Perspectives in Biology*, *11*(6), a034603. Advance online publication. DOI: 10.1101/cshperspect.a034603 PMID: 30910773

Chojnacka, K., Mikula, K., Izydorczyk, G., Skrzypczak, D., Witek-Krowiak, A., Moustakas, K., Ludwig, W., & Kułażyński, M. (2021). Improvements in drying technologies - Efficient solutions for cleaner production with higher energy efficiency and reduced emission. *Journal of Cleaner Production*, *320*, 128706. DOI: 10.1016/j.jclepro.2021.128706

Chong, P. S., Fung, M. L., Wong, K. H., & Lim, L. W. (2019). Therapeutic potential of Hericiumerinaceus for depressive disorder. *International Journal of Molecular Sciences*, *21*(1), 163. DOI: 10.3390/ijms21010163 PMID: 31881712

Chulze, S. N. (2010). Strategies to reduce mycotoxin levels in maize during storage: A review. *Food Additives & Contaminants. Part A, Chemistry, Analysis, Control, Exposure & Risk Assessment*, *27*(5), 651–657. DOI: 10.1080/19440040903573032 PMID: 20349375

Chun, S., Gopal, J., & Muthu, M. (2021). Antioxidant activity of mushroom extracts/polysaccharides —their antiviral properties and plausible antiCOVID-19 properties. In Antioxidants (Vol. 10, Issue 12). MDPI. DOI: 10.3390/antiox10121899

Chun, J., & Hartung, H.-P. (2010). Mechanism of action of oral fingolimod (FTY720) in multiple sclerosis. *Clinical Neuropharmacology*, *33*(2), 91–101. DOI: 10.1097/WNF.0b013e3181cbf825 PMID: 20061941

Chun, S. C., Paramasivan, M., & Chandrasekaran, M. (2018). Proline Accumulation Influenced by Osmotic Stress in Arbuscular Mycorrhizal Symbiotic Plants. *Frontiers in Microbiology*, *9*, 2525. Advance online publication. DOI: 10.3389/fmicb.2018.02525 PMID: 30459731

Chutima, R., & Lumyong, S. (2012). Production of indole-3-acetic acid by Thai native orchid-associated fungi. *Symbiosis*, *56*(1), 35–44. DOI: 10.1007/s13199-012-0158-2

Clapp, J. P., Fitter, A. H., & Young, J. P. W. (1999). Ribosomal small subunit sequence variation within spores of an arbuscular mycorrhizal fungus, *Scutellospora* sp. *Molecular Ecology*, *8*(6), 915–921. DOI: 10.1046/j.1365-294x.1999.00642.x PMID: 10434413

Cleland, J. (2013). World Population Growth; Past, Present and Future. *Environmental and Resource Economics*, *55*(4), 543–554. DOI: 10.1007/s10640-013-9675-6

Cobos, Á., & Díaz, O. (2023). "Superfoods": Reliability of the Information for Consumers Available on the Web. *Foods*, *12*(3), 546. Advance online publication. DOI: 10.3390/foods12030546 PMID: 36766074

Coelho, L. C. S., Mignoni, D. S. B., Silva, F. S. B., & Braga, M. R. (2019). Seed exudates of Sesbania virgata (Cav.) Pers. stimulate the asymbiotic phase of the arbuscular mycorrhizal fungus Gigaspora albida Becker & Hall. *Hoehnea*, *46*(1), e272018. DOI: 10.1590/2236-8906-27/2018

Coelho, M., Monteyne, A., Dirks, M., Finnigan, T., Stephens, F., & Wall, B. (2021). Daily mycoprotein consumption for 1 week does not affect insulin sensitivity or glycaemic control but modulates the plasma lipidome in healthy adults: A randomised controlled trial. *British Journal of Nutrition*, *125*(2), 147–160. DOI: 10.1017/S0007114520002524 PMID: 32660657

Colombo, R. P., Fernández Bidondo, L., Silvani, V. A., Carbonetto, M. B., Rascovan, N., Bompadre, M. J., Pérgola, M., Cuenca, G., & Godeas, A. M. (2014). Diversity of arbuscular mycorrhizal fungi in soil from the pampa ondulada, Argentina, Assessed by pyrosequencing and morphological techniques. *Canadian Journal of Microbiology*, *60*(12), 819–827. DOI: 10.1139/cjm-2014-0364 PMID: 25408171

Colot, H. V., Loros, J. J., & Dunlap, J. C. (2005). Temperature-modulated Alternative Splicing and Promoter Use in the Circadian Clock Gene frequency. *Molecular Biology of the Cell*, *16*(12), 5563–5571. DOI: 10.1091/mbc.e05-08-0756 PMID: 16195340

Čolović, R., Puvača, N., Cheli, F., Avantaggiato, G., Greco, D., Đuragić, O., Kos, J., & Pinotti, L. (2019). Decontamination of Mycotoxin-Contaminated Feedstuffs and Compound Feed. *Toxins*, *11*(11), 617. DOI: 10.3390/toxins11110617 PMID: 31731462

Conde-Hernández, L. A., Espinosa-Victoria, J. R., Trejo, A., & Guerrero-Beltrán, J. (2017). CO2-supercritical extraction, hydrodistillation and steam distillation of essential oil of rosemary (*Rosmarinus officinalis*). *Journal of Food Engineering*, *200*, 81–86. DOI: 10.1016/j.jfoodeng.2016.12.022

Cör, D., Knez, Ž., & Hrnčič, M. K. (2018). Antitumour, antimicrobial, antioxidant and antiacetylcholinesterase effect of Ganoderma lucidum terpenoids and polysaccharides: A review. *Molecules (Basel, Switzerland)*, *23*(3), 649. DOI: 10.3390/molecules23030649 PMID: 29534044

Cordeau, S., Triolet, M., Wayman, S., Steinberg, C., & Guillemin, J. (2016). Bioherbicides: Dead in the water? A review of the existing products for integrated weed management. *Crop Protection (Guildford, Surrey)*, *87*, 44–49. DOI: 10.1016/j.cropro.2016.04.016

Cornell, C., Kokkoris, V., Turcu, B., Dettman, J., Stefani, F., & Corradi, N. (2022). The arbuscular mycorrhizal fungus *Rhizophagus irregularis* harmonizes nuclear dynamics in the presence of distinct abiotic factors. *Fungal Genetics and Biology*, *158*, 103639. DOI: 10.1016/j.fgb.2021.103639 PMID: 34800644

Courty, P., Doidy, J., Garcia, K., Wipf, D., & Zimmermann, S. D. (2016). The transportome of mycorrhizal systems. *Molecular Mycorrhizal Symbiosis*, 239–256.

Cranenbrouck, S., Voets, L., Bivort, C., Renard, L., Strullu, D.-G., & Declerck, S. (2005). Methodologies for in Vitro Cultivation of Arbuscular Mycorrhizal Fungi with Root Organs. In Declerck, S., Fortin, J. A., & Strullu, D.-G. (Eds.), *In Vitro Culture of Mycorrhizas* (pp. 341–375). Springer Berlin Heidelberg., DOI: 10.1007/3-540-27331-X_18

Cui, J., Tian, S., Gu, Y., Wu, X., Wang, L., Wang, J., Chen, X., & Meng, Z. (2023). Toxicity effects of pesticides based on zebrafish (*Danio rerio*) models: Advances and perspectives. *Chemosphere*, *340*, 139825. DOI: 10.1016/j.chemosphere.2023.139825 PMID: 37586498

d'Errico, G., Aloj, V., Flematti, G. R., Sivasithamparam, K., Worth, C. M., Lombardi, N., Ritieni, A., Marra, R., Lorito, M., & Vinale, F. (2021). Metabolites of a Drechslera sp. endophyte with potential as biocontrol and bioremediation agent. *Natural Product Research*, *35*(22), 4508–4516. DOI: 10.1080/14786419.2020.1737058 PMID: 32159387

Daba, A. S., & Ezeronye, O. U. (2003). Anti-cancer effect of polysaccharides isolated from higher basidiomycetes mushrooms. *African Journal of Biotechnology*, *2*(12), 672–678. https://www.academicjournals.org/AJB. DOI: 10.5897/AJB2003.000-1123

Dai, H., Zhang, X., Zhao, B., Shi, J., Zhang, C., Wang, G., Yu, N., & Wang, E. (2022). Colonization of mutualistic mycorrhizal and parasitic blast fungi requires *OsRAM2*-regulated fatty acid biosynthesis in rice. *Molecular Plant-Microbe Interactions*, *35*(3), 178–186. DOI: 10.1094/MPMI-11-21-0270-R PMID: 34941378

Dalbanjan, N. P., Eelager, M. P., & Narasagoudr, S. S. (2024). Microbial protein sources: A comprehensive review on the potential usage of fungi and cyanobacteria in sustainable food systems. *Food and Humanity*, *3*, 100366. DOI: 10.1016/j.foohum.2024.100366

Danouche, M., Ferioun, M., Bahafid, W., & El Ghachtouli, N. (2021). Mycoremediation of azo dyes using *Cyberlindnera fabianii* yeast strain: Application of designs of experiments for decolorization optimization. *Water Environment Research*, *93*(8), 1402–1416. DOI: 10.1002/wer.1499 PMID: 33331006

Daou, R., Joubrane, K., Maroun, R. G., Rabbaa Khabbaz, L., Ismail, A., & El Khoury, A. (2021). Mycotoxins: Factors influencing production and control strategies. *AIMS Agriculture and Food*, *6*(1), 416–447. DOI: 10.3934/agrfood.2021025

Das, A. K., Nanda, P. K., Dandapat, P., Bandyopadhyay, S., Gullón, P., Sivaraman, G. K., McClements, D. J., Gullón, B., & Lorenzo, J. M. (2021). Edible mushrooms as functional ingredients for development of healthier and more sustainable Muscle Foods: A Flexitarian approach. *Molecules (Basel, Switzerland), 26*(9), 2463. DOI: 10.3390/molecules26092463 PMID: 33922630

Das, A., Chen, C. M., Mu, S. C., Yang, S. H., Ju, Y. M., & Li, S. C. (2022). Medicinal Components in Edible Mushrooms on Diabetes Mellitus Treatment. *Pharmaceutics, 14*(2), 436. Advance online publication. DOI: 10.3390/pharmaceutics14020436 PMID: 35214168

Das, A., & Sakthivel, K. M. (2020). Anthraquinone from Edible Fungi Pleurotus ostreatus Protects Human SH-SY5Y Neuroblastoma Cells Against 6-Hydroxydopamine-Induced Cell Death—Preclinical Validation of Gene Knockout Possibilities of PARK7, PINK1, and SNCA1 Using CRISPR SpCas9. *Applied Biochemistry and Biotechnology, 191*(2), 555–566. DOI: 10.1007/s12010-019-03188-7 PMID: 31820379

David, O. M., Olawusi, A. C., Oluwole, O. A., Adeola, P. O., & Odeyemi, A. T. (2023). Isolation, Molecular Characterization and Application of Aspergillus niger and Penicillium chrysogenum with Biofertilizer Potentials to Enhance Rice Growth. *Tropical Journal of Natural Product Research, 7*(4), 2790–2795. DOI: 10.26538/tjnpr/v7i4.20

De Castro, O., Avino, M., Di Maio, A., Menale, B., & Guida, M. (2018). Sanger and next generation sequencing in the characterisation of arbuscular mycorrhizal fungi (AMF) in *Pancratium maritimum* L. (Amaryllidaceae), a representative plant species of Mediterranean sand dunes. *Planta, 248*(6), 1443–1453. DOI: 10.1007/s00425-018-2981-z PMID: 30128603

De Girolamo, A., Lippolis, V., & Pascale, M. (2022). Overview of Recent Liquid Chromatography Mass Spectrometry-Based Methods for Natural Toxins Detection in Food Products. *Toxins, 14*(5), 328. DOI: 10.3390/toxins14050328 PMID: 35622576

de Jesus Oliveira, J., dos Passos, E. M., Alves, S. M., Sarmento, V. H. V., Bjerk, T. R., Cardoso, J. C., Blanco-Llamero, C., Souto, E. B., Severino, P., & da Costa Mendonça, M. (2024). Microemulsion of essential oil of *Citrus aurantium* var. *dulcis* for control of *Aleurocanthus woglumi* and evaluation of selectivity against *Aschersonia aleyrodis* and *Ceraeochrysa cornuta.Crop Protection (Guildford, Surrey), 178*, 106586. DOI: 10.1016/j.cropro.2024.106586

de Medeiros Barbosa, I., da Cruz Almeida, É. T., Gomes, A. C. A., & de Souza, E. L. (2020). Evidence on the induction of viable but non-culturable state in Listeria monocytogenes by *Origanum vulgare* L. and *Rosmarinus officinalis* L. essential oils in a meat-based broth. *Innovative Food Science & Emerging Technologies, 62*, 102351. DOI: 10.1016/j.ifset.2020.102351

de Moura Dickel, J. D., Carvalho, J. K., Silveira, M. A. D., Menegotto dos Santos, P., Rodrigues, M. L. F., Fagundes-Klen, M. R., Rosa, C. A., Johann, S., Buzanello, C. V., da Silva de Lucca, R. A., de Oliveira Santos, A. R., & da Rosa, M. F. (2023). Aspergillus sclerotiorum lipolytic activity and its application in bioremediation of high-fat dairy wastewater environments. *Environmental Science and Pollution Research International, 30*(13), 35517–35527. DOI: 10.1007/s11356-022-24669-z PMID: 36529799

De Santis, B., Debegnach, F., Gregori, E., Russo, S., Marchegiani, F., Moracci, G., & Brera, C. (2017). Development of a LC-MS/MS Method for the Multi-Mycotoxin Determination in Composite Cereal-Based Samples. *Toxins, 9*(5), 169. DOI: 10.3390/toxins9050169 PMID: 28524101

De Silva, D. D., Rapior, S., Hyde, K. D., & Bahkali, A. H. (2012). Medicinal mushrooms in prevention and control of diabetes mellitus. *Fungal Diversity*, *56*(1), 1–29. DOI: 10.1007/s13225-012-0187-4 PMID: 23097638

de Sousa, V. M. C., dos Santos, E. F., & Sgarbieri, V. C. (2011). The Importance of Prebiotics in Functional Foods and Clinical Practice. *Food and Nutrition Sciences*, *02*(02), 133–144. DOI: 10.4236/fns.2011.22019

De Souza, F. A., Kowalchuk, G. A., Leeflang, P., Van Veen, J. A., & Smit, E. (2004). PCR-Denaturing Gradient Gel Electrophoresis Profiling of Inter- and Intraspecies 18S rRNA Gene Sequence Heterogeneity Is an Accurate and Sensitive Method to Assess Species Diversity of Arbuscular Mycorrhizal Fungi of the Genus *Gigaspora.Applied and Environmental Microbiology*, *70*(3), 1413–1424. DOI: 10.1128/AEM.70.3.1413-1424.2004 PMID: 15006761

Demirhan, B., & Demirhan, B. E. (2023). Analysis of multi-mycotoxins in commonly consumed spices using the LC-MS/MS method for assessing food safety risks. *Microorganisms*, *11*(7), 1786. DOI: 10.3390/microorganisms11071786 PMID: 37512958

Demunshi, Y., & Chugh, A. (2010). Role of traditional knowledge in marine bioprospecting. *Biodiversity and Conservation*, *19*(11), 3015–3033. DOI: 10.1007/s10531-010-9879-9

Derbyshire, E.J. & Delange, J. (2021). Fungal protein–what is it and what is the health evidence? A systematic review focusing on mycoprotein.

Derbyshire, E. J. (2020). Flexitarian diets and health: A review of the evidence-based literature. *Frontiers in Nutrition*, *7*, 597. PMID: 28111625

Derbyshire, E. J., & Delange, J. (2021). Fungal protein–what is it and what is the health evidence? A systematic review focusing on mycoprotein. *Frontiers in Sustainable Food Systems*, *5*, 581682. DOI: 10.3389/fsufs.2021.581682

Derbyshire, E., & Ayoob, K.-T. (2019). Mycoprotein: Nutritional and health properties. *Nutrition Today*, *54*(1), 7–15. DOI: 10.1097/NT.0000000000000316

Desbarats, J. (2020). Pyridoxal 50-phosphate to mitigate immune dysregulation and coagulopathy in Covid-19, 2020050144.

Deshmukh, R., Khardenavis, A. A., & Purohit, H. J. (2016). Diverse metabolic capacities of fungi for bioremediation. *Indian Journal of Microbiology*, *56*(3), 247–264. DOI: 10.1007/s12088-016-0584-6 PMID: 27407289

Deveau, A., Bonito, G., Uehling, J., Paoletti, M., Becker, M., Bindschedler, S., Hacquard, S., Herve, V., Labbé, J., Lastovetsky, O. A., Mieszkin, S., Millet, L. J., Vajna, B., Junier, P., Bonfante, P., Krom, B. P., Olsson, S., van Elsas, J. D., & Wick, L. Y. (2018). Bacterial–fungal interactions: Ecology, mechanisms and challenges. *FEMS Microbiology Reviews*, *42*(3), 335–352. DOI: 10.1093/femsre/fuy008 PMID: 29471481

Devendrapandi, G., Liu, X., Balu, R., Ayyamperumal, R., Arasu, M. V., Lavanya, M., & Karthika, P. C. (2024). Innovative remediation strategies for persistent organic pollutants in soil and water: A comprehensive review. *Environmental Research*, *249*, 118404. DOI: 10.1016/j.envres.2024.118404 PMID: 38341071

Devi, P. V., Islam, J., Narzary, P., Sharma, D., & Sultana, F. (2024). Bioactive compounds, nutraceutical values and its application in food product development of oyster mushroom. *Journal of Future Foods*, *4*(4), 335–342. DOI: 10.1016/j.jfutfo.2023.11.005

Dey, A., Patel, S., & Rashid, M. M. (2020). Small-scale mushroom production unit for the upliftment of rural economy and women empowerment in India: A Review. *Current Journal of Applied Science and Technology*, *39*, 38–46. DOI: 10.9734/cjast/2020/v39i230495

Dhakal, A., Hashmi, M. F., & Sbar, E. (2023). Aflatoxin Toxicity. In *StatPearls*. StatPearls Publishing.

Dharumadurai Dhanasekaran, D. D., Subramaniyan Lawanya, S. L., Subhasish Saha, S. S., Nooruddin Thajuddin, N. T., & Annamalai Panneerselvam, A. P. (2011). Production of single cell protein from pineapple waste using yeast. https://www.cabidigitallibrary.org/doi/full/10.5555/20113172515

Di Francesco, A., Moret, E., Cignola, R., Garagozzo, L., Torelli, E., & Di Foggia, M. (2024). Yeasts volatile organic compounds (VOCs) as potential growth enhancers and molds biocontrol agents of mushrooms mycelia. *Fungal Biology*, *128*(4), 1859–1867. DOI: 10.1016/j.funbio.2024.05.007 PMID: 38876538

Diagne, N., Ngom, M., Djighaly, P. I., Fall, D., Hocher, V., & Svistoonoff, S. (2020). Roles of arbuscular mycorrhizal fungi on plant growth and performance: Importance in biotic and abiotic stressed regulation. *Diversity*, *12*(10), 370. DOI: 10.3390/d12100370

Dimopoulou, M., Kolonas, A., Mourtakos, S., Androutsos, O., & Gortzi, O. (2022). Nutritional Composition and Biological Properties of Sixteen Edible Mushroom Species. *Applied Sciences (Basel, Switzerland)*, *12*(16), 8074. Advance online publication. DOI: 10.3390/app12168074

Dinakarkumar, Y., Ramakrishnan, G., Gujjula, K. R., Vasu, V., Balamurugan, P., & Murali, G. (2024). Fungal bioremediation: An overview of the mechanisms, applications and future perspectives. *Environmental Chemistry and Ecotoxicology, 6*, 293–302. https://doi.org/https://doi.org/10.1016/j.enceco.2024.07.002

Dinakarkumar, Y., Gnanasekaran, R., Reddy, G. K., Vasu, V., Balamurugan, P., & Murali, G. (2024). *Fungal bioremediation: An overview of the mechanisms, applications and future perspectives*. Environmental Chemistry and Ecotoxicology.

Dinakarkumar, Y., Ramakrishnan, G., Gujjula, K. R., Vasu, V., Balamurugan, P., & Murali, G. (2024). Fungal bioremediation: An overview of the mechanisms, applications and future perspectives. *Environmental Chemistry and Ecotoxicology*, *6*, 293–302. DOI: 10.1016/j.enceco.2024.07.002

Di, Z., Chaoyang, L., Mengxi, Z., Yunlin, Z., Zhenggang, X., & Guiyan, Y. (2022). Curvularia coatesiae XK8, a Potential Bioadsorbent Material for Adsorbing Cd(II) and Sb(III) Compound Pollution: Characteristics and Effects. *Frontiers in Microbiology*, *12*(January), 1–13. DOI: 10.3389/fmicb.2021.816312 PMID: 35154041

Djonovic, S., Vargas, W. A., Kolomiets, M. V., Horndeski, M., Wiest, A., & Kenerley, C. M. (2007). A proteinaceous elicitor Sm1 from the beneficial fungus Trichoderma virens is required for induced systemic resistance in maize. *Plant Physiology*, *145*(3), 875–889. DOI: 10.1104/pp.107.103689 PMID: 17885089

Doidy, J., Grace, E., Kühn, C., Simon-Plas, F., Casieri, L., & Wipf, D. (2012). Sugar transporters in plants and in their interactions with fungi. *Trends in Plant Science*, *17*(7), 413–422. DOI: 10.1016/j.tplants.2012.03.009 PMID: 22513109

Domínguez-Núñez, J. A., Berrocal-Lobo, M., & Albanesi, A. S. (2019). *Ectomycorrhizal Fungi: Role as Biofertilizers in Forestry BT - Biofertilizers for Sustainable Agriculture and Environment* (B. Giri, R. Prasad, Q.-S. Wu, & A. Varma (eds.); pp. 67–82). Springer International Publishing. DOI: 10.1007/978-3-030-18933-4_4

Dong, L., Lu, Z., & Jiabao, L. (2022). *Alkali-resistant yeast and application thereof in production of single-cell protein by utilizing biogas slurry* (CN 113913309 A). https://lens.org/031-552-795-216-120

Doni, F., Zain, C. R. C. M., Isahak, A., Fathurrahman, F., Anhar, A., Mohamad, W. N. W., Yusoff, W. M. W., & Uphoff, N. (2018). A simple, efficient, and farmer-friendly Trichoderma-based biofertilizer evaluated with the SRI Rice Management System. *Organic Agriculture*, *8*(3), 207–223. DOI: 10.1007/s13165-017-0185-7

Doria, G., Conde, J., Veigas, B., Giestas, L., Almeida, C., Assunção, M., Rosa, J., & Baptista, P. V. (2012). Noble metal nanoparticles for biosensing applications. *Sensors (Basel)*, *12*(2), 1657–1687. DOI: 10.3390/s120201657 PMID: 22438731

Dorostkar, V., Afyuni, M., Khoshgoftarmanesh, A. H., & Rejali, F. (2014). Effects of salinity on exudates of Glomus species in wheat rhizosphere. *Advances in Environmental Biology*, •••, 219–226.

Douds, D. D.Jr, Lee, J., Rogers, L., Lohman, M. E., Pinzon, N., & Ganser, S. (2012). Utilization of inoculum of AM fungi produced on-farm for the production of Capsicum annuum: A summary of seven years of field trials on a conventional vegetable farm. *Biological Agriculture and Horticulture*, *28*(2), 129–145. DOI: 10.1080/01448765.2012.693362

Dövényi-Nagy, T., Rácz, C., Molnár, K., Bakó, K., Szláma, Z., Jóźwiak, Á., Farkas, Z., Pócsi, I., & Dobos, A. C. (2020). Pre-Harvest Modelling and Mitigation of Aflatoxins in Maize in a Changing Climatic Environment-A Review. *Toxins*, *12*(12), 768. DOI: 10.3390/toxins12120768 PMID: 33291729

Draoui, A., Ouinten, M., Gourine, N., & Yousfi, M. (2023). Acute anti-inflammatory activity of Artemisia campestris and *Citrus aurantium*: Effects of essential oil combinations. *Biocatalysis and Agricultural Biotechnology*, *53*, 102860. DOI: 10.1016/j.bcab.2023.102860

Drumm, F. C., Franco, D. S. P., Georgin, J., Grassi, P., Jahn, S. L., & Dotto, G. L. (2021). Macro-fungal (Agaricus bisporus) wastes as an adsorbent in the removal of the acid red 97 and crystal violet dyes from ideal colored effluents. *Environmental Science and Pollution Research International*, *28*(1), 405–415. DOI: 10.1007/s11356-020-10521-9 PMID: 32812157

Druzhinina, I. S., Seidl-Seiboth, V., Herrera-Estrella, A., Horwitz, B. A., Kenerley, C. M., Monte, E., Mukherjee, P. K., Zeilinger, S., Grigoriev, I. V., & Kubicek, C. P. (2011). Trichoderma: The genomics of opportunistic success. In *Nature Reviews Microbiology* (Vol. 9, Issue 10, pp. 749–759). DOI: 10.1038/nrmicro2637 du Jardin, P. (2015). Plant biostimulants: Definition, concept, main categories and regulation. *Scientia Horticulturae*, *196*, 3–14. https://doi.org/https://doi.org/10.1016/j.scienta.2015.09.021. DOI: 10.1016/j.scienta.2015.09.021

Dubey, D., Dhakal, B., Dhami, K., Sapkota, P., Rana, M., Poudel, N. S., & Aryal, L. (2019). Comparative study on effect of different substrates on yield performance of oyster mushroom. *Global Journal of Biology, Agriculture and Health Sciences*, •••, 14.

Dubey, S. C., Suresh, M., & Singh, B. (2007). Evaluation of Trichoderma species against Fusarium oxysporum F. Sp. ciceris for integrated management of chickpea wilt. *Biological Control*, *40*(1), 118–127. DOI: 10.1016/j.biocontrol.2006.06.006

Duc, N. H., Szentpéteri, V., Mayer, Z., & Posta, K. (2023). Distinct impact of arbuscular mycorrhizal isolates on tomato plant tolerance to drought combined with chronic and acute heat stress. *Plant Physiology and Biochemistry*, *201*, 107892. DOI: 10.1016/j.plaphy.2023.107892 PMID: 37490823

Duhamel, M., & Vandenkoornhuyse, P. (2013). Sustainable agriculture: Possible trajectories from mutualistic symbiosis and plant neodomestication. *Trends in Plant Science*, *18*(11), 597–600. DOI: 10.1016/j.tplants.2013.08.010 PMID: 24055138

Dunlap, J. C., & Loros, J. J. (2004). The *Neurospora* circadian system. *Journal of Biological Rhythms*, *19*(5), 414–424. DOI: 10.1177/0748730404269116 PMID: 15534321

Dunlap, J. C., & Loros, J. J. (2006). How fungi keep time: Circadian system in *Neurospora* and other fungi. *Current Opinion in Microbiology*, *9*(6), 579–587. DOI: 10.1016/j.mib.2006.10.008 PMID: 17064954

Dunlap, J. C., Loros, J. J., Colot, H. V., Mehra, A., Belden, W. J., Shi, M., Hong, C. I., Larrondo, L. F., Baker, C. L., Chen, C. H., Schwerdtfeger, C., Collopy, P. D., Gamsby, J. J., & Lambreghts, R. (2007). A circadian clock in *Neurospora*: How genes and proteins cooperate to produce a sustained, entrainable, and compensated biological oscillator with a period of about a day. *Cold Spring Harbor Symposia on Quantitative Biology*, *72*(1), 57–68. DOI: 10.1101/sqb.2007.72.072 PMID: 18522516

Dunlap, J. C., Loros, J. J., & DeCoursey, P. J. (2004). *Chronology: Biological timekeeping*. American Psychological Association.

Dunlop, M. V., Kilroe, S. P., Bowtell, J. L., Finnigan, T. J., Salmon, D. L., & Wall, B. T. (2017). Mycoprotein represents a bioavailable and insulinotropic non-animal-derived dietary protein source: A dose–response study. *British Journal of Nutrition*, *118*(9), 673–685. DOI: 10.1017/S0007114517002409 PMID: 29017627

Duo-Chuan, L. I., Chen, S., & Jing, L. U. (2005). Purification and partial characterization of two chitinases from the mycoparasitic fungus Talaromyces flavus. *Mycopathologia*, *159*(2), 223–229. DOI: 10.1007/s11046-004-9096-8 PMID: 15770448

Durrant, W., & Dong, X. (2004). Systemic acquired resistance. *Annual Review of Phytopathology*, *42*(1), 185–209. DOI: 10.1146/annurev.phyto.42.040803.140421 PMID: 15283665

Dutta, U., Jyote, S., Ghosh, K., Pal, S., Sengupta, S., & Laha, A. (2023). Bioremediation of contaminated environments through mycology: a review of current advancements and future prospects: a short review. *Journal of Advanced Zoology*, *44*(S-5), 2611–2616. DOI: 10.53555/jaz.v44iS5.3483

Du, X. J., Zhang, J. S., Yang, Y., Tang, Q. J., Jia, W., & Pan, Y. J. (2010). Purification, chemical modification and immunostimulating activity of polysaccharides from Tremella aurantialba fruit bodies. *Journal of Zhejiang University. Science. B.*, *11*(6), 437–442. DOI: 10.1631/jzus.B0900402 PMID: 20506575

Ebbisa, A. (2023). Arbuscular Mycorrhizal Fungi (AMF) in Optimizing Nutrient Bioavailability and Reducing Agrochemicals for Maintaining Sustainable Agroecosystems. In *Arbuscular Mycorrhizal Fungi in Agriculture - New Insights*. IntechOpen. DOI: 10.5772/intechopen.106995

Ebsa, G., Gizaw, B., Admassie, M., Desalegn, A., & Alemu, T. (2024). Screening, characterization and optimization of potential dichlorodiphenyl trichloroethane (DDT) degrading fungi. *Heliyon*, *10*(12), e33289. DOI: 10.1016/j.heliyon.2024.e33289 PMID: 39022069

Edwards, D. G., & Cummings, J. H. (2010). The protein quality of mycoprotein. Proceedings of the Nutrition Society, 69(OCE4), E331.

Edwards, D. G., & Cummings, J. H. (2010). Mycoprotein: A meat-free dietary protein with a low environmental footprint. *The Proceedings of the Nutrition Society*, *69*(3), 322–329.

Edwin, N. R., Fitzpatrick, A. H., Brennan, F., Abram, F., & O'Sullivan, O. (2024). An in-depth evaluation of metagenomic classifiers for soil microbiomes. *Environmental Microbiome*, *19*(1), 1–17. DOI: 10.1186/s40793-024-00561-w PMID: 38549112

Effiong, M. E., Umeokwochi, C. P., Afolabi, I. S., & Chinedu, S. N. (2024). Assessing the nutritional quality of Pleurotus ostreatus (oyster mushroom). *Frontiers in Nutrition*, *10*, 1279208. DOI: 10.3389/fnut.2023.1279208 PMID: 38292699

Egwim, E. C., Oyewole, O. A., & Yakubu, J. G. J. (2023). *Fungal bioremediation of pollutant*. Bentham., DOI: 10.2174/97898151234941230100009

Eisendle, M., Oberegger, H., Buttinger, R., Illmer, P., & Haas, H. (2004). Biosynthesis and uptake of Siderophores is controlled by the pacc-mediated Ambient-pH regulatory system in *Aspergillus nidulans*. *Eukaryotic Cell*, *3*(2), 561–563. DOI: 10.1128/EC.3.2.561-563.2004 PMID: 15075286

Ekanayake, M. S., & Manage, P. (2022). Mycoremediation Potential of Synthetic Textile Dyes by Aspergillus niger via Biosorption and Enzymatic Degradation: . *Environment and Natural Resources Journal*, *20*(3), 234–245.DOI: 10.32526/ennrj/20/202100171

El Ayari, T., Bouhdida, R., Ouzari, H. I., & El Menif, N. T. (2024). Bioremediation of petroleum refinery wastewater by fungal stains isolated from the fishing harbour of Bizerte (Mediterranean Sea). *Biodegradation*, *35*(5), 755–767. DOI: 10.1007/s10532-024-10083-4 PMID: 38687419

El Kharraf, S., El-Guendouz, S., Farah, A., Bennani, B., Mateus, M. C., El Hadrami, E. M., & Miguel, M. G. (2021). Hydrodistillation and simultaneous hydrodistillation-steam distillation of *Rosmarinus officinalis* and *Origanum compactum*: Antioxidant, anti-inflammatory, and antibacterial effect of the essential oils. *Industrial Crops and Products*, *168*, 113591. DOI: 10.1016/j.indcrop.2021.113591

El Khoury, Y., Noujeim, E., Ravlić, J., Oreste, M., Addante, R., Nemer, N., & Tarasco, E. (2020). The effect of entomopathogenic nematodes and fungi against four xylophagous pests. *Biocontrol Science and Technology*, *30*(9), 983–995. DOI: 10.1080/09583157.2020.1781059

El-Bondkly, A. M. A., & El-Gendy, M. M. A. A. (2022). Bioremoval of some heavy metals from aqueous solutions by two different indigenous fungi *Aspergillus sp.* AHM69 and *Penicillium sp.* AHM96 isolated from petroleum refining wastewater. *Heliyon*, *8*(7), e09854. Advance online publication. DOI: 10.1016/j.heliyon.2022.e09854 PMID: 35815132

El-Ghwas, D. E., Elkhateeb, W. A., & Daba, G. M. (2023). Fungi: The next generation of biofertilizers. *Environ Sci Arch*, *2*(1), 34–41. DOI: 10.5281/zenodo.7500741

El-Khawaga, H. A., Mustafa, A. E., El Khawaga, M. A., Mahfouz, A. Y., & Daigham, G. E. (2024). Biostimulating effect of endophytic Aspergillus flavus AUMC 16068 and its respective ex-polysaccharides in lead stress tolerance of Triticum aestivum plant. *Scientific Reports*, *14*(1), 11952. PMID: 38796501

El-Metwally, M. M., Mekawey, A. A. I., El-Halmouch, Y., & Naga, N. G. (2023). Symbiotic Relationships with Fungi: From Mutualism to Parasitism. In *Plant Mycobiome* (pp. 375–413). Springer International Publishing. DOI: 10.1007/978-3-031-28307-9_15

El-Saadony, M. T., Saad, A. M., Soliman, S. M., Salem, H. M., Ahmed, A. I., Mahmood, M., El-Tahan, A. M., Ebrahim, A. A. M., Abd El-Mageed, T. A., Negm, S. H., Selim, S., & Babalghith, A.

El-Sayed, R. A., Jebur, A. B., Kang, W., & El-Demerdash, F. M. (2022). An overview on the major mycotoxins in food products: Characteristics, toxicity, and analysis. *Journal of Future Foods*, *2*(2), 91–102. DOI: 10.1016/j.jfutfo.2022.03.002

Elumalai, K., Krishnappa, K., Pandiyan, J., Alharbi, N. S., Kadaikunnan, S., Khaled, J. M., Barnard, D. R., Vijayakumar, N., & Govindarajan, M. (2022). Characterization of secondary metabolites from Lamiaceae plant leaf essential oil: A novel perspective to combat medical and agricultural pests. *Physiological and Molecular Plant Pathology*, *117*, 101752. DOI: 10.1016/j.pmpp.2021.101752

Esaïe, T. (2015). Early effects of water stress on some biochemical and mineral parameters of mycorrhizal Vigna subterranea (L.) Verdc. (Fabaceae) cultivated in Cameroon http://www.innspub.net Vol. 7, No. 2, p. 21-35, 2015. *IJAAR, 7*, 21–35.

Espinoza-Sánchez, M. A., Arévalo-Niño, K., Quintero-Zapata, I., Castro-González, I., & Almaguer-Cantú, V. (2019). Cr(VI) adsorption from aqueous solution by fungal bioremediation based using Rhizopus sp. *Journal of Environmental Management*, *251*, 109595. https://doi.org/https://doi.org/10.1016/j.jenvman.2019.109595. DOI: 10.1016/j.jenvman.2019.109595 PMID: 31561145

Esterhuizen-Londt, M., Hertel, S., & Pflugmacher, S. (2017). Uptake and biotransformation of pure commercial microcystin-LR versus microcystin-LR from a natural cyanobacterial bloom extract in the aquatic fungus *Mucor hiemalis.Biotechnology Letters*, *39*(10), 1537–1545. DOI: 10.1007/s10529-017-2378-2 PMID: 28597371

Etesami, H. (2020). In Meena, R. S. (Ed.), *Enhanced Phosphorus Fertilizer Use Efficiency with Microorganisms BT -Nutrient Dynamics for Sustainable Crop Production* (pp. 215–245). Springer Singapore., DOI: 10.1007/978-981-13-8660-2_8

Etesami, H., & Glick, B. R. (2023). Exploring the potential: Can mycorrhizal fungi and hyphosphere silicate-solubilizing bacteria synergistically alleviate cadmium stress in plants? *Current Research in Biotechnology*, *6*, 100158. https://doi.org/https://doi.org/10.1016/j.crbiot.2023.100158. DOI: 10.1016/j.crbiot.2023.100158

Etesami, H., Jeong, B. R., & Glick, B. R. (2021). Contribution of Arbuscular Mycorrhizal Fungi, Phosphate–Solubilizing Bacteria, and Silicon to P Uptake by Plant. *Frontiers in Plant Science*, *12*, 699618. Advance online publication. DOI: 10.3389/fpls.2021.699618 PMID: 34276750

Eun Kang, J., Ciampi, A., & Hijri, M. (2020). SeSaMe: Metagenome Sequence Classification of Arbuscular Mycorrhizal Fungi-associated Microorganisms. *Genomics, Proteomics & Bioinformatics*, *18*(5), 601–612. DOI: 10.1016/j.gpb.2018.07.010 PMID: 33346086

European Biostimulants Industry Council (EBIC). (2012). *The European Biostimulants Industry Council*. EBIC.

European Commission. (2006). *Commission Regulation (EC) No 1881/2006 of 19 December 2006 setting maximum levels for certain contaminants in foodstuffs*. EUR-Lex. https://eur-lex.europa.eu/legal-content/EN/TXT/PDF/?uri=CELEX:02006R1881-20230326

Ey, J., Schömig, E., & Taubert, D. (2007). Dietary sources and antioxidant effects of ergothioneine. *Journal of Agricultural and Food Chemistry*, *55*(16), 6466–6474. DOI: 10.1021/jf071328f PMID: 17616140

Fadiji, A. E., Xiong, C., Egidi, E., & Singh, B. K. (2024). Formulation challenges associated with microbial biofertilizers in sustainable agriculture and paths forward. *Journal of Sustainable Agriculture and Environment*, *3*(3), e70006. Advance online publication. DOI: 10.1002/sae2.70006

Falk, S. P., Gadoury, D. M., Cortesi, P., Pearson, R. C., & Seem, R. C. (1995a). Parasitism of Uncinula necator cleistothecia by the mycoparasite Ampelomyces quisqualis. *Phytopathology*, *85*(7), 794–800. DOI: 10.1094/Phyto-85-794

Falk, S. P., Gadoury, D. M., Pearson, R. C., & Seem, R. C. (1995b). Partial control of grape powdery mildew by the mycoparasite *Ampelomyces quisqualis.Plant Disease*, *79*(5), 483–490. DOI: 10.1094/PD-79-0483

Fall, A. F., Nakabonge, G., Ssekandi, J., Founoune-Mboup, H., Apori, S. O., Ndiaye, A., Badji, A., & Ngom, K. (2022). Roles of Arbuscular Mycorrhizal Fungi on Soil Fertility: Contribution in the Improvement of Physical, Chemical, and Biological Properties of the Soil. *Frontiers in Fungal Biology*, *3*, 723892. Advance online publication. DOI: 10.3389/ffunb.2022.723892 PMID: 37746193

Fan, D., Ji, M., Wu, J., Chen, H., Jia, H., Zhang, X., Zhuang, X., & Kong, W. (2023). Grazing does not influence soil arbuscular mycorrhizal fungal diversity, but increases their interaction complexity with plants in dry grasslands on the Tibetan Plateau. *Ecological Indicators*, *148*, 110065. DOI: 10.1016/j.ecolind.2023.110065

Fandohan, P., Gbenou, J. D., Gnonlonfin, B., Hell, K., Marasas, W. F. O., & Wingfield, M. J. (2004). Effect of essential oils on the growth of *Fusarium verticillioides* and fumonisin contamination in corn. *Journal of Agricultural and Food Chemistry*, *52*(22), 6824–6829. DOI: 10.1021/jf040043p PMID: 15506822

Fan, Y., Lu, S., He, M., Yang, L., Hu, W., Yang, Z., Liu, X., Hui, D., Guo, J., & Yang, Y. (2021). Long-term throughfall exclusion decreases soil organic phosphorus associated with reduced plant roots and soil microbial biomass in a subtropical forest. *Geoderma, 404*, 115309. DOI: 10.1016/j.geoderma.2021.115309

FAO. (2003). Development of a framework for good agricultural practices. Committee on Agriculture. https://www.fao.org/3/Y8704e/Y8704e.htm

FAO. (2019). New standards to curb the global spread of plant pests and diseases.

Farag, M. A., Abib, B., Ayad, L., & Khattab, A. R. (2020). Sweet and bitter oranges: An updated comparative review of their bioactives, nutrition, food quality, therapeutic merits and biowaste valorization practices. *Food Chemistry, 331*, 127306. DOI: 10.1016/j.foodchem.2020.127306 PMID: 32593794

Faria, M. R., & Wraight, S. P. (2007). Mycoinsecticides and Mycoacaricides: A comprehensive list with worldwide coverage and international classification of formulation types. *Biological Control, 43*(3), 237–256. DOI: 10.1016/j.biocontrol.2007.08.001

Faria, M., & Wraight, S. P. (2001). Biological control of Bemisia tabaci with fungi. *Crop Protection (Guildford, Surrey), 20*(9), 767–778. DOI: 10.1016/S0261-2194(01)00110-7

Farraji, H., Zaman, N. Q., Tajuddin, R., & Faraji, H. (2016). Advantages and disadvantages of phytoremediation: A concise review. *Int J Env Tech Sci, 2*, 69–75.

Fasolin, L. H., Pereira, R. N., Pinheiro, A. C., Martins, J. T., Andrade, C. C. P., Ramos, O. L., & Vicente, A. A. (2019). Emergent food proteins–towards sustainability, health and innovation. *Food Research International, 2019*(125), 108586. DOI: 10.1016/j.foodres.2019.108586 PMID: 31554037

Feldmann, F., & Schneider, C. (2008). How to produce arbuscular mycorrhizal inoculum with desired characteristics. *Mycorrhiza Work in Horticulture and Forestry*, 305–322.

Feng, Z., Liu, X., Feng, G., Zhu, H., & Yao, Q. (2020). Linking lipid transfer with reduced arbuscule formation in tomato roots colonized by arbuscular mycorrhizal fungus under low pH stress. *Environmental Microbiology, 22*(3), 1036–1051. DOI: 10.1111/1462-2920.14810 PMID: 31608569

Fenu, G., & Malloci, F. M. (2021). Forecasting Plant and Crop Disease: An Explorative Study on Current Algorithms. *Big Data and Cognitive Computing, 5*(1), 2. Advance online publication. DOI: 10.3390/bdcc5010002

Fernández-Ríos, A., Laso, J., Hoehn, D., Amo-Setién, F. J., Abajas-Bustillo, R., Ortego, C., Fullana-i-Palmer, P., Bala, A., Batlle-Bayer, L., Balcells, M., Puig, R., Aldaco, R., & Margallo, M. (2022). A critical review of superfoods from a holistic nutritional and environmental approach. *Journal of Cleaner Production, 379*, 134491. DOI: 10.1016/j.jclepro.2022.134491

Ferreira, I. M. P. L. V. O., Pinho, O., Vieira, E., & Tavarela, J. G. (2010). Brewer's Saccharomyces yeast biomass: Characteristics and potential applications. *Trends in Food Science & Technology, 21*(2), 77–84. DOI: 10.1016/j.tifs.2009.10.008

Figlan, S., & Mwadzingeni, L. (2022). Breeding Tools for Assessing and Improving Resistance and Limiting Mycotoxin Production by *Fusarium graminearum* in Wheat. *Plants, 11*(15), 1933. DOI: 10.3390/plants11151933 PMID: 35893637

Finnigan, T., Mach, K., & Edlin, A. (2024). Mycoprotein: a healthy new protein with a low environmental impact. In *Sustainable protein sources* (pp. 539-566). Academic Press. https://doi.org/DOI: 10.1016/B978-0-323-91652-3.00011-3

Finnigan, T., Needham, L., & Abbott, C. (2017). Mycoprotein: a healthy new protein with a low environmental impact. In Nadathur, S. R., Wanasundara, J. P. D., & Scanlin, L. (Eds.), *Sustainable Protein Sources* (pp. 305–325). Academic Press. DOI: 10.1016/B978-0-12-802778-3.00019-6

Finnigan, T., Needham, L., & Pickering, K. (2019). Mycoprotein: A healthy new protein with a low environmental impact. *Current Developments in Nutrition*, *3*(6), nzz021. DOI: 10.1093/cdn/nzz021 PMID: 31187084

Firenzuoli, F., Gori, L., & Lombardo, G. (2008). The medicinal mushroom Agaricus blazei murrill: Review of literature and pharmaco-toxicological problems. In Evidence-based Complementary and Alternative Medicine (Vol. 5, Issue 1, pp. 3–15). DOI: 10.1093/ecam/nem007

Fischer, S., Hilger, T., Piepho, H.-P., Jordan, I., & Cadisch, G. (2019). Do we need more drought for better nutrition? The effect of precipitation on nutrient concentration in East African food crops. *The Science of the Total Environment*, *658*, 405–415. DOI: 10.1016/j.scitotenv.2018.12.181 PMID: 30579198

Flajs, D., & Peraica, M. (2009). Toxicological properties of citrinin. *Arhiv za Higijenu Rada i Toksikologiju*, *60*(4), 457–464. DOI: 10.2478/10004-1254-60-2009-1992 PMID: 20061247

Floss, D. S., Gomez, S. K., Park, H.-J., MacLean, A. M., Müller, L. M., Bhattarai, K. K., Lévesque-Tremblay, V., Maldonado-Mendoza, I. E., & Harrison, M. J. (2017). A transcriptional program for arbuscule degeneration during AM symbiosis is regulated by MYB1. *Current Biology*, *27*(8), 1206–1212. DOI: 10.1016/j.cub.2017.03.003 PMID: 28392110

Food and Agriculture Organization of the United Nations & World Health Organization. (2017). *Code of practice for the prevention and reduction of mycotoxin contamination in cereals (CXC 78-2017)*. https://www.fao.org/fao-who-codexalimentarius/sh-proxy/en/?lnk=1&url=https%253A%252F%252Fworkspace.fao.org%252Fsites%252Fcodex%252FStandards%252FCXC%2B78-2017%252FCXC_078e.pdf

Food and Agriculture Organization of the United Nations. (1995). *General standard for contaminants and toxins in food and feed (CXS 193-1995)*. https://www.fao.org/fileadmin/user_upload/agns/pdf/CXS_193e.pdf

Food and Agriculture Organization of the United Nations. (2023). *International Food Safety Authorities Network (INFOSAN)*. https://www.fao.org/sustainable-development-goals-helpdesk/champion/article-detail/international-food-safety-authorities-network-(infosan)/en

Food Standards Australia New Zealand. (2023). *Peanut, pistachio, and aflatoxins*. https://www.foodstandards.gov.au/sites/default/files/2023-11/Peanut%20pistachio%20and%20aflatoxins.pdf

Fouda, A. H., Hassan, S. E.-D., Eid, A. M., & Ewais, E. E.-D. (2015). Biotechnological applications of fungal endophytes associated with medicinal plant Asclepias sinaica (Bioss.). *Annals of Agricultural Science*, *60*(1), 95–104. DOI: 10.1016/j.aoas.2015.04.001

Fountain, J. C., Khera, P., Yang, L., Nayak, S. N., Scully, B. T., Lee, R. D., Chen, Z.-Y., Kemerait, R. C., Varshney, R. K., & Guo, B. (2015). Resistance to *Aspergillus flavus* in maize and peanut: Molecular biology, breeding, environmental stress, and future perspectives. *The Crop Journal*, *3*(3), 229–237. DOI: 10.1016/j.cj.2015.02.003

Frac, M., Hannula, S., Bełka, M., & Jedryczka, M. (2018). Fungal Biodiversity and Their Role in Soil Health. *Frontiers in Microbiology*, *9*, 707. Advance online publication. DOI: 10.3389/fmicb.2018.00707 PMID: 29755421

Franco Lucas, B., Götze, F., Vieira Costa, J. A., & Brunner, T. A. (2022). Consumer Perception Toward "Superfoods": A Segmentation Study. *Journal of International Food & Agribusiness Marketing*, *35*(5), 603–621. DOI: 10.1080/08974438.2022.2044955

Franken, P. (2012). The *Piriformospora indica* story: From laboratory to field. *Mycorrhiza*, *22*(8), 631–639. DOI: 10.1007/s00572-012-0459-x PMID: 22476582

Freire, J.C.P., Júnior, J.K.O., Silva, D.F., Sousa, J.P., Guerra, F.Q.S., de Oliveira Lima, E. (2017). Antifungal Activity of Essential Oils against *Candida albicans* Strains Isolated from Users of Dental Prostheses. *Evidence-Based Complementary and Alternative Medicine*. Article ID 7158756, 9 pages. https://doi.org/DOI: 10.1155/2017/7158756

Frey, S. D. (2007). Spatial Distribution of Soil Organisms. In *Soil Microbiology* (pp. 283–300). Ecology and Biochemistry.

Friedman, M. (2016). Mushroom Polysaccharides: Chemistry and Antiobesity, Antidiabetes, Anticancer, and Antibiotic Properties in Cells, Rodents, and Humans. *Foods*, *5*(4), 80. DOI: 10.3390/foods5040080 PMID: 28231175

Fryer, C. (2024). Mushrooms for Inflammation: Understanding their anti-inflammatory benefits. https://maoi.org/health/mushrooms-for-inflammation/

Furey, B., Slingerland, K., Bauter, M. R., Dunn, C., Goodman, R. E., & Koo, S. (2022). Safety evaluation of Fy Protein™ (Nutritional Fungi Protein), a macroingredient for human consumption. *Food and Chemical Toxicology*, *166*, 113005. DOI: 10.1016/j.fct.2022.113005

Fusconi, A., & Mucciarelli, M. (2018). How important is arbuscular mycorrhizal colonization in wetland and aquatic habitats? *Environmental and Experimental Botany*, *155*, 128–141. DOI: 10.1016/j.envexpbot.2018.06.016

Gajc-Wolska, J., Kowalczyk, K., Nowecka, M., Mazur, K., & Metera, A. (n.d.). *EFFECT OF ORGANIC-MINERAL FERTILIZERS ON THE YIELD AND QUALITY OF ENDIVE.*

Galea, V. J. (2021). Use of Stem Implanted Bioherbicide Capsules to Manage an Infestation of *Parkinsonia aculeata* in Northern Australia. *Plants*, *10*(9), 1909. DOI: 10.3390/plants10091909 PMID: 34579440

Ganeshpurkar, A., Rai, G., & Jain, A. (2010). Medicinal Mushrooms: Towards a New Horizon. *Pharmacognosy Reviews*, *4*(8), 127–135. DOI: 10.4103/0973-7847.70904 PMID: 22228952

Gao, L., Meng, J., Dai, W., Zhang, Z., Dong, H., Yuan, Q., Zhang, W., Liu, S., & Wu, X. (2023). Deciphering cell wall sensors enabling the construction of robust P. pastoris for single-cell protein production. *Biotechnology for Biofuels and Bioproducts, 16*(1), 178. DOI: 10.1186/s13068-023-02428-7

Gao, T., Qin, D., Zuo, S., Peng, Y., Xu, J., Yu, B., Song, H., & Dong, J. (2020). Decolorization and detoxification of triphenylmethane dyes by isolated endophytic fungus, Bjerkandera adusta SWUSI4 under non-nutritive conditions. *Bioresources and Bioprocessing, 7*(1), 53. DOI: 10.1186/s40643-020-00340-8

Gao, Y., Li, D., & Liu, Y. (2012). Production of single cell protein from soy molasses using Candida tropicalis. *Annals of Microbiology, 62*(3), 1165–1172. DOI: 10.1007/s13213-011-0356-9

Gardes, M., & Bruns, T. D. (1993). ITS primers with enhanced specificity for Basidiomycetes—Application to the identification of mycorrhizae and rusts. *Molecular Ecology, 2*(2), 113–118. DOI: 10.1111/j.1365-294X.1993.tb00005.x

Gardner, R. C., Howarth, A. J., Hahn, P., Brown-luedi, M., Shepherd, R. J., & Messing, J. (1981). The complete nucleotide sequence of an infectious clone of cauliflower mosaic virus by M13mp7 shotgun sequencing. *Nucleic Acids Research, 9*(12), 2871–2888. DOI: 10.1093/nar/9.12.2871 PMID: 6269062

Garzón, J. M., Rodríguez-Miranda, J. P., & Hernández-Gómez, C. (2017). Aporte de la biorremediación para solucionar problemas de contaminación y su relación con el desarrollo sostenible. *Universidad y salud, 19*(2), 309-318.

Gaur, P. K., Lau, H. P., Pestka, J. J., & Chu, F. S. (1981). Production and characterization of aflatoxin B2a antiserum. *Applied and Environmental Microbiology, 41*(2), 478–482. DOI: 10.1128/aem.41.2.478-482.1981 PMID: 7235694

Gavito, M. E., Jakobsen, I., Mikkelsen, T. N., & Mora, F. (2019). Direct evidence for modulation of photosynthesis by an arbuscular mycorrhiza-induced carbon sink strength. *The New Phytologist, 223*(2), 896–907. DOI: 10.1111/nph.15806 PMID: 30891762

Gegenbauer, C., Bellaire, A., Schintlmeister, A., Schmid, M. C., Kubicek, M., Voglmayr, H., Zotz, G., Richter, A., & Mayer, V. E. (2023). Exo- and endophytic fungi enable rapid transfer of nutrients from ant waste to orchid tissue. *The New Phytologist, 238*(5), 2210–2223. https://doi.org/https://doi.org/10.1111/nph.18761. DOI: 10.1111/nph.18761 PMID: 36683444

Genre, A., Chabaud, M., Timmers, T., Bonfante, P., & Barker, D. G. (2005). Arbuscular Mycorrhizal Fungi Elicit a Novel Intracellular Apparatus in Medicago truncatula Root Epidermal Cells before Infection [W]. *The Plant Cell, 17*(12), 3489–3499. DOI: 10.1105/tpc.105.035410 PMID: 16284314

Getahun, S., Kefale, H., & Gelaye, Y. (2024). Application of Precision Agriculture Technologies for Sustainable Crop Production and Environmental Sustainability: A Systematic Review. *The Scientific World Journal, 2024*(1). DOI: 10.1155/2024/2126734

Ghorbanpour, M., Omidvari, M., Abbaszadeh-Dahaji, P., Omidvar, R., & Kariman, K. (2018). Mechanisms underlying the protective effects of beneficial fungi against plant diseases. *Biological Control, 117*, 147–157. https://doi.org/https://doi.org/10.1016/j.biocontrol.2017.11.006. DOI: 10.1016/j.biocontrol.2017.11.006

Ghosh, D., Siddique Md, A., & Pal, D. (2024). *AI driven approach for Precision Agriculture in the Book 'AI in Agriculture for Sustainable and Economic Management.* Taylor and Francis.

Gianinazzi-Pearson, V., & Gianinazzi, S. (1988). Morphological Integration and Functional Compatibility Between Symbionts in Vesicular Arbuscular Endomycorrhizal Associations. In Scannerini, S., Smith, D., Bonfante-Fasolo, P., & Gianinazzi-Pearson, V. (Eds.), *Cell to Cell Signals in Plant, Animal and Microbial Symbiosis* (pp. 73–84). Springer Berlin Heidelberg. DOI: 10.1007/978-3-642-73154-9_5

Gianinazzi, S., Gollotte, A., Binet, M.-N., Van Tuinen, D., Redecker, D., & Wipf, D. (2010). Agroecology: The key role of arbuscular mycorrhizas in ecosystem services. *Mycorrhiza, 20*(8), 519–530. DOI: 10.1007/s00572-010-0333-3 PMID: 20697748

Gill, S. S., Gill, R., Trivedi, D. K., Anjum, N. A., Sharma, K. K., Ansari, M. W., Ansari, A. A., Johri, A. K., Prasad, R., Pereira, E., Varma, A., & Tuteja, N. (2016). Piriformospora indica: Potential and significance in plant stress tolerance. *Frontiers in Microbiology, 7*, 332. DOI: 10.3389/fmicb.2016.00332 PMID: 27047458

Giovannetti, M., Sbrana, C., Avio, L., Citernesi, A. S., & Logi, C. (1993). Differential hyphal morphogenesis in arbuscular mycorrhizal fungi during pre-infection stages. *The New Phytologist, 125*(3), 587–593. DOI: 10.1111/j.1469-8137.1993.tb03907.x PMID: 33874594

Giovannini, L., Sbrana, C., Avio, L., & Turrini, A. (2020). Diversity of a phosphate transporter gene among species and isolates of arbuscular mycorrhizal fungi. *FEMS Microbiology Letters, 367*(2), fnaa024. Advance online publication. DOI: 10.1093/femsle/fnaa024 PMID: 32043113

Gloer, J. B. (2007). Applications of fungal ecology in the search for new bioactive natural products. In Kubicek, C. P., & Druzhinina, I. S. (Eds.), *Environmental and microbial relationships* (pp. 257–283). Springer.

Gobbato, E., Marsh, J. F., Vernié, T., Wang, E., Maillet, F., Kim, J., Miller, J. B., Sun, J., Bano, S. A., Ratet, P., Mysore, K. S., Dénarié, J., Schultze, M., & Oldroyd, G. E. D. (2012). A GRAS-type transcription factor with a specific function in mycorrhizal signaling. *Current Biology, 22*(23), 2236–2241. DOI: 10.1016/j.cub.2012.09.044 PMID: 23122845

Gobbato, E., Wang, E., Higgins, G., Bano, S. A., Henry, C., Schultze, M., & Oldroyd, G. E. D. (2013). RAM1 and RAM2 function and expression during arbuscular mycorrhizal symbiosis and Aphanomyces euteiches colonization. *Plant Signaling & Behavior, 8*(10), e26049. DOI: 10.4161/psb.26049 PMID: 24270627

Godara, H., & Ramakrishna, W. (2023). Endophytes as nature's gift to plants to combat abiotic stresses. *Letters in Applied Microbiology, 76*(2), ovac067. Advance online publication. DOI: 10.1093/lambio/ovac067 PMID: 36763787

Gohel, N. M., Raghunandan, B. L., Patel, N. B., Parmar, H. V., & Raval, D. B. (2022). Role of fungal biocontrol agents for sustainable agriculture. *Fungal Biology*, •••, 577–606. DOI: 10.1007/978-981-16-8877-5_28

Goicoechea, N., & Antolín, M. C. (2017). Increased nutritional value in food crops. *Microbial Biotechnology*, *10*(5), 1004–1007. https://doi.org/https://doi.org/10.1111/1751-7915.12764. DOI: 10.1111/1751-7915.12764 PMID: 28696049

Goligar, N., Saadatmand, S., & Khavarinejad, R. A. (2023). Mycoremediation of lead and cadmium by lignocellulosic enzymes of *Pleurotus eryngii.AMB Express*, *13*(1), 127. DOI: 10.1186/s13568-023-01626-8 PMID: 37964138

Gomes, A. R., Freitas, A. C., Rocha-Santos, T. A. P., & Duarte, A. C. (2014). Bioactive compounds derived from echinoderms. *RSC Advances*, *4*(56), 29365–29382. DOI: 10.1039/C4RA03352C

González-Hernández, A. I., Suárez-Fernández, M. B., Pérez-Sánchez, R., Gómez-Sánchez, M. Á., & Morales-Corts, M. R. (2021). Compost Tea Induces Growth and Resistance against *Rhizoctonia solani* and *Phytophthora capsici* in Pepper. *Agronomy (Basel)*, *11*(4), 781. DOI: 10.3390/agronomy11040781

Gopalakrishnan, C., & Valluvaparidasan, V. (2009). Management of okra powdery mildew using Ampelomyces quisqualis. *Journal of Biological Control*, *23*(3), 325–327.

Gorin, M., Shabani, M., Votat, S., Lebrun, L., Mbokou, S. F., & Pontié, M. (2024). Application of fungal-based microbial fuel cells for biodegradation of pharmaceuticals: Comparative study of individual vs. mixed contaminant solutions. *Chemosphere*, *363*, 142849. DOI: 10.1016/j.chemosphere.2024.142849 PMID: 39009093

Gorissen, S. H., Horstman, A. M., Franssen, R., Crombag, J. J. R., Langer, H., Bierau, J., Respondek, F., & van Loon, L. J. C. (2016). Ingestion of wheat protein increases in vivo muscle protein synthesis rates in healthy older men in a randomized trial. *The Journal of Nutrition*, *146*(9), 1651–1659. DOI: 10.3945/jn.116.231340 PMID: 27440260

Gorte, O., Hollenbach, R., Papachristou, I., Steinweg, C., Silve, A., Frey, W., Syldatk, C., & Ochsenreither, K. (2020). Evaluation of Downstream Processing, Extraction, and Quantification Strategies for Single Cell Oil Produced by the Oleaginous Yeasts *Saitozyma podzolica* DSM 27192 and *Apiotrichum porosum* DSM 27194. *Frontiers in Bioengineering and Biotechnology*, *8*, 355. DOI: 10.3389/fbioe.2020.00355

Goryacheva, I. Y., Saeger, S. D., Eremin, S. A., & Peteghem, C. V. (2007). Immunochemical methods for rapid mycotoxin detection: Evolution from single to multiple analyte screening: A review. *Food Additives and Contaminants*, *24*(10), 1169–1183. DOI: 10.1080/02652030701557179 PMID: 17886190

Gothalwal, R. (2020). Bioremediation. In Managing Biological and Ecological Systems (pp. 359-378). CRC Press.

Gowtham, H. G., Hema, P., Murali, M., Shilpa, N., Nataraj, K., Basavaraj, G. L., Singh, S. B., Aiyaz, M., Udayashankar, A. C., & Amruthesh, K. N. (2024). Fungal Endophytes as Mitigators against Biotic and Abiotic Stresses in Crop Plants. *Journal of Fungi (Basel, Switzerland)*, *10*(2), 116. DOI: 10.3390/jof10020116 PMID: 38392787

Goyal, D., Misera, S., & Dantu, P. K. (2020). *Arbuscular mycorrhizae associations and role in mitigation of drought stress in plants. Agriculturally Important Fungi for Sustainable Agriculture* (Vol. 2). Functional Annotation for Crop Protection.

Grabsztunowicz, M., Stolarska, E., Tanwar, U. K., Arasimowicz-Jelonek, M., & Sobieszczuk-Nowicka, E. (2024). Heavy metal tolerance in crop plants: An in silico comprehensive analysis of Metal-tolerance gene family in barley. *Postepy Biochemii*, *70*(3), 358–370. DOI: 10.18388/pb.2021_559 PMID: 39365567

Greene, J. M., & McAdam, P. A. (1983). Nutritional Requirements and the Appropriate Use of Supplements. In *Nutrition in the middle and later years* (pp. 17–45). Butterworth-Heinemann., DOI: 10.1016/B978-0-7236-7046-9.50008-4

Green, S. (2003). A review of the potential for the use of bioherbicides to control forest weeds in the UK. *Forestry*, *76*(3), 285–298. DOI: 10.1093/forestry/76.3.285

Grimm, D., & Wösten, H. A. B. (2018). Mushroom cultivation in the circular economy. *Applied Microbiology and Biotechnology*, *102*(18), 7795–7803. DOI: 10.1007/s00253-018-9226-8 PMID: 30027491

Groden, E., & Lockwood, J. (1991). Effects of soil fungistasis on Beauveria bassiana and its relationship to disease incidence in the Colorado potato beetle, Leptinotarsa decemlineata, in Michigan and Rhode Island soils. *Journal of Invertebrate Pathology*, *57*(1), 7–16. DOI: 10.1016/0022-2011(91)90035-O

Grossmann, L., & Weiss, J. (2021). Alternative protein sources as technofunctional food ingredients. *Annual Review of Food Science and Technology*, *12*(1), 93–117. DOI: 10.1146/annurev-food-062520-093642 PMID: 33472014

Gruber-Dorninger, C., Jenkins, T., & Schatzmayr, G. (2019). Global mycotoxin occurrence in feed: A ten-year survey. *Toxins*, *11*(7), 375. DOI: 10.3390/toxins11070375 PMID: 31252650

Guan, Y., Lv, H., Wu, G., Chen, J., Wang, M., Zhang, M., Pang, H., Duan, Y., Wang, L., & Tan, Z. (2023). Effects of Lactic Acid Bacteria Reducing the Content of Harmful Fungi and Mycotoxins on the Quality of Mixed Fermented Feed. *Toxins*, *15*(3), 226. DOI: 10.3390/toxins15030226 PMID: 36977117

Guebsi, R., Mami, S., & Chokmani, K. (2024). Drones in Precision Agriculture: A Comprehensive Review of Applications, Technologies, and Challenges. *Drones (Basel)*, *8*(11), 686. DOI: 10.3390/drones8110686

Guohou, Q., Lanfu, W., Wanxiao, G., Yuansen, H., Yin, Z., Min, C., Chong, F., Guojie, Z., Zonghao, X., Hongwei, W., Yapeng, T., Guangyou, Z., & Xia, W. (2013). *Pichia pastoris for efficiently converting methanol to produce single cell protein and application of Pichia pastoris* (CN 103045494 A). https://lens.org/042-289-199-791-43X

Guo, J., Liu, X., Zhang, X., Wu, J., Chai, C., Ma, D., Chen, Q., Xiang, D., & Ge, W. (2019). Immobilized lignin peroxidase on Fe3O4@SiO2@polydopamine nanoparticles for degradation of organic pollutants. *International Journal of Biological Macromolecules*, *138*, 433–440. https://doi.org/https://doi.org/10.1016/j.ijbiomac.2019.07.105. DOI: 10.1016/j.ijbiomac.2019.07.105 PMID: 31325503

Gupta, S., & Pathak, B. (2020). *Chapter 6 - Mycoremediation of polycyclic aromatic hydrocarbons* (P. Singh, A. Kumar, & A. B. T.-A. of E. P. Borthakur (eds.); pp. 127–149). Elsevier. https://doi.org/https://doi.org/10.1016/B978-0-12-818095-2.00006-0

Gupta, P. (2023). *Bioremediation of Heavy Metal in Paper Mill Effluent*. Bentham., DOI: 10.2174/9789815123494123010005

Gupta, S. K., & Sharma, M. (2014). *Approaches and trends in plant disease management.* Scientific Publishers.

Gupta, S., Pathak, B., & Ravi, R. K. (2024). Biodegradation of naphthalene using biosurfactant producing *Fusarium proliferatum* WC416 isolated from refinery effluent. *Applied Biochemistry and Biotechnology*, *196*(5), 2549–2565. DOI: 10.1007/s12010-023-04364-6 PMID: 36689160

Gurikar, C., Shivaprasad, D. P., Sabillón, L., Gowda, N. A. N., & Siliveru, K. (2023). Impact of mycotoxins and their metabolites associated with food grains. *Grain & Oil Science and Technology*, *6*(1), 1–9. DOI: 10.1016/j.gaost.2022.10.001

Gutjahr, C., Banba, M., Croset, V., An, K., Miyao, A., & Paszkowski, U. (2008). Arbuscular mycorrhiza–specific signaling in rice transcends the common symbiosis signaling pathway. *The Plant Cell*, *20*(11), 2989–3005. DOI: 10.1105/tpc.108.062414 PMID: 19033527

Habschied, K., Krstanović, V., Zdunić, Z., Babić, J., Mastanjević, K., & Šarić, G. K. (2021). Mycotoxins Biocontrol Methods for Healthier Crops and Stored Products. *Journal of Fungi (Basel, Switzerland)*, *7*(5), 348. DOI: 10.3390/jof7050348 PMID: 33946920

Haider, M. (2021). Bioconversion of saw dust powder acid hydrolysis to single cell protein by the yeast Candida tropicalis. *Journal of Life and Bio Sciences Research*, *2*(02), 31–35. DOI: 10.38094/jlbsr2241

Hajiboland, R., Dashtebani, F., & Aliasgharzad, N. (2015). Physiological responses of halophytic C 4 grass Aeluropus littoralis to salinity and arbuscular mycorrhizal fungi colonization. *Photosynthetica*, *53*(4), 572–584. DOI: 10.1007/s11099-015-0131-4

Hajji, L., Hlaoua, W., Regaieg, H., & Horrigue-Raouani, N. (2016). Biocontrol potential of *Verticillium leptobactrum* and *Purpureocillium lilacinum* against Meloidogyne javanica and Globodera pallida on potato (Solanum tuberosum). *American Journal of Potato Research*, *94*(2), 178–183. DOI: 10.1007/s12230-016-9554-0

Hajlaoui, H., Mighri, H., Aouni, M., Gharsallah, N., & Kadri, A. (2016). Chemical composition and *in vitro* evaluation of antioxidant, antimicrobial, cytotoxicity and anti-acetylcholinesterase properties of Tunisian *Origanum majorana* L. essential oil. *Microbial Pathogenesis*, *95*, 86–94. DOI: 10.1016/j.micpath.2016.03.003 PMID: 26997648

Hamayun, M., Hussain, A., Khan, S. A., Kim, H.-Y., Khan, A. L., Waqas, M., Irshad, M., Iqbal, A., Rehman, G., Jan, S., & Lee, I.-J. (2017). Gibberellins Producing Endophytic Fungus Porostereum spadiceum AGH786 Rescues Growth of Salt Affected Soybean. *Frontiers in Microbiology*, *8*, 686. DOI: 10.3389/fmicb.2017.00686 PMID: 28473818

Hamayun, M., Khan, S. A., Khan, A. L., Rehman, G., Kim, Y.-H., Iqbal, I., Hussain, J., Sohn, E.-Y., & Lee, I.-J. (2010). Gibberellin production and plant growth promotion from pure cultures of *Cladosporium* sp. MH-6 isolated from cucumber (*Cucumis sativus* L.). *Mycologia*, *102*(5), 989–995. DOI: 10.3852/09-261 PMID: 20943499

Hamlen, R. A. (1979). Biological control of insects and mites on European greenhouse crops: Research and commercial implementation. *Proceedings of the Annual Meeting of the Florida State Horticultural Society*, *92*, 367–368.

Hammer, E. C., Pallon, J., Wallander, H., & Olsson, P. A. (2011). Tit for tat? A mycorrhizal fungus accumulates phosphorus under low plant carbon availability. *FEMS Microbiology Ecology*, *76*(2), 236–244. DOI: 10.1111/j.1574-6941.2011.01043.x PMID: 21223336

Han, N. S., Ahmad, W. A. N. W., & Ishak, W. W. (2016). Quality characteristics of Pleurotus sajor-caju powder: Study on nutritional compositions, functional properties and storage stability. *Sains Malaysiana*, *45*, 1617–1623.

Hanson, L. E., & Howell, C. R. (2002). Biocontrol efficacy and other characteristics of protoplast fusants between Trichoderma koningii and T. virens. *Mycological Research*, *106*(3), 321–328. DOI: 10.1017/S0953756202005592

Hanson, L. E., & Howell, C. R. (2004). Elicitors of Plant Defense Responses from Biocontrol Strains of Trichoderma viren. *Phytopathology*, *94*(2), 171–176. DOI: 10.1094/PHYTO.2004.94.2.171 PMID: 18943540

Han, Y., Feng, J., Han, M., & Zhu, B. (2020). Responses of arbuscular mycorrhizal fungi to nitrogen addition: A meta-analysis. *Global Change Biology*, *26*(12), 7229–7241. DOI: 10.1111/gcb.15369 PMID: 32981218

Hardoim, P. R., Van Overbeek, L. S., Berg, G., Pirttilä, A. M., Compant, S., Campisano, A., Döring, M., & Sessitsch, A. (2015). The Hidden World within Plants: Ecological and Evolutionary Considerations for Defining Functioning of Microbial Endophytes. *Microbiology and Molecular Biology Reviews : MMBR*, *79*(3), 293–320. DOI: 10.1128/MMBR.00050-14 PMID: 26136581

Haritash, A. K., & Kaushik, C. P. (2009). Biodegradation aspects of polycyclic aromatic hydrocarbons (PAHs): A review. *Journal of Hazardous Materials*, *169*(1-3), 1–15. DOI: 10.1016/j.jhazmat.2009.03.137 PMID: 19442441

Harman, G. E. (2006). Overview of mechanisms and uses of *Trichoderma* spp. *Phytopathology*, *96*(2), 190–194. DOI: 10.1094/PHYTO-96-0190 PMID: 18943924

Harman, G. E., Howell, C. R., Viterbo, A., Chet, I., & Lorito, M. (2004). Trichoderma species—Opportunistic, avirulent plant symbionts. *Nature Reviews. Microbiology*, *2*(1), 43–56. DOI: 10.1038/nrmicro797 PMID: 15035008

Harman, G. E., & Shoresh, M. (2007). The mechanisms and applications of symbiotic opportunistic plant symbionts. In Vurro, M., & Gressel, J. (Eds.), *Novel Biotechnologies for Biocontrol Agent Enhancement and Management. NATO Security through Science Series* (pp. 131–155). Springer., DOI: 10.1007/978-1-4020-5799-1_7

Harrach, B. D., Baltruschat, H., Barna, B., Fodor, J., & Kogel, K.-H. (2013). The Mutualistic Fungus *Piriformospora indica* Protects Barley Roots from a Loss of Antioxidant Capacity Caused by the Necrotrophic Pathogen *Fusarium culmorum.Molecular Plant-Microbe Interactions*, *26*(5), 599–605. DOI: 10.1094/MPMI-09-12-0216-R PMID: 23405867

Harrison, L. R., Colvin, B. M., Greene, J. T., Newman, L. E., & Cole, J. R.Jr. (1990). Pulmonary edema and hydrothorax in swine produced by fumonisin B1, a toxic metabolite of *Fusarium moniliforme.Journal of Veterinary Diagnostic Investigation : Official Publication of the American Association of Veterinary Laboratory Diagnosticians, Inc*, *2*(3), 217–221. DOI: 10.1177/104063879000200312 PMID: 2094448

Hartley, C. (1921). Damping-off in forest nurseries /. https://doi.org/DOI: 10.5962/bhl.title.108260

Hart, M. M., Reader, R. J., & Klironomos, J. N. (2003). Plant coexistence mediated by arbuscular mycorrhizal fungi. *Trends in Ecology & Evolution*, *18*(8), 418–423. DOI: 10.1016/S0169-5347(03)00127-7

Harwoko, H., Hartmann, R., Daletos, G., Ancheeva, E., Frank, M., Liu, Z., & Proksch, P. (2019). Biotransformation of Host Plant Flavonoids by the Fungal Endophyte *Epicoccum nigrum.ChemistrySelect*, *4*(45), 13054–13057. DOI: 10.1002/slct.201903168

Hasan, N. F. N., Wan Hanafi, W. N., Safian, M. F., & Zainal Ariffin, Z. (2024). Diversity of chlorpyrifos-degrading fungi isolated from chlorpyrifos-contaminated agricultural soil. [ScL]. *Science Letters*, *18*(1), 10–20.

Hasan, S., Ahmad, A., Purwar, A., Khan, N., Kundan, R., & Gupta, G. (2013). Production of extracellular enzymes in the entomopathogenic fungus *Verticillium lecanii.Bioinformation*, *9*(5), 238–242. DOI: 10.6026/97320630009238 PMID: 23515428

Hasan, S., Bhamra, A. K., Sil, K., Rajak, R. C., & Sandhu, S. S. (2002). Spore production of Metarhizium anisopliaem (ENT-12) by Solid State fermentation. *Journal of the Indian Botanical Society*, *8*, 85–88.

Haselwandter, K., Dobernigg, B., Beck, W., Jung, G., Cansier, A., & Winkelmann, G. (1992). Isolation and identification of hydroxamate siderophores of ericoid mycorrhizal fungi. *Biometals*, *5*(1), 51–56. DOI: 10.1007/BF01079698

Hashem, A., Alqarawi, A. A., Radhakrishnan, R., Al-Arjani, A. B. F., Aldehaish, H. A., Egamberdieva, D., & Abd_Allah, , E. F. (2018). Arbuscular mycorrhizal fungi regulate the oxidative system, hormones and ionic equilibrium to trigger salt stress tolerance in Cucumis sativus L. *Saudi Journal of Biological Sciences*, *25*(6), 1102–1114. DOI: 10.1016/j.sjbs.2018.03.009 PMID: 30174509

Hashempour-Baltork, F.. (2020). "Mycoproteins as Safe Meat Substitutes." Journal of Cleaner Production, vol. 253, Elsevier BV, Apr. 119958. Crossref, Liang, J., Xu, N. Nedele, A. K., Rigling, M., Zhu, L., and Zhang, Y. (2023).Upcycling of Soy Whey with Ischnoderma benzoinum toward Production of Bioflavors and Mycoprotein. *Journal of Agricultural and Food Chemistry*, *71*(23), 9070–9079.

Hashempour-Baltork, F., Hosseini, S. M., Assarehzadegan, M. A., Khosravi-Darani, K., & Hosseini, H. (2020). Safety assays and nutritional values of mycoprotein produced by Fusarium venenatum IR372C from date waste as substrate. *Journal of the Science of Food and Agriculture*, *100*(12), 4433–4441. DOI: 10.1002/jsfa.10483

Hawkins, N. J., Bass, C., Dixon, A., & Neve, P. (2019). The Evolutionary Origins of Pesticide Resistance. *Biological Reviews of the Cambridge Philosophical Society*, *94*(1), 135–155. DOI: 10.1111/brv.12440 PMID: 29971903

Helgason, T., Daniell, T. J., Husband, R., Fitter, A. H., & Young, J. P. W. (1998). Ploughing up the wood-wide web? *Nature*, *394*(6692), 431–431. DOI: 10.1038/28764 PMID: 9697763

Hell, K., Mutegi, C., & Fandohan, P. (2010). Aflatoxin control and prevention strategies in maize for Sub-Saharan Africa. *Julius-Kühn-Archiv*, (425), 534.

Herrera, M., Cavero, J., Franco-Luesma, S., Álvaro-Fuentes, J., Ariño, A., & Lorán, S. (2023). Mycotoxins and crop yield in maize as affected by irrigation management and tillage practices. *Agronomy (Basel)*, *13*(3), 798. DOI: 10.3390/agronomy13030798

Hettiarachchi, H., Caucci, S., & Schwärzel, K. (2020). Organic Waste Composting through Nexus Thinking.

Hewitt, O. H., & Shaikh, H. M. (2021). The Rhythm of Many: Biological Rhythms in the Marine Environment, From Macro-Scale Planktonic Ecosystems to Micro-Scale Holobionts. *Frontiers in Marine Science*, *8*, 1–21. DOI: 10.3389/fmars.2021.744169

Heydari, A., & Pessarakli, M. (2010). A review on biological control of fungal plant pathogens using microbial antagonists. *Journal of Biological Sciences (Faisalabad, Pakistan)*, *10*(4), 273–290. DOI: 10.3923/jbs.2010.273.290

Hiruma, K., Gerlach, N., Sacristán, S., Nakano, R. T., Hacquard, S., Kracher, B., & Schulze-Lefert, P. (2016). Root endophyte Colletotrichum tofieldiae confers plant fitness benefits that are phosphate status dependent. *Cell*, *165*(3), 464–474. DOI: 10.1016/j.cell.2016.02.028 PMID: 26997485

Hoang, D. T. T., Rashtbari, M., Anh, L. T., Wang, S., Tu, D. T., Hiep, N. V., & Razavi, B. S. (2022). Mutualistic interaction between arbuscular mycorrhiza fungi and soybean roots enhances drought resistant through regulating glucose exudation and rhizosphere expansion. *Soil Biology & Biochemistry*, *171*, 108728. DOI: 10.1016/j.soilbio.2022.108728

Hoque, R. A., Yadav, M., Yadav, H. S., & Boruah, R. (2023). Purification and characterization of novel manganese peroxidase from Trichoderma parestonica and its bio-conversion study of toxic arylamine. *Analytical Chemistry Letters*, *13*(6), 641–659. DOI: 10.1080/22297928.2023.2299257

Horbach, R., Navarro-Quesada, A. R., Knogge, W., & Deising, H. B. (2011). When and how to kill a plant cell: Infection strategies of plant pathogenic fungi. *Journal of Plant Physiology*, *168*(1), 51–62. DOI: 10.1016/j.jplph.2010.06.014 PMID: 20674079

Hossain, S., & Rahman, M. A. (2007). Essential fatty acid contents of Pleurotus ostreatus, Ganoderma lucidum and Agaricus bisporus. https://www.researchgate.net/publication/258050987

Hou, T., Sana, S. S., Li, H., Xing, Y., Nanda, A., Netala, V. R., & Zhang, Z. (2022). Essential oils and its antibacterial, antifungal and anti-oxidant activity applications: A review. *Food Bioscience*, *47*, 101716. DOI: 10.1016/j.fbio.2022.101716

Howell, C. R. (2003). Mechanisms Employed by Trichoderma Species in the Biological Control of Plant Diseases: The History and Evolution of Current Concepts. *Plant Disease*, *87*(1), 4–10. DOI: 10.1094/PDIS.2003.87.1.4 PMID: 30812698

Hsieh, Y. H., Wei, Y. H., Lo, J. C., Pan, H. Y., & Yang, S. Y. (2022). Arbuscular mycorrhizal symbiosis enhances tomato lateral root formation by modulating CEP2 peptide expression. *The New Phytologist, 235*(1), 292–305. DOI: 10.1111/nph.18128 PMID: 35358343

Huang, Y., Chen, X., Duan, N., Wu, S., Wang, Z., Wei, X., & Wang, Y. (2015). Selection and characterization of DNA aptamers against Staphylococcus aureus enterotoxin C1. *Food Chemistry, 166*, 623–629. DOI: 10.1016/j.foodchem.2014.06.039 PMID: 25053102

Huang, Y.-F., & Lo, S.-L. (2019). Predicting heating value of lignocellulosic biomass based on elemental analysis. *Energy, 191*, 116501. DOI: 10.1016/j.energy.2019.116501

Hu, K., Peris, A., Torán, J., Eljarrat, E., Sarrà, M., Blánquez, P., & Caminal, G. (2020). Exploring the degradation capability of *Trametes versicolor* on selected hydrophobic pesticides through setting sights simultaneously on culture broth and biological matrix. *Chemosphere, 250*, 126293. DOI: 10.1016/j.chemosphere.2020.126293 PMID: 32234621

Hu, M., & Palić, D. (2020). Micro- and nano-plastics activation of oxidative and inflammatory adverse outcome pathways. *Redox Biology, 37*, 101620. DOI: 10.1016/j.redox.2020.101620 PMID: 32863185

Huo, W. G., Chai, X. F., Wang, X. H., Batchelor, W. D., Kafle, A., & Gu, F. E. N. G. (2022). Indigenous arbuscular mycorrhizal fungi play a role in phosphorus depletion in organic manure amended high fertility soil. *Journal of Integrative Agriculture, 21*(10), 3051–3066.

Hussain, C. M., & Keçili, R. (2020). *Chapter 1 - Environmental pollution and environmental analysis* (C. M. Hussain & R. B. T.-M. E. A. T. for P. Keçili (eds.); pp. 1–36). Elsevier. https://doi.org/https://doi.org/10.1016/B978-0-12-816934-6.00001-1

Hu, Y., Bandara, A. R., Xu, J., Kakumyan, P., Hyde, K. D., & Mortimer, P. E. (2022). 'The use of agaricus subrufescens for rehabilitation of agricultural soils'. *Agronomy (Basel), 12*(9), 2034. Advance online publication. DOI: 10.3390/agronomy12092034

Hu, Y., Mortimer, P. E., Hyde, K. D., Kakumyan, P., & Thongklang, N. (2021). 'Mushroom cultivation for soil amendment and bioremediation'. *Circular Agricultural Systems, 1*(1), 1–14. DOI: 10.48130/CAS-2021-0011

Hu, Y., Yesilonis, I., & Szlavecz, K. (2021). Microbial and environmental controls on wood decomposition in deciduous forests of different ages. *Applied Soil Ecology, 166*, 103986. DOI: 10.1016/j.apsoil.2021.103986

IARC. (2012). Aflatoxins. [). Lyon, France: International Agency for Research on Cancer.]. *IARC Monographs on the Evaluation of Carcinogenic Risks to Humans, 100F*, 225–248.

ICAR-Directorate of Mushroom Research. Solan https://dmrsolan.icar.gov.in/Binding-Agent-mushroom-meat-Dr.Anuradha-1.pdf. Accessed 30.07.2024.

Ignoffo, C. M. (1981). The fungus Nomuraea rileyi as a microbial insecticide. In Burges, H. D. (Ed.), *Microbial Control of Pests and Plant Diseases* (pp. 513–538). Academic Press.

Ikan, C., Ben-Laouane, R., Ouhaddou, R., Anli, M., Boutasknit, A., Lahbouki, S., & Meddich, A. (2023). Interactions between arbuscular mycorrhizal fungus and indigenous compost improve salt stress tolerance in wheat (*Triticum durum*). *South African Journal of Botany*, *158*, 417–428. DOI: 10.1016/j.sajb.2023.05.038

Imran, A., Saadalla, M. J. A., Khan, S.-U., Mirza, M. S., Malik, K. A., & Hafeez, F. Y. (2014). Ochrobactrum sp. Pv2Z2 exhibits multiple traits of plant growth promotion, biodegradation and N-acyl-homoserine-lactone quorum sensing. *Annals of Microbiology*, *64*(4), 1797–1806. DOI: 10.1007/s13213-014-0824-0

Ina, K., Kataoka, T., & Ando, T. (2013). The use of lentinan for treating gastric cancer. *Anti-cancer Agents in Medicinal Chemistry*, *13*(5), 681–688. DOI: 10.2174/1871520611313050002 PMID: 23092289

Inderjit, S., Rawat, D., & Foy, C. L. (2004). Multifaceted approach to determine rice straw phytotoxicity. *Canadian Journal of Botany*, *82*(2), 168–176. DOI: 10.1139/b03-137

Iqbal, A., Shafi, M. I., Rafique, M., Jabeen, A., Asif, S., Zaman, M., & Jiang, L. (2023). Biofertilizers to improve soil health and crop yields. In *Sustainable Agriculture Reviews 61: Biochar to Improve Crop Production and Decrease Plant Stress under a Changing Climate* (pp. 247–272). Springer International Publishing.

Ishara, J. R., Sila, D. N., Kenji, G. M., & Buzera, A. K. (2018). Nutritional and functional properties of mushroom (Agaricus bisporus & Pleurotus ostreatus) and their blends with maize flour. *American Journal of Food Science and Technology*, *6*, 33–41. DOI: 10.12691/ajfst-6-1-6

Islam, M. N., Banik, M., Sura, S., Tucker, J. R., & Wang, X. (2022). Implications of Crop Rotation and Fungicide on *Fusarium* and Mycotoxin Spectra in Manitoba Barley, 2017-2019. *Toxins*, *14*(7), 463. DOI: 10.3390/toxins14070463 PMID: 35878201

Israilides, C., & Philippoussis, A. (2003). Bio-technologies of recycling agro-industrial wastes for the production of commercially important fungal polysaccharides and mushrooms. *Biotechnology & Genetic Engineering Reviews*, *20*(1), 247–260. DOI: 10.1080/02648725.2003.10648045 PMID: 14997854

Jaber, L. R., & Araj, S.-E. (2018). Interactions among endophytic fungal entomopathogens (Ascomycota: Hypocreales), the green peach aphid Myzus persicae Sulzer (Homoptera: Aphididae), and the aphid endoparasitoid Aphidius colemani Viereck (Hymenoptera: Braconidae). *Biological Control*, *116*, 53–61. DOI: 10.1016/j.biocontrol.2017.04.005

Jach, M. E., Serefko, A., Ziaja, M., & Kieliszek, M. (2022). Yeast protein as an easily accessible food source. *Metabolites*, *12*(1), 63. DOI: 10.3390/metabo12010063

Jackson, M. A., Dunlap, C. A., & Jaronski, S. T. (2009). Ecological considerations in producing and formulating fungal entomopathogens for use in insect biocontrol. *BioControl*, *55*(1), 129–145. DOI: 10.1007/s10526-009-9240-y

Jagdale, Y. D., Mahale, S. V., Zohra, B., Nayik, G. A., Dar, A. H., Khan, K. A., Abdi, G., & Karabagias, I. K. (2021). Nutritional Profile and Potential Health Benefits of Super Foods: A Review. *Sustainability (Basel)*, *13*(16), 9240. Advance online publication. DOI: 10.3390/su13169240

Jahromi, M. F., Liang, J. B., Rosfarizan, M., Goh, Y. M., Shokryazdan, P., & Ho, Y. W. (2011). Efficiency of rice straw lignocelluloses degradability by Aspergillus terreus ATCC 74135 in solid state fermentation. *African Journal of Biotechnology*, *10*, 4428–4435.

Jain, N., Bhatia, A., & Pathak, H. (2014). Emission of air pollutants from crop residue burning in India. *Aerosol and Air Quality Research*, *14*(1), 422–430. DOI: 10.4209/aaqr.2013.01.0031

Jain, N., Rana, I. S., Kanojiya, A., & Sandhu, S. S. (2008). Characterization of Beauveria bassiana strains based on protease and lipase activity and their role in pathogenicity. *Journal of Basic and Applied Mycology*, *I-II*, 18–22.

Jaklitsch, W. M. (2009). European species of Hypocrea Part I. The green-spored species. *Studies in Mycology*, *63*, 1–91. DOI: 10.3114/sim.2009.63.01 PMID: 19826500

Jakovljević, V. D., & Vrvić, M. M. (2018). Potential of pure and mixed cultures of *Cladosporium cladosporioides* and *Geotrichum candidum* for application in bioremediation and detergent industry. *Saudi Journal of Biological Sciences*, *25*(3), 529–536. DOI: 10.1016/j.sjbs.2016.01.020 PMID: 29686515

Jangir, P., Shekhawat, P. K., Bishnoi, A., Ram, H., & Soni, P. (2021). Role of Serendipita indica in enhancing drought tolerance in crops. *Physiological and Molecular Plant Pathology*, *116*, 101691. https://doi.org/https://doi.org/10.1016/j.pmpp.2021.101691. DOI: 10.1016/j.pmpp.2021.101691

Jan, R., Asaf, S., Numan, M., Lubna, , & Kim, K.-M. (2021). Plant Secondary Metabolite Biosynthesis and Transcriptional Regulation in Response to Biotic and Abiotic Stress Conditions. *Agronomy (Basel)*, *11*(5), 968. Advance online publication. DOI: 10.3390/agronomy11050968

Jansa, J., & Bukovská, P. (2024). In Parihar, M., Rakshit, A., Adholeya, A., & Chen, Y. (Eds.), *Role of Arbuscular Mycorrhizal Fungi in Nitrogen and Phosphorus Cycling Within Terrestrial Ecosystems BT - Arbuscular Mycorrhizal Fungi in Sustainable Agriculture: Nutrient and Crop Management* (pp. 143–166). Springer Nature Singapore., DOI: 10.1007/978-981-97-0300-5_7

Jasińska, A. (2018). Spent mushroom compost (SMC)–retrieved added value product closing loop in agricultural production.

Jayachandran, M., Xiao, J., & Xu, B. (2017). A critical review on health promoting benefits of edible mushrooms through gut microbiota. In International Journal of Molecular Sciences (Vol. 18, Issue 9). MDPI AG. DOI: 10.3390/ijms18091934

Jayaraman, S., Yadav, B., Dalal, R. C., Naorem, A., Sinha, N. K., & Rao, C. S. (2024). 'Mushroom Farming: A Review Focusing on Soil Health, Nutritional Security and Environmental Sustainability'. *Farming System*, *2*(3), 100098. DOI: 10.1016/j.farsys.2024.100098

Jayasuriya, W. J. A. B. N., Wanigatunge, C. A., Fernando, G. H., Abeytunga, D. T. U., & Suresh, T. S. (2015). Hypoglycaemic activity of culinary Pleurotus ostreatus and P. cystidiosus mushrooms in healthy volunteers and type 2 diabetic patients on diet control and the possible mechanisms of action. *Phytotherapy Research*, *29*(2), 303–309. DOI: 10.1002/ptr.5255 PMID: 25382404

Jedinak, A., Dudhgaonkar, S., Wu, Q. L., Simon, J., & Sliva, D. (2011). Anti-inflammatory activity of edible oyster mushroom is mediated through the inhibition of NF-κB and AP-1 signaling. *Nutrition Journal*, *10*(1), 1–10. DOI: 10.1186/1475-2891-10-52 PMID: 21575254

Jesus, E. da C., Liang, C., Quensen, J. F., Susilawati, E., Jackson, R. D., Balser, T. C., & Tiedje, J. M. (2016). Influence of corn, switchgrass, and prairie cropping systems on soil microbial communities in the upper Midwest of the United States. *Global Change Biology. Bioenergy*, *8*(2), 481–494. DOI: 10.1111/gcbb.12289

Jiang, X., Xiang, M., & Liu, X. (2017). Nematode-Trapping Fungi. *Microbiology spectrum, 5*(1), . https://doi.org/DOI: 10.1128/microbiolspec.FUNK-0022-2016

Jiang, S., Li, H., Zhang, L., Mu, W., Zhang, Y., Chen, T., Wu, J., Tang, H., Zheng, S., Liu, Y., Wu, Y., Luo, X., Xie, Y., & Ren, J. (2025). Generic Diagramming Platform (GDP): A comprehensive database of high-quality biomedical graphics. *Nucleic Acids Research*, *53*(D1), D1670–D1676. DOI: 10.1093/nar/gkae973

Jiang, Y., Xie, Q., Wang, W., Yang, J., Zhang, X., Yu, N., Zhou, Y., & Wang, E. (2018). Medicago AP2-domain transcription factor WRI5a is a master regulator of lipid biosynthesis and transfer during mycorrhizal symbiosis. *Molecular Plant*, *11*(11), 1344–1359. DOI: 10.1016/j.molp.2018.09.006 PMID: 30292683

Jibran, A., & Mol, J. (2011). *Pleurotus sajor-caju* protein: A potential biosorptive agent. *Advanced Biotechnology*, *11*, 25–27.

Ji, H., Zhang, M., Huang, C., Lin, W., Lu, Y., Wang, P., Dong, B., He, B., Wu, B., & Guo, L. (2024). The Impact of Piriformospora indica on plant heat and drought tolerance. *Frontiers in Plant Science*, *15*, 1479561. Advance online publication. DOI: 10.3389/fpls.2024.1479561 PMID: 39711589

Jin, R., Liao, H., Xuanming, L., Zheng, M., Xiong, X., Liu, X., Zhang, L., & Zhu, Y. (2012). Identification and characterization of a fungal strain with lignin and cellulose hydrolysis activities. *African Journal of Microbiological Research*, *6*(36), 6545–6550. DOI: 10.5897/AJMR12.476

Jobstvogt, N., Hanley, N., Hynes, S., Kenter, J., & Witte, U. (2014). Twenty thousand sterling under the sea: Estimating the value of protecting deep-sea biodiversity. *Ecological Economics*, *97*, 10–19. DOI: 10.1016/j.ecolecon.2013.10.019

Jogawat, A. (2019). Osmolytes and their Role in Abiotic Stress Tolerance in Plants. In *Molecular Plant Abiotic Stress* (pp. 91–104). Biology and Biotechnology. DOI: 10.1002/9781119463665.ch5

Jogawat, A., Vadassery, J., Verma, N., Oelmüller, R., Dua, M., Nevo, E., & Johri, A. K. (2016). PiHOG1, a stress regulator MAP kinase from the root endophyte fungus Piriformospora indica, confers salinity stress tolerance in rice plants. *Scientific Reports*, *6*(1), 36765. DOI: 10.1038/srep36765 PMID: 27849025

Johnson, L. (2008). Iron and siderophores in fungal–host interactions. *Mycological Research*, *112*(2), 170–183. DOI: 10.1016/j.mycres.2007.11.012 PMID: 18280720

Johnson, N. C., Wilson, G. W. T., Wilson, J. A., Miller, R. M., & Bowker, M. A. (2015). Mycorrhizal phenotypes and the l aw of the m inimum. *The New Phytologist*, *205*(4), 1473–1484. DOI: 10.1111/nph.13172 PMID: 25417818

Johnston, M. P. (2014). Secondary Data Analysis : A Method of which the Time Has Come. [QQML]. *Qualitatve and Quantative Methods in Libraryes*, *3*, 619–626.

Johri, A. K., Oelmüller, R., Dua, M., Yadav, V., Kumar, M., Tuteja, N., Varma, A., Bonfante, P., Persson, B. L., & Stroud, R. M. (2015). Fungal association and utilization of phosphate by plants: Success, limitations, and future prospects. *Frontiers in Microbiology*, *6*, 984. DOI: 10.3389/fmicb.2015.00984 PMID: 26528243

Jorgensen, H., & Olsson, L. (2006). Production of cellulases by Penicillium brasilianum IBT 20888: Effect of substrate on hydrolytic performance. *Enzyme and Microbial Technology*, *38*(3-4), 381–390. DOI: 10.1016/j.enzmictec.2005.06.018

Joseph, M. (2023). *Types Of Mushrooms From A to Z (With Photos) | Live Eat.* https://www.liveeatlearn.com/types-of-mushrooms-how-to-use-them/

Joseph, T. P., Chanda, W., Padhiar, A. A., & Batool, S. LiQun, S., Zhong, M. T., & Huang, M. (2018). A Preclinical Evaluation of the Antitumor Activities of Edible and Medicinal Mushrooms: A Molecular Insight. In Integrative Cancer Therapies (Vol. 17, Issue 2, pp. 200–209). SAGE Publications Inc. DOI: 10.1177/1534735417736861

Joshi, B. B., Bhatt, R. P., & Bahukhandi, D. (2010). Antagonistic and plant growth activity of Trichoderma isolates of Western Himalayas. *Journal of Environmental Biology*, *31*(6), 921–928. PMID: 21506476

Jovel, J., Patterson, J., Wang, W., Hotte, N., O'Keefe, S., Mitchel, T., Perry, T., Kao, D., Mason, A. L., Madsen, K. L., & Wong, G. K. S. (2016). Characterization of the Gut Microbiome Using 16S or Shotgun Metagenomics. *Frontiers in Microbiology*, *7*, 1–17. DOI: 10.3389/fmicb.2016.00459 PMID: 27148170

Juárez-Hernández, J., Castillo-Hernández, D., Pérez-Parada, C., Nava-Galicia, S., Cuervo-Parra, J. A., Surian-Cruz, E., Díaz-Godínez, G., Sánchez, C., & Bibbins-Martínez, M. (2021). Isolation of fungi from a textile industry effluent and the screening of their potential to degrade industrial dyes. *Journal of Fungi (Basel, Switzerland)*, *7*(10), 805. DOI: 10.3390/jof7100805 PMID: 34682227

Junaid, F. K. L. A., Khawaja, L. A., & Ali, S. (2020). Single cell proteins as a potential meat substitute: A critical review. *World Journal of Pharmaceutical Research*, *9*, 141–161. DOI: 10.20959/wjpr20202-16686

Juniper, S., & Abbott, L. (1993). Vesicular-arbuscular mycorrhizas and soil salinity. *Mycorrhiza*, *4*(2), 45–57. DOI: 10.1007/BF00204058

Jureczko, M., Przystaś, W., Krawczyk, T., Gonciarz, W., & Rudnicka, K. (2021). White-rot fungi-mediated biodegradation of cytostatic drugs-bleomycin and vincristine. *Journal of Hazardous Materials*, *407*, 124632. DOI: 10.1016/j.jhazmat.2020.124632 PMID: 33359974

Jyoti and Singh. D. P. (2016). Fungi as biocontrol agents in sustainable agriculture. In J. S. Singh & D. P. Singh (Eds.), Microbes and Environmental Management (pp. 172–194). Studium Press, New Delhi.

Kabaluk, J. T., & Ericsson, J. D. (2007). Metarhizium anisopliae seed treatment increases yield of Field corn when applied for Wireworm control. *Agronomy Journal*, *99*(5), 1377–1381. DOI: 10.2134/agronj2007.0017N

Kajula, M., Tejesvi, M. V., Kolehmainen, S., Mäkinen, A., Hokkanen, J., Mattila, S., & Pirttilä, A. M. (2010). The siderophore ferricrocin produced by specific foliar endophytic fungi in vitro. *Fungal Biology*, *114*(2–3), 248–254. DOI: 10.1016/j.funbio.2010.01.004 PMID: 20943135

Kalita, B., Boro, P., Talukdar, S., & Borah, L. (2024). Greenhouse Gas Emission Reduction by Crop Management and Climate-Smart Agriculture. In *Agricultural Greenhouse Gas Emissions: Problems and Solutions* (pp. 1–31). Springer Nature Singapore. DOI: 10.1007/978-981-97-7554-5_1

Kamal, N., Parshad, J., Saharan, B. S., Kayasth, M., Mudgal, V., Duhan, J. S., Mandal, B. S., & Sadh, P. K. (2023). Ecosystem protection through myco-remediation of chromium and arsenic. *Journal of Xenobiotics*, *13*(1), 159–171. DOI: 10.3390/jox13010013 PMID: 36976162

Kamle, M., Mahato, D. K., Devi, S., Lee, K. E., Kang, S. G., & Kumar, P. (2019). Fumonisins: Impact on Agriculture, Food, and Human Health and their Management Strategies. *Toxins*, *11*(6), 328. DOI: 10.3390/toxins11060328 PMID: 31181628

Kamle, M., Mahato, D. K., Gupta, A., Pandhi, S., Sharma, N., Sharma, B., Mishra, S., Arora, S., Selvakumar, R., Saurabh, V., Dhakane-Lad, J., Kumar, M., Barua, S., Kumar, A., Gamlath, S., & Kumar, P. (2022). Citrinin Mycotoxin Contamination in Food and Feed: Impact on Agriculture, Human Health, and Detection and Management Strategies. *Toxins*, *14*(2), 85. DOI: 10.3390/toxins14020085 PMID: 35202113

Kanabkaew, T., & Oanh, N. T. K. (2011). Development of spatial and temporal emission inventory for crop residue field burning. *Environmental Modeling and Assessment*, *16*(5), 453–464. DOI: 10.1007/s10666-010-9244-0

Kang, E., Li, Y., Zhang, X., Yan, Z., Zhang, W., Zhang, K., Yan, L., Wu, H., Li, M., Niu, Y., Yang, A., Wang, J., & Kang, X. (2022). Extreme drought decreases soil heterotrophic respiration but not methane flux by modifying the abundance of soil microbial functional groups in alpine peatland. *Catena*, *212*, 106043. DOI: 10.1016/j.catena.2022.106043

Kanipriya, M., Sugumaran, D., Pandiyarajan, V., Yogananda, B. S., Selvakumar, P., & Mohan, S. (2024, January). Analysis of Electromyography Signals for Control of Mechanical Prosthesis using Machine Learning Techniques. In 2024 2nd International Conference on Intelligent Data Communication Technologies and Internet of Things (IDCIoT) (pp. 1059-1067). IEEE.

Kapoor, R., & Bhatnagar, A. (2007). Attenuation of cadmium toxicity in mycorrhizal celery (Apium graveolens L.). *World Journal of Microbiology & Biotechnology*, *23*(8), 1083–1089. DOI: 10.1007/s11274-006-9337-8

Kargi, F., Shuler, M. L., Vashon, R., Seeley, H. W.Jr, Henry, A., & Austic, R. E. (1980). Continuous aerobic conversion of poultry waste into single-cell protein using a single reactor: Kinetic analysis and determination of optimal conditions. *Biotechnology and Bioengineering*, *22*(8), 1567–1600. DOI: 10.1002/bit.260220805

Karlovsky, P., Suman, M., Berthiller, F., De Meester, J., Eisenbrand, G., Perrin, I., Oswald, I. P., Speijers, G., Chiodini, A., Recker, T., & Dussort, P. (2016). Impact of food processing and detoxification treatments on mycotoxin contamination. *Mycotoxin Research*, *32*(4), 179–205. DOI: 10.1007/s12550-016-0257-7 PMID: 27554261

Kasana, R. C., Richa, S., Dhar, H., Dutt, S., & Gulati, A. (2008). A rapid and easy method for the detection of microbial cellulases on agar plates using Gram's Iodine. *Current Microbiology*, *57*(5), 503–507. DOI: 10.1007/s00284-008-9276-8

Katekar, V. P., Rao, A. B., & Sardeshpande, V. R. (2022). Review of the rose essential oil extraction by hydrodistillation: An investigation for the optimum operating condition for maximum yield. *Sustainable Chemistry and Pharmacy*, *29*, 100783. DOI: 10.1016/j.scp.2022.100783

Kaur, A., & Sahota, P. P. (2004). Studies on the role of fungal inocula in composting of wheat straw. *Asian J Microbiol Biotechnol Environ Exp Sci*, *6*, 671–673.

Kausar, H., Sariah, M., Saud, H. M., Alam, M. Z., & Ismail, M. R. (2010). Development of compatible lignocellulolytic fungal consortium for rapid composting of rice straw. *International Biodeterioration & Biodegradation*, *64*(7), 594–600. DOI: 10.1016/j.ibiod.2010.06.012

Kaya, C., Ashraf, M., Sonmez, O., Aydemir, S., Tuna, A., & Çullu, M. (2009). The influence of arbuscular mycorrhizal colonisation on key growth parameters and fruit yield of pepper plants grown at high salinity. *Scientia Horticulturae - SCI HORT-AMSTERDAM*, *121*, 1–6. DOI: 10.1016/j.scienta.2009.01.001

Keating, G., & Figgitt, D. (2003). Caspofungin: A review of its use in oesophageal candidiasis, invasive candidiasis and invasive aspergillosis. *Drugs*, *63*(20), 2235–2263. DOI: 10.2165/00003495-200363200-00008 PMID: 14498760

Keegan, R.-J. H., Lu, Z., Bogusz, J. M., Williams, J. E., & Holick, M. F. (2013). Photobiology of vitamin D in mushrooms and its bioavailability in humans. *Dermato-Endocrinology*, *5*(1), 165–176. DOI: 10.4161/derm.23321 PMID: 24494050

Keller, N. P., Turner, G., & Bennett, J. W. (2005). Fungal secondary metabolism - from biochemistry to genomics. *Nature Reviews. Microbiology*, *3*(12), 937–947. DOI: 10.1038/nrmicro1286 PMID: 16322742

Kenis, M., Hurley, B. P., Hajek, A. E., & Cock, M. J. (2017). Classical biological control of insect pests of trees: Facts and figures. *Biological Invasions*, *19*(11), 3401–3417. DOI: 10.1007/s10530-017-1414-4

Kerry, B. R., Crump, D. H., & Mullen, L. A. (1982). Studies of the cereal cyst-nematode, *Heterodera avenae* under continuous cereals, 1975–1978. II. Fungal parasitism of nematode females and eggs. *Annals of Applied Biology*, *100*(3), 489–499. DOI: 10.1111/j.1744-7348.1982.tb01415.x

Keskin, E., & Eyupoglu, O. E. (2023). Determination of mycotoxins by HPLC, LC-MS/MS and health risk assessment of the mycotoxins in bee products of Turkey. *Food Chemistry*, *400*, 134086. DOI: 10.1016/j.foodchem.2022.134086 PMID: 36075166

Keymer, A., Pimprikar, P., Wewer, V., Huber, C., Brands, M., Bucerius, S. L., Delaux, P.-M., Klingl, V., von Röpenack-Lahaye, E., & Wang, T. L. (2017). Lipid transfer from plants to arbuscular mycorrhiza fungi. *eLife*, *6*, e29107. DOI: 10.7554/eLife.29107 PMID: 28726631

Khalid, M., Hassani, D., Bilal, M., Liao, J., & Huang, D. (2017). Elevation of secondary metabolites synthesis in Brassica campestris ssp. Chinensis L. via exogenous inoculation of Piriformospora indica with appropriate fertilizer. *PLoS One*, *12*(5), e0177185. DOI: 10.1371/journal.pone.0177185 PMID: 28493970

Khan, A. L., Halo, B. A., Elyassi, A., Ali, S., Al-Hosni, K., Hussain, J., Al-Harrasi, A., & Lee, I.-J. (2016). Indole acetic acid and ACC deaminase from endophytic bacteria improves the growth of Solanum lycopersicum. *Electronic Journal of Biotechnology*, *21*, 58–64. DOI: 10.1016/j.ejbt.2016.02.001

Khan, A. L., Hamayun, M., Kang, S. M., Kim, Y. H., Joo, G. J., & Lee, I. J. (2014). Myrothecium verrucaria-mediated growth promotion of soybean plants and its potential to regulate plant endogenous hormones. *Biological Control*, *68*(1), 14–21. DOI: 10.1016/j.biocontrol.2013.10.004

Khan, A. L., Hamayun, M., Kang, S. M., Kim, Y. H., Jung, H. Y., Lee, J. H., & Lee, I. J. (2012). Endophytic fungal association via gibberellins and indole acetic acid can improve plant growth under abiotic stress: An example of Paecilomyces formosus LHL10. *BMC Microbiology*, *12*(1), 3. DOI: 10.1186/1471-2180-12-3 PMID: 22235902

Khan, A. L., Hamayun, M., Kim, Y.-H., Kang, S.-M., Lee, J.-H., & Lee, I.-J. (2011). Gibberellins producing endophytic Aspergillus fumigatus sp. LH02 influenced endogenous phytohormonal levels, isoflavonoids production and plant growth in salinity stress. *Process Biochemistry (Barking, London, England)*, *46*(2), 440–447. DOI: 10.1016/j.procbio.2010.09.013

Khan, A. R., Ullah, I., Waqas, M., Park, G.-S., Khan, A. L., Hong, S.-J., Ullah, R., Jung, B. K., Park, C. E., Ur-Rehman, S., Lee, I.-J., & Shin, J.-H. (2017). Host plant growth promotion and cadmium detoxification in Solanum nigrum, mediated by endophytic fungi. *Ecotoxicology and Environmental Safety*, *136*, 180–188. DOI: 10.1016/j.ecoenv.2016.03.014 PMID: 27931714

Khan, A., Murad, W., Salahuddin, , Ali, S., Shah, S. S., Halim, S. A., Khalid, A., Kashtoh, H., Khan, A., & Al-Harrasi, A. (2024). Contribution of mushroom farming to mitigating food scarcity: Current status, challenges and potential future prospects in Pakistan. *Heliyon*, *10*(23), e40362. DOI: 10.1016/j.heliyon.2024.e40362 PMID: 39660206

Khan, A., Panthari, D., Sharma, R. S., Punetha, A., Singh, A. V., & Upadhayay, V. K. (2023). *Biofertilizers: a microbial-assisted strategy to improve plant growth and soil health.*

Khanday, M., Bhat, R. A., Haq, S., Dervash, M. A., Bhatti, A. A., Nissa, M., & Mir, M. R. (2016). In Hakeem, K. R., Akhtar, J., & Sabir, M. (Eds.), *Arbuscular Mycorrhizal Fungi Boon for Plant Nutrition and Soil Health BT - Soil Science: Agricultural and Environmental Prospectives* (pp. 317–332). Springer International Publishing., DOI: 10.1007/978-3-319-34451-5_14

Khan, I., Aftab, M., Shakir, S., Ali, M., Qayyum, S., Rehman, M. U., Haleem, K. S., & Touseef, I. (2019). Mycoremediation of heavy metal (Cd and Cr)-polluted soil through indigenous metallotolerant fungal isolates. *Environmental Monitoring and Assessment*, *191*(9), 585. DOI: 10.1007/s10661-019-7769-5 PMID: 31440913

Khan, R. (2024). Mycotoxins in food: Occurrence, health implications, and control strategies-A comprehensive review. *Toxicon*, *248*, 108038. DOI: 10.1016/j.toxicon.2024.108038 PMID: 39047955

Khan, R., Anwar, F., & Ghazali, F. M. (2024). A comprehensive review of mycotoxins: Toxicology, detection, and effective mitigation approaches. *Heliyon*, *10*(8), e28361. DOI: 10.1016/j.heliyon.2024.e28361 PMID: 38628751

Khan, R., Pathak, B., & Fulekar, M. (2012). Microbial decolorization and degradation of synthetic dyes: A review. *Reviews in Environmental Science and Biotechnology*, *12*(1), 75–97. Advance online publication. DOI: 10.1007/s11157-012-9287-6

Khan, S. A., Hamayun, M., Khan, A. L., Lee, I. J., Shinwari, Z. K., Kamran, M., & Rehman, G. (2012). Isolation of plant growth promoting endophytic fungi from Phoma sp., native to saline soil. *African Journal of Biotechnology*, *11*(21), 4833–4842.

Khan, S. A., Hamayun, M., Yoon, H., Kim, H.-Y., Suh, S.-J., Hwang, S.-K., Kim, J.-M., Lee, I.-J., Choo, Y.-S., Yoon, U.-H., Kong, W.-S., Lee, B.-M., & Kim, J.-G. (2008). Plant growth promotion and Penicillium citrinum. *BMC Microbiology*, *8*(1), 231. DOI: 10.1186/1471-2180-8-231 PMID: 19099608

Khan, W., Zhu, Y., Khan, A., Zhao, L., Yang, Y.-M., Wang, N., Hao, M., Ma, Y., Nepal, J., Ullah, F., Rehman, M. M. U., Abrar, M., & Xiong, Y.-C. (2024). Above-and below-ground feedback loop of maize is jointly enhanced by plant growth-promoting rhizobacteria and arbuscular mycorrhizal fungi in drier soil. *The Science of the Total Environment*, *917*, 170417. DOI: 10.1016/j.scitotenv.2024.170417 PMID: 38280611

Khan, Y., Shah, S., & Hui, T. (2022). The Roles of Arbuscular Mycorrhizal Fungi in Influencing Plant Nutrients, Photosynthesis, and Metabolites of Cereal Crops—A Review. *Agronomy (Basel)*, *12*(9), 1–19. DOI: 10.3390/agronomy12092191

Khatua, S., Simal-Gandara, J., & Acharya, K. (2024). Myco-remediation of plastic pollution: Current knowledge and future prospects. *Biodegradation*, *35*(3), 249–279. DOI: 10.1007/s10532-023-10053-2 PMID: 37665521

Khoi, C. S., Chen, J. H., Lin, T. Y., Chiang, C. K., & Hung, K. Y. (2021). Ochratoxin A-Induced Nephrotoxicity: Up-to-Date Evidence. *International Journal of Molecular Sciences*, *22*(20), 11237. DOI: 10.3390/ijms222011237 PMID: 34681895

Kihal, A., Rodríguez-Prado, M., & Calsamiglia, S. (2022). The efficacy of mycotoxin binders to control mycotoxins in feeds and the potential risk of interactions with nutrient: A review. *Journal of Animal Science*, *100*(11), skac328. Advance online publication. DOI: 10.1093/jas/skac328 PMID: 36208465

Kim, J. J., Lee, M. H., Yoon, C. S., Kim, H. S., Yoo, J. K., & Kim, K. C. (2002). Control of cotton aphid and greenhouse whitefly with a fungal pathogen. Journal of Natural Institute of Agricultural Science and Technology, 7–14.

Kim, J. J., Goettel, M. S., & Gillespie, D. R. (2008). Evaluation of Lecanicillium longisporum, Vertalec® for simultaneous suppression of cotton aphid, Aphis gossypii, and cucumber powdery mildew, Sphaerotheca fuliginea, on potted cucumbers. *Biological Control*, *45*(3), 404–409. DOI: 10.1016/j.biocontrol.2008.02.003

Kim, K.-H., Kabir, E., & Jahan, S. A. (2017). Exposure to pesticides and the associated human health effects. *The Science of the Total Environment*, *575*, 525–535. DOI: 10.1016/j.scitotenv.2016.09.009 PMID: 27614863

Kiss, L. (2003). A review of fungal antagonists of powdery mildews and their potential as biocontrol agents. *Pest Management Science*, *59*(4), 475–483. DOI: 10.1002/ps.689 PMID: 12701710

Kjøller, R., & Rosendahl, S. (2001). Molecular diversity of glomalean (arbuscular mycorrhizal) fungi determined as distinct *Glomus* specific DNA sequences from roots of field grown peas fungi. *Mycological Research*, *105*(9), 1027–1032. DOI: 10.1016/S0953-7562(08)61964-4

Kobae, Y. (2019). Dynamic phosphate uptake in arbuscular mycorrhizal roots under field conditions. *Frontiers in Environmental Science*, *6*, 159. DOI: 10.3389/fenvs.2018.00159

Kochiieru, Y., Mankevičienė, A., Cesevičienė, J., Semaškienė, R., Ramanauskienė, J., Gorash, A., Janavičienė, S., & Venslovas, E. (2021). The impact of harvesting time on *Fusarium* mycotoxins in spring wheat grain and their interaction with grain quality. *Agronomy (Basel)*, *11*(4), 642. DOI: 10.3390/agronomy11040642

Koduah, P., Paul, F., & Dörr, J. M. (2017). Vitamin D in the prevention, prediction and treatment of neurodegenerative and neuroinflammatory diseases. *The EPMA Journal*, *8*(4), 313–325. DOI: 10.1007/s13167-017-0120-8 PMID: 29209434

Koide, R. T., & Kabir, Z. (2000). Extraradical hyphae of the mycorrhizal fungus Glomus intraradices can hydrolyse organic phosphate. *The New Phytologist*, *148*(3), 511–517.

Kolb, M., Bahadir, M., & Teichgräber, B. (2017). Determination of chemical oxygen demand (COD) using an alternative wet chemical method free of mercury and dichromate. *Water Research*, *122*, 645–654. DOI: 10.1016/j.watres.2017.06.034 PMID: 28649029

Kong, X., Jia, Y., Song, F., Tian, K., Lin, H., Bei, Z., Jia, X., Yao, B., Guo, P., & Tian, X. (2018). Insight into litter decomposition driven by nutrient demands of symbiosis system through the hypha bridge of arbuscular mycorrhizal fungi. *Environmental Science and Pollution Research International*, *25*(6), 5369–5378. DOI: 10.1007/s11356-017-0877-2 PMID: 29209973

Kong, Y., Zhang, L. L., Zhao, J., Zhang, Y. Y., Sun, B. G., & Chen, H. T. (2019). Isolation and identification of the umami peptides from shiitake mushroom by consecutive chromatography and LC-Q-TOF-MS. *Food Research International*, *121*, 463–470. DOI: 10.1016/j.foodres.2018.11.060 PMID: 31108770

Korpi, A., Järnberg, J., & Pasanen, A.-L. (2009). Microbial Volatile Organic Compounds. *Critical Reviews in Toxicology*, *39*(2), 139–193. DOI: 10.1080/10408440802291497 PMID: 19204852

Kos, J., Anić, M., Radić, B., Zadravec, M., Janić Hajnal, E., & Pleadin, J. (2023). Climate Change-A Global Threat Resulting in Increasing Mycotoxin Occurrence. *Foods*, *12*(14), 2704. DOI: 10.3390/foods12142704 PMID: 37509796

Kour, D., Khan, S. S., Ramniwas, S., Kumar, S., Rai, A. K., Rustagi, S., Chaubey, K. K., Singh, S., Yadav, A. N., & Ahluwalia, A. S. (2023a). Beneficial fungal communities for sustainable development: Present scenario and future challenges. *Journal of Applied Biology and Biotechnology*. Advance online publication. DOI: 10.7324/JABB.2024.148256

Kozarski, M., Klaus, A., Jakovljevic, D., Todorovic, N., Vunduk, J., Petrović, P., Niksic, M., Vrvic, M. M., & Van Griensven, L. (2015). Antioxidants of edible mushrooms. In Molecules (Vol. 20, Issue 10, pp. 19489–19525). MDPI AG. DOI: 10.3390/molecules201019489

Kozarski, M., Klaus, A., Niksic, M., Jakovljevic, D., Helsper, J. P. F. G., & Van Griensven, L. J. L. D. (2011). Antioxidative and immunomodulating activities of polysaccharide extracts of the medicinal mushrooms Agaricus bisporus, Agaricus brasiliensis, Ganoderma lucidum and Phellinus linteus. *Food Chemistry*, *129*(4), 1667–1675. DOI: 10.1016/j.foodchem.2011.06.029

Krause, J., & Tobin, G. (2013). Discovery, development, and regulation of natural products. In Marianna, K. (Ed.), *Using old solutions to new problems – natural drug discovery in the 21st century* (pp. 3–36). InTech. DOI: 10.5772/56424

Krishnamoorthy, R., Kim, C. G., Subramanian, P., Kim, K. Y., Selvakumar, G., & Sa, T. M. (2015). Arbuscular mycorrhizal fungi community structure, abundance and species richness changes in soil by different levels of heavy metal and metalloid concentration. *PLoS One*, *10*(6), e0128784. DOI: 10.1371/journal.pone.0128784 PMID: 26035444

Król, A., Mizerna, K., & Bożym, M. (2020). An assessment of pH-dependent release and mobility of heavy metals from metallurgical slag. *Journal of Hazardous Materials*, *384*, 121502. DOI: 10.1016/j.jhazmat.2019.121502 PMID: 31732354

Kryukov, A. A., Gorbunova, A. O., Machs, E. M., Mikhaylova, Y. V., Rodionov, A. V., Zhurbenko, P. M., & Yurkov, A. P. (2020). Perspectives of using Illumina MiSeq for identification of arbuscular mycorrhizal fungi. *Vavilovskii Zhurnal Genetiki i Selektsii*, *24*(2), 158–167. DOI: 10.18699/VJ19.38-o PMID: 33659795

Ksouri, S., Djebir, S., Bentorki, A. A., Gouri, A., Hadef, Y., & Benakhla, A. (2017). Antifungal activity of essential oils extract from *Origanum floribundum* Munby, *Rosmarinus officinalis* L. and *Thymus ciliatus* Desf. against *Candida albicans* isolated from bovine clinical mastitis. *Journal de Mycologie Médicale*, *27*(2), 245–249. DOI: 10.1016/j.mycmed.2017.03.004 PMID: 28454927

Kubicek, C. P., Mikus, M., Schuster, A., Schmoll, M., & Seiboth, B. (2009). Metabolic engineering strategies for the improvement of cellulase production by Hypocrea jecorina. *Biotechnology for Biofuels*, *2*(1), 19. DOI: 10.1186/1754-6834-2-19 PMID: 19723296

Kubista, M., Andrade, J. M., Bengtsson, M., Forootan, A., Jonák, J., Lind, K., Sindelka, R., Sjöback, R., Sjögreen, B., Strömbom, L., Ståhlberg, A., & Zoric, N. (2006). The real-time polymerase chain reaction. *Molecular Aspects of Medicine*, *27*(2-3), 95–125. DOI: 10.1016/j.mam.2005.12.007 PMID: 16460794

Kulshreshtha, S., Mathur, N., & Bhatnagar, P. (2014). Mushroom as a product and their role in mycoremediation. *AMB Express*, *4*(1), 1–7. DOI: 10.1186/s13568-014-0029-8 PMID: 24949264

Kumar, K. (2015). Role of edible mushroom as functional foods: A review. South Asian Journal of Food Technology and Environment, 01(3and4), 211–218. DOI: 10.46370/sajfte.2015.v01i03and04.02

Kumar, K., Mehra, R., Guiné, R. P. F., Lima, M. J., Kumar, N., Kaushik, R., Ahmed, N., Yadav, A. N., & Kumar, H. (2021). Edible Mushrooms: A Comprehensive Review on Bioactive Compounds with Health Benefits and Processing Aspects. Foods 2021, Vol. 10, Page 2996, 10(12), 2996. https://doi.org/ DOI: 10.3390/foods10122996

Kumar, A. V., B, P., Chandra, V., & Singam, P. (2024). In silico analysis and characterization of the F5H gene family in sorghum bicolor and its role in lignin production. *Plant Cell Biotechnology and Molecular Biology*, *25*(11-12), 84–106. DOI: 10.56557/pcbmb/2024/v25i11-128885

Kumar, A., Pathak, H., Bhadauria, S., & Sudan, J. (2021). Aflatoxin contamination in food crops: Causes, detection, and management: A review. *Food Production, Processing and Nutrition*, *3*(17), 17. Advance online publication. DOI: 10.1186/s43014-021-00064-y

Kumar, A., Saharan, B. S., Parshad, J., Gera, R., Choudhary, J., & Yadav, R. (2024). Revolutionizing Indian agriculture: The imperative of advanced biofertilizer technologies for sustainability. *Discover Agriculture*, *2*(1), 24. DOI: 10.1007/s44279-024-00037-y

Kumar, D., & Kalita, P. (2017). Reducing Postharvest Losses during Storage of Grain Crops to Strengthen Food Security in Developing Countries. *Foods*, *6*(1), 8. DOI: 10.3390/foods6010008 PMID: 28231087

Kumar, D., Singh, P. N., Willer, H., Sharma, S. K., Singh, U. B., Lagashetti, A. C., & Husain, R. (2023). Fungal Biofertilizers and Biopesticides and Their Roles in Sustainable Agriculture. In *Applied Mycology for Agriculture and Foods* (pp. 165–216). Apple Academic Press., DOI: 10.1201/9781003369868-9

Kumari, K. (2020). Mushrooms as source of dietary fiber and its medicinal value: A review article. *Journal of Pharmacognosy and Phytochemistry*, *9*, 2075–2078.

Kumari, M., Swarupa, P., Kesari, K. K., & Kumar, A. (2022). Microbial Inoculants as Plant Biostimulants: A Review on Risk Status. *Life (Chicago, Ill.)*, *13*(1), 12. DOI: 10.3390/life13010012 PMID: 36675961

Kumar, J., Ramlal, A., Mallick, D., & Mishra, V. (2021). An Overview of Some Biopesticides and Their Importance in Plant Protection for Commercial Acceptance. *Plants*, *10*(6), 1185. DOI: 10.3390/plants10061185 PMID: 34200860

Kumar, M. S., Reddy, G. C., Phogat, M., & Korav, S. (2018). Role of bio-fertilizers towards sustainable agricultural development: A review. *Journal of Pharmacognosy and Phytochemistry*, *7*(6), 1915–1921.

Kumar, P., Gupta, A., Mahato, D. K., Pandhi, S., Pandey, A. K., Kargwal, R., Mishra, S., Suhag, R., Sharma, N., Saurabh, V., Paul, V., Kumar, M., Selvakumar, R., Gamlath, S., Kamle, M., Enshasy, H. A. E., Mokhtar, J. A., & Harakeh, S. (2022). Aflatoxins in Cereals and Cereal-Based Products: Occurrence, Toxicity, Impact on Human Health, and Their Detoxification and Management Strategies. *Toxins*, *14*(10), 687. DOI: 10.3390/toxins14100687 PMID: 36287956

Kumar, R., Raj, T., Næss, G., Sørensen, M., & Dhawan, V. (2024). Opportunities and challenges in single-cell protein production using lignocellulosic material. *Biofuels, Bioproducts & Biorefining*, *18*(1), 310–321. DOI: 10.1002/bbb.2563

Kumar, V. (2017). *Mycoremediation: A Step Toward Cleaner Environment.*, DOI: 10.1007/978-3-319-68957-9_10

Kumar, V., & Dwivedi, S. K. (2021). Bioremediation mechanism and potential of copper by actively growing fungus Trichoderma lixii CR700 isolated from electroplating wastewater. *Journal of Environmental Management*, *277*, 111370. DOI: 10.1016/j.jenvman.2020.111370 PMID: 32979751

Kumar, V., Sharma, K., Kedam, N., Patel, A., Kate, T., & Rathnayake, U. (2024). A Comprehensive Review on Smart and Sustainable Agriculture Using IoT Technologies. *Smart Agricultural Technology*, *8*, 100487. Advance online publication. DOI: 10.1016/j.atech.2024.100487

Kumar, V., Singh, G., & Dwivedi, S. K. (2022). Arsenic and Nickel Stress Response, Their Bioremediation Potential and Mechanism of *Trichoderma lixii* Isolated from Electroplating Wastewater. *Geomicrobiology Journal*, *39*(3-5), 186–198. DOI: 10.1080/01490451.2021.1956648

Kurcz, A., Błażejak, S., Kot, A. M., Bzducha-Wróbel, A., & Kieliszek, M. (2018). Application of industrial wastes for the production of microbial single-cell protein by fodder yeast Candida utilis. *Waste and Biomass Valorization*, *9*(1), 57–64. DOI: 10.1007/s12649-016-9782-z

Kyslynska, A. S., Nadkernychna, O. V., Kopylov, Y. P., & Tsekhmister, H. V. (2023). The relation between mutualistic mycorrhiza and endophytic plant-fungus associations and their effect on host plants. *Agricultural Science and Practice*, *10*(1), 54–75. DOI: 10.15407/agrisp10.01.054

Lafhal, S., Bombarda, I., Dupuy, N., Jean, M., Ruiz, K., Vanloot, P., & Vanthuyne, N. (1610). 2020. Chiroptical fingerprints to characterize lavender and lavandin essential oils. *Journal of Chromatography. A*, *460568*. Advance online publication. DOI: 10.1016/j.chroma.2019.460568 PMID: 31586515

Lahbouki, S., Ech-chatir, L., Er-Raki, S., Outzourhit, A., & Meddich, A. (2022). Improving drought tolerance of Opuntia ficus-indica under field using Subsurface Water Retention Technology: Changes in physiological and biochemical parameters. *Canadian Journal of Soil Science, ja.*

Lahbouki, S., Anli, M., El Gabardi, S., Ait-El-Mokhtar, M., Ben-Laouane, R., Boutasknit, A., Ait-Rahou, Y., Outzourhit, A., Wahbi, S., Douira, A., & Meddich, A. (2022). Evaluation of arbuscular mycorrhizal fungi and vermicompost supplementation on growth, phenolic content and antioxidant activity of prickly pear cactus (*Opuntia ficus-indica*). *Plant Biosystems*, *156*(4), 882–892. DOI: 10.1080/11263504.2021.1947408

Lahbouki, S., Fernando, A. L., Rodrigues, C., Ben-Laouane, R., Ait-El-Mokhtar, M., Outzourhit, A., & Meddich, A. (2023). Effects of Humic Substances and Mycorrhizal Fungi on Drought-Stressed Cactus: Focus on Growth, Physiology, and Biochemistry. *Plants*, *12*(24), 4156. DOI: 10.3390/plants12244156 PMID: 38140483

Lahlali, R., & Hijri, M. (2010). Screening, identification and evaluation of potential biocontrol fungal endophytes against Rhizoctonia solani AG3 on potato plants: Biocontrol of Rhizoctonia solani using fungal endophytes. *FEMS Microbiology Letters*, *311*(2), 152–159. DOI: 10.1111/j.1574-6968.2010.02084.x PMID: 20738401

Lahlali, R., McGregor, L., Song, T., Gossen, B. D., Narisawa, K., & Peng, G. (2014). Heteroconium chaetospira Induces Resistance to Clubroot via Upregulation of Host Genes Involved in Jasmonic Acid, Ethylene, and Auxin Biosynthesis. *PLoS One*, *9*(4), e94144. DOI: 10.1371/journal.pone.0094144 PMID: 24714177

Lahmidi, N. A., Courty, P.-E., Brulé, D., Chatagnier, O., Arnould, C., Doidy, J., Berta, G., Lingua, G., Wipf, D., & Bonneau, L. (2016). Sugar exchanges in arbuscular mycorrhiza: *RiMST5* and *RiMST6*, two novel Rhizophagus irregularis monosaccharide transporters, are involved in both sugar uptake from the soil and from the plant partner. *Plant Physiology and Biochemistry*, *107*, 354–363. DOI: 10.1016/j.plaphy.2016.06.023 PMID: 27362299

Lainez-Cerón, E., Jiménez-Munguía, M. T., López-Malo, A., & Ramírez-Corona, N. (2021). Effect of process variables on heating profiles and extraction mechanisms during hydrodistillation of eucalyptus essential oil. *Heliyon*, *7*(10), e08234. DOI: 10.1016/j.heliyon.2021.e08234 PMID: 34754975

Lakhanpal, T. N., & Rana, M. (2005). Medicinal and nutraceutical genetic resources of mushrooms. *Plant Genetic Resources*, *3*(2), 288–303. DOI: 10.1079/PGR200581

Lakin-Thomas, P. L., Coté, G. G., & Brody, S. (1990). Circadian Rhythms in *Neurospora crassa*: Biochemistry and Genetics. *Critical Reviews in Microbiology*, *17*(5), 365–416. DOI: 10.3109/10408419009114762 PMID: 2147375

Landrigan, P. J., Fuller, R., Acosta, N. J., Adeyi, O., Arnold, R., Baldé, A. B., & Zhong, M. (2018). The Lancet Commission on pollution and health. *Lancet*, *391*(10119), 462–512. DOI: 10.1016/S0140-6736(17)32345-0 PMID: 29056410

Lanfranco, L., Fiorilli, V., Venice, F., & Bonfante, P. (2018). Strigolactones cross the kingdoms: Plants, fungi, and bacteria in the arbuscular mycorrhizal symbiosis. *Journal of Experimental Botany*, *69*(9), 2175–2188. DOI: 10.1093/jxb/erx432 PMID: 29309622

Lanfranco, L., Wyss, P., Marzachi, C., & Bonfante, P. (1995). Generation of RAPD-PCR primers for the identification of isolates of *Glomus mosseae*, an arbuscular mycorrhizal fungus. *Molecular Ecology*, *4*(1), 61–68. DOI: 10.1111/j.1365-294X.1995.tb00192.x PMID: 7711955

Lapeña, D., Kosa, G., Hansen, L. D., Mydland, L. T., Passoth, V., Horn, S. J., & Eijsink, V. G. (2020). Production and characterization of yeasts grown on media composed of spruce-derived sugars and protein hydrolysates from chicken by-products. *Microbial Cell Factories*, *19*(1), 1–14. DOI: 10.1186/s12934-020-1287-6

Larbi-Koranteng, S., Kankam, F., Adomako, J., & Abdulai, M. (2022). Role of Mycorrhizae in Crop Protection. In *Arbuscular Mycorrhizal Fungi in Agriculture-New Insights*. IntechOpen., DOI: 10.5772/intechopen.109020

Latef, A. (2011). Influence of arbuscular mycorrhizal fungi and copper on growth, accumulation of osmolyte, mineral nutrition and antioxidant enzyme activity of pepper (Capsicum annuum L.). *Mycorrhiza*, *21*(6), 495–503. DOI: 10.1007/s00572-010-0360-0 PMID: 21221660

Lauren, D. R., Jensen, D. J., & Smith, W. A. (2006). Mycotoxin contamination in graded fractions of maize (Zea mays) in New Zealand. *New Zealand Journal of Crop and Horticultural Science, 34*(1), 63–72. DOI: 10.1080/01140671.2006.9514389

Leal, M. C., Puga, J., Serôdio, J., Gomes, N. C. M., & Calado, R. (2012). Trends in the discovery of new marine natural products from invertebrates over the last two decades – where and what are we bioprospecting? *PLoS One, 7*(1), e30580. DOI: 10.1371/journal.pone.0030580 PMID: 22276216

Leal, P. L., Siqueira, J. O., & Stuermer, S. L. (2013). Switch of tropical Amazon forest to pasture affects taxonomic composition but not species abundance and diversity of arbuscular mycorrhizal fungal community. *Applied Soil Ecology, 71*, 72–80. DOI: 10.1016/j.apsoil.2013.05.010

Lee, D., Pan, J. H., Kim, D., Heo, W., Shin, E. C., Kim, Y. J., Shim, Y. Y., Reaney, M. J. T., Ko, S., Hong, S., Cho, H. T., Kim, T. G., Lee, K., & Kim, J. K. (2024). Mycoproteins and their health-promoting properties: Fusarium species and beyond. *Comprehensive Reviews in Food Science and Food Safety, 23*(3), e13365. DOI: 10.1111/1541-4337.13365

Lee, J. G., Lee, Y., Kim, C. S., & Han, S. B. (2021). Codex Alimentarius commission on ensuring food safety and promoting fair trade: Harmonization of standards between Korea and codex. *Food Science and Biotechnology, 30*(9), 1151–1170. DOI: 10.1007/s10068-021-00943-7 PMID: 34483698

Lee, K., Loros, J. J., & Dunlap, J. C. (2000). Interconnected Feedback Loops in the *Neurospora* Circadian System. *Science, 289*(5476), 107–110. DOI: 10.1126/science.289.5476.107 PMID: 10884222

Lee, S. J., Morse, D., & Hijri, M. (2019). Holobiont chronobiology: Mycorrhiza may be a key to linking aboveground and underground rhythms. *Mycorrhiza, 29*(5), 403–412. DOI: 10.1007/s00572-019-00903-4 PMID: 31190278

Lee, S.-J., Kong, M., Morse, D., & Hijri, M. (2018). Expression of putative circadian clock components in the arbuscular mycorrhizal fungus *Rhizoglomus irregulare.Mycorrhiza, 28*(5-6), 523–534. DOI: 10.1007/s00572-018-0843-y PMID: 29931403

Lee, W. Y., Park, E.-J., Ahn, J. K., & Ka, K.-H. (2009). Ergothioneine Contents in Fruiting Bodies and Their Enhancement in Mycelial Cultures by the Addition of Methionine. *Microbiology, 37*(1), 43. DOI: 10.1099/00221287-124-1-43 PMID: 23983506

Leger, R. J., Charnley, A. K., & Cooper, R. M. (1987). Characterization of cuticle-degrading proteases produced by the entomopathogen Metarhizium anisopliae. *Archives of Biochemistry and Biophysics, 253*(1), 221–232. DOI: 10.1016/0003-9861(87)90655-2 PMID: 3545084

Lehmann, J., & Schroth, G. (2002). Nutrient leaching. In *Trees, crops and soil fertility: concepts and research methods* (pp. 151–166). CABI Publishing., DOI: 10.1079/9780851995939.0151

Lemanceau, P., Alabouvette, C., & Meyer, J. M. (1986). Production of Fusarinine and Iron Assimilation by Pathogenic and Non-Pathogenic Fusarium. In Swinburne, T. R. (Ed.), *Iron, Siderophores, and Plant Diseases* (pp. 251–259). Springer US., DOI: 10.1007/978-1-4615-9480-2_27

Lengai, G. M. W., Muthomi, J. W., & Mbega, E. R. (2020). Phytochemical activity and role of botanical pesticides in pest management for sustainable agricultural crop production. *Scientific African*, *7*, e00239. DOI: 10.1016/j.sciaf.2019.e00239

Lenoir, I., Fontaine, J., Tisserant, B., Laruelle, F., & Lounès-Hadj Sahraoui, A. (2017). Beneficial contribution of the arbuscular mycorrhizal fungus, *Rhizophagus irregularis*, in the protection of *Medicago truncatula* roots against benzo[a]pyrene toxicity. *Mycorrhiza*, *27*(5), 465–476. DOI: 10.1007/s00572-017-0764-1 PMID: 28197735

Leroy, F., Beal, T., Gregorini, P., McAuliffe, G. A., & van Vliet, S. (2022). Nutritionism in a food policy context: The case of 'animal protein'. *Animal Production Science*, *62*(8), 712–720. DOI: 10.1071/AN21237

Leslie, J. F., Moretti, A., Mesterházy, Á., Ameye, M., Audenaert, K., Singh, P. K., Richard-Forget, F., Chulze, S. N., Ponte, E. M. D., Chala, A., Battilani, P., & Logrieco, A. F. (2021). Key Global Actions for Mycotoxin Management in Wheat and Other Small Grains. *Toxins*, *13*(10), 725. DOI: 10.3390/toxins13100725 PMID: 34679018

Leyva Salas, M., Mounier, J., Valence, F., Coton, M., Thierry, A., & Coton, E. (2017). Antifungal Microbial Agents for Food Biopreservation-A Review. *Microorganisms*, *5*(3), 37. DOI: 10.3390/microorganisms5030037 PMID: 28698479

Li, Y. P., Ahmadi, F., Kariman, K., & Lackner, M. (2024). Recent advances and challenges in single cell protein (SCP) technologies for food and feed production. *npj Science of Food, 8*(1), Article 66. https://doi.org/DOI: 10.1038/s41538-024-00299-2

Liang, H., Yang, M., Li, Q., Zhang, L., & Zhao, X. (2024). A comprehensive review of the main components of plant essential oils and the mechanisms responsible for the inhibitory effects on fungal growth and aflatoxin synthesis. *Innovative Food Science & Emerging Technologies*, *96*, 103747. DOI: 10.1016/j.ifset.2024.103747

Liao, H., Xu, C., Tan, S., Wei, Z., Ling, N., Yu, G., Raza, W., Zhang, R., Shen, Q., & Xu, Y. (2012). Production and characterization of acidophilic xylanolytic enzymes from Penicillium oxalicum GZ-2. *Bioresource Technology*, *123*, 117–124. DOI: 10.1016/j.biortech.2012.07.051

Li, H., Kilgallen, A. B., Münzel, T., Wolf, E., Lecour, S., Schulz, R., Daiber, A., & Van Laake, L. W. (2020). Influence of mental stress and environmental toxins on circadian clocks: Implications for redox regulation of the heart and cardioprotection. *British Journal of Pharmacology*, *177*(23), 5393–5412. DOI: 10.1111/bph.14949 PMID: 31833063

Li, K., Qiao, K., Xiong, J., Guo, H., & Zhang, Y. (2023). Nutritional Values and Bio-Functional Properties of Fungal Proteins: Applications in Foods as a Sustainable Source. *Foods (Basel, Switzerland)*, *12*(24), 4388. DOI: 10.3390/molecules28114388

Li, L., Feng, Y., Qi, F., & Hao, R. (2023). Research Progress of Piriformospora indica in Improving Plant Growth and Stress Resistance to Plant. *Journal of Fungi (Basel, Switzerland)*, *9*(10), 965. DOI: 10.3390/jof9100965 PMID: 37888222

Li, M., Yu, L., Zhao, J., Zhang, H., Chen, W., Zhai, Q., & Tian, F. (2021). Role of dietary edible mushrooms in the modulation of gut microbiota. *Journal of Functional Foods*, *83*, 104538. DOI: 10.1016/j.jff.2021.104538

Li, N., Alfiky, A., Wang, W., Islam, M., Nourollahi, K., Liu, X., & Kang, S. (2018). Volatile Compound-Mediated Recognition and Inhibition between *Trichoderma* Biocontrol Agents and *Fusarium oxysporum. Frontiers in Microbiology*, *9*, 2614. DOI: 10.3389/fmicb.2018.02614 PMID: 30455673

Lindequist, U., Niedermeyer, T. H. J., & Jülich, W. D. (2005). The pharmacological potential of mushrooms. In Evidence-based Complementary and Alternative Medicine (Vol. 2, Issue 3, pp. 285–299). https://doi.org/DOI: 10.1093/ecam/neh107

Linder, T. (2023). Beyond Agriculture— How Microorganisms Can Revolutionize Global Food Production. *ACS Food Science & Technology*, *3*(7), 1144–1152. DOI: 10.1021/acsfoodscitech.3c00099

Linder, T. (2024). What next for mycoprotein? *Current Opinion in Food Science*, *101199*, 101199. Advance online publication. DOI: 10.1016/j.cofs.2024.101199

Lin, G., Tanguay, A., Guertin, C., Todorova, S., & Brodeur, J. (2017). A new method for loading predatory mites with entomopathogenic fungi for biological control of their prey. *Biological Control*, *115*, 105–111. DOI: 10.1016/j.biocontrol.2017.09.012

Lin, P., Zhang, M., Wang, M., Li, Y., Liu, J., & Chen, Y. (2021). Inoculation with arbuscular mycorrhizal fungus modulates defense-related genes expression in banana seedlings susceptible to wilt disease. *Plant Signaling & Behavior*, *16*(5), 1884782. DOI: 10.1080/15592324.2021.1884782 PMID: 33793381

Liu, A., Wang, B., & Hamel, C. (2004). Arbuscular mycorrhiza colonization and development at suboptimal root zone temperature. *Mycorrhiza*, *14*(2), 93–101. DOI: 10.1007/s00572-003-0242-9 PMID: 12748840

Liu, C. Y., Guo, X. N., Wu, X. L., Dai, F. J., & Wu, Q. S. (2022). The comprehensive effects of Rhizophagus intraradices and P on root system architecture and P transportation in Citrus limon L. *Agriculture*, *12*(3), 317. DOI: 10.3390/agriculture12030317

Liu, J., Sun, L., Zhang, N., Zhang, J., Guo, J., Li, C., & Qi, D. (2016). Effects of nutrients in substrates of different grains on aflatoxin B1 production by *Aspergillus flavus.BioMed Research International*, *2016*(1), 7232858. PMID: 27294129

Liu, K., Tian, J., Xiang, M., & Liu, X. (2012). How carnivorous fungi use three-celled constricting rings to trap nematodes. *Protein & Cell*, *3*(5), 325–328. DOI: 10.1007/s13238-012-2031-8 PMID: 22528749

Liu, L., Li, Y., Li, S., Hu, N., He, Y., Pong, R., Lin, D., Lu, L., & Law, M. (2012). Comparison of Next-Generation Sequencing Systems. *Journal of Biomedicine & Biotechnology*, *2012*, 1–11. PMID: 22829749

Liu, M., Zhang, X., Luan, H., Zhang, Y., Xu, W., Feng, W., & Song, P. (2024). Bioenzymatic detoxification of mycotoxins. *Frontiers in Microbiology*, *15*, 1434987. DOI: 10.3389/fmicb.2024.1434987 PMID: 39091297

Liu, N., Liu, C., Dudaš, T. N., Loc, M. Č., Bagi, F. F., & Van der Fels-Klerx, H. J. (2021). Improved aflatoxins and fumonisins forecasting models for maize (PREMA and PREFUM), using combined mechanistic and Bayesian Network modeling—Serbia as a case study. *Frontiers in Microbiology*, *12*, 643604. DOI: 10.3389/fmicb.2021.643604 PMID: 33967981

Liu, X., Ashforth, E., Ren, B., Song, F., Dai, H., Liu, M., Wang, J., Xie, Q., & Zhang, L. (2010). Bioprospecting microbial natural product libraries from the marine environment for drug discovery. *The Journal of Antibiotics*, *63*(8), 415–422. DOI: 10.1038/ja.2010.56 PMID: 20606699

Liu, Y. (2003). Molecular Mechanisms of Entrainment in the *Neurospora* Circadian Clock. *Journal of Biological Rhythms*, *18*(3), 195–205. DOI: 10.1177/0748730403018003002 PMID: 12828277

Liu, Y., & Bell-Pedersen, D. (2006). Circadian Rhythms in *Neurospora crassa* and Other Filamentous Fungi. *Eukaryotic Cell*, *5*(8), 1184–1193. DOI: 10.1128/EC.00133-06 PMID: 16896204

Liu, Y., Feng, S., & Liu, H. (2024). Measurements of Drying and Wetting Gas Diffusion Coefficients and Gas Permeability of Unsaturated Soils Using a New Flexible-Wall Device. *Journal of Geotechnical and Geoenvironmental Engineering*, *150*(11), 6024006. DOI: 10.1061/JGGEFK.GTENG-12475

Liu, Z., Zhang, G., Zhang, Y., Jin, Q., Zhao, J., & Li, J. (2016). Factors controlling mycotoxin contamination in maize and food in the Hebei province, China. *Agronomy for Sustainable Development*, *36*(39), 39. Advance online publication. DOI: 10.1007/s13593-016-0374-x

Liuzzi, G. M., Petraglia, T., Latronico, T., Crescenzi, A., & Rossano, R. (2023). Antioxidant Compounds from Edible Mushrooms as Potential Candidates for Treating Age-Related Neurodegenerative Diseases. *Nutrients*, *15*(8), 1913. DOI: 10.3390/nu15081913 PMID: 37111131

Li, X., Ma, W., Ma, Z., Zhang, Q., & Li, H. (2021). The Occurrence and Contamination Level of Ochratoxin A in Plant and Animal-Derived Food Commodities. *Molecules (Basel, Switzerland)*, *26*(22), 6928. DOI: 10.3390/molecules26226928 PMID: 34834020

Li, X., Song, C., Kang, X., Chen, F., Li, A., Wang, Y., Zou, J., Yin, J., Li, Y., Sun, Z., Ma, X., & Liu, J. (2024). Assembly and functional profile of rhizosphere microbial community during the *Salix viminalis*-AMF remediation of polycyclic aromatic hydrocarbon polluted soils. *Journal of Environmental Management*, *370*, 122503. DOI: 10.1016/j.jenvman.2024.122503 PMID: 39299104

Li, Z., Forester, S., Jennings-Dobbs, E., & Heber, D. (2023). Perspective: A Comprehensive Evaluation of Data Quality in Nutrient Databases. *Advances in Nutrition*, *14*(3), 379–391. DOI: 10.1016/j.advnut.2023.02.005 PMID: 36849083

Lo, C. T. (1997). Biological control of turf diseases using Trichoderma harzianum. Plant Protection. *Bulletin.*, *39*, 207–225.

Loi, M., Fanelli, F., Liuzzi, V. C., Logrieco, A. F., & Mulè, G. (2017). Mycotoxin biotransformation by native and commercial enzymes: Present and future perspectives. *Toxins*, *9*(4), 111. DOI: 10.3390/toxins9040111 PMID: 28338601

Loi, M., Logrieco, A. F., Pusztahelyi, T., Leiter, É., Hornok, L., & Pócsi, I. (2023). Advanced mycotoxin control and decontamination techniques in view of an increased aflatoxin risk in Europe due to climate change. *Frontiers in Microbiology*, *13*, 1085891. DOI: 10.3389/fmicb.2022.1085891 PMID: 36762096

Lombardi, L. M., & Brody, S. (2005). Circadian rhythms in *Neurospora crassa*: Clock gene homologues in fungi. *Fungal Genetics and Biology*, *42*(11), 887–892. DOI: 10.1016/j.fgb.2005.06.005 PMID: 16154782

Lonchamp, J., Stewart, K., Munialo, C. D., Evans, L., Akintoye, M., Gordon, S., Clegg, P. S., Willoughby, N., & Euston, S. R. (2022). Mycoprotein as novel functional ingredient: Mapping of functionality, composition and structure throughout the Quorn fermentation process. *Food Chemistry*, *396*, 133736. DOI: 10.1016/j.foodchem.2022.133736

Lone, R., Shuab, R., Khan, S., Ahmad, J., & Koul, K. K. (2017). Arbuscular Mycorrhizal Fungi for Sustainable Agriculture. In Kumar, V., Kumar, M., Sharma, S., & Prasad, R. (Eds.), *Probiotics and Plant Health* (pp. 553–577). Springer Singapore., DOI: 10.1007/978-981-10-3473-2_25

López-Bucio, J., Pelagio-Flores, R., & Herrera-Estrella, A. (2015). Trichoderma as biostimulant: Exploiting the multilevel properties of a plant beneficial fungus. *Scientia Horticulturae*, *196*, 109–123. DOI: 10.1016/j.scienta.2015.08.043

Ludwig-Müller, J. (2010). Hormonal responses in host plants triggered by arbuscular mycorrhizal fungi. Arbuscular mycorrhizas: physiology and function, 169-190

Luginbuehl, L., & Oldroyd, G. E. D. (2016). Calcium signaling and transcriptional regulation in arbuscular mycorrhizal symbiosis. *Molecular Mycorrhizal Symbiosis*, 125–140.

Luginbuehl, L. H., Menard, G. N., Kurup, S., Van Erp, H., Radhakrishnan, G. V., Breakspear, A., Oldroyd, G. E. D., & Eastmond, P. J. (2017). Fatty acids in arbuscular mycorrhizal fungi are synthesized by the host plant. *Science*, *356*(6343), 1175–1178. DOI: 10.1126/science.aan0081 PMID: 28596311

Lugtenberg, B. J. J., Caradus, J. R., & Johnson, L. J. (2016). Fungal endophytes for sustainable crop production. *FEMS Microbiology Ecology*, *92*(12), fiw194. Advance online publication. DOI: 10.1093/femsec/fiw194 PMID: 27624083

Lukić, B., Panico, A., Huguenot, D., Fabbricino, M., van Hullebusch, E. D., & Esposito, G. (2017). A review on the efficiency of landfarming integrated with composting as a soil remediation treatment. *Environmental Technology Reviews*, *6*(1), 94–116. DOI: 10.1080/21622515.2017.1310310

Lumini, E., Orgiazzi, A., Borriello, R., Bonfante, P., & Bianciotto, V. (2010). Disclosing arbuscular mycorrhizal fungal biodiversity in soil through a land-use gradient using a pyrosequencing approach. *Environmental Microbiology*, *12*(8), 2165–2179. DOI: 10.1111/j.1462-2920.2009.02099.x PMID: 21966911

Lundell, T. K., Makela, M. R., & Hilden, K. (2010). Lignin-modifying enzymes in filamentous basidiomycetes—Ecological, functional and phylogenetic review. *Journal of Basic Microbiology*, *50*(1), 5–20. DOI: 10.1002/jobm.200900338

Lutsky, I. I., & Mor, N. (1981). Alimentary toxic aleukia (septic angina, endemic panmyelotoxicosis, alimentary hemorrhagic aleukia): T-2 toxin-induced intoxication of cats. *American Journal of Pathology*, *104*(2), 189. PMID: 6973281

Luz, C., Ferrer, J., Mañes, J., & Meca, G. (2018). Toxicity reduction of ochratoxin A by lactic acid bacteria. *Food and Chemical Toxicology, 112*, 60–66. DOI: 10.1016/j.fct.2017.12.030 PMID: 29274433

Łysakowska, P., Sobota, A., & Wirkijowska, A. (2023). Medicinal Mushrooms: Their Bioactive Components, Nutritional Value and Application in Functional Food Production—A Review. In Molecules (Vol. 28, Issue 14). Multidisciplinary Digital Publishing Institute (MDPI). DOI: 10.3390/molecules28145393

Macalady, J. L., Hamilton, T. L., Grettenberger, C. L., Jones, D. S., Tsao, L. E., & Burgos, W. D. (2013). Energy, ecology and the distribution of microbial life. *Philosophical Transactions of the Royal Society of London. Series B, Biological Sciences, 368*(1622), 20120383. DOI: 10.1098/rstb.2012.0383 PMID: 23754819

Madaan, K., Sharma, S., & Kalia, A. (2024). Effect of selenium and zinc biofortification on the biochemical parameters of Pleurotus spp. under submerged and solid-state fermentation. *Journal of Trace Elements in Medicine and Biology, 82*, 127365. DOI: 10.1016/j.jtemb.2023.127365 PMID: 38171269

Madouh, T. A., & Quoreshi, A. M. (2023). The function of arbuscular mycorrhizal fungi associated with drought stress resistance in native plants of arid desert ecosystems: A review. *Diversity, 15*(3), 391. DOI: 10.3390/d15030391

Magkourilou, E., Bell, C. A., Daniell, T. J., & Field, K. J. (2024). The functionality of arbuscular mycorrhizal networks across scales of experimental complexity and ecological relevance. *Functional Ecology*, •••, 1–16. DOI: 10.1111/1365-2435.14618

Mahadev Madgule, N. (2023). Vinayaka, Yeshwant M. Sonkhaskar, Dhiren Ramanbhai Patel, R. Karthikeyan, P. Selvakumar,(2023). Mechanical properties and microstructure of activated TIG welded similar joints of Inconel alloys by desirability approaches. *Materials Today: Proceedings, 77*, 528–533. DOI: 10.1016/j.matpr.2022.12.250

Mahato, D. K., Devi, S., Pandhi, S., Sharma, B., Maurya, K. K., Mishra, S., Dhawan, K., Selvakumar, R., Kamle, M., Mishra, A. K., & Kumar, P. (2021a). Occurrence, Impact on Agriculture, Human Health, and Management Strategies of Zearalenone in Food and Feed: A Review. *Toxins, 13*(2), 92. DOI: 10.3390/toxins13020092 PMID: 33530606

Mahato, D. K., Kamle, M., Sharma, B., Pandhi, S., Devi, S., Dhawan, K., Selvakumar, R., Mishra, D., Kumar, A., Arora, S., Singh, N. A., & Kumar, P. (2021b). Patulin in food: A mycotoxin concern for human health and its management strategies. *Toxicon, 198*, 12–23. DOI: 10.1016/j.toxicon.2021.04.027 PMID: 33933519

Mahuku, G., Nzioki, H. S., Mutegi, C., Kanampiu, F., Narrod, C., & Makumbi, D. (2019). Pre-harvest management is a critical practice for minimizing aflatoxin contamination of maize. *Food Control, 96*, 219–226. DOI: 10.1016/j.foodcont.2018.08.032 PMID: 30713368

Ma, J., Janoušková, M., Li, Y., Yu, X., Yan, Y., Zou, Z., & He, C. (2015). Impact of arbuscular mycorrhizal fungi (AMF) on cucumber growth and phosphorus uptake under cold stress. *Functional Plant Biology, 42*(12), 1158–1167. DOI: 10.1071/FP15106 PMID: 32480753

Ma, J., Sun, Y., Meng, D., Zhou, Z., Zhang, Y., & Yang, R. (2023). Yeast proteins: The novel and sustainable alternative protein in food applications. *Trends in Food Science & Technology*, *135*, 190–201. DOI: 10.1016/j.tifs.2023.04.003

Majumder, R., Miatur, S., Saha, A., & Hossain, S. (2024). Mycoprotein: Production and nutritional aspects: A review. *Sustainable Food Technology*, *2*(1), 81–91. DOI: 10.1039/D3FB00169E

Makhuvele, R., Naidu, K., Gbashi, S., Thipe, V. C., Adebo, O. A., & Njobeh, P. B. (2020). The use of plant extracts and their phytochemicals for control of toxigenic fungi and mycotoxins. *Heliyon*, *6*(10), e05291. DOI: 10.1016/j.heliyon.2020.e05291 PMID: 33134582

Makrane, H., Aziz, M., Mekhfi, H., Ziyyat, A., Legssyer, A., Melhaoui, A., Berrabah, M., Bnouham, M., Alem, C., Elombo, F. K., Gressier, B., Desjeux, J. F., & Eto, B. (2019). Myorelaxant Activity of essential oil from *Origanum majorana* L. on rat and rabbit. *Journal of Ethnopharmacology*, *228*, 40–49. DOI: 10.1016/j.jep.2018.08.036 PMID: 30205180

Ma, L., Liu, Z., Kong, Z., Wang, M., Li, T., Zhu, H., Wan, Q., Liu, D., & Shen, Q. (2021). Functional characterization of a novel copper-dependent lytic polysaccharide monooxygenase TgAA11 from Trichoderma guizhouense NJAU 4742 in the oxidative degradation of chitin. *Carbohydrate Polymers*, *258*, 117708. https://doi.org/https://doi.org/10.1016/j.carbpol.2021.117708. DOI: 10.1016/j.carbpol.2021.117708 PMID: 33593577

Malafatti-Picca, L., de Barros Chaves, M. R., de Castro, A. M., Valoni, É., de Oliveira, V. M., Marsaioli, A. J., de Franceschi de Angelis, D., & Attili-Angelis, D. (2019). Hydrocarbon-associated substrates reveal promising fungi for poly (ethylene terephthalate)(PET) depolymerization. *Brazilian Journal of Microbiology : Publication of the Brazilian Society for Microbiology*, *50*(3), 633–648. DOI: 10.1007/s42770-019-00093-3 PMID: 31175657

Malekzad, H., Sahandi Zangabad, P., Mirshekari, H., Karimi, M., & Hamblin, M. R. (2017). Noble metal nanoparticles in biosensors: Recent studies and applications. *Nanotechnology Reviews*, *6*(3), 301–329. DOI: 10.1515/ntrev-2016-0014 PMID: 29335674

Malik, M. A., Jan, Y., Al-Keridis, L. A., Haq, A., Ahmad, J., Adnan, M., Alshammari, N., Ashraf, S. A., & Panda, B. P. (2022). Effect of Vitamin-D-Enriched Edible Mushrooms on Vitamin D Status, Bone Health and Expression of CYP2R1, CYP27B1 and VDR Gene in Wistar Rats. *Journal of Fungi (Basel, Switzerland)*, *8*(8), 864. Advance online publication. DOI: 10.3390/jof8080864 PMID: 36012852

Malviya, D., Singh, P., Singh, U. B., Paul, S., Kumar Bisen, P., Rai, J. P., Verma, R. L., Fiyaz, R. A., Kumar, A., Kumari, P., Dei, S., Ahmed, M. R., Bagyaraj, D. J., & Singh, H. V. (2023). Arbuscular mycorrhizal fungi-mediated activation of plant defense responses in direct seeded rice (Oryza sativa L.) against root-knot nematode Meloidogyne graminicola. *Frontiers in Microbiology*, *14*, 1104490. Advance online publication. DOI: 10.3389/fmicb.2023.1104490 PMID: 37200920

Manathunga, K. K., Gunasekara, N. W., Meegahakumbura, M. K., Ratnaweera, P. B., Faraj, T. K., & Wanasinghe, D. N. (2024). Exploring Endophytic Fungi as Natural Antagonists against Fungal Pathogens of Food Crops. *Journal of Fungi (Basel, Switzerland)*, *10*(9), 606. DOI: 10.3390/jof10090606 PMID: 39330366

Manjula, N. G. Tajunnisa, Mamani, V., Meghana, C. A., & Mayegowda, S. B. (2023). Fungal-Based Synthesis to Generate Nanoparticles for Nanobioremediation. In *Green Nanoremediation: Sustainable Management of Environmental Pollution* (pp. 83-108). Cham: Springer International Publishing.

Manoussopoulos, Y., Mantzoukas, S., Lagogiannis, I., Goudoudaki, S., & Kambouris, M. (2019). Effects of Three Strawberry Entomopathogenic Fungi on the Prefeeding Behavior of the Aphid Myzus persicae. *Journal of Insect Behavior*, *32*(2), 99–108. DOI: 10.1007/s10905-019-09709-w

Mantzoukas, S., & Eliopoulos, P. A. (2020). Endophytic Entomopathogenic Fungi: A Valuable Biological Control Tool against Plant Pests. *Applied Sciences (Basel, Switzerland)*, *10*(1), 360. DOI: 10.3390/app10010360

Marete, G. N., Kanja, L. W., Mbaria, J. M., Okumu, M. O., Ateku, P. A., Korhonen, H., & Joutsjoki, V. (2020). Effects of the use of good agricultural practices on aflatoxin levels in maize grown in Nandi County, Kenya. *Sci*, *2*(4), 85. DOI: 10.3390/sci2040085

Maria, D. L. W. T. A., & Sebastian, G. M. J. (2023). *Single cell protein from thermophilic fungi* (US 11707078 B2). U.S. Patent. https://lens.org/142-801-579-684-076

Mariod, A.A. (2016). Effect of Essential Oils on Organoleptic (Smell, Taste, and Texture) Properties of Food. *Essential Oils in Food Preservation, Flavor and Safety*. Elsevier. 131-137. https://doi.org/DOI: 10.1016/B978-0-12-416641-7.00013-4

Marleau, J., Dalpé, Y., St-Arnaud, M., & Hijri, M. (2011). Spore development and nuclear inheritance in arbuscular mycorrhizal fungi. *BMC Evolutionary Biology*, *11*(1), 51. DOI: 10.1186/1471-2148-11-51 PMID: 21349193

Martinez, C., Blanc, F., Le Claire, E., Besnard, O., Nicole, M., & Baccou, J. C. (2001). Salicylic acid and ethylene pathways are differentially activated in melon cotyledons by active or heat-denatured cellulase from Trichoderma longibrachiatum. *Plant Physiology*, *127*(1), 334–344. DOI: 10.1104/pp.127.1.334 PMID: 11553761

Martin, F. M., & van der Heijden, M. G. A. (2024). The mycorrhizal symbiosis: Research frontiers in genomics, ecology, and agricultural application. *The New Phytologist*, *242*(4), 1486–1506. DOI: 10.1111/nph.19541 PMID: 38297461

Martins, A., Vieira, H., Gaspar, H., & Santos, S. (2014). Marketed marine natural products in the pharmaceutical and cosmeceutical industries: Tips for success. *Marine Drugs*, *12*(2), 1066–1101. DOI: 10.3390/md12021066 PMID: 24549205

Masanga, J. O., Matheka, J. M., Omer, R. A., Ommeh, S. C., Monda, E. O., & Alakonya, A. E. (2015). Downregulation of transcription factor aflR in Aspergillus flavus confers reduction to aflatoxin accumulation in transgenic maize with alteration of host plant architecture. *Plant Cell Reports*, *34*(8), 1379–1387. DOI: 10.1007/s00299-015-1794-9 PMID: 25895735

Masrahi, A. S., Alasmari, A., Shahin, M. G., Qumsani, A. T., Oraby, H. F., & Awad-Allah, M. M. (2023). Role of arbuscular mycorrhizal fungi and phosphate solubilizing bacteria in improving yield, yield components, and nutrients uptake of barley under salinity soil. *Agriculture*, *13*(3), 537. DOI: 10.3390/agriculture13030537

Masyita, A., Mustika Sari, R., Dwi Astuti, A., Yasir, B., Rahma Rumata, N., Emran, T. B., Nainu, F., & Simal-Gandara, J. (2022). Terpenes and terpenoids as main bioactive compounds of essential oils, their roles in human health and potential application as natural food preservatives. *Food Chemistry: X*, *13*, 100217. DOI: 10.1016/j.fochx.2022.100217 PMID: 35498985

Mathew, S. O., Sandhu, S. S., & Rajak, R. C. (1998). Bioactivity of Nomuraea rileyi against Spilosoma obliqua: Effect of dosage, temperature, and relative humidity. *Journal of the Indian Botanical Society*, *77*, 23–25.

Matrella, R., Monaci, L., Milillo, M. A., Palmisano, F., & Tantillo, M. G. (2006). Ochratoxin A determination in paired kidneys and muscle samples from swines slaughtered in southern Italy. *Food Control*, *17*(2), 114–117. DOI: 10.1016/j.foodcont.2004.08.008

Matsumoto, M., Takahashi, Y., Fujimoto, H., Shindo, K., Okada, K., & Ikeda, H. (2015). *Cladosporium* cladosporioides promotes plant defense responses via secreted elicitors. *Bioscience, Biotechnology, and Biochemistry*, *79*(1), 149–157. DOI: 10.1080/09168451.2014.949272 PMID: 25175163

Matz, C., Webb, J. S., Schupp, P. J., Phang, S. Y., Penesyan, A., Egan, S., Steinberg, P., & Kjelleberg, S. (2008). Marine biofilm bacteria evade eukaryotic predation by targeted chemical defense. *PLoS One*, *3*(7), e2744. DOI: 10.1371/journal.pone.0002744 PMID: 18648491

Mayans, B., Camacho-Arévalo, R., García-Delgado, C., Antón-Herrero, R., Escolástico, C., Segura, M. L., & Eymar, E. (2021). An assessment of *Pleurotus ostreatus* to remove sulfonamides, and its role as a biofilter based on its own spent mushroom substrate. *Environmental Science and Pollution Research International*, *28*(6), 7032–7042. DOI: 10.1007/s11356-020-11078-3 PMID: 33025433

Mayer, A. M. S., Glaser, K. B., Cuevas, C., Jacobs, R. S., Kem, W., Little, R. D., McIntosh, J. M., Newman, D. J., Potts, B. C., & Shuster, D. E. (2010). The odyssey of marine pharmaceuticals: A current pipeline perspective. *Trends in Pharmacological Sciences*, *31*(6), 255–265. DOI: 10.1016/j.tips.2010.02.005 PMID: 20363514

Maza, M., Pajot, H. F., Amoroso, M. J., & Yasem, M. G. (2014). Post-harvest sugarcane residue degradation by autochthonous fungi. *International Biodeterioration & Biodegradation*, *87*, 18–25. DOI: 10.1016/j.ibiod.2013.10.020

McFarland, J. W., Ruess, R. W., Kielland, K., Pregitzer, K., Hendrick, R., & Allen, M. (n.d.). *Cross-Ecosystem Comparisons of In Situ Plant Uptake of Amino Acid-N and NH4+.*

Meddich, A. (2023). Biostimulants for resilient agriculture—Improving plant tolerance to abiotic stress: A concise review. *Gesunde Pflanzen*, *75*(4), 709–727. DOI: 10.1007/s10343-022-00784-2

Meena, K. K., Kumar, P., Sorty, A. M., Bitla, U., & Pathak, H. (2022). Ecology of Arbuscular Mycorrhizae and Influence on Drought Tolerance in Crop Plants. In *Microbial BioTechnology for Sustainable Agriculture* (Vol. 1, pp. 261–285). Springer Nature Singapore. DOI: 10.1007/978-981-16-4843-4_8

Mello, I. S., Targanski, S., Pietro-Souza, W., Stachack, F. F. F., Terezo, A. J., & Soares, M. A. (2020). Endophytic bacteria stimulate mercury phytoremediation by modulating its bioaccumulation and volatilization. *Ecotoxicology and Environmental Safety*, *202*, 110818. DOI: 10.1016/j.ecoenv.2020.110818 PMID: 32590206

Mendoza-Soto, A. B., Rodríguez-Corral, A. Z., Bojórquez-López, A., Cervantes-Rojo, M., Castro-Martínez, C., & Lopez-Meyer, M. (2022). Arbuscular mycorrhizal symbiosis leads to differential regulation of genes and mirnas associated with the cell wall in tomato leaves. *Biology (Basel)*, *11*(6), 854. DOI: 10.3390/biology11060854 PMID: 35741375

Menezes-Oliveira, V. B., Scott-Fordsmand, J. J., Soares, A. M. V. M., & Amorim, M. J. B. (2013). Effects of temperature and copper pollution on soil community—Extreme temperature events can lead to community extinction. *Environmental Toxicology and Chemistry*, *32*(12), 2678–2685. DOI: 10.1002/etc.2345 PMID: 23939831

Meng, J., Liu, S., Gao, L., Hong, K., Liu, S., & Wu, X. (2023). Economical production of Pichia pastoris single cell protein from methanol at industrial pilot scale. *Microbial Cell Factories*, *22*(1), 198. DOI: 10.1186/s12934-023-02198-9

Menkis, A., Vasiliauskas, R., Taylor, A. F. S., Stenlid, J., & Finlay, R. (2007). Afforestation of abandoned farmland with conifer seedlings inoculated with three ectomycorrhizal fungi—Impact on plant performance and ectomycorrhizal community. *Mycorrhiza*, *17*(4), 337–348. DOI: 10.1007/s00572-007-0110-0 PMID: 17277941

Mesterhazy, A., Szieberth, D., Tóth, E. T., Nagy, Z., Szabó, B., Herczig, B., Bors, I., & Tóth, B. (2022). The role of preharvest natural infection and toxin contamination in food and feed safety in maize, South-East Hungary, 2014–2021. *Journal of Fungi (Basel, Switzerland)*, *8*(10), 1104. DOI: 10.3390/jof8101104 PMID: 36294669

Mészáros, E., Bodor, A., Kovács, E., Papp, S., Kovács, K., Perei, K., & Feigl, G. (2023). Impacts of Plastics on Plant Development: Recent Advances and Future Research Directions. *Plants*, *12*(18), 3282. Advance online publication. DOI: 10.3390/plants12183282 PMID: 37765446

Metoui, N., Gargouri, S., Amri, I., Fezzani, T., Jamoussi, B., & Hamrouni, L. (2015). Activity antifungal of the essential oils; aqueous and ethanol extracts from *Citrus aurantium* L. *Natural Product Research*, *29*(23), 2238–2241. DOI: 10.1080/14786419.2015.1007136 PMID: 26207731

Meyer, V., Basenko, E. Y., Benz, J. P., Braus, G. H., Caddick, M. X., Csukai, M., De Vries, R., Frisvad, D., & Gunde-Cimerman, N. (2020). Growing a Circular Economy with Fungal Biotechnology: A White Paper. *Fungal Biology and Biotechnology*, *7*(1), 1–23. DOI: 10.1186/s40694-020-00095-z PMID: 32280481

Meyling, N. V., & Eilenberg, J. (2007). Ecology of the entomopathogenic fungi Beauveria bassiana and Metarhizium anisopliae in temperate agroecosystems: Potential for conservation biological control. *Biological Control*, *43*(2), 145–155. DOI: 10.1016/j.biocontrol.2007.07.007

Mia, R., Selim, M., Shamim, A. M., Chowdhury, M., Sultana, S., Armin, M., Hossain, M., Akter, R., Dey, S., & Naznin, H. (2019). Review on various types of pollution problem in textile dyeing & printing industries of Bangladesh and recommandation for mitigation. *Journal of Textile Engineering & Fashion Technology*, *5*(4). Advance online publication. DOI: 10.15406/jteft.2019.05.00205

Mikkelsen, B. L., Rosendahl, S., & Jakobsen, I. (2008). Underground resource allocation between individual networks of mycorrhizal fungi. *The New Phytologist*, *180*(4), 890–898. DOI: 10.1111/j.1469-8137.2008.02623.x PMID: 18801003

Milagres, A. M. F., Lacis, L. S., & Prade, R. A. (1993). Characterization of xylanase production by a local isolate of Penicillium janthinellum. *Enzyme and Microbial Technology*, *15*(3), 248–253. DOI: 10.1016/0141-0229(93)90145-R

Ministry of Health. Labour and Welfare, Japan. (2020). *Standards and criteria for food under the Food Sanitation Act (abstract).* Retrieved from https://www.mhlw.go.jp/english/topics/importedfoods/19/appendix1.html

Mir, R. A., Mantoo, A. G., Sofi, Z. A., Bhat, D. A., Bashir, A., & Bashir, S. (2023). Types of Environmental Pollution and Its Effects on the Environment and Society. In *Geospatial Analytics for Environmental Pollution Modeling: Analysis, Control and Management* (pp. 1–31). Springer Nature Switzerland. DOI: 10.1007/978-3-031-45300-7_1

Mirshad, P. P., & Puthur, J. T. (2016). Arbuscular mycorrhizal association enhances drought tolerance potential of promising bioenergy grass (Saccharum arundinaceum retz.). *Environmental Monitoring and Assessment*, *188*(7), 425. DOI: 10.1007/s10661-016-5428-7 PMID: 27329476

Mishra, D., Rajvir, S., Mishra, U., & Kumar, S. S. (2013). Role of bio-fertilizer in organic agriculture: A review. *Research Journal of Recent Sciences ISSN*, *2277*, 2502.

Mishra, S., Singh, A., Keswani, C., Saxena, A., Sarma, B. K., & Singh, H. B. (2015). Harnessing Plant-Microbe Interactions for Enhanced Protection Against Phytopathogens. In Arora, N. K. (Ed.), *Plant Microbes Symbiosis: Applied Facets* (pp. 111–125). Springer India., DOI: 10.1007/978-81-322-2068-8_5

Mnkandla, S. M., & Otomo, P. V. (2021). Effectiveness of mycofiltration for removal of contaminants from water: A systematic review protocol. *Environmental Evidence*, *10*(1), 1–8. DOI: 10.1186/s13750-021-00232-0

Mohamed, H., & Saad, A. (2009). The biocontrol of postharvest disease (Botryodiplodia theobromae) of guava (Psidium guajava L.) by the application of yeast strains. *Postharvest Biology and Technology*, *53*(3), 123–130. DOI: 10.1016/j.postharvbio.2009.04.001

Mohammadi-Sichani, M., Mazaheri Assadi, M., Farazmand, A., Kianirad, M., Ahadi, A. M., & Hadian-Ghahderijani, H. (2019). Ability of Agaricus bisporus, Pleurotus ostreatus and Ganoderma lucidum compost in biodegradation of petroleum hydrocarbon-contaminated soil. *International Journal of Environmental Science and Technology*, *16*(5), 2313–2320. DOI: 10.1007/s13762-017-1636-0

Mohapatra, D., Rath, S. K., & Mohapatra, P. K. (2022). Soil fungi for bioremediation of pesticide toxicants: A perspective. *Geomicrobiology Journal*, *39*(3-5), 352–372. DOI: 10.1080/01490451.2021.2019855

Molfetta, M., Morais, E. G., Barreira, L., Bruno, G. L., Porcelli, F., Dugat-Bony, E., Bonnarme, P., & Minervini, F. (2022). Protein sources alternative to meat: State of the art and involvement of fermentation. *Food*, *11*(14), 2065. DOI: 10.3390/foods11142065 PMID: 35885308

Molinski, T. F., Dalisay, D. S., Lievens, S. L., & Saludes, J. P. (2009). Drug development from marine natural products. *National Review*, *8*(1), 69–85. PMID: 19096380

Monaco, C., Perello, A., Alippi, H. E., & Pasquare, A. O. (1991). Trichoderma spp.: A biocontrol agent of Fusarium spp. and Sclerotium rolfsii by seed treatment. *Advances in Horticultural Science*, *5*(3), 92–95.

Monteyne, A. J., Coelho, M. O., Porter, C., Abdelrahman, D. R., Jameson, T. S. O., Jackman, S. R., Blackwell, J. R., Finnigan, T. J. A., Stephens, F. B., Dirks, M. L., & Wall, B. T. (2020). Mycoprotein ingestion stimulates protein synthesis rates to a greater extent than milk protein in rested and exercised skeletal muscle of healthy young men: A randomized controlled trial. *The American Journal of Clinical Nutrition*, *112*(2), 318–333. DOI: 10.1093/ajcn/nqaa092 PMID: 32438401

Mora, C., Tittensor, D. P., Adl, S., Simpson, A. G. B., & Worm, B. (2011). How many species are there on Earth and in the ocean? *PLoS Biology*, *9*(8), 1–8. DOI: 10.1371/journal.pbio.1001127 PMID: 21886479

Morales, D., Rutckeviski, R., Villalva, M., Abreu, H., Soler-Rivas, C., Santoyo, S., Iacomini, M., & Smiderle, F. R. (2020). Isolation and comparison of α- and β-D-glucans from shiitake mushrooms (Lentinula edodes) with different biological activities. *Carbohydrate Polymers*, *229*, 115521. DOI: 10.1016/j.carbpol.2019.115521 PMID: 31826486

Morales-Payan, J. P., & Stall, W. M. (n.d.). *PAPAYA (CARICA PAPAYA) RESPONSE TO FOLIAR TREATMENTS WITH ORGANIC COMPLEXES OF PEPTIDES AND AMINO ACIDS. 14.*

Morales-Vargas, A. T., López-Ramírez, V., Álvarez-Mejía, C., & Vázquez-Martínez, J. (2024). Endophytic Fungi for Crops Adaptation to Abiotic Stresses. *Microorganisms*, *12*(7), 1357. DOI: 10.3390/microorganisms12071357 PMID: 39065124

Morath, S. U., Hung, R., & Bennett, J. W. (2012). Fungal volatile organic compounds: A review with emphasis on their biotechnological potential. *Fungal Biology Reviews*, *26*(2–3), 73–83. DOI: 10.1016/j.fbr.2012.07.001

Morgan, M. R. (1989). Mycotoxin immunoassays (with special reference to ELISAs). *Tetrahedron*, *45*(8), 2237–2249. DOI: 10.1016/S0040-4020(01)83429-3

Mouhamadou, B., Faure, M., Sage, L., Marçais, J., Souard, F., & Geremia, R. A. (2013). Potential of autochthonous fungal strains isolated from contaminated soils for degradation of polychlorinated biphenyls. *Fungal Biology*, *117*(4), 268–274. DOI: 10.1016/j.funbio.2013.02.004 PMID: 23622721

Mudaliar, S., Agrawal, K., & Verma, P. (2023). Fungi: a veritable tool for refractory pollutant remediation. *Clean Technologies Toward a Sustainable Future*, 283.

Muhonja, C. N., Makonde, H., Magoma, G., & Imbuga, M. (2018). Biodegradability of polyethylene by bacteria and fungi from Dandora dumpsite Nairobi-Kenya. *PLoS One*, *13*(7), e0198446. DOI: 10.1371/journal.pone.0198446 PMID: 29979708

Muhorakeye, M. C., Namikoye, E. S., Khamis, F. M., Wanjohi, W., & Akutse, K. S. (2024). Biostimulant and antagonistic potential of endophytic fungi against fusarium wilt pathogen of tomato Fusarium oxysporum f. sp. lycopersici. *Scientific Reports*, *14*(1), 15365. DOI: 10.1038/s41598-024-66101-1 PMID: 38965302

Mukherjee, P. K., Horwitz, B. A., & Kenerley, C. M. (2012). Secondary metabolism in Trichoderma—A genomic perspective. *Microbiology (Reading, England)*, *158*(Pt 1), 35–45. DOI: 10.1099/mic.0.053629-0 PMID: 21998165

Mukherjee, P. K., & Raghu, K. (1997). Effect of temperature on antagonistic and biocontrol potential of Trichoderma sp. on Sclerotium rolfsii. *Mycopathologia, 139*(3), 151–155. DOI: 10.1023/A:1006868009184 PMID: 16283454

Mukhopadhyay, R., & Kumar, D. (2020). Trichoderma: A beneficial antifungal agent and insights into its mechanism of biocontrol potential. *Egyptian Journal of Biological Pest Control*, *30*(1), 133. DOI: 10.1186/s41938-020-00333-x

Mukhopadhyay, R., & Pan, S. K. (2012). Isolation and selection of some antagonistic Trichoderma species from different new alluvial zones of Nadia district, West Bengal. *Journal of the Botanical Society of Bengal*, *66*(2), 149–152.

Mulyadi, & Jiang, L. (2023). Combined application of arbuscular mycorrhizal fungi (AMF) and nitrogen fertilizer alters the physicochemical soil properties, nitrogen uptake, and rice yield in a polybag experiment. *Agriculture, 13*(7), 1364.

Munkvold, G. P., Proctor, R. H., & Moretti, A. (2021). Mycotoxin production in Fusarium according to contemporary species concepts. *Annual Review of Phytopathology*, *59*(1), 373–402. DOI: 10.1146/annurev-phyto-020620-102825 PMID: 34077240

Münzel, T., Hahad, O., Daiber, A., & Landrigan, P. J. (2023). Soil and water pollution and human health: What should cardiologists worry about? *Cardiovascular Research*, *119*(2), 440–449. DOI: 10.1093/cvr/cvac082 PMID: 35772469

Murphy, P. A., Hendrich, S., Landgren, C., & Bryant, C. M. (2006). Food mycotoxins: An update. *Journal of Food Science*, *71*(5), R51–R65. DOI: 10.1111/j.1750-3841.2006.00052.x

Mustafa, F., Chopra, H., Baig, A. A., Avula, S. K., Kumari, S., Mohanta, T. K., Saravanan, M., Mishra, A. K., Sharma, N., & Mohanta, Y. K. (2022). Edible Mushrooms as Novel Myco-Therapeutics: Effects on Lipid Level, Obesity, and BMI. In *Journal of Fungi* (Vol. 8, Issue 2). MDPI. https://doi.org/DOI: 10.3390/jof8020211

Musyoka, D. M., Njeru, E. M., Nyamwange, M. M., & Maingi, J. M. (2020). Arbuscular mycorrhizal fungi and Bradyrhizobium co-inoculation enhances nitrogen fixation and growth of green grams (*Vigna radiata* L.) under water stress. *Journal of Plant Nutrition*, *43*(7), 1036–1047. DOI: 10.1080/01904167.2020.1711940

Mutungi, C., Muthoni, F., Bekunda, M., Gaspar, A., Kabula, E., & Abass, A. (2019). Physical quality of maize grain harvested and stored by smallholder farmers in the Northern highlands of Tanzania: Effects of harvesting and pre-storage handling practices in two marginally contrasting agro-locations. *Journal of Stored Products Research*, *84*, 101517. DOI: 10.1016/j.jspr.2019.101517

Mwampashi, L. L., Magubika, A. J., Ringo, J. F., Theonest, D. J., Tryphone, G. M., Chilagane, L. A., & Nassary, E. K. (2024). Exploring agro-ecological significance, knowledge gaps, and research priorities in arbuscular mycorrhizal fungi. *Frontiers in Microbiology*, *15*, 1491861. Advance online publication. DOI: 10.3389/fmicb.2024.1491861 PMID: 39552643

Mwangi, R. W., Macharia, J. M., Wagara, I. N., & Bence, R. L. (2022). The antioxidant potential of different edible and medicinal mushrooms. *Biomedicine and Pharmacotherapy*, *147*, 112621. DOI: 10.1016/j.biopha.2022.112621 PMID: 35026489

Nabi, S. U., Raja, W. H., Kumawat, K. L., Mir, J. I., Sharma, O. C., Singh, D. B., & Sheikh, M. A. (2017). Post-harvest diseases of temperate fruits and their management strategies—A review. *International Journal of Pure and Applied Biosciences*, *5*(3), 885–898. DOI: 10.18782/2320-7051.2981

Nadimi, M., Beaudet, D., Forget, L., Hijri, M., & Lang, B. F. (2012). Group I Intron–Mediated Trans-splicing in Mitochondria of *Gigaspora rosea* and a Robust Phylogenetic Affiliation of Arbuscular Mycorrhizal Fungi with Mortierellales. *Molecular Biology and Evolution*, *29*(9), 2199–2210. DOI: 10.1093/molbev/mss088 PMID: 22411852

Nagajyoti, P. C., Lee, K. D., & Sreekanth, T. V. M. (2010). Heavy metals, occurrence and toxicity for plants: A review. *Environmental Chemistry Letters*, *8*(3), 199–216. DOI: 10.1007/s10311-010-0297-8

Nagy, B., Măicăneanu, A., Indolean, C., Mânzatu, C., Silaghi-Dumitrescu, L., & Majdik, C. (2014). Comparative study of Cd (II) biosorption on cultivated *Agaricus bisporus* and wild *Lactarius piperatus* based biocomposites. Linear and nonlinear equilibrium modelling and kinetics. *Journal of the Taiwan Institute of Chemical Engineers*, *45*(3), 921–929. DOI: 10.1016/j.jtice.2013.08.013

Nalage, D., Khedkar, G., Kalyankar, A., Sarkate, A., Ghodke, S., Bedre, V. B., & Khedkar, C. D. (2016). Single cell proteins. *Encyclopedia of food and health, 4*, 790-794. http://dx.doi.org/DOI: 10.1016/B978-0-12-384947-2.00628-0

Nandi, S., Sikder, R., Rapior, S., Arnould, S., Simal-Gandara, J., & Acharya, K. (2024). A review for cancer treatment with mushroom metabolites through targeting mitochondrial signaling pathway: In vitro and in vivo evaluations, clinical studies and future prospects for mycomedicine. *Fitoterapia*, *172*, 105681. DOI: 10.1016/j.fitote.2023.105681 PMID: 37743029

Narayan Paudel, P., Satyal, P., Satyal, R., Setzer, W. N., Gyawali, R. 2022. molecules Chemical Composition, Enantiomeric Distribution, Antimicrobial and Antioxidant Activities of *Origanum majorana* L. Essential Oil from Nepal. https://doi.org/DOI: 10.3390/molecules27186136

Nasseri, A. T., Rasoul-Amini, S., Morowvat, M. H., & Ghasemi, Y. (2011). Single cell protein: Production and process. *American Journal of Food Technology*, *6*(2), 103–116. DOI: 10.3923/ajft.2011.103.116

Nassimi, Z., & Taheri, P. (2017). Endophytic fungus *Piriformospora indica* induced systemic resistance against rice sheath blight via affecting hydrogen peroxide and antioxidants. *Biocontrol Science and Technology*, *27*(2), 252–267. DOI: 10.1080/09583157.2016.1277690

Nasslahsen, B., Prin, Y., Ferhout, H., Smouni, A., & Duponnois, R. (2022). Management of Plant Beneficial Fungal Endophytes to Improve the Performance of Agroecological Practices. *Journal of Fungi (Basel, Switzerland)*, *8*(10), 1087. DOI: 10.3390/jof8101087 PMID: 36294652

Nath, A., Chattopadhyay, A., & Joshi, S. R. (2015). Biological Activity of Endophytic Fungi of Rauwolfia serpentina Benth: An Ethnomedicinal Plant Used in Folk Medicines in Northeast India. *Proceedings of the National Academy of Sciences. India. Section B, Biological Sciences*, *85*(1), 233–240. DOI: 10.1007/s40011-013-0184-8

Navarro, A., Elia, A., Conversa, G., Pasquale, C., & Mastrorilli, M. (2012). Potted mycorrhizal carnation plants and saline stress: Growth, quality and nutritional plant responses. *Scientia Horticulturae*, *140*, 131–139. DOI: 10.1016/j.scienta.2012.03.016

Navarro, J. M., & Morte, A. (2024). Arbuscular Mycorrhizal Fungi as Biofertilizers to Increase the Plant Quality of Sour-Orange Seedlings. *Agronomy (Basel), 14*(1), 230. DOI: 10.3390/agronomy14010230

Nayak, S. K., & Mishra, B. B. (2020). *Frontiers in Soil and Environmental Microbiology* (Nayak, S. K., & Mishra, B. B., Eds.). CRC Press., DOI: 10.1201/9780429485794

Nayeem, M., Chauhan, K., Khan, S., Rattu, G., Dhaka, R. K., & Sidduqui, H. (2017). Optimization of low-cost substrate for the production of single cell protein using Kluyveromyces marxianus. *Pharma Innov J, 6*(8).

Ndeko, A. B., Founoune-Mboup, H., Kane, A., & Cournac, L. (2022). Arbuscular mycorrhizal fungi alleviate the negative effect of temperature stress in millet lines with contrasting soil aggregation potential. *Gesunde Pflanzen, 74*(1), 53–67. DOI: 10.1007/s10343-021-00588-w

Nega, A. (2014). Review on concepts in biological control of plant pathogens. *Journal of Biology, Agriculture and Healthcare, 4*(27), 33–54.

Neme, K., & Mohammed, A. (2017). Mycotoxin occurrence in grains and the role of postharvest management as a mitigation strategy: A review. *Food Control, 78*, 412–425. DOI: 10.1016/j.foodcont.2017.03.012

Németh, M. Z., Pintye, A., Horváth, Á. N., Vági, P., Kovács, G. M., Gorfer, M., & Kiss, L. (2019). Green Fluorescent Protein Transformation Sheds More Light on a Widespread Mycoparasitic Interaction. *Phytopathology, 109*(8), 1404–1416. DOI: 10.1094/PHYTO-01-19-0013-R PMID: 30900938

Ng, Z. Y., Kee, P. E., Abdullah, R., Lan, J. C. W., Ling, T. C., Jiang, J. J., Lim, J. W., & Khoo, K. S. (2024). Conversion of lignocellulosic biomass waste into mycoprotein: Current status and future directions for sustainable protein production. *Biomass Conversion and Biorefinery*, •••, 1–27. DOI: 10.1007/s13399-024-06130-y

Niemi, K., Vuorinen, T., & Ernstsen, A. (2002). Ectomycorrhizal fungi and exogenous auxins influence root and mycorrhiza formation of Scots pine hypocotyl cuttings in vitro. *Tree Physiology, 22*(17), 1231–1239. DOI: 10.1093/treephys/22.17.1231 PMID: 12464576

Nirmalkar, V. K., Singh, S., Tiwari, R. S., Said, P. P., & Kaushik, D. K. (2017). Field efficacy of Trichoderma harzianum and rhizobium against wilt complex of chickpea. *International Journal of Current Microbiology and Applied Sciences, 6*(7), 1421–1429. DOI: 10.20546/ijcmas.2017.607.170

Nizamani, M. M., Hughes, A. C., Zhang, H. L., & Wang, Y. (2024). Revolutionizing agriculture with nanotechnology: Innovative approaches in fungal disease management and plant health monitoring. *The Science of the Total Environment, 928*, 172473. DOI: 10.1016/j.scitotenv.2024.172473 PMID: 38615773

Nji, Q. N., Babalola, O. O., Ekwomadu, T. I., Nleya, N., & Mwanza, M. (2022). Six Main Contributing Factors to High Levels of Mycotoxin Contamination in African Foods. *Toxins, 14*(5), 318. DOI: 10.3390/toxins14050318 PMID: 35622564

Nødvig, C. S., Nielsen, J. B., Kogle, M. E., & Mortensen, U. H. (2015). A CRISPR-Cas9 System for Genetic Engineering of Filamentous Fungi. *PLoS One, 10*(7), e0133085. DOI: 10.1371/journal.pone.0133085 PMID: 26177455

Noman, E., Al-Gheethi, A., Talip, B. A., Mohamed, R., & Kassim, A. H. (2021). Decolourization of dye wastewater by A Malaysian isolate of *Aspergillus iizukae* 605EAN strain: A biokinetic, mechanism and microstructure study. *International Journal of Environmental Analytical Chemistry, 101*(11), 1592–1615. DOI: 10.1080/03067319.2019.1686146

Noma, T., & Strickler, K. (2000). Effects of Beauveria bassiana on Lygus hesperus (Hemiptera: Miridae) feeding and oviposition. *Environmental Entomology, 29*(2), 394–402. DOI: 10.1093/ee/29.2.394

Nouh, F. A. A., & Abdel-Azeem, A. M. (2025). Climatic changes and drought against forest fungi. In *Forest Fungi* (pp. 47–62). Academic Press. DOI: 10.1016/B978-0-443-18870-1.00008-1

Nuccio, E. E., Blazewicz, S. J., Lafler, M., Campbell, A. N., Kakouridis, A., Kimbrel, J. A., Wollard, J., Vyshenska, D., Riley, R., Tomatsu, A., Hestrin, R., Malmstrom, R. R., Firestone, M., & Pett-Ridge, J. (2022). HT-SIP: A semi-automated stable isotope probing pipeline identifies cross-kingdom interactions in the hyphosphere of arbuscular mycorrhizal fungi. *Microbiome, 10*(1), 1–20. DOI: 10.1186/s40168-022-01391-z PMID: 36434737

Núñez del Prado, E., Iannacone, J., & Gómez, H. (2008). Effect of two Entomopathogenic fungi in controlling Aleurodicus cocois (Curtis, 1846) (Hemiptera: Aleyrodidae). *Chilean Journal of Agricultural Research, 68*(1), 21–30. DOI: 10.4067/S0718-58392008000100003

Nurhaslina, C. R., Andi Bacho, S., & Mustapa, A. N. (2022). Review on drying methods for herbal plants. *Materials Today: Proceedings, 63*, S122–S139. DOI: 10.1016/j.matpr.2022.02.052

Nurzyńska-Wierdak, R., & Walasek-Janusz, M. (2025). Chemical Composition, Biological Activity, and Potential Uses of Oregano (*Origanum vulgare* L.) and Oregano Essential Oil. *Pharmaceuticals (Basel, Switzerland), 18*(2), 267. DOI: 10.3390/ph18020267 PMID: 40006079

Nyanhongo, G. S., Gűbitz, G., Sukyai, P., Leitner, C., Haltrich, D., & Ludwig, R. (2007). Oxidoreductases from *Trametes* spp. in biotechnology: A wealth of catalytic activity. *Food Technology and Biotechnology, 45*(3), 250–268.

Nyong, T. M., & Njenga, P. K. (2017). Comparative Analysis of Agaricus bisporus Under Different Substrates And Agaricus bitorquis initiation, Incubation, And Casing Period. *IOSR Journal of Pharmacy and Biological Sciences, 12*, 82–88.

O., Elrys, A. S., El-Tarabily, K. A., and Abu Qamar, S. F. (2022). Plant growth-promoting microorganisms as biocontrol agents of plant diseases: Mechanisms, challenges and future perspectives. *Frontiers in Plant Science, 13*. https://www.frontiersin.org/journals/plant-science/articles/10.3389/fpls.2022.923880

Ocón, A., Hampp, R., & Requena, N. (2007). Trehalose turnover during abiotic stress in arbuscular mycorrhizal fungi. *The New Phytologist, 174*(4), 879–891. DOI: 10.1111/j.1469-8137.2007.02048.x PMID: 17504469

Ogbonna, A. I., Nwadiaro, P. O., Chuku, A., Ogbonna, C. I. C., & Onwuliri, F. C. (2015). Cellulase induction in three Aspergillus species isolated from Artemisia annua L. plantation soil using different cellulose substrates. *J Appl Biotechnol, 3*(1), 63–74. DOI: 10.5296/jab.v3i1.6738

Okamura, T., Ogata, T., Minamoto, N., Takeno, T., Noda, H., Fukuda, S., & Ohsugi, M. (2001). Characteristics of wine produced by mushroom fermentation. *Bioscience, Biotechnology, and Biochemistry, 65*(7), 1596–1600. DOI: 10.1271/bbb.65.1596 PMID: 11515544

Okeudo-Cogan, M. C., Murray, B. S., Ettelaie, R., Connell, S. D., Radford, S. J., Micklethwaite, S., & Sarkar, A. (2023). Understanding the microstructure of a functional meat analogue: Demystifying interactions between fungal hyphae and egg white protein. *Food Hydrocolloids, 140*, 108606. DOI: 10.1016/j.foodhyd.2023.108606

Okori, F., Cherotich, S., Baidhe, E., Komakech, A. J., & Banadda, N. (2022). Grain hermetic storage and post-harvest loss reduction in sub-saharan Africa: Effects on grain damage, weight loss, germination, insect infestation, and mold and mycotoxin contamination. *Journal of Biosystems Engineering, 47*(1), 48–68. DOI: 10.1007/s42853-022-00130-4

Okuda, Y. (2022). Sustainability perspectives for future continuity of mushroom production: The bright and dark sides. *Frontiers in Sustainable Food Systems, 6*, 1026508. Advance online publication. DOI: 10.3389/fsufs.2022.1026508

Oliveira, I. S., da Silva, A. G.Junior, de Andrade, C. A. S., & Oliveira, M. D. L. (2019). Biosensors for early detection of fungi spoilage and toxigenic and mycotoxins in food. *Current Opinion in Food Science, 29*, 64–79. DOI: 10.1016/j.cofs.2019.08.004

Onji, Y., Aoki, Y., Tani, N., Umebayashi, K., Kitada, Y., & Dohi, Y. (1998). Direct analysis of several Fusarium mycotoxins in cereals by capillary gas chromatography–mass spectrometry. *Journal of Chromatography. A, 815*(1), 59–65. DOI: 10.1016/S0021-9673(98)00357-4 PMID: 9718707

Onwosi, C. O., Igbokwe, V. C., Odimba, J. N., Eke, I. E., Nwankwoala, M. O., Iroh, I. N., & Ezeogu, L. I. (2017a). 'Composting Technology in Waste Stabilization: On the Methods, Challenges and Future Prospects'. *Journal of Environmental Management, 190*, 140–157. DOI: 10.1016/j.jenvman.2016.12.051 PMID: 28040590

Onwude, D. I., Chen, G., Eke-emezie, N., Kabutey, A., Khaled, A. Y., & Sturm, B. (2020). Recent Advances in Reducing Food Losses in the Supply Chain of Fresh Agricultural Produce. *Processes (Basel, Switzerland), 8*(11), 1431. DOI: 10.3390/pr8111431

Onyeaka, H., Anumudu, C. K., Okpe, C., Okafor, A., Ihenetu, F., Miri, T., ... & Anyogu, A. (2022). Single cell protein for foods and feeds: A review of trends. *The Open Microbiology Journal, 16*(1), e187428582206160. https://, 2022, 16, e187428582206160DOI: 10.2174/18742858-v16-e2206160

Oreste, M., Bubici, G., Poliseno, M., & Tarasco, E. (2016). Effect of *Beauveria bassiana* and *Metarhizium anisopliae* on the *Trialeurodes vaporariorum-Encarsia Formosa* system. *Journal of Pest Science, 89*(1), 153–160. DOI: 10.1007/s10340-015-0660-4

Org, I., Elkhateeb, W. A., Daba, G. M., & Elkhateeb, W. A. (2023). Vitamin D Enriched Edible Mushrooms: A. [Review]. *Journal of Pharmacy and Drug Development, 2*(1). Advance online publication. DOI: 10.58489/2836-2322/013

Oro, L., Feliziani, E., Ciani, M., Romanazzi, G., & Comitini, F. (2018). Volatile organic compounds from Wickerhamomyces anomalus, Metschnikowia pulcherrima and Saccharomyces cerevisiae inhibit growth of decay causing fungi and control postharvest diseases of strawberries. *International Journal of Food Microbiology*, *265*, 18–22. DOI: 10.1016/j.ijfoodmicro.2017.10.027 PMID: 29107842

Ortas, I., Rafique, M., & Çekiç, F. Ö. (2021). Do Mycorrhizal Fungi enable plants to cope with abiotic stresses by overcoming the detrimental effects of salinity and improving drought tolerance? *Symbiotic Soil Microorganisms: Biology and Applications*, 391–428.

Ortellado, L. E., Villalba, L. L., Zapata, P. D., & Fonseca, M. I. (2022). In Udayanga, D., Bhatt, P., Manamgoda, D., & Saez, J. M. (Eds.), *Mycoremediation of Wastewater by Fungal Lipases BT - Mycoremediation Protocols* (pp. 213–219). Springer US., DOI: 10.1007/978-1-0716-2006-9_18

Ortiz-Urquiza, A., Luo, Z., & Keyhani, N. O. (2015). Improving mycoinsecticides for insect biological control. *Applied Microbiology and Biotechnology*, *99*(3), 1057–1068. DOI: 10.1007/s00253-014-6270-x PMID: 25503318

Ospina, J. D., Tovar, C. D. G., Flores, J. C. M., & Orozco, M. S. S. (2016). Relationship between refractive index and thymol concentration in essential oils of Lippia origanoides Kunth. *Chilean Journal of Agricultural & Animal Science*, *32*(2), 127–133. DOI: 10.4067/S0719-38902016000200006

Ouedraogo, N., Savadogo, A., Somda, M. K., Zongo, C., & Traore, A. S. (2017). Effect of mineral salts and nitrogen source on yeast (Candida utilis NOY1) biomass production using tubers wastes. *African Journal of Biotechnology*, *16*(8), 359–365. DOI: 10.5897/AJB2016.15801

Ouhaddou, R., Meddich, A., Ikan, C., Lahlali, R., Ait Barka, E., Hajirezaei, M. R., Duponnois, R., & Baslam, M. (2023). Enhancing Maize Productivity and Soil Health under Salt Stress through Physiological Adaptation and Metabolic Regulation Using Indigenous Biostimulants. *Plants*, *12*(21), 3703. DOI: 10.3390/plants12213703 PMID: 37960059

Ovchinnikova, E., Chiasson, D., Wen, Z., Wu, Y., Tahaei, H., Smith, P. M., & Kaiser, B. N. (2023). Arbuscular-Mycorrhizal Symbiosis in Medicago Regulated by the Transcription Factor *MtbHLHm1*; 1 and the Ammonium Facilitator Protein *MtAMF1*; 3. *International Journal of Molecular Sciences*, *24*(18), 14263. DOI: 10.3390/ijms241814263 PMID: 37762569

Owen, D., Williams, A. P., Griffith, G. W., & Withers, P. J. A. (2015). Use of commercial bio-inoculants to increase agricultural production through improved phosphrous acquisition. *Applied Soil Ecology*, *86*, 41–54. https://doi.org/https://doi.org/10.1016/j.apsoil.2014.09.012. DOI: 10.1016/j.apsoil.2014.09.012

Owiny, A. A., & Dusengemungu, L. (2024). Mycorrhizae in mine wasteland reclamation. *Heliyon*, *10*(13), e33141. https://doi.org/https://doi.org/10.1016/j.heliyon.2024.e33141. DOI: 10.1016/j.heliyon.2024.e33141 PMID: 39035525

Ownley, B. H., Griffin, M. R., Klingeman, W. E., Gwinn, K. D., Moulton, J. K., & Pereira, R. M. (2008). Beauveria bassiana: Endophytic colonization and plant disease control. *Journal of Invertebrate Pathology*, *98*(3), 267–270. DOI: 10.1016/j.jip.2008.01.010 PMID: 18442830

Ownley, B. H., & Windham, M. (2007). Biological control of plant pathogens. In Trigiano, R. N., Windham, M. T., & Windham, A. S. (Eds.), *Plant pathology: Concepts and laboratory exercises* (2nd ed., pp. 423–436). CRC Press.

Pahalvi, H. N., Rafiya, L., Rashid, S., Nisar, B., & Kamili, A. N. (2021). *Chemical Fertilizers and Their Impact on Soil Health BT - Microbiota and Biofertilizers, Vol 2: Ecofriendly Tools for Reclamation of Degraded Soil Environs* (G. H. Dar, R. A. Bhat, M. A. Mehmood, and K. R. Hakeem (eds.); pp. 1–20). Springer International Publishing. DOI: 10.1007/978-3-030-61010-4_1

Palanivel, T. M., Pracejus, B., & Novo, L. A. (2023). Bioremediation of copper using indigenous fungi *Aspergillus* species isolated from an abandoned copper mine soil. *Chemosphere*, *314*, 137688. DOI: 10.1016/j.chemosphere.2022.137688 PMID: 36584825

Pandey, D., Kehri, H. K., Zoomi, I., Akhtar, O., & Singh, A. K. (2019). *Mycorrhizal Fungi: Biodiversity, Ecological Significance, and Industrial Applications BT - Recent Advancement in White Biotechnology Through Fungi: Volume 1: Diversity and Enzymes Perspectives* (A. N. Yadav, S. Mishra, S. Singh, and A. Gupta (eds.); pp. 181–199). Springer International Publishing. DOI: 10.1007/978-3-030-10480-1_5

Pandey, A. K., Samota, M. K., Kumar, A., Silva, A. S., & Dubey, N. K. (2023). Fungal mycotoxins in food commodities: Present status and future concerns. *Frontiers in Sustainable Food Systems*, *7*, 1162595. Advance online publication. DOI: 10.3389/fsufs.2023.1162595

Pant, K. K., Naik, J., Barthakur, S., & Chandra, V. (2025). High-temperature stress in wheat (*Triticum aestivum L.*): Unfolding the impacts, tolerance and methods to mitigate the detrimental effects. *Cereal Research Communications*. Advance online publication. DOI: 10.1007/s42976-025-00634-7

Park, H. J. (2022). Current Uses of Mushrooms in Cancer Treatment and Their Anticancer Mechanisms. *International Journal of Molecular Sciences*, *23*(18), 10502. DOI: 10.3390/ijms231810502 PMID: 36142412

Parrado, J., Bautista, J., Romero, E. J., García-Martínez, A. M., Friaza, V., & Tejada, M. (2008). Production of a carob enzymatic extract: Potential use as a biofertilizer. *Bioresource Technology*, *99*(7), 2312–2318. DOI: 10.1016/j.biortech.2007.05.029 PMID: 17601731

Patel, T. K., Anand, R., Singh, A. P., Shankar, J., & Tiwary, B. N. (2014). Evaluation of aflatoxin B1 biosynthesis in *A. flavus* isolates from central India and identification of atoxigenic isolates. *Biotechnology and Bioprocess Engineering; BBE*, *19*(6), 1105–1113. DOI: 10.1007/s12257-014-0464-z

Pathak, A. P., Rathod, M., & Gudda, R. (2015). Chapter Cultivation by Optimizing Substrate Dependent Growth of Selected Oyster Mushroom Species under Environmentally Controlled Conditions: A Review. *Biotechnology of Mushroom*, 24.

Patninan, K., Sermwittayawong, D., Nuinamwong, S., Khimmaktong, W., Noipha, K., & Nongporn, H.-T. (2023). Purified Polysaccharides Extracted from Gray Oyster Mushroom (Pleurotus sajor-caju (Fr.) Sing.) Stimulate Glucose Uptake in C2C12 Myotubes through the activation of AMP-Activated Protein Kinase (AMPK) and Glucose Transporter 1 (GLUT1) Proteins. *Sains Malaysiana*, *52*(10), 2815–2827. DOI: 10.17576/jsm-2023-5210-07

Paudel, S., & States, S. L. (2023). Urban green spaces and sustainability: Exploring the ecosystem services and disservices of grassy lawns versus floral meadows. *Urban Forestry & Urban Greening*, *84*, 127932. https://doi.org/https://doi.org/10.1016/j.ufug.2023.127932. DOI: 10.1016/j.ufug.2023.127932

Paul, B., Das, D., Aich, T., & Pal, D. (2024). Plant Based Biocoagulants from Cucurbita pepo and Cicer arietinum for Improving Water Quality. *International Journal of Agriculture Environment and Biotechnology*, *17*(1), 29–36. DOI: 10.30954/0974-1712.01.2024.5

Pavan, J., Vireesha, P., Chaitanya, M., Raja, G. P., & Priya, M. K. (2024). Biopesticides in Indian agriculture: Advances, challenges, and pathways to sustainable crop protection. *International Journal of Advanced Biochemistry Research*, *8*(8S), 936–943. DOI: 10.33545/26174693.2024.v8.i8Sm.1966

Pedranzani, H., Rodríguez-Rivera, M., Gutiérrez, M., Porcel, R., Hause, B., & Ruiz-Lozano, J. M. (2016). Arbuscular mycorrhizal symbiosis regulates physiology and performance of Digitaria eriantha plants subjected to abiotic stresses by modulating antioxidant and jasmonate levels. *Mycorrhiza*, *26*(2), 141–152. DOI: 10.1007/s00572-015-0653-4 PMID: 26184604

Pell, J. K., Hannam, J. J., & Steinkraus, D. C. (2010). Conservation biological control using fungal entomopathogens. *BioControl*, *55*(1), 187–198. DOI: 10.1007/s10526-009-9245-6

Peng, L., Tang, R., Wang, G., Ma, R., Li, Y., Li, G., & Yuan, J. (2023). 'Effect of Aeration Rate, Aeration Pattern, and Turning Frequency on Maturity and Gaseous Emissions during Kitchen Waste Composting'. *Environmental Technology & Innovation*, *29*, 102997. DOI: 10.1016/j.eti.2022.102997

Peng, Z., Zulfiqar, T., Yang, H., Wang, M., & Zhang, F. (2024). Effect of Arbuscular Mycorrhizal Fungi (AMF) on photosynthetic characteristics of cotton seedlings under saline-alkali stress. *Scientific Reports*, *14*(1), 8633. DOI: 10.1038/s41598-024-58979-8 PMID: 38622240

Pepe, A., Giovannetti, M., & Sbrana, C. (2016). Different levels of hyphal self-incompatibility modulate interconnectedness of mycorrhizal networks in three arbuscular mycorrhizal fungi within the Glomeraceae. *Mycorrhiza*, *26*(4), 325–332. DOI: 10.1007/s00572-015-0671-2 PMID: 26630971

Peraica, M., Radic, B., Lucic, A., & Pavlovic, M. (1999). Toxic effects of mycotoxins in humans. *Bulletin of the World Health Organization*, *77*(9), 754–766. PMID: 10534900

Pereira, P., Nesci, A., Castillo, C., & Etcheverry, M. (2010). Impact of bacterial biological control agents on fumonisin B1 content and Fusarium verticillioides infection of field-grown maize. *Biological Control*, *53*(3), 258–266. DOI: 10.1016/j.biocontrol.2010.02.001

Pérez-Chávez, A. M., Mayer, L., & Albertó, E. (2019). 'Mushroom Cultivation and Biogas Production: A Sustainable Reuse of Organic Resources'. *Energy for Sustainable Development*, *50*, 50–60. DOI: 10.1016/j.esd.2019.03.002

Pergola, M., De Falco, E., Belliggiano, A., & Ievoli, C. (2024). The Most Relevant Socio-Economic Aspects of Medicinal and Aromatic Plants through a Literature Review. *Agriculture*, *14*(3), 405. DOI: 10.3390/agriculture14030405

Perrone, G., Ferrara, M., Medina, A., Pascale, M., & Magan, N. (2020). Toxigenic Fungi and Mycotoxins in a Climate Change Scenario: Ecology, Genomics, Distribution, Prediction and Prevention of the Risk. *Microorganisms*, *8*(10), 1496. DOI: 10.3390/microorganisms8101496 PMID: 33003323

Pervez, M. T., Hasnain, M. J. U., Abbas, S. H., Moustafa, M. F., Aslam, N., & Shah, S. S. M. (2022). A Comprehensive Review of Performance of Next-Generation Sequencing Platforms. *BioMed Research International*, *2022*(1), 1–12. DOI: 10.1155/2022/3457806 PMID: 36212714

Pestka, J. J. (2010). Deoxynivalenol: Mechanisms of action, human exposure, and toxicological relevance. *Archives of Toxicology*, *84*(9), 663–679. DOI: 10.1007/s00204-010-0579-8 PMID: 20798930

Pestka, J. J., & Smolinski, A. T. (2005). Deoxynivalenol: Toxicology and potential effects on humans. *Journal of Toxicology and Environmental Health. Part B, Critical Reviews*, *8*(1), 39–69. DOI: 10.1080/10937400590889458 PMID: 15762554

PESTKA, J., & ABOUZIED, M. (1995). Immunological assays for mycotoxin detection: Immunoassay applications to food analysis. *Food technology (Chicago)*, *49*(2), 120–128.

Pfohl-Leszkowicz, A., & Manderville, R. A. (2007). Ochratoxin A: An overview on toxicity and carcinogenicity in animals and humans. *Molecular Nutrition & Food Research*, *51*(1), 61–99. DOI: 10.1002/mnfr.200600137 PMID: 17195275

Philip, B., Behiry, S. I., Salem, M. Z. M., Amer, M. A., El-Samra, I. A., Abdelkhalek, A., & Heflish, A. (2024). Trichoderma afroharzianum TRI07 metabolites inhibit Alternaria alternata growth and induce tomato defense-related enzymes. *Scientific Reports*, *14*(1), 1874. DOI: 10.1038/s41598-024-52301-2 PMID: 38253713

Picciotti, U., Araujo Dalbon, V., Ciancio, A., Colagiero, M., Cozzi, G., De Bellis, L., Finetti-Sialer, M. M., Greco, D., Ippolito, A., Lahbib, N., Logrieco, A. F., López-Llorca, L. V., Lopez-Moya, F., Luvisi, A., Mincuzzi, A., Molina-Acevedo, J. P., Pazzani, C., Scortichini, M., Scrascia, M., & Porcelli, F. (2023). "Ectomosphere": Insects and microorganism interactions. *Microorganisms*, *11*(2), 440. DOI: 10.3390/microorganisms11020440 PMID: 36838405

Pichon, V., & Combès, A. (2016). Selective tools for the solid-phase extraction of Ochratoxin A from various complex samples: Immunosorbents, oligosorbents, and molecularly imprinted polymers. *Analytical and Bioanalytical Chemistry*, *408*(25), 6983–6999. DOI: 10.1007/s00216-016-9886-0 PMID: 27585915

Pickova, D., Ostry, V., & Malir, F. (2021). A recent overview of producers and important dietary sources of aflatoxins. *Toxins*, *13*(3), 186. DOI: 10.3390/toxins13030186 PMID: 33802572

Pillaca-Pullo, O. S., Lopes, A. M., & Estela-Escalante, W. D. (2024). Reusing wastewater from Coffea arabica processing to produce single-cell protein using Candida sorboxylosa: Optimizing of culture conditions. *Biotechnology Progress*, *40*(1), e3393. DOI: 10.1002/btpr.3393

Pinos, N. Q., Berbara, R. L. L., Tavares, O. C. H., & García, A. C. (2019). Different Structures in Humic Substances Lead to Impaired Germination but Increased Protection against Saline Stress in Corn. *Communications in Soil Science and Plant Analysis*, *50*(17), 2209–2225. DOI: 10.1080/00103624.2019.1659294

Pinto, Y., & Bhatt, A. S. (2024). Sequencing-based analysis of microbiomes. *Nature Reviews. Genetics*, *25*(12), 829–845. DOI: 10.1038/s41576-024-00746-6 PMID: 38918544

Pirinçci, Ş. Ş., Ertekin, Ö., Laguna, D. E., Özen, F. Ş., Öztürk, Z. Z., & Öztürk, S. (2018). Label-free QCM immunosensor for the detection of ochratoxin A. *Sensors (Basel)*, *18*(4), 1161. DOI: 10.3390/s18041161 PMID: 29641432

Pittendrigh, C. S. (1960). Circadian Rhythms and the Circadian Organization of Living Systems. *Cold Spring Harbor Symposia on Quantitative Biology*, *25*(0), 159–184. DOI: 10.1101/SQB.1960.025.01.015 PMID: 13736116

Pobiega, K., Sękul, J., Pakulska, A., Latoszewska, M., Michońska, A., Korzeniowska, Z., Macherzyńska, Z., Pląder, M., Duda, W., Szafraniuk, J., Kufel, A., Dominiak, Ł., Lis, Z., Kłusek, E., Kozicka, E., Wierzbicka, A., Trusińska, M., Rybak, K., Kot, A. M., & Nowacka, M. (2024). Fungal proteins: Sources, production and purification methods, industrial applications, and future perspectives. *Applied Sciences (Basel, Switzerland)*, *14*(14), 6259. DOI: 10.3390/app14146259

Podkowa, A., Kryczyk-Poprawa, A., Opoka, W., & Muszyńska, B. (2021). Culinary–medicinal mushrooms: a review of organic compounds and bioelements with antioxidant activity. In European Food Research and Technology (Vol. 247, Issue 3, pp. 513–533). Springer Science and Business Media Deutschland GmbH. DOI: 10.1007/s00217-020-03646-1

Polak-Śliwińska, M., & Paszczyk, B. (2021). Trichothecenes in Food and Feed, Relevance to Human and Animal Health and Methods of Detection: A Systematic Review. *Molecules (Basel, Switzerland)*, *26*(2), 454. DOI: 10.3390/molecules26020454 PMID: 33467103

Pons, S., Fournier, S., Chervin, C., Bécard, G., Rochange, S., Frei Dit Frey, N., & Puech Pagès, V. (2020). Phytohormone production by the arbuscular mycorrhizal fungus Rhizophagus irregularis. *PLoS One*, *15*(10), e0240886. DOI: 10.1371/journal.pone.0240886 PMID: 33064769

Pooja, P., Tallapragada, S., Lamba, A., & Punia, S. (2024). Role played by arbuscular mycorrhizal fungi in amelioration of salinity stress: A review. *Plant and Soil*, •••, 1–16.

Popkin, B. M., & Ng, S. W. (2022). The nutrition transition to a stage of high obesity and noncommunicable disease prevalence dominated by ultra-processed foods is not inevitable. *Obesity Reviews*, *23*(1), e13366. https://doi.org/https://doi.org/10.1111/obr.13366. DOI: 10.1111/obr.13366 PMID: 34632692

Porcel, R., Aroca, R., Azcon, R., & Ruiz-Lozano, J. M. (2016). Regulation of cation transporter genes by the arbuscular mycorrhizal symbiosis in rice plants subjected to salinity suggests improved salt tolerance due to reduced Na+ root-to-shoot distribution. *Mycorrhiza*, *26*(7), 673–684. DOI: 10.1007/s00572-016-0704-5 PMID: 27113587

Porras-Soriano, A., Soriano-Martín, M. L., Porras-Piedra, A., & Azcón, R. (2009). Arbuscular mycorrhizal fungi increased growth, nutrient uptake and tolerance to salinity in olive trees under nursery conditions. *Journal of Plant Physiology*, *166*(13), 1350–1359. https://doi.org/https://doi.org/10.1016/j.jplph.2009.02.010. DOI: 10.1016/j.jplph.2009.02.010 PMID: 19342122

Pourseyed, S. H., Tavassoli, M., Bernousi, I., & Mardani, K. (2010). Metarhizium anisopliae (Ascomycota: Hypocreales): an effective alternative to chemical acaricides against different developmental stages of fowl tick Argas persicus (Acari: Argasidae). *Veterinary Parasitology*, *172*(3-4), 305–310. DOI: 10.1016/j.vetpar.2010.05.014 PMID: 20541868

Powlson, D. S., Gregory, P. J., Whalley, W. R., & Quinton, J. N. (2014). Soil management in relation to sustainable agriculture and ecosystem services. *Food Policy*, *36*, 572–587.

Pozdnyakova, N. N., Balandina, S. A., Dubrovskaya, E. V., Golubev, C. N., & Turkovskaya, O. V. (2018, January). Ligninolytic basidiomycetes as promising organisms for the mycoremediation of PAH-contaminated Environments. [). IOP Publishing.]. *IOP Conference Series. Earth and Environmental Science*, *107*, 012071. DOI: 10.1088/1755-1315/107/1/012071

Pradhan, J., Sarma, A., Kalita, S., Talukdar, R., & Tayung, K. (2023). Plant Growth-Promoting Effect of Seed-Borne Endophytic Fungi Isolated from Senna Alata (L.) Roxb. *Proceedings of the National Academy of Sciences. India. Section B, Biological Sciences*, *93*(4), 901–908. DOI: 10.1007/s40011-023-01483-2

Prado, M. R. V., Weber, O. L. D. S., Moraes, M. F. D., Santos, C. L. R. D., Santos Tunes, M., & Ramos, F. T. (2016). Humic Substances on Soybeans Grown Under Water Stress. *Communications in Soil Science and Plant Analysis*, *47*(21), 2405–2413. DOI: 10.1080/00103624.2016.1243715

Prajapati, S. K., Rawat, V., Singh, D., Saroj, S., Ravikant, D., & Verma, S. (2023). Mushroom Cultivation: A Sustainable Approach to Future Agriculture to Ensure Quality Food and Nutritional Security of the Current Population in India. *International Journal of Multidisciplinary Research and Growth Evaluation.*, *4*, 697–705.

Preece, C., Verbruggen, E., Liu, L., Weedon, J. T., & Peñuelas, J. (2019). Effects of past and current drought on the composition and diversity of soil microbial communities. *Soil Biology & Biochemistry*, *131*, 28–39. DOI: 10.1016/j.soilbio.2018.12.022

Prins, T. W., Tudzynski, P., von Tiedemann, A., Tudzynski, B., ten Have, A., Hansen, M. E., & van Kan, J. A. L. (2000). Infection strategies of Botrytis cinerea and related necrotrophic pathogens. *Fungal Genetics and Biology*, *30*(2), 1–23. DOI: 10.1006/fgbi.2000.1224

Priya, A., Manab Sarma, P., & Lal, B. (2016). Isolation and characterization of Candida vishwanathii strain TERI MS1 for degradation of petroleum hydrocarbons in marine environment. *Desalination and Water Treatment*, *57*(46), 22099–22106. DOI: 10.1080/19443994.2015.1134351

Priyashantha, A. K. H., Karunarathna, S. C., Lu, L., & Tibpromma, S. (2023). Fungal Endophytes: An Alternative Biocontrol Agent against Phytopathogenic Fungi. *Encyclopedia*, *3*(2), 759–780. DOI: 10.3390/encyclopedia3020055

Proksch, P., Edrada, R. A., & Ebel, R. (2002). Drugs from the seas – current status and microbiological implications. *Applied Microbiology and Biotechnology*, *59*(2–3), 125–134. PMID: 12111137

Pumplin, N., Zhang, X., Noar, R. D., & Harrison, M. J. (2012). Polar localization of a symbiosis-specific phosphate transporter is mediated by a transient reorientation of secretion. *Proceedings of the National Academy of Sciences of the United States of America*, *109*(11), E665–E672. DOI: 10.1073/pnas.1110215109 PMID: 22355114

Purchase, D., Scholes, L. N. L., Revitt, D. M., & Shutes, R. B. E. (2009). Effects of temperature on metal tolerance and the accumulation of Zn and Pb by metal-tolerant fungi isolated from urban runoff treatment wetlands. *Journal of Applied Microbiology*, *106*(4), 1163–1174. DOI: 10.1111/j.1365-2672.2008.04082.x PMID: 19187150

Purnomo, A. S., Putra, S. R., Shimizu, K., & Kondo, R. (2014). Biodegradation of heptachlor and heptachlor epoxide-contaminated soils by white-rot fungal inocula. *Environmental Science and Pollution Research International*, *21*(19), 11305–11312. DOI: 10.1007/s11356-014-3026-1 PMID: 24840358

Püschel, D., Bitterlich, M., Rydlová, J., & Jansa, J. (2020). Facilitation of plant water uptake by an arbuscular mycorrhizal fungus: A Gordian knot of roots and hyphae. *Mycorrhiza*, *30*(2), 299–313. DOI: 10.1007/s00572-020-00949-9 PMID: 32253570

Pusztahelyi, T., Holb, I. J., & Pócsi, I. (2015). Secondary metabolites in fungus-plant interactions. *Frontiers in Plant Science*, *6*. Advance online publication. DOI: 10.3389/fpls.2015.00573 PMID: 26300892

Qadeer, A., Wakeel, A., Cheema, S. A., & Sanaullah, M. (2022). Interactive effects of salinity and drought stresses on soil respiration and microbial activities. *Pakistan Journal of Agricultural Sciences*, *59*(1), 35–42. DOI: 10.21162/PAKJAS/22.1263

Qiao, M., Hong, C., Jiao, Y., Hou, S., & Gao, H. (2024). Impacts of drought on photosynthesis in major food crops and the related mechanisms of plant responses to drought. *Plants*, *13*(13), 1808. DOI: 10.3390/plants13131808 PMID: 38999648

Qin, H., Lu, K., Strong, P. J., Xu, Q., Wu, Q., Xu, Z., Xu, J., & Wang, H. (2015). Long-term fertilizer application effects on the soil, root arbuscular mycorrhizal fungi and community composition in rotation agriculture. *Applied Soil Ecology*, *89*, 35–43. DOI: 10.1016/j.apsoil.2015.01.008

Qin, Y., Zhang, W., Feng, Z., Feng, G., Zhu, H., & Yao, Q. (2022). Arbuscular mycorrhizal fungus differentially regulates P mobilizing bacterial community and abundance in rhizosphere and hyphosphere. *Applied Soil Ecology*, *170*, 104294. DOI: 10.1016/j.apsoil.2021.104294

Qiu, K., Wang, S., Duan, F., Sang, Z., Wei, S., Liu, H., & Tan, H. (2024). Rosemary: Unrevealing an old aromatic crop as a new source of promising functional food additive—A review. *Comprehensive Reviews in Food Science and Food Safety*, *23*(1), e13273. DOI: 10.1111/1541-4337.13273 PMID: 38284599

Qi, Y., Zhang, R., Zhang, M., Wen, Q., & Shen, J. (2021). Effects of exogenous ascorbic acid on the mycelia growth and primordia formation of *Pleurotus ostreatus.Journal of Basic Microbiology*, *61*(8), 736–744. DOI: 10.1002/jobm.202100143 PMID: 34252217

Quintela, S. (2020). Mycotoxins in beverages: occurrence, regulation, economic impact and cost-effectiveness of preventive and removal methods. In *Safety issues in beverage production* (pp. 147–186). Academic Press. DOI: 10.1016/B978-0-12-816679-6.00005-X

Quiroga, G., Erice, G., Aroca, R., Chaumont, F., & Ruiz-Lozano, J. M. (2017). Enhanced drought stress tolerance by the arbuscular mycorrhizal symbiosis in a drought-sensitive maize cultivar is related to a broader and differential regulation of host plant aquaporins than in a drought-tolerant cultivar. *Frontiers in Plant Science*, *8*, 1056. DOI: 10.3389/fpls.2017.01056 PMID: 28674550

Rabea, E. I., Badawy, M. E., Stevens, C. V., Smagghe, G., & Steurbaut, W. (2003). Chitosan as antimicrobial agent: Applications and mode of action. *Biomacromolecules*, *4*(6), 1457–1465. DOI: 10.1021/bm034130m PMID: 14606868

Radhakrishnan, R., Kang, S.-M., Baek, I.-Y., & Lee, I.-J. (2014). Characterization of plant growth-promoting traits of *Penicillium* species against the effects of high soil salinity and root disease. *Journal of Plant Interactions*, *9*(1), 754–762. DOI: 10.1080/17429145.2014.930524

Radujković, D., Verbruggen, E., Sigurdsson, B. D., Leblans, N. I. W., Janssens, I. A., Vicca, S., & Weedon, J. T. (2018). Prolonged exposure does not increase soil microbial community compositional response to warming along geothermal gradients. *FEMS Microbiology Ecology*, *94*(2), fix174. DOI: 10.1093/femsec/fix174 PMID: 29228354

Raeisi Vanani, A., Sheikhi Shahrivar, F., Nouri, A., & Sepehri, M. (2024). Exploring rice tolerance to salinity and drought stresses through Piriformospora indica inoculation: Understanding physiological and metabolic adaptations. *Frontiers in Plant Science*, *15*, 1428631. Advance online publication. DOI: 10.3389/fpls.2024.1428631 PMID: 39385986

Rages, A. A., Haider, M. M., & Aydin, M. (2021). Alkaline hydrolysis of olive fruits wastes for the production of single cell protein by *Candida lipolytica.Biocatalysis and Agricultural Biotechnology*, *33*, 101999. DOI: 10.1016/j.bcab.2021.101999

Rai, M., Ingle, A., Gaikwad, S., Gupta, I., Yadav, A., Gade, A., & Duran, N. (2018). Fungi: myconanofactory, mycoremediation and medicine. In *Fungi* (pp. 201-219). CRC Press. DOI: 10.1201/9781315369471-10

Rai, M., Zimowska, B., Shinde, S., & Tres, M. V. (2021). Bioherbicidal potential of different species of Phoma: Opportunities and challenges. *Applied Microbiology and Biotechnology*, *105*(8), 3009–3018. DOI: 10.1007/s00253-021-11234-w PMID: 33770245

Rai, N., Rai, S. P., & Sarma, B. K. (2021). Prospects for Abiotic Stress Tolerance in Crops Utilizing Phyto- and Bio-Stimulants. *Frontiers in Sustainable Food Systems*, *5*, 754853. DOI: 10.3389/fsufs.2021.754853

Raina, S. A., Yahmed, N. Ben, Bhat, R. A., & Dervash, M. A. (2020). Mycoremediation: a sustainable tool for abating environmental pollution. *Bioremediation and Biotechnology: Sustainable Approaches to Pollution Degradation*, 269–291.

Rai, R., & Vijayakumar, B. S. (2023). Myco-Remediation of Textile Dyes Via Biosorption by Aspergillus tamarii Isolated from Domestic Wastewater. *Water, Air, and Soil Pollution*, *234*(8), 542. DOI: 10.1007/s11270-023-06535-x

Raja, B.-L., Soufian, L., Salma, T., Wissal, B., Ouissame, R., Said, W., Cherkaoui, E. M., Marouane, B., & Abdelilah, M. (2022). Use of biostimulants to improve salinity tolerance in cereals. In *Sustainable Remedies for Abiotic Stress in Cereals* (pp. 471–517). Springer. DOI: 10.1007/978-981-19-5121-3_19

Rajput, P., Benjwal, S., & Pandey, R. (2025). A Comprehensive Review on the Role of Bioremediation in Heavy Metal Contamination. *Nature Environment & Pollution Technology, 24.*

Ramakrishnan, T., Mohan Gift, M. D., Chitradevi, S., Jegan, R., Hency Jose, P. S., Nagaraja, H. N., & Hailegiorgis, S. M. (2022). Study of numerous resins used in polymer matrix composite materials. *Advances in Materials Science and Engineering*, *2022*(1), 1088926.

Raman, J., Jang, K. Y., Oh, Y. L., Oh, M., Im, J.-H., Lakshmanan, H., & Sabaratnam, V. (2020). Cultivation and nutritional value of prominent Pleurotusspp: An overview. *Mycobiology*, *49*(1), 1–14. DOI: 10.1080/12298093.2020.1835142 PMID: 33536808

Ramudingana, P., Makhado, N., Kamutando, C. N., Thantsha, M. S., & Mamphogoro, T. P. (2025). Fungal Biocontrol Agents in the Management of Postharvest Losses of Fresh Produce-A Comprehensive Review. *Journal of Fungi (Basel, Switzerland)*, *11*(1), 82. DOI: 10.3390/jof11010082 PMID: 39852501

Rana, R. K., Monga, S., & Kaur, T. (2024). Scaling up Mushroom Production for Sustainable Entrepreneurship.

Rangel, D. E., Braga, G. U., Anderson, A. J., & Roberts, D. W. (2005). Variability in conidial thermotolerance of Metarhizium anisopliae isolates from different geographic origins. *Journal of Invertebrate Pathology*, *88*(2), 116–125. DOI: 10.1016/j.jip.2004.11.007 PMID: 15766928

Rasheed, H. A., Rehman, A., Li, C., Bai, M., Karim, A., Dai, J., Cui, H., & Lin, L. (2024). Fabrication of *Citrus bergamia* essential oil-loaded sodium caseinate/peach gum nanocomplexes: Physicochemical, spectral, and structural characterization. *International Journal of Biological Macromolecules*, *260*, 129475. DOI: 10.1016/j.ijbiomac.2024.129475 PMID: 38262830

Raut J.K., Adhikari M.K. (2021) Mushroom: a true super food. Comprehensive insights in vegetables of Nepal. Nepal academy of science and technology (NAST), 3323: 201-234.

Raut, J. K. (2019). Current Status, Challenges, and Prospects of Mushroom Industry in Nepal. *Journal of Agricultural Economics*, *4*(4), 154–160. DOI: 10.11648/j.ijae.20190404.13

Ravindra, P. (2000). Value-added food: Single cell protein. *Biotechnology Advances*, *18*(6), 459–479. DOI: 10.1016/S0734-9750(00)00045-8

Ray, P., Pal, D., & Pandey, S. (2023). Influence of Sustainable Biocoagulants Trigonella foenumgraecum and Moringa oleifera for Improving Water Potability. [IJTSRD]. *International Journal of Trend in Scientific Research and Development*, *7*, 1239–1246.

Raziq, A., Lateef, M., Ullah, A., Ullah, H., & Khan, M. W. (2020). Single cell protein (SCP) production and potential substrates: A comprehensive review. [PAB]. *Pure and Applied Biology*, *9*(3), 1743–1754. DOI: 10.19045/bspab.2020.90185

Redecker, D., Hijri, I., & Wiemken, A. (2003). Molecular Identification of Arbuscular Mycorrhizal Fungi in Roots: Perspectives and Problems. *Folia Geobotanica*, *38*(2), 113–124. DOI: 10.1007/BF02803144

Redecker, D., Schüßler, A., Stockinger, H., Stürmer, S. L., Morton, J. B., & Walker, C. (2013). An evidence-based consensus for the classification of arbuscular mycorrhizal fungi (Glomeromycota). *Mycorrhiza*, *23*(7), 515–531. DOI: 10.1007/s00572-013-0486-y PMID: 23558516

Redman, R. S., Sheehan, K. B., Stout, R. G., Rodriguez, R. J., & Henson, J. M. (2001). Thermotolerance generated by plant/fungal symbiosis. *Science, 298*(5598), 1581–1583. DOI: 10.1126/science.1078055 PMID: 12446900

Reese, E. T., & Mandels, M. (1963). Enzymatic hydrolysis of cellulose and its derivatives. In Whistler, R. L. (Ed.), *Methods carbohydrate chemistry* (3rd ed., pp. 139–143). Academic Press.

Reihani, S. F. S., & Khosravi-Darani, K. (2018). Mycoprotein production from date waste using Fusarium venenatum in a submerged culture. *Applied Food Biotechnology*, *5*(4), 243–352.

Reihani, S. F. S., & Khosravi-Darani, K. (2019). Influencing factors on single-cell protein production by submerged fermentation: A review. *Electronic Journal of Biotechnology*, *37*, 34–40. DOI: 10.1016/j.ejbt.2018.11.005

Reinhard, H., & Zimmerli, B. (1999). Reversed-phase liquid chromatographic behavior of the mycotoxins citrinin and ochratoxin A. *Journal of Chromatography. A*, *862*(2), 147–159. DOI: 10.1016/S0021-9673(99)00929-2 PMID: 10596972

Reis, A. (2024). 18 Different Types Of Edible Mushrooms And How To Use Them. https://cookingchew.com/different-types-of-edible-mushrooms-and-how-to-use-them.html

Reis, F. S., Hu, Y., & Fernandes, T. H. (2022). Mushrooms as future generation healthy foods.

Renshaw, J. C., Robson, G. D., Trinci, A. P. J., Wiebe, M. G., Livens, F. R., Collison, D., & Taylor, R. J. (2002). Fungal siderophores: Structures, functions and applications. *Mycological Research*, *106*(10), 1123–1142. DOI: 10.1017/S0953756202006548

Rétif, F., Kunz, C., Calabro, K., Duval, C., Prado, S., Bailly, C., & Baudouin, E. (2023). Seed fungal endophytes as biostimulants and biocontrol agents to improve seed performance. *Frontiers in Plant Science*, *14*, 1260292. DOI: 10.3389/fpls.2023.1260292 PMID: 37941673

Reverchon, F., María del Ortega-Larrocea, P., & Pérez-Moreno, J. (2010). Saprophytic fungal communities change in diversity and species composition across a volcanic soil chronosequence at Sierra del Chichinautzin, Mexico. *Annals of Microbiology*, *60*(2), 217–226. DOI: 10.1007/s13213-010-0030-7

Rhodes, C. J. (2014). Mycoremediation (bioremediation with fungi) – growing mushrooms to clean the earth. *Chemical Speciation and Bioavailability*, *26*(3), 196–198. DOI: 10.3184/095422914X14047407349335

Ribeiro, G. O., Rodrigues, L. D. A. P., Santos, T. B. S. D., Alves, J. P. S., Oliveira, R. S., Nery, T. B. R., Barbosa, J. D. V., & Soares, M. B. P. (2023). Innovations and developments in single cell protein: Bibliometric review and patents analysis. *Frontiers in Microbiology*, *13*, 1093464. DOI: 10.3389/fmicb.2022.1093464

Ribeiro-Santos, R., Carvalho-Costa, D., Cavaleiro, C., Costa, H. S., Albuquerque, T. G., Castilho, M. C., Ramos, F., Melo, N. R., & Sanches-Silva, A. (2015). A novel insight on an ancient aromatic plant: The rosemary (*Rosmarinus officinalis* L.). *Trends in Food Science & Technology*, *45*(2), 355–368. DOI: 10.1016/j.tifs.2015.07.015

Rillig, M. C., Wright, S. F., Nichols, K. A., Schmidt, W. F., & Torn, M. S. (2001). Large contribution of arbuscular mycorrhizal fungi to soil carbon pools in tropical forest soils. *Plant and Soil, 233*(2), 167–177. DOI: 10.1023/A:1010364221169

Ringot, D., Chango, A., Schneider, Y. J., & Larondelle, Y. (2006). Toxicokinetics and toxicodynamics of ochratoxin A, an update. *Chemico-Biological Interactions, 159*(1), 18–46. DOI: 10.1016/j.cbi.2005.10.106 PMID: 16293235

Ripa, F. A., Cao, W., Tong, S., & Sun, J. (2019). Assessment of Plant Growth Promoting and Abiotic Stress Tolerance Properties of Wheat Endophytic Fungi. *BioMed Research International, 2019*, 1–12. DOI: 10.1155/2019/6105865 PMID: 31032353

Ristaino, J. B., Anderson, P. K., Bebber, D. P., Brauman, K. A., Cunniffe, N. J., Fedoroff, N. V., & Wei, Q. (2021). The persistent threat of emerging plant disease pandemics to global food security. *Proceedings of the National Academy of Sciences of the United States of America, 118*(23), e2022239118. DOI: 10.1073/pnas.2022239118 PMID: 34021073

Ritala, A., Häkkinen, S. T., Toivari, M., & Wiebe, M. G. (2017). Single cell protein—state-of-the-art, industrial landscape and patents 2001–2016. *Frontiers in microbiology, 8*, 2009. https://doi.org/DOI: 10.3389/fmicb.2017.02009

Robinson Boyer, L., Brain, P., Xu, X., & Jeffries, P. (2014). Inoculation of drought-stressed strawberry with a mixed inoculum of two arbuscular mycorrhizal fungi: Effects on population dynamics of fungal species in roots and consequential plant tolerance to water deficiency. *Mycorrhiza, 25*(3), 215–227. Advance online publication. DOI: 10.1007/s00572-014-0603-6 PMID: 25186649

Robinson, B., Winans, K., Kendall, A., Dlott, J., & Dlott, F. (2019). 'A Life Cycle Assessment of Agaricus Bisporus Mushroom Production in the USA'. *The International Journal of Life Cycle Assessment, 24*(3), 456–467. DOI: 10.1007/s11367-018-1456-6

Rodrigues, W. P., Silva, J. R., Ferreira, L. S., Machado Filho, J. A., Figueiredo, F. A., Ferraz, T. M., Bernado, W. P., Bezerra, L. B. S., de Abreu, D. P., & Cespom, L. (2018). Stomatal and photochemical limitations of photosynthesis in coffee (*Coffea* spp.) plants subjected to elevated temperatures. *Crop & Pasture Science, 69*(3), 317–325. DOI: 10.1071/CP17044

Rodríguez-Carrasco, Y., Berrada, H., Font, G., & Mañes, J. (2012). Multi-mycotoxin analysis in wheat semolina using an acetonitrile-based extraction procedure and gas chromatography–tandem mass spectrometry. *Journal of Chromatography. A, 1270*, 28–40. DOI: 10.1016/j.chroma.2012.10.061 PMID: 23182289

Rodriguez, R. J., & Redman, R. S. (1997). *Fungal Life-Styles and Ecosystem Dynamics: Biological Aspects of Plant Pathogens*. Plant Endophytes and Saprophytes., DOI: 10.1016/S0065-2296(08)60073-7

Ronaghi, M. (2001). Pyrosequencing Sheds Light on DNA Sequencing. *Genome Research, 11*(1), 3–11. DOI: 10.1101/gr.150601 PMID: 11156611

Rose, M. T., Patti, A. F., Little, K. R., Brown, A. L., Jackson, W. R., & Cavagnaro, T. R. (2014). A Meta-Analysis and Review of Plant-Growth Response to Humic Substances. In *Advances in Agronomy* (Vol. 124, pp. 37–89). Elsevier., DOI: 10.1016/B978-0-12-800138-7.00002-4

Rosendahl, S. (2008). Communities, populations and individuals of arbuscular mycorrhizal fungi. *The New Phytologist*, *178*(2), 253–266. DOI: 10.1111/j.1469-8137.2008.02378.x PMID: 18248587

Roth, R., & Paszkowski, U. (2017). Plant carbon nourishment of arbuscular mycorrhizal fungi. *Current Opinion in Plant Biology*, *39*, 50–56. DOI: 10.1016/j.pbi.2017.05.008 PMID: 28601651

Rouphael, Y., & Colla, G. (2020). Editorial: Biostimulants in Agriculture. In *Frontiers in Plant Science* (Vol. 11). Frontiers Media S.A. DOI: 10.3389/fpls.2020.00040

Rouphael, Y., Franken, P., Schneider, C., Schwarz, D., Giovannetti, M., Agnolucci, M., De Pascale, S., Bonini, P., & Colla, G. (2015). Arbuscular mycorrhizal fungi act as biostimulants in horticultural crops. *Scientia Horticulturae*, *196*, 91–108. https://doi.org/https://doi.org/10.1016/j.scienta.2015.09.002. DOI: 10.1016/j.scienta.2015.09.002

Roy, H. E., & Cottrell, T. E. (2008). Forgotten natural enemies: Interactions between coccinellids and insect-parasitic fungi. *European Journal of Entomology*, *105*(3), 391–398. DOI: 10.14411/eje.2008.049

Rudakiya, D. M., Patel, D. H., & Gupte, A. (2020). Exploiting the potential of metal and solvent tolerant laccase from Tricholoma giganteum AGDR1 for the removal of pesticides. *International Journal of Biological Macromolecules*, *144*, 586–595. https://doi.org/https://doi.org/10.1016/j.ijbiomac.2019.12.068. DOI: 10.1016/j.ijbiomac.2019.12.068 PMID: 31830449

Rudresh, D. L., Shivaprakash, M. K., & Prasad, R. D. (2005). Tricalcium phosphate solubilizing abilities of Aspergillus niger and Penicillium sp. in maize rhizosphere. *Brazilian Journal of Microbiology : Publication of the Brazilian Society for Microbiology*, *36*(1), 9–12. DOI: 10.1590/S1517-83822005000100002

Ruiz-Lozano, J. M., Azcon, R., & Gomez, M. (1995). Effects of arbuscular-mycorrhizal glomus species on drought tolerance: Physiological and nutritional plant responses. *Applied and Environmental Microbiology*, *61*(2), 456–460. DOI: 10.1128/aem.61.2.456-460.1995 PMID: 16534929

Ruscasso, F., Bezus, B., Garmendia, G., Vero, S., Curutchet, G., Cavello, I., & Cavalitto, S. (2021). Debaryomyces hansenii F39A as biosorbent for textile dye removal. *Revista Argentina de Microbiologia*, *53*(3), 257–265. https://doi.org/https://doi.org/10.1016/j.ram.2020.10.004. DOI: 10.1016/j.ram.2020.10.004 PMID: 33454152

Russell, J. R., Huang, J., Anand, P., Kucera, K., Sandoval, A. G., Dantzler, K. W., Hickman, D. S., Jee, J., Kimovec, F. M., Koppstein, D., Marks, D. H., Mittermiller, P. A., Núñez, S. J., Santiago, M., Townes, M. A., Vishnevetsky, M., Williams, N. E., Vargas, M. P. N., Boulanger, L.-A., & Strobel, S. A. (2011). Biodegradation of polyester polyurethane by endophytic fungi. *Applied and Environmental Microbiology*, *77*(17), 6076–6084. DOI: 10.1128/AEM.00521-11 PMID: 21764951

Russo, F., Ceci, A., Pinzari, F., Siciliano, A., Guida, M., Malusà, E., Tartanus, M., Miszczak, A., Maggi, O., & Persiani, A. M. (2019). Bioremediation of dichlorodiphenyltrichloroethane (DDT)-contaminated agricultural soils: Potential of two autochthonous saprotrophic fungal strains. *Applied and Environmental Microbiology*, *85*(21), e01720–e19. DOI: 10.1128/AEM.01720-19 PMID: 31444208

Ruzzi, M., & Aroca, R. (2015). Plant growth-promoting rhizobacteria act as biostimulants in horticulture. *Scientia Horticulturae*, *196*, 124–134. DOI: 10.1016/j.scienta.2015.08.042

Saad, M. M. G., Saad, M. A., Saad, B. S., Zakaria, F. A., Husain, A.-R. A., & Abdelgaleil, S. A. M. (2022). Bioremediation and microbial-assisted phytoremediation of heavy metals by endophytic Fusarium species isolated from Convolvulus arvensis. *Bioremediation Journal*, *28*(2), 202–212. DOI: 10.1080/10889868.2022.2138256

Sabaratnam, V., Kah-Hui, W., Naidu, M., & David, P. (2013). Neuronal health - can culinary and medicinal mushrooms help? *Journal of Traditional and Complementary Medicine*, *3*(1), 62–68. DOI: 10.4103/2225-4110.106549 PMID: 24716157

Sadiq, S., Inam, H. M., Ahmad, I., Ahad, K., & Rashid, A. (2015). Bioremediation potential of white rot fungi, *Pleurotus* spp against organochlorines. *Journal of Bioremediation & Biodegradation*, *6*(308), 2.

Sadiq, S., Mahmood-ul-Hassan, M., Ahad, K., & Nazir, S. (2018). Bioremediation of hexachlorocyclohexane (HCH) in soil using spent mushroom compost of Pleurotus ostreatus. *Bioremediation Journal*, *22*(3–4), 126–135. DOI: 10.1080/10889868.2018.1516615

Saeed, Q., Xiukang, W., Haider, F. U., Kučerik, J., Mumtaz, M. Z., Holatko, J., Naseem, M., Kintl, A., Ejaz, M., Naveed, M., Brtnicky, M., & Mustafa, A. (2021). Rhizosphere Bacteria in Plant Growth Promotion, Biocontrol, and Bioremediation of Contaminated Sites: A Comprehensive Review of Effects and Mechanisms. *International Journal of Molecular Sciences*, *22*(19), 10529. Advance online publication. DOI: 10.3390/ijms221910529 PMID: 34638870

Sagar, P. R., Raol, G. G., Prajapati, D., Chaudhari, H. G., & Kiri, B. D. (2024). Statistical Evaluation for the formulation of low-cost Single cell protein production in Food waste medium by Pichia occidentalis: A sustainable valorization approach. *Waste and Biomass Valorization*, *15*(6), 3457–3475. DOI: 10.1007/s12649-023-02382-3

Sakshi, S., Singh, S. K., & Haritash, A. K. (2019). Polycyclic aromatic hydrocarbons: Soil pollution and remediation. *International Journal of Environmental Science and Technology*, *16*(10), 6489–6512. DOI: 10.1007/s13762-019-02414-3

Saleem, H., Zaidi, S. J., Ismail, A. F., & Goh, P. S. (2022). Advances of nanomaterials for air pollution remediation and their impacts on the environment. *Chemosphere*, *287*, 132083. https://doi.org/https://doi.org/10.1016/j.chemosphere.2021.132083. DOI: 10.1016/j.chemosphere.2021.132083 PMID: 34488054

Saleh, M. H., Rashedi, I., & Keating, A. (2017). Immunonodulatory properties of Coriuolus versicolor: The role of Polysaccharopeptide. *Frontiers in Immunology*, *8*, 1087. DOI: 10.3389/fimmu.2017.01087 PMID: 28932226

Salgado, C. L., Muñoz, R., Blanco, A., & Lienqueo, M. E. (2021). Valorization and upgrading of the nutritional value of seaweed and seaweed waste using the marine fungi Paradendryphiella salina to produce mycoprotein. *Algal Research*, *53*, 102135. DOI: 10.1016/j.algal.2020.102135

Salichos, L., & Rokas, A. (2010). The diversity and evolution of circadian clock proteins in fungi. *Mycologia*, *102*(2), 269–278. DOI: 10.3852/09-073 PMID: 20361495

Salwan, R., Sharma, A., Kaur, R., Sharma, R., & Sharma, V. (2023). Signaling in arbuscular mycorrhizal association. In *The Chemical Dialogue Between Plants and Beneficial Microorganisms* (pp. 127–135). Elsevier. DOI: 10.1016/B978-0-323-91734-6.00005-3

Samanta, S. K., Singh, O. V., & Jain, R. K. (2002). Polycyclic aromatic hydrocarbons: Environmental pollution and bioremediation. *Trends in Biotechnology*, *20*(6), 243–248. DOI: 10.1016/S0167-7799(02)01943-1 PMID: 12007492

Sande, D., Oliveira, G. P., Moura, M. A. F. E., Martins, B. A., Lima, M. T. N. S., & Takahashi, J. A. (2019). Edible mushrooms as a ubiquitous source of essential fatty acids. *Food Research International*, *125*, 108524. DOI: 10.1016/j.foodres.2019.108524 PMID: 31554069

Sandhu, S. S., Rajak, R. C., & Hasija, S. K. (2000). Potential of entomopathogens for the biological management of medically important pests: Progress and prospect. Glimpses in Plant Sciences, 110–117.

Sandhu, D. K., & Arora, D. S. (1985). Laccase production by Polyporus sanguineus under different nutritional and environmental conditions. *Experientia*, *41*(3), 355–356. DOI: 10.1007/BF02004501

Sandhu, S. S. (1995). Effect of physical factors on germination of entomopathogenic fungus Beauveria bassiana conidia. *National Academy Science Letters*, *18*(1–2), 1–5.

Sandhu, S. S., Rajak, R. C., & Agarwal, G. P. (1993). Studies on prolonged storage of Beauveria bassiana conidia: Effects of temperature and relative humidity on conidial viability and virulence against chickpea borer Helicoverpa armigera. *Biocontrol Science and Technology*, *3*(1), 47–53. DOI: 10.1080/09583159309355258

Sandhu, S. S., Sharma, A. K., Beniwal, V., Goel, G., Batra, P., Kumar, A., Jaglan, S., Sharma, A. K., & Malhotra, S. (2012). Myco-biocontrol of insect pests: Factors involved, mechanism, and regulation. *Journal of Pathogens*, *126819*, 1–10. Advance online publication. DOI: 10.1155/2012/126819 PMID: 22567344

Sangeeta, S. D., Ramniwas, S., Mugabi, R., Uddin, J., & Nayik, G. A. (2024, August 25). Revolutionizing Mushroom processing: Innovative techniques and technologies. *Food Chemistry: X*, *23*, 101774. DOI: 10.1016/j.fochx.2024.101774 PMID: 39280230

Sangwan, S., & Prasanna, R. (2022). Mycorrhizae helper bacteria: Unlocking their potential as bioenhancers of plant–arbuscular mycorrhizal fungal associations. *Microbial Ecology*, *84*(1), 1–10. DOI: 10.1007/s00248-021-01831-7 PMID: 34417849

Sankar, N. R., Kumar, V. K., Shailaja, R., Saritha, K., & Naidu, N. V. (2011). Single cell protein production by *Trichoderma harzianum* using waste banana peel. http://idosi.org/ijmr/ijmr2(1)11/13a.pdf

Santos, A. R., Carreiró, F., Freitas, A., Barros, S., Brites, C., Ramos, F., & Sanches Silva, A. (2022). Mycotoxins Contamination in Rice: Analytical Methods, Occurrence and Detoxification Strategies. *Toxins*, *14*(9), 647. DOI: 10.3390/toxins14090647 PMID: 36136585

Santos, A., Bustamante, M. A., Tortosa, G., Moral, R., & Bernal, M. P. (2016). 'Gaseous Emissions and Process Development during Composting of Pig Slurry: The Influence of the Proportion of Cotton Gin Waste'. *Journal of Cleaner Production*, *112*, 81–90. DOI: 10.1016/j.jclepro.2015.08.084

Saravanan, A., Senthil Kumar, P., Jeevanantham, S., Karishma, S., Tajsabreen, B., Yaashikaa, P. R., & Reshma, B. (2021). Effective water/wastewater treatment methodologies for toxic pollutants removal: Processes and applications towards sustainable development. *Chemosphere, 280*, 130595. https://doi .org/https://doi.org/10.1016/j.chemosphere.2021.130595. DOI: 10.1016/j.chemosphere.2021.130595 PMID: 33940449

Saravanan, K., Rajendran, M., & Kathirvel, P. (2023). Utilization of Water Hyacinth (Eichhornia crassipes) as a Feasible Substrate for the Production of Single Cell Protein Using Saccharomyces cerevisiae (Baker's Yeast). *Waste and Biomass Valorization, 14*(12), 4231–4242. DOI: 10.1007/s12649-023-02138-z

Sarrocco, S., & Vannacci, G. (2018). Preharvest application of beneficial fungi as a strategy to prevent mycotoxin contamination. *World Mycotoxin Journal, 11*(1), 227–238. DOI: 10.3920/WMJ2017.2276

Sartori, M. V., Nesci, A. V., Castillo, C., & Etcheverry, M. G. (2013). Biological control of fumonisins production in maize at field level.

Saslis-Lagoudakis, C. H., Savolainen, V., Williamson, E. M., Forest, F., Wagstaff, S. J., Baral, S. R., Watson, M. F., Pendry, C. A., & Hawkins, J. A. (2012). Phylogenies reveal predictive power of traditional medicine in bioprospecting. *Proceedings of the National Academy of Sciences of the United States of America, 109*(39), 15835–15840. DOI: 10.1073/pnas.1202242109 PMID: 22984175

Sato, T., Ezawa, T., Cheng, W., & Tawaraya, K. (2015). Release of acid phosphatase from extraradical hyphae of arbuscular mycorrhizal fungus Rhizophagus clarus. *Soil Science and Plant Nutrition, 61*(2), 269–274. DOI: 10.1080/00380768.2014.993298

Sawangwan, T., Wansanit, W., Pattani, L., & Noysang, C. (2018). Study of prebiotic properties from edible mushroom extraction. *Agriculture and Natural Resources (Bangkok), 52*(6), 519–524. DOI: 10.1016/j.anres.2018.11.020

Schena, L., Nigro, F., Soleti Ligorio, V., Yaseen, T., Ippolito, A., & El Ghaouth, A. (2005). Biocontrol activity of Bio-Coat and Biocure against postharvest rots of table grapes and sweet cherries. *Acta Horticulturae*, (682), 2115–2120. DOI: 10.17660/ActaHortic.2005.682.288

Schindler, D. W., Carpenter, S. R., Chapra, S. C., Hecky, R. E., & Orihel, D. M. (2016). Reducing phosphorus to curb lake eutrophication is a success. *Environmental Science & Technology, 50*(17), 8923–8929. DOI: 10.1021/acs.est.6b02204 PMID: 27494041

Schonbeck, F., & Dehne, H. W. (1977). Damage to mycorrhizal and nonmycorrhizal cotton seedlings by Thielaviopsis basicola. *The Plant Disease Reporter, 61*(4), 266–267.

Schrader, J., Schilling, M., Holtmann, D., Sell, D., Villela Filho, M., Marx, A., & Vorholt, J. A. (2009). Methanol-based industrial biotechnology: Current status and future perspectives of methylotrophic bacteria. *Trends in Biotechnology, 27*(2), 107–115. DOI: 10.1016/j.tibtech.2008.10.009

Schrank, A., & Vainstein, M. H. (2010). Metarhizium anisopliae enzymes and toxins. *Toxicon: official journal of the International Society on Toxinology, 56*(7), 1267–1274. https://doi.org/DOI: 10.1016/j. toxicon.2010.03.008

Schreinemachers, P., & Tipraqsa, P. (2012). Agricultural pesticides and land use intensification in high, middle and low income countries. *Food Policy*, *37*(6), 616–626. DOI: 10.1016/j.foodpol.2012.06.003

Schulz, E., & Oslage, H. J. (1976). Composition and nutritive value of single-cell protein (SCP). *Animal Feed Science and Technology*, *1*(1), 9–24. DOI: 10.1016/0377-8401(76)90003-1

Schulz, K., Pöhlmann, C., Dietrich, R., Märtlbauer, E., & Elßner, T. (2019). Electrochemical biochip assays based on anti-idiotypic antibodies for rapid and automated on-site detection of low molecular weight toxins. *Frontiers in Chemistry*, *7*, 31. DOI: 10.3389/fchem.2019.00031 PMID: 30775361

Scott, P. M. (1995). Mycotoxin methodology. *Food Additives and Contaminants*, *12*(3), 395–403. DOI: 10.1080/02652039509374321 PMID: 7664934

Šegvić Klarić, M., Cvetnić, Z., Pepeljnjak, S., & Kosalec, I. (2009). Co-occurrence of aflatoxins, ochratoxin A, fumonisins, and zearalenone in cereals and feed, determined by competitive direct enzyme-linked immunosorbent assay and thin-layer chromatography. *Arhiv za Higijenu Rada i Toksikologiju*, *60*(4), 427–433. DOI: 10.2478/10004-1254-60-2009-1975 PMID: 20061243

Selvakumar, P. (2024). Climate Change. In Sofia Idris, I. G. I. (Ed.), *Global Challenges for the Environment and Climate Change* (pp. 100–110). Global., DOI: 10.4018/979-8-3693-2845-3.ch006

Selvakumar, P. (2024). Reasons for Biodiversity Loss. In Ashok Kumar Rathoure, I. G. I. (Ed.), *Impact of Societal Development and Infrastructure on Biodiversity Decline* (pp. 37–49). Global., DOI: 10.4018/979-8-3693-6950-0.ch004

Selvakumar, P.. (2024). The Role of Corrosion Inhibitors. In Thakur, A., & Kumar, A. (Eds.), *Innovations in Nanomaterials-Based Corrosion Inhibitors* (pp. 130–141). IGI Global., DOI: 10.4018/979-8-3693-3088-3.ch006

Selvakumar, P., Mariappan, P., Shunmugam, S., Paul, R., & Jabez, S. (2024). Removal of chemicals from effluent using photobioreactor technology to improve environmental and health impacts. *Zeitschrift für Physikalische Chemie*, *2024*(7), 1295–1302. Advance online publication. DOI: 10.1515/zpch-2023-0450

Selvakumar, P., Palanisamy, S. K., Cinthaikinian, S., Palanisamy, V., Mariappan, R., & Selvakumar, P. K. (2024). Biosensors and its diverse applications in healthcare systems. *Zeitschrift für Physikalische Chemie*, *2024*. Advance online publication. DOI: 10.1515/zpch-2023-0406

Sery, D. J.-M., Van Tuinen, D., Drain, A., Mounier, A., & Zézé, A. (2018). The genus Rhizophagus dominates arbuscular mycorrhizal fungi communities in contrasted cassava field soils in Côte d'Ivoire. *Rhizosphere*, *7*, 8–17. DOI: 10.1016/j.rhisph.2018.06.007

Seyyedi-Mansour, S., Carpena, M., Donn, P., Barciela, P., Perez-Vazquez, A., Echave, J., Pereira, A. G., & Prieto, M. A. (2024). Citrus Seed Waste and Circular Bioeconomy: Insights on Nutritional Profile, Health Benefits, and Application as Food Ingredient. *Applied Sciences (Basel, Switzerland)*, *14*(20), 9463. DOI: 10.3390/app14209463

Shaffique, S., Khan, M. A., Wani, S. H., Pande, A., Imran, M., Kang, S.-M., Rahim, W., Khan, S. A., Bhatta, D., Kwon, E.-H., & Lee, I.-J. (2022). A Review on the Role of Endophytes and Plant Growth Promoting Rhizobacteria in Mitigating Heat Stress in Plants. *Microorganisms*, *10*(7), 1286. DOI: 10.3390/microorganisms10071286 PMID: 35889005

Shaher, H., Naushin, F., Hasan, M., & Bagyaraj, D. J. (2024). Synergistic impact of Rhizobium and arbuscular mycorrhizal fungi on lentil plant tolerance to heavy metal-rich fly ash amended soil. *Discover Plants*, *1*(1), 9. DOI: 10.1007/s44372-024-00010-5

Shahrajabian, M. H., Chaski, C., Polyzos, N., & Petropoulos, S. A. (2021). Biostimulants Application: A Low Input Cropping Management Tool for Sustainable Farming of Vegetables. *Biomolecules*, *11*(5), 698. Advance online publication. DOI: 10.3390/biom11050698 PMID: 34067181

Shamim, M. Z., Mishra, A. K., Kausar, T., Mahanta, S., Sarma, B., Kumar, V., Mishra, P. K., Panda, J., Baek, K. H., & Mohanta, Y. K. (2023). Exploring Edible Mushrooms for Diabetes: Unveiling Their Role in Prevention and Treatment. *Molecules (Basel, Switzerland)*, *28*(6), 2837. DOI: 10.3390/molecules28062837 PMID: 36985818

Shanmugapriya, S., Manivannan, G., Selvakumar, G., & Sivakumar, N. (2019). *Extracellular Fungal Peroxidases and Laccases for Waste Treatment: Recent Improvement BT - Recent Advancement in White Biotechnology Through Fungi: Volume 3: Perspective for Sustainable Environments* (A. N. Yadav, S. Singh, S. Mishra, & A. Gupta (eds.); pp. 153–187). Springer International Publishing. DOI: 10.1007/978-3-030-25506-0_6

Sharif, M., Zafar, M. H., Aqib, A. I., Saeed, M., Farag, M. R., & Alagawany, M. (2021). Single cell protein: Sources, mechanism of production, nutritional value and its uses in aquaculture nutrition. *Aquaculture (Amsterdam, Netherlands)*, *531*, 735885. DOI: 10.1016/j.aquaculture.2020.735885

Sharma, M., Kansal, R., & Singh, D. (2018). Endophytic Microorganisms: Their Role in Plant Growth and Crop Improvement. In *Crop Improvement Through Microbial Biotechnology* (pp. 391–413). Elsevier. DOI: 10.1016/B978-0-444-63987-5.00020-7

Sharma, A. K., Sharma, M., Sharma, A. K., Sharma, M., & Sharma, M. (2023). Mapping the impact of environmental pollutants on human health and environment: A systematic review and meta-analysis. *Journal of Geochemical Exploration*, *255*, 107325. DOI: 10.1016/j.gexplo.2023.107325

Sharma, A., Sharma, R., Arora, A., Shah, R., Singh, A., Pranaw, K., & Nain, L. (2014). Insights into rapid composting of paddy straw augmented with efficient microorganism consortium. *International Journal of Recycling of Organic Waste in Agriculture*, *3*(2), 54. DOI: 10.1007/s40093-014-0054-2

Sharma, I., Raina, A., Choudhary, M., Apra, , Kaul, S., & Dhar, M. K. (2023). Fungal endophyte bioinoculants as a green alternative towards sustainable agriculture. *Heliyon*, *9*(9), e19487. DOI: 10.1016/j.heliyon.2023.e19487 PMID: 37662754

Sharma, K., Gupta, S., Thokchom, S. D., Jangir, P., & Kapoor, R. (2021). Arbuscular mycorrhiza-mediated regulation of polyamines and aquaporins during abiotic stress: Deep insights on the recondite players. *Frontiers in Plant Science*, *12*, 642101. DOI: 10.3389/fpls.2021.642101 PMID: 34220878

Sharma, N., & Tripathi, A. (2008). Effects of *Citrus sinensis* (L) Obseck epecarp essential oil on growth and morphogenesis of *Aspergillus niger* (L)." Van Tieghem. *Microbiological Research, 163*(3), 337–344. DOI: 10.1016/j.micres.2006.06.009 PMID: 16870411

Sharma, R., Garg, P., Kumar, P., Bhatia, S. K., & Kulshrestha, S. (2020). Microbial fermentation and its role in quality improvement of fermented foods. *Fermentation (Basel, Switzerland), 6*(4), 106. DOI: 10.3390/fermentation6040106

Sharma, R., & Sharma, P. (2021). Fungal entomopathogens: A systematic review. *Egyptian Journal of Biological Pest Control, 31*(1), 57. DOI: 10.1186/s41938-021-00404-7

Sharmeen, J. B., Mahomoodally, F. M., Zengin, G., & Maggi, F. (2021). Essential oils as natural sources of fragrance compounds for cosmetics and cosmeceuticals. *Molecules (Basel, Switzerland), 26*(3), 666. DOI: 10.3390/molecules26030666 PMID: 33514008

Sheikh, A. A., Loveleen, L., Pant, K. K., & Chandra, V. (2022). Cherries: A Treasure Of Bioactive Compounds With Promising Health Promoting Functional Potential. *IJFANS International Journal of Food and Nutritional Sciences, 11*(8), 794–821.

Shimazu, M. (2004). A novel technique to inoculate conidia of entomopathogenic fungi and its application for investigation of susceptibility of the Japanese pine sawyer, Monochamus alternatus, to Beauveria bassiana. *Applied Entomology and Zoology, 39*(3), 485–490. DOI: 10.1303/aez.2004.485

Shin, H.-J., Ro, H.-S., Kawauchi, M., & Honda, Y. (2025). Review on mushroom mycelium-based products and their production process: From upstream to downstream. *Bioresources and Bioprocessing, 12*(1), 3. DOI: 10.1186/s40643-024-00836-7 PMID: 39794674

Shirur, M. (2022). Leveraging government policies and schemes for promoting mushroom entrepreneurship in India. *Mushroom Research, 30*(2).

Shivakumar, C. K., Thippeswamy, B., & Krishnappa, M. (2014). Studies on Heavy Metals Detoxification Biomarkers in Fungal Consortia. *Caribbean Journal of Sciences and Technology, 2*(01 SE-Research Article), 496–502. https://www.caribjscitech.com/index.php/cjst/article/view/62

Shukla, K. P., Singh, N. K., & Sharma, S. (2010). Bioremediation: Developments, current practices and perspectives. *Genetic Engineering and Biotechnology Journal, 3*, 1–20.

Sible, C. N., Seebauer, J. R., & Below, F. E. (2021). Plant Biostimulants: A Categorical Review, Their Implications for Row Crop Production, and Relation to Soil Health Indicators. *Agronomy (Basel), 11*(7), 1297. DOI: 10.3390/agronomy11071297

Siddeeg, S. M., Tahoon, M. A., Mnif, W., & Ben Rebah, F. (2020). Iron Oxide/Chitosan Magnetic Nanocomposite Immobilized Manganese Peroxidase for Decolorization of Textile Wastewater. In *Processes* (Vol. 8, Issue 1). DOI: 10.3390/pr8010005

Siegrist, M., Michel, F., & Hartmann, C. (2024). The shift from meat to plant-based proteins: Consumers and public policy. *Current Opinion in Food Science, 58*, 101182. DOI: 10.1016/j.cofs.2024.101182

Simon, L., Lalonde, M., & Bruns, T. D. (1992). Specific amplification of 18S fungal ribosomal genes from vesicular-arbuscular endomycorrhizal fungi colonizing roots. *Applied and Environmental Microbiology*, *58*(1), 291–295. DOI: 10.1128/aem.58.1.291-295.1992 PMID: 1339260

Simon, L., Lévesque, R. C., & Lalonde, M. (1993). Identification of endomycorrhizal fungi colonizing roots by fluorescent single-strand conformation polymorphism-polymerase chain reaction. *Applied and Environmental Microbiology*, *59*(12), 4211–4215. DOI: 10.1128/aem.59.12.4211-4215.1993 PMID: 8285712

Singh, A., Kumari, R., & Yadav, A. N. (2021). Fungal secondary metabolites for bioremediation of hazardous heavy metals. *Recent trends in mycological research: volume 2: environmental and industrial perspective*, 65-98.

Singh, M., Vijay, B., Kamal, S., & Wakchaure, G. C. (2011). Mushrooms Cultivation, Marketing, and Consumption Editors.

Singh, N., Kumar, A., & Sharma, B. (2019). *Role of Fungal Enzymes for Bioremediation of Hazardous Chemicals BT - Recent Advancement in White Biotechnology Through Fungi: Volume 3: Perspective for Sustainable Environments* (A. N. Yadav, S. Singh, S. Mishra, & A. Gupta (eds.); pp. 237–256). Springer International Publishing. DOI: 10.1007/978-3-030-25506-0_9

Singh, A., & Sarkar, S. (2021). Effect of excess use of fertilizers on insect/pest infestation in *Oryza sativa*: A review. *Journal of Entomology and Zoology Studies*, *9*(5), 246–252. DOI: 10.22271/j.ento.2021.v9.i5c.8834

Singh, D., Raina, T. K., & Singh, J. (2017). Entomopathogenic fungi: An effective biocontrol agent for management of insect populations naturally. *Journal of Pharmaceutical Sciences and Research.*, *9*(6), 830–839.

Singh, H. (2006). *Mycoremediation: fungal bioremediation*. John Wiley & Sons. DOI: 10.1002/0470050594

Singh, H. B. (2014). Management of plant pathogens with microorganisms. *Proceedings of the Indian National Science Academy. Part A, Physical Sciences*, *80*(2), 443–454. DOI: 10.16943/ptinsa/2014/v80i2/55120

Singh, V. (2024). Soil Pollution. In *Textbook of Environment and Ecology* (pp. 267–274). Springer Nature Singapore. DOI: 10.1007/978-981-99-8846-4_18

Singh, V., & Chawla, S. (2012). Cultural Practices: An Eco-friendly Innovative Approach in Plant Disease Management. DOI: 10.13140/RG.2.1.4216.2327

Singh, Y., & Sidhu, H. S. (2014). Management of Cereal Crop Residues for Sustainable Rice-Wheat Production System in the Indo-Gangetic Plains of India. *Proceedings of the Indian National Science Academy. Part A, Physical Sciences*, *80*(1), 95–114. DOI: 10.16943/ptinsa/2014/v80i1/55089

Sirtori, C. R., & Lammi, C. (2023). Challenges and Opportunities for Mycoprotein-Based Food Products. *The American Journal of Clinical Nutrition*, *118*(1), 5–7. DOI: 10.1016/j.ajcnut.2023.05.019

Sishodia, R. P., Ray, R. L., & Singh, S. K. (2020). Applications of Remote Sensing in Precision Agriculture: A Review. *Remote Sensing (Basel)*, *12*(19), 3136. DOI: 10.3390/rs12193136

Slaughter, G. (2006). Current advances in biosensor design and fabrication. *Encyclopedia of Analytical Chemistry: Applications, Theory and Instrumentation*, 1-25.

Smetana, S., Mathys, A., Knoch, A., & Heinz, V. (2015). Meat alternatives: Life cycle assessment of most known meat substitutes. *The International Journal of Life Cycle Assessment*, *20*(9), 1254–1267. DOI: 10.1007/s11367-015-0931-6

Smith, S. E. R. D. J. (2008). *Mycorrhizal Symbiosis* (3rd ed.). Academic Press, London.

Smith, J. M., Whiteside, M. D., & Jones, M. D. (2020). Rapid nitrogen loss from ectomycorrhizal pine germinants signaled by their fungal symbiont. *Mycorrhiza*, *30*(4), 407–417. DOI: 10.1007/s00572-020-00959-7 PMID: 32363468

Smith, S. E., & Read, D. J. (2010). *Mycorrhizal symbiosis*. Academic press.

Smith, S. E., & Smith, F. A. (2011). Roles of arbuscular mycorrhizas in plant nutrition and growth: New paradigms from cellular to ecosystem scales. *Annual Review of Plant Biology*, *62*(1), 227–250. DOI: 10.1146/annurev-arplant-042110-103846 PMID: 21391813

Snini, S. P., & Mathieu, F. (2020). Biocontrol Agents and Natural Compounds Against Mycotoxinogenic Fungi. *Toxins*, *12*(6), 353. DOI: 10.3390/toxins12060353 PMID: 32481521

Soares, P. R. S., Birolli, W. G., Ferreira, I. M., & Porto, A. L. M. (2021). Biodegradation pathway of the organophosphate pesticides chlorpyrifos, methyl parathion and profenofos by the marine-derived fungus *Aspergillus sydowii* CBMAI 935 and its potential for methylation reactions of phenolic compounds. *Marine Pollution Bulletin*, *166*, 112185. DOI: 10.1016/j.marpolbul.2021.112185 PMID: 33640600

Sodaneath, H., Lee, J. I., Yang, S. O., Jung, H., Ryu, H. W., & Cho, K. S. (2017). Decolorization of textile dyes in an air-lift bioreactor inoculated with *Bjerkandera adusta* OBR105. *Journal of Environmental Science and Health. Part A, Toxic/Hazardous Substances & Environmental Engineering*, *52*(11), 1099–1111. DOI: 10.1080/10934529.2017.1340753 PMID: 28763254

Solaiman, Z. (2024). Role of AMF in Organic Matter Decomposition, Carbon Sequestration and Climate Change Mitigation. In *Arbuscular mycorrhizal fungi in sustainable agriculture: Nutrient and crop management* (pp. 131–141). Springer Nature Singapore. DOI: 10.1007/978-981-97-0300-5_6

Solanki, M. K., Kashyap, B. K., Solanki, A. C., Malviya, M. K., & Surapathrudu, K. (2019). *Helpful Linkages of Trichoderma s in the process of Mycoremediation and Mycorestoration. Plant Health Under Biotic Stress* (Vol. 2). Microbial Interactions.

Somasundaram, J., Sinha, N. K., & Ram, C. (2020). 'No-Till Farming and Conservation Agriculture in South Asia–Issues, Challenges, Prospects, and Benefits'. *Critical Reviews in Plant Sciences*, *39*(3), 236–279. DOI: 10.1080/07352689.2020.1782069

Song, Y. Y., Zeng, R. Sen, Xu, J. F., Li, J., Shen, X., & Yihdego, W. G. (2010). Interplant Communication of Tomato Plants through Underground Common Mycorrhizal Networks. *PLOS ONE*, *5*(10), e13324-. DOI: 10.1371/journal.pone.0013324

Son, Y., Martínez, C. E., & Kao-Kniffin, J. (2024). Three important roles and chemical properties of glomalin-related soil protein. *Frontiers in Soil Science*, *4*, 1418072. DOI: 10.3389/fsoil.2024.1418072

Sośnicka, A., Kózka, B., Makarova, K., Giebułtowicz, J., Klimaszewska, M., & Turło, J. (2022). Optimization of white-rot fungi mycelial culture components for bioremediation of pharmaceutical-derived pollutants. *Water (Basel), 14*(9), 1374. DOI: 10.3390/w14091374

Sota, T., Kagata, H., Ando, Y., Utsumi, S., & Osono, T. (2014). Metagenomic Approach Yields Insights into Fungal Diversity and Functioning. In *Species Diversity and Community Structure*, 1–23. DOI: 10.1007/978-4-431-54261-2_1

Souto, A. L., Sylvestre, M., Tölke, E. D., Tavares, J. F., Barbosa-Filho, J. M., & Cebrián-Torrejón, G. (2021). Plant-Derived Pesticides as an Alternative to Pest Management and Sustainable Agricultural Production: Prospects, Applications and Challenges. *Molecules (Basel, Switzerland), 26*(16), 4835. DOI: 10.3390/molecules26164835 PMID: 34443421

Spalvins, K., Geiba, Z., Kusnere, Z., & Blumberga, D. (2020). Waste Cooking Oil as Substrate for Single Cell Protein Production by Yeast Yarrowia lipolytica. *Environmental and Climate Technologies, 24*(3), 457–469. DOI: 10.2478/rtuect-2020-0116

Spalvins, K., Ivanovs, K., & Blumberga, D. (2018). Single cell protein production from waste biomass: Review of various agricultural by-products. https://doi.org/DOI: 10.15159/AR.18.129

Srikanth, M., Sandeep, T. S. R. S., Sucharitha, K., & Godi, S. (2022). Biodegradation of plastic polymers by fungi: A brief review. *Bioresources and Bioprocessing, 9*(1), 42. DOI: 10.1186/s40643-022-00532-4 PMID: 38647755

Stanker, L. H., Merrill, P., Scotcher, M. C., & Cheng, L. W. (2008). Development and partial characterization of high-affinity monoclonal antibodies for botulinum toxin type A and their use in analysis of milk by sandwich ELISA. *Journal of Immunological Methods, 336*(1), 1–8. DOI: 10.1016/j.jim.2008.03.003 PMID: 18452945

Starck, C., Cassettari, T., Wright, J., Petocz, P., Beckett, E., & Fayet-Moore, F. (2024). Mushrooms: A food-based solution to vitamin D deficiency to include in dietary guidelines. *Frontiers in Nutrition, 11*, 1384273. DOI: 10.3389/fnut.2024.1384273 PMID: 38660061

Stefani, F., Bencherif, K., Sabourin, S., Hadj-Sahraoui, A. L., Banchini, C., Séguin, S., & Dalpé, Y. (2020). Taxonomic assignment of arbuscular mycorrhizal fungi in an 18S metagenomic dataset: A case study with saltcedar (*Tamarix aphylla*). *Mycorrhiza, 30*(2-3), 243–255. DOI: 10.1007/s00572-020-00946-y PMID: 32180012

Stein, E., Molitor, A., Kogel, K. H., & Waller, F. (2008). Systemic resistance in Arabidopsis conferred by the mycorrhizal fungus Piriformospora indica requires jasmonic acid signaling and the cytoplasmic function of NPR1. *Plant & Cell Physiology, 49*(11), 1747–1751. DOI: 10.1093/pcp/pcn147 PMID: 18842596

Stępniewska, Z., & Kuźniar, A. (2013). Endophytic microorganisms—Promising applications in bioremediation of greenhouse gases. *Applied Microbiology and Biotechnology, 97*(22), 9589–9596. DOI: 10.1007/s00253-013-5235-9 PMID: 24048641

Stover, H. J., Thorn, R. G., Bowles, J. M., Bernards, M. A., & Jacobs, C. R. (2012). Arbuscular mycorrhizal fungi and vascular plant species abundance and community structure in tallgrass prairies with varying agricultural disturbance histories. *Applied Soil Ecology*, *60*, 61–70. DOI: 10.1016/j.apsoil.2012.02.016

Strong, P. J., Self, R., Allikian, K., Szewczyk, E., Speight, R., O'Hara, I., & Harrison, M. D. (2022). Filamentous fungi for future functional food and feed. *Current Opinion in Biotechnology*, *76*, 102729. DOI: 10.1016/j.copbio.2022.102729 PMID: 35525176

Stuart, M., Lapworth, D., Crane, E., & Hart, A. (2012). Review of risk from potential emerging contaminants in UK groundwater. *The Science of the Total Environment*, *416*, 1–21. DOI: 10.1016/j.scitotenv.2011.11.072 PMID: 22209399

Sturtewagen, L., De Soete, W., Dewulf, J., Lachat, C., Lauryssen, S., Heirman, B., Rossi, F., & Schaubroeck, T. (2016). Resource use profile and nutritional value assessment of a typical Belgian meal catered or home-cooked, with pork or QuornTM as a protein source. *Journal of Cleaner Production*, *112*, 196–204. DOI: 10.1016/j.jclepro.2015.09.006

Sudha, G., Vadivukkarasi, S., Shree, R. B. I., & Lakshmanan, P. (2012). Antioxidant activity of various extracts from an edible mushroom Pleurotus. *Food Science and Biotechnology*, *21*(3), 661–668. DOI: 10.1007/s10068-012-0086-1 PMID: 30263280

Suman, G., Nupur, M., Anuradha, S., & Pradeep, B. (2015). Single cell protein production: A review. *International Journal of Current Microbiology and Applied Sciences*, *4*(9), 251–262.

Sun, W., & Shahrajabian, M. H. (2023). The Application of Arbuscular Mycorrhizal Fungi as Microbial Biostimulant, Sustainable Approaches in Modern Agriculture. *Plants*, *12*(17), 3101. Advance online publication. DOI: 10.3390/plants12173101 PMID: 37687348

Suzuki, T., Kurisu, M., Nose, N., & Watanabe, A. (1981). Determination of butenolide (Fusarium toxin) by gas chromatography with an electron-capture detector. Food Hygiene and Safety Science (Shokuhin Eiseigaku Zasshi), 22(3), 197-202_1.

Szekeres, A., Leitgeb, B., Kredics, L., Antal, Z., Hatvani, L., Manczinger, L., & Vágvölgyi, C. (2005). Peptaibols and related peptaibiotics of Trichoderma. A review. *Acta Microbiologica et Immunologica Hungarica*, *52*(2), 137–168. DOI: 10.1556/AMicr.52.2005.2.2 PMID: 16003936

Tadesse Mawcha, K., Malinga, L., Muir, D., Ge, J., & Ndolo, D. (2024). Recent Advances in Biopesticide Research and Development: A Focus on Microbial: A Review. *F1000 Research*, *13*, 1071. DOI: 10.12688/f1000research.154392.3 PMID: 39512238

Tadesse Mawcha, K., Malinga, L., Muir, D., Ge, J., & Ndolo, D. (2025). Recent Advances in Biopesticide Research and Development with a Focus on Microbials. *F1000 Research*, *13*, 1071. DOI: 10.12688/f1000research.154392.4 PMID: 39512238

Tagkouli, D., Kaliora, A., Bekiaris, G., Koutrotsios, G., Christea, M., Zervakis, G. I., & Kalogeropoulos, N. (2020). Free amino acids in three Pleurotus species cultivated on agricultural and agro-industrial by-products. *Molecules (Basel, Switzerland)*, *25*(17), 4015. DOI: 10.3390/molecules25174015 PMID: 32887476

Taheur, F. B., Kouidhi, B., Al Qurashi, Y. M. A., Salah-Abbès, J. B., & Chaieb, K. (2019). Biotechnology of mycotoxins detoxification using microorganisms and enzymes. *Toxicon, 160*, 12–22. DOI: 10.1016/j.toxicon.2019.02.001 PMID: 30772433

Takio, N., Yadav, M., & Yadav, H. S. (2021). Catalase-mediated remediation of environmental pollutants and potential application – a review. *Biocatalysis and Biotransformation, 39*(6), 389–407. DOI: 10.1080/10242422.2021.1932838

Tamasloukht, B., Kuhn, B., Becard, G., & Franken, P. (2000). RNA accumulation patterns of arbuscular mycorrhizal fungi during presymbiotic stages. *Programs, Abstratcts and Papers of the Third International Congress on Symbiosis.*

Tamasloukht, M., Séjalon-Delmas, N., Kluever, A., Jauneau, A., Roux, C., Bécard, G., & Franken, P. (2003). Root Factors Induce Mitochondrial-Related Gene Expression and Fungal Respiration during the Developmental Switch from Asymbiosis to Presymbiosis in the Arbuscular Mycorrhizal Fungus *Gigaspora rosea.Plant Physiology, 131*(3), 1468–1478. DOI: 10.1104/pp.012898 PMID: 12644696

Tanaya, K., Kumari, A., Singh, A. K., & Singh, D. (2024). Bioremediation: An Economical Approach for Treatment of Textile Dye Effluents. *Water, Air, and Soil Pollution, 235*(8), 516. DOI: 10.1007/s11270-024-07287-y

Tang, H., Hassan, M. U., Feng, L., Nawaz, M., Shah, A. N., Qari, S. H., & Miao, J. (2022). The critical role of arbuscular mycorrhizal fungi to improve drought tolerance and nitrogen use efficiency in crops. *Frontiers in Plant Science, 13*, 919166. DOI: 10.3389/fpls.2022.919166 PMID: 35873982

Tang, X., Li, X., Li, P., Zhang, Q., Li, R., Zhang, W., Ding, X., Lei, J., & Zhang, Z. (2014). Development and application of an immunoaffinity column enzyme immunoassay for mycotoxin zearalenone in complicated samples. *PLoS One, 9*(1), e85606. DOI: 10.1371/journal.pone.0085606 PMID: 24465616

Tao, C., Wang, Z., Liu, S., Lv, N., Deng, X., Xiong, W., Shen, Z., Zhang, N., Geisen, S., Li, R., Shen, Q., & Kowalchuk, G. A. (2023). Additive fungal interactions drive biocontrol of Fusarium wilt disease. *The New Phytologist, 238*(3), 1198–1214. DOI: 10.1111/nph.18793 PMID: 36740577

Tarigan, S. I., Toth, S., Szalai, M., Kiss, J., Turoczi, G., & Toepfer, S. (2022). Biological control properties of microbial plant biostimulants. A review. *Biocontrol Science and Technology, 32*(12), 1351–1371. DOI: 10.1080/09583157.2022.2129589

Tarman, K., Lindequist, U., Wende, K., Porzel, A., Arnold, N., & Wessjohann, L. A. (2011). Isolation of a new natural product and cytotoxic and antimicrobial activities of extracts from fungi of indonesian marine habitats. *Marine Drugs, 9*(3), 294–306. DOI: 10.3390/md9030294 PMID: 21556160

Tedersoo, L., Sánchez-Ramírez, S., Kõljalg, U., Bahram, M., Döring, M., Schigel, D., May, T., Ryberg, M., & Abarenkov, K. (2018). High-level classification of the Fungi and a tool for evolutionary ecological analyses. *Fungal Diversity, 90*(1), 135–159. DOI: 10.1007/s13225-018-0401-0

Terrasan, C. R., Temer, B., Duarte, M. C., & Carmona, E. C. (2010). Production of xylanolytic enzymes by Penicillium janczewskii. *Bioresource Technology, 101*(11), 4139–4143. DOI: 10.1016/j.biortech.2010.01.011

Thacharodi, A., Hassan, S., Singh, T., Mandal, R., Chinnadurai, J., Khan, H. A., Hussain, M. A., Brindhadevi, K., & Pugazhendhi, A. (2023). Bioremediation of polycyclic aromatic hydrocarbons: An updated microbiological review. *Chemosphere*, *328*, 138498. DOI: 10.1016/j.chemosphere.2023.138498 PMID: 36996919

Thakur, R., Rajak, R., & Sandhu, S. S. (2005). Biochemical and molecular characteristics of indigenous strains of the entomopathogenic fungus Beauveria bassiana of Central India. *Biocontrol Science and Technology*, *15*(7), 733–744. DOI: 10.1080/09583150500136584

Thakur, R., & Sandhu, S. S. (2010). Distribution, occurrence and natural invertebrate hosts of indigenous entomopathogenic fungi of Central India. *Indian Journal of Microbiology*, *50*(1), 89–96. DOI: 10.1007/s12088-010-0007-z PMID: 23100813

Thambugala, K. M., Daranagama, D. A., Phillips, A. J. L., Kannangara, S. D., & Promputtha, I. (2020). Fungi vs. Fungi in Biocontrol: An Overview of Fungal Antagonists Applied Against Fungal Plant Pathogens. *Frontiers in Cellular and Infection Microbiology*, *10*, 604923. DOI: 10.3389/fcimb.2020.604923 PMID: 33330142

Thepbandit, W., & Athinuwat, D. (2024). Rhizosphere Microorganisms Supply Availability of Soil Nutrients and Induce Plant Defense. *Microorganisms*, *12*(3), 558. Advance online publication. DOI: 10.3390/microorganisms12030558 PMID: 38543610

Thines, E., Anke, H., & Weber, R. W. S. (2004). Fungal secondary metabolites as inhibitors of infection-related morphogenesis in phytopathogenic fungi. *Mycological Research*, *108*(1), 14–25. DOI: 10.1017/S0953756203008943 PMID: 15035501

Thirkell, T. J., Charters, M. D., Elliott, A. J., Sait, S. M., & Field, K. J. (2017). Are mycorrhizal fungi our sustainable saviours? Considerations for achieving food security. *Journal of Ecology*, *105*(4), 921–929. DOI: 10.1111/1365-2745.12788

Thumkasem, N., On-mee, T., Chittapun, S., Pornpukdeewattana, S., Ketudat-Cairns, M., Thongprajukaew, K., Antimanon, S., & Charoenrat, T. (2023). Rhodotorula paludigena CM33 cultivation process development for high β-carotene single cell protein production. *Biocatalysis and Agricultural Biotechnology*, *54*, 102926. DOI: 10.1016/j.bcab.2023.102926

Tian, X., Liu, X. Q., Liu, X. R., & Li, Q. S., Abd_Allah, E. F., & Wu, Q. S. (. (2023). Mycorrhizal cucumber with *Diversispora versiformis* has active heat stress tolerance by up-regulating expression of both *CsHsp70s* and *CsPIPs* genes. *Scientia Horticulturae*, *319*, 112194. DOI: 10.1016/j.scienta.2023.112194

Tilman, D., Cassman, K. G., Matson, P. A., Naylor, R., & Polasky, S. (2002). Agricultural sustainability and intensive production practices. *Nature*, *418*(6898), 671–677. DOI: 10.1038/nature01014 PMID: 12167873

Ting, A. S. Y., Mah, S. W., & Tee, C. S. (2012). Evaluating the feasibility of induced host resistance by endophytic isolate Penicillium citrinum BTF08 as a control mechanism for Fusarium wilt in banana plantlets. *Biological Control*, *61*(2), 155–159. DOI: 10.1016/j.biocontrol.2012.01.010

Tisserant, E., Malbreil, M., Kuo, A., Kohler, A., Symeonidi, A., Balestrini, R., & Martin, F. (2013). Genome of an arbuscular mycorrhizal fungus provides insight into the oldest plant symbiosis. *Proceedings of the National Academy of Sciences of the United States of America, 110*(50), 20117–20122.

Tiwari, D., Hussain, A., Tiwari, S. K., Ahmed, S., Sultan, M. W., & Ahamed, M. I. (2023). *Removal of Heavy Metals using Microbial Bioremediation*. Bentham., DOI: 10.2174/9789815123494123010004

Tiwari, P., & Park, K. I. (2024). Advanced Fungal Biotechnologies in Accomplishing Sustainable Development Goals (SDGs): What Do We Know and What Comes Next? *Journal of Fungi (Basel, Switzerland), 10*(7), 506. DOI: 10.3390/jof10070506 PMID: 39057391

Tjamos, E. C., Papavizas, G. C., & Cook, R. J. (1992). *Biological control of plant diseases: Progress and challenges for the future*. Plenum Press. DOI: 10.1007/978-1-4757-9468-7

Toleikiene, M., Arlauskiene, A., Fliessbach, A., Iqbal, S., Sarunaite, L., & Kadziuliene, Z. (2020). The decomposition of standardised organic materials in loam and clay loam arable soils during a non-vegetation period. *Soil and Water Research, 15*(3), 181–190. DOI: 10.17221/31/2019-SWR

Tongpool, R., & Pongpat, P. (2013). Analysis of shiitake environmental performance via life cycle assessment. *International Journal of Environmental Sciences and Development, 4*, 552–557. DOI: 10.7763/IJESD.2013.V4.412

Toppo, P., Subba, R., Roy, K., Mukherjee, S., & Mathur, P. (2023). Elucidating the Strategies for Isolation of Endophytic Fungi and Their Functional Attributes for the Regulation of Plant Growth and Resilience to Stress. *Journal of Plant Growth Regulation, 42*(3), 1342–1363. DOI: 10.1007/s00344-022-10638-w

Tothill, I. J. W. M. J. (2011). Biosensors and nanomaterials and their application for mycotoxin determination. *World Mycotoxin Journal, 4*(4), 361–374. DOI: 10.3920/WMJ2011.1318

Traversari, S., Cacini, S., Galieni, A., Nesi, B., Nicastro, N., & Pane, C. (2021). Precision Agriculture Digital Technologies for Sustainable Fungal Disease Management of Ornamental Plants. *Sustainability (Basel), 13*(7), 3707. DOI: 10.3390/su13073707

Tubiello, F. N. 2014. 'Agriculture, Forestry, and Other Land Use Emissions by Sources and Removals by Sinks: 1990-2011 Analysis'. DOI: 10.13140/2.1.4143.4245

Tudi, M., Ruan, H. D., Wang, L., Lyu, J., Sadler, R., Connell, D., & Chu, C. (2021). Agriculture Development, Pesticide Application and Its Impact on the Environment. *International Journal of Environmental Research and Public Health, 18*(3), 1112. DOI: 10.3390/ijerph18031112 PMID: 33513796

Tuomela, M., Vikman, M., Hatakka, A., & Itavaava, M. (2000). Biodegradation of lignin in a compost environment: A review. *Bioresource Technology, 72*(2), 169–183. DOI: 10.1016/S0960-8524(99)00104-2

Turnbull, W. H., Leeds, A. R., & Edwards, G. D. (1992). Mycoprotein reduces blood lipids in free-living subjects. *The American Journal of Clinical Nutrition, 55*(2), 415–419. DOI: 10.1093/ajcn/55.2.415 PMID: 1734679

Turnbull, W., Ar, L., & Edwards, G. (1990). Effect of mycoprotein on blood lipids. *The American Journal of Clinical Nutrition, 52*(4), 646–650. DOI: 10.1093/ajcn/52.4.646

Turner, N. W., Subrahmanyam, S., & Piletsky, S. A. (2009). Analytical methods for determination of mycotoxins: A review. *Analytica Chimica Acta*, *632*(2), 168–180. DOI: 10.1016/j.aca.2008.11.010 PMID: 19110091

Tyśkiewicz, R., Nowak, A., Ozimek, E., & Jaroszuk-Ściseł, J. (2022). Trichoderma: The Current Status of Its Application in Agriculture for the Biocontrol of Fungal Phytopathogens and Stimulation of Plant Growth. *International Journal of Molecular Sciences*, *23*(4), 2329. Advance online publication. DOI: 10.3390/ijms23042329 PMID: 35216444

U.S. Department of Agriculture, Foreign Agricultural Service. (2018). *China releases standard for maximum levels of mycotoxins in foods*. https://apps.fas.usda.gov/newgainapi/api/report/downloadreportbyfilename?filename=China%20Releases%20Standard%20for%20Maximum%20Levels%20of%20Mycotoxins%20in%20Foods%20_Beijing_China%20-%20Peoples%20Republic%20of_5-9-2018.pdf

U.S. Food and Drug Administration. (2000). *Guidance for industry: Action levels for poisonous or deleterious substances in human food and animal feed*. https://www.fda.gov/regulatory-information/search-fda-guidance-documents/guidance-industry-action-levels-poisonous-or-deleterious-substances-human-food-and-animal-feed

Uffelman, C. N., Chan, N. I., Davis, E. M., Wang, Y., McGowan, B. S., & Campbell, W. W. (2023). An Assessment of Mushroom Consumption on Cardiometabolic Disease Risk Factors and Morbidities in Humans: A Systematic Review. In Nutrients (Vol. 15, Issue 5). MDPI. DOI: 10.3390/nu15051079

Ugalde, U. O., & Castrillo, J. I. (2002). Single cell proteins from fungi and yeasts. In *Applied mycology and biotechnology* (Vol. 2, pp. 123–149). Elsevier., https://www.sciencedirect.com/science/article/pii/S1874533402800089

Ukaogo, P. O., Ewuzie, U., & Onwuka, C. V. (2020). Environmental pollution: causes, effects, and the remedies. In *Microorganisms for sustainable environment and health* (pp. 419–429). Elsevier. DOI: 10.1016/B978-0-12-819001-2.00021-8

United Nations Environment Programme (UNEP). (2023). Annual Report 2023. https://www.unep.org/resources/annual-report-2023

United Nations, Department of Economic and Social Affairs, Population Division (2017). *World Population Prospects: The 2017 Revision, Key Findings and Advance Tables.* Working Paper No. ESA/P/WP/248.

Uppala, S. S., Bowen, K. L., & Woods, F. M. (2013). Pre-harvest aflatoxin contamination and soluble sugars of peanut. *Peanut Science*, *40*(1), 40–51. DOI: 10.3146/PS12-9.1

Urbain, P., Singler, F., Ihorst, G., Biesalski, H. K., & Bertz, H. (2011). Bioavailability of vitamin D2 from UV-B-irradiated button mushrooms in healthy adults deficient in serum 25-hydroxyvitamin D: a randomized controlled trial. European Journal of Clinical Nutrition 2011 65:8, 65(8), 965–971. DOI: 10.1038/ejcn.2011.53

Urík, M., Čerňanský, S., Ševc, J., Šimonovičová, A., & Littera, P. (2007). Biovolatilization of arsenic by different fungal strains. *Water, Air, and Soil Pollution*, *186*(1-4), 337–342. DOI: 10.1007/s11270-007-9489-7

Usuki, F., & Narisawa, K. (2007). A mutualistic symbiosis between a dark septate endophytic fungus, *Heteroconium chaetospira,* and a nonmycorrhizal plant, Chinese cabbage. *Mycologia*, *99*(2), 175–184. DOI: 10.1080/15572536.2007.11832577 PMID: 17682770

Văcar, C. L., Covaci, E., Chakraborty, S., Li, B., Weindorf, D. C., Fren iu, T., Pârvu, M., & Podar, D. (2021). Heavy Metal-Resistant Filamentous Fungi as Potential Mercury Bioremediators. *Journal of Fungi (Basel, Switzerland)*, *7*(5), 386. Advance online publication. DOI: 10.3390/jof7050386 PMID: 34069296

Vadassery, J., Ritter, C., Venus, Y., Camehl, I., Varma, A., Shahollari, B., Novák, O., Strnad, M., Ludwig-Müller, J., & Oelmüller, R. (2008). The Role of Auxins and Cytokinins in the Mutualistic Interaction Between *Arabidopsis* and *Piriformospora indica.Molecular Plant-Microbe Interactions*, *21*(10), 1371–1383. DOI: 10.1094/MPMI-21-10-1371 PMID: 18785832

Vadassery, J., Tripathi, S., Prasad, R., Varma, A., & Oelmüller, R. (2009). Monodehydroascorbate reductase 2 and dehydroascorbate reductase 5 are crucial for a mutualistic interaction between Piriformospora indica and Arabidopsis. *Journal of Plant Physiology*, *166*(12), 1263–1274. DOI: 10.1016/j.jplph.2008.12.016 PMID: 19386380

Vaksmaa, A., Guerrero-Cruz, S., Ghosh, P., Zeghal, E., Hernando-Morales, V., & Niemann, H. (2023). Role of fungi in bioremediation of emerging pollutants. *Frontiers in Marine Science*, *10*, 1070905. DOI: 10.3389/fmars.2023.1070905

Valente, S., Piombo, E., Schroeckh, V., Meloni, G. R., Heinekamp, T., Brakhage, A. A., & Spadaro, D. (2021). CRISPR-Cas9-based discovery of the verrucosidin biosynthesis gene cluster in Penicillium polonicum. *Frontiers in Microbiology*, *12*, 660871. DOI: 10.3389/fmicb.2021.660871 PMID: 34093475

Valverde, M. E., Hernández-Pérez, T., & Paredes-López, O. (2015). Edible mushrooms: Improving human health and promoting quality life. *International Journal of Microbiology*, *376387*, 1–14. Advance online publication. DOI: 10.1155/2015/376387 PMID: 25685150

Van Dongen, P. W., & de Groot, A. N. (1995). History of ergot alkaloids from ergotism to ergometrine. *European Journal of Obstetrics, Gynecology, and Reproductive Biology*, *60*(2), 109–116. DOI: 10.1016/0028-2243(95)02104-Z PMID: 7641960

Van Lexmond, M. B., Bonmatin, J.-M., Goulson, D., & Noome, D. A. (2015). Worldwide integrated assessment on systemic pesticides: Global collapse of the entomofauna: exploring the role of systemic insecticides. *Environmental Science and Pollution Research International*, *22*(1), 1–4. DOI: 10.1007/s11356-014-3220-1 PMID: 25146115

Van Loo, E. J., Caputo, V., & Lusk, J. L. (2020). Consumer Preferences for Farm-raised Meat, Lab-grown Meat, and Plant-based Meat Alternatives: Does Information or Brand Matter? *Food Policy*, *95*, 101931. DOI: 10.1016/j.foodpol.2020.101931

Van Oosten, M. J., Pepe, O., De Pascale, S., Silletti, S., & Maggio, A. (2017). The role of biostimulants and bioeffectors as alleviators of abiotic stress in crop plants. *Chemical and Biological Technologies in Agriculture*, *4*(1), 5. DOI: 10.1186/s40538-017-0089-5

Van Tuinen, D., Jacquot, E., Zhao, B., Gollotte, A., & Gianinazzi Pearson, V. (1998). Characterization of root colonization profiles by a microcosm community of arbuscular mycorrhizal fungi using 25S rDNA-targeted nested PCR. *Molecular Ecology*, *7*(7), 879–887. DOI: 10.1046/j.1365-294x.1998.00410.x PMID: 9691489

Varavinit, S., Srithongkum, P., De-Eknamkul, C., Assavanig, A., & Charoensiri, K. (1996). Production of Single Cell Protein from Cassava Starch in Air-lift Fermenter by Cephalosporium eichhorniae. *Stärke*, *48*(10), 379–382. DOI: 10.1002/star.19960481007

Varghese, R., & Ray, J. G. (2023). Sustainable agriculture of tropical spices: Arbuscular mycorrhizal fungi as an ecotechnological tool-a critical review. *Journal of Applied Research on Medicinal and Aromatic Plants*, *36*, 100507. DOI: 10.1016/j.jarmap.2023.100507

Varma, A., Savita Verma, , Sudha, , Sahay, N., Bütehorn, B., & Franken, P. (1999). *Piriformospora indica*, a Cultivable Plant-Growth-Promoting Root Endophyte. *Applied and Environmental Microbiology*, *65*(6), 2741–2744. DOI: 10.1128/AEM.65.6.2741-2744.1999 PMID: 10347070

Varma, A., Sherameti, I., Tripathi, S., Prasad, R., Das, A., Sharma, M., Bakshi, M., Johnson, J. M., Bhardwaj, S., Arora, M., Rastogi, K., Agrawal, A., Kharkwal, A. C., Talukdar, S., Bagde, U. S., Bisaria, V. S., Upadhyaya, C. P., Won, P. S., Chen, Y., & Oelmüller, R. (2012). 13 The Symbiotic Fungus Piriformospora indica: Review. In *Fungal Associations* (pp. 231–254). Springer Berlin Heidelberg., DOI: 10.1007/978-3-642-30826-0_13

Vasilachi, I. C., Stoleru, V., & Gavrilescu, M. (2023). Analysis of Heavy Metal Impacts on Cereal Crop Growth and Development in Contaminated Soils. *Agriculture*, *13*(10), 1983. Advance online publication. DOI: 10.3390/agriculture13101983

Vassilev, N., & Mendes, G. de O. (2024). Soil Fungi in Sustainable Agriculture. *Microorganisms*, *12*(1), 163. DOI: 10.3390/microorganisms12010163 PMID: 38257989

Vega, F. E., Goettel, M. S., Blackwell, M., Chandler, D., Jackson, M. A., Keller, S., Koike, M., Maniania, N. K., Monzón, A., Ownley, B. H., Pell, J. K., Rangel, D. E., & Roy, H. E. (2009). Fungal entomopathogens: New insights on their ecology. *Fungal Ecology*, *2*(4), 149–159. DOI: 10.1016/j.funeco.2009.05.001

Vega, F. E., Posada, F., Aime, M. C., Pava-Ripoll, M., Infante, F., & Rehner, S. A. (2008). Entomopathogenic fungal endophytes. *Biological Control*, *46*(1), 72–82. DOI: 10.1016/j.biocontrol.2008.01.008

Venkatesh, N., & Keller, N. P. (2019). Mycotoxins in conversation with bacteria and fungi. *Frontiers in Microbiology*, *10*, 403. DOI: 10.3389/fmicb.2019.00403 PMID: 30941105

Venthodika, A., Chhikara, N., Mann, S., Garg, M. K., Sofi, S. A., & Panghal, A. (2021). Bioactive compounds of Aegle marmelos L., medicinal values and its food applications: A critical review. *Phytotherapy Research*, *35*(4), 1887–1907. DOI: 10.1002/ptr.6934 PMID: 33159390

Venturella, G., Ferraro, V., Cirlincione, F., & Gargano, M. L. (2021). Medicinal Mushrooms: Bioactive Compounds, Use, and Clinical Trials. *International Journal of Molecular Sciences*, *2*(2), 634. DOI: 10.3390/ijms22020634 PMID: 33435246

Verma, A., Shameem, N., Jatav, H. S., Sathyanarayana, E., Parray, J. A., Poczai, P., & Sayyed, R. Z. (2022). Fungal Endophytes to Combat Biotic and Abiotic Stresses for Climate-Smart and Sustainable Agriculture. *Frontiers in Plant Science, 13*, 953836. Advance online publication. DOI: 10.3389/fpls.2022.953836 PMID: 35865289

Vestergaard, S., Nielsen, C., Harding, S., & Eilenberg, J. (2002). First field trials to control Melolontha melolontha with Beauveria brongniartii in christmas trees in Denmark. *IOBC/WPRS Bulletin, 25*(7), 51–58.

Viegas, C., Nurme, J., Piecková, E., & Viegas, S. (2018). Sterigmatocystin in foodstuffs and feed: Aspects to consider. *Mycology, 11*(2), 91–104. DOI: 10.1080/21501203.2018.1492980 PMID: 32923018

Vierheilig, H. (2004). Further root colonization by arbuscular mycorrhizal fungi in already mycorrhizal plants is suppressed after a critical level of root colonization. *Journal of Plant Physiology, 161*(3), 339–341. DOI: 10.1078/0176-1617-01097 PMID: 15077632

Vijayakumar, G., Rajkumar, M., Rajiv Chandar, N., Selvakumar, P., & Duraisamy, R. (2024). Muniyandi Rajkumar, Rajiv Chandar N, Selvakumar P and Ramesh Duraisamy, (2024). Environmentally friendly TDS removal from waste water by electrochemical ion exchange batch-type recirculation (EIR) technique. *Environmental Science. Water Research & Technology, 10*(4), 826–835. DOI: 10.1039/D3EW00793F

Vinale, F., Sivasithamparam, K., Ghisalberti, E. L., Marra, R., Woo, S. L., & Lorito, M. (2008b). Trichoderma-plant-pathogen interactions. *Soil Biology & Biochemistry, 40*(1), 1–10. DOI: 10.1016/j.soilbio.2007.07.002

Vinale, F., Sivasithamparam, K., Ghisalberti, E., Marra, R., Barbetti, M., Li, H., Woo, S. L., & Lorito, M. (2008a). A novel role for Trichoderma secondary metabolites in the interactions with plants. *Physiological and Molecular Plant Pathology, 72*(1-3), 80–86. DOI: 10.1016/j.pmpp.2008.05.005

Vipotnik, Z., Michelin, M., & Tavares, T. (2022). Biodegradation of chrysene and benzo[a]pyrene and removal of metals from naturally contaminated soil by isolated Trametes versicolor strain and laccase produced thereof. *Environmental Technology & Innovation, 28*, 102737. https://doi.org/https://doi.org/10.1016/j.eti.2022.102737. DOI: 10.1016/j.eti.2022.102737

Voets, L., Dupré de Boulois, H., Renard, L., Strullu, D.-G., & Declerck, S. (2005). Development of an autotrophic culture system for the in vitro mycorrhization of potato plantlets. *FEMS Microbiology Letters, 248*(1), 111–118. DOI: 10.1016/j.femsle.2005.05.025 PMID: 15941625

Wahab, A., Muhammad, M., Munir, A., Abdi, G., Zaman, W., Ayaz, A., & Reddy, S. P. P. (2023). Role of arbuscular mycorrhizal fungi in regulating growth, enhancing productivity, and potentially influencing ecosystems under abiotic and biotic stresses. *Plants, 12*(17), 3102. DOI: 10.3390/plants12173102 PMID: 37687353

Walker, C. (1995). AM or VAM: What's in a Word? In Varma, A., & Hock, B. (Eds.), *Mycorrhiza: Structure, Function, Molecular Biology and Biotechnology* (pp. 25–26). Springer Berlin Heidelberg., DOI: 10.1007/978-3-662-08897-5_2

Waller, F., Achatz, B., Baltruschat, H., Fodor, J., Becker, K., Fischer, M., Heier, T., Hückelhoven, R., Neumann, C., Von Wettstein, D., Franken, P., & Kogel, K.-H. (2005). The endophytic fungus Piriformospora indica reprograms barley to salt-stress tolerance, disease resistance, and higher yield. *Proceedings of the National Academy of Sciences of the United States of America*, *102*(38), 13386–13391. DOI: 10.1073/pnas.0504423102 PMID: 16174735

Wang, M., & Zhao, R. (2023). A review on nutritional advantages of edible mushrooms and its industrialization development situation in protein meat sanalogues. In Journal of Future Foods (Vol. 3, Issue 1, pp. 1–7). Beijing Academy of Food Sciences. DOI: 10.1016/j.jfutfo.2022.09.001

Wang, B., & Qiu, Y. L. (2006). Phylogenetic distribution and evolution of mycorrhizas in land plants. *Mycorrhiza*, *16*(5), 299–363. DOI: 10.1007/s00572-005-0033-6 PMID: 16845554

Wang, B., Yeun, L. H., Xue, J., Liu, Y., Ané, J., & Qiu, Y. (2010). Presence of three mycorrhizal genes in the common ancestor of land plants suggests a key role of mycorrhizas in the colonization of land by plants. *The New Phytologist*, *186*(2), 514–525. DOI: 10.1111/j.1469-8137.2009.03137.x PMID: 20059702

Wang, C., & St Leger, R. J. (2005). Developmental and transcriptional responses to host and nonhost cuticles by the specific locust pathogen Metarhizium anisopliae var. acridum. *Eukaryotic Cell*, *4*(5), 937–947. DOI: 10.1128/EC.4.5.937-947.2005 PMID: 15879528

Wang, D., Jin, S., Lu, Q., & Chen, Y. (2023). Advances and Challenges in CRISPR/Cas-Based Fungal Genome Engineering for Secondary Metabolite Production: A Review. *Journal of Fungi (Basel, Switzerland)*, *9*(3), 362. DOI: 10.3390/jof9030362 PMID: 36983530

Wang, E., Yu, N., Bano, S. A., Liu, C., Miller, A. J., Cousins, D., Zhang, X., Ratet, P., Tadege, M., Mysore, K. S., Downie, J. A., Murray, J. D., Oldroyd, G. E. D., & Schultze, M. (2014). A H+-ATPase that energizes nutrient uptake during mycorrhizal symbioses in rice and *Medicago truncatula.The Plant Cell*, *26*(4), 1818–1830. DOI: 10.1105/tpc.113.120527 PMID: 24781115

Wang, F., Li, X., Li, Y., Han, J., Chen, Y., Zeng, J., Su, M., Zhuo, J., Ren, H., Liu, H., Hou, L., Fan, Y., Yan, X., Song, S., Zhao, J., Jin, D., Zhang, M., & Pei, Y. (2021). Arabidopsis P4 ATPase-mediated cell detoxification confers resistance to *Fusarium graminearum* and *Verticillium dahliae.Nature Communications*, *12*(1), 6426. DOI: 10.1038/s41467-021-26727-5 PMID: 34741039

Wang, F., Sun, Y., & Shi, Z. (2019). Arbuscular mycorrhiza enhances biomass production and salt tolerance of sweet sorghum. *Microorganisms*, *7*(9), 289. DOI: 10.3390/microorganisms7090289 PMID: 31450847

Wang, G., Jin, Z., George, T. S., Feng, G., & Zhang, L. (2023). Arbuscular mycorrhizal fungi enhance plant phosphorus uptake through stimulating hyphosphere soil microbiome functional profiles for phosphorus turnover. *The New Phytologist*, *238*(6), 2578–2593. DOI: 10.1111/nph.18772 PMID: 36694293

Wang, H., Zhao, R., Zhao, D., Liu, S., Fu, J., Zhang, Y., Dai, N., Song, D., & Ding, H. (2022). Microbial-Mediated Emissions of Greenhouse Gas from Farmland Soils: A Review. *Processes (Basel, Switzerland)*, *10*(11), 2361. DOI: 10.3390/pr10112361

Wang, J., Yu, J., Yang, Q., McDermott, J., Scott, A., Vukovich, M., Lagrois, R., Gong, Q., Greenleaf, W., Eisenstein, M., Ferguson, B. S., & Soh, H. T. (2017). Multiparameter particle display (MPPD): A quantitative screening method for the discovery of highly specific aptamers. *Angewandte Chemie*, *129*(3), 762–765. DOI: 10.1002/ange.201608880 PMID: 27933702

Wang, X., Selvam, A., & Wong, J. W. C. (2016). Influence of Lime on Struvite Formation and Nitrogen Conservation during Food Waste Composting. *Bioresource Technology*, *217*, 227–232. DOI: 10.1016/j.biortech.2016.02.117 PMID: 27020123

Wang, Y. J., He, X. H., Meng, L. L., Zou, Y. N., & Wu, Q. S. (2023). Extraradical mycorrhizal hyphae promote soil carbon sequestration through difficultly extractable glomalin-related soil protein in response to soil water stress. *Microbial Ecology*, *86*(2), 1023–1034. DOI: 10.1007/s00248-022-02153-y PMID: 36471016

Wang, Y., Liu, Z., Hao, X., Wang, Z., Wang, Z., Liu, S., Tao, C., Wang, D., Wang, B., Shen, Z., Shen, Q., & Li, R. (2023). Biodiversity of the beneficial soil-borne fungi steered by Trichoderma-amended biofertilizers stimulates plant production. *NPJ Biofilms and Microbiomes*, *9*(1), 46. DOI: 10.1038/s41522-023-00416-1 PMID: 37407614

Wang, Y., Zhang, B., Chen, N., Wang, C., Feng, S., & Xu, H. (2018). Combined bioremediation of soil co-contaminated with cadmium and endosulfan by *Pleurotus eryngii* and *Coprinus comatus.Journal of Soils and Sediments*, *18*(6), 2136–2147. DOI: 10.1007/s11368-017-1762-9

Wani, B. A., Bodha, R. H., & Wani, A. H. (2010). Nutritional and medicinal importance of mushrooms. In Journal of Medicinal Plants Research (Vol. 4, Issue 24, pp. 2598–2604). Academic Journals. DOI: 10.5897/JMPR09.565

Wani, K. I., Zehra, A., Choudhary, S., Naeem, M., Khan, M. M. A., Castroverde, C. D. M., & Aftab, T. (2020). Mechanistic insights into strigolactone biosynthesis, signaling, and regulation during plant growth and development. *Journal of Plant Growth Regulation*, ●●●, 1–17.

Wanxiao, G., Gang, Y., Xiaohui, R., Yuansen, H., Hongwei, W., Yanchao, F., Guanzhang, B., Jun, X., Zhihong, L., Jingmei, H., Min, M., Qian, X., Yapei, D., Yanbing, C., & Congle, Y. (2015). Pichia pastoris for producing methanol protein and lipase at same time and application thereof (CN 105132301 A). https://lens.org/056-903-998-744-571

Ward, O. P., Qin, W. M., Dhanjoon, J., Ye, J., & Singh, A. (2005). Physiology and biotechnology of Aspergillus. *Advances in Applied Microbiology*, *58*, 1–75. DOI: 10.1016/S0065-2164(05)58001-8

Wei, X., Xie, B., Wan, C., Song, R., Zhong, W., Xin, S., & Song, K. (2024). Enhancing Soil Health and Plant Growth through Microbial Fertilizers: Mechanisms, Benefits, and Sustainable Agricultural Practices. In *Agronomy* (Vol. 14, Issue 3). DOI: 10.3390/agronomy14030609

Wei, D., Wei, S., Peng, A., Yang, C., & Chen, C. (2022). Responses of bacterial communities in soils under winter wheat to nightly warming and nitrogen addition. *Agronomy (Basel)*, *12*(7), 1616. DOI: 10.3390/agronomy12071616

Weinberg, E. D. (1999). The Role of Iron In Protozoan and Fungal Infectious Diseases. *The Journal of Eukaryotic Microbiology*, *46*(3), 231–238. DOI: 10.1111/j.1550-7408.1999.tb05119.x PMID: 10377984

Weindling, R. (1941). Experimental consideration of the mold-toxins of Gliocladium and Trichoderma. *Phytopathology*, *31*(11), 991–1003.

Wei, X., Zhang, W., Zulfiqar, F., Zhang, C., & Chen, J. (2022). Ericoid mycorrhizal fungi as biostimulants for improving propagation and production of ericaceous plants. *Frontiers in Plant Science*, *13*, 1027390. Advance online publication. DOI: 10.3389/fpls.2022.1027390 PMID: 36466284

Wei, Y., Wang, L., & Liu, J. (2023). The diabetogenic effects of pesticides: Evidence based on epidemiological and toxicological studies. *Environmental Pollution*, *331*, 121927. DOI: 10.1016/j.envpol.2023.121927 PMID: 37268216

Wei, Y., Yang, H., Hu, J., Li, H., Zhao, Z., Wu, Y., Li, J., Zhou, Y., Yang, K., & Yang, H. (2023). Trichoderma harzianum inoculation promotes sweet sorghum growth in the saline soil by modulating rhizosphere available nutrients and bacterial community. *Frontiers in Plant Science*, *14*, 1258131. Advance online publication. DOI: 10.3389/fpls.2023.1258131 PMID: 37771481

Weng, W., Yan, J., Zhou, M., Yao, X., Gao, A., Ma, C., Cheng, J., & Ruan, J. (2022). Roles of Arbuscular mycorrhizal Fungi as a Biocontrol Agent in the Control of Plant Diseases. *Microorganisms*, *10*(7), 1266. Advance online publication. DOI: 10.3390/microorganisms10071266 PMID: 35888985

Whitelaw, M. A. (2000). Growth promotion of plants inoculated with phosphate-solubilizing fungi. *Advances in Agronomy*, *69*, 99–151. DOI: 10.1016/S0065-2113(08)60948-7

White, P. J., & Brown, P. H. (2010). Plant nutrition for sustainable development and global health. *Annals of Botany*, *105*(7), 1073–1080. DOI: 10.1093/aob/mcq085 PMID: 20430785

Whittaker, J. A., Johnson, R. I., Finnigan, T. J., Avery, S. V., & Dyer, P. S. (2020). The Biotechnology of Quorn Mycoprotein: Past, Present and Future Challenges. In *Grand challenges fungal biotechnology* (pp. 59–79). Springer. DOI: 10.1007/978-3-030-29541-7_3

Wiebe, M. (2002). Myco-protein from Fusarium venenatum: A well-established product for human consumption. *Applied Microbiology and Biotechnology*, *58*(4), 421–427. DOI: 10.1007/s00253-002-0931-x

Wiebe, M. G. (2004). Quorn™ mycoprotein–Overview of a successful fungal product. *The Mycologist*, *18*(1), 17–20. DOI: 10.1017/S0269915X04001089

Wiest, A., Grzegorski, D., Xu, B. W., Goulard, C., Rebuffat, S., Ebbole, D. J., Bodo, B., & Kenerley, C. (2002). Identification of peptaibols from Trichoderma virens and cloning of a peptaibol synthetase. *The Journal of Biological Chemistry*, *277*(23), 20862–20868. DOI: 10.1074/jbc.M201654200 PMID: 11909873

Wild, C. P., & Gong, Y. Y. (2010). Mycotoxins and human disease: A largely ignored global health issue. *Carcinogenesis*, *31*(1), 71–82. DOI: 10.1093/carcin/bgp264 PMID: 19875698

Wildman, H. G. (1999). *Pharmaceutical bioprospecting and its relationship to the conservation and utilization of bioresources*. IUPAC.

Williams, J. H., Phillips, T. D., Jolly, P. E., Stiles, J. K., Jolly, C. M., & Aggarwal, D. (2004). Human aflatoxicosis in developing countries: A review of toxicology, exposure, potential health consequences, and interventions. *The American Journal of Clinical Nutrition*, *80*(5), 1106–1122. DOI: 10.1093/ajcn/80.5.1106 PMID: 15531656

Willing, C. E., Wan, J., Yeam, J. J., Cessna, A. M., & Peay, K. G. (2024). Arbuscular mycorrhizal fungi equalize differences in plant fitness and facilitate plant species coexistence through niche differentiation. *Nature Ecology & Evolution*, *8*(11), 2058–2071. DOI: 10.1038/s41559-024-02526-1 PMID: 39251818

Wilson, G. W. T., Rice, C. W., Rillig, M. C., Springer, A., & Hartnett, D. C. (2009). Soil aggregation and carbon sequestration are tightly correlated with the abundance of arbuscular mycorrhizal fungi: Results from long-term field experiments. *Ecology Letters*, *12*(5), 452–461. DOI: 10.1111/j.1461-0248.2009.01303.x PMID: 19320689

Wisniewski, M., Droby, S., Norelli, J., Liu, J., & Schena, L. (2016). Alternative management technologies for postharvest disease control: The journey from simplicity to complexity. *Postharvest Biology and Technology*, *122*, 3–10. DOI: 10.1016/j.postharvbio.2016.05.012

Woo, S. L., Scala, F., Ruocco, M., & Lorito, M. (2006). The molecular biology of the interactions between Trichoderma spp., phytopathogenic fungi, and plants. *Phytopathology*, *96*(2), 181–185. DOI: 10.1094/PHYTO-96-0181 PMID: 18943922

Wraight, S. P., Jackson, M. A., & Kokc, S. L. (2001). Production, stabilization and formulation of fungal biocontrol agents. Fungi as biocontrol agents: progress, problems and potential, 253-287. DOI: 10.1079/9780851993560.0253

Wraight, S., Carruthers, R., Jaronski, S., Bradley, C., Garza, C., & Galaini-Wraight, S. (2000). Evaluation of the Entomopathogenic fungi Beauveria bassiana and Paecilomyces fumosoroseus for microbial control of the Silverleaf Whitefly, Bemisia argentifolii. *Biological Control*, *17*(3), 203–217. DOI: 10.1006/bcon.1999.0799

Wu, F. (2015). Global impacts of aflatoxin in maize: Trade and human health. *World Mycotoxin Journal*, *8*(2), 137–142. DOI: 10.3920/WMJ2014.1737

Wu, F. (2022). Mycotoxin risks are lower in biotech corn. *Current Opinion in Biotechnology*, *78*, 102792. DOI: 10.1016/j.copbio.2022.102792 PMID: 36088737

Wu, J., Hu, J., Zhao, S., He, M., Hu, G., Ge, X., & Peng, N. (2018). Single-cell Protein and Xylitol Production by a Novel Yeast Strain Candida intermedia FL023 from Lignocellulosic Hydrolysates and Xylose. *Applied Biochemistry and Biotechnology*, *185*(1), 163–178. DOI: 10.1007/s12010-017-2644-8

Wu, J., & Yu, H. Q. (2007). Biosorption of 2, 4-dichlorophenol by immobilized white-rot fungus Phanerochaete chrysosporium from aqueous solutions. *Bioresource Technology*, *98*(2), 253–259. DOI: 10.1016/j.biortech.2006.01.018 PMID: 16517157

Wu, Q.-S., Srivastava, A. K., & Zou, Y.-N. (2013). AMF-induced tolerance to drought stress in citrus: A review. *Scientia Horticulturae*, *164*, 77–87. https://doi.org/https://doi.org/10.1016/j.scienta.2013.09.010. DOI: 10.1016/j.scienta.2013.09.010

Xiao, P., & Kondo, R. (2020). Potency of Phlebia species of white rot fungi for the aerobic degradation, transformation and mineralization of lindane. *journal of microbiology, 58*, 395-404.

Xiao, J., Hu, C., Chen, Y., Yang, B., & Hua, J. (2014). Effects of low magnesium and an arbuscular mycorrhizal fungus on the growth, magnesium distribution and photosynthesis of two citrus cultivars. *Scientia Horticulturae*, *177*, 14–20. DOI: 10.1016/j.scienta.2014.07.016

Xiao, P., Mori, T., Kamei, I., & Kondo, R. (2011). Metabolism of organochlorine pesticide heptachlor and its metabolite heptachlor epoxide by white rot fungi, belonging to genus *Phlebia.FEMS Microbiology Letters*, *314*(2), 140–146. DOI: 10.1111/j.1574-6968.2010.02152.x PMID: 21087297

Xiao, Z., Zhao, Q., Li, W., Gao, L., & Liu, G. (2023). Strain improvement of Trichoderma harzianum for enhanced biocontrol capacity: Strategies and prospects. *Frontiers in Microbiology*, *14*, 1146210. Advance online publication. DOI: 10.3389/fmicb.2023.1146210 PMID: 37125207

Xie, D., Zhou, Y., Wang, Y., Xu, L., Yang, C., Zhang, G., Zhang, H., Du, Q., & Jin, P. (2024). Insight into sensory characteristics and volatile differences of freeze-dried *Citrus aurantium* L. var. amara Engl. flower essential oil-casein phospholipid nanocomposites sustained-release scented tea by combining GC-E-nose and SBSE-GC-MS. *Food Control*, *166*, 110714. DOI: 10.1016/j.foodcont.2024.110714

Xie, Y., Wang, F., Wang, K., Yue, H., & Lan, X. (2020). Responses of bacterial phoD gene abundance and diversity to crop rotation and feedbacks to phosphorus uptake in wheat. *Applied Soil Ecology*, *154*, 103604. DOI: 10.1016/j.apsoil.2020.103604

Xie, Y., Yang, F., Tan, W., Li, X., Jiao, C., Huang, R., & Yang, B. B. (2016). The anti-cancer components of Ganoderma lucidum possesses cardiovascular protective effect by regulating circular RNA expression. *Oncoscience*, *3*(7–8), 203–207. DOI: 10.18632/oncoscience.316 PMID: 27713910

Xin, W., & Le, G. (2023). *Pichia pastoris with multi-carbon-source utilization and high-protein synthesis and application of Pichia pastoris* (CN 116478838 A). https://lens.org/122-059-031-651-763

Xing, F., Liu, X., Wang, L., Selvaraj, J. N., Jin, N., Wang, Y., Zhao, Y., & Liu, Y. (2017). Distribution and variation of fungi and major mycotoxins in pre-and post-nature drying maize in North China Plain. *Food Control*, *80*, 244–251. DOI: 10.1016/j.foodcont.2017.03.055

Xing, H.-S., & Sun, P.-F., LI, F., LI, W., & Guo, S.-X. (2019). Effects of arbuscular mycorrhizal fungi on heat tolerance of *Lavandula angustifolia.Junwu Xuebao*, *38*(5), 698–706.

Xue, L., Klinnawee, L., Zhou, Y., Saridis, G., Vijayakumar, V., Brands, M., Dörmann, P., Gigolashvili, T., Turck, F., & Bucher, M. (2018). *AP2* transcription factor *CBX1* with a specific function in symbiotic exchange of nutrients in mycorrhizal Lotus japonicus. *Proceedings of the National Academy of Sciences of the United States of America*, *115*(39), E9239–E9246. DOI: 10.1073/pnas.1812275115 PMID: 30209216

Xu, F., Chen, P., Li, H., Qiao, S., Wang, J., Wang, Y., Wang, X., Wu, B., Liu, H., Wang, C., & Xu, H. (2021). Comparative transcriptome analysis reveals the differential response to cadmium stress of two *Pleurotus* fungi: *Pleurotus cornucopiae* and *Pleurotus ostreatus.Journal of Hazardous Materials*, *416*, 125814. DOI: 10.1016/j.jhazmat.2021.125814 PMID: 33866290

Xu, G., Pan, J., Rehman, M., Li, X., Cao, S., Wang, C., Wang, X., Chen, C., Nie, J., Wang, M., Li, R., Li, R., & Chen, P. (2024). Arbuscular mycorrhizal fungi-mediated drought stress tolerance in kenaf (*Hibiscus cannabinus* L.): A mechanistic approach. *Plant Growth Regulation*, *103*(3), 1–22. DOI: 10.1007/s10725-024-01150-0

Xu, J. (2006). Microbial ecology in the age of genomics and metagenomics: Concepts, tools, and recent advances. *Molecular Ecology*, *15*(7), 1713–1731. DOI: 10.1111/j.1365-294X.2006.02882.x PMID: 16689892

Xu, L., Zhang, Z., Zhang, Q., & Li, P. (2016). Mycotoxin determination in foods using advanced sensors based on antibodies or aptamers. *Toxins*, *8*(8), 239. DOI: 10.3390/toxins8080239 PMID: 27529281

Xu, Q., Zhang, T., Niu, Y., Mukherjee, S., Abou-Elwafa, S. F., Nguyen, N. S. H., al Aboud, N. M., Wang, Y., Pu, M., Zhang, Y., Tran, H. T., Almazroui, M., Hooda, P. S., Bolan, N. S., Rinklebe, J., & Shaheen, S. M. (2024). A comprehensive review of agricultural waste utilization through sustainable conversion techniques, with a focus on the additives effect on the fate of phosphorus and toxic elements during composting process. *The Science of the Total Environment*, *942*, 173567. DOI: 10.1016/j.scitotenv.2024.173567 PMID: 38848918

Xu, X., Wang, X., Cleary, M., Wang, P., Lu, N., Sun, Y., & Rönnberg, J. (2020). Slope position rather than thinning intensity affects arbuscular mycorrhizal fungi (AMF) community in Chinese fir plantations. *Forests*, *11*(3), 273. DOI: 10.3390/f11030273

Yadav, A., & Yadav, K. (2024). Challenges and Opportunities in Biofertilizer Commercialization. *SVOA Microbiology, 5*(1), 01–14. DOI: 10.58624/SVOAMB.2024.05.037

Yadav, J. S. S., Bezawada, J., Ajila, C. M., Yan, S., Tyagi, R. D., & Surampalli, R. Y. (2014). Mixed culture of Kluyveromyces marxianus and Candida krusei for single-cell protein production and organic load removal from whey. *Bioresource Technology*, *164*, 119–127. DOI: 10.1016/j.biortech.2014.04.069

Yadav, J. S. S., Yan, S., Ajila, C. M., Bezawada, J., Tyagi, R. D., & Surampalli, R. Y. (2016). Food-grade single-cell protein production, characterization and ultrafiltration recovery of residual fermented whey proteins from whey. *Food and Bioproducts Processing*, *99*, 156–165. DOI: 10.1016/j.fbp.2016.04.012

Yadav, P., Rai, S., Singh, M., & Mishra, V. (2021). Mycoremediation of environmental pollutants: A review with special emphasis on mushrooms. *Environmental Sustainability*, *4*(4), 605–618. Advance online publication. DOI: 10.1007/s42398-021-00197-3

Yahfoufi, N., Alsadi, N., Jambi, M., & Matar, C. (2018). The Immunomodulatory and Anti-Inflammatory Role of Polyphenols. Nutrients 2018, Vol. 10, Page 1618, 10(11), 1618. DOI: 10.3390/nu10111618

Yakhin, O. I., Lubyanov, A. A., Yakhin, I. A., & Brown, P. H. (2017). Biostimulants in Plant Science: A Global Perspective. *Frontiers in Plant Science*, *7*. Advance online publication. DOI: 10.3389/fpls.2016.02049 PMID: 28184225

Yang, B., Ma, H.-Y., Wang, X.-M., Jia, Y., Hu, J., Li, X., & Dai, C.-C. (2014). Improvement of nitrogen accumulation and metabolism in rice (Oryza sativa L.) by the endophyte Phomopsis liquidambari. *Plant Physiology and Biochemistry*, *82*, 172–182. DOI: 10.1016/j.plaphy.2014.06.002 PMID: 24972305

Yang, B., Wang, X.-M., Ma, H.-Y., Yang, T., Jia, Y., Zhou, J., & Dai, C.-C. (2015). Fungal endophyte Phomopsis liquidambari affects nitrogen transformation processes and related microorganisms in the rice rhizosphere. *Frontiers in Microbiology*, *6*. Advance online publication. DOI: 10.3389/fmicb.2015.00982 PMID: 26441912

Yang, H., Fang, C., Li, Y., Wu, Y., Fransson, P., Rillig, M. C., & Weih, M. (2022). Temporal complementarity between roots and mycorrhizal fungi drives wheat nitrogen use efficiency. *The New Phytologist*, *236*(3), 1168–1181. DOI: 10.1111/nph.18419 PMID: 35927946

Yang, H., Wang, Y., Zhang, Z., Yan, R., & Zhu, D. (2014). Whole-Genome Shotgun Assembly and Analysis of the Genome of *Shiraia* sp. Strain Slf14, a Novel Endophytic Fungus Producing Huperzine A and Hypocrellin A. *Genome Announcements*, *2*(1), e00011–e00014. DOI: 10.1128/genomeA.00011-14 PMID: 24503982

Yang, J., Yue, H. R., Pan, L. Y., Feng, J. X., Zhao, S., Suwannarangsee, S., Champreda, V., Liu, C.-G., & Zhao, X. Q. (2023). Fungal strain improvement for efficient cellulase production and lignocellulosic biorefinery: Current status and future prospects. *Bioresource Technology*, *385*, 129449. DOI: 10.1016/j.biortech.2023.129449 PMID: 37406833

Yang, Y., Han, X., Liang, Y., Ghosh, A., Chen, J., & Tang, M. (2015). The combined effects of arbuscular mycorrhizal fungi (AMF) and lead (Pb) stress on Pb accumulation, plant growth parameters, photosynthesis, and antioxidant enzymes in Robinia pseudoacacia L. *PLoS One*, *10*(12), e0145726. DOI: 10.1371/journal.pone.0145726 PMID: 26698576

Yang, Y., Luo, W., Xu, J., Guan, P., Chang, L., Wu, X., & Wu, D. (2022). Fallow Land Enhances Carbon Sequestration in Glomalin and Soil Aggregates Through Regulating Diversity and Network Complexity of Arbuscular Mycorrhizal Fungi Under Climate Change in Relatively High-Latitude Regions. *Frontiers in Microbiology*, *13*, 930622. Advance online publication. DOI: 10.3389/fmicb.2022.930622 PMID: 35859742

Yan, J., Han, B., Gui, X., Wang, G., Xu, L., Yan, Y., Madzak, C., Pan, D., Wang, Y., Zha, G., & Jiao, L. (2018). Engineering Yarrowia lipolytica to simultaneously produce lipase and single cell protein from agro-industrial wastes for feed. *Scientific Reports*, *8*(1), 758. DOI: 10.1038/s41598-018-19238-9

Yao, K. Y., Zhang, T. Z., Wang, H. F., & Liu, J. X. (2018). Upgrading of by-product from beverage industry through solid-state fermentation with Candida utilis and Bacillus subtilis. *Letters in Applied Microbiology*, *67*(6), 557–563. DOI: 10.1111/lam.13078

Yao, X., Guo, H., Zhang, K., Zhao, M., Ruan, J., & Chen, J. (2023a). Trichoderma and its role in biological control of plant fungal and nematode disease. *Frontiers in Microbiology*, *14*, 1160551. Advance online publication. DOI: 10.3389/fmicb.2023.1160551 PMID: 37206337

Yarzábal Rodríguez, L. A., Álvarez Gutiérrez, P. E., Gunde-Cimerman, N., Ciancas Jiménez, J. C., Gutiérrez-Cepeda, A., Ocaña, A. M. F., & Batista-García, R. A. (2024). Exploring extremophilic fungi in soil mycobiome for sustainable agriculture amid global change. *Nature Communications*, *15*(1), 6951. DOI: 10.1038/s41467-024-51223-x PMID: 39138171

Yarzábal, L. A., & Chica, E. J. (2021). Microbial-Based Technologies for Improving Smallholder Agriculture in the Ecuadorian Andes: Current Situation, Challenges, and Prospects. *Frontiers in Sustainable Food Systems*, *5*, 617444. Advance online publication. DOI: 10.3389/fsufs.2021.617444

Yehia, R. S. (2023). Highlighting the potential for crude oil bioremediation of locally isolated Cunninghamella echinulata and Mucor circinelloides. *Brazilian Journal of Microbiology : Publication of the Brazilian Society for Microbiology*, *54*(3), 1969–1981. DOI: 10.1007/s42770-023-01008-z PMID: 37249816

Yin, Z. W., Fan, B. Q., & Ren, P. (2011). Isolation and identification of a cellulose degrading fungus Y5 and its capability of degrading wheat straw. *Huan Jing Ke Xue*, *32*, 247–252.

Young, J. C., & Games, D. E. (1994). Analysis of Fusarium mycotoxins by gas chromatography-Fourier transform infrared spectroscopy. *Journal of Chromatography. A*, *663*(2), 211–218. DOI: 10.1016/0021-9673(94)85247-2 PMID: 8173667

Yousufi, M. K. (2012). Impact of pH on the single cell protein produced on okara-wheat grit substrates using Rhizopus oligosporus and Aspergillus oryzae. *IOSR J. Environ. Sci. Toxicol. Food Technol*, *1*, 32-35. ISSN: 2319-2402, ISBN: 2319-2399. Volume 1, Issue 2 (Sep-Oct. 2012), PP 32-35 www.iosrjournals.org

Yuan, M. M., Kakouridis, A., Starr, E., Nguyen, N., Shi, S., Pett-Ridge, J., Nuccio, E., Zhou, J., & Firestone, M. (2021). Fungal-Bacterial Cooccurrence Patterns Differ between Arbuscular Mycorrhizal Fungi and Nonmycorrhizal Fungi across Soil Niches. *mBio*, *12*(2), 1–19. DOI: 10.1128/mBio.03509-20 PMID: 33879589

Yuan, Y., Feng, H., Wang, L., Li, Z., Shi, Y., Zhao, L., Feng, Z., & Zhu, H. (2017). Potential of Endophytic Fungi Isolated from Cotton Roots for Biological Control against Verticillium Wilt Disease. *PLoS One*, *12*(1), e0170557. DOI: 10.1371/journal.pone.0170557 PMID: 28107448

Yuan, Y., & Sun, Y. (2024). Practices, Challenges, and Future of Digital Transformation in Smallholder Agriculture: Insights from a Literature Review. *Agriculture*, *14*(12), 2193. DOI: 10.3390/agriculture14122193

Yu, Q., Guo, M., Zhang, B., Wu, H., Zhang, Y., & Zhang, L. (2020). Analysis of nutritional composition in 23 kinds of edible fungi. *Journal of Food Quality*, *2020*, 8821315. DOI: 10.1155/2020/8821315

Yusuf, Ç., & Peker, K. (2009). Benefit/cost analysis of mushroom production for diversification of income in developing countries. https://www.researchgate.net/publication/267956799

Yuvaraj, M., & Ramasamy, M. (2020). Role of Fungi in Agriculture. In Mahyar Mirmajlessi, S., & Radhakrishnan, R. (Eds.), *Biostimulants in Plant Science*. IntechOpen., DOI: 10.5772/intechopen.89718

Zadravec, M., Markov, K., Lešić, T., Frece, J., Petrović, D., & Pleadin, J. (2022). Biocontrol methods in avoidance and downsizing of mycotoxin contamination of food crops. *Processes (Basel, Switzerland)*, *10*(4), 655. DOI: 10.3390/pr10040655

Zahuri, A. A., Wan Mohtar, W. H. M., Hanafiah, Z. M., Abdul Patah, M. F., Show, P.-L., Gafforov, Y., & Wan-Mohtar, W. A. A. Q. I. (2024). Mycoremediation of Industrial Textile Wastewater Using Ganoderma lucidum Pellets and Activated Dolomite in Batch Bioreactor. *Molecular Biotechnology*. Advance online publication. DOI: 10.1007/s12033-023-01035-z PMID: 38286973

Zain, M. E. (2011). Impact of mycotoxins on humans and animals. *Journal of Saudi Chemical Society*, *15*(2), 129–144. DOI: 10.1016/j.jscs.2010.06.006

Zaki, A. G., Hasanien, Y. A., & Abdel-Razek, A. S. (2022). Biosorption optimization of lead(II) and cadmium(II) ions by two novel nanosilica-immobilized fungal mutants. *Journal of Applied Microbiology*, *133*(2), 987–1000. DOI: 10.1111/jam.15624 PMID: 35578549

Zaman, M. U., Pandit, B. A., Kumar, R., & Sofi, T. A. (2023). Mycoremediation: An Innovative and Sustainable Approach for Environmental Restoration. *International Journal for Research in Applied Science and Engineering Technology*, *11*(6), 2407–2413. DOI: 10.22214/ijraset.2023.53736

Zaynab, M., Fatima, M., Abbas, S., Sharif, Y., Umair, M., Zafar, M. H., & Bahadar, K. (2018). Role of secondary metabolites in plant defense against pathogens. *Microbial Pathogenesis*, *124*, 198–202. DOI: 10.1016/j.micpath.2018.08.034 PMID: 30145251

Zeilinger, S., Gruber, S., Bansal, R., & Mukherjee, P. K. (2016). Secondary metabolism in Trichoderma – Chemistry meets genomics. *Fungal Biology Reviews*, *30*(2), 74–90. DOI: 10.1016/j.fbr.2016.05.001

Zeng, G. M., Yu, H. Y., Huang, H. L., Chen, Y. N., & Huang, G. H. (2006). Laccase activities of a soil fungus Penicilluim simplicissimum in relation to lignin degradation. *World Journal of Microbiology & Biotechnology*, *22*(4), 317–324. DOI: 10.1007/s11274-005-9025-0

Zeng, H., Yu, L., Liu, P., Wang, Z., Chen, Y., & Wang, J. (2021). Nitrogen fertilization has a stronger influence than cropping pattern on AMF community in maize/soybean strip intercropping systems. *Applied Soil Ecology*, *167*, 104034. DOI: 10.1016/j.apsoil.2021.104034

Zhang, Y., Venkitasamy, C., Pan, Z., & Wang, W. (2013). Recent developments on umami ingredients of edible mushrooms - A review. In Trends in Food Science and Technology (Vol. 33, Issue 2, pp. 78–92). DOI: 10.1016/j.tifs.2013.08.002

Zhang, Y., Zhang, K., & Hyde, K. (2014). The ecology of Nematophagous fungi in natural environments. Fungal Diversity Research Series, 211-229. DOI: 10.1007/978-94-017-8730-7_4

Zhang, C., Guan, X., Yu, S., Zhou, J., & Chen, J. (2023). Production of Meat Alternatives Using Live Cells, Cultures and Plant Proteins. *Current Opinion in Food Science*, *43*, 43–52. DOI: 10.1016/j.cofs.2021.11.002

Zhang, C., Jiang, J., Li, J., Zhang, J., Zhang, X., & Wang, H. (2023). Long transportation duration affects nutrient composition, mycotoxins and microbial community in whole-plant corn silage. *Frontiers in Veterinary Science*, *10*, 1189358. DOI: 10.3389/fvets.2023.1189358 PMID: 37275604

Zhang, D., Satake, M., Fukuzawa, S., Sugahara, K., Niitsu, A., Shirai, T., & Tachibana, K. (2012). Two new indole alkaloids, 2-(3,3-dimethylprop-1-ene)-costaclavine and 2-(3,3-dimethylprop-1-ene)-epicostaclavine, from the marine-derived fungus Aspergillus fumigatus. *Journal of Natural Medicines*, *66*(1), 222–226. DOI: 10.1007/s11418-011-0565-3 PMID: 21792727

Zhang, F., Jia-Dong, H. E., Qiu-Dan, N. I., Qiang-Sheng, W. U., & Ying-Ning, Z. O. U. (2018). Enhancement of drought tolerance in trifoliate orange by mycorrhiza: Changes in root sucrose and proline metabolisms. *Notulae Botanicae Horti Agrobotanici Cluj-Napoca*, *46*(1), 270–276. DOI: 10.15835/nbha46110983

Zhang, F., Labourel, A., Haon, M., Kemppainen, M., Da Silva Machado, E., Brouilly, N., Veneault-Fourrey, C., Kohler, A., Rosso, M., Pardo, A., Henrissat, B., Berrin, J.-G., & Martin, F. (2022). The ectomycorrhizal basidiomycete *Laccaria bicolo*r releases a GH28 polygalacturonase that plays a key role in symbiosis establishment. *The New Phytologist*, *233*(6), 2534–2547. DOI: 10.1111/nph.17940 PMID: 34942023

Zhang, J., Diao, F., Hao, B., Xu, L., Jia, B., Hou, Y., Ding, S., & Guo, W. (2023). Multiomics reveals Claroideoglomus etunicatum regulates plant hormone signal transduction, photosynthesis and La compartmentalization in maize to promote growth under La stress. *Ecotoxicology and Environmental Safety*, *262*, 115128. DOI: 10.1016/j.ecoenv.2023.115128 PMID: 37315361

Zhang, L., Chen, F. X., Zeng, Z., Xu, M., Sun, F., Yang, L., Bi, X., Lin, Y., Gao, Y. J., Hao, H. X., Yi, W., Li, M., & Xie, Y. (2021). Advances in Metagenomics and Its Application in Environmental Microorganisms. *Frontiers in Microbiology*, *12*, 1–15. DOI: 10.3389/fmicb.2021.766364 PMID: 34975791

Zhang, L., Xu, X., Li, Y., Zhang, F., & Shen, Q. (2018). *Mortierella hyalina* increases the effectiveness of nitrogen fertilization by improving the soil nitrogen status. *Applied Microbiology and Biotechnology*, *102*(3), 1459–1469. DOI: 10.1007/s00253-017-8647-9 PMID: 29274059

Zhang, Q., Zhang, J., Yang, L., Zhang, L., Jiang, D., Chen, W., & Li, G. (2014). Diversity and biocontrol potential of endophytic fungi in Brassica napus. *Biological Control*, *72*, 98–108. DOI: 10.1016/j.biocontrol.2014.02.018

Zhang, T., Dou, W., Zhang, X., Zhao, Y., Zhang, Y., Jiang, L., & Sui, X. (2021). The development history and recent updates on soy protein-based meat alternatives. *Trends in Food Science & Technology*, *109*, 702–710. DOI: 10.1016/j.tifs.2021.01.060

Zhang, X., Pumplin, N., Ivanov, S., & Harrison, M. J. (2015). EXO70I is required for development of a sub-domain of the periarbuscular membrane during arbuscular mycorrhizal symbiosis. *Current Biology*, *25*(16), 2189–2195. DOI: 10.1016/j.cub.2015.06.075 PMID: 26234213

Zhang, Y., Arends, J. B. A., Van de Wiele, T., & Boon, N. (2011). Bioreactor technology in marine microbiology: From design to future application. *Biotechnology Advances*, *29*(3), 312–321. DOI: 10.1016/j.biotechadv.2011.01.004 PMID: 21251973

Zhang, Z., Chen, X., & Gao, L. (2024). New strategy for the biosynthesis of alternative feed protein: Single-cell protein production from straw-based biomass. *Global Change Biology. Bioenergy*, *16*(2), e13120. DOI: 10.1111/gcbb.13120

Zhanying, L., Jing, Y., & Jinna, C. (2022). Candida utilis and application thereof (CN 114836331 A). https://lens.org/194-618-495-505-094

Zhao, J., Zhou, L., Wang, J., Shan, T., Zhong, L., Liu, X., & Gao, X. (n.d.). *Endophytic fungi for producing bioactive compounds originally from their host plants.*

Zhao, S., Gao, Q., Rong, C., Wang, S., Zhao, Z., Liu, Y., & Xu, J. (2020). Immunomodulatory effects of edible and medicinal mushrooms and their bioactive immunoregulatory products. In Journal of Fungi (Vol. 6, Issue 4, pp. 1–37). MDPI AG. DOI: 10.3390/jof6040269

Zhao, S., Zhang, Q., Xu, Z., and Zhang, Y. (2015). Effects of *Rhizopus stolonifer* on soil microbial community and organic matter decomposition. Journal of Environmental Sciences, 37, 92 99. https://doi.org/DOI: 10.1016/j.jes.2015.04.002

Zhao, J., Xie, X., Jiang, Y., Li, J., Fu, Q., Qiu, Y., Fu, X., Yao, Z., Dai, Z., Qiu, Y., & Chen, H. (2024). Effects of simulated warming on soil microbial community diversity and composition across diverse ecosystems. *The Science of the Total Environment*, *911*, 168793. DOI: 10.1016/j.scitotenv.2023.168793 PMID: 37996030

Zhao, R., Guo, W., Bi, N., Guo, J., Wang, L., Zhao, J., & Zhang, J. (2015). Arbuscular mycorrhizal fungi affect the growth, nutrient uptake and water status of maize (Zea mays L.) grown in two types of coal mine spoils under drought stress. *Applied Soil Ecology*, *88*, 41–49. https://doi.org/https://doi.org/10.1016/j.apsoil.2014.11.016. DOI: 10.1016/j.apsoil.2014.11.016

Zhao, W., Chen, Z., Yang, X., Sheng, L., Mao, H., & Zhu, S. (2023). Metagenomics reveal arbuscular mycorrhizal fungi altering functional gene expression of rhizosphere microbial community to enhance *Iris tectorum*'s resistance to Cr stress. *The Science of the Total Environment*, *895*, 164970. DOI: 10.1016/j.scitotenv.2023.164970 PMID: 37343864

Zhao, Z., Chen, L., & Xiao, Y. (2021). The combined use of arbuscular mycorrhizal fungi, biochar and nitrogen fertilizer is most beneficial to cultivate *Cichorium intybus* L. in Cd-contaminated soil. *Ecotoxicology and Environmental Safety*, *217*, 112154. DOI: 10.1016/j.ecoenv.2021.112154 PMID: 33901784

Zheng, F. L., Wang, Y. J., & Hashem, A., Abd_Allah, E. F., & Wu, Q. S. (. (2024). Mycorrhizae with *Funneliformis mosseae* regulate the trehalose synthesis and sucrose cleavage for enhancing drought tolerance in trifoliate orange. *Scientia Horticulturae*, *337*, 113486. DOI: 10.1016/j.scienta.2024.113486

Zheng, M. M., Wang, C., Li, W. X., Guo, L., Cai, Z. J., Wang, B. R., Chen, J., & Shen, R. F. (2021). Changes of acid and alkaline phosphatase activities in long-term chemical fertilization are driven by the similar soil properties and associated microbial community composition in acidic soil. *European Journal of Soil Biology*, *104*, 103312. DOI: 10.1016/j.ejsobi.2021.103312

Zheng, S., Yang, M., & Yang, Z. (2005). Biomass production of yeast isolate from salad oil manufacturing wastewater. *Bioresource Technology*, *96*(10), 1183–1187. DOI: 10.1016/j.biortech.2004.09.022

Zhong, L., Carere, J., Lu, Z., Lu, F., & Zhou, T. (2018). Patulin in Apples and Apple-Based Food Products: The Burdens and the Mitigation Strategies. *Toxins*, *10*(11), 475. DOI: 10.3390/toxins10110475 PMID: 30445713

Zhou, K., & Yu, L. (2006). Total phenolic contents and antioxidant properties of commonly consumed vegetables grown in Colorado. *Lebensmittel-Wissenschaft + Technologie*, *39*(10), 1155–1162. DOI: 10.1016/j.lwt.2005.07.015

Zhou, Y. H., Liu, Y. J., Wang, Y. D., & Yang, C. X. (2022). *Claroideoglomus etunicatum* improved the growth and saline–alkaline tolerance of *Potentilla anserina* by altering physiological and biochemical properties. *Biocell*, *46*(8), 1967–1978. DOI: 10.32604/biocell.2022.019304

Zhu, B., Gao, T., Zhang, D., Ding, K., Li, C., & Ma, F. (2022). Functions of arbuscular mycorrhizal fungi in horticultural crops. *Scientia Horticulturae*, *303*, 111219. DOI: 10.1016/j.scienta.2022.111219

Zhu, C., Tian, G., Luo, G., Kong, Y., Guo, J., Wang, M., & Shen, Q. (2018). N-fertilizer-driven association between the arbuscular mycorrhizal fungal community and diazotrophic community impacts wheat yield. *Agriculture, Ecosystems & Environment*, *254*, 191–201. DOI: 10.1016/j.agee.2017.11.029

Zhu, N. (2007). Effect of low initial C/N ratio on aerobic composting of swine manure with rice straw. *Bioresource Technology*, *98*(1), 9–13. DOI: 10.1016/j.biortech.2005.12.003

Zhuo, R., Zhang, J., Yu, H., Ma, F., & Zhang, X. (2019). The roles of *Pleurotus ostreatus* HAUCC 162 laccase isoenzymes in decolorization of synthetic dyes and the transformation pathways. *Chemosphere*, *234*, 733–745. DOI: 10.1016/j.chemosphere.2019.06.113 PMID: 31234090

Zia, R., Nawaz, M. S., Siddique, M. J., Hakim, S., & Imran, A. (2021). Plant survival under drought stress: Implications, adaptive responses, and integrated rhizosphere management strategy for stress mitigation. *Microbiological Research*, *242*, 126626. https://doi.org/https://doi.org/10.1016/j.micres.2020.126626. DOI: 10.1016/j.micres.2020.126626 PMID: 33189069

Zięba, P., Kała, K., Włodarczyk, A., Szewczyk, A., Kunicki, E., Sękara, A., & Muszyńska, B. (2020). Selenium and Zinc Biofortification of Pleurotus eryngii Mycelium and Fruiting Bodies as a Tool for Controlling Their Biological Activity. *Molecules (Basel, Switzerland)*, *25*(4), 889. DOI: 10.3390/molecules25040889 PMID: 32079328

Zilly, A., dos Santos Bazanella, G. C., Helm, C. V., Araújo, C. A. V., de Souza, C. G. M., Bracht, A., & Peralta, R. M. (2012). Solid-state bioconversion of passion fruit waste by white-rot fungi for production of oxidative and hydrolytic enzymes. *Food and Bioprocess Technology*, *5*(5), 1573–1580. DOI: 10.1007/s11947-011-0532-8

Zingales, V., Taroncher, M., Martino, P. A., Ruiz, M. J., & Caloni, F. (2022). Climate Change and Effects on Molds and Mycotoxins. *Toxins*, *14*(7), 445. DOI: 10.3390/toxins14070445 PMID: 35878185

Zolotova, N., Kosyreva, A., Dzhalilova, D., Fokichev, N., & Makarova, O. (2022). Harmful effects of the microplastic pollution on animal health: A literature review. *PeerJ*, *10*, e13503. DOI: 10.7717/peerj.13503 PMID: 35722253

Zong, K., Huang, J., Nara, K., Chen, Y., Shen, Z., & Lian, C. (2015). Inoculation of ectomycorrhizal fungi contributes to the survival of tree seedlings in a copper mine tailing. *Journal of Forest Research*, *20*(6), 493–500. DOI: 10.1007/s10310-015-0506-1

Zulfiqar, F., Moosa, A., Ali, H. M., Bermejo, N. F., & Munné-Bosch, S. (2024). Biostimulants: A sufficiently effective tool for sustainable agriculture in the era of climate change? *Plant Physiology and Biochemistry*, *211*, 108699. DOI: 10.1016/j.plaphy.2024.108699 PMID: 38749375

About the Contributors

Arun Kumar Kashyap is currently working as an assistant professor of Biotechnology at Govt. E Raghavendra Rao PG. Science College, Bilaspur (C.G.). he has pursuit his doctoral degree from Jawaharlal Nehru University (JNU), New Delhi in Plant Fungal interaction. Dr Kashyap has more 7 year experience of working the mycological solutions for better future. Dr Kashyap has developed a genetically modified fungus which after interaction with plants confers resistance to various abiotic and biotic responses. Dr kashyap is actively engaged in finding different fungal varieties which have ability to provide solutions for current problems and which re potential source of industrial raw material. Dr Kashyap have 15 Research article in various international peer revived journal and presented his work to 6 national and international conferences. Dr Kashyap have also edited 3 books published by national publishers based in india. Dr Kashyap has also uthored 15 Chapters in different edited books published by International and national publishers.

Abhimnyu Jogawat, has been recognised as a to 2% scientist in the year 2023 by Stanford based on his publication in that year. Dr Jogawat is currently working as a post doctoral fellow at the National Institute of Plant Genome Research, New Delhi in the project related to plant fungal relationship. Before that he has successfully completed 3 projects related to similar area. Dr Jogawat has completed his doctoral degree from JNU, New Delhi. He has published 19 research article in reputed international journal. He has authored 6 book chapter in different edited book. He is a recipient of best paper award and fellow of many academic societies.

***

A. U. Akash is currently pursuing his Ph.D. at the Division of Plant Pathology, Indian Agricultural Research Institute (IARI), New Delhi, India. He completed his master's degree in agriculture with a specialization in Plant Pathology from Professor Jayashankar Telangana State Agricultural University, Hyderabad, Telangana, and holds a bachelor's degree in Agriculture from Tamil Nadu Agricultural University, Coimbatore, Tamil Nadu. He has contributed to over 10 publications, focusing on agronomy and plant pathology. His current research interest lies in enhancing plant health through sustainable and innovative approaches, with a commitment to advancing practices that support long-term agricultural resilience and productivity.

Vikas Chandra has obtained his B.Sc. and M.Sc. degree in Biotechnology from Guru Ghasidas Vishwavidyalaya, Bilaspur, Chhattisgarh. He has obtained his Ph.D. degree from University of Calcutta, Kolkata. Dr. Vikas is currently working as an Assistant Professor at Department of Biotechnology, Guru Ghasidas Vishwavidyalaya, Bilaspur, Chhattisgarh.

Abhishek Dadsena is currently persuing his M.Sc. in Biotechnology from Guru Ghasidas Vishwavidyalaya, Bilaspur, Chhattisgarh.

Joydeb Dey has pursued his PhD from the Department of Life Science & Bioinformatics, Assam University, Silchar, under the supervision of Dr. Himanshu Kishore Prasad, Associate Professor, Assam University. His thesis is entitled 'Arginine catabolism and its role in pathogenesis of *C. albicans*'. His work was focused on functional characterization and purification of three potential Ureohydrolase superfamily members from *C. albicans*.

Salah Er-Raki is actually a Full Professor of Applied physics at Faculty of Sciences and Techniques, Cadi Ayyad University, Morocco. He received his PhD in "crop water requirements estimates through combining the modeling, in situ measurements and remote sensing" in 2007. His main research interests include the application of multispectral remote sensing observations (optical, radar and thermal data) in land-surface-atmosphere models for high spatio-temporal resolution monitoring of crops water needs especially in arid and semi-arid regions. He also contributed in developing a software named « SAMIR, Satellite monitoring of irrigation", which such basic concepts are based on his thesis work. He also participates in several international experiments (e.g. Yaqui-2006-2008 in Mexico) and he was a co-leader of the joint international laboratory "Télédétection et Ressources en Eau en Méditerranée semi-Aride", LMI-TREMA. Salah Er-Raki has published more than 130 papers in international peer-reviewed journals, which have been cited more than 4000 times and brought him an H-index of 35. He serves as a Guest editor, editorial boards and reviewer of many international scientific journals with high impact factor among them are: Agricultural Water Management, Remote Sensing of Environment, remote sensing, Agricultural and Forest Meteorology, Computers and Electronics in Agriculture, Theoretical and Applied Climatology, Soft Computing, irrigation science, water resources management,... He has attended and serves as chair sessions in many international congresses and symposiums. He supervised and co-supervised 8 PhD students and he coordinates several research projects (e.g. PHC Maghreb, PRIMA, H2020 RISE action,...). He is awarded as a Morocco's top researchers (2016- 2020) in Agricultural Sciences ranked by Essential Science Indicators (ESI) and Web of Sciences, in the Second edition of "Research Excellence Awards", organized by Clarivate and CNRST, June 2021. He also awarded as International Young Scientist Award, by the Foundation of Innovative Research, Noval Research Foundation and International Research and Academic Excellence Awards (IRAEA - 2021), August 2022.

Amira Hachana National Institute of Agricultural Research, Tunisia Research interests: Agronomic and environmental performances of field crop systems Precision agriculture (microbial bio-fertilizer application) Modeling plant-environment interactions Remote sensing application on agriculture Soil-microbe-plant interactions Soil and crop mapping Contact: E-mail: amira.hachana.inrat@gmail.com Phone: +216 58 629 429

Amjada. S. Khan is currently pursuing her Ph.D. at the Division of Plant Pathology, Indian Agricultural Research Institute (IARI), New Delhi, India. She completed her bachelor's degree in Agriculture from Kerala Agricultural University and master's degree in agriculture with a specialization in Plant Pathology from ICAR-IARI, New Delhi. Her research contributes to sustainable agriculture and food security by addressing critical challenges in crop protection using advanced molecular tools. Her focus is on bridging fundamental research with practical agricultural solutions.

Khushboo Kumari is currently pursuing a PhD at Mahatma Gandhi Central University, Motihari, Bihar, India. She has completed her master's degree from Patna Science College, Patna University, Bihar, India. She has completed her bachelor's degree from Patna Science College, Bihar, India. She has a keen research interest in microbiology and sustainable agricultural practices.

Soufiane Lahbouki is a researcher specializing in biotechnology and plant physiology, with a focus on sustainable agriculture, mycorrhizal fungi, and drought stress resilience. His work explores plant-microbe interactions, soil fertility, and strategies to enhance crop tolerance to abiotic stress, with multiple publications in high-impact journals.

Pooja Malave is an Assistant Professor in the Departement of Biotechnology, Dr D.Y Patil College of arts, commerce and Science Pimpri Pune. PhD student at Krishna Vishwa Vidyapeeth Malkapur Karad.

Abdelilah Meddich is a professor at Cadi Ayyad University, Marrakech, specializing in plant biotechnology and physiology. His research focuses on plant-microbe interactions, particularly arbuscular mycorrhizal fungi (AMF), and their role in improving plant growth, nutrient uptake, and stress tolerance under abiotic conditions.

P. Selvakumar is currently an Associate Professor in the Department of Science and Humanities at Nehru Institute of Technology, Coimbatore, Tamil Nadu, India. Since 2023, he has served as the Department IQAC Coordinator at the institute. Prior to joining Nehru Institute of Technology, he held positions as an Assistant Professor in the Department of Science and Humanities various engineering colleges in Coimbatore, Tamil Nadu, India, spanning from 2011 to 2023. Dr.P. Selvakumar's academic and research interests lie primarily in the field of phytochemistry. He has made significant contributions to his field, with an impressive record that includes 30 journal publications, 12 patents, and 20 conference papers. Additionally, he has organized 5 seminars and conferences, further demonstrating his commitment to academic and scientific exchange. Throughout his career, Dr. P.Selvakumar has been recognized for his dedication and excellence in teaching, having been honored with several awards, including the Best Teacher Awards. He holds a prominent position in the academic community, contributing actively to research.

Tarun Kumar Patel is a researcher in the Department of Biotechnology, Sant Guru Ghasidas Govt. P.G. College, Kurud, Dhamtari Chhattisgarh.

Himanshu Prasad is a faculty member in the Department of Life Science and Bioinformatics at Assam University, Silchar, Assam. His research interests encompass antifungal mechanisms and microbial biotechnology.

Anjali Roy is currently a PhD research scholar at the Department of Life Science & Bioinformatics, Assam University, Silchar. Her field of research is associated with microbial biotechnology, dealing with yeasts. She received a bachelor's degree with honors in botany from Gurucharan University, Silchar and a master's degree in botany as well from the Department of Life Science & Bioinformatics, Assam University.

Gurucharan Sahu has obtained his B.Sc. and M.Sc. degree in Biotechnology from Guru Ghasidas Vishwavidyalaya, Bilaspur, Chhattisgarh.

Lalit Kumar Sahu, have completed her Master's degree in Biotechnology. She was born on March 1, 2001, in Bilaspur, Chhattisgarh, and has always been passionate about pursuing innovative and sustainable solutions in science. Her research focuses on the green synthesis of silver nanoparticles, which combines traditional knowledge with modern scientific approaches. Her academic journey has been deeply rooted in exploring environmentally friendly methods to address global challenges. She is dedicated to contributing to the scientific community through her work and aspire to create sustainable solutions that can benefit society and the environment.

Sujan Samanta is a M.Sc. student in the Department of Biosciences

Dipali Saxena, an accomplished Assistant Professor, boasts over 10 years of academic experience in Food Sciences and Technology. She earned her Ph.D. in Food Technology from Jiwaji University, Gwalior, specializing in Nutraceuticals and functional foods. Graduating as the university topper and receiving the MSc. Gold Medal in Food Technology from Jiwaji University, Gwalior, Madhya Pradesh, highlights her academic excellence. Currently, Dr. Saxena serves as the Senior Faculty and Assistant Professor in the Department of Food Nutrition at Shri Vaishnav Institute of Home Science and Shri Vaishnav Institute of Agriculture, part of Shri Vaishnav Vidyapeeth Vishwavidyalaya, Indore. Additionally, she holds positions as a Member of the Board of Studies (BOS), Faculty of Studies (FOS), and various committees including Admission, Cultural, Anti-ragging, ABC credit, and Library Committees at Shri Vaishnav Vidyapeeth Vishwavidyalaya, Indore.

Naira Sbbar is a Ph.D. candidate specializing in plant biotechnology and physiology, focusing on sustainable agriculture, mycorrhizal fungi, and plant stress resilience. She is affiliated with Cadi Ayyad University, Marrakech, Morocco. Her research explores plant-microbe interactions, soil fertility, and strategies to enhance crop tolerance to abiotic stress, with an emphasis on the role of arbuscular mycorrhizal fungi (AMF) in improving plant growth under challenging environmental conditions.

Sharmistha Sharma is currently a PhD. research scholar at the Department of Life Science & Bioinformatics, Assam University, Silchar. Her interest lies in the field of research associated with microbial biotechnology, that deals primarily with fungi. She received a bachelor's degree with honors in botany from Gurucharan University, Silchar and a master's degree in botany as well from the Department of Life Science & Bioinformatics, Assam University.

Durgeshwer Singh is working as an Assistant Professor in the Department of Botany at Mahatma Gandhi Central University, Motihari, Bihar, India. He completed his PhD from the Defence Laboratory, Defence Research and Development Organization (DRDO) Jodhpur. Dr. Singh did his M.Sc. in Botany from Jai Narain Vyas University, Jodhpur. With over a decade of research experience, Dr. Singh specializes in nanobiotechnology, microbiology and bioremediation. He has made significant contributions to the development of nanosilver containing dressing for wound healing applications, microbial interactions, and environmental remediation. Dr. Singh has published more than 15 research papers in peer-reviewed international journals and has presented his work at various international and national conferences.

He also has a patent on chitin-based dressing containing nanosilver. His research has led to innovative advancements in nanotechnology-based biomedical applications and bioremediation strategies.

Neha Singh is currently pursuing Ph.D. at the division of Plant Pathology, Indian Agricultural Research Institute (IARI), New Delhi, India. She has completed her master's degree in agriculture with a specialisation in plant Pathology from Navsari Agricultural University, Navsari, Gujarat and holds a bachelor's degree in Agriculture from Rajendra Prasad Central Agricultural University, PUSA, Bihar. Her current research interest lies in enhancing plant growth using sustainable approaches.

Khirabdhi Tanaya is a research scholar at Department of Botany, Mahatma Gandhi Central University, Bihar, India. She has obtained her master's degree with Gold medal at university level from Mahatma Gandhi Central University, Motihari, Bihar. She has done her Bachelor's Degree from Shailabala Women's College, Cuttack, Odisha. She has research interest in Microbiology and Bioremediation.

Nikita Verma, completed her Master's degree in Biotechnology, with her dissertation focusing on Rational Structure-Based Inhibitor Design Against Sortase C from Enterococcus faecalis. She is deeply interested in exploring computational and structural biology to address key challenges in antimicrobial resistance. Her research aims to contribute to the development of novel therapeutic strategies and expand the understanding of target-based drug design.

Harsha Yadav has obtained her B.Sc. and M.Sc. degree in Biotechnology from Guru Ghasidas Vishwavidyalaya, Bilaspur, Chhattisgarh.

Index

A

Abiotic stress 4, 19, 22, 26, 27, 30, 39, 41, 66, 73, 74, 75, 77, 82, 83, 91, 92, 93, 94, 98, 99, 100, 106, 112, 118, 120, 122, 123, 125, 230
aflatoxin 323, 378, 379, 380, 384, 385, 386, 388, 391, 392, 393, 397, 399, 400, 402, 403, 404, 409, 410, 411, 412, 413, 417, 419, 423, 424, 428, 429, 431, 432, 433, 436, 437
Agricultural Wastes 152, 167, 261, 262, 263, 269, 270, 274, 277
agriculture 1, 2, 3, 4, 6, 7, 8, 10, 11, 13, 16, 17, 18, 19, 20, 21, 22, 23, 24, 25, 26, 27, 31, 36, 37, 38, 40, 41, 43, 45, 46, 48, 53, 56, 64, 66, 68, 74, 85, 86, 87, 89, 91, 92, 93, 94, 96, 97, 98, 100, 102, 104, 108, 109, 113, 114, 117, 119, 120, 121, 131, 132, 135, 138, 140, 141, 142, 143, 144, 146, 147, 148, 160, 161, 162, 166, 174, 177, 185, 186, 201, 227, 238, 254, 263, 270, 271, 274, 276, 277, 278, 281, 286, 293, 310, 320, 321, 324, 325, 327, 328, 329, 330, 331, 333, 336, 337, 338, 339, 343, 345, 358, 360, 361, 372, 374, 375, 379, 390, 398, 400, 401, 403, 404, 407, 417, 429, 430, 432
Alternative protein 164, 175, 186, 190, 192, 193, 196, 201, 205
AMF Diversity 74, 75, 76, 77, 78
Animal feed 157, 164, 175, 176, 182, 183, 188, 190, 378, 379, 382, 383, 395, 396, 404, 407, 408, 427, 428, 430
Antagonistic fungi 50
Antifungal potency 320
Approaches 33, 37, 43, 56, 61, 64, 119, 181, 212, 216, 217, 219, 244, 251, 252, 258, 262, 266, 267, 271, 273, 274, 280, 291, 299, 324, 329, 331, 332, 333, 335, 336, 337, 338, 339, 340, 341, 342, 343, 345, 347, 361, 387, 393, 397, 407, 425, 427, 428, 433
Arbuscular Mycorrhizal Fungi 1, 2, 3, 6, 8, 9, 11, 17, 19, 20, 21, 22, 23, 26, 27, 31, 32, 33, 34, 39, 40, 41, 42, 43, 44, 73, 74, 86, 87, 88, 89, 90, 91, 92, 93, 94, 95, 96, 98, 99, 102, 103, 115, 117, 119, 136, 137, 138, 143, 144, 145, 146, 147, 148, 149, 209, 216, 219, 223, 224, 227, 228, 229, 230, 231, 232, 233, 234, 338, 340

B

Bioactive 3, 7, 27, 36, 57, 59, 69, 100, 101, 121, 142, 151, 154, 155, 160, 162, 166, 182, 198, 204, 206, 262, 268, 269, 270, 271, 310, 311, 319, 323, 327, 328, 329, 330, 331, 332, 334, 337, 338, 339, 340, 341, 343, 344, 345, 346, 349, 350, 353, 354, 355, 356, 367, 368, 372, 376, 426
Biodegradation 115, 238, 243, 244, 245, 247, 248, 249, 252, 254, 255, 256, 258, 260, 261, 262, 270, 271, 272, 273, 274, 275, 276, 277, 278, 279, 280, 281, 284, 288, 296, 298, 299, 301, 302, 304, 305, 336, 342
Biofortified 151, 157
Breaking Down 2, 7, 27, 32, 36, 49, 54, 101, 110, 134, 237, 244, 245, 249, 250, 251, 261, 264, 266, 270, 289, 335, 340, 342, 344, 395

C

Chronic diseases 190, 356
Circadian Clock 212, 224, 228, 229, 230, 231, 232
Climate Change 1, 2, 5, 11, 12, 13, 16, 17, 18, 23, 39, 60, 73, 74, 94, 109, 121, 138, 141, 191, 271, 274, 275, 277, 280, 285, 340, 348, 350, 378, 379, 386, 387, 402, 403, 404, 405, 427, 434
Compost 7, 85, 89, 145, 152, 157, 227, 243, 258, 259, 270, 271, 272, 281, 296, 359, 371
Crop improvement 119, 122, 144

D

Defense response 97, 106, 122
detection methods 378, 390, 396, 421, 422
Discovering Novel Bioactive Compounds 327, 330, 332
Disease Resistance 1, 2, 9, 10, 16, 27, 120, 122, 140, 330

E

Eco-friendly 4, 8, 16, 19, 26, 45, 56, 70, 86, 98, 101, 131, 132, 133, 136, 141, 190, 201, 237, 238, 252, 262, 284, 287, 291, 297, 334, 335, 341, 344
Edible 108, 137, 144, 151, 152, 154, 155, 159, 160, 161, 162, 183, 203, 243, 273, 349, 351, 352, 354, 356, 363, 364, 365, 368, 369, 371, 372, 373, 374, 375, 376
Endophyte 22, 28, 100, 101, 103, 106, 108, 111, 113, 115, 120, 121, 145, 286, 300
environmental pollution 25, 97, 169, 237, 257, 258, 262, 263, 271, 273, 285, 302, 303, 304, 306, 325, 335
essential oils 155, 309, 310, 311, 312, 313, 314, 315,

316, 317, 318, 319, 320, 322, 323, 324, 325

F

fat 151, 153, 154, 174, 178, 183, 191, 195, 196, 198, 199, 250, 255, 351, 354
food safety 181, 186, 194, 201, 203, 324, 368, 377, 379, 380, 383, 384, 387, 390, 391, 392, 395, 396, 399, 400, 402, 409, 417, 422, 429, 430, 431, 432
Food security 11, 13, 86, 101, 131, 135, 149, 158, 163, 164, 166, 168, 177, 182, 190, 193, 278, 321, 325, 328, 332, 339, 343, 358, 361, 364, 367, 378, 379, 384, 398, 399, 402, 407, 408
Functional foods 153, 161, 190, 354, 362, 365, 369, 371
Fungal biocontrol agents 45, 46, 47, 48, 50, 58, 59, 61, 64, 69, 71
Fungal Biodegradation 261, 262, 270, 271, 272, 273, 274, 275, 276, 277, 278, 298
Fungal Bioprospecting 327, 328, 337, 338, 339, 340, 341, 342, 343, 344, 345
Fungal Biostimulants 25, 26, 27, 28, 30, 32, 33, 35, 36, 37, 38, 97, 99, 105, 106, 108, 112, 123, 124, 125
Fungal Diversity 19, 72, 73, 74, 88, 119, 233, 263, 327, 328, 331, 332, 337, 369

G

Gene 10, 13, 16, 18, 37, 66, 73, 78, 79, 80, 83, 86, 89, 101, 111, 113, 124, 127, 144, 145, 173, 212, 213, 215, 216, 218, 219, 221, 222, 223, 224, 225, 226, 228, 231, 233, 234, 235, 262, 273, 329, 331, 334, 337, 339, 372, 427, 428, 436

H

Health hazard 108
Heat Shock Proteins 4, 5, 12, 73, 107

I

Immunomodulation 154, 354

L

Laccase 245, 249, 252, 254, 258, 260, 261, 280, 281, 289, 307, 356
Life cycle assessment 190, 206, 275, 374, 375

M

meat 152, 158, 161, 162, 165, 174, 177, 180, 182, 184, 185, 190, 191, 192, 193, 194, 195, 197, 198, 199, 200, 201, 203, 204, 205, 206, 207, 269, 322, 335, 365, 375, 418
medicinal and aromatic plants 94, 311, 324
Metabolites 4, 10, 16, 22, 30, 33, 37, 42, 48, 50, 54, 57, 61, 71, 85, 99, 100, 101, 104, 107, 111, 116, 119, 124, 126, 133, 136, 155, 164, 166, 168, 185, 198, 230, 250, 261, 270, 291, 298, 300, 305, 322, 324, 327, 328, 329, 330, 332, 334, 337, 339, 341, 343, 368, 373, 377, 378, 387, 388, 401, 407, 408, 426, 427, 430
minimum inhibitory concentration 316
mitigation strategies 377, 378, 379, 397, 398, 405, 407
Mushroom 151, 152, 153, 154, 155, 156, 157, 158, 159, 160, 161, 162, 203, 259, 298, 302, 303, 349, 350, 351, 352, 353, 354, 356, 357, 358, 359, 360, 361, 362, 363, 364, 365, 367, 368, 369, 370, 371, 373, 374, 375, 376
Myco-fertilizers 131, 132, 133, 134, 135, 138, 139, 140, 141, 142
Mycoprotein 164, 166, 167, 174, 177, 178, 179, 180, 183, 184, 185, 186, 187, 188, 189, 190, 191, 192, 193, 194, 195, 196, 197, 198, 199, 200, 201, 203, 204, 205, 206
mycoproteins 165, 166, 174, 176, 179, 186, 191, 192, 193, 195, 196, 197, 198, 199, 200, 201, 202, 205
Mycoremediation 109, 237, 238, 242, 243, 244, 245, 247, 249, 250, 251, 252, 253, 254, 255, 256, 257, 258, 260, 283, 284, 286, 287, 288, 291, 295, 296, 297, 298, 299, 300, 302, 304, 305, 307, 341, 342, 344, 350
Mycorrhizal 1, 2, 3, 6, 8, 9, 11, 17, 19, 20, 21, 22, 23, 25, 26, 27, 28, 31, 32, 33, 34, 35, 36, 39, 40, 41, 42, 43, 44, 70, 73, 74, 81, 82, 85, 86, 87, 88, 89, 90, 91, 92, 93, 94, 95, 96, 98, 99, 100, 102, 103, 107, 115, 117, 119, 132, 133, 134, 135, 136, 137, 138, 139, 140, 141, 142, 143, 144, 145, 146, 147, 148, 149, 209, 210, 211, 214, 216, 217, 218, 219, 223, 224, 225, 227, 228, 229, 230, 231, 232, 233, 234, 287, 298, 330, 336, 338, 340
Mycosorption 248, 252
Mycotoxin Contamination 148, 377, 378, 379, 384, 385, 386, 387, 390, 391, 392, 393, 394, 397, 398, 399, 400, 401, 403, 404, 407, 408, 417, 423, 425, 426, 427, 428, 432, 433, 434, 437
Mycotoxins 174, 329, 377, 378, 379, 381, 382, 383, 384, 385, 387, 388, 389, 390, 391, 392, 393, 395, 396, 397, 398, 399, 400, 401, 402, 403, 404, 405,

407, 408, 409, 410, 411, 412, 413, 414, 417, 418, 420, 421, 422, 423, 424, 425, 426, 427, 428, 429, 430, 431, 432, 433, 434, 435, 436, 437

N

Nutraceuticals 153, 163, 164, 190, 354, 365
Nutrient Uptake 1, 5, 6, 7, 8, 9, 10, 12, 13, 16, 25, 26, 27, 28, 30, 31, 33, 36, 42, 44, 74, 76, 77, 83, 85, 95, 98, 100, 106, 108, 123, 126, 127, 132, 137, 138, 139, 140, 142, 143, 144, 227, 239, 272, 328, 330, 336, 338, 340
Nutrient use efficiency 12, 17, 25, 73, 98, 112, 122
Nutrition 30, 42, 44, 75, 88, 92, 99, 101, 105, 126, 138, 146, 148, 149, 152, 157, 158, 159, 160, 162, 164, 175, 176, 177, 178, 181, 182, 184, 185, 188, 189, 190, 203, 204, 205, 206, 210, 243, 244, 284, 286, 322, 350, 364, 368, 369, 371, 372, 373, 375, 402, 431, 434, 436
Nutritional Status 34, 107, 127, 152, 158, 195, 423

O

Osmotic balance 31, 73, 77, 78, 81, 83

P

Phytopathogen control 61
phytopathogenic fungi 54, 68, 71, 119
Plant growth 2, 3, 7, 8, 9, 15, 17, 19, 21, 22, 25, 27, 28, 30, 31, 32, 33, 35, 39, 42, 43, 50, 54, 55, 57, 65, 66, 73, 74, 76, 78, 81, 82, 86, 88, 90, 95, 98, 99, 100, 101, 103, 104, 105, 106, 108, 111, 113, 115, 116, 118, 119, 120, 122, 123, 124, 125, 127, 131, 132, 133, 134, 135, 136, 138, 139, 141, 143, 146, 147, 149, 217, 221, 227, 241, 330, 336, 339, 340, 417
Plant resilience 3, 5, 6, 11, 16, 18, 26, 27, 74, 76, 86, 137
Pollution 11, 25, 92, 97, 120, 132, 152, 158, 159, 169, 230, 237, 238, 239, 240, 242, 252, 253, 255, 256, 257, 258, 259, 260, 261, 262, 263, 271, 273, 274, 283, 284, 285, 286, 288, 290, 291, 296, 297, 298, 299, 302, 303, 304, 305, 306, 325, 335, 336, 341, 364
protein 12, 26, 50, 63, 67, 68, 77, 78, 80, 81, 82, 84, 92, 94, 95, 98, 102, 104, 107, 111, 123, 126, 138, 151, 152, 153, 154, 157, 158, 159, 161, 163, 164, 165, 166, 167, 168, 169, 170, 171, 173, 174, 175, 176, 177, 178, 180, 181, 182, 183, 184, 185, 186, 187, 188, 189, 190, 191, 192, 193, 194, 195, 196, 197, 198, 199, 200, 201, 203, 204, 205, 206, 207, 210, 221, 223, 224, 234, 235, 238, 250, 252, 264, 269, 270, 301, 334, 335, 337, 340, 349, 351, 352, 354, 356, 360, 365, 375, 422, 430
Protein quality 166, 180, 190, 195, 204

R

regulatory frameworks 14, 17, 38, 262, 273, 275, 329, 343, 377, 378, 384, 390, 391, 392, 397, 398
Regulatory measures 378
Root System 4, 9, 27, 91, 137, 138, 211, 217, 219, 221, 243

S

Sequencing 111, 209, 211, 212, 214, 215, 216, 217, 219, 221, 222, 223, 225, 226, 227, 228, 229, 231, 232, 264, 328, 329, 331, 337, 338, 340, 342
Single cell protein 164, 165, 167, 168, 174, 181, 183, 184, 185, 186, 187, 188, 189, 190
soil and water pollutants 287, 291
Soil health 1, 2, 6, 8, 9, 10, 11, 13, 15, 16, 18, 20, 25, 27, 45, 61, 92, 97, 98, 99, 119, 122, 131, 132, 133, 135, 139, 140, 141, 142, 146, 147, 149, 227, 239, 240, 242, 263, 270, 273, 274, 276, 277, 330, 339, 340, 358, 359, 371
Soil Microbiome 218, 222, 233
Spent mushroom substrate 152, 157, 302, 350, 359
Stress tolerance 3, 4, 7, 8, 12, 13, 15, 16, 19, 22, 27, 32, 33, 73, 75, 81, 83, 86, 87, 89, 93, 94, 95, 97, 98, 99, 105, 106, 114, 115, 118, 120, 121, 123, 125, 135, 137, 145, 223, 230, 272, 273, 275, 298, 330, 340
Superfood 151, 153, 154, 157, 158, 159, 349, 350
Sustainability 2, 3, 11, 13, 14, 17, 18, 19, 20, 21, 22, 25, 26, 37, 45, 46, 74, 81, 105, 109, 110, 121, 131, 132, 135, 141, 149, 164, 177, 180, 181, 182, 183, 184, 200, 201, 203, 204, 237, 238, 242, 252, 258, 260, 262, 266, 267, 269, 270, 271, 272, 274, 275, 276, 277, 309, 311, 328, 329, 331, 332, 335, 337, 338, 339, 344, 358, 359, 361, 362, 364, 370, 371, 373, 408
Sustainable 1, 2, 3, 6, 7, 8, 10, 11, 13, 15, 16, 18, 19, 20, 21, 22, 23, 24, 25, 26, 27, 31, 32, 36, 38, 40, 43, 44, 45, 46, 47, 53, 56, 61, 64, 66, 73, 74, 75, 81, 85, 86, 87, 89, 91, 93, 94, 97, 98, 100, 101, 102, 105, 106, 108, 109, 112, 113, 114, 117, 118, 119, 120, 121, 131, 132, 133, 138, 140, 141, 142, 143, 144, 145, 146, 148, 149, 151, 152, 158, 160, 162, 163, 164, 166, 167, 168, 176, 177, 180, 181, 182, 184, 186, 187, 188, 190, 191, 192, 193, 198,

199, 201, 202, 204, 205, 209, 238, 245, 247, 258, 259, 261, 262, 263, 265, 266, 267, 269, 270, 271, 273, 274, 275, 276, 277, 278, 284, 286, 296, 297, 298, 299, 302, 303, 306, 307, 310, 311, 320, 321, 323, 325, 328, 329, 330, 331, 332, 335, 336, 338, 339, 340, 341, 342, 343, 344, 345, 349, 350, 358, 359, 360, 361, 364, 367, 369, 372, 373, 374, 376, 400, 403, 426

Sustainable agriculture 1, 2, 3, 6, 7, 8, 11, 13, 16, 18, 19, 20, 21, 22, 23, 24, 25, 26, 27, 31, 36, 40, 45, 46, 53, 64, 66, 74, 85, 86, 87, 89, 91, 94, 98, 100, 102, 108, 109, 113, 114, 117, 120, 121, 131, 132, 138, 140, 141, 142, 143, 144, 146, 148, 166, 263, 270, 321, 338, 339, 361, 372

Sustainable protein 163, 164, 167, 168, 176, 182, 184, 187, 190, 192, 193, 198, 205

Symbiotic relationship 3, 75, 79, 80, 81, 100, 106, 121, 133, 134, 136

T

Technique 58, 70, 173, 215, 216, 225, 235, 238, 281, 298, 313, 318, 348, 387, 388, 389, 390, 420

U

Urbanization 237, 297

W

Water Absorption 4, 6, 73, 85, 135, 137, 138, 141, 338

www.ingramcontent.com/pod-product-compliance
Lightning Source LLC
LaVergne TN
LVHW080619300926
849162LV00005B/1209
* 9 7 9 8 3 6 9 3 8 2 0 1 1 *